Improve Your Grade.

Access included with each new book.

Get help with **exam prep.**
REGISTER NOW!

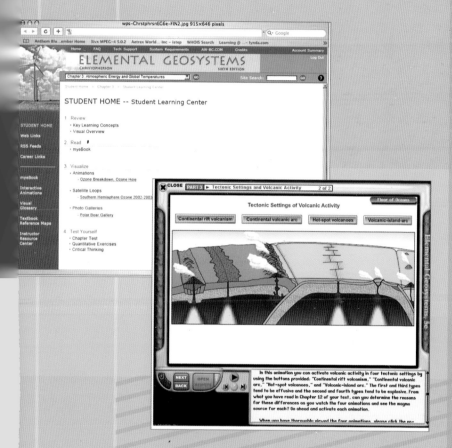

Registration will let you:

- Prepare for exams using online quizzes
- Grasp difficult concepts with tutorial animations and videos
- Learn at your pace using self-study media
- Access and annotate your text online 24/7 using myeBook

PEARSON
mygeoscience place

www.mygeoscienceplace.com

TO REGISTER

1. Go to www.mygeoscienceplace.com.
2. Click "Register."
3. Follow on-screen instructions to create your login name and password.

Your Access Code is:

Note: If there is no silver foil covering the access code, it may already have been redeemed, and therefore may no longer be valid. In that case, you can purchase online access using a major credit card. To do so, go to www.mygeoscienceplace.com and click on "Buy Access," and follow the on-screen instructions.

TO LOG IN

1. Go to www.mygeoscienceplace.com.
2. Click "Log In."
3. Pick your book cover.
4. Enter your login name and password.

Hint:
Remember to bookmark the site after you log in.

Technical Support:
http://247pearsoned.custhelp.com

Stop!

Did you register?
Don't miss out!

Turn back one page
to register and get help
preparing for exams...
whenever and wherever
you need it.

www.mygeoscienceplace.com

Elemental
Geosystems

Combination of satellite *Terra* MODIS sensor image and *GOES* satellite image produces a true-color view of North and South from 35,000 km (22,000 mi) in space. *[Courtesy of MODIS Science Team, Goddard Space Flight Center, NASA and NOAA.]*

The Arkansas River in western Kansas disappears as groundwater from the High Plains Aquifer is drawn down by groundwater pumping for central-pivot irrigation, dropping more than 30 m (100 ft) since 1950 (see Chapter 6). Such an influent condition occurs when the water table drops below the riverbed, effectively drying it out. The Arkansas River, sixth among US rivers in length, still flows along the surface to the west of this region and to the east and south of the High Plains Aquifer where it continues as a tributary to the Mississippi River. [Photo by Bobbé Christopherson, July 2008.]

Sixth Edition

Elemental
Geosystems

Robert W. Christopherson

Prentice Hall
New York Boston San Francisco
London Toronto Sydney Tokyo Singapore Mad
Mexico City Munich Paris Cape Town Hong Kong Mon

Library of Congress Cataloging-in-Publication Data

Christopherson, Robert W.
  Elemental geosystems / Robert W. Christopherson. — 6th ed.
    p. cm.
  Includes bibliographical references and index.
  ISBN-13: 978-0-321-59521-8
  ISBN-10: 0-321-59521-1
  1. Physical geography. I. Title.
  GB54.5.C47 2010
  910'.02—dc22

2008049468

Geography Acquisitions Editor: *Christian Botting*
Geography Project Manager: *Tim Flem*
Editor in Chief, Chemistry and Geosciences: *Nicole Folchetti*
Editorial Assistant: *Haifa Mahabir*
Managing Editor, Chemistry and Geosciences:
  Gina M. Cheselka
Project Manager, Production: *Shari Toron*
Production Management: *GGS Higher Education Resources,*
  A division of PreMedia Global, Inc.
Production Editors: *Cindy Miller, Suganya Karuppasamy*
Production Assistant to the Author: *Bobbé Christopherson*
Art Director: *Maureen Eide*
Art Project Manager: *Connie Long*
Art Studio: *Precision Graphics;*
  Project Manager - Holly Birch

Showcased Art: *David Fierstein*
Interior and Cover Design: *Wee Design*
Cover Designer: *Bobbé Christopherson, Tamara Newman*
Copy Editor: *Pamela Rockwell*
Proofreader: *Jeff Georgeson*
Composition: *GGS Higher Education Resources, A division of*
  PreMedia Global, Inc.
Senior Operations Supervisor: *Alan Fischer*
Marketing Manager: *Amy Porubsky*
Marketing Assistant: *Keri Parcells*
Senior Production Supervisor, Media: *Liz Winer*
Media Project Manager: *Natasha Wolfe*

Front and Back Cover Photos: *Spearville Wind Energy Facility, Kansas; false cirrus clouds, jet contrails at dawn; storm clouds sweep past Jan Mayen Island, Norwegian Sea; Green River Wyoming landscape; Bison females and calves in the Big Basin Prairie Reserve; all photos by* Bobbé Christopherson.
Dedication Page Quote: *B. Kingsolver, Small Wonder (New York: Harper Collins Publishers, 2002), p. 39.*

© 2010, 2007, 2004, 2001, 1998, 1995 Pearson Education, Inc.
Pearson Education, Inc.
Upper Saddle River, NJ 07458

Printed in the United States of America
10 9 8 7 6 5 4 3

ISBN-13: 978-0-321-59521-8
ISBN-10: 0-321-59521-1

Prentice Hall
is an imprint of

www.pearsonhighered.com

To all the students and teachers of Earth,
our home planet, and a sustainable future.

For the polar bears and all the grand children who
will enjoy them, may there be sea ice.

The land still provides our genesis,
however we might like to forget
that our food comes from dank,
muddy Earth, that the oxygen in our lungs was
recently inside a leaf, and that every newspaper or
book we may pick up is made from the hearts of
trees that died for the sake of our imagined lives.
What you hold in your hands right now,
beneath these words,
is consecrated air and time and sunlight.

—Barbara Kingsolver

About our Sustainability Initiatives

This book is carefully crafted to minimize environmental impact. The materials used to manufacture this book originated from sources committed to responsible forestry practices. The paper is FSC® certified. The binding, cover, and paper come from facilities that minimize waste, energy consumption, and the use of harmful chemicals.

Pearson closes the loop by recycling every out-of-date text returned to our warehouse. We pulp the books, and the pulp is used to produce items such as paper coffee cups and shopping bags. In addition, Pearson aims to become the first carbon neutral educational publishing company.

The future holds great promise for reducing our impact on Earth's environment, and Pearson is proud to be leading the way. We strive to publish the best books with the most up-to-date and accurate content, and to do so in ways that minimize our impact on Earth.

MIX
Paper from
responsible sources
FSC
www.fsc.org
FSC® C084269

Prentice Hall
is an imprint of

Brief Contents

Contents

PART II Water, Weather, and Climate Systems 142

5 Atmospheric Water and Weather 144

PART III Earth's Changing
Landscape
Systems   262

8 The Dynamic Planet   264

9 Tectonics, Earthquakes, and Volcanoes   296

10 Weathering, Karst Landscapes, and Mass Movement   336

11 River Systems and Landforms 364

12 Wind Processes and Desert Landscapes 400

13 The Oceans, Coastal Processes, and Landforms 428

14 Glacial and Periglacial Landscapes   460

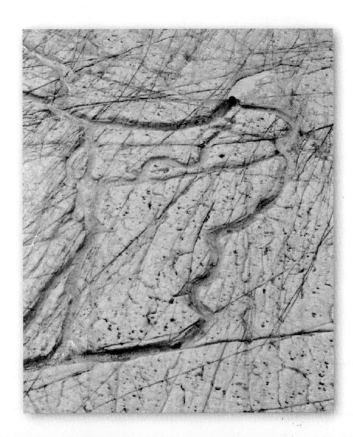

Preface

Welcome to the Sixth Edition of *Elemental Geosystems*! Much has happened since the last edition. The International Polar Year (IPY; http://www.ipy.org/) began in 2007 and concluded March 2009. The Intergovernmental Panel on Climate Change (IPCC; http://www.ipcc.ch/) completed its Fourth Assessment Report and published four supporting volumes of its consensus science regarding human-forced climate change and is a source for several new figures in our text. Armed with the spatial analysis tools of geographic science, physical geographers are well equipped to participate in a planetary understanding of environmental conditions.

In every chapter, *Elemental Geosystems* presents updated science, information, and Internet URLs to enhance your understanding of Earth systems. The currency of *Elemental Geosystems* makes it an important resource for students and teachers alike. The dramatic cover photo captures this quality as global winds and wind-generated energy resources relate to content in several chapters; an applied topic of physical geography. By mid–2008 worldwide wind-generated energy rose more than 1000% since 1998, surpassing 100,000 million watts (100 gigawatts) of installed electrical capacity—more about this in Chapter 4.

The goal of physical geography is to explain the spatial dimension of Earth's dynamic systems—its energy, air, water, weather, climate, tectonics, landforms, rocks, soils, plants, ecosystems, and biomes. Understanding human-Earth relations is part of the challenge of physical geography—to create a holistic (or complete) view of the planet and its inhabitants.

This edition builds on the widespread success of the first five editions in the United States, Canada, and elsewhere, as well as its companion texts, *Geosystems*, now in its seventh edition, and *Geosystems, Canadian Edition*, second edition. Students and teachers continue to communicate their appreciation of the up-to-date content, systems organization, scientific accuracy, clarity of the summary and review sections, and overall relevancy. Welcome to physical geography!

New to the Sixth Edition

There is not room in this Preface to list all the changes and additions to *Elemental Geosystems*, 6/e. However, here is a brief sample. Begin by examining all 17 chapter-opening and 4 part-opening photos and images and read their captions—all but two are new. You see the devastation of an EF-5 tornado that swept away Greensburg, Kansas, with comparison photos made one year after the disaster; the midnight Sun in the Arctic; aerial and close-up views of a wind farm; dramatic weather systems and clouds passing over a volcano in the North Atlantic; Hurricanes Katrina (2005), Gustav, and Ike (both 2008); a 2003 and 2007 photo comparison of the severe water-level drop of Lake Mead at Hoover Dam; a palm-oak hammock of trees in Florida; aerial view of the Green River landscape in Wyoming; detail of red sandstone and lichen in the desert; close-up of Niagara River plunging over the escarpment; Monument Valley scene; beach scene in Mexico; a glacier as it calves into the sea; a Kansas prairie bison herd; among others. These are content-specific and begin the learning process from the start of the chapter. Throughout the text there are hundreds of new photos and satellite images, many used in compound figures with arts and maps.

2007 was the second warmest in the climate record since 1880. 2005 broke many records climatologically as the warmest year, 2006 was fifth warmest, as shown in the updated global average temperature graph in Figure 7.28a and accompanying text. Chapter 1 discusses the extinction of 67% of all harlequin frog species as an example of applied systems analysis in this time of rising temperatures and climate change impacts.

In the Arctic region, there is coverage of the new records for land and water temperatures and sea-ice melt that were set over the past five years. A record low sea-ice extent occurred in 2007; 2008 sea-ice melt was nearly as bad—featured in News Report 3.1 and Chapter 14 in Figure 14.29c and d and text. The rate of ice melt in the high latitudes of both hemispheres is accelerating, an important component to a rising sea level.

Chapter 5 includes an exclusive discussion of derechos—straight-line (non-tornado) winds that annually cause so much damage. Also, the weather coverage includes record-breaking tornadoes in intensity and numbers. Figure 5.40 presents a world tropical cyclone map with note of bizarre, first-ever hurricanes that hit Spain and Brazil, the mid-Pacific and Persian Gulf. A rare late-season category 4 Hurricane Paloma struck the Caribbean in November 2008. The 2005 Atlantic tropical storm season set many records and was followed by the 2008 hurricanes that continued the assault of the Gulf Coast, Florida, and the Atlantic Coast. Chapter 13 illustrates with photos and satellite images the impact on coastal regions and barrier islands from these storms, such as the loss of the Chandeleur Islands in the Gulf of Mexico (Figure 13.15) and Galveston Island and Bolivar Peninsula, Texas (Figure 13.16).

New content and photos enhance every chapter in PART III covering "Earth's Changing Landscape Systems." The new Ocean Drilling Program research ship is profiled. In Figure 9.33, a new Pu'u O'o crater aerial photo at Kīlauea volcano, Hawai'i, from summer 2007 is compared to an aerial photo at roughly the same location and angle in 1999. Pu'u O'o had huge collapses on the west side in 2007, then filling with lava, then collapsing again. In Chapter 9, updated discussions cover the 2008 Chengdu, China, earthquake.

Chapter 10 features new headings "Salt-Crystal Growth (Salt weathering)," and "Dissolution of Carbonates," replacing

"Crystalization" and "Carbonation and Solution," respectively. Included is a newly redone world karst map and new weathering photo examples.

Updated News Report 11.3 "What Once Was Bayou Lafourche," is a poignant story of a hurricane levee compound that is slightly west of where Hurricane Katrina made landfall and that experienced a direct hit by Hurricane Gustav in 2008. Expanded coverage of the Mississippi Delta and photos of Bonnet Carré spillway in New Orleans and river flood avoidance structures are in Focus Study 11.1. *Elemental Geosystems* takes the approach that the 2005 storm was named Katrina but the disaster was of human origins due to poorly conceived planning and flawed engineering. A new aerial photo from August 30, 2005, shows the Inner Harbor Navigation Canal levee break. A new section "Engineering Failures in New Orleans, 2005," with satellite before-and-after *Landsat* images of the city with a detailed locator map illustrates the catastrophe.

In Chapter 12, Focus Study 12.1 has the latest information on the troubled Colorado River system, a subject discussed in all my books since the first edition. This continuing focus study accurately forecasted the present crisis in advance and covers this new predicament of record-low flows since 2002. Note the tie-in to the Chapter 6 opening photo comparison showing Lake Mead levels at Hoover Dam. In Chapter 12, a new sediment-accumulation-model art is added to the deflation-model art for desert pavement formation, representing both hypotheses.

Focus Study 14.1 has expanded coverage of Dome C, Antarctica, and the ice-core analysis that now stretches the climate record back to 800,000 years ago, confirming present levels of greenhouse gases in the atmosphere as the highest in the entire record. News Report 14.3 is new coverage of the increase in meltponds in Greenland and the climatic causes and implications. New mass balance data through 2006 is the latest available along with comparitive photos of the ice losses.

In Chapter 15 the latest soil science in the 2006 edition of *Keys to Soil Taxonomy*, 10/e, is referenced and the new "M" designation for anthropogenic soil structures is described. In Chapter 16, USDA 1990 map of Hardiness Zones is compared to a 2006 proposed revised map, along with a differences map, as climate change takes its spatial toll on this key planting guide with zones migrating northward. Also, we track the future climate shifts in the Midwest by tracking the climate of Illinois from the present to 2090. The endangered status of polar bears and the USGS forecast of losing two-thirds of their population by 2050 is covered as plant and animal extinctions increase.

The text culminates with Chapter 17, "Earth and the Human Denominator," a unique capstone chapter that summarizes physical geography as an important discipline in understanding Earth's present status and possible future. An updated image of the International Space Station as it dramatically expands is Figure 17.1. All population references are updated through mid-2008. *Elemental Geosystems* ends with a Casinni-Huygens satellite image of Saturn and its rings backlit by the Sun from a perspective not seen from Earth. A small dot is visible between the rings, it's Earth 1.23 billion km distant—such a perspective! This chapter is sure to stimulate further thought and discussion, dealing as it does with the most profound issue of our time, Earth's stewardship.

Systems Organization Makes *Elemental Geosystems* Flow

Each section of this book is organized around the flow of energy, materials, and information. *Elemental Geosystems* presents subjects in the same sequence in which they occur in nature. In this way, you and your teacher logically progress through topics that unfold according to the flow of individual systems, or in accord with time and the flow of events. See Figure 1.8 in the text for an illustration of this systems organization.

For flexibility, *Elemental Geosystems* is divided into four parts, each containing chapters that link content in logical groupings. The diagram in Figure 1.9 illustrates our part structure. A quick check of the Table of Contents and this illustration shows you the order of chapters within these four parts, which are summarized with photos on the back cover.

Elemental Geosystems Is a Text That Teaches

Three heading levels are used throughout the text and precise topic sentences begin each paragraph to help you outline and review material. **Boldface** words are defined where they first appear in the text. The Glossary collects these terms and concepts alphabetically, with a chapter-number reference. The complete Glossary appears on our *Student Learning Center* that accompanies this text at www.mygeoscienceplace.com. *Italics* are used in the text to emphasize other words and phrases of importance. Every figure has a title that summarizes the caption. Also, in the introduction to each chapter, a feature "**In this chapter:**" gives you an overview.

An important continuing feature is a list of *Key Learning Concepts* that opens each chapter, stating what you should be able to do upon completing the chapter. These objectives are keyed to the main headings in the chapter. Note at the beginning of each chapter, after the key learning concepts list, is a sample of "*Student Learning Center Tools*" described later in this Preface.

At the end of each chapter is a unique *Summary and Review* section that corresponds to the *Key Learning Concepts*, which includes a narrative review defining the bold-faced terms, a key terms list with page numbers, and specific review questions for that concept. You can conveniently review each concept, test your understanding with review questions, and check key terms in the glossary.

A *Network and Critical Thinking Tools* section ends each chapter, challenging you to take the next step with information from the chapter, in many cases leading you to the *Student Learning Center*. The key learning concepts help you determine what you want to learn, the text guides you in developing information and more questions, the summary and review assesses what you have learned and what more you might want to know about the subject, and this critical thinking section provokes action and application.

Focus Study essays, some completely revised and several new to this edition, provide additional explanation of key topics. News Reports relate topics of special interest. See the Table of Contents to sample titles of these important and timely features.

We live on a planet served by the Internet and its World Wide Web, a resource that weaves threads of information from around the globe into a vast fabric. The fact that we have Internet access into almost all the compartments aboard Spaceship Earth is clearly evident in *Elemental Geosystems*. You will find more than 200 URLs (Internet addresses) in the body of the text (printed in blue and boldface) and find an additional 300 links on the *Student Learning Center*. Given the fluid nature of the Internet, URLs were rechecked at press time for accuracy. If some URLs have changed since publication, you can most likely find the new location using elements of the old address. These Internet links begin with a revised Table 1.1 presenting the URLs for major geography organizations.

The *Elemental Geosystems* Learning/Teaching Package

The sixth edition provides a complete physical geography program for you and your teacher.

For You, the Student:

- The *Elemental Geosystems* **Student Learning Center** provides online resources for each chapter. Click on the book cover at **www.mygeoscienceplace.com** for access. Once you are in our Learning Center, bookmark the URL address, find reviewing tools, self-tests and quizzes, critical thinking items, 67 figure and content animations, 4 photo galleries and 9 movie and video clips, and 18 satellite loops. Among many learning assets are a satellite feature of the entire 2005 Atlantic tropical storm season and sea-surface temperatures in a dramatic 6-month movie, and an animated map that portrays global temperature data from 1891–2006 that you view as a movie. Also, *Career Link* essays feature geographers and other scientists in a variety of professional fields practicing their spatial analysis craft to give you an idea of possible career opportunities. For teachers of previous book editions, the Learning Center includes all the contents of the CD-ROM that accompanied those texts.
- Included with the Student Learning Center is ***myeBook***, an integrated, full-featured electronic version of the book that allows you to add personal notes and highlights to the pages and see any comments made by the instructor. The search function allows for full-text searching of book content.
- *Student Study Guide*, Sixth Edition (ISBN: 0321595785), by Robert Christopherson and Charlie Thomsen of American River College. The *Study Guide* includes additional learning objectives, a complete chapter

outline, critical thinking exercises, problems, and short essay work using actual figures from the text, as well as a self-test with answer key included.
- *Encounter Geosystems* (ISBN: 0321636996), by Charlie Thomsen of American River College and Robert Christopherson. This exciting new publication contains interactive explorations keyed to each chapter of the textbook. You are directed to special files using *Google Earth*™ Mapping Service where you literally fly over landscapes, see layers of information on maps, and measure and analyze data.

For You the Teacher:

Elemental Geosystems is designed to give you flexibility in presenting your course. The text is comprehensive in that it is true to each scientific discipline from which it draws subject matter. This diversity is a strength of physical geography, yet makes it difficult to cover the entire book in a school term. *Elemental Geosystems* is organized to help you customize your presentation. You should feel free to use the text based on your specialty or emphasis, rearranging parts and chapters as desired. The four-part structure of chapters, systems organization within each chapter, and focus study and news report features, all assist you in sampling some chapters while covering others in greater depth. The following materials are available to assist you—have a great class!

- ***Instructor Resource Center on DVD*** (ISBN: 0321595661), contains everything you need for efficient course preparation. Find all your digital resources in one, well-organized, easy-to-access place. Included on this DVD are:

  Figures—JPEGs of all illustrations and, new to this edition, all the photos from the book.
  Animations—All the animations ready to use in your classroom presentations.
  PowerPoint™—Pre-authored slides outline the concepts of each chapter with embedded art for use in the classroom, or to customize with other items for your own presentation needs.
  TestGen—TestGen software featuring a thorough and revised *Test Bank*, with questions and answers. Edit existing questions or add your own.
  Electronic files—The complete *Instructor Resource Manual* and *Test Bank* in one easily accessible location.

- ***Instructor Resource Manual***, Sixth Edition (by Robert Christopherson and Charlie Thomsen). The *Instructor Resource Manual*, intended as a resource for both new and experienced teachers, includes lecture outlines and key terms, additional source materials, teaching tips, complete annotation of chapter review questions, and a list of overhead transparencies. Digitial files are available on the *Instructor Resource Center on DVD*.
- ***Test Bank*** by Robert Christopherson and Charlie Thomsen. This collaboration has produced the most extensive and fully revised *Test Bank* available in

physical geography. This test bank employs **TestGen**—a computerized test generator that lets you view and edit *Test Bank* questions, transfer questions to tests, and print customized formats. Visit the Pearson Prentice Hall Instructor Resource Center catalog at **www.pearsonhighered.com** to download in MAC or PC format. Files also available on the *Instructor Resource Center on DVD*.

- *Overhead Transparencies* (ISBN: 0321595777) includes approximately 200 illustrations from the text on 200 transparencies—all enlarged for excellent classroom visibility.
- **Online Course Management Systems**: Pearson Prentice Hall offers content specific tools for *Elemental Geosystems*, Sixth Edition, ready for easy importation into the *BlackBoard*, *WebCT*, and *CourseCompass* course management system platforms. Using these you can easily post your syllabus, communicate with students on-line or off-line, administer quizzes, record student results, and track their progress. Please call your local Pearson Prentice Hall representative for more details.
- *Applied Physical Geography—Geosystems in the Laboratory*, Seventh Edition (ISBN: 0136011810) by Robert Christopherson and Charlie Thomsen of American River College. The new seventh edition is the result of a careful revision. Twenty lab exercises, divided into logical sections, allow flexibility in presentation. Each exercise comes with a list of learning concepts. We integrate *Google Earth*™ Mapping Service links to KMZ files into the exercises that allow students to actually experience and manipulate topographic maps in DEM relief as they work problems. Our manual comes with its own complete glossary, stereolenses for viewing photo stereopairs and stereomaps in the manual. A complete **Solutions and Answers Manual**, 7/e, is available to teachers (ISBN: 0136011802).

Acknowledgments

As in all past editions, I give recognition to my family—for all they have endured since the first edition of *Elemental Geosystems'* goals—especially the next generation: Chavon, Bryce, Payton, Brock, Trevor, Blake, Chase, Téyenna, and our newest Cade. When I look into our grandchildren's faces it tells me why we need to work toward a sustainable future, one for the children.

I give special gratitude to all the students over 30 years at American River College for defining the importance of Earth's future, for their questions, and their enthusiasm. My thanks go to the many authors and scientists who published research, articles and books that enrich my work. To all the colleagues who served as reviewers on one or more editions of each book, or who offered helpful suggestions in conversations at our national and regional geography meetings. And, although unnamed here, to all the correspondence received from students and teachers from across the globe who shared with me through e-mail, FAX, and phone—a continuing appreciated dialogue. I am grateful to all for the generosity of ideas and sacrifice of time. Here is a master list of all our reviewers of the *Geosystems* books, including those who reviewed the student animations.

Ted J. Alsop, *Utah State University*
Ward Barrett, *University of Minnesota*
Steve Bass, *Mesa Community College*
Stefan Becker, *University of Wisconsin–Oshkosh*
Daniel Bedford, *Weber State University*
David Berner, *Normandale Community College*
Franco Biondi, *University of Nevada, Reno*
Peter D. Blanken, *University of Colorado, Boulder*
Patricia Boudinot, *George Mason University*
David R. Butler, *Southwest Texas State University*
Mary-Louise Byrne, *Wilfred Laurier University*
Ian A. Campbell, *University of Alberta–Edmonton*
Randall S. Cerveny, *Arizona State University*
Fred Chambers, *University of Colorado, Boulder*
Muncel Chang, *Butte College* Emeritus
Andrew Comrie, *University of Arizona*
C. Mark Cowell, *Indiana State University*
Richard A. Crooker, *Kutztown University*
Armando M. da Silva, *Towson State University*
Dirk H. de Boer, *University of Saskatchewan*
Dennis Dahms, *University of Northern Iowa*
Mario P. Delisio, *Boise State University*
Joseph R. Desloges, *University of Toronto*
Lee R. Dexter, *Northern Arizona University*
Don W. Duckson, Jr., *Frostburg State University*
Christopher H. Exline, *University of Nevada–Reno*
Michael M. Folsom, *Eastern Washington University*
Mark Francek, *Central Michigan University*
Glen Fredlund, *University of Wisconsin–Milwaukee*
William Garcia, *University of North Carolina–Charlotte*
Doug Goodin, *Kansas State University*
David E. Greenland, *University of North Carolina–Chapel Hill*
Duane Griffin, *Bucknell University*
John W. Hall, *Louisiana State University–Shreveport*
Barry N. Haack, *George Mason University*
Roy Haggerty, *Oregon State University*
Vern Harnapp, *University of Akron*
John Harrington, *Kansas State University*
Blake Harrison, *Southern Connecticut University*
Jason "Jake" Haugland, *University of Colorado, Boulder*
Gail Hobbs, *Pierce College*
Thomas W. Holder, *University of Georgia*
David A. Howarth, *University of Louisville*
Patricia G. Humbertson, *Youngstown State University*
David W. Icenogle, *Auburn University*
Philip L. Jackson, *Oregon State University*
J. Peter Johnson, Jr., *Carleton University*
Guy King, *California State University–Chico*
Ronald G. Knapp, *SUNY–The College at New Paltz*
Peter W. Knightes, *Central Texas College*
Thomas Krabacher, *California State University–Sacramento*
Richard Kurzhals, *Grand Rapids Junior College*
Hsiang-te Kung, *University of Memphis*
Steve Ladochy, *California State University, Los Angeles*
Charles W. Lafon, *Texas A & M University*

Paul R. Larson, *Southern Utah University*
Robert D. Larson, *Southwest Texas State University*
Elena Lioubimtseva, *Grand Valley State University*
Joyce Lundberg, *Carleton University*
W. Andrew Marcus, *Montana State University*
Nadine Martin, *University of Arizona*
Elliot G. McIntire, *California State University, Northridge*
Norman Meek, *California State University, San Bernardino*
Leigh W. Mintz, *California State University–Hayward, Emeritus*
Sherry Morea-Oaks, Boulder, CO
Patrick Moss, *University of Wisconsin, Madison*
Lawrence C. Nkemdirim, *University of Calgary*
Andrew Oliphant, *San Francisco State University*
John E. Oliver, *Indiana State University*
Bradley M. Opdyke, *Michigan State University*
Richard L. Orndorff, *University of Nevada, Las Vegas*
Patrick Pease, *East Carolina University*
James Penn, *Southeastern Louisiana University*
Greg Pope, *Montclair State University*
Robin J. Rapai, *University of North Dakota*
Philip D. Renner, *American River College* Emeritus
William C. Rense, *Shippensburg University*
Leslie Rigg, Northern Illinois University
Dar Roberts, *University of California–Santa Barbara*
Wolf Roder, *University of Cincinnati*
Robert Rohli, *Louisiana State University*
Bill Russell, *L.A. Pierce College*
Dorothy Sack, *Ohio University*
Randall Schaetzl, *Michigan State University*
Glenn R. Sebastian, *University of South Alabama*
Daniel A. Selwa, *U.S.C. Coastal Carolina College*
Peter Siska, *Austin Peay State University*
Lee Slater, *Rutgers University*
Thomas W. Small, *Frostburg State University*
Daniel J. Smith, *University of Victoria*
Stephen J. Stadler, *Oklahoma State University*
Michael Talbot, *Pima Community College*
Paul E. Todhunter, *University of North Dakota*
Susanna T.Y. Tong, *University of Cincinnati*
Liem Tran, *Florida Atlantic University*
Suzanne Traub-Metlay, *Front Range Community College*
Alice V. Turkington, *University of Kentucky*
David Weide, *University of Nevada–Las Vegas*
Thomas B. Williams, *Western Illinois University*
Brenton M. Yarnal, *Pennsylvania State University*
Catherine H. Yansa, *Michigan State University*
Stephen R. Yool, *University of Arizona*
Susie Zeigler-Svatek, *University of Minnesota*

I extend my continuing gratitude to the editorial, production, and sales staff of Pearson Prentice Hall. Thanks to Pearson Science President Paul Corey for his leadership. Thanks to former editor Dan Kaveney and new editor Christian Botting for their support. To work with such professionals and feel a friendship is appreciated. And thanks to all the staff for allowing us to participate in the entire publishing process. Warm recognition to Charlie Thomsen for his creative collaboration on the lab manual and ancillaries. And thanks to Project Manager Tim Flem who maintains organizational control with such skill, for all his coordination effort and help; and to Pearson support staff Connie Long and Shari Toron. To Marketing Manager Amy Porubsky and the many sales representatives who spend months in the field communicating the *Elemental Geosystems* approach, as always, thanks and safe travels to them—may the wind be at their backs!

My thanks to our publishing partner Production Editor Cindy Miller of GGS Higher Education Resources for such friendship, managerial ability, and sustaining care. And to Suganya Karuppasamy of GGS for her skills and oversight of pages, intense precision, and dedicated work. To Pam Rockwell and her skills as a copyeditor I give an author's thanks. Thanks to the art and design team at Pearson Prentice Hall for our cover design using Bobbé's photo and for their beautiful text design. And, to David Fierstein's guidance and artistic skill revising many of the arts in this edition.

My continuing collaboration with my photographer, expedition partner, and production assistant, Bobbé Christopherson, is at the heart of my life's work. You see 379 of her photographs in the sixth edition, including the cover photo and the title page photo, all carefully made to illustrate physical geography principles. Please visit the four photo galleries at the Student Learning Center and learn more from her ability with a camera. She processes satellite imagery and all photos, completes many of the figure layouts, obtains permissions, and assists in keeping me organized. Bobbé's insightful photographs contribute greatly to each edition. And, she continues as my best friend, soul mate, and companion.

Physical geography teaches us a holistic view of the intricate supporting web that is Earth's environment and our place in it. Dramatic changes that demand our understanding are occurring in many human-Earth relations, as we alter physical, chemical, and biological systems. Our attention to climate change science is in recognition to the linkages between all planetary systems and the impacts we are experiencing. The threatened polar bears become our early warning system, our miner's Canary. All things considered, this is a critical time for you to be enrolled in a physical geography course! The best to you in your studies—and ***carpe diem!***

Robert W. Christopherson
P. O. Box 128
Lincoln, California 95648–0128
E-mail: bobobbe@aol.com
Web site: **www.mygeoscienceplace.com**

Greensburg, Kansas, essentially leveled by an EF-5 tornado on the Enhanced Fujita Scale (see Chapter 5 for more on this). Winds exceeded 330 kmph (205 mph) in the 2.7-km (1.7-mi) wide funnel that locked onto the ground along a 35-km (22-mi) track, visible on the *IKONOS* image. The May 4, 2007, twister destroyed 95% of all public, commercial, and private structures in this town.

2008 photos one-year later, inset clockwise from upper left: aerial of Greensburg from the east; surviving trees, stripped and pruned by the tornado, slowly resprout; block after block of vacant lots; lots still for sale, steps to nowhere, empty foundations; FEMA-ville still in use; the 5.4.7 Art Center built to LEED "green" standards by University of Kansas architecture students (www.547artscenter.org). Greensburg is in slow recovery (www.greensburggreentown.org). The town is attempting to rebuild "green," meaning with highest energy efficiency and using renewable energy sources. *[Image by satellite IKONOS May 14, courtesy of MJ Harden/GeoEye. All rights reserved. All photos by Bobbé Christopherson.]*

Essentials of Geography

KEY LEARNING CONCEPTS

After reading the chapter, you should be able to:

- **Define** geography and physical geography in particular.

- **Describe** systems analysis and open and closed systems and **relate** these concepts to Earth systems.

- **Explain** Earth's reference grid: latitude and longitude and latitudinal geographic zones and time.

- **Define** cartography and mapping basics: map scale and map projections.

- **Describe** remote sensing and **explain** geographic information systems (GIS) as tools used in geographic analysis.

STUDENT LEARNING CENTER TOOLS

The Elemental Geosystems Student Learning Center provides on-line resources for this chapter by clicking on the book cover at **www.mygeoscienceplace.com**. Once you are in our Learning Center, bookmark the URL address. For this chapter find reviewing tools, eBook links, self-tests and quizzes, critical thinking items, and other helpful features. Some highlights:

▶ Career Link about an astronaut with 1272 hours in orbit who made two of the images in the chapter.

▶ "Finding Polaris" illustration to help you find the North Star.

▶ Textbook reference maps and Visual Glossary.

▶ More than 30 additional URLs along with Quick Links to all the URLs in the chapter.

Welcome to the sixth edition of *Elemental Geosystems*! Much has happened since the last edition, including the International Polar Year (IPY, http://www.ipy.org/), a two-year global science research effort in the high latitudes that ended March 2009. This global interdisciplinary research, exploration, and discovery effort combined many scientific fields that are at the core of physical geography and this text.

The Intergovernmental Panel on Climate Change (IPCC, http://www.ipcc.ch/) completed its Fourth Assessment Report and published four volumes of supporting science on human-forced climate change. In every chapter, *Elemental Geosystems* presents updated science and information to enhance your understanding of changing Earth systems. The IPCC was corecipient of the 2007 Nobel Peace Prize for their 20 years of scientific exploration of this global issue and its myriad impacts. The highest global air temperatures for meteorological stations in the instrumental or proxy record were registered in 2005; 2007 tied for second highest.

Headlines crackle with news of an earthquake in China; an unseasonal typhoon in the Philippines and in the Caribbean in 2008; record floods in one region as record droughts plague others; melting and cracking of the Petermann glacier, the largest on Greenland's north coast; or the politics of hazard forecasting and perception, among many stories—all stimulate the public's interest in physical geography. Scientists and governments conduct research to understand the worldwide impact of global changes in weather, climate, water resources, and the landscape.

The UN Environment Programme (UNEP, http://www.unep.org/) reported data in 2005 from 15 of the largest financial institutions that estimated weather-related damage losses (drought, floods, hail, tornadoes, derechos, tropical systems, storm surges, blizzard and ice storms, and wildfires) will exceed $1 trillion by A.D. 2040 (adjusted to current dollars)—up from $210 billion in 2005 (Hurricane Katrina alone produced $125 billion in losses). The 2005 tropical storm/hurricane season when added to 2008 easily exceeds $250 billion in damages. Global climate change is driving this increasing total, at a pace that is accelerating.

Society is facing challenging times as we in physical geography work to understand how Earth's systems operate and then forecast changes in these systems to determine what future environments we may expect. The scope of this challenge was outlined by scientist Jack Williams:

> By the end of the 21st century, large portions of the Earth's surface may experience climates not found at present, and some 20th-century climates may disappear. . . . Novel climates are projected to develop primarily in the tropics and subtropics.*

Physical geography is the one course you can take that explains the operations of Earth systems from a spatial perspective and gives you the tools to understand what is happening. Related to these increasing hazards, former Canadian Minister of the Environment David Anderson stated:

> Current preoccupation is with terrorism, but in the long term climate change will outweigh terrorism as an issue for the international community. Terrorism will come and go; it has in the past . . . and it's very important. But climate change is going to make some very fundamental changes to human existence on the planet.†

Why do all of these conditions occur? How are these events different from past experience? Why do the environment and rates of global change vary from the equator to the midlatitudes, between deserts and polar regions? How does solar energy influence the distribution of trees, soils, climates, and lifestyles? How does energy produce the patterns of wind, weather, and ocean currents? How do natural systems affect human populations? Why did the countries of the world join in the IPY to study the polar regions? In this book, we explore those questions and more through geography's unique perspective. Welcome to an exploration of physical geography!

We live in an extraordinary era of **Earth systems science**. This science contributes to our emerging view of Earth as a complete entity—an interacting set of physical, chemical, and biological systems that produce a whole Earth. Physical geography is at the heart of Earth systems science as we answer the *spatial* questions concerning Earth's physical systems and their interaction with living things. The need for this spatial, integrative approach is apparent when we consider the need to forecast events such as the December 2004 magnitude 9.3 earthquake that struck Sumatra, triggering a massive tsunami (seismic sea wave) that radiated across the Indian Ocean, killing more than 170,000 people (Figure 1.1), or the October 2005 quake in the Kashmir region of Pakistan that killed 83,000 people, or the Chengdu quake in August 2008 that took 70,000 lives.

In this chapter: Our study of *Elemental Geosystems*—Earth systems—begins in this chapter with a look at the science of physical geography and the geographic tools we use. Physical geography is key to studying entire Earth systems because of its integrative spatial approach.

Physical geographers analyze systems to study the environment. Therefore, we discuss systems and the feedback mechanisms that influence system operations. We then consider location, a key theme of geographic inquiry—the latitude, longitude, and time coordinates that inscribe Earth's surface—and the new technologies in use to measure them. The study of longitude and a universal time system provide us with interesting insights into geography. Next, we examine maps as critical tools that geographers use to portray physical and cultural information. This chapter concludes with an overview of the technology that is adding exciting dimensions to geography: remote sensing from space and computer-based geographic information systems (GIS).

---

*Jack Williams et al., "Projected distributions of novel and disappearing climates by A.D. 2100," *Proceedings of the National Academy of Sciences*, April 3, 2007.

†Environmental News Network, February 6, 2004, http://www.enn.com/.

Location

Absolute and relative location on Earth. Location answers the question *"Where?"* The specific planetary address of a location. This freeway sign is posted on Interstate 5 in Oregon telling drivers their position of Earth.

Region

Areas having uniform characteristics; how they form and change; their relation to other regions. The American Midwest is a distinct region of grasslands and prairie, corn, soybean, and feed crops, and huge regional feedlots and animal processing factories. Vast distances, diminishing precipitation, and lowering groundwater conditions emerge as issues.

Human–Earth Relationships

Humans and the environment: resource exploitation, hazard perception, and environmental pollution and modification. A coal-fired power plant at Barentsburg on Spitsbergen Island in the Arctic Ocean adds greenhouse gases and pollution to the atmosphere. A wind farm of comparable mega-watt output can produce pollution-free energy that is based on abundant winds.

Geographic Science

Place

Tangible and intangible living and nonliving characteristics that make each place unique. No two places on Earth are exactly alike. The Bay of Fundy in the Canadian Maritimes, separates New Brunswick and Nova Scotia grading into the Bay of Maine toward the south. The area has a rich history, farm and pasturelands, and the highest average tides in the world – up to 17 m (55.7 ft) in height. The tidal bore current in the photo is moving at a rapid 12 kmph (7.5 mph) speed!

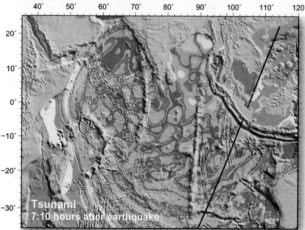

Movement

Communication, movement, circulation, migration, and diffusion across Earth's surface. Global interdependence links all regions and places. The radiating waves of energy in an Indian Ocean tsunami generated by a magnitude 9.3 earthquake off the coast of Sumatra, Indonesia, on December 26, 2004. Damage and loss of life exceeding 170,000 people, resulted when waves hit distant shores.

FIGURE 1.1 Five themes of geographic science.
Five fundamental themes in geography defined—location, place, movement, regions, and human–Earth relationships. [Regions, Human–Earth relationships, Place photos by Bobbé Christopherson; Location photo by author; and Movement GFO-Satellite image courtesy of NOAA, Laboratory for Satellite Altimetry.]

The Science of Geography

Geography (from *geo*, "Earth," and *graphein*, "to write") is the science that studies the relationships among natural systems, geographic areas, society, cultural activities, and the interdependence of all of these *over space*. The term **spatial** refers to the nature and character of physical space, its measurement, and the distribution of things within it.

For example, think of your own route to the classroom or library today or the way you get to work and how you used your knowledge of street patterns, traffic trouble spots, one-way streets, parking spaces, or bike rack locations to minimize walking distance. All these are spatial considerations. Humans are spatial actors. We profoundly influence vast areas because of our mobility and access to energy and technology. In turn, Earth's systems influence our activities in a most obvious way—these systems give us life.

To help in this definition, we separate geographic science into five spatial themes: **location, region, human–Earth relationships, movement**, and **place**, each illustrated and defined in Figure 1.1 (p. 3). *Geosystems* draws on each theme. For a listing of some important geography organizations and their URLs, see Table 1.1.

Geographic Analysis

Within these five geographic themes, a *method* rather than a specific body of knowledge governs geography. The method is **spatial analysis**. Using this method, geography synthesizes (brings together) topics from many fields, integrating information to form a whole-Earth concept. Geographers view phenomena in spaces and areas having distinctive characteristics. The language of geography reflects this spatial view: *space, territory, zone, pattern, distribution, place, location, region, sphere, province,* and *distance*. Geographers analyze the differences and similarities among places and locations.

Process, a set of actions or mechanisms that operate in some special order, is central to geographic analysis. As examples in *Elemental Geosystems*, numerous processes are involved in Earth's vast water–atmosphere–weather system, or in continental crust movements and earthquake occurrences, or in ecosystem functions. Geographers use spatial analysis to examine how Earth's processes interact over space or area.

Therefore, **physical geography** is *the spatial analysis of all the physical elements and processes that make up the environment: energy, air, water, weather, climate, landforms, soils, animals, plants, microorganisms, and Earth itself*. We add to this the oldest theme in the geographic tradition, that of human activity. As a science, physical geographers employ the **scientific method**. Focus Study 1.1 explains this important process of science.

The Geographic Continuum

Geography is eclectic, integrating a wide range of subject matter from diverse fields; virtually any subject can be examined geographically. Figure 1.2 (p. 7) shows a continuous distribution—a continuum—along which the content of geography is arranged. Disciplines in the physical and life sciences are at one end, and those in the human and cultural sciences are at the other. As the figure shows, geography draws from many subject areas.

The continuum in Figure 1.2 reflects a basic duality, or split, within geography—*physical geography* versus *human/cultural geography*. Society parallels this duality. Humans sometimes think of themselves as exempt from physical Earth processes—like actors not paying attention to their stage, props, and lighting. We all depend on Earth's systems to provide oxygen, water, nutrients, energy, and materials to support life. The growing complexity of the human–Earth relation requires that we shift our study of geographic processes, and perhaps our link to Earth systems, toward the center of the continuum in Figure 1.2 to attain a more *holistic*, or balanced, perspective—such is the thrust of *Elemental Geosystems*.

Table 1.1

A Few Geography Organizations	URL Addresses
American Geographical Society	http://www.amergeog.org/
Association of American Geographers*	http://www.aag.org/
National Council for Geographic Education	http://www.ncge.org/
National Geographic Society	http://www.nationalgeographic.com/
Canadian Association of Geographers	http://www.cag-acg.ca/en/
Royal Canadian Geographical Society	http://www.rcgs.org/
Institute of Australian Geographers	http://www.iag.org.au/
Australian Geography Teachers Association	http://www.agta.asn.au/
European Geography Association	http://www.egea.eu/
Royal Geographical Society, Institute of British Geographers	http://www.rgs.org/
Complete global listing of geography organizations	http://www.geoggeol.fau.edu/Links/GeographicalSites.htm

* Includes nine regional divisions: East Lakes, Great Plains/Rocky Mountain, Middle Atlantic, Middle States, New England–St. Lawrence Valley, Pacific Coast Regional, Southeastern, Southwestern, West Lakes.

The Scientific Method

The term *scientific method* may have an aura of complexity that it should not. The scientific method is simply the application of common sense in an organized and objective manner. A scientist observes, makes a general statement to summarize the observations, formulates a hypothesis, conducts experiments to test the hypothesis, and develops a theory and governing scientific laws. Sir Isaac Newton (1642–1727) developed this method of discovering the patterns of nature, although the term *scientific method* was applied later.

Scientists are curious about nature and appreciate the challenge of problem solving. Society depends on the discoveries of science to solve mysteries, find solutions, and foster progress. *Complexity* dominates nature, making several outcomes possible as a system operates. Science serves an important function in reducing such uncertainty. Yet, the more knowledge we have, the more the uncertainty and awareness of other possible scenarios (outcomes and events) increase. This in turn demands more precise and aggressive science.

A problem arises when scientific uncertainty fuels an antiscience viewpoint in society. However, defining uncertainty is a by-product of scientific discovery. As we become aware of the scientific principles of complexity and chaos in natural and human-made systems, the need becomes stronger for critical thinking and the scientific method in all aspects of life.

Steps in the Scientific Method

Follow the scientific method illustration in Figure 1.1.1 as you read. The scientific method begins with our perception of the real world and a determination of what we know, what we want to know, and the many unanswered questions that exist. Scientists who study the physical environment turn to nature for clues that they can observe and measure. They figure out what data are needed and begin to collect those data. Then, they analyze these observations and data to identify patterns that may be present. This search for patterns requires *inductive reasoning*, or the process of drawing generalizations from specific facts. This step is important in modern Earth systems science, in which the goal is to understand *a whole functioning Earth*, rather than isolated, small compartments of information. Such understanding allows the scientist to construct models that simulate general operations of Earth systems.

If patterns are discovered, the researcher may formulate a *hypothesis*—a formal generalization of a principle. Examples include the planetesimal hypothesis, nuclear-winter hypothesis, and moisture-benefits-from-hurricanes hypothesis. Further observations are related to the general principles established by the hypothesis. Additional data gathering may support or refute the hypothesis, or predictions made according to it may prove accurate or inaccurate. All these findings provide feedback for adjusting data collection and model building and for refining the hypothesis statement. Verification of the hypothesis after exhaustive testing may lead to its elevation to the status of a *theory*.

A theory is constructed on the basis of several extensively tested hypotheses. Theories represent truly broad general principles—unifying concepts that tie together the laws that govern nature (for example, the theory of relativity, theory of evolution, atomic theory, Big Bang theory, stratospheric ozone depletion theory, or plate tectonics theory). A theory is a powerful device with which to understand both the order and chaos (disorder) in nature. Using a theory allows predictions to be made about things not yet known, the effects of which can be tested and verified or disproved. The value of a theory is the continued observation, testing, understanding, and pursuit of knowledge that the theory stimulates. A general theory reinforces our perception of the real world, acting as positive feedback.

Pure science does not make value judgments. Instead, pure science provides people and their institutions with objective information on which to base their own value judgments. Social and political judgments about the applications of science are increasingly critical as Earth's natural systems respond to the impact of modern civilization. Jane Lubchenco in her 1997 American Association for the Advancement of Science (AAAS) Presidential Address stated:

> Science alone does not hold the power to achieve the goal of greater sustainability, but scientific knowledge and wisdom are needed to help inform decisions that will enable society to move toward that end.

The Method and Applied Science

The growing awareness that human activity is producing global change places increasing pressure on scientists to participate in decision making. Numerous editorials in scientific journals have called for such an application of *applied science*. F. Sherwood Rowland and Mario Molina provide an example, discussed in Chapter 2, for they first proposed a hypothesis that certain human-made chemicals caused damaging reactions in the stratosphere. They hypothesized in 1974—along with Paul Crutzen, who also contributed to these discoveries—that chlorine-containing products such as chlorofluorocarbons (CFCs), commonly used as aerosol propellants, refrigerants, foaming agents, and cleaning solvents, were depleting our protective stratospheric ozone (O_3) layer.

Subsequently, surface, atmosphere, and satellite measurements confirmed the photochemical reactions, provided data to map the real losses that were occurring, and led to international treaties and agreements to ban the chemical culprits. In 1995, the Royal Swedish Academy of Sciences awarded these three scientists the Nobel Prize for their pioneering work. Such successful applied science is strengthening society's resolve and promoting treaties to ban problem chemicals and practices and introduce sustainable substitutes.

(continued)

Focus Study 1.1 *(continued)*

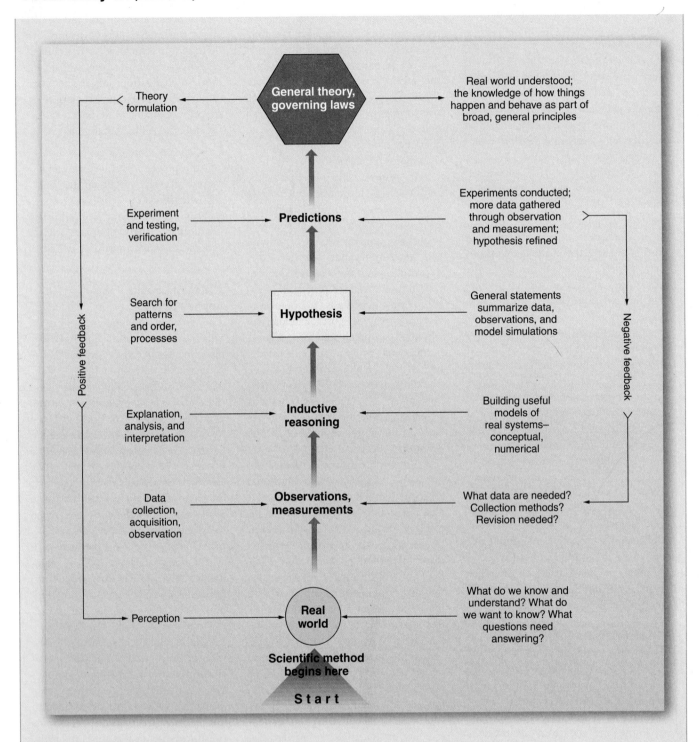

FIGURE 1.1.1 Scientific method flow chart.
The scientific method process: from perceptions to observations, reasoning, hypothesis, predictions, and possibly to general theory and natural laws.

Earth Systems Concepts

The word *system* pervades our lives daily: "Check the car's cooling system"; "How does the grading system work?"; "There is a weather system approaching." Systems of many kinds surround us. *Systems analysis* began with studies of energy and temperature (thermodynamics) in the nineteenth century and was fostered in engineering during World War II. Today, geographers use systems methodology as an analytical tool. In this textbook's 4 parts and 17 chapters, content is organized along logical flow paths consistent with nature and systems thinking.

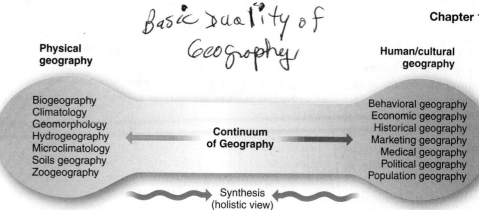

Basic Duality of Geography (handwritten)

Physical geography

Biogeography
Climatology
Geomorphology
Hydrogeography
Microclimatology
Soils geography
Zoogeography

Continuum of Geography

Human/cultural geography

Behavioral geography
Economic geography
Historical geography
Marketing geography
Medical geography
Political geography
Population geography

Synthesis (holistic view)

FIGURE 1.2 The content of geography.
Geography derives subject matter from many different sciences. The focus of this book is physical geography, but we also integrate some human and cultural components. Movement toward the middle of the continuum suggests a synthesis of Earth topics and human topics.

Systems Theory

Simply stated, a **system** is any ordered, interrelated set of things and their attributes, linked by flows of energy and matter, as distinct from the surrounding environment outside the system. The elements within a system may be arranged in a series or interwoven with one another. A system comprises any number of subsystems. Within Earth's systems, both matter and energy are stored and retrieved and energy is transformed from one type to another. (Remember: *Matter* is mass that assumes a physical shape and occupies space; *energy* is a capacity to change the motion of, or to do work on, matter.)

Open System Systems in nature are generally not self-contained: Inputs of energy and matter flow into the system, and outputs of energy and matter flow from the system. Such a system is an **open system**. Earth is an open system *in terms of energy*, for solar energy enters freely and heat energy goes back into space. Most natural systems are open in terms of energy. Figure 1.3 illustrates an open system and presents the inputs and outputs of an automobile as an example.

Closed System A system that is shut off from the surrounding environment so that it is self-contained is a **closed system**. Although such closed systems are rarely found in nature, Earth is essentially a closed system *in terms of physical matter and resources*—air, water, and material resources. The only exceptions are the slow escape of lightweight gases (such as hydrogen) from the atmosphere into space and the input of frequent but tiny meteors and cosmic and meteoric dust. The fact that Earth is a closed material system makes recycling efforts inevitable if we want a sustainable global economy.

System Example Figure 1.4 illustrates a simple open-flow system, using plant photosynthesis as an example. In photosynthesis, plants use an energy input (certain wavelengths of sunlight) and material inputs of water, nutrients, and carbon dioxide. The photosynthetic process converts these inputs to stored chemical energy in the form of plant sugars (carbohydrates). The process also releases an output from the plant system: the oxygen we breathe.

Reversing the process shown in Figure 1.4, plants derive energy for their operations from respiration. In *respiration*, the plant consumes inputs of chemical energy

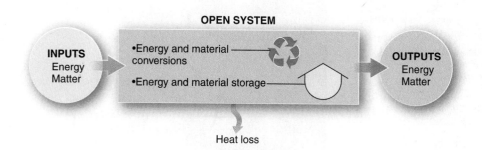

OPEN SYSTEM

INPUTS
Energy
Matter

• Energy and material conversions
• Energy and material storage

OUTPUTS
Energy
Matter

Heat loss

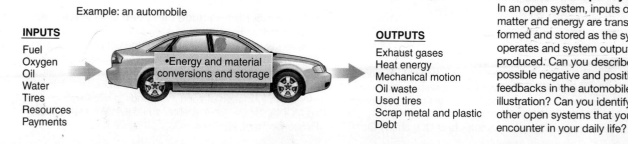

Example: an automobile

INPUTS
Fuel
Oxygen
Oil
Water
Tires
Resources
Payments

• Energy and material conversions and storage

OUTPUTS
Exhaust gases
Heat energy
Mechanical motion
Oil waste
Used tires
Scrap metal and plastic
Debt

FIGURE 1.3 An open system.
In an open system, inputs of matter and energy are transformed and stored as the system operates and system outputs are produced. Can you describe possible negative and positive feedbacks in the automobile illustration? Can you identify other open systems that you encounter in your daily life?

(carbohydrates) and oxygen and releases outputs of carbon dioxide, water, and heat energy into the environment. Thus, a plant acts as an open system. (Photosynthesis and respiration processes are discussed further in Chapter 16.)

System Feedback As a system operates, it generates outputs that influence its own operations. These outputs function as "information" that is returned to various points in the system via pathways as **feedback loops**. Feedback information can guide, and sometimes control, further system operations. In the plant's photosynthetic system (see Figure 1.4), any increase or decrease in daylength (sunlight availability), carbon dioxide, or water will produce feedback that causes specific responses in the plant. For example, decreasing the water input will slow the growth process; increasing daylength will increase the growth process, within limits.

If the feedback *information* discourages response in the system, it is **negative feedback**. *Further production in the system decreases the growth of the system.* Such negative feedback causes self-regulation in a natural system, stabilizing the system. Think of a weight-loss diet—the human body is an open system. When you stand on the scales, the good or bad news reported about your weight acts as negative feedback to you. This, in turn, provides you with guidance (negative feedback) as to how inputs such as food Calories need to be adjusted, or how exercise and increased metabolism are needed to burn excess Calories.

If feedback *information* encourages increased response in the system, it is **positive feedback**. *Further production in the system stimulates the growth of the system.* In finance, a compound-interest-bearing account provides an example: The larger the account becomes, the more interest it earns, thus the larger the account becomes, and so on. Unchecked positive feedback in a system can create a runaway ("snowballing") condition. In natural systems, such unchecked growth will reach a critical limit, leading to instability, disruption, or death of organisms. Global climate change creates an example of positive feedback in Figure 1.5 showing meltponds in West Greenland.

As explained in News Report 14.3, Chapter 14, meltponds are darker and reflect less sunlight (have a lower albedo) and therefore absorb more solar energy, which in turn melts more ice, which forms more meltponds, and so forth. Thus, a positive feedback system is in operation, further enhancing the effects of higher temperatures and warming trends. Based on satellite images and aerial surveys, scientists determined a spatial increase in meltpond distribution of 400% since 2001 across the Arctic. Partially in response to this increased melt rate, and to warmer air and sea-surface temperatures, the extent of Arctic Ocean sea ice was the smallest in the entire satellite record during 2007 and again in 2008, and ice losses from Greenland continue to accelerate.

Think of the numerous devastating wildfires that occurred globally in the last half decade, including area-record wildfires in the western United States, as examples of positive feedback. As the fires burned, they dried wet shrubs and green wood around the fire, thus providing more fuel for combustion. The greater the fire, the greater becomes the availability of fuel, and thus more fire is possible—a positive feedback for the fire. Control of the input of flammable fuel and oxygen is the key to extinguishing such fires. Knowing this process allows

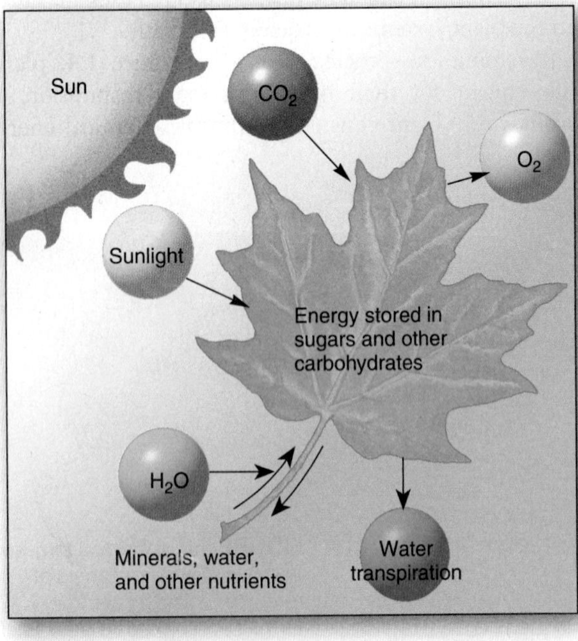

FIGURE 1.4 A leaf is a natural open system.
Simple open system of a plant leaf. Plants take light, carbon dioxide (CO_2), water (H_2O), and nutrients as system inputs to produce outputs of oxygen and carbohydrates (sugars) through the photosynthesis process. Plant respiration is simply a reverse of this process, producing outputs of CO_2 and H_2O.

FIGURE 1.5 High-latitude meltpond increase results from positive feedback.
Meltponds on the western region of the Greenland ice sheet in 2003. The darker, water-saturated surface layer from melt and blue-colored meltponds indicate change in surface energy budgets on the ice, signaling significant increase in temperatures. The formation of such meltponds is accelerating. [Image by MODIS sensor, *Terra* satellite, courtesy of MODIS Rapid Response Team, NASA/GSFC.]

us to design strategies for controlling fuel availability, regulating excessive landscaping in vulnerable urban areas, and practicing "control burns."

Can you describe possible negative and positive feedback in the automobile illustration in Figure 1.3? For instance, how would the inputs and outputs be affected if tailpipe exhaust emission standards are weakened or strengthened? What if the automobile is operated at high elevation? Or if the car is a sports utility vehicle (SUV) that weighs 2300 kg (5070 lb)? Or if the price of gasoline increases or decreases as it did in 2008? Take a moment to assess the effect of such changes on automobile system operations.

System Equilibrium Most systems maintain structure and character over time. An energy and material system that remains balanced over time, in which conditions are constant or recur, is in a *steady-state condition*. When the rates of inputs and outputs in the system are equal and the amounts of energy and matter in storage within the system are constant (or more realistically, as they fluctuate around a stable average, such as a person's body weight), the system is in **steady-state equilibrium**.

However, a steady-state system may demonstrate a changing trend over time, a condition described as **dynamic equilibrium**. These changing trends of either increasing or decreasing system operations may appear gradual. Figure 1.6 illustrates these two equilibrium conditions, steady-state and dynamic.

Examples of dynamic equilibrium include long-term climatic changes and the present pattern of increasing temperatures in the atmosphere and ocean or the ongoing loss of barrier islands along the East and Gulf coastlines. The present rate of species extinction exhibits a downward trend in numbers of living species—for example, Hawai'i has had 114 species extinctions since 1980, nearly half the US total.

Note that systems try to maintain their functional operations; they tend to resist abrupt change. However, a system may reach a **threshold**, or *tipping point*, at which it can no longer maintain its character, so it lurches to a new operational level. This abrupt change places the system in a *metastable equilibrium*. An example of such a condition is a landscape, such as a hillside or coastal bluff, that adjusts after a sudden landslide. A new equilibrium is eventually achieved among slope, materials, and energy over time.

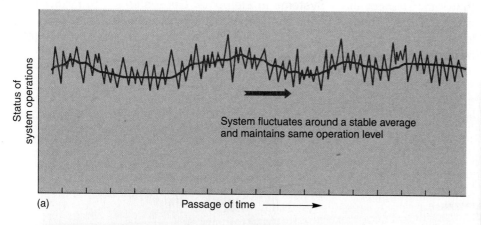

(a)   Passage of time ⟶

System fluctuates around a stable average and maintains same operation level

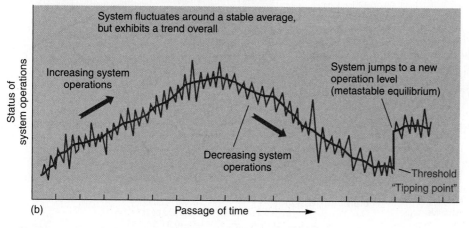

System fluctuates around a stable average, but exhibits a trend overall

Increasing system operations

System jumps to a new operation level (metastable equilibrium)

Decreasing system operations

Threshold "Tipping point"

(b)   Passage of time ⟶

Coastal systems pass a threshold.

FIGURE 1.6 System equilibria: steady-state and dynamic.
(a) A steady-state equilibrium over time; system operations fluctuate around a stable average. Other systems are in a condition of dynamic equilibrium, with an increasing or decreasing operational trend as in (b). Rather than changing gradually, some systems may reach a *threshold* where system operations lurch (change abruptly) to a new set of relations—a "tipping point." Landslides along the Pacific Coast south of San Francisco destroyed homes and property as a threshold was reached and the bluffs collapsed. [Inset photo by Sam Sargent/Liaison Agency Inc.]

This threshold concept raises concern in the scientific community, especially if some natural systems reach their threshold limits, such as the capacity of the oceans to absorb heat energy or the ability of the atmosphere to absorb carbon dioxide. The relatively sudden collapse of ice shelves surrounding a portion of Antarctica—such as the Wilkins ice shelf in 2008—or the sudden crack-up of the Ward Hunt in 2003, Ayles in 2005, and six other nearby ice shelves in 2008 (on the north coast of Ellesmere Island, Canada) serve as examples of systems at a threshold, changing to a new status, that of disintegration (more on this in Chapter 14). The bleaching (death) of living coral reefs worldwide accelerated dramatically from 1997 to 2008. Warming conditions and some pollution in the ocean led to such a threshold and coral system collapse. News Report 1.1 (p. 14) looks at a tipping point reached by a majority of the harlequin frog species of tropical Central and South America. In this report, note the power of systems analysis to explain what is causing these extinctions.

Mount Pinatubo—Global System Impact A particularly dramatic example of interactions between volcanic eruptions and Earth systems illustrates the strength of spatial analysis in physical geography and the systems organization of this textbook. Mount Pinatubo in the Philippines erupted violently in 1991, injecting 15–20 million tons of ash and sulfuric acid mist into the upper atmosphere (Figure 1.7). The smoke and ash soared 40 km (25 mi) into the atmosphere and spread out 500 km wide (310 mi) in its mushroom-cloud top. This was the second-greatest eruption during the twentieth century; Mount Katmai in Alaska (1912) was the only one greater. The eruption materials from Mount Pinatubo affected Earth systems in several ways, which are noted on the map.

As you progress through this book, you will see the story of Mount Pinatubo and its implications woven through eight chapters: Chapter 1 (systems theory), Chapter 3 (effects on energy budgets in the atmosphere), Chapter 4 (satellite images of the spread of debris by atmospheric winds), Chapter 7 (temporary effect on global atmospheric temperatures), Chapters 8 and 9 (volcanic processes), Chapter 14 (past climatic effects of volcanoes), and Chapter 16 (effects on net photosynthesis). This illustrates the value of a systems approach; instead of simply describing the eruption, we see the linkages and global impacts of such a volcanic explosion.

Systems in *Elemental Geosystems* To further illustrate the conceptual systems organization used in this textbook, notice how chapters and portions of chapters

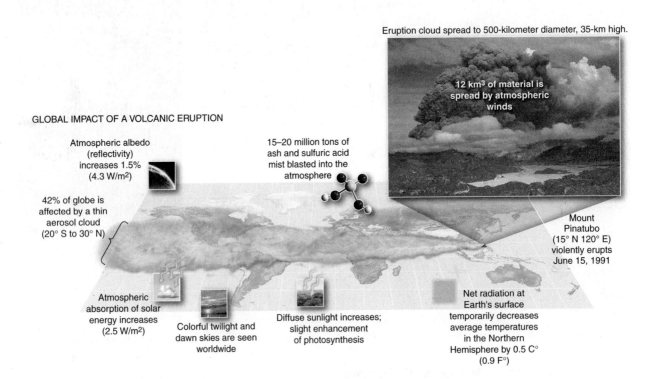

FIGURE 1.7 The eruption of Mount Pinatubo.
The 1991 Mount Pinatubo eruption widely affected the Earth–atmosphere system. Geographers and other scientists use the latest technology to study how such eruptions affect the atmosphere's dynamic equilibrium. For a summary of the impacts from this eruption, refer to the discussion of volcanoes in Chapter 9. [Photo by Van Cappellen/REA/SABA.]

are presented in simple sequences organized around the flow of energy, materials, and information (Figure 1.8). In this way, a logical progression is followed through topics that unfold according to the flow within individual systems or with time and the flow of events.

The sequence (Figure 1.8a)—*input* (components and driving force), *actions* (movements, processes, and storage changes), *outputs* (results and consequences), and *human impacts/impact on humans* (measure of relevance)—is seen in PART I as an example. The Sun begins Chapter 2;

Systems in *Elemental Geosystems*

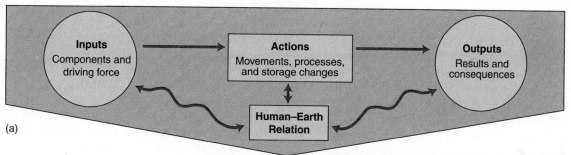

Examples of systems organization in text:

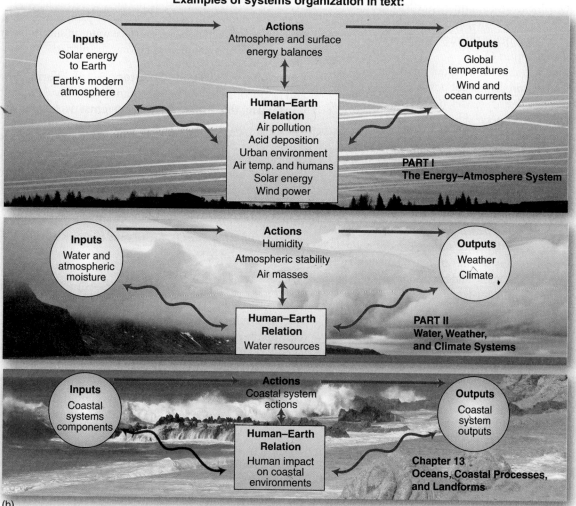

FIGURE 1.8 The systems in *Elemental Geosystems*.
(a) Chapters, sections, and topics are organized around simple flow systems, or around time and the flow of events, at various scales. (b) As an example, here are PARTS I and II. Chapter 13, which examines coastal processes, provides a chapter example. Note the outline headings used in Chapter 13: "Coastal System Components," "Coastal System Actions," "Coastal System Outputs," and "Human Impact on Coastal Environments"—along the systems flow shown in (a).

the energy flows across space to the top of the atmosphere and through the atmosphere to the surface and surface energy budgets (Chapters 2 and 3). Then we look at the outputs of temperature (Chapter 3) and winds and ocean currents (Chapter 4). Note the same logical systems flow in the other three parts and in each chapter of this text—PART II and Chapter 13 also are shown, as examples.

Models of Systems   A **model** is a simplified, idealized representation of part of the real world. Models are designed with varying degrees of generalization. You may have built or drawn a model of something in your life that was a simplification of the real thing—such as a model airplane or model house. Physical geographers use simple system models to demonstrate complex associations in the environment.

The simplicity of a model makes a system easier to understand and to simulate in experiments (for example, the leaf model in Figure 1.4). A good example is a model of the *hydrologic system*, which models Earth's entire water system, its related energy flows, and atmosphere, surface, and subsurface environments through which water moves (see Figure 6.1 in Chapter 6).

Adjusting the variables in a model produces differing conditions and allows predictions of possible system operations. A general circulation model (GCM) of the atmosphere (see Figure 7.28), operated by the Goddard Institute, accurately predicted the effect of Mount Pinatubo's ash on the atmosphere—the lowering and subsequent recovery of global air temperatures. However, predictions are only as good as the assumptions and accuracy built into the model. It is best to view a model for what it is—a simplification to help us understand complex processes.

We discuss many *system models* in this text, including the hydrologic system, water balance, surface energy budgets, earthquakes and faulting as outputs of Earth systems, river drainage basins, glacier mass budgets, soil profiles, and various ecosystems. Computer-based models are in use to study most natural systems—from the upper atmosphere, to climate change, to Earth's interior. Each of Earth's four "spheres" represents such a model of Earth's systems.

Earth's Four "Spheres"

Earth's surface is a vast area of 500 million km^2 (193 million mi^2) where four immense open systems interact. Figure 1.9 shows a simple model of three **abiotic** (nonliving) systems overlapping to form the realm of the **biotic** (living) system. The abiotic spheres are the *atmosphere, hydrosphere,* and *lithosphere.* The biotic sphere is the *biosphere.* Because these four systems are not independent units in nature, their boundaries are really transition zones rather than sharp barriers. As noted in the figure, these four spheres form the part structure in which chapters are grouped in this book. The various arrows in the figure show that content in each part and among chapters interrelates as we build our discussion of Earth's abiotic and biotic systems.

Atmosphere (PART I, Chapters 2–4)   The **atmosphere** is a thin, gaseous veil surrounding Earth, held to the planet by the force of gravity. Formed by gases arising from within Earth's crust and interior and the exhalations of all life over time, the lower atmosphere is unique in the Solar System. It is a combination of nitrogen, oxygen, argon, carbon dioxide, water vapor, and trace gases.

Hydrosphere (PART II, Chapters 5–7)   Earth's waters exist in the atmosphere, on the surface, and in the crust near the surface and, collectively, form the **hydrosphere.** That portion of the hydrosphere that is frozen is the **cryosphere**—ice sheets, ice caps and fields, glaciers, ice shelves, sea ice, and subsurface ground ice and frozen ground (permafrost)—it and related topics are covered in all four parts of the book. Water of the hydrosphere exists in all three states: liquid, solid (the frozen cryosphere), and gaseous (water vapor). Water occurs in two general chemical conditions, fresh and saline (salty); it exhibits important heat-storage properties; and it is an extraordinary solvent. Water is the medium of life. Among the planets in the Solar System, only Earth possesses surface water in such quantity, adding to Earth's uniqueness among the planets.

Lithosphere (PART III, Chapters 8–14)   Earth's crust and a portion of the upper mantle directly below the crust form the **lithosphere.** The crust is quite brittle compared with the layers deep beneath the surface, which move slowly in response to an uneven distribution of heat energy and pressure. In a broad sense, the term *lithosphere* sometimes refers to the entire solid planet. The soil layer sometimes is referred to as the *edaphosphere* and generally covers Earth's land surfaces. In this text, soils represent a bridge between the lithosphere and biosphere and are featured in the first chapter in PART IV, "Biogeography Systems."

Biosphere (PART IV, Chapters 15–16)   The intricate, interconnected web that links all organisms with their physical environment is the **biosphere,** sometimes called the **ecosphere.** The biosphere is the area in which physical and chemical factors form the context of life and which exists in the overlap among the abiotic spheres, extending from the seafloor, and even from the upper layers of the crustal rock, to about 8 km (5 mi) into the atmosphere. Life is sustainable within these natural limits. In turn, life processes have powerfully shaped the other three spheres through various interactive processes. The biosphere evolves, reorganizes itself at times, faces extinctions, and manages to flourish. Earth's biosphere is the only one known in the Solar System; thus, life as we know it is unique to Earth. Chapter 17 offers a concluding capstone for the text.

A Spherical Planet

We all have heard that some people believed Earth to be flat. Yet the roundness or *sphericity* of Earth is not as modern a concept as many think. For instance, the Greek

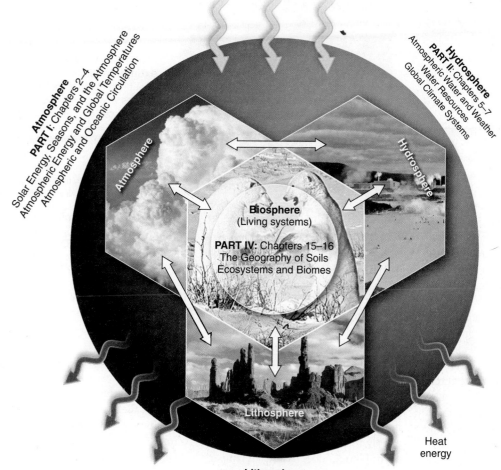

Elemental Geosystems:
Our "Sphere of Contents"

Solar energy

Atmosphere
PART I: Chapters 2–4
Solar Energy, Seasons, and the Atmosphere
Atmospheric Energy and Global Temperatures
Atmospheric and Oceanic Circulation

Hydrosphere
PART II: Chapters 5–7
Atmospheric Water and Weather
Water Resources
Global Climate Systems

Atmosphere

Hydrosphere

Biosphere
(Living systems)
PART IV: Chapters 15–16
The Geography of Soils
Ecosystems and Biomes

Lithosphere

Heat
energy

Lithosphere
PART III: Chapters 8–14
The Dynamic Planet
Earthquakes and Volcanoes
Weathering, Karst Landscapes, and Mass Movement
River Systems and Landforms
Wind Processes and Desert Landscapes
The Oceans, Coastal Processes, and Landforms
Glacial and Periglacial Landscapes

FIGURE 1.9 Earth's four spheres.
Each sphere is a model for the four "parts" used to organize *Elemental Geosystems*: our "sphere of contents."
PART I—Atmosphere: The Energy–Atmosphere System (Dramatic clouds.)
PART II—Hydrosphere: Water, Weather, and Climate Systems (Geothermal Blue Lagoon, Svartsengi geothermal area Iceland.)
PART III—Lithosphere: Earth's Changing Landscape Systems (300-m-tall Totem Pole, Monument Valley, Navajo Tribal Park, Arizona.)
PART IV—Biosphere: Biogeography Systems (Fighting polar bears, near Churchill, Manitoba.)
[Lithosphere photo by author; all others by Bobbé Christopherson.]

mathematician and philosopher Pythagoras (ca. 580–500) determined through observation that Earth is round. We do not know what Pythagoras observed to conclude that Earth is a sphere, but we can guess.

He might have noticed ships sailing beyond the horizon and apparently sinking below the surface, only to arrive back at port with dry decks. Perhaps he noticed Earth's curved shadow cast on the lunar surface during an

eclipse of the Moon. He might have deduced that the Sun and Moon are not just disks but are spherical, and that Earth must be a sphere as well.

Earth's sphericity was generally accepted by the educated populace as early as the first century A.D. Christopher Columbus, for example, knew he was sailing around a sphere in 1492; that is why he expected to reach the East Indies. In 1687, Sir Isaac Newton postulated that the

NEWS REPORT 1.1

Frogs at a Tipping Point—Check the System

Global climate change is pushing systems to their threshold. Many amphibian species are threatened by these changes. The "Declining Amphibian Population Task Force," founded in 1991 and merged with the IUCN (World Conservation Union) Amphibian Group in 2006, is a response to this growing crisis. Scientific understanding of frog population declines from habitat loss and pollution, and now rising temperatures, is emerging.

The key to unraveling what was happening to harlequin frog (*Atelopus varius*, Figure 1.1.1) and golden toad (*Bufo periglenes*) species in the Monteverde Cloud Forest Reserve, Costa Rica, and elsewhere in Central and South America, was an examination of climate change and how this alters conditions for the transmission of diseases and pathogens. Research had to uncover complex synergistic interactions (combined actions that magnify results) in physical and biological systems. One infectious agent, the *chytrid fungus* (*Bactrachochytrium dendrobactidis*), was spreading as thermal-optimum conditions were reached in its mid-elevation range (1000 to 2400 m; 3280–7875 ft) in the mountain-cloud forest. Here is how system operations and feedbacks produced this species failure.

Human-forced climate change is increasing the temperature of the ocean and atmosphere. Higher temperatures cause higher evaporation rates. However, even though humidity (water-vapor content of the air) is higher, relative humidity is lower because the air can absorb more, which affects the condensation level (see Chapter 5). As the warm, moist air moves onshore and reaches the mountains, the air mass lifts and cools

and condensation occurs at higher elevations than before. Increasing levels of air pollution provide more condensation nuclei for cloud droplets to form (see Chapters 2 and 5).

Increased clouds affect the daily temperature range: Clouds at night, acting like insulation, raise nighttime minimum temperatures, whereas clouds during the day act as reflectors, sending sunlight back into space and lowering daytime maximum temperatures (see Chapter 3). Nighttime minimum temperatures are increasing more rapidly than daytime maximums. The chytrid fungus does best when temperatures are between 17°C and 25°C, stops growing at 28°C, and dies at 30°C. The cloud cover keeps the daytime maximum below 25°C (77°F) at the leaf-litter and moss-ground cover microscale; without cloud cover, the surface reaches 30°C. In these more favorable conditions the disease pathogen flourishes.

Harlequin frogs have moist, porous skin that the fungus penetrates, killing the frog. Between 1986 and 2006, approximately 67% of the 110 known species of harlequin frogs went extinct—extinction is forever. The IUCN Red List classifies the remaining population as "Critically Endangered."

From this example can you see the power of systems analysis? Global warming intensifies the hydrologic cycle, raises air and water temperatures, and increases evaporation, all of which changes the nature of cloud formations and alters daily temperature patterns. These changes impact ecological systems by increasing the range and number of infectious disease pathogens, which attack species and ultimately end their existence—follow the system interactions. A summary of the research by J. Alan Pounds and his team of scientists concluded with a sobering thought,

> The powerful synergy between pathogen transmission and climate change should give us cause for concern about human health in a warmer world. . . . The frogs are sending an alarm call to all concerned about the future of biodiversity and the need to protect the greatest of all open-access resources—the atmosphere.[*]

[*]A. Blaustein and A. Dobson, "A message from the frogs," summary of J. A. Ponds et al., "Widespread amphibian extinctions from epidemic disease driven by global warming," *Nature* 439 (January 12, 2006): 144.

round Earth, along with the other planets, could not be perfectly spherical. Newton reasoned that Earth is slightly misshapen by its spinning, making it bulge through the equator and flatten at the poles.

Earth's equatorial bulge and its polar oblateness (slight flattening) are universally accepted and confirmed by satellite observations. **Geodesy** is the science that attempts to determine Earth's shape and size by surveys and mathematical

0

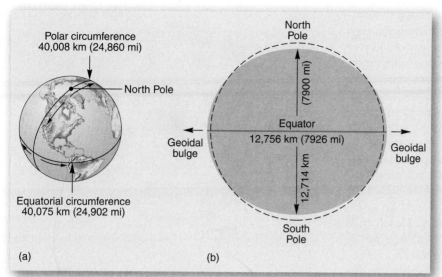

FIGURE 1.10 Earth's dimensions.
Earth's equatorial and polar circumference
(a) and diameter (b). The dashed line is a
perfect circle for reference to Earth's geoid.

calculations. This modern era of precise Earth measurement is the "geoidal epoch" because Earth is considered a **geoid**, meaning literally "the shape of Earth is uniquely Earth-shaped."

Today's satellite-based measurements are so precise that scientists know how Earth's shape is changing in ways that are unnoticeable to us. Following the last ice age and the transfer of much water from the cryosphere to the hydrosphere, Earth's polar oblateness decreased. This reversed in 1997 and the oblateness is now increasing, with the mass redistribution of water as lower-latitude subpolar glaciers melt and the ocean warms. In other words, Earth's equatorial girth is expanding slightly. Also, Earth's shape changes slightly with the seasons as the crust compresses a little during each hemisphere's winter. Such is the excitement in modern geodesy!

Figure 1.10 gives Earth's average polar and equatorial circumferences and diameters. The Greek geographer,

astronomer, and librarian Eratosthenes (ca. 276–195 B.C.) first measured Earth's polar circumference more than 2200 years ago. The ingenious reasoning by which he arrived at his calculation follows.

Measuring Earth in 247 B.C. Foremost among early geographers, Eratosthenes served as the librarian of Alexandria in Egypt during the third century B.C. He was in a position of scientific leadership, for Alexandria's library was the finest in the ancient world. Among his achievements was the calculation of Earth's polar circumference to a high level of accuracy, quite a feat for 247 B.C. Here's how he did it. As you read, follow along on Figure 1.11.

Travelers told Eratosthenes that on June 21 they saw the Sun's rays shine directly to the bottom of a well at Syene, the location of present-day Aswan, Egypt. This meant that the Sun was directly overhead on that day. North of Syene

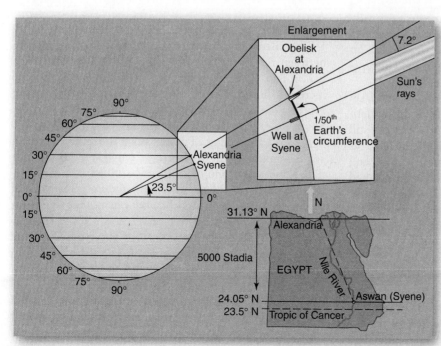

FIGURE 1.11 Eratosthenes' calculation.
Eratosthenes' work teaches the value of observing carefully and integrating all observations with previous learning. Calculating Earth's circumference required application of his knowledge of Earth–Sun relationships, geometry, and geography.

in Alexandria, Eratosthenes knew from his own observations that the Sun's rays never were directly overhead, even at noon on June 21. This day is the longest day of the year and the day on which the noon Sun is at its northernmost position in the sky. He knew objects in Alexandria, unlike objects in Syene, always cast a noontime shadow. Using his knowledge of geometry, Eratosthenes conducted an experiment to explain the different observations.

In Alexandria at noon on June 21, he measured the angle of a shadow cast by an obelisk (a perpendicular column). Knowing the height of the obelisk and measuring the length of the Sun's shadow from its base, he solved the triangle for the angle of the Sun's rays, which he determined to be 7.2° off from directly overhead. However, at Syene on the same day, the angle of the Sun's rays was 0° from a perpendicular, as it was directly overhead.

Geometric principles told Eratosthenes that the distance on the ground between Alexandria and Syene formed an arc of Earth's circumference equal to the angle of the Sun's rays at Alexandria. Since 7.2° is roughly 1/50 of the 360° (360° ÷ 7.2° = 50) in Earth's total circumference, the distance between Alexandria and Syene must represent approximately 1/50 of Earth's total polar circumference.

Next, Eratosthenes determined the surface distance between the two cities as 5000 stadia. A *stadium*, a Greek unit of measure, equals approximately 185 m (607 ft). He then multiplied 5000 stadia by 50 to determine that Earth's polar circumference is about 250,000 stadia. Eratosthenes' calculations convert to roughly 46,250 km (28,738 mi), which is remarkably close to the correct value of 40,008 km (24,860 mi) for Earth's polar circumference. Not bad for 247 B.C.! (Several values can be used for the distance of a Greek stadium—the 185 m used here represents an average value.)

Location and Time on Earth

An essential for geographic science is an internationally accepted, coordinated grid system to determine location on Earth. The terms *latitude* and *longitude* were in use on maps as early as the first century A.D., with the concepts themselves dating back to Eratosthenes and others.

The geographer, astronomer, and mathematician Ptolemy (ca. A.D. 90–168) contributed greatly to modern maps, and many of his terms and configurations are still used today. Ptolemy divided the circle into 360 degrees (360°), with each degree comprising 60 minutes (60') and each minute including 60 seconds (60") in a manner adapted from the ancient Babylonians. He located places using these degrees, minutes, and seconds. However, the precise length of a degree of latitude and a degree of longitude remained unresolved for the next 17 centuries.

Latitude

Latitude *is an angular distance north or south of the equator*, measured from the center of Earth (Figure 1.12a). On a map or globe, the lines designating these angles of

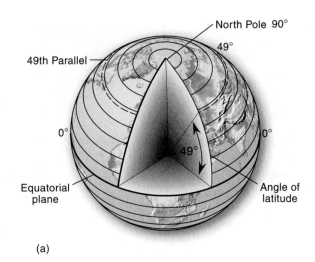

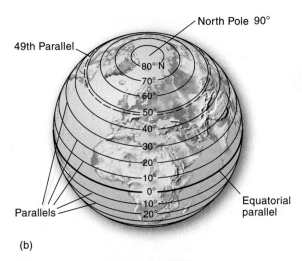

FIGURE 1.12 Parallels of latitude.
(a) Latitude is measured in degrees north or south of the equator, which is 0°. Each pole is at 90°. Note the measurement of 49° latitude. (b) These angles of latitude determine parallels along Earth's surface. Do you know your present latitude?

latitude run east and west, parallel to the equator (Figure 1.12b). Because Earth's equator divides the distance between the North Pole and the South Pole exactly in half, it is assigned the value of 0° latitude. Thus, latitude increases in value from the equator northward to the North Pole, at 90° north latitude, and southward to the South Pole, at 90° south latitude.

A line connecting all points along the same latitudinal angle is a **parallel**. In the figure, an angle of 49° north latitude is measured, and by connecting all points at this latitude we have the 49th parallel. Thus, *latitude* is the name of the angle (49° north latitude), *parallel* names the line (49th parallel), and both indicate distance north of the equator.

Latitude is readily determined by observing *fixed celestial objects* such as the Sun or the stars, a method dating to ancient times. During daylight hours, the angle of the Sun above the horizon indicates the observer's latitude, after adjustment is made for the season and for the time of day. Because Polaris (the North Star) is almost directly overhead at the North Pole, persons anywhere in the

Northern Hemisphere can determine their latitude at night simply by sighting Polaris and measuring its angle above the local horizon. The angle of elevation of Polaris above the horizon equals the latitude of the observation point.

In the Southern Hemisphere, Polaris cannot be seen, because it is below the horizon. Instead, latitude measurement south of the equator is accomplished by sighting on a constellation that points to a celestial location above the South Pole. This indicator constellation is the Southern Cross (*Crux Australis*).

Latitudinal Geographic Zones Natural environments differ dramatically from the equator to the poles. These differences result from the amount of solar energy received, which varies by latitude and season of the year. As a convenience, geographers identify *latitudinal geographic zones* as regions with fairly consistent qualities.

Figure 1.13 portrays these zones, their locations, and their names: *equatorial* and *tropical*, *subtropical*, *midlatitude*, *subarctic* or *subantarctic*, and *arctic* or *antarctic*. "Lower latitudes" are those nearer the equator, whereas "higher latitudes" are those nearer the poles. These generalized latitudinal zones are useful for reference and comparison, but they do not have rigid boundaries, as one zone transitions into another.

The *Tropic of Cancer* (about 23.5° north parallel) and the *Tropic of Capricorn* (about 23.5° south parallel) are the most extreme northern and southern parallels that experience perpendicular (directly overhead) rays of the Sun at local noon. When the Sun arrives overhead at these tropics, it marks the first day of summer in each hemisphere. (The tropics are discussed further in Chapter 2.) The *Arctic Circle* (about 66.5° north parallel) and the *Antarctic Circle* (about 66.5° south parallel) are the parallels farthest from the poles that still experience 24 uninterrupted hours of night (including dawn and twilight) during local winter or of day during local summer.

Longitude

Longitude *is an angular distance east or west of a point on Earth's surface*, measured from the center of Earth (Figure 1.14a). On a map or globe, the lines designating these angles of longitude run north and south at right angles (90°) to the equator and to all parallels. A line connecting all points along the same longitude is a **meridian**. In the figure a longitudinal angle of 60° E is measured.

Thus, *longitude* is the name of the angle, *meridian* names the line, and both indicate distance east or west of an arbitrary **prime meridian** (Figure 1.14b). Earth's prime meridian (0°) passes through the old Royal Observatory at Greenwich, England, as set by treaty—the *Greenwich prime meridian*.

We noted that latitude is easily determined by sighting the Sun, North Star, or Southern Cross, but a method of accurately determining longitude, especially at sea, remained a major difficulty until the mid-1700s. In addition, a specific determination of longitude was needed before world standard time could be established. The interesting geographic quest for longitude and time is the topic of Focus Study 1.2.

Today, latitude, longitude, elevation, and Earth measurements are accurately calibrated using a handheld instrument that reads radio signals from satellites, a technology known as the **Global Positioning System (GPS)**. News Report 1.2 discusses the dramatic spatial applications of GPS.

Great Circles and Small Circles

Great circles and small circles are important concepts that help summarize latitude and longitude (Figure 1.15). A **great circle** is any circle of Earth's circumference whose center coincides with the center of Earth. An infinite number of great circles can be drawn on Earth. Every

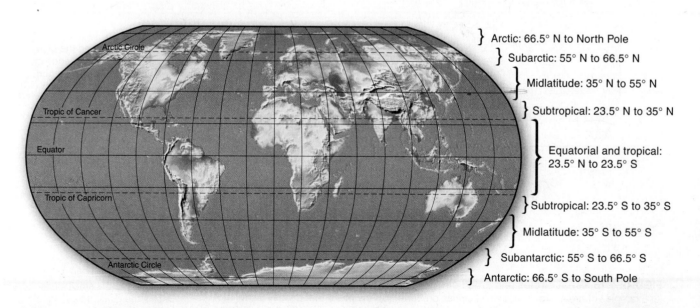

FIGURE 1.13 Latitudinal geographic zones.
Geographic zones are generalizations that characterize various regions by latitude. Think of these as transitioning into one another over broad areas.

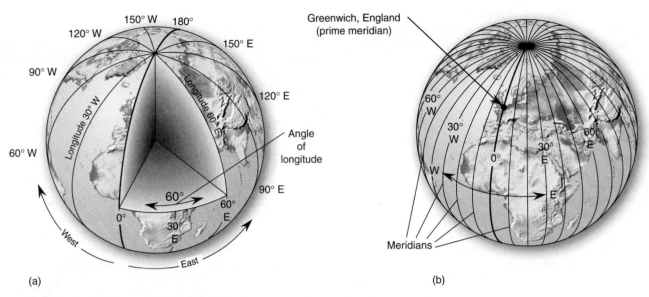

(a)

(b)

FIGURE 1.14 Meridians of longitude.

(a) Longitude is measured in degrees east or west of a 0° starting line, the prime meridian. Note the measurement of 60°E longitude. (b) Angles of longitude measured from this prime meridian determine other meridians. The prime meridian is drawn from the North Pole, through the Royal Observatory in Greenwich, England, to the South Pole. North America is west of Greenwich; therefore, it is in the Western Hemisphere. Do you know your present longitude?

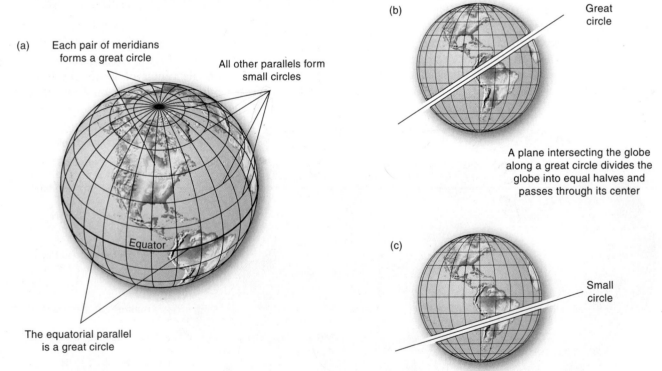

FIGURE 1.15 Great circles and small circles.

(a) Examples of great circles and small circles on Earth. (b) Any plane that divides Earth into equal halves will intersect the globe along a great circle; this great circle is a full circumference of the globe and is the shortest distance between any two surface points. (c) Any plane that splits the globe into unequal portions will intersect the globe along a small circle.

FOCUS STUDY 1.2

The Timely Search for Longitude

Unlike latitude, longitude cannot be determined readily from fixed celestial bodies. Determining longitude is particularly critical at sea, where no landmarks are visible. The problem is Earth's rotation, which constantly changes the apparent position of the Sun and stars.

In his historical novel *Shogun*, author James Clavell expressed the frustration of the longitude problem through his pilot, Blackthorn: "Find how to fix longitude and you're the richest man in the world. . . . The Queen, God bless her, 'll give you ten thousand pound and a dukedom for the answer to the riddle. . . . Out of sight of land you're always lost, lad."*

In the early 1600s, Galileo explained that longitude could be measured by using two clocks. Any point on Earth takes 24 hours to travel around the full 360° of one rotation (1 day). If you divide 360° by 24 hours, you find that any point on Earth travels through 15° of longitude every hour. Thus, if there were a way to measure time accurately at sea, a comparison of two clocks could give a value for longitude.

One clock would indicate the time back at home port (Figure 1.2.1). The other clock would be reset at local noon each day, as determined by the highest Sun position in the sky (solar zenith). The time difference then would indicate the longitudinal difference traveled: 1 hour for each 15° of longitude. The principle was sound; all that was needed were accurate

*From J. Clavell, *Shogun*, Copyright © 1975 by James Clavell (New York: Delacorte Press, a division of Dell Publishing Group, Inc.), p. 10.

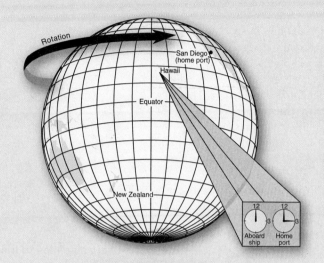

FIGURE 1.2.1
Clock times determine longitude.
Using two clocks on a ship to determine longitude. For example, if the shipboard clock reads local noon and the clock set for home port reads 3:00 P.M., ship time is 3 hours earlier than home time. Therefore, calculating 3 hours at 15° per hour puts the ship at 45° west longitude from home port.

clocks. Unfortunately, the pendulum clock invented by Christian Huygens in 1656 did not work on the rolling deck of a ship at sea!

In 1707, the British lost four ships and 2000 men in a sea tragedy that was blamed specifically on the longitude problem. In response, Parliament passed an act in 1714—"Publik Reward . . . to Discover the Longitude at Sea"—and authorized a prize worth more than $2 million in today's dollars to the first successful inventor of an accurate seafaring clock. The Board of Longitude was established to judge any devices submitted.

John Harrison, a self-taught country clockmaker, began work on the problem in 1728 and finally produced his brilliant marine chronometer, known as Number 4, in 1760. The clock was tested on a voyage to Jamaica in 1761. When taken ashore and compared to land-based longitude, Harrison's ingenious Number 4 was only 5 seconds slow, an error that translates to only 1.25', or 2.3 km (1.4 mi), well within Parliament's standard. After many delays, Harrison

finally received most of the prize money in his last years of life.

With his marine clocks, John Harrison tested the waters of space–time. He succeeded, against all odds, in using the fourth—temporal—dimension to link points on the three-dimensional globe. He wrested the world's whereabouts from the stars, and locked the secret in a pocket watch.[†]

From that time it was possible to determine longitude accurately on land and sea, as long as everyone agreed upon a meridian to use as a reference for time comparisons—the Royal Observatory in Greenwich, England. In this modern era of atomic clocks and satellites in mathematically precise orbits, we have far greater accuracy available for the determination of longitude on Earth's surface and a basis for precise navigation.

[†]From Dava Sobel, *Longitude, The Story of a Lone Genius Who Solved the Greatest Scientific Problem of His Time* (New York: Walker and Co., 1995), p. 175.

meridian is one-half of a great circle that passes through the poles. On flat maps, airline and shipping routes appear to arch their way across oceans and landmasses. These are *great circle routes*, the shortest distance between two points on Earth (discussion with Figure 1.22 is just ahead).

Only one parallel is a great circle—the *equatorial parallel*. All other parallels diminish in length toward the poles and, along with any other non-great circles that one might draw, constitute **small circles**. These circles have centers that do not coincide with Earth's center.

NEWS REPORT 1.2

GPS: A Personal Locator

The Global Positioning System (GPS) comprises 24 orbiting satellites, in six orbital planes, that transmit navigational signals for Earth-bound use (backup GPS satellites are in orbital storage as replacements). Originally devised in the 1970s by the Department of Defense for military purposes, the present system is commercially available worldwide.

A receiver, some about the size of a pocket radio or smaller, senses signals from three or more satellites at the same time and triangulates the receiver's location. The key is the time lag the signals take, travelling at the speed of light (Figure 1.2.1). The receiver reports a calculatation of latitude and longitude within 10-m accuracy (33 ft) and elevation within 15 m (49 ft). *Differential GPS (DGPS)* increases accuracy by comparing readings with another base station (reference receiver) for a differential correction (Figure 1.2.2).

The GPS satellite measurements help refine our knowledge of Earth's exact shape and how this shape is changing. GPS is useful for diverse applications, such as ocean navigation, land surveying, tracking small changes in Earth's crust, managing the movement of fleets of trucks, mining and resource mapping, tracking wildlife migration and behavior, police and security work, and environmental planning. Measuring and mapping an ominous bulge on an Oregon volcano is another present-day application. GPS also is useful to the backpacker and sportsperson.

To measure and study earthquakes in southern California, such as the 2008 Chino Hills quake, the Jet Propulsion Laboratory (JPL) operates the GPS Observation Office that monitors a network of 250 seismic instruments—the Southern California Integrated GPS Network (SCIGN, http://www.scign.org/). This system can record fault movement as small as 1 mm (0.04 in.).

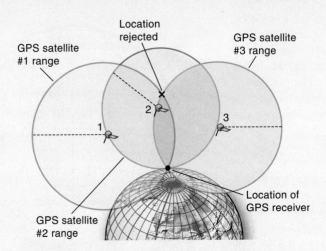

FIGURE 1.2.1 **Using satellites to triangulate location through GPS.**
Imagine a ranging sphere around each of three GPS satellites. These spheres intersect at two points, one easily rejected because it is some distance above Earth and the other at the true location of the GPS receiver. The three spheres can only intersect at two points. The key is gauging the distance between the satellites and the GPS receiver—signals travel at the speed of light.

FIGURE 1.2.2 **GPS in action.**
A GPS unit in operation for surveying and location analysis in an environmental study.
[Photo courtesy of Trimble Navigation Ltd., Sunnyvale, California.]

Commercial airlines use GPS to improve accuracy of routes flown and thus increase fuel efficiency. Scientists used GPS to accurately determine the height of Mount Everest in the Himalayan Mountains—now 8850 m compared to the former measurement of 8848 m (29,035 ft; 29,028 ft) (Figure 1.2.3). GPS measurements of Mount Kilimanjaro lowered its summit from 5895 m to 5892 m (119,340 ft; 19,330 ft).

In "precision agriculture," farmers use GPS to determine crop yields on specific parts of their farms. A detailed plot map is made to guide the farmer in fertilizing, proper seed distribution, irrigation applications, or other tasks as needed. A computer and GPS unit on board the farm equipment guides the work. This is the science of *variable-rate technology*, made possible by GPS. One example of this application, among others, is the establishment of Precision Agriculture (http://www.precisionag.com/) by the California Polytechnic State University, San Luis Obispo, California State University,

FIGURE 1.2.3 **GPS on Mount Everest.**
A Trimble 4800 global positioning system (GPS) installation establishes Earth's highest benchmark. The installation is only 18 m (60 ft) from the 8850-m (29,035-ft) summit of Mount Everest. Scientists calculated this new measurement of the mountain's height and announced the results in 1999. [Photo and GPS installation by explorer, climber Wally Berg, May 20, 1998.]

Fresno, and the University of California, Davis. A full, downloadable curriculum with supporting faculty is posted on their web site.

The importance of GPS to geography is obvious; this precise technology reduces the need to maintain ground control points for location, mapping, and spatial analysis. Instead, geographers working in the field can determine their position accurately as they work. Boundaries and data points in a study area are easily determined and entered into a data base, reducing the need for traditional surveys. Additional frequencies were added in 2003 and 2006, which increased accuracy significantly. Due to this and myriad other applications, GPS sales are exceeding $10 billion a year. (For a GPS overview, see http://www.colorado.edu/geography/gcraft/notes/gps/gps_f.html.)

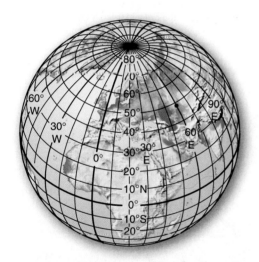

FIGURE 1.16 Earth's coordinate grid system.
Latitude and parallels and longitude and meridians allow all places on Earth to be precisely located. The location noted is at 49°N latitude by 60°E longitude.

Figure 1.16 combines latitude and parallels with longitude and meridians to illustrate Earth's complete coordinate grid system. Note the dot that marks our measurement of 49° N by 60° E, a location in western Kazakhstan. Next time you look at a world globe, follow the parallel and meridian that converge on your location.

Prime Meridian and Standard Time

Coordination of international trade, airline schedules, business and agricultural activities, and daily living depends on a worldwide time system. Today, we take for granted standard time zones and an agreed-upon prime meridian, but such a standard is a relatively recent development.

Setting time was not a great problem in small European countries, most of which are less than 15° wide. But in North America, which spans more than 90° of longitude (the equivalent of six 15° time zones), the problem was serious. In 1870, railroad travelers going from Maine to San Francisco made 22 adjustments to their watches to stay consistent with local time. In Canada, Sir Sanford Fleming led the fight for standard time and for an international agreement on a prime meridian.

Twenty-seven countries attended the 1884 International Meridian Conference in Washington, DC. Before that year, most nations used their own national capital as a prime meridian for their land maps, whereas more than 70% of the world's merchant ships were using Greenwich as a prime meridian on marine charts. After lengthy debate at the conference, most participating nations chose the highly respected Royal Observatory at Greenwich, London, England, as the place for the prime meridian of 0° longitude for all maps. Thus, a world standard was set—**Greenwich Mean Time (GMT)**—and a consistent Universal Time was established. (See http://www.gmt2000.co.uk/meridian/place/plco0a1.htm.)

The basis of time is that Earth revolves 360° every 24 hours, or 15° per hour (360° ÷ 24 = 15°). Thus, a time zone of 1 hour is established for each 15° increment of longitude, or 7.5° on either side of a *central meridian*. Assuming it is 9:00 P.M. in Greenwich, then it is 4:00 P.M. in Baltimore (+5 hr), 3:00 P.M. in Oklahoma City (+6 hr), 2:00 P.M. in Salt Lake City (+7 hr), 1:00 P.M. in Seattle and Los Angeles (+8 hr), noon in Anchorage (+9 hr), and 11:00 A.M. in Honolulu (+10 hr). To the east, it is midnight in Ar Riyāḍ, Saudi Arabia (−3 hr). (The designation A.M. is for *ante meridiem*, "before noon," whereas P.M. is for *post meridiem*, "after noon." A 24-hour clock avoids the use of these designations: 3 P.M. is stated as 15:00 hours; 3 A.M. is 3:00 hours.)

As you can see from the modern international time zones in Figure 1.17, national boundaries and political considerations distort time boundaries. For example, China spans four time zones, but its government decided to keep the entire country operating at the same time. Thus, in some parts of China clocks are several hours off from what the Sun is doing. In the United States, parts of Florida and west Texas are in the Central time zone.

International Date Line An important corollary of the prime meridian is the 180° meridian on the opposite side of the planet. This meridian is the **International Date Line** and marks the place where each day officially begins (at 12:01 A.M.). From this "line" the new day sweeps westward. This *westward* movement of time is created by Earth's turning *eastward* on its axis.

At the International Date Line, the west side of the line is always one day ahead of the east side. No matter what time of day it is when the line is crossed, the calendar changes a day (Figure 1.18). Note in the illustration the departures from the IDL and the 180° meridian; this deviation is due to local administrative and political preferences. The political control of island nations extends only 3.5 nautical miles (4 mi) offshore. So in 1995 when

Kiribati moved the IDL designation to the east of its islands in order to experience "first dawn" on Earth, the political move did not apply to the ocean between its territorial waters and the 180th meridian.

Locating the date line in the sparsely populated Pacific Ocean minimizes most local confusion. However, early explorers before the date-line concept were "lost," on the calendar. For example, Magellan's crew returned from the first circumnavigation of Earth in 1522, confident from their ship's log that the day of their arrival home was Wednesday, September 7. They were shocked when informed by insistent local residents that it was actually Thursday, September 8. Of course, without an International Date Line they had no idea that they must advance a day somewhere when sailing around the world in a westward direction. Imagine the confusion as the crew accounted for each day in their log!

Coordinated Universal Time The **Coordinated Universal Time (UTC)*** time-signal system replaced GMT universal time in 1972 and became the legal reference

*UTC is in use because agreement was not reached on whether to use the English word order, CUT, or the French order, TUC. UTC was the compromise and is recommended for all timekeeping applications; use of the term GMT is discouraged.

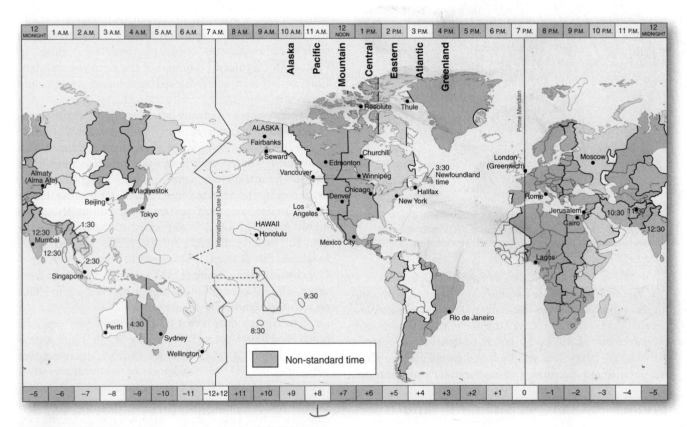

FIGURE 1.17 Modern international standard time zones.
Numbers along the bottom of the map indicate how many hours each zone is earlier (plus sign) or later (minus sign) than the Coordinated Universal Time (UTC) standard at the prime meridian. The United States has five time zones; Canada is divided into six. [Adapted from Standard Time Zone Chart of the World, Defense Mapping Agency, Bethesda, Maryland. See **http://aa.usno.navy.mil/faq/docs/world_tzones.html**.]

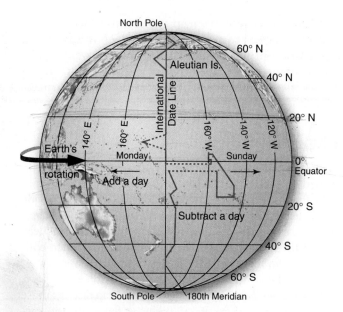

FIGURE 1.18 International Date Line.
International Date Line location, approximately along the 180th meridian (see also Figure 1.17). Note that it is officially one day later west of the IDL.

for official time in all countries. Although the prime meridian still runs through Greenwich, UTC is based on average time calculations collected by the BIPM near Paris and broadcast worldwide (the International Bureau of Weights and Measures, see http://www.bipm.fr/). You might still see official UTC referred to as GMT or Zulu time.

Regular vibrations (natural frequency) of cesium atoms in primary standard clocks measure the length of a second and UTC—accuracies now range down to 15 decimal places, or nanoseconds of time. Time and Frequency Services of the National Institute for Standards and Technology (NIST), US Department of Commerce, operates several of the most advanced clocks that report to BIPM. (For more on time, call 303-499-7111 or 808-335-4363, or see http://nist.time.gov/ for UTC.) In Canada, the Institute for Measurement Standards, National Research Council Canada, participates in determining UTC (English, 613-745-1578; French, 613-745-9426; for more see http://www.nrc.ca/).

Daylight Saving Time In 70 countries, mainly in the temperate latitudes, time is set ahead 1 hour in the spring and set back 1 hour in the fall—a practice known as **daylight saving time**. The idea to extend daylight for early evening activities (at the expense of daylight in the morning) was first proposed by Benjamin Franklin. It was not adopted until World War I and again in World War II, when Great Britain, Australia, Germany, Canada, and the United States used the practice to save energy (1 less hour of artificial lighting needed).

Time "springs forward" 1 hour the second Sunday in March and "falls back" an hour the first Sunday in November, except in a few places that do not use daylight saving time (Hawai'i, Arizona, and Saskatchewan). In Europe, the last Sundays in March and October are used to begin and end what is called "summer time." (See http://webexhibits.org/daylightsaving/).

Maps, Scales, and Projections

The earliest known graphic maps date to 2300 B.C., when the Babylonians used clay tablets to record information about the region of the Tigris and Euphrates rivers (the area of modern-day Iraq). Today, the making of maps and charts is a specialized science as well as an art, blending aspects of geography, engineering, mathematics, graphics, computer science, and artistic specialties. Mapmaking is similar in ways to architecture, in which aesthetics and utility combine to produce a useful product.

A **map** is a generalized view of an area, usually some portion of Earth's surface, as seen from above and greatly reduced in size. **Cartography** is that part of geography that involves mapmaking. Maps are critical tools with which geographers depict spatial information and analyze spatial relationships.

We all use maps at some time to visualize our location and our relationship to other places, or maybe to plan a trip, or to coordinate commercial and economic activities. Have you found yourself looking at a map, planning real and imagined adventures to far-distant places? Maps are wonderful tools! Understanding a few basics about maps is essential to our study of physical geography.

Map Scales

Architects, toy designers, and mapmakers have something in common: They all create scale models. They reduce real things and places to the more convenient scale of a drawing; a model car, train, or plane; a diagram; or a map. An architect renders a blueprint of a structure to guide the building contractors, selecting a scale so that 1 cm (or inch) on the drawing represents so many meters (or feet) on the proposed building. Often, the drawing is 1/50 to 1/100 of real size.

The cartographer does the same thing in preparing a map. **Scale** is the ratio of the image on a map to the real world; it relates a unit on the map to a similar unit on the ground. A 1:1 scale means that a centimeter on the map represents a centimeter on the ground (although this is an impractical map scale, for the map is as large as the area mapped!). A more appropriate scale for a local map is 1:24,000, in which 1 unit on the map represents 24,000 identical units on the ground.

Cartographers express map scale as a representative fraction, a graphic scale, or a written scale (Figure 1.19). A *representative fraction* (RF, or *fractional scale*) is expressed with either a colon or a slash, as in 1:125,000 or 1/125,000. No actual units of measurement are mentioned because any unit is applicable as long as both parts of the fraction are in the same unit: 1 cm to 125,00 cm, 1 in. to 125,000 in., or even 1 arm length to 125,000 arm lengths, and so on.

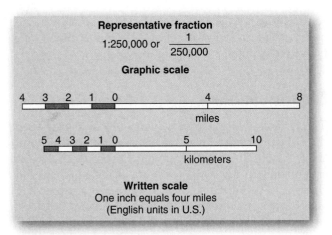

Representative fraction

$$1{:}250{,}000 \text{ or } \frac{1}{250{,}000}$$

Graphic scale

4 3 2 1 0   4   8

miles

5 4 3 2 1 0   5   10

kilometers

Written scale
One inch equals four miles
(English units in U.S.)

FIGURE 1.19 Map scale.
Three common expressions of map scale—representative fraction, graphic scale, and written scale.

A *graphic scale*, or bar scale, is a bar graph with units to allow measurement of distances on the map. An important advantage of a graphic scale is that if the map is enlarged or reduced, the graphic scale enlarges or reduces along with the map. In contrast, written and fractional scales become incorrect with enlargement or reduction. As an example, you can shrink a map from 1:24,000 to 1:63,360, but the scale will still say "1 in. to 2000 ft," instead of the new correct scale of 1 in. to 5280 ft (1 mi).

Scales are *small*, *medium*, and *large*, depending on the ratio described. In relative terms, a scale of 1:24,000 is a large scale, whereas a scale of 1:50,000,000 is a small scale. The greater the denominator in a fractional scale (or the number on the right in a ratio expression), the smaller the scale and the more abstract the map is in relation to what is being mapped. Examples of selected representative fractions and written scales are listed in Table 1.2 for small-, medium-, and large-scale maps.

Table 1.2 Sample Representative Fractions and Written Scales for Small-, Medium-, and Large-Scale Maps

System	Scale Size	Representative Fraction	Written Scale
English	Small	1:3,168,000	1 in. = 50 mi
		1:1,000,000	1 in. = 16 mi
		1:250,000	1 in. = 4 mi
	Medium	1:125,000	1 in. = 2 mi
		1:63,360	1 in. = 1 mi
		(or 1:62,500)	
	Large	1:24,000	1 in. = 2000 ft

System		Representative Fraction	Written Scale
Metric		1:1,000,000	1 cm = 10.0 km
		1:25,000	1 cm = 0.25 km
		1:10,000	1 cm = 0.10 km

If there is a globe or map available in your library or classroom, check to see the scale at which it was drawn. See if you can find examples of representative, graphic, and written scales on wall maps, highway maps, and in atlases. In general, do you think a world globe is a small- or a large-scale map of Earth's surface?

Map Projections

A globe is not always a helpful map representation of Earth. When you go on a trip, you need more detailed information than a globe can provide. Consequently, to provide local detail, cartographers prepare large-scale flat maps, which are two-dimensional representations (scale models) of our three-dimensional Earth. Unfortunately, such conversion from three dimensions to two causes distortion.

A globe is the only true representation of *distance*, *direction*, *area*, *shape*, and *proximity*. A flat map distorts those properties. Therefore, in preparing a flat map, the cartographer must decide which characteristic to preserve, which to distort, and how much distortion is acceptable. To understand this problem, consider these important properties of a *globe*:

- Parallels always are parallel to each other, always are evenly spaced along meridians, and always decrease in length toward the poles.
- Meridians converge at both poles and are evenly spaced along any individual parallel.
- The distance between meridians decreases toward poles, with the spacing between meridians at the 60th parallel equal to one-half the equatorial spacing.
- Parallels and meridians always cross each other at right angles.

The problem is that all these qualities cannot be reproduced on a flat surface. Simply taking a globe apart and laying it flat on a table illustrates the challenge faced by cartographers (Figure 1.20). You can see the empty spaces that open up between the sections, or gores, of the globe. This reduction of the spherical Earth to a flat surface is a **map projection**. Thus, no flat map projection of Earth can ever have all the features of a globe. Flat maps always possess some degree of distortion—much less for large-scale maps representing a few kilometers; much more for small-scale maps covering individual countries, continents, or the entire world.

Properties of Projections *The best projection is always determined by its intended use.* The major decisions in selecting a map projection involve the properties of **equal area** (equivalence) and **true shape** (conformality). If a cartographer selects equal area as the desired trait—for example, for a map showing the distribution of world climates—then true shape must be sacrificed by *stretching* and *shearing*, which allow parallels and meridians to cross at other than right angles. On an equal-area map, a coin covers the same amount of surface area no matter where

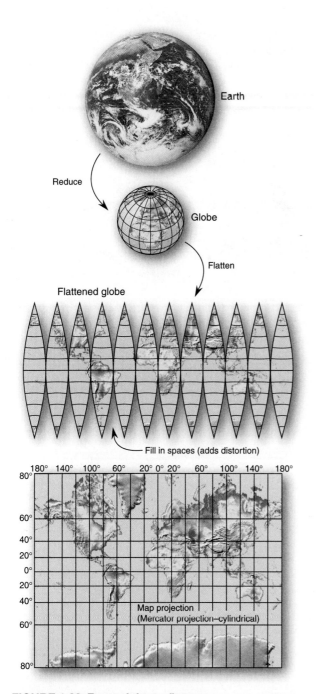

FIGURE 1.20 From globe to flat map.
Conversion of the globe to a flat map projection requires decisions about which properties to preserve and what amount of distortion is acceptable. [Astronaut photo from *Apollo 17* 1972, NASA.]

you place it on the map. In contrast, if a cartographer selects the property of true shape, such as for a map used for navigational purposes, then equal area must be sacrificed, and the scale will actually change from one region of the map to another.

The Nature and Classes of Projections

Despite the fact that modern cartographic technology uses mathematical constructions and computer-assisted graphics, the word *projection* is still used. The term comes from times past, when geographers actually projected the shadow of a

wire-skeleton globe onto a geometric surface. The wires represented parallels, meridians, and outlines of the continents. A light source then casts a shadow pattern of latitude and longitude lines from the globe onto various geometric surfaces, such as a *cylinder*, *plane*, or *cone*.

Figure 1.21 illustrates the general classes of map projections and the perspectives from which they are generated. The classes shown include the cylindrical, planar (or azimuthal), and conic. Another class of projections, which cannot be derived from this physical-perspective approach, is the nonperspective oval shape. Still other projections are derived from purely mathematical calculations.

With projections, the contact line or contact point between the wire globe and the projection surface—called a *standard line* or *standard point*—is *the only place where all globe properties are preserved*. Thus, a *standard parallel* or *standard meridian* is a standard line true to scale along its entire length without any distortion. Areas away from this critical tangent line or point become increasingly distorted. Consequently, this area of optimum spatial properties should be centered on the region of interest.

The commonly used **Mercator projection** (from Gerardus Mercator, A.D. 1569) is a cylindrical projection (Figure 1.21a). The Mercator is a true-shape projection, with meridians appearing as equally spaced straight lines and parallels appearing as straight lines that are spaced closer together near the equator. The poles are infinitely stretched, with the 84th north parallel and 84th south parallel fixed at the same length as that of the equator. Note in Figure 1.20 and Figure 1.21a that the Mercator projection is cut around the 80th parallel in each hemisphere because of the severe distortion at higher latitudes.

Unfortunately, Mercator classroom maps present false notions of the size (area) of midlatitude and poleward landmasses. A dramatic example on the Mercator projection is Greenland, which looks bigger than all of South America. In reality, Greenland is an island only one-eighth the size of South America and is actually 20% smaller than Argentina alone.

The advantage of the Mercator projection is that a line of constant direction, known as a **rhumb line**, is straight and thus facilitates plotting directions between two points (see Figure 1.22). Thus, the Mercator projection is useful in navigation and is the standard for nautical charts prepared by the National Ocean Service since 1910 (formerly US Coast and Geodetic Survey).

The *gnomonic*, or *planar*, *projection* in Figure 1.21b is generated by projecting a light source at the center of a globe onto a plane that is tangent to (touching) the globe's surface. The resulting severe distortion prevents showing a full hemisphere on one projection. However, a valuable feature is derived: All great circle routes, which are the shortest distance between two points on Earth's surface, are projected as straight lines (Figure 1.22a). The great circle routes plotted on a gnomonic projection then can be transferred to a true-direction projection, such as the Mercator, for determination of precise compass headings (Figure 1.22b). Rhumb lines cross successive meridians at

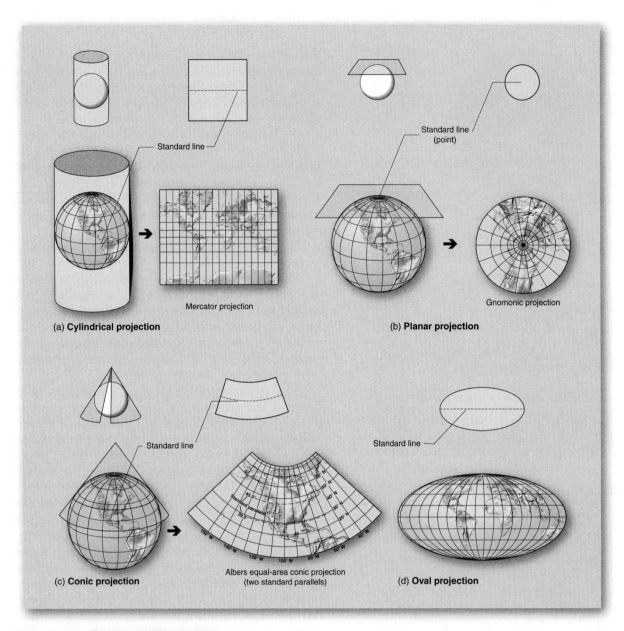

FIGURE 1.21 Map projection classes.
Four general classes and perspectives of map projections—cylindrical, planar, conic, and oval projections.

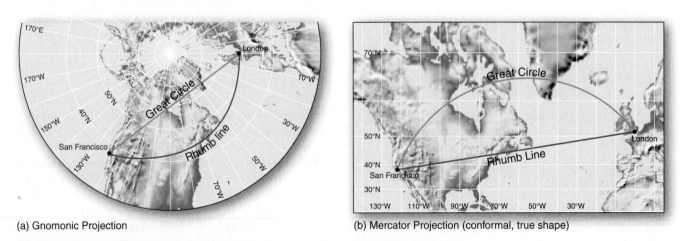

(a) Gnomonic Projection

(b) Mercator Projection (conformal, true shape)

FIGURE 1.22 Determining great circle routes.
Comparison of rhumb line and great-circle routes from San Francisco to London on gnomonic (a)
and Mercator (b) projections. All great circle routes are the shortest distance between any two points.

the same angle, appearing as a straight line only on a Mercator projection.

For more information on maps used in this text and standard map symbols, turn to Appendix A, "Maps in This Text and Topographic Maps." Topographic maps are essential tools of landscape analysis. Geographers, other scientists, travelers, and anyone visiting the outdoors may use topographic maps. US Geological Survey (USGS) topographic maps appear in several chapters of this text. Perhaps you have used a "topo" map in planning a hike. *The National Map* that the USGS calls "The Nation's Topographic Map for the 21st Century" is discussed in the appendix and is available on the Internet. Go to **http://nationalmap.gov/** and click through "The National Map Viewer" and find your campus and your home town; the topographic shading shows you the texture of the landscape.

Remote Sensing and GIS

Geographers probe, analyze, and map our home planet through remote sensing and geographic information systems (GIS). These technologies enhance our understanding of Earth. Geographers use remote-sensing data to study—among many subjects—humid and arid lands, natural and economic vegetation, snow and ice, Earth energy budgets, seasonal variation of atmospheric and oceanic circulation, sea-level measurements, atmospheric chemistry, weather, geologic features and events, changes in the timing of seasons, and the human activities that produce global change.

Remote Sensing

In this era of observations from orbit outside the atmosphere, from aircraft within it, and from remote submersibles in the oceans, scientists obtain a wide array of remotely sensed data (Figure 1.23). Remote sensing is nothing new to humans—we do it with our eyes all the time. When we scan the environment with our eyes, we are sensing the shape, size, and color of objects from a distance, registering energy from the visible wavelength portion of the electromagnetic spectrum. Similarly, when a camera views the wavelengths for which its film or sensor is designed (visible light or infrared), it remotely senses energy that is reflected or emitted from a scene.

Our eyes and cameras are familiar means of obtaining **remote-sensing** information about a distant subject without having physical contact. Aerial photographs have been used for years to improve the accuracy of surface maps faster and more cheaply than can be done by on-site surveys.

Remote sensors on satellites, the International Space Station, and other craft sense a broader range of wavelengths than can our eyes. They can be designed to "see" wavelengths shorter than visible light (ultraviolet) and wavelengths longer than visible light (infrared and microwave radar).

Following launch, satellites can be set in several types of orbits: geostationary, polar, and Sun-synchronous,

illustrated in Figure 1.24. Figure 1.24a shows a geostationary (geosynchronous) orbit, typically 35,790 km (22,239 mi) altitude, where satellites pace Earth's rotation speed and therefore remain "parked" above a specific location, usually the equator, which provides a full hemisphere and is good for recording weather; an example is *GOES*.

Figure 1.24b shows a polar-orbiting satellite typically residing between 200 km and 1000 km (124 mi–621 mi) altitude and passing over every portion of the globe as Earth rotates beneath it, completing each orbit in 90 minutes. Such satellites are good for distant regions such as the poles, for environmental studies, and for spy surveillance; examples include *Terra* and *Aqua*.

Figure 1.24c demonstrates a Sun-synchronous orbit, shifting its 600-km to 800-km (373-mi to 497-mi) altitude, near-polar track about 1° per day (365 days in a year, 360° in a circle), providing a surface below in constant sunlight for visible-wavelength images or in constant darkness for long-wave radiation images.

Satellites do not take conventional-film photographs. Rather, they record *images* that are transmitted to Earth-based receivers in a way similar to television satellite transmissions or a digital camera. A scene is scanned and broken down into pixels (*pic*ture *el*ements), each identified by coordinates named *lines* (horizontal rows) and *samples* (vertical columns). For example, a grid of 6000 lines and 7000 samples forms 42,000,000 pixels, providing great detail. The large amount of data needed to produce a single image requires computer processing and data storage at ground stations.

Digital data are processed in many ways to enhance their utility: simulated natural color, "false" color to highlight a particular feature, enhanced contrast, signal filtering, and different levels of sampling and resolution. Active and passive are two types of remote-sensing systems.

Active Remote Sensing Active systems direct a beam of energy at a surface and analyze the energy reflected back. An example is *radar* (*ra*dio *d*etection *a*nd *r*anging). A radar transmitter emits short bursts of energy that have relatively long wavelengths (0.3 to 10 m) toward the subject terrain, penetrating clouds and darkness. Energy reflected back to a radar receiver for analysis is known as *backscatter*. An example is the computer image of wind and sea-surface patterns over the Pacific in Figure 4.6a, developed from 150,000 radar-derived measurements made on a single day by the *Seasat* satellite.

In addition, NASA sent imaging radar systems into orbit on several Space Shuttles. The subjects of study included oceanography, landforms and geology, and biogeography. Shuttle missions in 1994 by *Endeavour* and *Atlantis* marked dramatic contributions to Earth observations using radar and other sensors to study stratospheric ozone, weather, volcanic activity, earthquakes, and water resources, among many subjects.

One Space Shuttle mission in September 1994 was appropriately loaded with radar sensors to study volcanoes. Only 8 hours after launch, the Kliuchevskoi Volcano

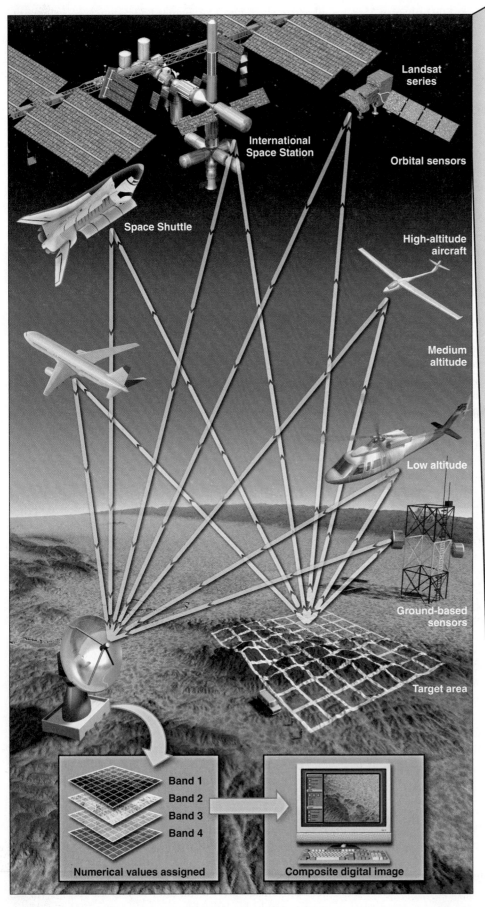

A sample of orbital platforms

Envisat: ESA environment-monitoring satellite; 10 sensors, including a next-generation radar.

ERBS: Earth Radiation Budget Satellite.

GOES: Weather monitoring and forecasting.

JASON: Measures sea level heights.

Landsat: First in 1972 to *Landsat-7* in 1999, millions of images for Earth systems science and global change.

NOAA: First in 1978 through *NOAA-15, -16,* -17, and -18 now operating, global data, short- and long-term weather forecasts.

RADARSAT-1, -2: Synthetic Aperature Radar in near polar orbit operated by Canada Space Agency.

SciSat-1: Analyzes trace gases, thin clouds, atmospheric aerosols with Arctic focus.

SeaStar: Carries the SeaWiFS (Sea-viewing Wide Field-of-View instrument) to observe Earth's oceans and microscopic marine plants.

Terra and *Aqua:* Environmental change, error-free surface images, cloud properties, through five instrument packages.

TOMS-EP*:* Total Ozone Mapping Spectrometer, monitoring stratospheric ozone, similar instruments on *NIMBUS-7* and *Meteor-3.*

TOPEX/POSEIDON: Measures sea level heights.

TRMM: Tropical Rainfall Measuring Mission, includes lightning detection and global energy budget measurements.

UARS: Since 1991 measuring atmospheric chemistry and ozone layer changes.

For more info see:
http://www.nasa.gov/centers/goddard/mission/index.html

Landsat series

International Space Station

Orbital sensors

Space Shuttle

High-altitude aircraft

Medium altitude

Low altitude

Ground-based sensors

Target area

Band 1
Band 2
Band 3
Band 4

Numerical values assigned

Composite digital image

FIGURE 1.23 Remote-sensing technologies.
Remote sensing from orbiting spacecraft, aircraft in the atmosphere, and ground-based sensors measures and monitors Earth's systems. Computers process data to produce digital images for analysis. Many of the physical systems discussed in this text are studied using these technologies. A sample of remote-sensing platforms is in the margin. A Space Shuttle is shown in its inverted orbital flight mode (illustration not to scale).

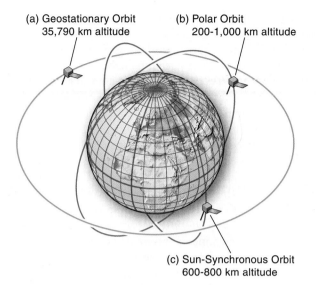

(a) Geostationary Orbit
35,790 km altitude

(b) Polar Orbit
200-1,000 km altitude

(c) Sun-Synchronous Orbit
600-800 km altitude

FIGURE 1.24 Three satellite orbital paths.
(a) Geostationary (geosynchronous) orbit, typically 35,790 km altitude. (b) Polar orbiting satellites typically are between 200–1000 km altitude. (c) Sun-synchronous orbits shift their near-polar track at 600 to 800 km altitude.

on the Kamchatka Peninsula of Russia erupted unexpectedly. Previously, this volcano had erupted in 1737 and 1945. The Shuttle radar was able to see through ash and smoke and expose lava flows and the volcanic eruption in dramatic images (Figure 1.25). Astronaut Mission Specialist Dr. Thomas Jones operated the radar and camera to make the image and photo in the figure. A Career Link profiles this astronaut on the Student Learning Center web site. Chapter 9 discusses volcanic processes.

Canada operates *RADARSAT–1, –2* and *SCISAT–1* and the Japanese have *JERS–1* that image in radar wavelengths (Figure 1.26). The European Space Agency (ESA) now operates two Earth resource satellites (*ERS–1* and *–2*). They work in tandem, producing a spectacular 10-cm (3.9-in) resolution. Imaging the same area at different times, pairs of images produce a digital, three-dimensional data set.

ESA's environment-monitoring satellite, *Envisat*, went into service in 2002. In addition to a more sophisticated radar than *ERS*, *Envisat* carries nine passive sensor packages, making it a multitalented craft. It monitors ocean

(a) Visible light

(b) Radar

(c)

FIGURE 1.25 A volcanic eruption seen from orbit.
(a) Photograph (passive, visible light) and (b) image (active, radar) of the eruption of the Kliuchevskoi Volcano on the Kamchatka Peninsula of Siberia, Russia, as captured by the Space Shuttle *Endeavour*, September 1994. (c) *Terra* image of the Kamchatka Peninsula made June 26, 2002. Snow cover highlights the many active volcanoes. [(a) Photo and (b) image courtesy of JPL/NASA; (c) *Terra* image, June 26, 2002, from MODIS Land Rapid Response Team, NASA/GSFC.]

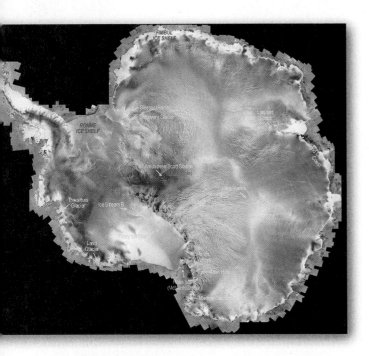

FIGURE 1.26 *RADARSAT–1* **composite of Antarctica.**
A mosaic of *RADARSAT–1* images made over an 18-day
period. Resolution can be as fine as 8 m (26 ft) with this
microwave Synthetic Aperture Radar (SAR) system. The platform is placed in a near-polar orbit, approximately 800 km (497
mi) above the Earth. [Image courtesy of Canadian Space Agency.]

temperatures, sea level, and wave patterns; polar ice;
forests; biological activity in the oceans and on land; cloud
heights; atmospheric ozone and pollution; and carbon
dioxide concentrations (see http://envisat.esa.int/).

Passive Remote Sensing Passive remote-sensing systems record energy radiated from a surface, particularly
visible light and infrared. Our own eyes are passive
remote sensors, as was the *Apollo 17* astronaut camera that
made the picture of Earth on the back cover of this book
from a distance of 37,000 km (23,000 mi).

Passive remote sensors on five *Landsat* satellites,
launched by the United States, provide a variety of data,
as shown in images of the Appalachian Mountains in
Chapter 16, river deltas and New Orleans before and after
Katrina in Chapter 11, and an Icelandic ice cap in Chapter
14. Two *Landsats* remain operational, *Landsat–5* and *–7*.
See http://landsat.gsfc.nasa.gov/ and http://geo.arc.
nasa.gov/sge/landsat/landsat.html for more information and additional links.

A NASA, National Oceanic and Atmospheric Administration (NOAA, see http://www.noaa.gov/), and U.K.
joint venture, the *Polar Operational Environmental Satellites*
(POES), completes nearly polar orbits 14 times a day,
carrying the *advanced very high resolution radiometer*
(AVHRR). These are in Sun-synchronous orbits aboard
NOAA-15 and *NOAA-17* for morning hours, and *NOAA-*
16 and *NOAA-18* for afternoon hours. These sensors are
sensitive to visible and infrared wavelengths. They are
used for weather analysis and forecasting, sea-surface
temperature measurement, ozone concentrations, and

temperatures in the stratosphere, among many other
functions. See http://www.oso.noaa.gov/poesstatus/.

Key to NASA's Earth Observing System (EOS) is
satellite *Terra*, which began beaming back data and images
in 2000 (see http://terra.nasa.gov/), followed by another
satellite in the series called *Aqua*. Five instrument packages observe Earth systems in detail, exploring the atmosphere, landscapes, oceans, environmental change, and
climate, among other abilities. Of more than 100 remote-
sensing images presented throughout this text, watch for
those from the *Terra* satellite. The image of Earth on the
half-title page of this book includes a cloud snapshot from
GOES added to *Terra* MODIS images in 2000.

The *Geostationary Operational Environmental Satellites*,
known as *GOES*, became operational in late 1994, providing frequent infrared and visible images—the ones you
see on television weather reports. Geostationary satellites
stay in semipermanent positions because they keep pace
with Earth's rotational speed at their altitude of
35,400 km (22,000 mi).

GOES-9, launched in 1995, is now above 160° E in a
backup orbital mode. *GOES-10*, active in 1998, was moved
in December 2006 to monitor South America from above
60° W. *GOES-12* sits above 75° W longitude to monitor
central and eastern North America and the western
Atlantic (Figure 1.27). Think of these satellites as hovering
over these meridians for continuous coverage, using visual
wavelengths for daylight hours and infrared for nigthtime
views. *GOES-11* operates above 135° W longitude for
western North America and eastern Pacific Ocean weather.

FIRST GOES-12
VISIBLE IMAGE
AUGUST 17, 2001 18:00 UTC (2:00 PM EDT)

FIGURE 1.27 *GOES-12* **first image.**
GOES-12 demonstrates in its first image excellent image
quality from its 37,500 km (23,300 mi) orbital post. This satellite
"hovers" above 75° W longitude for central and eastern Atlantic
coverage. [Image courtesy of NOAA.]

GOES-13, launched in May 2006, is parked in orbit to become next in line when needed. For the future, *GOES-O*, launched in July 2007, will become *GOES-14*, and *GOES-P*, after a successful 2008 launch, will be operational as *GOES-15*. (See the Geostationary Satellite Server at http://www.goes.noaa.gov/. The *GOES* Project Science appears at http://rsd.gsfc.nasa.gov/goes/ or see http://www.ghcc.msfc.nasa.gov/GOES/.)

Other satellites used for weather include Japan's *MTSAT-1R* (Multi-Functional Transport Satellite, 2005) and *MTSAT-2* (2006), which now serve weather and communication needs over the western Pacific (*GOES-9* serves as backup). China's *Feng Yun-2B* and *Feng Yun-1D* above 105°E, cover the Far East, and *METEOSAT-7*, above 58°E, covers Europe and Africa and is operated by the European Space Agency. (See the Remote Sensing Virtual Library at http://virtual.vtt.fi/space/rsvlib/ for remote-sensing links; click on "Satellite Data" for specific coverage.)

One of two commercial systems includes the three French satellites (1, 2, and 4) called *SPOT* (Systeme Probatoire d'Observation de la Terre) that resolve objects on Earth down to 10 to 20 m (33 to 66 ft), depending on which sensors are used. Other commercial systems are operated by GeoEye, formerly Space Imaging, which was acquired by ORBIMAGE. Its satellites offer 0.41-m to 4-m resolution from *Ikonos–2*, *OrbView–2*, *–3*, and *–5*, and its new *GeoEye-1* satellite, all in Sun-synchronous orbits at 680 km (420 mi) altitude (http://www.geoeye.com/). See the Chapter 1 opening photo for an example of an image from *Ikonos–2*.

Exciting new developments are occurring all the time, such as combining spectroscopy (the study of electromagnetic radiation through absorption features; see Chapter 2), remote-sensing technology, and geographic information systems (discussed next) to produce *hyperspectral images* displayed as a *datacube* with the actual spatial information (surface *X–Y* plane) and spectral analysis (in depth as a *Z* plane) presented together. There are many applications of this technology for analyzing Earth systems and features.

The USGS *EROS* (Earth Resources Observation and Science) Data Center, near Sioux Falls, South Dakota, is the national repository for geophysical and geospatial data—remote-sensing imagery, photos, maps, the Global Land Inventory System, NASA's EOS Gateway, and more (Figure 1.28). *EROS* is where we gather all the data and imagery in one place that will guide future generations about Earth systems operations and what conditions were like in this era.

For a sampling of global remote-sensing coverage from a variety of satellite platforms, go to the USGS Global Visualization Viewer, select the satellite, click on the map for a location, and then watch (http://glovis.usgs.gov/).

Geographic Information Systems (GIS)

Techniques like remote sensing, which are important tools for acquiring large volumes of spatial data, are most valuable when that data can be stored, processed, and retrieved in

FIGURE 1.28 EROS Data Center (EDS).
USGS operates the EROS (Earth Resources Observation Systems) Data Center, outside Sioux Falls, South Dakota. This center operates one of the largest computer complexes dedicated to science (see http://edc.usgs.gov/). [Photo by Bobbé Christopherson.]

useful ways. A **geographic information system (GIS)** is a computer-based, data-processing tool for gathering, manipulating, and analyzing geographic information. Today's sophisticated computer systems allow the integration of geographic information from direct surveys (on-the-ground mapping) and remote sensing in complex ways never before possible. Through a GIS, Earth and human phenomena are analyzed over time. This can include the study and forecasting of diseases and epidemics such as West Nile virus, SARS, and the vicious Ebola virus, to name a few recent examples. Examples are shown in Figure 1.29b and c.

The beginning component for any GIS is a coordinate system such as latitude–longitude, which establishes reference points against which to position data. The coordinate system is digitized, along with all areas, points, and lines. Remotely sensed imagery and data are then added on the coordinate system.

A GIS is capable of analyzing patterns and relationships within a single data plane, such as the floodplain or soil layer in Figure 1.29a. The GIS also can generate an overlay analysis where two or more data planes interact. When the layers are combined, the resulting synthesis—a *composite overlay*—is a valuable product, ready for use in analyzing complex problems. A research study may follow specific points or areas through the complex of overlay planes. The utility of a GIS compared with that of a fixed map is the ability to manipulate the variables for analysis and to constantly change the map—the map is alive!

Before the advent of computers, an environmental-impact analysis required someone to gather data and painstakingly hand-produce overlays of information to determine positive and negative impacts of a project or event. Today, this layered information is handled by a

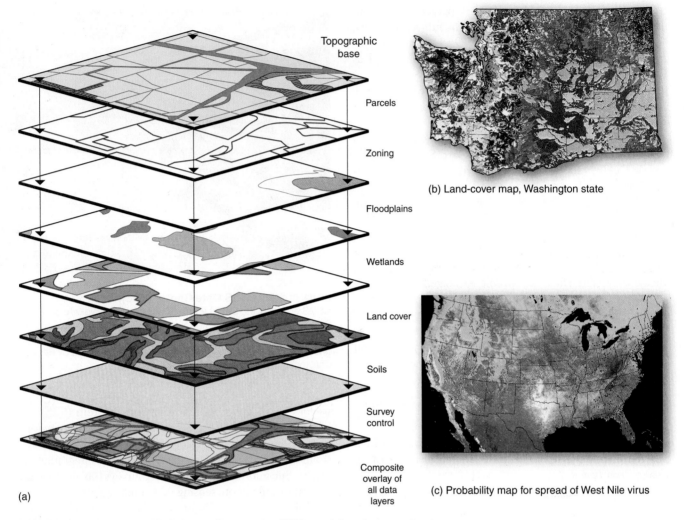

Topographic base

Parcels

Zoning

Floodplains

Wetlands

Land cover

Soils

Survey control

Composite overlay of all data layers

(b) Land-cover map, Washington state

(c) Probability map for spread of West Nile virus

FIGURE 1.29 A geographic information system (GIS) model and map output.
(a) Layered spatial data in a geographic information system (GIS) format. (b) This comprehensive land-cover map is an important component of a statewide GIS analysis for Washington state. The goal is to evaluate the protection of species and biodiversity. (c) A NASA GIS study uses satellite-derived temperature, humidity, and vegetation variables in a GIS overlay of environmental conditions for a probability map of the spread of West Nile virus. Various colors indicate temperatures and temperature ranges, moisture or dryness, and vegetation as an indicator of humidity. [(a) After USGS. (b) GIS map courtesy of Dr. Kelly M. Cassidy, *Gap Analysis of Washington State*, v. 5, Map 3, University of Washington, 1997. (c) Composite image from *NOAA-15* AVHRR data, NASA/GSFC.]

computer-driven GIS, which assesses the complex inter-connections among different components. In this way, subtle changes in one element of a landscape may be identified as having a powerful impact elsewhere.

GIS is a rapidly expanding career field in many sectors of the economy. Regardless of your academic major, the ability to analyze data spatially is important. The annual ESRI International Users Conference, whose attendance is approaching 15,000 people, is where GIS professionals share the latest applications of the GIS technology (see **http://www.esri.com/**). We are in the midst of a GIS revolution. Potent new careers are emerging in business, marketing, and academia. For a list of current trends in GIS, simply enter this topic in your search engine; for an alphabetized GIS resources list, check out **http://www.geo.ed.ac.uk/home/giswww.html** and **http://erg.usgs.gov/isb/pubs/gis_poster/**.

GIS degree programs are available at many colleges and universities. GIS curriculum and certificate programs are now available at many community colleges. For example, at American River College, Sacramento, California, a curriculum offering 15 GIS courses drives programs leading to a GIS certificate or an A.S. degree. To look at this program, go to **http://www.arc.losrios.edu/~earthsci/** and click "GIS."

A consortium of three universities forms the National Center for Geographic Information and Analysis (NCGIA) for GIS education, research, outreach, and model generation. These GIS centers are: Department of Geography, University of California Santa Barbara, Santa Barbara, CA 93106; Department of Surveying and Engineering, University of Maine, Orono, ME 04669; and State University of New York–Buffalo, Buffalo, NY 14260. (See **http://www.ncgia.ucsb.edu/**, **http://www.ncgia.maine.edu/**, or **http://www.geog.buffalo.edu/ncgia**.)

Summary and Review— Essentials of Geography

To assist you, here is a review in handy summary form of the Key Learning Concepts listed on this chapter's title page. Each concept review concludes with a list of the key terms from the chapter, their page numbers, and review questions. Such summary and review sections follow each chapter in the book.

■ *Define* geography and physical geography in particular.

Geography brings together disciplines from the physical and life sciences with the cultural and human sciences to attain a holistic view of Earth—physical geography is an essential aspect of the emerging **Earth systems science**. **Geography** is a science of method, a special way of analyzing phenomena over space; **spatial** refers to the nature and character of physical space. Geography integrates a wide range of subject matter, and geographic education recognizes five major themes: **location**, **region**, **human–Earth relationships**, **movement**, and **place**. Geography's method is **spatial analysis**, used to study the interdependence among geographic areas, natural systems, society, and cultural activities over geographic space. **Process**—that is, analyzing a set of actions or mechanisms that operate in some special order—is central to geographic synthesis.

   Physical geography applies spatial analysis to all the physical elements and processes that make up the environment: energy, air, water, weather, climate, landforms, soils, animals, plants, microorganisms, and Earth itself. Understanding the complex relations among these elements is important to human survival because Earth's physical systems and human society are so intertwined. The development of hypotheses and theories about the Universe, Earth, and life involves the **scientific method**.

   Earth systems science (p. 2)
   geography (p. 4)
   spatial (p. 4)
   location (p. 4)
   region (p. 4)
   human–Earth relationships (p. 4)
   movement (p. 4)
   place (p. 4)
   spatial analysis (p. 4)
   process (p. 4)
   physical geography (p. 4)
   scientific method (p. 4)

1. What is unique about the science of geography? On the basis of information in this chapter, define physical geography and review the geographic approach.
2. In general terms, how might a physical geographer analyze water pollution in the Great Lakes?
3. Assess your geographic literacy by examining atlases and maps. What types of maps have you used? Do you know what projections they employed? Do you know the names and locations of the four oceans, seven continents, and most individual countries? Can you identify the new countries that have emerged since 1990?
4. Suggest a representative example for each of the five geographic themes and use that theme in a sentence.
5. Have you made decisions today that involve geographic concepts discussed within the five themes presented? Explain briefly.

■ *Describe* systems analysis and open and closed systems and *relate* those concepts to Earth systems.

A **system** is any ordered, related set of things and their attributes, as distinct from their surrounding environment. Systems analysis is an important organizational and analytical tool used by geographers. Earth is an **open system** in terms of energy, receiving energy from the Sun, but it is essentially a **closed system** in terms of matter and physical resources.

   As a system operates, "information" is returned to various points in the system via pathways as **feedback loops**. If the feedback information discourages response in the system, it is **negative feedback**. (Further production in the system decreases the growth of the system.) If feedback information encourages response in the system, it is **positive feedback**. (Further production in the system stimulates the growth of the system.) When the rates of inputs and outputs in the system are equal and the amounts of energy and matter in storage within the system are constant (or as they fluctuate around a stable average), the system is in **steady-state equilibrium**. A system that demonstrates a steady increase or decrease in system operations—a trend over time—is in **dynamic equilibrium**. A **threshold** is the moment at which a system can no longer maintain its character, so it lurches to a new operational level, which may not be compatible with previous conditions. Geographers often construct a simplified **model** of natural systems to better understand them.

   Four immense open systems powerfully interact at Earth's surface: three nonliving **abiotic** systems (**atmosphere**, **hydrosphere** [including the **cryosphere**], and **lithosphere**) and a living **biotic** system (**biosphere**, or **ecosphere**).

   system (p. 7)
   open system (p. 7)
   closed system (p. 7)
   feedback loops (p. 8)
   negative feedback (p. 8)
   positive feedback (p. 8)
   steady-state equilibrium (p. 9)
   dynamic equilibrium (p. 9)
   threshold (p. 9)
   model (p. 12)
   abiotic (p. 12)
   biotic (p. 12)
   atmosphere (p. 12)
   hydrosphere (p. 12)
   cryosphere (p. 12)
   lithosphere (p. 12)
   biosphere (p. 12)
   ecosphere (p. 12)

6. Define systems theory as an organizational strategy. What are open systems, closed systems, and negative feedback? When is a system in a steady-state equilibrium condition? What type of system (open or closed) is the human body? A lake? A wheat plant?
7. Describe Earth as a system in terms of both energy and matter; use simple diagrams to illustrate your description.
8. What are the three abiotic spheres (nonliving) that make up Earth's environment? Relate these to the biotic (living) sphere: the biosphere.

■ *Explain* Earth's reference grid: latitude, longitude and latitudinal geographic zones and time.

The science that studies Earth's shape and size is **geodesy**. Earth bulges slightly through the equator and is oblate (flattened) at the poles, producing a misshapen spheroid called a **geoid**. Absolute location on Earth is described with a specific reference grid of **parallels** of **latitude** (measuring distances north and south of the equator) and **meridians** of **longitude** (measuring distances east and west of a prime meridian). A historic breakthrough in navigation and timekeeping occurred with the establishment of an international **prime meridian** (0° through Greenwich, England) and the invention of precise chronometers that enabled accurate measurement of longitude. Latitude, longitude, and elevation are accurately calibrated using a handheld **global positioning system (GPS)** instrument that reads radio signals from satellites. A **great circle** is any circle of Earth's circumference whose center coincides with the center of Earth. Great circle routes are the shortest distance between two points on Earth. **Small circles** are those whose centers do not coincide with Earth's center.

The prime meridian provided the basis for **Greenwich Mean Time (GMT)**, the world's first universal time system. A corollary of the prime meridian is the 180° meridian, the **International Date Line**, which marks the place where each day officially begins. Today, **Coordinated Universal Time (UTC)** is the worldwide standard and the basis for international time zones. **Daylight saving time** is a seasonal change of clocks by 1 hour in summer months.

> geodesy (p. 14)
> geoid (p. 15)
> latitude (p. 16)
> parallel (p. 16)
> longitude (p. 17)
> meridian (p. 17)
> prime meridian (p. 17)
> Global Positioning System (GPS) (p. 17)
> great circle (p. 17)
> small circles (p. 19)
> Greenwich Mean Time (GMT) (p. 21)
> International Date Line (p. 22)
> Coordinated Universal Time (UTC) (p. 22)
> daylight saving time (p. 23)

9. Draw a simple sketch describing Earth's shape and size.

10. How did Eratosthenes use Sun angles to figure out that the 5000-stadia distance between Alexandria and Syene was 1/50 of Earth's circumference? Once he knew this fraction of Earth's circumference, how did he calculate the length of Earth's circumference?

11. What are the latitude and longitude coordinates (in degrees, minutes, and seconds) of your present location? Where can you find this information?

12. Define *latitude* and *parallel* and define *longitude* and *meridian* using a simple sketch with labels.

13. Define a great circle, great circle routes, and a small circle. In terms of these concepts, describe the equator, other parallels, and meridians.

14. Identify the various latitudinal geographic zones that roughly subdivide Earth's surface. In which zone do you live?

15. What does timekeeping have to do with longitude? Explain this relationship. How is Coordinated Universal Time (UTC) determined on Earth?

16. What and where is the prime meridian? How was the location originally selected? Describe the meridian that is opposite the prime meridian on Earth's surface.

17. What is GPS and how does it assist you in finding location and elevation on Earth? Give a couple of examples where it was utilized to correct heights for some famous mountains.

■ *Define* cartography and mapping basics: map scale and map projections.

A **map** is a generalized view of an area, usually some portion of Earth's surface, as seen from above, and greatly reduced in size. The science and art of mapmaking is called **cartography**. Geographers use maps for the spatial portrayal of Earth's physical systems. **Scale** is the ratio of the image on a map to the real world; it relates a unit on the map to an identical unit on the ground. Cartographers create **map projections** for specific purposes, selecting the best compromise of projection for each application. Compromise is always necessary because Earth's round, three-dimensional surface cannot be exactly duplicated on a flat, two-dimensional map. **Equal area** (equivalence), **true shape** (conformality), true direction, and true distance are all considerations in selecting a projection. The **Mercator projection** is in the cylindrical class of projection; it has true-shape qualities and straight lines that show constant direction. **Rhumb lines** are lines of constant direction and appear as straight lines on the Mercator.

> map (p. 23)
> cartography (p. 23)
> scale (p. 23)
> map projection (p. 24)
> equal area (p. 24)
> true shape (p. 24)
> Mercator projection (p. 25)
> rhumb line (p. 25)

18. Define cartography. Explain why it is an integrative discipline.

19. What is map scale? In what three ways may it be expressed on a map?

20. State whether each of the following ratios is a large scale, medium scale, or small scale: 1:3,168,000; 1:24,000; 1:250,000.

21. Describe the differences between the characteristics of a globe and those that result when a flat map is prepared.

22. What type of map projection is used in Figure 1.13? In Figure 1.17? (See Appendix A.)

■ *Describe* remote sensing and *explain* geographic information system (GIS) methodology as a tool used in geographic analysis.

Orbital and aerial **remote sensing** analyzes the operation of Earth's systems. Satellites do not take photographs, but record images that are transmitted to Earth-based receivers. Satellite images are recorded in digital form for later processing, enhancement, and generation. Aerial photographs have been used for years to improve the accuracy of surface maps.

The mountain of data already collected has led to the development of **geographic information system (GIS)** technology. Computers process geographic information from direct surveys and remote sensing in complex ways never before possible. GIS methodology is an important step in better understanding Earth's systems and is a vital career opportunity for geographers.

The science of physical geography is in a unique position to synthesize the spatial, environmental, and human aspects of our increasingly complex relationship with our home planet—Earth.

remote sensing (p. 27)
geographic information system (GIS) (p. 31)

23. What is remote sensing? What are you viewing when you observe a weather satellite image on TV or in the newspaper? Explain.

24. Describe the *Terra*, *Landsat*, *GOES*, and *NOAA* satellites and explain them using several examples.

25. If you were in charge of planning for development of a large tract of land, how would GIS methodologies assist you? How might planning and zoning be affected if a portion of the tract in the GIS is a floodplain or prime agricultural land?

NetWork and Critical Thinking Tools

A. Select a location (for example, your campus, home, workplace, a public place, or a city) and determine the following: latitude, longitude, and elevation. Describe the resources you used to gather this geographic information. Have you ever used a GPS unit to determine these aspects of your location?

B. Let's say there is a world globe in the library or geography department that is 61 cm (24 in.) in diameter. We know that Earth has an equatorial diameter of 12,756 km (7926 mi), so the scale of the globe is the ratio of 61 cm to 12,756 km. We divide Earth's actual diameter by the globe's diameter (12,756 km ÷ 61 cm) and determine that 1 cm of the globe's diameter equals about _____ cm of Earth's actual diameter. Thus, the representative fraction for the globe is expressed in centimeters as 1/ _____. (Hint: 1 km = 1000 m, 1 m = 100 cm; therefore, Earth's diameter of 12,756 km represents 1,275,600,000 cm.)

C. The Chapter 1 opening feature shows the aftermath of the May 4, 2007, EF-05 monster tornado that leveled Greenburg, Kansas, at the time it occurred and one-year later. As you take time examing the aerial photo and orbital image, think through each of the five themes of geographic science detailed in Figure 1.1 and describe how each theme might be implemented to analyze this storm and its destruction—how the theme applies. Use your Internet search browser to explore the tornado and turn ahead to the related section in Chapter 5 and go to some of the sources listed. Be sure to include any URLs for the sites you visit.

Part I   The Energy–Atmosphere System

Sunrise illuminates false-cirrus clouds seeded by jet contrails. Research is progressing on the impacts of such human-induced clouds in the energy-atmosphere system and the dimming of insolation from the Sun. Such is the role of physical geography and spatial analysis of Earth's systems. *[Photo by Bobbé Christopherson.]*

Our planet and our lives are powered by radiant energy from the star that is closest to Earth—the Sun. For more than 4.6 billion years, solar energy has traveled across interplanetary space to Earth, where a small portion of the solar output is intercepted. Because of Earth's curvature, the energy at the top of the atmosphere is unevenly distributed, creating imbalances from the equator to each pole—the equatorial region experiences energy surpluses; the polar regions experience energy deficits. Also, the pulse of seasonal change varies the distribution of energy during the year.

Earth's atmosphere acts as an efficient filter, absorbing most harmful radiation, charged particles, and space debris so that they do not reach Earth's surface. In the lower atmosphere, the uneven-ness of daily energy receipt empowers atmospheric and surface energy budgets, giving rise to global patterns of temperature and circulation of wind and ocean currents. Each of us depends on many systems that are set into motion by energy from the Sun. These systems are the subject of PART I.

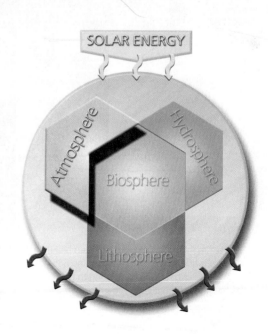

The Midnight Sun seen from a research ship in the Arctic Ocean. The camera points due north, viewing the Sun's rays from across the North Pole some 885 km (550 mi) distant (see the inset illustration). This seemed so strange at the time to have natural lighting that felt like about 3 P.M. (15:00 hours), yet have our watches clearly show it to be midnight. [Photo by Bobbé Christopherson.]

The inset illustration shows:

23.5° N
Tropic of Cancer
Equator
Tropic of Capricorn
Antarctic Circle
S 23.5°
Summer (June) solstice

Chapter 2

Solar Energy, Seasons, and the Atmosphere

KEY LEARNING CONCEPTS

After reading the chapter, you should be able to:

- *Distinguish* among galaxies, stars, and planets and *locate* Earth.

- *Describe* the Sun's operation and *explain* the characteristics of the solar wind and the electromagnetic spectrum of radiant energy.

- *Define* solar altitude, solar declination, and daylength and *describe* the annual variability of each—Earth's seasonality.

- *Construct* a general model of the atmosphere based on composition, temperature, and function and *diagram* this model in a simple sketch.

- *Describe* conditions within the stratosphere; specifically, *review* the function and status of the ozonosphere (ozone layer).

- *Distinguish* between natural and anthropogenic variable gases and materials in the lower atmosphere and *describe* the sources and effects of air pollution and acid deposition.

STUDENT LEARNING CENTER TOOLS

The *Elemental Geosystems* Student Learning Center provides on-line resources for this chapter by clicking on the book cover at **www.mygeoscienceplace.com**. Once you are in our Learning Center, bookmark the URL address. For this chapter find reviewing tools, eBook links, self-tests and quizzes, critical thinking items, figure animations, photo galleries, and satellite loops. Some highlights:

▶ Thinking spatially reviews key figures in the chapter with your interaction.

▶ Animation on the formation of our Solar System.

▶ Earth–Sun relations and the march of the seasons in an interactive animation.

▶ Operation of stratospheric ozone animation, its depletion, and a guiding tutorial.

▶ More than 10 additional URLs along with Quick Links to all the URLs in the chapter.

The Universe is populated with billions of galaxies. One of these is our own Milky Way Galaxy, consisting of billions of stars. Among these stars is an average yellow star we call the Sun, although the dramatic *SOHO* satellite image in Figure 2.2 seems anything but average! Our Sun radiates energy in all directions and upon its family of orbiting planets. Of special interest to us is the solar energy that falls on the third planet, our immediate home.

In this chapter: Solar energy at the top of the atmosphere sets into motion the winds, weather systems, and ocean currents that greatly influence our lives. This solar energy input to the atmosphere plus Earth's tilt, orientation, revolution, and rotation produce daily, seasonal, and annual patterns of changing daylength and Sun angles.

Earth's atmosphere is a unique reservoir of gases—the product of nearly 5 billion years of development. We all participate in the atmosphere with each breath we take. We examine the atmosphere's structure, composition, and function. Our study also must include the spatial aspects of human-induced problems that affect the atmosphere, such as the stratospheric ozone problem, air pollution, and acid deposition.

The Solar System, ① Sun, and Earth

Our Solar System is located on a remote, trailing edge of the **Milky Way Galaxy,** a flattened, disk-shaped mass estimated to contain nearly 400 billion stars. From our Earth-bound perspective in the Milky Way, the galaxy appears to stretch across the night sky like a narrow band of hazy light. On a clear night the unaided eye can see only a few thousand of these billions of stars (Figure 2.1a, b). Have you ever observed the Milky Way Galaxy in the night sky?

According to prevailing theory, our Solar System condensed from a large, slowly rotating, collapsing cloud of dust and gas called a *nebula*. **Gravity,** the mutual attracting force exerted by the mass of an object upon all other objects, was the key organizing force in this condensing solar nebula. The beginnings of the formation of the Sun and its Solar System are estimated to have occurred more than 4.6 billion years ago.

The concept that suns condense from nebular clouds and planetesimals form in orbits around their central masses is the **planetesimal hypothesis,** or *dust-cloud hypothesis.* The planets appear to accrete (growth by accumulation) from dust, gases, and icy comets that are drawn by gravity into collision and coalescence. Astronomers are observing this formation process under way in other parts of the galaxy. By 2008, astronomers had identified more than 300 planets orbiting other stars.

⑥ Dimensions, Distances, and Earth's Orbit

The **speed of light** is 299,792 kmps (186,282 mps), which is about 9.5 trillion km or nearly 6 trillion mi per year. This tremendous distance that light travels in a year

is known as a *light-year* and is used as a unit of measurement for the vast universe.

For spatial comparison, our Moon is an average distance of 384,400 km (238,866 mi) from Earth, or about 1.28 seconds in terms of light speed. Our entire Solar System is approximately 11 hours in diameter, measured by light speed. In contrast, the Milky Way is about 100,000 light-years from side to side, and the known Universe that is observable from Earth stretches approximately 12 billion light-years in all directions. (See a Solar System simulator at **http://space.jpl.nasa.gov/.**)

Earth's orbit around the Sun is presently elliptical—an oval-shaped path. Earth's average distance from the Sun is approximately 150 million km (93 million mi), which means that light reaches Earth from the Sun in an average of 8 minutes and 20 seconds. Earth is at **perihelion** (its closest position to the Sun) on January 3 and at **aphelion** (its farthest position from the Sun) on July 4 (Figure 2.1d).

② Solar Energy: From Sun to Earth

Our Sun is unique to us and yet commonplace in our galaxy. It is the ultimate energy source for most life processes in our biosphere. The Sun is the only object in the entire Solar System that produces thermonuclear energy. Clearly, the Sun is the dominant object in our region of space.

The solar mass produces tremendous pressure and high temperatures deep in its dense interior. Under these conditions, the Sun's abundant atoms of hydrogen are forced together, and pairs of hydrogen nuclei are fused (joined) in the process of **fusion,** liberating enormous quantities of energy. A sunny day can seem so peaceful, certainly belying the violence proceeding on the Sun. The Sun's principal outputs consist of the solar wind and radiant energy in portions of the electromagnetic spectrum. Let us trace each of these emissions across space to Earth.

⑦ Solar Wind

The Sun constantly emits ionized (electrically charged) particles (principally, electrons and protons) that surge outward in all directions from the Sun's surface. The term **solar wind** was first applied to this phenomenon in 1958. The solar wind grows stronger during periods of increased sunspot activity. **Sunspots** are large magnetic storms that reveal solar activity (Figure 2.2). These surface disturbances produce flares and prominences. In addition, outbursts of charged material referred to as *coronal mass ejections* contribute to the solar wind flow of material to space.

A regular cycle exists for sunspot occurrences, averaging 11 years from maximum to maximum; however, the cycle may vary from 7 to 17 years. The 1990–1991 maximum was the most intense ever observed. A sunspot minimum occurred in 1997, an intense maximum in 2001, a minimum in 2007, and a forecasted maximum in 2012, or before, roughly maintains the average. (For more on the sunspot cycle, see **http://solarscience.msfc.nasa.gov/** and for the

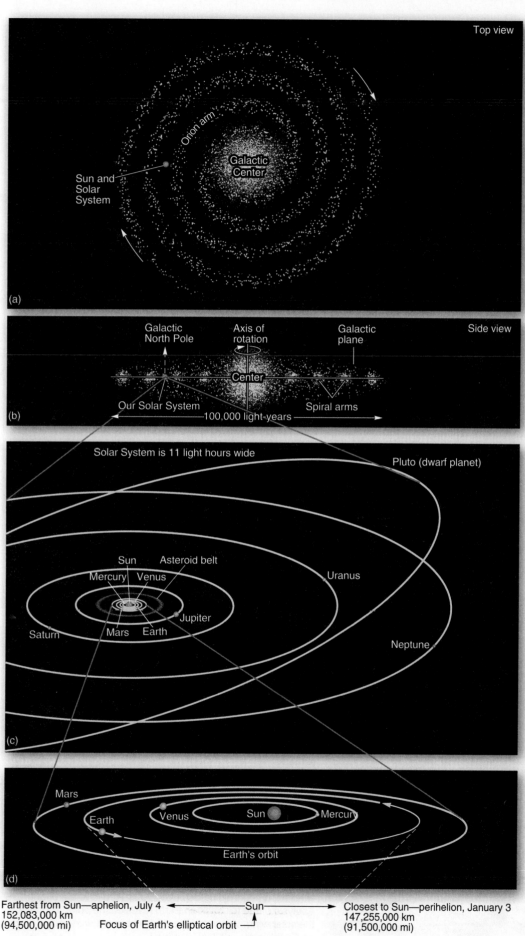

(a) Top view

Orion arm

Galactic Center

Sun and Solar System

(b) Side view

Galactic North Pole

Axis of rotation

Galactic plane

Center

Our Solar System

Spiral arms

100,000 light-years

(c) Solar System is 11 light hours wide

Pluto (dwarf planet)

Sun   Asteroid belt

Mercury   Venus

Uranus

Jupiter

Saturn   Mars   Earth

Neptune

(d) Mars

Earth

Venus

Sun

Mercury

Earth's orbit

Farthest from Sun—aphelion, July 4
152,083,000 km
(94,500,000 mi)

Sun

Focus of Earth's elliptical orbit

Closest to Sun—perihelion, January 3
147,255,000 km
(91,500,000 mi)

ANIMATION

Nebular Hypothesis

FIGURE 2.1 Milky Way Galaxy, Solar System, and Earth's orbit.
The Milky Way Galaxy viewed from above (a) and cross-section side view (b). Our Solar System of eight planets and asteroids is some 30,000 light-years from the center of the Galaxy. In 2006, Pluto was reclassified a "dwarf planet" and part of the Kuiper Asteroid Belt. All of the planets except Pluto have orbits closely aligned to the plane of the ecliptic (c). The four inner terrestrial planets and the structure of Earth's elliptical orbit, illustrating perihelion (closest) and aphelion (farthest) positions during the year, are shown in (d).

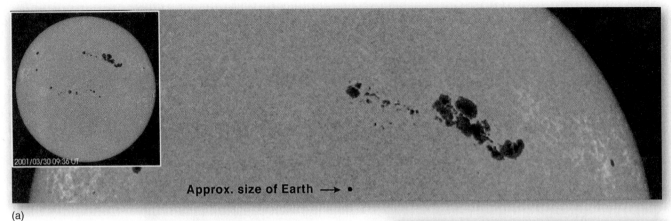

(a)

FIGURE 2.2 Images of the Sun and sunspots.
(a) The Sun and a large sunspot group (visible dark patches) in a recent active cycle, imaged by the MDI (Michelson Doppler Imager) instrument aboard satellite *SOHO*. This group was the source of numerous flares and coronal mass ejections, including the largest flare in recent decades. The area within the sunspot group is more than 13 times the entire surface area of Earth (i.e., 13 times 500 million km^2)—Earth is shown for scale. (b) A dramatic Sun captured by instruments aboard the *SOHO* satellite. A twirling prominence rises into the Sun's corona, and other prominences are also visible. [(a) Image courtesy of *SOHO/EIT* (Solar and Heliospheric Observatory/Extreme Ultraviolet Imaging Telescope) Consortium; (b) Image courtesy of *SOHO/MDI* Consortium—*MDI* is from Stanford-Lockheed/Martin Institute for Space Research. *SOHO* is an international project of cooperation between the European Space Agency and NASA. See **http://sohowww.nascom.nasa.gov/**.]

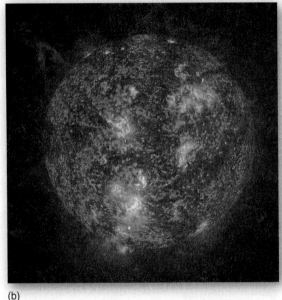

(b)

latest space weather, see **http://www.spaceweather.com/** or **http://www.swpc.noaa.gov/**).

Earth's outer defense against the charged particles of the solar wind is the **magnetosphere**, which is a magnetic field surrounding Earth, generated by dynamo-like motions within our planet. The magnetosphere deflects the solar wind toward both poles, so that only a small portion of it enters the atmosphere.

Because the solar wind does not reach Earth's surface, research on this phenomenon must be conducted in space. On July 20, 1969, the *Apollo XI* astronauts deployed a solar wind experiment on the lunar surface that exhibited particle impacts that confirmed the presence and character of the solar wind (Figure 2.3).

The solar wind creates **auroras** in the upper atmosphere when absorbed energy is radiated as light energy of varying

FIGURE 2.3 Astronaut and solar wind experiment.
Without a protective atmosphere, the lunar surface receives charged particles of the solar wind and all of the Sun's electromagnetic radiation. The unrolled sheet of foil is a solar wind experiment being deployed by an *Apollo XI* astronaut. Analysis of the foil back on Earth revealed the composition of the solar wind. Why wouldn't this experiment work if deployed on Earth's surface? [NASA photo.]

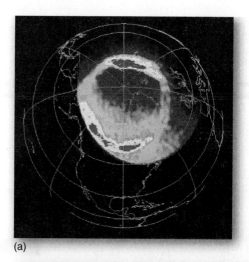

(a)

(b)

FIGURE 2.4 Auroras from space and from the ground in Alaska.
(a) *Polar* satellite false-color image of auroral halo over Earth's North Pole region from April 1996—the UVI sensor is able to capture the aurora on the day and night sides of Earth. (b) Surface view of the aurora borealis in the night sky over Alaska. [(a) Image from the Ultraviolet Imager (UVI) aboard *Polar* satellite courtesy of NASA; (b) photo by Johnny Johnson/Tony Stone Images, Inc.]

colors. These lighting effects are the *aurora borealis* (northern lights) and *aurora australis* (southern lights). The auroras generally are visible poleward of 65° latitude when the solar wind is active. More intense outbursts produce auroras at lower latitudes. They appear as folded sheets of green, yellow, blue, and red light that undulate across the skies of high latitudes, as shown in Figure 2.4. Research continues as to possible links between the sunspot cycle and patterns of drought and periods of wetness—the sunspot–weather connection. (For auroral activity, see http://www.gi.alaska.edu/.)

Electromagnetic Spectrum of Radiant Energy

The key solar input to life is electromagnetic energy. Solar radiation occupies a portion of the **electromagnetic spectrum** of radiant energy. This radiant energy travels at the

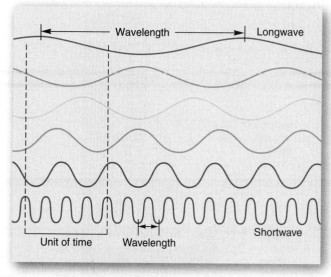

FIGURE 2.5 Wavelength and frequency.
Wavelength and frequency describe the electromagnetic spectrum. These are two ways of defining the same phenomenon—electromagnetic wave motion. Shorter wavelengths are higher in frequency, whereas longer wavelengths are lower in frequency.

speed of light to Earth. The total spectrum of this radiant energy is made up of different wavelengths. A **wavelength** is the distance between corresponding points on any two successive waves (Figure 2.5). The number of waves passing a fixed point in 1 second is the *frequency*.

The Sun emits radiant energy composed of 8% ultraviolet, X-ray, and gamma ray wavelengths; 47% visible light wavelengths; and 45% infrared wavelengths. A portion of the electromagnetic spectrum is illustrated in Figure 2.6; note the wavelengths at which various phenomena and human applications of energy occur.

An important physical law states that all objects radiate energy in wavelengths related to their individual surface temperatures: the hotter the object, the shorter the wavelengths emitted. This law holds true for the Sun and Earth. Figure 2.7 shows that the hot Sun radiates shorter wavelength energy, concentrated around 0.5 μm (micrometer).

The Sun's surface temperature is about 6000 K (6273°C, 11,459°F), and its emission curve is similar to that predicted for an idealized 6000 K surface, or *blackbody radiator* (shown in Figure 2.7). An ideal blackbody absorbs all the radiant energy that it receives and subsequently emits all that it receives. A hotter object like the Sun emits a much greater amount of energy per unit area of its surface than does a similar area of a cooler object like Earth. Shorter wavelength emissions are dominant at these higher temperatures.

Although cooler than the Sun, Earth radiates nearly all that it absorbs and acts as a blackbody (also shown in Figure 2.7). Because Earth is a cooler radiating body, longer wavelengths are emitted as radiation mostly in the infrared portion of the spectrum, centered around 10.0 μm. Some of the atmospheric gases vary in their response to radiation received, being transparent to some while absorbing others.

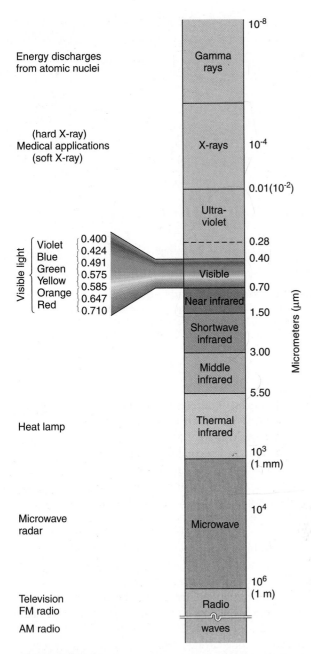

FIGURE 2.6 A portion of the electromagnetic spectrum of radiant energy.
Shorter wavelengths are toward the top of the spectrum in this figure; longer wavelengths are toward the bottom.

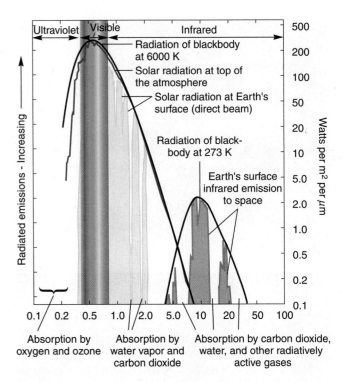

FIGURE 2.7 Solar and terrestrial energy distribution by wavelength.
The solar output peaks in shorter wavelengths of visible light in relation to its higher surface temperature, whereas Earth's emissions are concentrated in the infrared portion of the spectrum in relation to its lower surface temperature. Ideal blackbody curves for a radiating body like the Sun (hotter) and Earth (cooler) illustrate the concept. The dropouts in the plot lines of solar and terrestrial radiation represent absorption bands of water vapor, water, carbon dioxide, oxygen, ozone (O_3), and other gases. The graph is not drawn to scale. [Adapted from W. D. Sellers, *Physical Climatology* (Chicago: University of Chicago Press, 1965), p. 20. Used by permission.]

ANIMATION Electromagnetic Spectrum and Plants

the outer boundary of Earth's energy system and provides a useful point at which to assess the arriving solar radiation before it is diminished by passage through the atmosphere.

The Solar Constant Earth's distance from the Sun results in its interception of only one two-billionth of the Sun's total energy output. Nevertheless, this tiny fraction of the Sun's overall output represents an enormous amount of energy input into Earth's systems. **Insolation** refers to *in*coming *sol*ar radi*ation*.

Knowing the amount of insolation intercepted by Earth is important to climatologists and other scientists. The **solar constant** is the average value of insolation received at the thermopause when Earth is at its average distance from the Sun. That value of the solar constant is

Figure 2.8 illustrates the flows of energy into and out of Earth systems. To summarize, the solar spectrum is *shortwave radiation* that peaks in the visible wavelengths, and Earth's spectrum is *longwave radiation* concentrated in infrared wavelengths. In Chapter 3, we see that Earth, clouds, sky, ground, and all things that are terrestrial are *cool-body* radiators.

Incoming Energy at the Top of the Atmosphere

The region at the top of the atmosphere, approximately 480 km (300 mi) above Earth's surface, is the **thermopause**. It is

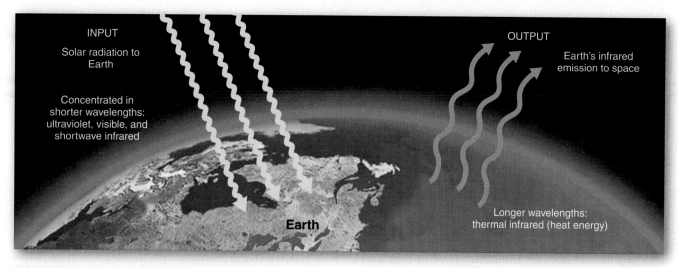

FIGURE 2.8 Earth's energy budget simplified.
Solar radiation is concentrated in shorter wavelengths. Earth emits longer wavelengths of infrared to the atmosphere and eventually to space.

Surface area receiving insolation

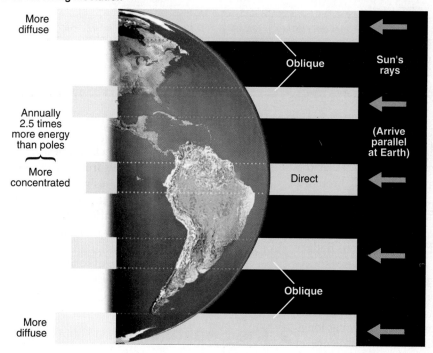

FIGURE 2.9 Insolation receipts and Earth's curved surface.
Solar insolation angles determine the concentration of energy received at each latitude. Lower latitudes receive more concentrated energy from a more direct solar beam. Higher latitudes receive slanting (oblique) rays and more diffuse energy. Note the area covered by identical columns of solar energy arriving at Earth's surface at higher latitudes (more diffuse, larger area covered) and at lower latitudes (more concentrated, smaller area covered).

1372 watts per square meter (W/m^2).* As we follow insolation through the atmosphere to Earth's surface in Chapter 3, we see the value of the solar constant reduced by half or more through reflection, scattering, and absorption of shortwave radiation.

*A *watt* is equal to 1 joule (a unit of energy) per second and is the standard unit of power in the International System of Units (SI). (See the conversion tables in Appendix C of this text for more information on measurement conversions.) In nonmetric *calorie* heat units, the solar constant is expressed as approximately 2 calories per square centimeter per minute, or 2 *langleys* per minute (a langley being 1 cal/cm²). A calorie is the amount of energy required to raise the temperature of 1 gram of water (at 15°C) 1 degree Celsius and is equal to 4.184 joules.

Uneven Distribution of Insolation Earth's curved surface presents a continuous varying angle to the incoming parallel rays of insolation (Figure 2.9). The latitudinal variation in the angle of solar rays results in an uneven global distribution of insolation. The only point receiving insolation perpendicular to the surface (from directly overhead) is the **subsolar point**. During the year this point occurs at lower latitudes between the tropics (about 23.5° north and 23.5° south latitudes), where the energy received is more concentrated. All other places away from the subsolar point receive insolation at an angle less than 90° and thus experience more diffuse energy; this effect is pronounced at higher latitudes.

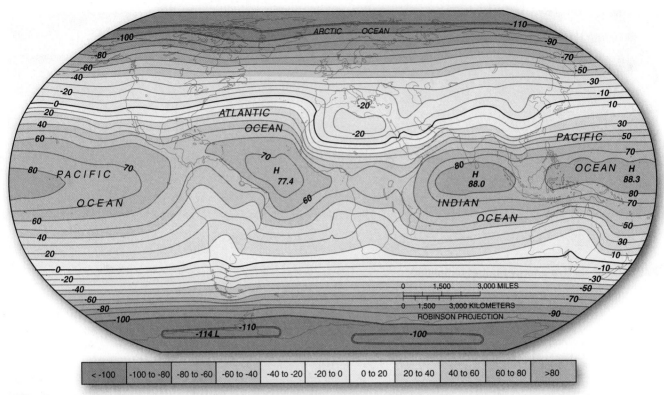

FIGURE 2.10 Daily net radiation patterns at top of atmosphere.
Average daily net radiation flows for a 9-year period (1979 to 1987) measured at the top of the atmosphere by the Earth Radiation Budget (ERB) instrument in orbit. Units are watts per square meter (W/m²).
[Map courtesy of Dr. H. Lee Kyle, Goddard Space Flight Center, NASA.]

The thermopause above the equatorial region receives 2.5 times more insolation annually than the thermopause above the poles. Of lesser importance is the fact that the lower-angle solar rays toward the poles must pass through a greater thickness of atmosphere, resulting in greater losses of energy due to scattering, absorption, and reflection.

Global Net Radiation   The Earth Radiation Budget (ERB) instrument aboard several satellites measures shortwave and longwave flows of energy at the top of the atmosphere. ERB sensors collected the data used to develop the map in Figure 2.10. This map shows *net radiation*, or the balance between incoming shortwave and outgoing longwave radiation. Note the latitudinal energy imbalance in net radiation—positive values in lower latitudes and negative values toward the poles.

In middle and high latitudes, approximately poleward of 36° north and south latitudes, net radiation is negative. This happens because Earth's climate system loses more energy to space than it gains from the Sun as measured at the top of the atmosphere. In the lower atmosphere, these polar energy deficits are offset by flows of heat energy from tropical energy surpluses, as we see in Chapters 3 and 4. The atmosphere and ocean form a giant heat engine, driven by differences in energy from place to place and causing major circulations within the lower atmosphere and in the ocean. The largest net radiation values are above the tropical oceans along a narrow equatorial zone, averaging 80 W/m². Net radiation minimums are lowest over Antarctica.

Of interest is the −20 W/m² area over the Sahara desert region, where usually clear skies—which permit large longwave radiation losses from Earth's surface—and reflective (light-colored) surfaces work together to reduce net radiation values measured at the thermopause. Clouds in the lower atmosphere also affect net radiation patterns at the top of the atmosphere by reflecting greater amounts of shortwave energy to space.

Having examined the flow of solar energy to Earth, let us now look at the nature of seasonal change as it affects these energy receipts.

The Seasons

The periodic rhythm of warmth and cold, dawn and daylight, twilight and night have fascinated humans for centuries. In fact, many ancient societies demonstrated a greater awareness of seasonal change than modern peoples and formally commemorated natural energy rhythms with festivals and monuments. Presently, warm-season length is on the increase worldwide in response to global temperature change.

Seasonality

Seasonality refers to both the seasonal variation of the Sun's position above the horizon and the changing daylengths during the year. Seasonal variations are a response to changes in the Sun's **altitude**, or the angle between the horizon and the

FIGURE 2.11 Subsolar point and pen in the sand.
We stuck this pen in sand at 22° N latitude. The local noon Sun was directly overhead, placing us at the subsolar point. Note there is no shadow since the Sun's altitude is 90° above the horizon. It is June 7 when the perpendicular rays of the Sun pass this latitude on their way to the solstice and the Tropic of Cancer. [Photo by Bobbé Christopherson.]

Sun. At sunrise or sunset, the Sun is at the horizon, so its altitude is 0°. If during the day the Sun reaches halfway between the horizon and directly overhead, it is at 45°, and if directly overhead it is at 90° altitude (called the *zenith*). The Sun is directly overhead only at the subsolar point, as we demonstrate in the photo made at 22° N latitude in Figure 2.11.

The Sun's **declination** is the latitude location of the subsolar point. Declination annually migrates through 47° of latitude between the *Tropic of Cancer* at about 23.5° N and the *Tropic of Capricorn* at about 23.5° S latitude. The subsolar point does not reach the US or Canadian mainland. Other than Hawai'i, all other states and provinces are too far north.

Seasonality also means a changing **daylength**, or duration of exposure. Daylength varies during the year depending on latitude. The equator always receives equal hours of day and night: If you live in Ecuador, Kenya, or Singapore, every day and night is 12 hours long, year-round. People living along 40° N latitude (Philadelphia, Denver, Madrid, Beijing), or 40° S latitude (Buenos Aires, Capetown, Melbourne), experience about 6 hours' difference in daylight between winter (9 hours) and summer (15 hours). At 50° N or S latitude (Winnipeg, Paris, Falkland, or Malvinas, Islands), people experience almost 8 hours of annual daylength variation.

Reasons for Seasons

Seasons result from variations in the Sun's altitude above the horizon, declination latitude, and daylength. These in turn are created by several physical factors that operate

Table 2.1 Five Reasons for Seasons

Factor	Description
Revolution	Orbit around the Sun; requires 365.24 days to complete at 107,280 kmph (66,660 mph)
Rotation	Earth turning on its axis; takes approximately 24 hours to complete at 1675 kmph (1041 mph) at the equator
Tilt	Axis is aligned at a 23.5° angle from a perpendicular to the plane of the ecliptic (the plane of Earth's orbit)
Axial parallelism	Remains in a fixed alignment, with Polaris directly overhead at the North Pole throughout the year
Sphericity	Appears as an oblate spheroid to the Sun's parallel rays; the geoid

together: Earth's revolution in orbit around the Sun, its daily rotation on its axis, its tilted axis, the unchanging orientation of its axis, and its sphericity. We now briefly describe these factors, which are summarized in Table 2.1. Of course, the essential ingredient is having a single source of radiant energy—the Sun.

Revolution The structure of Earth's orbit and **revolution** about the Sun are shown in Figure 2.1 and Figure 2.12. Note the distinction between orbital *revolution* and *rotation*—the spinning of Earth on its axis. At an average distance from the Sun of 150 million km (93 million mi), Earth completes its annual orbit in 365.24 days at speeds averaging 107,280 kmph (66,660 mph) in a counterclockwise direction when viewed from above Earth's North Pole. This number is based on a *tropical year*, measured from equinox to equinox, or the elapsed time between two crossings of the equator by the Sun.

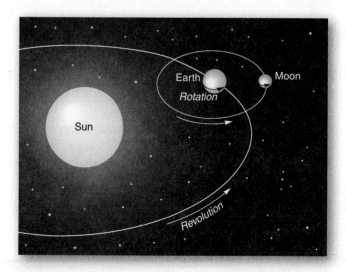

FIGURE 2.12 Earth's revolution and rotation.
Earth's *revolution* about the Sun and *rotation* on its axis, as viewed from above Earth's orbit. Note that the Moon's rotation on its axis and revolution about Earth are counterclockwise as well.

Rotation Earth's **rotation**, or turning, is a complex motion that averages 24 hours in duration. A true day varies slightly from 24 hours, but by international agreement a day is considered to be exactly 24 hours (86,400 seconds) in length. Rotation determines daylength, causes the apparent deflection of winds and ocean currents, and produces the twice-daily rise and fall of the ocean tides in relation to the gravitational pull of the Sun and Moon.

Earth rotates about its *axis*, an imaginary line extending through the planet from the geographic North Pole to the South Pole. When viewed from above the North Pole, Earth rotates counterclockwise around this axis; viewed from above the equator it moves west to east, or eastward. This west-to-east rotation creates the Sun's *apparent* daily journey from east to west, even though the Sun remains in a fixed position in the center of the Solar System. Similarly, the Moon revolves around Earth and rotates counterclockwise on its axis (Figure 2.12).

Although every point on Earth takes the same 24 hours to complete one rotation, the linear velocity of rotation at any point on Earth's surface varies dramatically with latitude. The equator is 40,075 km (24,902 mi) long; therefore, rotational velocity at the equator must be approximately 1675 kmph (1041 mph) to cover that distance in one day. At 60° latitude, a parallel is only half the length of the equator, or 20,038 km (12,451 mi) long, so the rotational velocity there is 838 kmph (521 mph). At the poles, the velocity is 0. (This variation in rotational velocity establishes the effect of the Coriolis force, discussed in Chapter 4.)

Earth's rotation produces the diurnal (daily) pattern of day and night. The dividing line between day and night is the **circle of illumination** (as illustrated in Figure 2.14). Because this day-night dividing circle of illumination intersects the equator, *daylength at the equator is always evenly divided*—12 hours of day and 12 hours of night. All other latitudes experience uneven daylength through the seasons, except for 2 days a year, on the equinoxes.

Tilt of Earth's Axis To understand Earth's *axial tilt*, imagine Earth's elliptical orbit about the Sun as a plane, with half of the Sun and Earth above the plane and half below. (It may help to envision two spheres floating in water, where the water's surface forms a plane.) This flat surface is the **plane of the ecliptic**. Now imagine a perpendicular line passing through the plane. From this perpendicular, Earth's axis is tilted about 23.5°. It forms close to a 66.5° angle from the plane itself (Figure 2.13). The axis through Earth's two poles points to near the star *Polaris*, which is, appropriately, the *North Star*.

This text uses "about" with the tilt angle just described because Earth's axial tilt changes over a complex 41,000-year cycle (see Figure 14.27). The axial tilt roughly ranges between 22° and 24.5° from a perpendicular to the plane of the ecliptic. The present tilt is 23° 27' (or 66° 33' from the plane). In decimal numbers, 23° 27' is approximately 23.45°. For convenience, this is rounded off to a 23.5° tilt (or 66.5° from the plane) in our discussion. Scientific

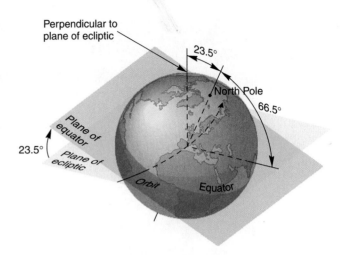

FIGURE 2.13 The plane of Earth's orbit—the ecliptic—and Earth's axial tilt.
Note that the plane of the equator is inclined at about 23.5° to the plane of the ecliptic.

evidence shows that the angle of tilt is lessening in its 41,000-year cycle.

Axial Parallelism Throughout our annual journey around the Sun, Earth's axis *maintains the same alignment* relative to the plane of the ecliptic and to Polaris and the other stars. You can see this alignment in Figure 2.14. Note that in each position shown of Earth revolving about the Sun, the axis is oriented identically, or parallel to itself. This is known as **axial parallelism**.

Annual March of the Seasons

The combined effect of all these physical factors is the annual march of the seasons on Earth. Daylength is the most evident way of sensing changes in season at latitudes away from the equator. The extremes of daylength occur in December and June. The times around December 21 and June 21 are the *solstices*. They mark the times of the year when the Sun's declination places it directly over one of the two *tropics*, parallels of latitude that represent the Sun's farthest northerly or southerly position. ("Tropic" is from the Latin *tropicus*, meaning a turn or change, so a tropic latitude is where the Sun's declination appears to stand still briefly—Sun stance, or *sol stice*; then it turns and heads toward the other tropic.) Table 2.2 presents the key seasonal anniversary dates, their names, and the subsolar point location (declination).

Figure 2.14 demonstrates the annual march of the seasons and illustrates Earth's relationship to the Sun during each month of the year. Let us begin with December (the globe to the far right in the figure). On December 21 or 22, at the **winter solstice** (literally, "winter Sun stance") or **December solstice**, the circle of illumination excludes the North Pole region from sunlight and includes the South Pole region. The subsolar point is at 23.5° S latitude, a parallel known as the **Tropic of Capricorn**, at the moment of the solstice. The Northern Hemisphere is tilted away from

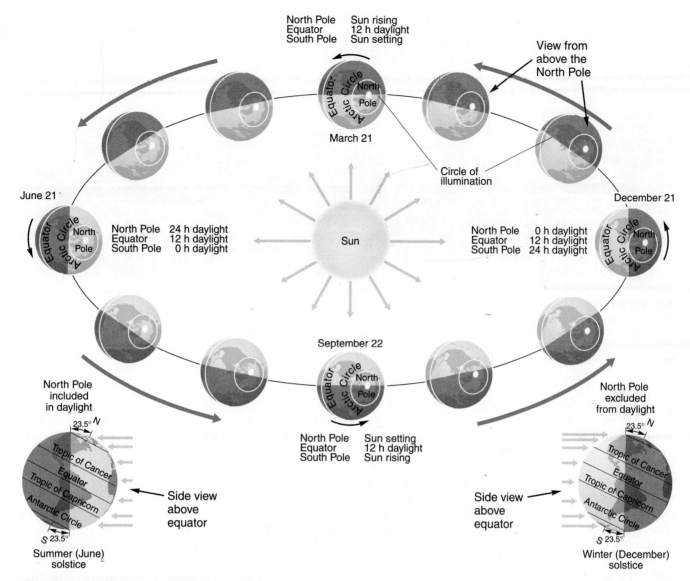

FIGURE 2.14 Annual march of the seasons.
Annual march of the seasons as Earth revolves about the Sun. Shading indicates the changing position of the circle of illumination in the Northern Hemisphere (viewed from above the North Pole) and in both Northern and Southern Hemispheres (viewed from above the equator). Note the hours of daylight for the equator and the poles. To follow the text, begin on the right side at December 21 and move counterclockwise.

Earth-Sun Relations, Seasons

these more direct rays of sunlight, thereby creating a lower angle for the incoming solar rays and a more diffuse pattern of insolation, thus causing our northern winter. Examine the photo of Earth on the back cover of this book. During what month do you think it was taken?

From 66.5° N latitude to 90° N (the North Pole), the Sun remains below the horizon the entire day. This latitude

(66.5° N) marks the *Arctic Circle*, the southernmost parallel (in the Northern Hemisphere) that experiences a 24-hour period of darkness. Twilight and dawn provide some lighting for more than a month at the beginning and end of the Arctic night. During the next 3 months, daylength and solar angles gradually increase in the Northern Hemisphere as Earth completes one-fourth of its orbit.

Table 2.2	Annual March of the Seasons	
Approximate Date	**Northern Hemisphere Name**	**Location of the Subsolar Point**
December 21–22	Winter solstice (December solstice)	23.5° S latitude (Tropic of Capricorn)
March 20–21	Vernal equinox (March equinox)	0° (equator)
June 20–21	Summer solstice (June solstice)	23.5° N latitude (Tropic of Cancer)
September 22–23	Autumnal equinox (September equinox)	0° (equator)

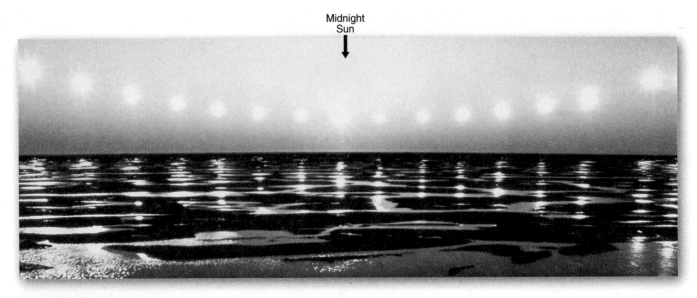

FIGURE 2.15 The Midnight Sun.
The Midnight Sun north of the Arctic Circle captured in a series of 18 exposures on the same piece of film. The camera is facing due north. Midnight is the exposure showing the Sun closest to the horizon. The photographer removed the lens cap at regular intervals to make the multiple exposures. [Photo by Gary Braasch/Tony Stone Images.]

The **vernal equinox**, or **March equinox**, occurs on March 20 or 21. At that time, the circle of illumination passes through both poles, so that all locations on Earth experience a 12-hour day and a 12-hour night. Those living around 40° N latitude (New York, Denver) have gained 3 hours of daylight since the December solstice. At the North Pole, the Sun peeks above the horizon for the first time since the previous September; at the South Pole, the Sun is setting, marking the beginning of a period of twilight, then a long, dark winter.

From March, the seasons move on to June 20 or 21 and the **summer solstice**, or **June solstice**. The subsolar point has now shifted from the equator to 23.5° N latitude, the **Tropic of Cancer**. Because the circle of illumination now includes the North Pole region, everything north of the Arctic Circle receives 24 hours of daylight—the "Midnight Sun" (Figure 2.15; chapter-opening photo). In contrast, the region from the *Antarctic Circle* to the South Pole (66.5°–90° S latitude) is in darkness the entire day, although an extended period of twilight and dawn provides some lighting at the beginning and end of the Antarctic night.

Note that the orientation of Earth's axis has remained fixed relative to the heavens and parallel to its previous position 6 months earlier. The Northern Hemisphere in June is tilted toward the Sun, receives higher Sun angles, and experiences longer days—and therefore more insolation—than the Southern Hemisphere. Those living at 40° N latitude now receive more than 15 hours of sunlight a day, which is 6 hours more than in December.

September 22 or 23 is the time of the **autumnal equinox**, or **September equinox**, when Earth's orientation is such that the circle of illumination again passes through both poles so that all parts of the globe receive a 12-hour day and a 12-hour night. The subsolar point has returned to the equator, with days growing shorter to the north and longer to the south. Researchers stationed at the South Pole see the disk of the Sun just rising, ending their 6 months of night. At the North Pole, a period of twilight transitioning into darkness begins. In the Northern Hemisphere, fall arrives, a time of many colorful changes in the landscape.

Seasonal Observations   For most of you reading this text in the Northern Hemisphere, the point of sunrise migrates along the horizon from the southeast in December to the northeast in June. The point of sunset migrates from the southwest to the northwest during those same times. The Sun's altitude at local noon at 40° N latitude migrates from a 26° angle above the horizon at the winter (December) solstice to a 73° angle above the horizon at the summer (June) solstice—a range of 47° (Figure 2.16). Think back over the past year. What seasonal changes have you observed in Sun angles, vegetation, temperatures, and weather?

Atmospheric Composition, Temperature, and Function

Insolation cascades through the atmosphere toward Earth's surface, powering the physical systems in the atmosphere, in the oceans, and on land—varying through the seasonal changes just discussed. Along the way the atmosphere works as an efficient filter, removing harmful radiation from sunlight. Let's now examine this unique atmosphere—its composition, temperature, and function.

The modern atmosphere probably is the fourth general atmosphere in Earth's history. This modern atmosphere is a gaseous mixture of ancient origin, the sum of all

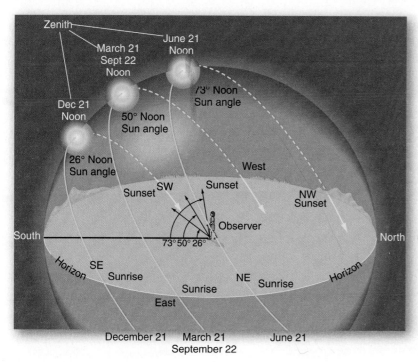

FIGURE 2.16 Seasonal observations—sunrise, noon, and sunset through the year. Seasonal observations at 40° N latitude for the December solstice, March equinox, June solstice, and September equinox. The Sun's altitude increases from 26° in December to 73° above the horizon in June—a difference of 47°. See the NOAA/ESRL Sunrise/Sunset Calculator at **http://www.srrb.noaa.gov/highlights/sunrise/sunrise.html.**

the exhalations and inhalations of life on Earth throughout time. The principal substance of this atmosphere is air, the medium of life as well as a major industrial and chemical raw material. *Air* is a simple additive mixture of gases that is naturally odorless, colorless, tasteless, and formless, blended so thoroughly that it behaves as if it were a single gas.

In his book *The Lives of a Cell*, the late physician and self-styled "biology-watcher" Lewis Thomas compared the atmosphere of Earth to an enormous cell membrane. The membrane around a cell regulates the interactions between the cell's delicate inner workings and the potentially disruptive outer environment. Each cell membrane is selective as to what it will and will not allow to pass through. The modern atmosphere acts as Earth's protective membrane, as Thomas described so vividly (Figure 2.17).

As a practical matter, we consider the top of our atmosphere to be around 480 km (300 mi) above Earth's surface, the same altitude we used for measuring the solar constant and initial insolation receipt. Beyond that altitude, the atmosphere is rarefied (nearly a vacuum) and is called the **exosphere**, which means "outer sphere." It contains scarce, lightweight hydrogen and helium atoms, weakly bound by gravity as far as 32,000 km (20,000 mi) from Earth.

Atmospheric Profile

Earth's modern atmosphere is arranged in a series of imperfectly shaped concentric "shells" or "spheres" that grade into one another, all bound to the planet by gravity. As critical as the atmosphere is to us, it represents only the thinnest envelope, amounting to less than one-millionth of Earth's total mass. Important to the following discussion is

Figure 2.18, a vertical cross section, or side view, of Earth's atmosphere. We simplify our discussion of the atmosphere by looking at three aspects: its *composition, temperature*, and *function*. These categories are shown along the left side of Figure 2.18a.

This 480-km atmospheric profile exerts its weight, pressing downward under the pull of gravity. Air molecules create air pressure through their motion, size, and number. Pressure is exerted on all surfaces in contact with the air. The weight (force over a unit area) of the atmosphere, or **air pressure**, pushes in on all of us. Fortunately, that same pressure also exists inside us, pushing outward; otherwise we would be crushed by the mass of air around us.

The atmosphere exerts an average force of approximately 1 kg/cm^2 (14.7 lb/in.^{2}) at sea level. Under the influence of gravity, air is compressed and therefore denser near Earth's surface; it thins rapidly with increasing altitude (Figure 2.19a). Consequently, over half the total mass of the atmosphere is compressed below 5500 m (18,000 ft), 75% below 10,700 m (35,100 ft), and 90% is compressed below 16,000 m (52,500 ft). All but 0.1% of the atmosphere exists within an altitude of 50 km (31 mi), as shown in the pressure profile in Figure 2.19b (percentage column is farthest to the right).

Few people are aware that in routine air travel they are sitting above 80% of the total atmospheric volume! To better understand this pressure profile, imagine skydiving from a high-altitude balloon 33 km (20 mi) above Earth. What would you experience? How fast would you fall? What sounds would you hear? See News Report 2.1 (p. 56) to read about someone who did this very thing.

At sea level, the atmosphere exerts a pressure of 1013.2 mb (*millibar*, force per square meter of surface area) or 29.92 in. of mercury, as measured by a *barometer*. In

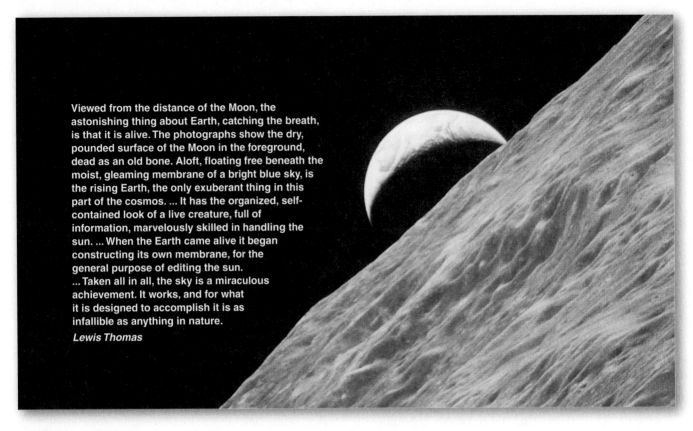

FIGURE 2.17 Earthrise.
Earthrise over the ". . . dead as an old bone . . ." lunar surface.
[NASA photo. Quotation from "The World's Biggest Membrane" from *The Lives of a Cell* by Lewis
Thomas. Copyright © 1973 by the Massachusetts Medical Society. Originally published in the *New
England Journal of Medicine.*]

Canada and other countries, normal air pressure is expressed as 101.32 kPa (*kilopascal*; 1 kPa = 10 mb). For more on air pressure, the instruments that measure it, and the role air pressure plays in generating winds, see Chapter 4.

Atmospheric Composition Criterion

Using chemical *composition* as a criterion, we find that the atmosphere divides into two broad regions, the *heterosphere* (80 to 480 km altitude) and the *homosphere* (Earth's surface to 80 km altitude). As you read, note that we follow the same path that incoming solar radiation travels through the atmosphere to Earth's surface.

Heterosphere   The **heterosphere** is the outer atmosphere in terms of composition. It begins at about 80 km (50 mi) altitude and extends outward to the transition to the exosphere and interplanetary space (Figure 2.18). The International Space Station and most Space Shuttle missions orbit in the upper heterosphere.

As the prefix *hetero-* implies, this region is not uniform—its gases are *not evenly mixed*. This distribution is quite different from the nicely blended gases we breathe near Earth's surface, in the homosphere. Gases in the heterosphere are distributed in distinct layers that are sorted by gravity according to their atomic weight, with the lightest elements (hydrogen and helium) at the margins of outer space and the heavier elements (oxygen and nitrogen) dominant in the lower heterosphere. Less than 0.001% of the atmosphere's mass is in the heterosphere.

Homosphere   Between the heterosphere and Earth's surface is the other compositional shell of the atmosphere, the **homosphere**. This region extends from the surface to an altitude of 80 km (50 mi). Even though the atmosphere rapidly changes density in the homosphere, decreasing with increasing altitude, the blend (proportion) of gases is nearly uniform throughout the homosphere. The only exceptions are the concentration of ozone (O_3) in the "ozone layer," from 19 to 50 km (12 to 31 mi), and the variations in water vapor, carbon dioxide, pollutants, and some trace chemicals in the lowest portion of the atmosphere.

The stable mixture of gases throughout the homosphere evolved slowly. The present proportion, which includes oxygen, was attained approximately 500 million years ago. Figure 2.20a presents by volume the stable ingredients that constitute dry, clean air in the homosphere. Earth's atmosphere is sampled at the Mauna Loa Observatory, Hawai'i, or MLO site (http://www.mlo.noaa.gov/). MLO is on the north slope of Mauna Loa at 3394 m (11,135 ft) elevation (shown in Figure 2.20b). The continuous measurement of carbon dioxide at this station is important.

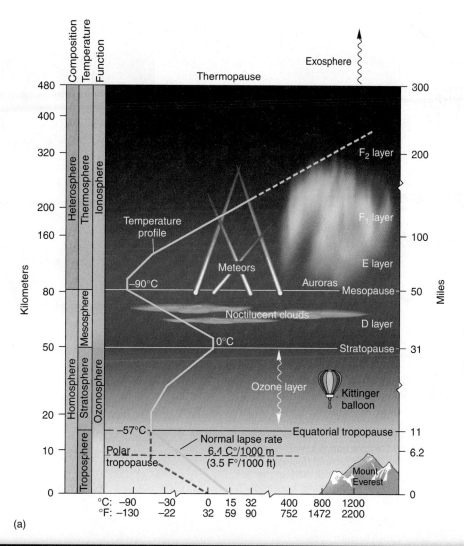

(a)

(b)

FIGURE 2.18 Modern atmosphere profile.
(a) An integrated chart of our modern atmosphere. Columns show division of the atmosphere by composition, temperature, and function. The chart spans from Earth's surface to the thermopause at 480 km (300 mi). (The small balloon shows the height achieved by Kittinger, discussed in News Report 2.1.) (b) Space Shuttle astronauts captured a dramatic sunset through various atmospheric layers across the "edge" of our planet—*Earth's limb*. A silhouetted cumulonimbus thunderhead cloud is seen rising to the tropopause. [Space Shuttle photo from NASA.]

The homosphere is a vast reservoir of relatively inert *nitrogen*, originating principally from volcanic sources. Nitrogen is a key element of life, yet we exhale all the nitrogen that we inhale. This apparent contradiction is explained by the fact that nitrogen is absorbed into our bodies not from the air we breathe, but through compounds in food. In the soil, nitrogen is bound to these compounds by nitrogen-fixing bacteria, and it is returned to the atmosphere by denitrifying bacteria that remove nitrogen from organic materials. The nitrogen cycle discussion is in Chapter 16.

Oxygen, a by-product of photosynthesis, also is essential for life processes. Slight spatial variations occur in the

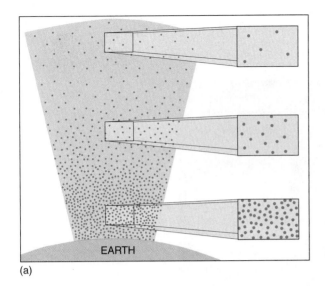

(a)

FIGURE 2.19 Density decreases with altitude.
(a) The atmosphere, denser near Earth's surface, rapidly decreases in density with altitude. The difference in density is easily measured because air exerts its weight as pressure. Have you experienced pressure changes that you could feel on your eardrums? How high above sea level were you at the time? (b) Pressure profile of the atmosphere. Note the rapid decrease in atmospheric pressure with altitude and that about 90% of the atmospheric mass resides in the troposphere (far-right column).

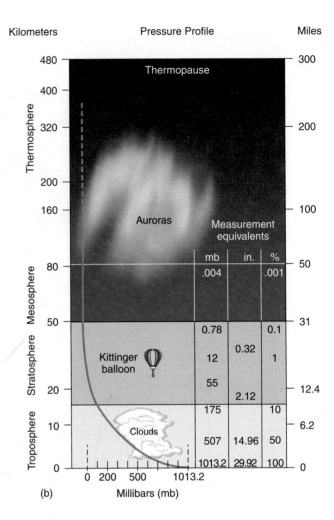

(b)

percentage of oxygen in the atmosphere because of variations in photosynthetic rates with latitude, seasonal changes, and the lag time as atmospheric circulation slowly mixes the air. Although it forms about one-fifth of the atmosphere, oxygen forms compounds that compose about half of Earth's crust. Oxygen readily reacts with many elements to form these materials. Both nitrogen and oxygen reserves in the atmosphere are so extensive that, at present, they far exceed human capabilities to disrupt or deplete them.

The gas *argon*, constituting about 1% of the homosphere, is completely inert (an unreactive "noble" gas) and therefore is unusable in life processes. Argon is a residue from the radioactive decay of an isotope (form) of potassium called potassium-40 (symbolized ^{40}K). Because industry has found uses for inert argon (in light bulbs, welding, and some lasers), it is extracted or "mined" from the atmosphere, in addition to nitrogen and oxygen.

Carbon dioxide is a natural by-product of life processes. It is essentially a stable atmospheric component, qualifying it for inclusion in Figure 2.20. Although its present percentage in the atmosphere is small at 0.039% (388.60 ppm, May 2008), CO_2 is important in maintaining global temperatures. Its percentage has been increasing over the past 200 years as a result of human activities, principally the combustion of fossil fuels.

According to ice-core records, CO_2 is higher now than at any time in the past 800,000 years. This increase appears to be accelerating. From 1990 to 1999, CO_2 emissions rose at an average of 1.1% per year; compare this to the average increase since 2000 of 3.1% per year—this equals a 2 to 3 ppm per year increase (Figure 2.21, p. 57).* A distinct climatic threshold is approaching at 450 ppm, sometime in the decade of the 2020s. Beyond this tipping point there would be irreversible ice-sheet and species losses. Dr. James Hansen, a NASA scientist at Goddard Institute for Space Sciences, said in March 2008:

> If humanity wishes to preserve a planet similar to that on which civilization developed, paleoclimatic evidence and ongoing climate change suggest that CO_2 will need to be reduced to at most 350 ppm [the average in 1987].

If this rate of increase continues unabated in business-as-usual fashion for the rest of the century, this concentration could reach 1400 ppm CO_2, according to forecasts. Scientists are telling us to avoid this by reducing CO_2 emissions to avoid crossing the *450 threshold*. Chapter 7

*Proceedings of the National Academy of Sciences, May 22, 2007, v. 104, n. 24, p. 10288.

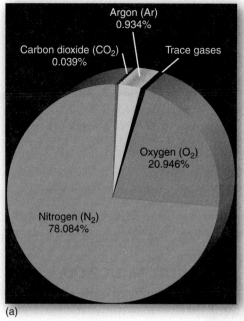

(a)

(b)

FIGURE 2.20 Composition of the homosphere and its measurement.
(a) Stable components of the modern atmosphere (percentage concentration by volume).
(b) The air sampling tower at Mauna Loa Observatory, operational since 1957, benefits from its remoteness in the middle of the Pacific Ocean, ideal for monitoring atmospheric composition far from pollution sources. A marine inversion layer keeps volcanic emissions from nearby Kīlauea volcano and dust at a minimum. (Average carbon dioxide concentration measured in May 2008 at Mauna Loa, Hawai'i; see **http://www.esrl.noaa.gov/gmd/ccgg/ trends/co2_data_mlo.html.**)

discusses the implications of CO_2 increases for global warming and climate change.

Atmospheric Temperature Criterion

Shifting to temperature as a criterion, the atmosphere has four distinct zones—the *thermosphere, mesosphere, stratosphere,* and *troposphere*—from thermopause to Earth's surface (labeled in Figure 2.18).

Thermosphere We define the **thermosphere** ("heat sphere") as roughly corresponding to the heterosphere (80 km out to 480 km, or 50–300 mi). The upper limit of the thermosphere is the *thermopause* (the suffix *-pause* means "to change"). During periods of a less active Sun

(fewer sunspots and coronal bursts), the thermopause may lower in altitude from the average 480 km (300 mi) to only 250 km altitude (155 mi). An active Sun causes the outer atmosphere to swell to an altitude of 550 km (340 mi), where it can create frictional drag on satellites in low orbit.

The temperature profile in Figure 2.18 (yellow curve) shows that temperatures rise sharply in the thermosphere, to 1200°C (2200°F) and higher. Despite such high temperatures, the thermosphere is not "hot" in the way you might expect. Temperature and heat are different concepts. The intense solar radiation in this portion of the atmosphere excites individual molecules (principally nitrogen and oxygen) to high levels of vibration. This **kinetic energy**, the energy of motion, is the vibrational energy that we measure as *temperature.*

NEWS REPORT 2.1

Falling through the Atmosphere—The Highest Skydive

Imagine a small, unpressurized compartment large enough for only one person, floating at 31.3 km (19.5 mi) altitude, carried to such height by a helium-filled balloon. Even though it is daytime, the sky is almost black, for this is above all but 1.0% of the atmospheric mass. The air pressure at 31,300 m (102,800 ft) is barely measurable—in aircraft testing this altitude is used as the beginning of space. The year is 1961. (See the position of this balloon in Figures 2.18a and 2.19b.)

Captain Joseph Kittinger, Jr., an Air Force officer, stood looking out of the opening in his capsule. He paused for the view of Earth's curved horizon, the New Mexico landscape more than 31 km below, and the dark heavens above, and then leaped into the stratospheric void (Figure 2.1.1). What do you think he would initially hear and feel? At what speed would he fall? What would the temperature be? How long would he fall before opening his parachute?

He heard nothing, no rushing sound. The fabric of his pressure suit did not flutter, for there was not enough air to create friction against the fibers. Because of these factors he had no sensation of movement until he glanced back at the balloon and capsule. He saw them retreat rapidly from his motionless perception.

His speed was remarkable. In the lower atmosphere, objects fall at

FIGURE 2.1.1 Stratospheric leap into history.
Moments into Captain Joseph Kittinger's historic exploration of the atmosphere, captured by a remotely triggered camera. The clouds are more than 26,000 m (85,000 ft) below him. He carries an instrument pack on his seat, his main chute, and pure oxygen for his breathing mask. [Volkmar Wentzel/NGS Image Collection, used by permission from the National Geographic Society, *National Geographic Magazine*, December 1960, p. 855. All rights reserved.]

"terminal velocity," reaching about 200 kmph (125 mph) in a few seconds. However, in the rarefied stratosphere, the lack of air resistance permits incredible velocity. Captain Kittinger quickly accelerated to 988 kmph (614 mph), slightly less than the speed of sound at sea level.

His free fall took him through the stratosphere and its ozone layer. As he encountered denser layers of atmospheric gases, his free fall was slowed by frictional drag. As he flew past 15,240 m (50,000 ft), he had slowed to 400 kmph (250 mph). He dropped

into the lower atmosphere, finally falling below regular airplane altitudes. His free fall lasted 4 minutes and 25 seconds, slowing to terminal velocity and the opening of his main chute at 5500 m (18,000 ft). He safely drifted to Earth's surface. This remarkable 13-minute, 35-second voyage through 99% of the atmospheric mass remains a record to this day. The Smithsonian National Air and Space Museum in Washington, DC, features Captain Kittinger's flight suit and the photo in Figure 2.1.1 in a special display to honor him.

However, the actual heat involved is very small. The reason is that the density of the molecules is so low that little actual *heat*—the flow of kinetic energy from one body to another because of a temperature difference between them—is produced. Heating in the atmosphere near Earth's surface is different because the greater number of molecules in the denser atmosphere transmit their kinetic energy as **sensible heat**, meaning that we can measure its temperature. (Density, temperature, and heat capacity determine the sensible heat of a substance.)

Mesosphere The **mesosphere** is the area from 50 to 80 km (30 to 50 mi) above Earth and is the highest in altitude of the three temperature regions within the homosphere. As Figure 2.18 shows, the mesosphere's outer

boundary, the *mesopause*, is the coldest portion of the atmosphere, averaging −90°C (−130°F), although that temperature may vary considerably (25 to 30 C°, or 45 to 54 F°). Very low pressures (low density of molecules) exist in the mesosphere, as shown on Figure 2.19b.

Stratosphere The **stratosphere** extends from 18 to 50 km (11 to 31 mi) from Earth's surface. Temperatures increase with altitude throughout the stratosphere, from −57°C (−70°F) at 18 km (tropopause), warming to 0°C (32°F) at 50 km, the altitude of the stratosphere's outer boundary, the *stratopause*. This is the location of the ozone layer.

Troposphere The **troposphere** is the final layer encountered by incoming solar radiation as it surges

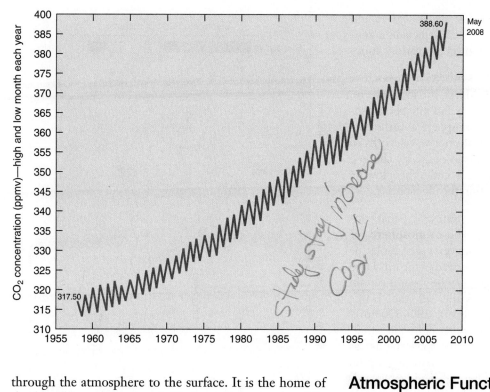

FIGURE 2.21 Carbon dioxide concentration increase 1958–2008.
Fifty-one years of CO_2 data measured at the Mauna Loa Observatory through May 2008. The highest and lowest average months are plotted for each year on the graph. [Data posted at **ftp://ftp.cmdl. noaa.gov/ccg/ co2/trends/ co2_mm_mlo.txt.**]

through the atmosphere to the surface. It is the home of the biosphere, the atmospheric layer that supports life, and the region of principal weather activity.

The troposphere contains approximately 90% of the total mass of the atmosphere and the bulk of all water vapor, clouds, weather, air pollution, and life forms. The *tropopause*, its upper limit, is defined by an average temperature of −57°C (−70°F), but its exact altitude varies with the season, latitude, and surface temperatures and pressures. Near the equator, because of intense heating from the surface, the tropopause occurs at 18 km (11 mi); in the middle latitudes, it occurs at 12 km (8 mi); and at the North and South Poles, it is only 8 km (5 mi) or less above Earth's surface.

Figure 2.22 illustrates the normal temperature profile within the troposphere during daytime. As the graph shows, temperatures decrease rapidly with increasing altitude at an average of 6.4 C° per kilometer (3.5 F° per 1000 feet), a rate known as the **normal lapse rate**. This temperature plot also appears in Figure 2.18.

The normal lapse rate is an average. The **environmental lapse rate** is the actual lapse rate at any particular time and place, which may deviate considerably because of local weather conditions. This variation in temperature gradient in the lower troposphere is central to our discussion of weather processes (Chapter 5).

In the stratosphere, the marked warming with increasing altitude causes the tropopause to act like a lid, essentially preventing whatever is in the cooler (denser) air below from mixing into the warmer (less dense) stratosphere. However, the tropopause may be disrupted above the midlatitudes wherever jet streams produce vertical turbulence and an interchange between the troposphere and the stratosphere (Chapter 4). Also, hurricanes occasionally inject moisture above this temperature-inversion layer at the tropopause.

Atmospheric Function Criterion

Looking at our final atmospheric criterion of *function*, we find that the atmosphere has two specific zones that remove most of the harmful wavelengths of incoming

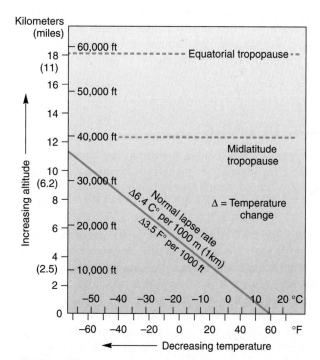

FIGURE 2.22 The temperature profile of the troposphere.
During daytime, temperature decreases at a rate known as the *normal lapse rate*. Scientists use a concept called the *standard atmosphere* as an accepted description of air temperature and pressure changes with altitude. The values in this profile are part of the standard atmosphere. Note the approximate locations of the equatorial and midlatitude tropopauses.

solar radiation and charged particles: the *ionosphere* and the *ozonosphere* (*ozone layer*). Figure 2.23 depicts in a general way the absorption of radiation by the various functional layers of the atmosphere.

Ionosphere   The outer functional layer, the **ionosphere**, extends throughout the thermosphere and into the mesosphere below (Figure 2.18). The ionosphere absorbs cosmic rays, gamma rays, X-rays, and shorter wavelengths of ultraviolet radiation, changing atoms to positively charged ions and giving the ionosphere its name. The glowing auroral lights occur principally within the ionosphere.

Ozonosphere   That portion of the stratosphere that contains an increased level of ozone is the **ozonosphere**, or **ozone layer**. Ozone is a highly reactive oxygen molecule made up of three oxygen atoms, O_3, instead of the usual two atoms, O_2, that make up oxygen gas. Ozone absorbs certain wavelengths of ultraviolet (principally, all the UVC, 100–290 nm, and about 90% of UVB, 290–320 nm).* Ozone absorbs wavelengths of ultraviolet light and subsequently radiates this energy at longer wavelengths, as infrared radiation. This process converts most harmful ultraviolet radiation, effectively "filtering" it and safeguarding life at Earth's surface. UVA, at 320–400 nm, makes up about 98% of all UV radiation that reaches Earth's surface.

The ozone layer is presumed to have been relatively stable over the past several hundred million years (allowing for daily and seasonal fluctuations). Today, however, it is in a state of continuous change. Focus Study 2.1 presents an analysis of the crisis in this critical portion of our atmosphere. Fortunately, international treaties to prevent further losses appear to be working.

⑤ Variable Atmospheric Components

The troposphere contains natural and human-caused variable gases, particles, and other chemicals. The spatial aspects of these variables are important applied topics in physical geography and the study of Earth's atmosphere.

Air pollution is not a new problem. Romans complained over 2000 years ago about the foul air of their cities. Filling Roman air was the stench of open sewers, smoke from fires, and fumes from ceramic-making kilns and smelters (furnaces) that converted ores into metals. In human experience, cities are always the place where the environment's natural ability to process and recycle waste is most taxed. Regulations to curb human-caused air pollution have met with great success, although much remains to be done. Before we discuss these topics, let's examine some natural sources of air pollution.

*Nanometer (nm) = one-millionth of a millimeter; 1 nm = 10^{-9}. For comparison, a micrometer, or micron (μm), = one-millionth of a meter; 1 μm = 10^{-6} m. A millimeter (mm) = one-thousandth of a meter; 1 mm = 10^{-3}.

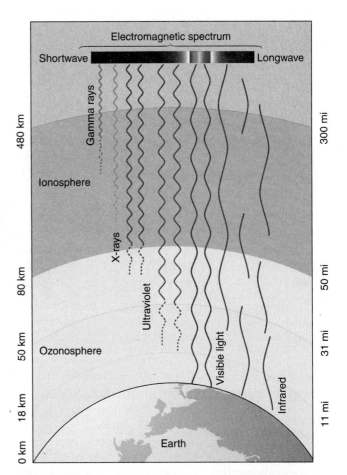

FIGURE 2.23 The atmosphere protects Earth's surface. As solar energy passes through the atmosphere, the shortest wavelengths are absorbed. Only a fraction of the ultraviolet radiation, and most of the visible light and infrared, reaches Earth's surface. When they are above these protective layers, astronauts must wear a spacesuit to survive; the suit duplicates this filtering process.

ANIMATION   Ozone Breakdown, Ozone Hole

Natural Sources

Natural air pollution sources produce a greater quantity of pollutants—nitrogen oxides, carbon monoxide, hydrocarbons from plants and trees, and carbon dioxide—than do human-made sources. Table 2.3 lists some of these

(text continued on page 62)

Table 2.3	Sources of Natural Variable Gases and Materials
Sources	**Contribution**
Volcanoes	Sulfur oxides, particulates
Forest fires	Carbon monoxide and dioxide, nitrogen oxides, particulates
Plants	Hydrocarbons, pollens
Decaying plants	Methane, hydrogen sulfides
Soil	Dust and viruses
Ocean	Salt spray and particulates

FOCUS STUDY 2.1

Stratospheric Ozone Losses: A Worldwide Health Hazard

Consider:

- The stratospheric ozone above Antarctica during the months of Antarctic spring (September to November) suffered its most severe loss in 2006, covering an area 3 times larger than the United States. Each year, the "ozone hole" (actually a thinning) widens and deepens, even though the rate of accumulation of offending chemicals has slowed.

- As stratospheric ozone flows into the depleted region from lower latitudes following the Antarctic spring, the protective layer thins over southern South America, southern Africa, Australia, and New Zealand. In Ushuaia, Argentina, surface UVB intensity increased 236% compared to normal levels during this time.

- An international scientific consensus confirmed previous assessments of the anthropogenic, or human-caused, disruption of the ozone layer—chlorine atoms and chlorine monoxide molecules in the stratosphere are of human origin. (See *Scientific Assessment of Ozone Depletion*, by NASA, NOAA, United Nations Environment Programme, and World Meteorological Organization.)

- At Earth's opposite pole, a similar ozone depletion over the Arctic annually exceeds 30%, with the loss rising to as high as 45%. The Canadian and US governments report an "ultraviolet index" to help the public protect themselves (http://www.ec.gc.ca/ozone/).

- The present generation of first-nations people in the Arctic region are receiving more than a 30% greater lifetime UV dose than prior generations,

increasing incidences of skin cancer, cornea damage, cataracts, immune system suppression, skin aging, and sunburn, among other disorders.

- Increased ultraviolet radiation is affecting atmospheric chemistry, biological systems, oceanic phytoplankton (small, photosynthetic organisms that form the basis of the ocean's primary food production), fisheries, crop yields, and human skin, eye tissues, and immunity.

More ultraviolet radiation than ever before is breaking through Earth's protective ozone layer. What is happening in the stratosphere everywhere and in the polar regions specifically? Why is this happening, and how are people and their governments responding? What effects does this have on you personally?

Monitoring Earth's Fragile Safety Screen

A sample from the ozone layer's densest part (at 29 km, or 18 mi, altitude) contains only 1 part ozone per 4 million parts of air—compressed to surface pressure, it would be only 3 mm thick. Yet, this rarefied layer was in steady-state equilibrium for several hundred million years, absorbing intense ultraviolet

radiation and permitting life to proceed safely on Earth.

The total ozone mapping spectrometer (TOMS) began operations in 1978 aboard various satellites. A September 2006 image by the OMI (Ozone Monitoring Instrument) on NASA's *Aura* satellite is in Figure 2.1.1 and shows the all-time record ozone depletion. These measures mean that the ozone column over Antarctica was virtually gone. *Aura* and its OMI package became operational in 2004. (See http://www.nasa.gov/mission_pages/aura/main/index.html.)

Ozone Losses Explained

What is causing the decline in stratospheric ozone? In 1974, two atmospheric chemists, F. Sherwood Rowland and Mario Molina, hypothesized that some synthetic chemicals were releasing chlorine atoms that decompose ozone. These **chlorofluorocarbons, or CFCs,** are synthetic molecules of chlorine, fluorine, and carbon. (See Rowland and Molina's report: "Stratospheric sink for chlorofluoromethanes: Chlorine atom catalyzed destruction of ozone," *Nature* 249 [1974]: 810.)

CFCs are stable (inert) under conditions at Earth's surface and they possess remarkable heat properties. Both qualities made them valuable as

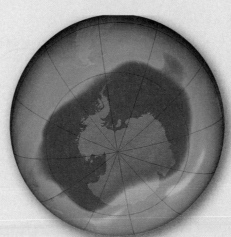

FIGURE 2.1.1 The Antarctic ozone hole.
This OMI image for September 24, 2006, shows the largest ozone hole observed, nearly 29.5 million km^2 (11.4 million mi^2). Measurements dropped below 100 Dobson units in the region of blues and purples; the greens, yellows, and reds denote more ozone. The ozone "hole" has grown larger since 1979. [OMI instrument aboard *Aqua* courtesy of GSFC/NASA.]

Total Ozone (Dobson Units)
110 220 330 440 550

(continued)

Focus Study 2.1 *(continued)*

propellants in aerosol sprays and as refrigerants. Also, some 45% of CFCs were solvents in the electronics industry and used as foaming agents. Being inert, CFC molecules do not dissolve in water and do not break down in biological processes. (In contrast, chlorine compounds derived from volcanic eruptions and the ocean are water-soluble and rarely reach the stratosphere.)

Researchers Rowland and Molina hypothesized that stable CFC molecules slowly migrate into the stratosphere, where intense ultraviolet radiation splits them, freeing chlorine (Cl) atoms. These Cl atoms produce a complex set of reactions that break up ozone molecules (O_3) and leave oxygen gas molecules (O_2) in their place. Oxygen gas molecules are transparent to ultraviolet radiation. The effect is severe, for a single chlorine atom decomposes more than 100,000 ozone molecules. The long residence time of chlorine atoms in the ozone layer (40 to 100 years) is likely to produce long-term consequences through this century from the chlorine already in place. More than 22 million metric tons (24 million tons) of CFCs were sold worldwide and subsequently released into the atmosphere. (See http://www.epa.gov/ozone/index.html.)

Political Realities—An International Response and the Future

As the science became established, sales and production of CFCs declined until a March 1981 presidential order permitted the export and sales of banned products. Sales increased and hit a new peak in 1987 at 1.2 million metric tons (1.32 million tons), when an international agreement went into effect halting further sales growth.

Chemical manufacturers once claimed that no hard evidence existed to prove the ozone-depletion model, and they successfully delayed remedial action for 15 years. Today, with extensive scientific evidence and verification of losses, even the CFC manufacturers admit that the problem is serious. The *Montreal Protocol on Substances That Deplete the Ozone Layer* (1987), as amended five times in 1990, 1992, 1995, 1997, and 1999, aims to

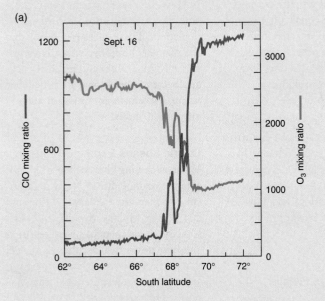

(a)

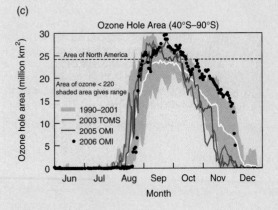

(b)

(c)

SATELLITE

Southern Hemisphere Ozone 2002–2003

FIGURE 2.1.2 Chemical evidence of ozone damage by humans.

(a) The negative correlation between chlorine monoxide and O_3 over the Antarctic continent poleward of 68° S latitude. The data were collected from flights during September 1987 at stratospheric altitudes. (b) Total ozone column data from 1990 to 2006 illustrate the seasonal loss. (c) The area of ozone hole depletion generally peaks in late September, reaching a record low 85 Dobson units in 2006. [(a) Data from NASA; (b and c) TOMS aboard various satellites and OMI on *Aqua*, collected data courtesy of GSFC/NASA.]

reduce and eliminate CFC damage to the ozone layer. With 189 signatory countries, the protocol is regarded as probably the most successful international agreement in history. (See the United Nations Ozone Secretariat at http://ozone.unep.org/ and the Montreal Protocol at http://www.ec.gc.ca/international/multilat/ozone1_e.htm.)

CFC sales continue their decline and all production of harmful CFCs will cease by 2010, although there is concern over some of the substitute compounds and a robust black market for banned CFCs, rivaling drug trafficking at some ports. If the protocol is fully enforced, it is estimated that the stratosphere will return to more normal conditions in a century.

The world without this treaty might prove challenging. Imagine possible stratospheric ozone losses of 50% in the midlatitudes, producing 1.5 million more cases of malignant melanoma (a skin cancer). We owe much to Doctors Rowland and Molina. Mario Molina stated, "It was frustrating for many years, but it really paid off with the Protocol, which was a marvelous example of what the international community can

do working together. We can see from atmospheric measurements that it is already working" [*Nature* 389 (September 18, 1997): 219]. For their work, Rowland, Molina, and another colleague, Paul Crutzen, received the 1995 Nobel Prize for Chemistry. In making the award, the Royal Swedish Academy of Sciences said, "By explaining the chemical mechanism that affects the thickness of the ozone layer, the three researchers have contributed to our salvation from a global environmental problem that could have catastrophic consequences."

Ozone Losses over the Poles

How do Northern Hemisphere CFCs become concentrated over the South Pole? Evidently, chlorine freed in the Northern Hemisphere midlatitudes concentrates over Antarctica through the work of atmospheric winds. Persistent cold temperatures over the South Pole and the presence of thin, icy clouds in the stratosphere promote development of ozone losses. Over the North Pole, conditions are more changeable, so the depletion area is smaller, although growing each year.

Polar stratospheric clouds (*PSCs*) are thin clouds that are important cata-

lysts in the release of chlorine for ozone-depleting reactions. During the long, cold, winter months, a tight circulation pattern forms over the Antarctic continent—the polar vortex. Figure 2.1.2a graphs the negative correlation between ClO presence and O_3 losses over the Antarctic continent within the polar vortex. Figure 2.1.2b graphs the progressive loss of ozone during the year to its usual September low, and Figure 2.1.2c presents satellite data on the areal size of the ozone-depleted region during the year.

Atmospheric science is active in Antarctica. The US scientific base, Palmer Station (64° 46′ S) on Anvers Island, Antarctic Peninsula, experiences a 285% increase in surface UVB radiation above normal levels in the fall. Nearby, at Ukraine's Akademik Vernadsky Station, the former British Faraday Base in the Argentine Islands (65° 9′ S), one of the original ozone spectrophotometer instruments is actively measuring ultraviolet radiation (Figure 2.1.3). Ukraine scientists are studying UVA (320–400 nm) and UVB (290–320 nm) wavelengths using a variety of instruments, including *biodosimeters* that use biological material to measure radiation impact.

(a)

(b)

FIGURE 2.1.3 Atmospheric science in Antarctica.
(a) Ukraine's Vernadsky Station operates year-round performing valuable scientific research. (b) This ozone spectrophotometer instrument measures the ultraviolet component of insolation arriving through the opening in the science station's roof. [Photos by Bobbé Christopherson.]

(continued)

Focus Study 2.1 *(continued)*

UV Index Helps Save Your Skin

Newspaper and television weather reports regularly include the *UV Index* (UVI) in forecasts, as reported by the National Weather Service (NWS) and the Environmental Protection Agency (EPA) since 1994. A revision in 2004 aligned the UVI with guidelines adopted by the World Health Organization and World Meteorological Organization.

Note that the 2004 UVI scale does not rely on "burn times." People were misinterpreting time-to-burn estimates thinking there was some safe level of exposure, or safe tanning time—which was the wrong message. Table 2.1.1 presents the revised UVI. Fair-skinned people (melano-compromised) must exercise more caution than moderate to darker skin types (melano-competent and melano-protected), but all skin types should feel warned, especially for babies and children.

As stratospheric ozone levels continue to thin, and eventually stabilize, surface exposure to cancer-causing radiation climbs. The public now is alerted to take extra precautions in the form of sunscreens, hats, and sunglasses. Remember, damage is cumulative and it may be decades before you experience the ill effects triggered by this summer's sunburn. (For more information, contact the American Cancer Society, 800-227-2345, http://www.cancer.org/; or write the American Academy of Dermatology, P.O. Box 681069, Schaumburg, IL 60168-1069.)

Table 2.1.1 UV Index (EPA, NWS, WHO, WMO)		
Exposure Risk Category	**UVI Range**	**Comments**
Low	less than 2	Low danger for average person. Wear sunglasses on bright days. Watch out for reflection off snow.
Moderate	3–5	Take covering precautions, such as sunglasses, sunscreen, hats, protective clothing, and stay in shade during midday hours.
High	6–7	Use sunscreens with SPF of 15 or higher. Reduce time in the Sun between 11 A.M. and 4 P.M. Use protections mentioned above.
Very High	8–10	Minimize Sun exposure 10 A.M. to 4 P.M. Use sunscreens with SPF ratings of over 15. Use protections mentioned above.
Extreme	11+	Unprotected skin is at risk of burn. Sunscreen application every 2 hours if out-of-doors. Avoid direct Sun exposure during midday hours. Use protections mentioned above.

See http://www.epa.gov/sunwise/uvindex.html for a UVI forecast for your location.

natural sources and the substances they contribute to the air. However, any attempt to diminish the impact of human-made air pollution through a comparison with natural sources is irrelevant, for we have evolved with and adapted to the natural ingredients in the air and not to what we have introduced. We have not evolved in relation to the comparatively recent concentrations of *anthropogenic* (human-caused) contaminants in our metropolitan regions.

A dramatic natural source of pollution was the 1991 eruption of Mount Pinatubo in the Philippines (15° N 120° E), perhaps the last century's largest eruption. This event injected between 15 and 20 million tons of sulfur dioxide (SO_2) into the stratosphere. The spread of these emissions is shown in a sequence of satellite images that begins Chapter 4.

Devastating wildfires on several continents, including widespread fires across the western United States, produce natural air pollution as each year's fire season breaks the previous year's record. Soot, ash, and gases darken skies and damage health in affected regions. Wind patterns spread the pollution from the fires to nearby cities, closing airports and forcing evacuations to avoid the health-related dangers. Wildfire smoke contains particulate matter (dust, smoke, soot, ash), nitrogen oxides, carbon monoxide, and volatile organic compounds.

Scientists established the connection between climate change/global warming and wildfire occurrence, meaning this source of smoke is on the increase as is the release of enormous quantities of carbon dioxide to the atmosphere. One sample of the findings related to climate-change impacts and the western United States:

> Higher spring and summer temperatures and earlier snowmelt are extending the wildfire season and increasing the intensity of wildfires in the western US. . . . large wildfire activity increased suddenly and markedly in the mid-1980s, with higher large-wildfire frequency . . . longer wildfire durations, and longer wildfire seasons.*

*A. L. Westerling et al., "Warming and earlier spring increase western US forest wildfire activity," *Science* 313, 5789 (August 18, 2006): 940.

For example, in drought-plagued southern California, dozens of wildfires across the region burned more than 600,000 acres (243,000 ha) in October 2007 (Figure 2.24a). In 2008, California experienced nearly 3000 wildfires, a record in its state history, consuming millions of acres (Figure 2.24b). Each year across the country, wildfire acreage breaks the record of the previous year—2008 more than doubled the acreage lost in 2001 (see http:// www.usgs.gov/hazards/wildfires/).

Natural Factors That Affect Air Pollution

The problems resulting from both natural and human-made atmospheric contaminants are made worse by several important natural factors. Among these, wind, local and regional landscape characteristics, and temperature inversions in the troposphere dominate.

Wind Winds gather and move pollutants from one area to another, sometimes reducing the concentration of pollution in one location while increasing it in another. Wind can produce dramatic episodes of dust movement. (Dust is defined as

(a)

(b)

particles less than 62 μm, or 0.0025 in., in diameter.) Traveling on prevailing winds, dust from Africa contributes to the soils of South America and Europe. Such movement is confirmed by chemical analysis, frequently employed by scientists to track dust to its source area (Figure 2.25).

Such air movements make the atmosphere's condition an international issue. Indeed, prevailing winds transport air pollution from the United States to Canada, causing much complaint and negotiation between the two governments. Pollution in North America is tracked to Europe, adding to European air pollution problems. In Europe, the cross-boundary drift of pollution is a major issue because of the proximity of countries. This issue led in part to Europe's unification and the European Union (EU).

Arctic haze is a term from the 1950s, when pilots noticed decreased visibility, either horizontally or at a slant angle from their aircraft. Since there is no heavy industry at these latitudes and sparse population, this seasonal haze (concentration of microscopic particles and air pollution that diminishes air clarity) was a mystery at first. A remarkable attribute of industrialization in the Northern Hemisphere is this haze across the Arctic region. Simply, winds of atmospheric circulation transport pollution to sites far distant from points of origin. There is no comparable haze over the Antarctic continent. Based on this analysis, can you think of why Antarctica lacks such a condition?

Local and Regional Landscapes Local and regional landscapes are another important factor in air pollution. Surrounding mountains and hills can form barriers to air movement or can direct pollutants from one area to another. Some of the worst incidents have resulted when local landscapes have trapped and concentrated air pollution.

Places such as Iceland and Hawai'i have their own natural pollution with which to deal. During periods of sustained volcanic activity at Kīlauea, some 2000 metric tons (2200 tons) of sulfur dioxide are produced a day. Concentrations are sometimes high enough to merit broadcast warnings about health concerns, losses to agriculture, and other economic impacts from volcanic smog and acid rain. Hawaiians coined the word *vog* to describe their *vol*canic sm*og*.

Temperature Inversion Vertical temperature and atmospheric density distribution in the troposphere also can worsen pollution conditions. A **temperature inversion**

FIGURE 2.24 California wildfires fill the atmosphere with smoke.
(a) Wildfires in drought- and high-temperature-plagued portions of California, October 23, 2007. More than 500,000 acres (117,000 ha) burned and the fires continued through November. The smoke contains particulate matter (dust, smoke, soot, ash), nitrogen oxides, carbon monoxide, and volatile organic compounds. (b) Sierra Nevada mountain wildfires move along uncontrolled during July 2008. Recent California wildfires consumed millions of acres of drought-damaged forest and chaparral. [(a) *Terra* MODIS sensor, courtesy of the MODIS Land Rapid Response Team, NASA/GSFC; (b) aerial photo by Bobbé Christopherson.]

FIGURE 2.25 Natural variable dust in the atmosphere.
(a) A dust storm in central Nevada. (b) A *Terra* orbital view of a sinuous windblown dust plume flowing northward from Africa, across the Mediterranean Sea, toward Turkey. The Nile Delta appears obscured by the dust storm. [(a) Photo by author; (b) *Terra* MODIS image from 5/13/2001 courtesy of the MODIS Land Rapid Response Team NASA/GSFC.]

(a)   (b)

occurs when the normal temperature decrease with altitude (normal lapse rate) begins to *increase* at some altitude. This can happen at any point from ground level to several thousand meters. Figure 2.26 compares a normal temperature profile with that of a temperature inversion. The normal profile (Figure 2.26a) permits warmer (less dense) air at the surface to rise, ventilating the valley and moderating surface pollution. But a warm air inversion (Figure 2.26b) prevents the rise of cooler (denser) air beneath, halting the vertical mixing of pollutants with other atmospheric gases. Thus, instead of being carried away, pollutants are trapped under the *inversion layer*.

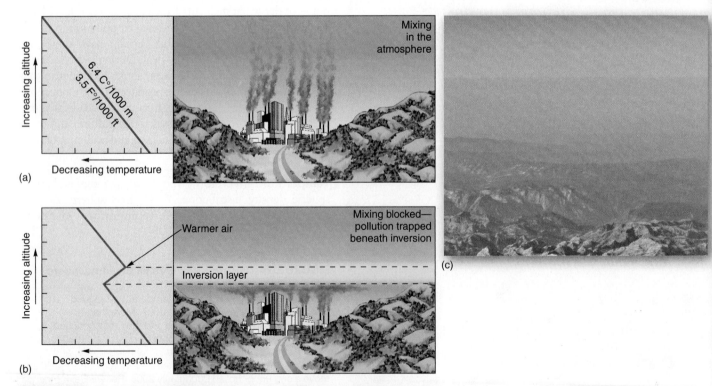

FIGURE 2.26 Normal and inverted temperature profiles.
(a) A comparison of a normal temperature profile in the atmosphere with (b) a temperature inversion in the lower atmosphere. Note how the warmer air layer prevents mixing of the denser (cooler) air below the inversion, thereby trapping pollution. (c) An inversion layer is visible in the morning hours over a valley. [Photo by Bobbé Christopherson.]

Anthropogenic Pollution in the Atmosphere

Anthropogenic, or human-caused, air pollution remains most prevalent in urbanized regions. Approximately 2% of annual deaths in the United States are attributable to air pollution. Comparable risks are identified in Canada, Europe, Mexico, Asia—especially China and India—and elsewhere.

The human population is moving to cities and is therefore coming into increasing contact with air pollution. In 2008, more than 50% of the world's population was living in metropolitan regions, some one-third with unhealthful levels of air pollution. This represents a potentially massive public health issue in this century.

Table 2.4 lists the names, chemical symbols, principal sources, and impacts of variable anthropogenic components in the air. The first eight pollutants in the table result from

Table 2.4	Anthropogenic Gases and Materials in the Lower Atmosphere		
Name	**Symbol**	**Sources**	**Description and Effects of Criteria Pollutants (Main sources listed for US)**
Carbon monoxide	CO	Incomplete combustion of fuels	Odorless, colorless, tasteless gas. Toxicity; affinity for hemoglobin. Displaces O_2 in bloodstream; 50 to 100 ppm causes headaches and vision and judgment losses. *Source:* 78% transportation
Nitrogen oxides	NO_x (NO, NO_2)	High temperature/ pressure combustion	Reddish-brown choking gas. Inflames respiratory system, destroys lung tissue. Damages plants; 3 to 5 ppm is dangerous. *Source:* 55% transportation
Volatile organic compounds	VOC	Incomplete combustion of fossil fuels such as gasoline; cleaning and paint solvents	Prime agents of ozone formation. *Sources:* 45% industrial, 47% transportation
Ozone	O_3	Photochemical reactions	Highly reactive, unstable gas. Oxidizes surfaces, dries rubber and elastic. Damages plants at 0.01 to 0.09 ppm. Agricultural losses at 0.1 ppm; 0.3 to 1.0 ppm irritates eyes, nose, throat
Peroxyacetyl nitrates	PAN	Photochemical reactions	Produced by NO + VOC photochemistry. No human health effects. Major damage to plants, forests, crops
Sulfur oxides	SO_x (SO_2, SO_3)	Combustion of sulfur-containing fuels	Colorless; irritating smell 0.1 to 1 ppm; impairs breathing, taste threshold. Human asthma, bronchitis, emphysema. Leads to acid deposition. *Source:* 85% fuel combustion
Particulate matter	PM	Dust, dirt, soot, salt, metals, organics; agriculture, construction, roads	Complex mixture of solid and aerosol particles. Dust, smoke, and haze affect visibility. Various health effects: bronchitis, pulmonary function. PM_{10} negative health effects established by researchers. *Sources:* 42% industrial, 33% fuel combustion, 25% transportation
Carbon dioxide	CO_2	Complete combustion, mainly from fossil-fuel consumption	Principal greenhouse gas. Atmospheric concentration increasing; 64% of greenhouse warming effect
Methane	CH_4	Organic processes	Secondary greenhouse gas. Atmospheric concentration increasing; 19% of greenhouse warming effect
Water vapor	H_2O vapor	Combustion processes, steam	See Chapter 7 for more on the role of water vapor in the atmosphere.

combustion of fossil fuels in transportation (specifically automobiles, including light trucks) and at stationary sources such as power plants and factories. Overall, automobiles contribute more than 60% of United States and 50% of Canadian human-caused air pollution. Reducing air pollution from the transportation sector involves known technologies and strategies, monetary savings for the consumer, and significant health benefits.

Stationary sources, such as electric power plants and industrial plants that use fossil fuels, contribute the most sulfur oxides and particulates. For this reason, concentrations of these substances are focused in the Northern Hemisphere and the industrial, developed countries.

The last three gases shown in Table 2.4 are discussed elsewhere in this text: Water vapor is examined with water and weather (Chapter 5); carbon dioxide and methane are covered with greenhouse gases and climate (Chapters 3 and 7).

Photochemical Smog Pollution Photochemical smog was not generally experienced in the past, but developed with the advent of the automobile. Today, it is the major component of anthropogenic air pollution. **Photochemical smog** results from the interaction of sunlight and the combustion products in automobile and light truck exhaust (nitrogen oxides and VOCs). Although the term *smog*—a combination of the words *smoke* and *fog*—is a misnomer, it is generally used to describe this phenomenon. Smog is responsible for the hazy sky and reduced sunlight in many of our cities and was center

stage as a concern during the 2008 Olympic Summer Games held in Beijing, where air pollution sieges are common.

Figure 2.27 summarizes how *nitrogen dioxide* in car exhaust converts into major air pollutants—*ozone*, **peroxyacetyl nitrates (PAN)**, and *nitric acid*. In the photochemistry reaction in the illustration, ultraviolet radiation liberates atomic oxygen (O) and a nitric oxide (NO) molecule from the NO_2. The free oxygen atom combines with an oxygen molecule, O_2, to form the oxidant ozone, O_3. In addition, the nitric oxide (NO) molecule reacts with VOCs to create PAN, which produces no known health effect in humans but is particularly damaging to plants, including both agricultural crops and forests.

Ozone is the primary ingredient in photochemical smog. (This is the same gas that is beneficial to us in the stratosphere in absorbing ultraviolet radiation.) The reactivity of ozone is a health threat, for it damages biological tissues. For several reasons, children are at greatest risk from ozone pollution—one in four children in US cities is at risk of developing health problems from ozone pollution. This ratio is significant; it means that more than 12 million children are vulnerable in those cities with the worst polluted air (Los Angeles, New York City, Atlanta, Houston, and Detroit).

Industrial Smog and Sulfur Oxides Over the past 300 years, except in some developing countries, coal has slowly replaced wood as the basic fuel used by society. The

FIGURE 2.27 Photochemical reactions.
The interaction of automobile exhaust (NO_2, VOCs, CO) and ultraviolet radiation in sunlight causes photochemical reactions. To the left, note the formation of nitric acid and acid deposition.

Solar radiation

Ultraviolet radiation

H_2O (water)

HNO_3 (nitric acid)

CO
VOC
NO_2
NO_2 (nitrogen dioxide)
CO
VOC
NO_2
CO (carbon monoxide)
NO_2
VOC
NO_2
CO
VOC (Volatile organic compounds)

O_2 (molecular oxygen)
O (atomic oxygen)
O_3 (ozone)

NO + VOC (nitric oxide)
PAN (peroxyacetyl nitrates)

Photochemical smog

Acid deposition

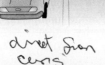

times more potent than the larger $PM_{2.5}$ and PM_{10} particles because they can get into smaller channels in lung tissue and cause scarring, abnormal thickening, and damage called *fibrosis*. Asthma prevalence has nearly doubled since 1980 in the United States. Representatives of pollution sources dispute these findings and the concern generated by such studies.

We are now contributing significantly to the creation of the **anthropogenic atmosphere**, a tentative label for Earth's next (fifth) atmosphere. The urban air we breathe today may be just a preview. What is the air quality like where you live, work, and go to college? How might you find out its status?

Benefits of the Clean Air Act

This case study remains compelling despite the period of time it covers. The findings apply to the present situation in many ways and the actions taken by society. Concentration of many air pollutants declined over the past several decades because of Clean Air Act (CAA) legislation (1970, 1977, 1990), resulting in the saving of trillions of dollars in avoided health, economic, and environmental losses. Despite this success, air pollution controls are subject to a continuing political pressure and a weakening of some standards and proposals that will increase pollution emissions.

Since 1970 and the CAA, there were significant reductions in atmospheric concentrations of carbon monoxide (−45%), nitrogen oxides (−22%), volatile organic compounds (−48%), PM_{10} particulates (−75%), sulfur oxides (−52%; see Focus Study 2.2, Figure 2.2.1), and lead (−98%). Prior to the CAA, lead was added to gasoline, emitted in the exhaust, dispersed over great distances, and settled in living tissues, especially in children. These all represent remarkable reductions—a successful linking of science and public policy.

To be justified, abatement (mitigation and prevention) costs must not exceed the financial benefits derived from reducing pollution damage. Compliance with the CAA affected patterns of industrial production, employment, and capital investment. Although these expenditures must be viewed as investments that generated benefits, the dislocation and job loss in some regions were difficult—reductions in high-sulfur coal mining and cutbacks in polluting industries such as steel, for example.

In 1990, Congress requested the EPA to answer the following question: How do the overall health, welfare, ecological, and economic benefits of CAA programs compare with the costs of these programs? In response, the EPA performed an exhaustive cost-benefit analysis and published a draft report in 1996 and a final report in 1997. *The Benefits of the Clean Air Act, 1970 to 1990* (Office of Policy, Planning, and Evaluation, US EPA) calculated the following:

- The *total direct cost* to implement the CAA for all federal, state, and local rules from 1970 to 1990 was

FIGURE 2.28 Typical industrial smog.
Pollution generated by industry and coal-fired electrical generation differs from that produced by transportation. Industrial pollution has high concentrations of sulfur oxides, particulates, and carbon dioxide. This coal-fired power plant at Barentsburg, Spitsbergen, part of Norway, lacks scrubbers to reduce stack emissions. [Photo by Bobbé Christopherson.]

Industrial Revolution required high-grade energy to run machines. The air pollution associated with coal-burning industries is known as **industrial smog** (Figure 2.28). The term *smog* was coined by a London physician at the beginning of the twentieth century to describe the combination of fog and smoke containing sulfur gases (sulfur is an impurity in fossil fuels).

Once in the atmosphere, **sulfur dioxide** (SO_2) reacts with oxygen (O) to form sulfur trioxide (SO_3), which is highly reactive and, in the presence of water or water vapor, forms **sulfate aerosols**, tiny particles about 0.1 to 1 μm in diameter. Sulfuric acid (H_2SO_4) can form even in moderately polluted air, at normal temperatures. Sulfur dioxide–laden air is dangerous to health, corrodes metals, and deteriorates stone building materials at accelerated rates. Sulfuric acid deposition, added to nitric acid deposition, has increased in severity since it was first described in the 1970s. Focus Study 2.2 discusses this vital atmospheric issue.

In the United States, coal-burning electric utilities and steel manufacturing are the main sources of sulfur dioxide, principally in the East and Midwest. And, because of the prevailing movement of air masses, they are the main sources of sulfur dioxide in adjacent Canadian regions. As much as 70% of Canadian sulfur dioxide is initiated within the United States.

 Particulates A diverse mixture of fine particles, both solid and aerosol, that impact human health is **particulate matter (PM)**. Haze, smoke, and dust are visible reminders of particulate material in the air we breathe. PM_{10}, particulate smaller than 10 microns (10 μm) in diameter, was designated a matter for concern in 1987. $PM_{2.5}$ is currently being debated as an appropriate standard for human health. New studies are implicating even smaller particles, known as *ultrafines*, which are $PM_{0.1}$. These are many

$523 billion (in 1990-value dollars). This cost was borne by businesses, consumers, and government entities.

- The estimate of *direct monetized benefits* from the CAA from 1970 to 1990 falls in a range from $5.6 to $49.4 trillion, with a central mean of *$22.2 trillion.*
- Therefore, the *net financial benefit* of the CAA is *$21.7 trillion*—a ratio of 42 to 1 of benefits over costs. "The finding is overwhelming. The benefits far exceed the costs of the CAA in the first 20 years," said Richard Morgenstern, associate administrator for policy planning and evaluation at the EPA.

The benefits to society, directly and indirectly, have been widespread across the entire population: improved health and environment, less lead to harm children, lowered cancer rates, less acid deposition, and an estimated 206,000 fewer deaths related to air pollution in 1990 alone, among many benefits described. Further, these benefits continued during a period (1970–2008) in which the US population grew by a third to over 303 million and the economy expanded by nearly 200%.

As you reflect on this chapter—the path of insolation through the atmosphere to Earth's surface, our modern

FOCUS STUDY 2.2

Acid Deposition: A Continuing Blight on the Landscape

Acid deposition is a major environmental problem in some areas of the United States, Canada, Europe, and Asia. Such deposition is most familiar as "acid rain," but it also occurs as "acid snow" and in dry form as dust or aerosols. (Aerosols are tiny liquid droplets or solid particles.) In addition, winds can carry the acid-producing chemicals many kilometers from their sources before they settle on the landscape, where they enter streams and lakes as runoff and groundwater flows.

Acid deposition is causally linked to serious problems: declining fish populations and fish kills in the northeastern United States, southeastern Canada, Sweden, and Norway; widespread forest damage in these same places and Germany; widespread changes in soil chemistry; and damage to buildings, sculptures, and historic artifacts. In New Hampshire's Hubbard Brook Experimental Forest (http://www.hubbardbrook.org/), a study covering 1960 to the present found half the nutrient calcium and magnesium base cations (see Chapter 15) were leached from the soil. Excess acids are the cause of the decline.

Despite scientific agreement about the problem, which the US General Accounting Office calls a "combination of meteorological, chemical, and biological phenomena," corrective action was delayed by its complexity and politics.

The acidity of precipitation is measured on the pH scale, which expresses the relative abundance of free hydrogen ions (H^+) in a solution. Free hydrogen ions in a solution are what make an acid corrosive, for they easily combine with other ions. The pH scale is logarithmic: Each whole number represents a 10-fold change. A pH of 7.0 is neutral (neither acidic nor basic). Values less than 7.0 are increasingly *acidic*, and values greater than 7.0 are increasingly *basic*, or *alkaline*. (A pH scale for soil acidity and alkalinity is portrayed graphically in Chapter 15.)

Natural precipitation dissolves carbon dioxide from the atmosphere to form carbonic acid. This process releases hydrogen ions and produces an average pH reading for precipitation of 5.65. The normal range for precipitation is 5.3–6.0. Thus, normal precipitation is always slightly acidic.

Some anthropogenic gases are converted to acids in the atmosphere and then are removed by wet and dry deposition processes. Specifically, nitrogen and sulfur oxides released in the combustion of fossil fuels can produce (HNO_3) and sulfuric acid (H_2SO_4) in the atmosphere.

Acid Precipitation Damage

Precipitation as acidic as pH 2.0 has fallen in the eastern United States, Scandinavia, and Europe. By comparison, vinegar and lemon juice register

slightly less than pH 3.0. Aquatic plant and animal life perishes when lakes drop below pH 4.8.

More than 50,000 lakes and some 100,000 km (62,000 mi) of streams in the United States and Canada are at a pH level below normal (i.e., below pH 5.3), with several hundred lakes incapable of supporting any aquatic life— 15% of the lakes in New England and 41% in the Adirondack Mountains. More than 7000 lakes in the Sudbury, Ontario, region suffer acid damage. Acid deposition causes the release of aluminum and magnesium from clay minerals in the soil, and both of these are harmful to fish and plant communities. (Environment Canada's acid rain web page is at http://www.ec.gc.ca/acidrain/.)

Also, relatively harmless mercury deposits in lake-bottom sediments convert in acidified lake waters into highly toxic *methylmercury*, which is deadly to aquatic life. Local health advisories in two provinces and 22 US states are regularly issued to warn those who fish of the methylmercury problem. Mercury atoms rapidly bond with carbon and move through biological systems as an *organometallic compound*.

Damage to forests results from the rearrangement of soil nutrients, the death of soil microorganisms, and an aluminum-induced calcium deficiency that is currently under investigation. The most advanced impact is seen in forests in Europe, especially in eastern Europe, principally because of its long history of burning coal and the density of industrial activity. In Germany and Poland, up to 50% of the forests are dead or damaged; in Switzerland, 30% are afflicted.

In the United States, regional-scale decline in forest cover is significant, especially red spruce and sugar maples. In some maples, aluminum is collecting around rootlets; in spruce, acid fogs and rains leach calcium from needles directly. Affected trees are susceptible to winter cold, insects, and droughts. In New England, some stands of spruce are as much as 75% affected, as evidenced through analysis of tree-growth rings, which become narrower in adverse growing years. Acid-laden cloud cover damages trees at higher elevations in the Appalachians.

Government estimates of damage in the United States, Canada, and Europe exceed $50 billion annually.

Because wind and weather patterns are international, efforts at reducing acidic deposition also must be international in scope. Figure 2.2.1 maps the reduction in sulfate deposition between the 1990–1994 and 1996–2000 periods. However, this progress is only a beginning. According to a study in *BioScience*, power plants and other sources must cut sulfur dioxide emissions 80% beyond the Clean Air Act mandate to truly reverse damage trends.

Ten leading acid-deposition researchers reported in an extensive study in *BioScience*, March 2001:

Model calculations suggest that the greater the reduction in atmospheric sulfur deposition, the greater the magnitude and rate of chemical recovery. Less aggressive proposals for controls of sulfur emissions will result in slower chemical and biological recovery and in delays in regaining the services of a fully functional ecosystem. North America and Europe are in the midst of a large-scale experiment. Sulfuric and nitric acids have acidified soils, lakes, and streams, thereby stressing or killing terrestrial and aquatic biota.*

Acid deposition is an issue of global spatial significance for which science is providing strong incentives for action. Reductions in troublesome emissions are closely tied to energy conservation and therefore directly related to production of greenhouse gases and global warming concerns—thus, linking these environmental issues.

*C. T. Driscoll et al., "Acidic deposition in the Northeastern United States: Sources and inputs, ecosystems effects, and management strategies," *BioScience* 51 (March 2001): 195.

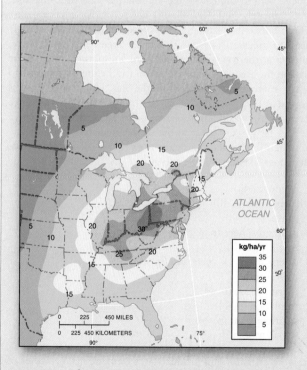 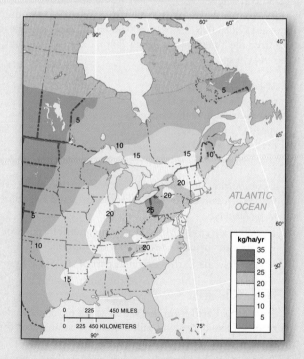

FIGURE 2.2.1 Improvement in sulfate wet deposition rate.
Spatial portrayal of annual sulfate (principally SO$_4$) wet deposition on the landscape, (a) 1990 to 1994 and (b) 1996 to 2000, in kilograms per hectare per year. The Canadian Environmental Protection Act, the Canada–United States Border Air Quality Strategy, and the US Clean Air Act regulations lowered levels of emissions that add acid to the environment. [Maps from *Cleaner Air through Cooperation, Canada–United States, Progress under the Air Quality Agreement—2003, Environment Canada, 2003*. Reproduced with the permission of the Minister of Public Works and Government Services Canada.]

atmosphere, the treaties to protect stratospheric ozone, and the EPA study of benefits from the CAA—the positive steps to protect natural systems should be refreshing. Society knew what to do, took action, and reaped enormous economic and health benefits. An important role for physical geographers is to explain these global impacts through spatial analysis and to guide an informed citizenry toward a better understanding.

Summary and Review—Solar Energy, Seasons, and the Atmosphere

■ *Distinguish* among galaxies, stars, and planets and *locate* Earth.

Our Solar System—Sun and eight planets—is located on a remote, trailing edge of the **Milky Way Galaxy**, a flattened, disk-shaped mass estimated to contain up to 400 billion stars. **Gravity**, the mutual attracting force exerted by the mass of an object upon all other objects, is an organizing force in the Universe. The process of suns (stars) condensing from nebular clouds with planetesimals (protoplanets) forming in orbits around their central masses is the **planetesimal hypothesis**.

The Solar System, planets, and Earth began to condense from a nebular cloud of dust, gas, debris, and icy comets approximately 4.6 billion years ago. Distances in space are so vast that the **speed of light** (300,000 kmps, or 186,000 mps, which is about 9.5 trillion km, or nearly 6 trillion mi, per year) is used to express distance.

In its orbit, Earth is at **perihelion** (its closest position to the Sun) during our Northern Hemisphere winter (January 3 at 147,255,000 km, or 91,500,000 mi). It is at **aphelion** (its farthest position from the Sun) during our Northern Hemisphere summer (July 4 at 152,083,000 km, or 94,500,000 mi). Earth's average distance from the Sun is approximately 8 minutes and 20 seconds in terms of light speed.

Milky Way Galaxy (p. 40)
gravity (p. 40)
planetesimal hypothesis (p. 40)
speed of light (p. 40)
perihelion (p. 40)
aphelion (p. 40)

1. Describe the Sun's status among stars in the Milky Way Galaxy. Describe the Sun's location, size, and relationship to its planets.
2. If you have seen the Milky Way at night, briefly describe it. Use specifics from the text in your description.
3. Briefly describe Earth's origin as part of the Solar System.
4. Compare the eight planets of the Solar System and their distances from the Sun.
5. How far is Earth from the Sun in terms of light speed? In terms of kilometers and miles? Relate this distance to the shape of Earth's orbit during the year.
6. Briefly describe the relationship among these concepts: Universe, Milky Way Galaxy, Solar System, Sun, and Planet Earth.
7. What is the role of gravity, and the gravitational force, in the Solar System and in planetary formation?

■ *Describe* the Sun's operation and *explain* the characteristics of the solar wind and the electromagnetic spectrum of radiant energy.

The **fusion** process—hydrogen atoms forced together under tremendous temperature and pressure in the Sun's interior—generates incredible quantities of energy. Solar energy in the form of charged particles of **solar wind** travels out in all directions from disturbances on the Sun. The Sun's most conspicuous features are large **sunspots**, caused by magnetic disturbances. Solar wind is deflected by Earth's **magnetosphere**, producing various effects in the upper atmosphere, including spectacular **auroras**, the northern and southern lights that surge across the skies at higher latitudes. Another effect of the solar wind in the atmosphere is its possible influence on weather.

The **electromagnetic spectrum** of radiant energy travels outward in all directions from the Sun. The total spectrum of this radiant energy is made up of different **wavelengths**—the distance between corresponding points on any two successive waves of radiant energy. Eventually, some of this radiant energy reaches Earth's surface.

Electromagnetic radiation from the Sun passes through Earth's magnetic field to the top of the atmosphere—the **thermopause**, at approximately 500 km (300 mi) altitude. Solar radiation that reaches a horizontal plane at Earth is **insolation**, a term specifically applied to radiation arriving at Earth's surface and atmosphere. Insolation at the top of the atmosphere is expressed as the **solar constant**: the average insolation received at the thermopause when Earth is at its average distance from the Sun. The solar constant is measured as 1372 W/m^2 (2.0 $cal/cm^2/min$); 2 langleys/min. The place receiving maximum insolation is the **subsolar point**, where solar rays are perpendicular to the surface (radiating from directly overhead). All other locations away from the subsolar point receive slanting rays and more diffuse energy.

fusion (p. 40)
solar wind (p. 40)
sunspots (p. 40)
magnetosphere (p. 42)
auroras (p. 42)
electromagnetic spectrum (p. 43)
wavelength (p. 43)
thermopause (p. 44)
insolation (p. 44)
solar constant (p. 44)
subsolar point (p. 45)

8. How does the Sun produce such tremendous quantities of energy?
9. What is the sunspot cycle? What stage in the cycle is expected for 2012?
10. Describe Earth's magnetosphere and its effects on the solar wind and the electromagnetic spectrum.
11. Compare the filtering aspects of the atmosphere to the astronaut's protective spacesuit shown in Figure 2.3, specifically in regard to the solar wind.

Chapter 2 Solar Energy, Seasons, and the Atmosphere **71**

12. Describe the various segments of the electromagnetic spectrum, from shortest to longest wavelength. What wavelengths are mainly produced by the Sun? Which are principally radiated by Earth to space?

13. What is the solar constant? Why is it important to know?

14. Select 0° or 90° latitude on Figure 2.10 and compare the approximate amount of energy received. What differences do you observe? On the map where do you find the largest net radiation? The lowest net radiation?

15. If Earth were flat and oriented perpendicularly to incoming solar radiation (insolation), what would be the latitudinal distribution of solar energy at the top of the atmosphere?

■ *Define* solar altitude, solar declination, and daylength and *describe* the annual variability of each—Earth's seasonality.

The angle between the Sun and the horizon is the Sun's **altitude**. The Sun's **declination** is the latitude of the subsolar point. Declination annually migrates through 47° of latitude, moving between the *Tropic of Cancer* at 23.5° N (June) and the *Tropic of Capricorn* at 23.5° S (December) latitude. Seasonality means an annual change in the Sun's altitude and changing **daylength**, or duration of exposure.

Earth's distinct seasons are produced by interactions of **revolution** (annual orbit about the Sun), **rotation** (turning on the axis), *axial tilt* (about 23.5° from a perpendicular to the **plane of the ecliptic**—an imaginary plane touching all points of Earth's orbit), **axial parallelism** (the parallel alignment of the axis throughout the year), and sphericity. Earth rotates about its *axis*, an imaginary line extending through the planet from the geographic North Pole to the South Pole. As it rotates, the traveling boundary that divides daylight and darkness is called the **circle of illumination**.

On December 21 or 22, at the moment of the **winter solstice** ("winter Sun stance"), or **December solstice**, the circle of illumination excludes the North Pole but includes the South Pole. The subsolar point is at 23.5° S latitude, the parallel called the **Tropic of Capricorn**. The moment of the **vernal equinox**, or **March equinox**, occurs on March 20 or 21. At that time, the circle of illumination passes through both poles so that all locations on Earth experience a 12-hour day and a 12-hour night.

June 20 or 21 is the moment of the **summer solstice**, or **June solstice**. The subsolar point now has shifted from the equator to 23.5° N latitude, the **Tropic of Cancer**. Because the circle of illumination now includes the North Polar region, everything north of the Arctic Circle receives 24 hours of daylight—the "midnight Sun." September 22 or 23 is the time of the **autumnal equinox**, or **September equinox**, when Earth's orientation is such that the circle of illumination again passes through both poles so that all parts of the globe experience a 12-hour day and a 12-hour night.

altitude (p. 46)
declination (p. 47)
daylength (p. 47)
revolution (p. 47)
rotation (p. 48)
circle of illumination (p. 48)
plane of the ecliptic (p. 48)

axial parallelism (p. 48)
winter solstice, December solstice (p. 48)
Tropic of Capricorn (p. 48)
vernal equinox, March equinox (p. 50)
summer solstice, June solstice (p. 50)
Tropic of Cancer (p. 50)
autumnal equinox, September equinox (p. 50)

16. Contrast the concept of seasonality between the equator and the polar regions.

17. The concept of seasonality refers to two specific observations. How do these two aspects of seasonality change during a year at 0° latitude? At 40°? At 90°?

18. Differentiate between the Sun's altitude and its declination at Earth's surface.

19. For the latitude at which you live, how does daylength vary during the year? How does the Sun's altitude vary? Does your local newspaper publish a weather calendar containing such information?

20. List the five physical factors that operate together to produce seasons.

21. Describe Earth's revolution and rotation and differentiate between them.

22. Define Earth's present tilt relative to its orbit about the Sun.

■ *Construct* a general model of the atmosphere based on composition, temperature, and function and *diagram* this model in a simple sketch.

Our modern atmosphere is a gaseous mixture so evenly mixed it behaves as if it were a single gas. It is naturally odorless, colorless, tasteless, and formless. The principal substance of this atmosphere is air—the medium of life.

Above 480 km (300 mi) altitude, the atmosphere is rarefied (nearly a vacuum) and is called the **exosphere**, which means "outer sphere." The weight (force over a unit area) of the atmosphere, exerted on all surfaces, is **air pressure**. It decreases rapidly with altitude.

By *composition*, we divide the atmosphere into the **heterosphere**, extending from 480 km to 80 km, and the **homosphere**, extending from 80 km to Earth's surface. Within the heterosphere and using *temperature* as a criterion, we identify the **thermosphere**. Its upper limit, called the *thermopause*, is at approximately 480 km altitude. **Kinetic energy**, the energy of motion, is the vibrational energy that we measure and call temperature. However, the actual heat produced in the thermosphere is very small. The density of the molecules is so low that little actual *heat*, the flow of kinetic energy from one body to another because of a temperature difference between them, is produced. Nearer Earth's surface the greater number of molecules in the denser atmosphere transmit their kinetic energy as **sensible heat**, meaning that we can feel it.

The homosphere includes the **mesosphere, stratosphere,** and **troposphere**, as defined by temperature criteria. The normal temperature profile within the troposphere during the daytime decreases rapidly with increasing altitude at an average of 6.4 C° per km (3.5 F° per 1000 ft), a rate known as the **normal lapse rate**. The top of the troposphere is wherever a temperature of −57°C (−70°F) is recorded, a transition known as the *tropopause*. The actual lapse rate at any particular time and place may deviate considerably because of local weather conditions and is called the **environmental lapse rate**.

We distinguish a region in the heterosphere by its *function*. The **ionosphere** absorbs cosmic rays, gamma rays, X-rays, and shorter wavelengths of ultraviolet radiation and converts them into kinetic energy. A functional region within the stratosphere is the **ozonosphere**, or **ozone layer**, which absorbs life-threatening ultraviolet radiation, subsequently raising the temperature of the stratosphere.

exosphere (p. 51)
air pressure (p. 51)
heterosphere (p. 52)
homosphere (p. 52)
thermosphere (p. 55)
kinetic energy (p. 55)
sensible heat (p. 56)
mesosphere (p. 56)
stratosphere (p. 56)
troposphere (p. 56)
normal lapse rate (p. 57)
environmental lapse rate (p. 57)
ionosphere (p. 58)
ozonosphere, ozone layer (p. 58)

23. What is air? Where did the components in Earth's present atmosphere originate?
24. In view of the analogy by the late Lewis Thomas, characterize the various functions the atmosphere performs that protect the surface environment.
25. What three distinct criteria are employed in dividing the atmosphere for study?
26. Describe the overall temperature profile of the atmosphere and list the four layers defined by temperature.
27. Describe the two divisions of the atmosphere on the basis of composition.
28. What are the two primary functional layers of the atmosphere and what does each do?

■ *Describe* conditions within the stratosphere; specifically, *review* the function and status of the ozonosphere (ozone layer).

The overall reduction of the stratospheric ozonosphere, or ozone layer, during the past several decades represents a hazard for society and many natural systems and is caused by chemicals introduced into the atmosphere by humans. Since World War II, quantities of human-made **chlorofluorocarbons (CFCs)** and bromine-containing compounds have made their way into the stratosphere. The increased ultraviolet light at those altitudes breaks down these stable chemical compounds, thus freeing chlorine and bromine atoms. These atoms act as catalysts in reactions that destroy ozone molecules.

chlorofluorocarbons, CFCs (p. 59)

29. Why is stratospheric ozone (O_3) so important? Describe the effects created by increases in ultraviolet light reaching the surface.
30. Summarize the ozone predicament and describe trends and any treaties that intend to protect the ozone layer.
31. Evaluate Crutzen, Rowland, and Molina's use of the scientific method in investigating stratospheric ozone depletion.

■ *Distinguish* between natural and anthropogenic variable gases and materials in the lower atmosphere and *describe* the sources and effects of air pollution and acid deposition.

Within the troposphere, both natural and human-caused variable gases, particles, and other chemicals are part of the atmosphere. We coevolved with natural "pollution" and thus are adapted to it. But we are not adapted to cope with our own anthropogenic pollution. It constitutes a major health threat, particularly where people are concentrated in cities.

Vertical temperature and atmospheric density distribution in the troposphere also can worsen pollution conditions. A **temperature inversion** occurs when the normal temperature decrease with altitude (normal lapse rate) reverses. In other words, temperature begins to increase at some altitude.

Photochemical smog results from the interaction of sunlight and the products of automobile exhaust, the single largest contributor to pollution that produces smog. Car exhaust, containing *nitrogen dioxide* and *volatile organic compounds (VOCs)*, in the presence of ultraviolet light in sunlight converts into major air pollutants—*ozone*, **peroxyacetyl nitrates (PAN)**, and *nitric acid*. The principal photochemical by-products include *ozone* (O_3), which causes negative health effects, oxidizes surfaces, and kills or damages plants; and PAN, which produce no known health effects in humans but are particularly damaging to plants, including both agricultural crops and forests.

The distribution of human-produced **industrial smog** over North America, Europe, and Asia is related to transportation and electrical production. Such characteristic pollution contains **sulfur dioxide**. Sulfur dioxide in the atmosphere reacts to produce **sulfate aerosols**, which produce sulfuric acid (H_2SO_4) deposition and affect the Earth energy budget by scattering and reflecting solar energy. **Particulate matter (PM)** is a diverse mixture of fine particles, both solid and aerosol, that impact human health.

Energy conservation and efficiency and reducing emissions are essential strategies for reducing air pollution. Earth's next atmosphere most accurately may be described as the **anthropogenic atmosphere** (human-influenced atmosphere).

temperature inversion (p. 63)
photochemical smog (p. 66)
peroxyacetyl nitrates (PAN) (p. 66)
industrial smog (p. 67)
sulfur dioxide (p. 67)
sulfate aerosols (p. 67)
particulate matter (PM) (p. 67)
anthropogenic atmosphere (p. 67)

32. Why are anthropogenic gases more significant to human health than those produced from natural sources?
33. In what ways does a temperature inversion worsen an air pollution episode? Why?
34. What is the difference between industrial smog and photochemical smog?
35. Describe the relationship between automobiles and the production of ozone and PAN in city air. What are the principal negative impacts of these gases?
36. How are sulfur impurities in fossil fuels related to the formation of acid in the atmosphere and acid deposition on the land?
37. In summary, what are the results from the first 20 years under Clean Air Act regulations? In your opinion, do you see viable arguments for its repeal?

NetWork and Critical Thinking Tools

A. A scientific study, *The Benefits of the Clean Air Act, 1970 to 1990* (Office of Policy, Planning, and Evaluation, US EPA, October 1997), determined that the Clean Air Act provided the American people with benefits. Health, welfare, ecological, and economic benefits rated 42 to 1 over the costs of CAA implementation (estimated at $21.7 trillion in net benefits compared to $523 billion in costs). Do you think this is significant? Should this be part of the debate about weakening or strengthening of the Clean Air Act? Do you think similar benefits might result from other regulations such as the Clean Water Act or groundwater protection measures, or action on global climate change?

B. Relative to item A: In your opinion, why is the public generally unaware of these widespread benefits? What are the difficulties in informing the public? Why is there such media attention given to antiscience and nonscientific opinions? Take a moment and brainstorm recommendations for action, education, and public awareness on these issues.

C. To determine the total ozone column at your present location, go to http://toms.gsfc.nasa.gov/teacher/ozone_overhead. html, "What was the total ozone column at your house?" Select a point on the map or enter your latitude and longitude and the date you want to check. The ozone column is measured by the Ozone Monitoring Instrument (OMI) sensor aboard the *Aqua* satellite and is mainly sensitive to stratospheric ozone. (Note also the limitations listed on the extent of data availability.) For several different dates, when do the lowest values occur? The highest values occur? Briefly explain what your results mean. How do you interpret the values found?

D. Go to the Student Learning Center for this chapter and open the animation "Earth–Sun Relations, Seasons." After reviewing the first two screens and narrations you find an animation of the march of the seasons as Earth revolves around the Sun during a year. Note the animation begins at the March Equinox. Activate the animation and listen to the narration as it stops at each of the four anniversary dates. Next, using your mouse and the slider below the animation, click the marker and move it to today's date as you work with this. Describe what you see in the Earth–Sun relation, circle of illumination, sunlight at the poles or lack of sunlight. Lastly, set the slider on the date of the final exam for this physical geography or Earth science class. Note the distance you must move Earth to get from today to the final exam day. Remember, on the voyage of revolution you are traveling at 107,280 kmph (66,660 mph)—quite a trip together.

Jet condensation trails stimulate high-cirrus cloud formation. Inset photos and image show thin, newer contrails in contrast to older contrails widening over time as they develop. Note the density of contrails over the Midwest and Great Lakes region in the satellite image. Studies found that such contrails affect daily temperature ranges between daytime maximum and nighttime minimum—another connection between transportation systems and Earth's atmospheric energy budget. *[Photos by Bobbé Christopherson; inset image Terra MODIS satellite courtesy of MODIS Land Rapid Response Team, NASA/GSFC.]*

3

Atmospheric Energy and Global Temperatures

KEY LEARNING CONCEPTS

After reading the chapter, you should be able to:

- **Identify** the pathways of solar energy through the troposphere to Earth's surface: transmission, scattering, diffuse radiation, refraction, albedo (reflectivity), conduction, convection, and advection.

- **Describe** the greenhouse effect and the patterns of global net radiation and surface energy balances.

- **Review** the temperature concepts and temperature controls that produce global temperature patterns.

- **Interpret** the pattern of Earth's temperatures for January and July and annual temperature ranges.

- **Contrast** wind chill and heat index and **determine** human response to these apparent temperature effects.

- **Portray** typical urban heat island conditions and **contrast** the microclimatology of urban areas with that of surrounding rural environments.

STUDENT LEARNING CENTER TOOLS

The Elemental Geosystems Student Learning Center provides on-line resources for this chapter by clicking on the book cover at **www.mygeoscienceplace.com**. Once you are in our Learning Center, bookmark the URL address. For this chapter you will find reviewing tools, eBook links, self-tests and quizzes, critical thinking items, figure animations, photo galleries, and satellite loops. Some highlights:

- ▶ Narrated map animations of albedo, shortwave radiation, net radiation, and latent and sensible heat fluxes.

- ▶ Earth–atmosphere energy balance animation and Earth's greenhouse effect animation

- ▶ Narrated map animations of surface and sea-surface temperatures.

- ▶ More than 10 additional URLs along with Quick Links to all the URLs in the chapter.

arth's biosphere pulses with flows of energy. Think for a moment of your own life and activities—all reflect atmosphere and surface energy patterns, air temperatures, the seasons, climate, and the daily weather we experience. Find a nice spot outdoors to read sections of this chapter so you can actually observe the concepts happening as you learn—depending on local weather and seasons, of course.

In this chapter: This chapter examines the cascade of energy *input* through the atmosphere to Earth's surface as insolation is absorbed and redirected along various pathways in the troposphere. Reflected light and emitted longwave energy from the atmosphere and surface environment counter the insolation gain. An important *output* produced by this atmosphere and surface energy system is the pattern of global temperatures. Global temperature patterns presently are changing in a warming trend that is affecting us all. Also, you learn what apparent temperature means as wind and humidity alter the temperatures we sense as the wind chill or heat index. Finally, the cities in which we live alter surface energy characteristics, so the temperatures and climates of our urban areas differ from those of surrounding rural areas.

Energy Essentials

In a photograph of Earth taken from space, the pattern of surface response to incoming insolation is clearly visible (see Earth's photo on the back cover of this text). Land and water surfaces, clouds, and atmospheric gases and dust intercept solar energy. The flows of energy are manifest in swirling weather patterns, powerful oceanic currents, and the varied distribution of vegetation. Specific energy patterns differ for deserts, oceans, mountaintops, rain forests, and ice-covered landscapes. In addition, clear or cloudy weather may mean a 75% difference in the amount of energy reaching the surface, because clouds reflect incoming energy.

Energy Pathways and Principles

Earth's atmosphere and surface are heated by solar energy, which is unevenly distributed by latitude and which fluctuates seasonally. Figure 3.1 is a simplified flow diagram of shortwave and longwave radiation in the Earth–atmosphere system. You will find it helpful to refer to this figure, and the more detailed energy balance illustration in Figure 3.10, as you read through the following section. We first look at some important pathways and principles for insolation as it passes through the atmosphere to Earth's surface.

Transmission refers to the passage of shortwave and longwave energy through either the atmosphere or water. Our budget of atmospheric energy comprises shortwave radiation *inputs* (ultraviolet light, visible light, and near-infrared wavelengths) and longwave radiation *outputs* (thermal infrared) that pass through the atmosphere by transmission.

Insolation Input *Insolation* is the single energy input driving the Earth–atmosphere system. The world map in Figure 3.2 shows the distribution of average annual solar energy at Earth's surface. It includes all radiation arriving at Earth's surface, both direct and diffuse (or downward scattered).

Several patterns are notable on the map. Insolation decreases poleward from about 25° latitude in both the Northern and Southern Hemispheres. Consistent daylength and high Sun altitude produce average annual values of 180–200 W/m^2 throughout the equatorial and tropical latitudes. In general, greater insolation of 240–280 W/m^2 occurs in low-latitude deserts worldwide because of frequently cloudless skies. Note the energy pattern in the cloudless subtropical deserts in both hemispheres (for example, the Sonoran, Sahara, Arabian, Gobi, Atacama, Namib, Kalahari, and Australian deserts).

Scattering (Diffuse Radiation) Insolation encounters an increasing density of atmospheric gases as it travels

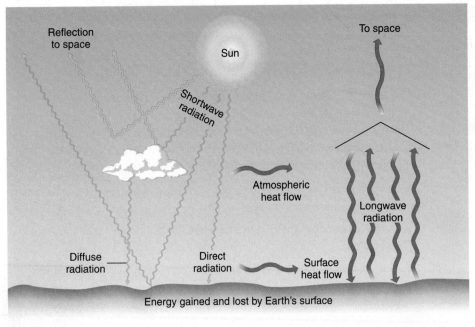

Energy gained and lost by Earth's surface

Global Warming, Climate Change

Earth-Atmosphere Energy Balance

FIGURE 3.1 Energy gained and lost by Earth's surface and atmosphere.
Simplified view of the Earth–atmosphere energy system—circuits include incoming shortwave insolation, reflected shortwave radiation, and outgoing longwave radiation.

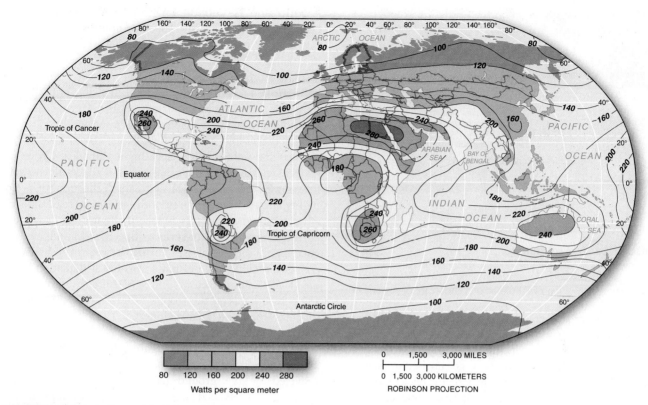

FIGURE 3.2 Insolation at Earth's surface.
Average annual solar radiation receipt on a horizontal surface at ground level in watts per square meter (100 W/m² = 75 kcal/cm²/yr). [After M. I. Budyko, *The Heat Balance of the Earth's Surface*, Washington, DC: US Department of Commerce, 1958, p. 99.]

Global Shortwave Radiation

toward the surface. Gas molecules redirect radiation, changing the direction of the light's movement *without altering its wavelengths*. This phenomenon is **scattering** and represents 7% of Earth's reflectivity (see Figure 3.10). Dust particles, pollutants, ice, cloud droplets, and water vapor produce further scattering. Air pollution—smog and haze—produces scattering and an almost white sky because the larger particles associated with air pollution act to scatter all wavelengths of the visible light.

How effectively the Sun's rays reach the surface depends on their angle and the thickness of the atmosphere they must pass through. Direct rays experience less scattering and absorption than do low, oblique-angle rays that must travel farther through the atmosphere. Figure 3.10 shows that some incoming insolation is diffused by clouds and atmosphere and is transmitted to Earth as **diffuse radiation**, the downward component of scattered light. For instance, on a cloudy day this light is multidirectional and thus casts shadowless light on the ground.

Refraction. As insolation enters the atmosphere, it passes from one medium to another, from virtually empty space into atmospheric gases, or from air into water. This transition subjects the insolation to a change of speed, which also shifts its direction, a bending action called **refraction**. In the same way, a crystal or prism refracts light passing through it, bending different wavelengths to different angles, separating the light into its component

colors to display the spectrum. A rainbow is created when visible light passes through myriad raindrops and is refracted and reflected toward the observer at a precise angle (see Figure 3.3). Another example of refraction is a *mirage*, an image that appears near the horizon where

FIGURE 3.3 A rainbow.
Raindrops, and in this photo moisture droplets from the Niagara River, refract and reflect light to produce a primary rainbow. Note that the color order in the primary rainbow has the shortest wavelengths on the inside of the bow and the longest wavelengths on the outside of the bow. In the secondary bow note the color sequence is reversed because of an extra angle of reflection within each moisture droplet. [Photo by Bobbé Christopherson.]

FIGURE 3.4 Sun refraction.
The distorted appearance of the Sun, nearing sunset over the ocean, is produced by refraction of the Sun's image in the atmosphere. Have you ever noticed this effect? [Photo by author.]

light waves are refracted by layers of air of differing temperatures (and resulting differing densities) on a hot day.

The Sun's image is refracted in its passage from space through the atmosphere, and so, at sunrise, we see the Sun about 4 minutes before it actually peeks over the horizon. Similarly, the Sun actually sets at sunset, but its image is refracted over the horizon for about 4 minutes afterward. The distortion of the setting Sun in Figure 3.4 is a product of refraction.

Albedo and Reflection A portion of arriving energy bounces directly back into space without being converted into heat or performing any work. The reflective quality of a surface is its **albedo**. **Albedo** is the percentage of insolation that is reflected. This returned energy is **reflection**, a term that applies to both visible and ultraviolet light. Albedo is an important control over the amount of insolation that is available to heat a surface.

In the visible wavelengths, darker colors have lower albedos and lighter colors have higher albedos. On water surfaces, the angle of the solar rays also affects albedo values; lower angles produce a greater reflection than do higher angles. In addition, smooth surfaces increase albedo, whereas rougher surfaces reduce it. Figure 3.5 illustrates albedo values for various surfaces. Specific locations experience highly variable albedo during the year in response to changes in cloud and ground cover. Earth Radiation Budget (ERB) orbiting sensors measure average albedos from 19%–38% between the tropics (23.5° N to 23.5° S) to as high as 80% in the polar regions.

Earth and its atmosphere reflect 31% of all insolation when averaged over a year. Figure 3.10 shows that Earth's average albedo is a combination of light reflected by clouds, reflected by the ground (combined land and oceanic surfaces), and reflected and scattered by the atmosphere. By comparison, a full Moon, which is bright enough to read by under clear, night skies, has only a 6%–8% albedo value. Thus, with *earthshine* being four times brighter than

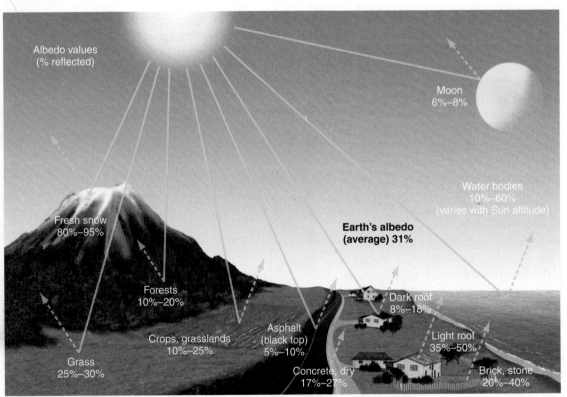

FIGURE 3.5 Various albedo values.
Selected albedos for different surfaces. Generally, light surfaces are more reflective and thus have higher albedo values. Darker surfaces are less reflective. [Data from M. I. Budyko, *The Heat Balance of the Earth's Surface*, Washington, DC: US Department of Commerce, 1958, p. 36.]

Global Albedo Values, Global Shortwave Radiation

moonlight (four times the albedo), and with Earth being four times larger than the Moon, it is no surprise that astronauts report how startling our planet looks from space.

Clouds and the Atmosphere's Albedo

An unpredictable factor in the tropospheric energy budget, and therefore in refining climatic models, is the role of clouds. Clouds reflect insolation and thus cool Earth's surface. **Cloud-albedo forcing** occurs when clouds cause an increase in albedo. Yet clouds act as insulation, trapping longwave radiation from Earth and raising minimum temperatures. **Cloud-greenhouse forcing** occurs when clouds act to increase greenhouse warming. Figure 3.6 illustrates the general effects of clouds on shortwave radiation and longwave radiation.

Other mechanisms affect atmospheric albedo and therefore atmospheric and surface energy budgets. Industrialization is producing a haze of pollution that is increasing the reflectivity of the atmosphere, including sulfate aerosols, soot and fly ash, and black carbon. Emissions of sulfur dioxide and subsequent chemical reactions in the atmosphere form *sulfate aerosols*. These aerosols act as an insolation-reflecting haze in clear-sky conditions.

Scientists are actively analyzing such pollution-dimmed sunlight reductions. The pollution causes both an atmospheric warming through absorption by the pollutants and a surface cooling through reduction in insolation reaching ground and water surfaces. An overall term has emerged that describes this decline in insolation making it to Earth's surface—**global dimming**. In determining present temperature increases in global warming, this solar dimming is perhaps forcing an underestimation of the actual amount of warming.

Absorption

Absorption is the *assimilation* of radiation and its *conversion* from one form to another. Insolation (both direct and diffuse) that is not part of the 31% reflected from Earth's surfaces is absorbed and converted into longwave radiation or is used by plants in photosynthesis. The temperature of the absorbing surface is raised in the process, causing that surface to radiate more total energy at shorter wavelengths. In addition to absorption by land and water surfaces, absorption also occurs in atmospheric gases, dust, clouds, and stratospheric ozone.

Conduction, Convection, and Advection

There are several means of transferring heat energy in a system. **Conduction** is the molecule-to-molecule transfer of heat energy as it diffuses through a substance. As molecules warm, their vibration increases, causing collisions that produce motion in neighboring molecules, thus transferring heat from warmer to cooler materials. Different materials (gases, liquids, and solids) conduct sensible heat directionally from areas of higher temperature to those of lower temperature. This heat flow transfers energy through matter at varying rates, depending on the conductivity of the material. Earth's land surface is a better conductor than air; moist air is a slightly better conductor than dry air.

Gases and liquids transfer energy through **convection** movements, in which their physical mixing involves a strong vertical motion. In the atmosphere or bodies of water, warmer (less dense) masses tend to rise and cooler (denser) masses tend to sink, establishing patterns of convection. When a lateral (horizontal) motion is dominant, the term **advection** applies. Sensible heat is transported physically through the medium in these ways. You commonly experience such energy flows in the kitchen: Energy is conducted through the handle of a pan or boiling water bubbles in the saucepan in convective motions (Figure 3.7); a convection oven uses a fan to circulate heated air to uniformly cook food.

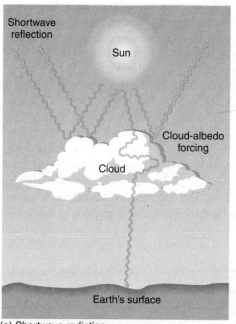

(a) Shortwave radiation

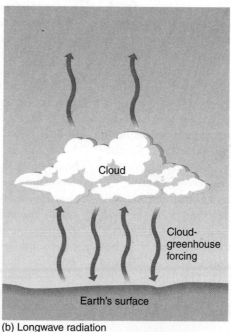

(b) Longwave radiation

FIGURE 3.6 The effects of clouds on shortwave and longwave radiation.
(a) Shortwave radiation is reflected and scattered by clouds; a high percentage is returned to space. (b) Longwave radiation emitted by Earth is absorbed and radiated by clouds; some infrared energy is radiated to space and some back toward the surface.

FIGURE 3.7 Heat energy transfer processes.
A pan of water heating on the stove illustrates heat transfer. Infrared energy *radiates* from the burner to the saucepan and the air. Energy *conducts* through the molecules of the pan and the handle. The water physically mixes, carrying heat energy by *convection*. The energy in the water and handle is measurable as *sensible heat*. The vapor leaving the surface of the water contains the *latent heat* absorbed in the change of water to water vapor.

In physical geography, we find many examples of these energy transfer processes:

- conduction (surface energy budgets, temperature differences between land and water bodies, the heating of surfaces and overlying air, soil temperatures);
- convection (atmospheric and oceanic circulation, air mass movements and weather systems, internal motions deep within Planet Earth that produce a magnetic field and movements in the crust); and
- advection (horizontal movement of winds from land to sea and back, fog that forms and moves to another area).

The Greenhouse Effect and Atmospheric Warming

Previously, in Chapter 2, we characterized Earth as a cool-body radiator, emitting energy in longer wavelengths from its surface and atmosphere back toward space. However, some of this longwave radiation is absorbed by carbon dioxide, water vapor, methane, chlorofluorocarbons (CFCs), and other gases in the lower atmosphere and is then radiated to Earth, thus delaying heat loss to space. This counterradiation process is an important factor in warming the lower atmosphere. The approximate similarity between this process and the way a greenhouse operates gives the process its name—the **greenhouse effect**.

In a greenhouse, the glass is transparent to shortwave insolation, allowing light to pass through to the soil, plants, and materials inside. The absorbed energy is then radiated as longwave radiation back toward the glass, but the glass effectively traps both the longer wavelengths and the warmed air inside the greenhouse. Thus, the glass acts as a one-way filter, allowing the light in but not allowing the heat out. The same process also can be observed in a car parked in direct sunlight.

In the atmosphere, the greenhouse analogy does not completely apply because the longwave radiation is not trapped as in a greenhouse. Rather, its passage to space is *delayed* as the heat radiates between Earth's surface and certain gases and particulates in the atmosphere. Today's increasing carbon dioxide concentration is causing more longwave radiation absorption in the lower atmosphere, thus forcing a warming trend and disruption of the Earth–atmosphere energy system, according to the scientific consensus.

Clouds and Earth's "Greenhouse" Clouds affect the heating of the lower atmosphere in several ways, depending on cloud type. Not only is the percentage of cloud cover important, but the cloud type, height, and thickness (water content and density) also have an effect. High-altitude, ice-crystal clouds reflect insolation with albedos of about 50%, whereas thick, lower cloud cover reflects about 90% of incoming insolation.

To understand the actual effects on the atmosphere's energy budget, however, we must consider both transmission of shortwave and longwave radiation and cloud type. Figure 3.8a portrays the *cloud-greenhouse forcing* caused by high clouds (warming, because their greenhouse effects exceed their albedo effects), and Figure 3.8b portrays the *cloud-albedo forcing* produced by lower, thicker clouds (cooling, because albedo effects exceed greenhouse effects). Understanding the nature of global cloud cover is crucial in refining computer models that forecast global climate change.

Jet contrails (*condensation trails*) produce high cirrus clouds, sometimes called *false cirrus clouds* (see PART I panorama photo), in Figure 3.8c. The chapter-opening photos and satellite image illustrate how cloud development is stimulated by aircraft exhaust. Scientists suspected that these clouds affected atmospheric and surface temperatures. As shown in Figure 3.8a, these high-altitude ice-crystal clouds are efficient at reducing outgoing longwave radiation and therefore contribute to global heating. These contrail-seeded cirrus have a higher density and higher albedo than normal cirrus clouds, thus reducing some surface insolation, leading to cooling.

The tragedy that struck the World Trade Center, and humanity, on 9/11/2001 inadvertently provided researchers with a chance to study these effects. Following 9/11, there was a 3-day grounding of all commercial airline traffic and therefore no contrails across the United States. Weather data from 4000 stations over 30 years were compared with data during the 3-day shutdown. *Diurnal temperature range* (DTR)—the difference between daytime maximum and nighttime minimum temperatures—increased during the 3 days. Specifically, maximum temperatures were more responsive to the missing clouds than minimums. Later study of synoptic conditions (air pressure, humidity, air masses data—all gathered at a specific time), analysis of Canadian air space, a check on conditions during adjacent 3-day periods, and a look at the few military flights that took place during the halt of air travel served to verify the initial findings and strengthen

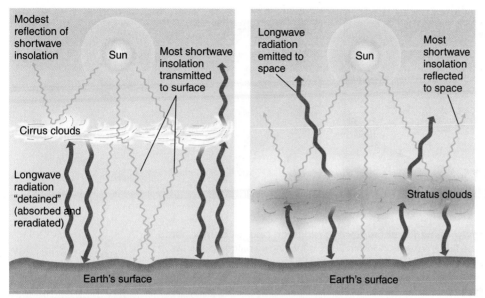

(a) High clouds: net greenhouse forcing and atmospheric warming

(b) Low clouds: net albedo forcing and atmospheric cooling

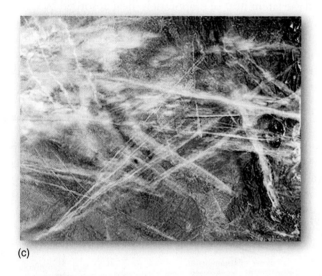

(c)

FIGURE 3.8 Energy effects of two cloud types.
(a) High, ice-crystal clouds (called *cirrus*) transmit most of the insolation. However, they absorb and delay losses of outgoing longwave infrared, producing a greater greenhouse forcing and a net warming of Earth. (b) Low, thick clouds (*stratus*) reflect most of the incoming insolation and radiate longwave to space, producing a greater albedo forcing and a net cooling of Earth. (c) Jet-airliner exhaust triggers cirrus cloud development. Newer, thinner contrails spread to form cloud cover. [(c) Astronaut photo May 15, 2002, courtesy of NASA-JSC.]

the conclusion.* The elimination of contrail-stimulated clouds produced an increase in DTR—slightly higher afternoon and slightly lower nighttime temperature readings across the United States. The largest effects were in those regions where the greatest number of flights occur in the fall. This ongoing study is helpful in the spatial analysis of possible impacts of future aircraft design on the Earth–atmosphere energy budget.

To better understand the role of natural clouds as well as jet contrails, NASA operates the CERES program, an acronym for Clouds and Earth's Radiant Energy System. CERES sensors aboard satellites began operating in 1997 (*TRMM*) and on the newest in 2000 (*Terra*). Data collected

by CERES sensors cover both reflected and emitted radiation. Figure 3.9 shows the first global monthly images from satellite *Terra* for March 2000 in W/m² flux (meaning energy "flow"). In Figure 3.9a, lighter regions indicate where more sunlight is reflected into space than is absorbed—for example, by lighter-colored land surfaces such as deserts or by cloud cover such as over tropical lands. Green and blue areas illustrate where less light was reflected.

In Figure 3.9b orange and red indicate regions where more heat was absorbed and emitted to space, whereas less heat energy is escaping in the blue and purple areas. Those blue regions of lower longwave emissions over tropical lands are due to tall, thick clouds along the equatorial convergence (Amazon, equatorial Africa, and Indonesia); these clouds also caused higher shortwave reflection. Subtropical desert regions exhibit greater longwave radiation emissions owing to the presence of little cloud cover and greater radiative energy losses from surfaces that have absorbed a lot of energy, as we saw in Figure 3.2.

*See David J. Travis et al., "Contrails reduce daily temperature range," *Nature* 418 (August 8, 2002): 601, from the Department of Geography and Geology, University of Wisconsin–Whitewater; and, D. J. Travis, A. M. Carleton, and R. G. Lauritsen, "Regional variations in US diurnal temperature range for the 9/11–14/2001 aircraft groundings: Evidence of jet contrail influence on climate," *Journal of Climate* 17 (March 1, 2004): 1123–34.

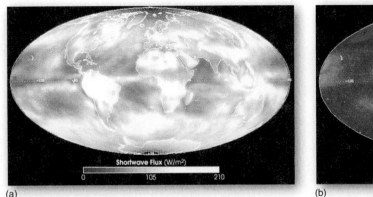
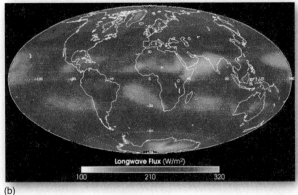

FIGURE 3.9 Shortwave and longwave images show Earth's radiation budget components.
The CERES sensors aboard *Terra* made these portraits in March 2000, capturing (a) outgoing shortwave
energy flux reflected from clouds, land, and water—Earth's albedo—and (b) longwave energy flux emitted
by surfaces back to space. The scale beneath each image displays value gradations in watts per square
meter. [Images courtesy of CERES Instrument Team, Langley Research Center, NASA.]

 Global
Shortwave

Earth–Atmosphere Energy Balance

The Earth–atmosphere energy system naturally balances
in a steady-state equilibrium. The natural energy balance
occurs through energy transfers that are radiative and
nonradiative. Radiative transfer is by longwave radiation
between the surface and the atmosphere. Nonradiative

(physical motion) transfers include convection, conduction,
and the latent heat of evaporation (heat absorbed by water
vapor when water evaporates).

Figure 3.10 summarizes the Earth–atmosphere radiation balance. It brings together all the elements discussed to
this point in the chapter by following 100% of energy
through the system. To summarize: Of 100% of the solar

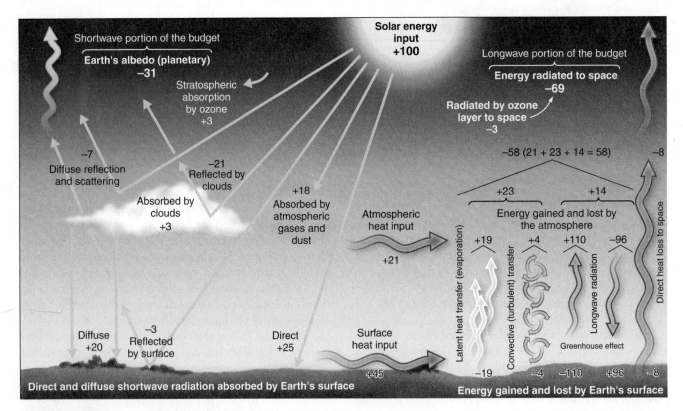

FIGURE 3.10 Earth–atmosphere energy budget.
Solar energy cascades through the lower atmosphere (left-hand portion of the illustration), where
it is absorbed, reflected, and scattered. Clouds, atmosphere, and the surface reflect 31% of this
insolation back to space. Atmospheric gases and dust and Earth's surface absorb energy and
radiate longwave radiation (right-hand portion of the illustration). Over time, Earth emits, on average, 69% of incoming energy to space. When added to Earth's average albedo (31%, reflected
energy), this equals the total energy input from the Sun.

 Global Warming,
Climate Change

Earth-Atmosphere
Energy Balance

energy arriving, Earth's average albedo reflects 31% into space; Earth eventually emits the longwave radiation part of the budget into space: 21% (atmosphere heating) + 45% (surface heating) + 3% (ozone emission) = 69% (net outgoing longwave).

Figure 3.11 summarizes this radiation balance for all shortwave and longwave energy by latitude. In the equatorial zone, energy surpluses dominate, for in those areas more energy is received than is lost. The solar angle is high, with consistent daylength. However, deficits exist in the polar regions, where more energy is lost than gained. At the poles, the Sun is extremely low in the sky, snow and ice surfaces are light and reflective, and for 6 months during the year no insolation is received.

The imbalance of net radiation from *tropical surpluses* to the *polar deficits* drives a vast global circulation of both energy and mass. The meridional (north–south) transfer agents are winds, ocean currents, dynamic weather systems, and related phenomena. As you go about your daily activities, use these dynamic natural systems as reminders of the constant flow of solar energy.

Dramatic examples of such energy and mass transfers are tropical cyclones—hurricanes and typhoons. Forming in the tropics, these powerful storms mature and migrate to higher latitudes, carrying with them energy, water, and water vapor. Scientists established in 2005 a connection between increasing sea-surface and air temperatures and the increasing intensity and power dissipation of tropical storms and the hurricanes into which they develop.

Energy at Earth's Surface

Solar energy is the principal heat source at Earth's surface. The direct and diffuse radiation and longwave radiation arriving daily at the ground surface are of great interest to geographers. Examine the inputs, outputs, and related terms along the ground surface in Figure 3.10.

Daily Radiation Patterns

The fluctuating daily pattern of incoming shortwave energy and outgoing longwave energy at the surface, and resultant air temperature, are shown in Figure 3.12. This graph represents conditions for bare soil on a cloudless day in the middle latitudes. Incoming energy arrives during daylight, beginning at sunrise, peaking at noon, and ending at sunset.

The shape and height of this insolation curve varies with season and latitude. The highest trend for such a curve occurs at the time of the summer solstice (around June 21 in the Northern Hemisphere and December 21 in the Southern Hemisphere). The temperature plot also responds to seasons and variations in input. Within a 24-hour day, air temperature generally peaks between 3:00 and 4:00 P.M. and dips to its lowest point right at or slightly after sunrise.

The relationship between the insolation curve and the air temperature curve on the graph is interesting—they do not align; there is a *lag*. The warmest time of day occurs not at the moment of maximum insolation, but at the

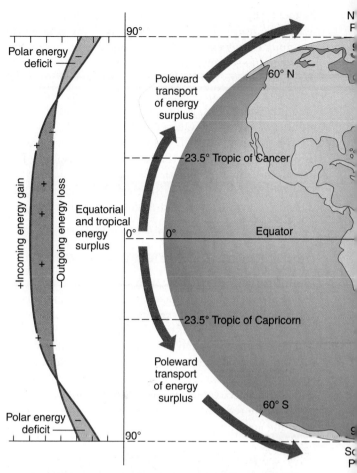

FIGURE 3.11 Energy budget by latitude.
Earth's energy surpluses and deficits by latitude produce poleward transport of energy and mass in each hemisphere—atmospheric circulation and ocean currents. Outside of the tropics, atmospheric winds dominate the transport of energy toward each pole.

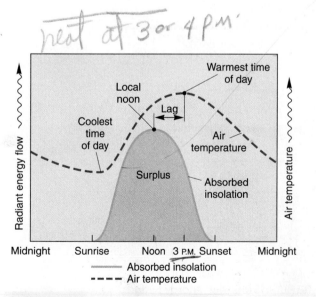

FIGURE 3.12 Daily radiation curves.
Sample radiation curves for a typical day show the changes in insolation (orange) and air temperature (dashed). Comparing the curves demonstrates a lag between local noon (the insolation peak for the day) and the warmest time of day.

moment when a *maximum of insolation is absorbed* and emitted to the atmosphere from the ground. As long as the incoming energy exceeds the outgoing energy, air temperature continues to increase, not peaking until the incoming energy begins to diminish in the afternoon as the Sun's altitude decreases. In contrast, if you have ever gone camping in the mountains, you no doubt experienced the coldest time of day with a wake-up chill at sunrise!

The annual pattern of insolation and air temperature exhibits a similar lag. For the Northern Hemisphere, January is usually the coldest month, occurring after the December solstice and the shortest days. Similarly, the warmest months of July and August occur after the June solstice and the longest days.

Simplified Surface Energy Balance

Earth's surface is supplied with energy that daily and seasonally varies. The climate at or near Earth's surface is generally termed the *boundary layer climate*. **Microclimatology** is the study of this portion of the atmosphere. The following discussion is more meaningful if you keep in mind an actual surface—perhaps a park, a front yard, or a place on campus.

The surface receives visible light and longwave radiation, and it reflects visible light and radiates longwave radiation according to this basic scheme:

$$+ \, SW\downarrow \; - \, SW\uparrow \; + \, LW\downarrow \; - \, LW\uparrow \; = \; NET \, R$$
(Insolation) (Reflection) (Infrared) (Infrared) (Net Radiation)

We use SW = shortwave and LW = longwave (you may come across other symbols in the microclimatology literature, such as K for shortwave, L for longwave, and Q* for

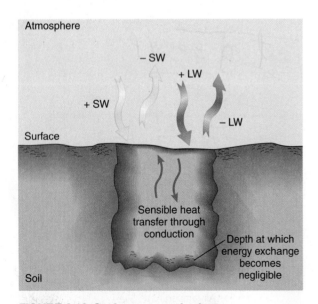

FIGURE 3.13 Surface energy budget.
Idealized input and output energy budget components for a surface and a soil column. Sensible heat transfer in the soil is through conduction, predominantly downward during the day or in summer and toward the surface at night or in winter (SW = shortwave, LW = longwave)

NET R). Along the surface, as illustrated in Figure 3.13, are the components of a surface energy balance. The column of soil continues to a depth at which energy exchange with surrounding materials, or with the surface, becomes negligible, usually less than a meter.

Energy moving toward the surface is regarded as positive (a gain), and energy moving away from the surface is considered negative (a loss). The amount of insolation arriving at the surface ($+SW\downarrow$) varies with season, cloudiness, and latitude. The albedo dictates the amount of insolation reflected at the surface ($-SW\uparrow$). Darker and rougher surfaces reflect less energy, absorb more energy, and therefore produce more net radiation. Lighter and smoother surfaces reflect more shortwave energy, leading to less net radiation.

Surface albedo values ($-SW\uparrow$) dictate the amount of insolation reflected, and therefore not absorbed, at the surface. As an example, imagine a snow-covered landscape as a surface energy system. Snow has high reflectivity, sending sunlight back to space and reducing the amount of insolation absorbed. Consequently, surfaces and air temperatures are lower, and thus the snow cover does not melt. This is a system with *positive feedback*. Remember from Chapter 1 that positive feedback increases response in the system: more cooling—more snow—increasing albedo—more cooling—which leads to colder, drier air—then less snow—and so on.

Adding and subtracting the heat flow at the surface completes the calculation of net radiation (NET R), or the balance of all radiation at Earth's surface—shortwave and longwave.

   Global Net Radiation

Net Radiation   **Net radiation (NET R)**, the net radiation (of all wavelengths) available at Earth's surface, is the final outcome of the entire radiation-balance process discussed in this chapter. NET R is expended from a nonvegetated surface through three pathways:

- *Turbulent sensible heat transfer*, the quantity of sensible heat energy physically transferred from air to surface, and surface to air, through convection and conduction.
- *Latent heat of evaporation*, the heat energy that is stored in water vapor as water evaporates. Large quantities of latent heat are absorbed into water vapor during its change of state from liquid to gas.
- *Ground heating and cooling*, energy that flows into and out of the ground surface by conduction.

Understanding the NET R balance is essential to solar energy technologies that concentrate shortwave and longwave energy for use. Solar energy offers great potential worldwide and will see expanded use in the near future. Focus Study 3.1 briefly reviews solar energy strategies.

Variation in the expenditure of NET R among *sensible heat* (heat we can feel), *latent heat*, and ground heating and

(text continued on page 87)

FOCUS STUDY 3.1

Solar Energy Collection and Concentration

Consider the following:

- The insolation receipt in just 35 minutes at the surface of the United States exceeds the amount of energy derived from the burning of fossil fuels (coal, oil, natural gas) in a year.
- An average building in the United States receives 6 to 10 times more energy from the Sun hitting its exterior than is required to heat the inside.
- Old photographs of homes in southern California and Florida show solar (flat-plate) water heaters on many rooftops. Early twentieth-century newspaper ads and merchandise catalogues featured the Climax solar water heater (1905) and the Day and Night water heater (1909). Low-priced natural gas and oil displaced many of these applied solar energy technologies.

Today, installed solar water heating serves more than 50 million households worldwide.

- Installed photovoltaic solar-electric capacity is doubling every 24 months and is the fastest-growing energy source in the world, rising to over 9000 MW installed through 2006.
- Sunlight is direct and widely available, and solar power installations are decentralized, labor-intensive, and renewable. Sunlight has been collected for centuries through various technologies. Yet it is underutilized.

Rural villages in developing countries could benefit greatly from the simplest, most cost-effective solar application—the *solar-panel cooker*. For example, women in Kenya walk many kilometers collecting fuelwood for cooking fires (Figure 3.1.1a). Each village and refugee camp is surrounded

by impoverished land, stripped bare of wood. Using simple cardboard solar cookers, villagers are able to cook meals and sanitize their drinking water right at home without scavenging for wood (Figure 3.1.1b). These devices are simple yet efficient, reaching temperatures between 105° and 135°C (220° and 275°F) (Figure 3.1.1c). (See Solar Cookers International at http://solarcooking.org/.)

In such countries, the pressing need is for decentralized energy sources, appropriate in scale to everyday needs, such as cooking, heating water, and pasteurization. *Net per capita (per person) cost for solar cookers is far less than for centralized electrical production*, regardless of fuel source.

Collecting and Concentrating Solar Energy

Any surface that receives light from the Sun is a *solar collector*—remember our discussion of NET R in this chapter.

(a)

(b)

(c)

FIGURE 3.1.1 The solar-cooking solution.
(a) Five women haul firewood many miles to the Dadaab refugee camp in northeastern Kenya. (b) Kenyan women in training to use their solar panel cookers, which do not require scavenging the countryside for scarce fuelwood. These simple cookers collect and focus direct and diffuse insolation using a shiny surface and an enclosed box or cooking bag that traps infrared radiation (c). This is a small-scale, efficient application of the greenhouse effect. Construction is easy, using cardboard components. Temperatures easily exceed 105°C and can reach 135°C for baking, boiling, purifying water, and sterilizing instruments. [Photos (a) and (b) by Solar Cookers International, Sacramento, California; (c) Bobbé Christopherson.]

(continued)

Focus Study 3.1 *(continued)*

But the diffuse nature of solar energy received at the surface requires that it be collected, concentrated, transformed, and stored to be useful. Space heating is the simplest application. Windows that are carefully designed and placed allow sunlight to shine into a building, where it is absorbed and converted into sensible heat. Here we have an everyday application of the greenhouse effect.

A *passive solar system* captures heat energy and stores it in a "thermal mass," such as water-filled tanks, adobe, tile, or concrete. An *active solar system* involves heating water or air in a collector and then pumping it through a plumbing system to a tank, where it can provide hot water for direct use or for space heating.

Solar energy systems can generate heat energy of an appropriate scale for approximately half the present domestic applications in the United States (space heating and water heating). In marginal climates, solar-assisted water and space heating is feasible as a backup; even in New England, Northern Plains states, and southern Canada, solar collection systems prove effective. (See the National Solar Radiation Data Base at http://www.nrel.gov/rredc/.)

Focusing (concentrating) mirrors, such as Fresnel lenses, or parabolic (curved surface) troughs and dishes can be used to attain very high temperatures to heat water or other heat-storing fluids. Kramer Junction, California, about 225 km (140 mi) northeast of Los Angeles, has the world's largest operating solar electric-generating facility, with a capacity of 150 MW (megawatts; 150 million watts). Long troughs of computer-guided curved mirrors concentrate sunlight to create temperatures of 390°C (735°F) in vacuum-sealed tubes filled with synthetic oil. The heated oil is then used to heat water; the heated water produces steam that rotates turbines to generate cost-effective electricity. The facility converts 23% of the sunlight it receives into electricity during peak hours (Figure 3.1.2a).

The National Renewable Energy Laboratory (NREL, http://www.nrel.gov) and the National Center for Photovoltaics (at http://www.nrel.gov/ncpv/) were established in 1974 to coordinate solar energy research, development, and testing in partnership with private industry. NREL's headquarters building in Golden, Colorado, features the latest in passive and active solar design, although Germany (installations) and Japan (production) continue to be the global market leaders.

Electricity Directly from Sunlight

Producing electricity by *photovoltaic cells* (PVs) is a technology that has been used in spacecraft since 1958. Familiar to us all are the solar cells in pocket calculators (over 100 million units now in use). When sunlight shines upon a semiconductor material in these cells, it stimulates a flow of electrons (an electrical current) in the cell. PV cells are arranged in modules that can be assembled in large arrays. At NREL's Outdoor Test Facility, successful tests continue on arrays of prototype solar cells. Some of these panels have been in operation for more than 25 years (Figure 3.1.2b).

(a)

(b)

(c)

FIGURE 3.1.2 Solar thermal and photovoltaic energy production.
(a) Kramer Junction 150-MW solar-thermal energy installation in southern California. (b) NREL Outdoor Test Facility in Golden, Colorado, where a variety of photovoltaic cell arrays successfully convert sunlight directly into electricity. (c) A residence with a 7380-W rooftop solar photovoltaic array generates electricity. Excess electricity is fed to the grid for credits that more than offset 100% of the home's electric bill. [Photos by Bobbé Christopherson.]

The efficiency of these cells has improved to the level that they are generally cost-competitive, especially if government policies and subsidies were to be balanced evenly among all energy sources. The residential installation in Figure 3.1.2c features 36 panels, generating 205 W each, 7380 W total, at a 21.5% conversion efficiency. People are able to run their electric meters in reverse and supply electricity to the power grid. In experiments, NREL developed high-efficiency concentrator solar cells that broke the 40% conversion efficiency barrier. (See the Department of Energy's "Photovoltaic Home Page" at http://www.eere.energy.gov/solar/.)

The Promise of Solar Energy

Solar energy is a wise choice for the future. It is directly available to the consumer; it is based on a renewable energy source of an appropriate scale for end-use needs, such as water and space heating and cooling (it matches them well); and most solar strategies are labor-intensive (in contrast to capital-intensive, centralized, nonrenewable power production). Solar is preferable to further development of our decreasing fossil-fuel reserves, further increases in oil imports and tanker spills, investment in foreign military operations, or adding more nuclear power with its many problems.

Whether or not the alternative path of solar energy is followed comes down to political control. The technology is ready for installation and is cost-effective when all direct and indirect costs are considered for other energy sources. NREL asserts in its mission statement that the laboratory wants to "lead the nation toward a sustainable energy future by developing renewable energy technologies, improving energy efficiency, advancing related science, and engineering commercialization."

cooling produces the variety of environments we experience in nature. Figure 3.14a illustrates a desert environment, in which few clouds, sparse plants, and little water result in high sensible heat production and hot ground. Figure 3.14b shows the area near Pitt Meadows, British Columbia, featuring a moist, vegetated environment of irrigated mixed orchard and rye grass. Here, high amounts of energy are expended for latent heat of evaporation, with more moderate amounts spent as sensible heat.

Now that we have examined the flow of solar energy across space, through the atmosphere, and to Earth's surface, we turn our attention to important outputs of this system: the pattern of global temperatures and, in the next chapter, the flow of atmospheric and oceanic circulation.

Temperature Concepts and Controls

Air temperature has a remarkable influence on our lives. What temperature is it now (both indoors and outdoors) as you read these words? How is temperature measured and what does the value mean? How is today's air temperature controlling your plans for the day? Our bodies sense temperature and subjectively judge comfort, reacting to changing temperatures with predictable responses. A variety of temperature regimes worldwide affect cultures, decision making, and resources consumed. We read of heat waves and cold spells affecting people, crops, events, and energy consumption.

Heat and temperature are not the same. *Heat* is a form of energy that flows from one system or object to another

(a)

(b)

FIGURE 3.14 Net radiation comparison and the landscape.
(a) In the California desert east of Los Angeles (at about 35° N), the daily expenditures of net radiation produce a desert landscape of high temperature and sparse vegetation. (b) Irrigated blueberry orchards are characteristic of the agricultural activity near Pitt Meadows, British Columbia (at about 49° N), which has a moist environment and moderate temperatures. [Photos by author.]

Global Latent Heat Flux

Global Sensible Heat

because the two are at different temperatures. **Temperature** is a measure of the average kinetic energy (motion) of individual molecules in matter. We feel the effect of temperature as the *sensible heat* transfer from warmer objects to cooler objects. For instance, when you jump into a cool lake you can sense the heat transfer from your skin to the water as kinetic energy leaves your body and flows to the water; a chill develops.

Temperature and heat are related because changes in temperature are caused by the absorption or emission (gain or loss) of heat energy. The term *heat energy* is frequently used to describe energy that is added to or removed from a system or substance.

Temperature Scales

The temperature at which all atomic and molecular motion in matter completely stops is 0° *absolute temperature*, commonly, "*absolute zero.*" Its value on the different temperature-measuring scales is −273° Celsius (C), −459.4° Fahrenheit (F), and 0 Kelvin (K). Figure 3.15 compares these three scales. Formulas for converting between Celsius, SI (Système International), and English units are in Appendix C of this text.

The Fahrenheit scale places the melting point of ice at 32°F, with 180 subdivisions to the boiling point of water at 212°F. The scale is named for its developer, Daniel G. Fahrenheit, a German physicist (1686–1736). Note that there is only one melting point for ice, but there are many freezing points for water, ranging from 32°F down to −40°F, depending on its purity, volume, and certain conditions in the atmosphere.

About a year after the adoption of the Fahrenheit scale, Swedish astronomer Anders Celsius (1701–1744) developed the Celsius scale (formerly called centigrade). He placed the melting point of ice at 0°C and the boiling temperature of water at sea level at 100°C, dividing his scale into 100 degrees using a decimal system.

British physicist Lord Kelvin (born William Thomson, 1824–1907) proposed the Kelvin scale in 1848. Science uses this scale because it starts at absolute zero temperature and thus readings are proportional to the actual kinetic energy in a material. The Kelvin scale's melting point for ice is 273 K, and its boiling point of water is 373 K.

Most countries use the Celsius scale to express temperature. The United States remains the only major country still using the Fahrenheit scale. This textbook presents Celsius (with Fahrenheit equivalents in parentheses) throughout to help bridge this transitional era in the United States. The continuing pressure of the scientific community, not to mention the rest of the world, makes adoption of Celsius and SI units inevitable for the United States.

Measuring Temperature

A *mercury thermometer* or *alcohol thermometer* in a sealed glass tube measures outdoor temperatures. Fahrenheit invented both the alcohol and mercury thermometers. Cold climates demand alcohol thermometers because alcohol freezes at −112°C (−170°F), whereas mercury freezes at −39°C

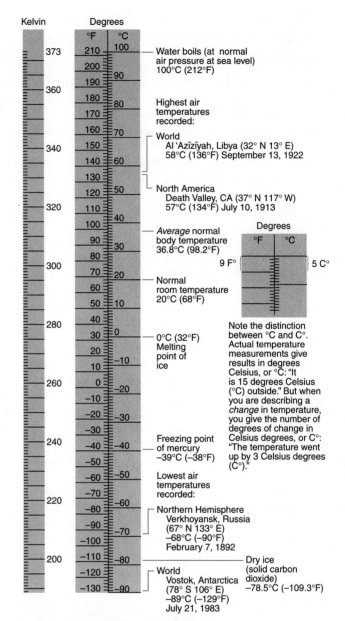

FIGURE 3.15 Temperature scales.
Scales for expressing temperature in Kelvin (K) and degrees Celsius (°C) and Fahrenheit (°F).

(−38.2°F) . The principle of these thermometers is simple: When fluids are heated, they expand; upon cooling, they contract. A thermometer stores fluid in a small reservoir at one end and is marked with calibrations to measure the expansion or contraction of the fluid, which reflects the temperature of the thermometer's environment.

Thermometers for standardized official readings are placed outdoors in small shelters that are white (for high albedo) and louvered (for ventilation) to avoid overheating of the instruments. In increasing use are *thermisters* that measure temperature by sensing the electrical resistance of a semiconducting material. A thermister is in the shelter in Figure 3.16. Weather shelters are placed at least 1.2–1.8 m (4–6 ft) above a surface, usually on turf; in the United States 1.2 m (4 ft) is standard. Official temperature measurements are made in the shade to avoid the effect of direct insolation.

Temperature readings are taken daily, sometimes hourly, at more than 16,000 weather stations worldwide. One of the goals of the Global Climate Observing System (GCOS, **http://www.wmo.ch/pages/prog/gcos/**) is to establish a reference network of one station per 250,000 km² (95,800 mi²). (See the World Meteorological Organization at **http://www.wmo.ch/**.)

Daily minimum-maximum readings are averaged to get the *daily mean temperature*. The *monthly mean temperature* is the total of daily mean temperatures for the month divided by the number of days in the month. An *annual temperature range* expresses the difference between the lowest and highest monthly mean temperatures for a given year.

Principal Temperature Controls

The interaction of complex control mechanisms produces Earth's temperature patterns. These principal influences upon temperatures include latitude, altitude/elevation, cloud cover, and land-water heating differences.

Latitude   Insolation is the single most important influence on temperature variations. Figure 2.9 showed how insolation intensity decreases as we move away from the subsolar point, a point that migrates between the Tropic of Cancer and the Tropic of Capricorn—that is, between about 23.5° N and 23.5° S—during the year. In addition, daylength and Sun angle change throughout the year, increasing in seasonal effect with increasing latitude. The five cities graphed in Figure 3.17 demonstrate the effects of latitudinal position. From equator to poles, Earth ranges from continually warm, to seasonally variable, to continually cold.

FIGURE 3.16 Instrument shelter.
This standard thermometer instrument shelter is white (for high albedo) and louvered (for ventilation), replacing the traditional cotton-region shelter (also known as a Stevenson screen). [Photo by Bobbé Christopherson.]

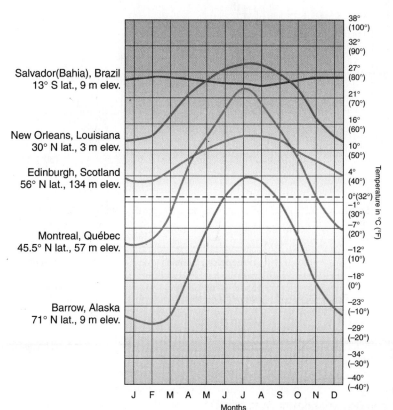

FIGURE 3.17 Latitude affects temperatures.
A comparison of five cities from near the equator to north of the Arctic Circle demonstrates changing seasonality and increasing differences between average minimum and maximum temperatures.

Altitude/Elevation Within the troposphere, temperatures decrease with increasing altitude above Earth's surface. (Recall that the *normal lapse rate* of temperature change with altitude is 6.4 C°/1000 m, or 3.5 F°/1000 ft; see Figure 2.22.) The density of the atmosphere also diminishes with increasing altitude. In fact, the density of the atmosphere at an elevation of 5500 m (18,000 ft) is about half that at sea level. As the atmosphere thins, there is a loss in its ability to absorb and radiate sensible heat. Thus, worldwide, mountainous areas experience lower temperatures than do regions nearer sea level, even at similar latitudes. Note that "altitude" refers to airborne objects above Earth's surface, whereas "elevation" usually refers to the height of a point on Earth's surface above some plane of reference.

The consequences are that, at higher elevations, average air temperatures are lower, nighttime cooling increases, and the temperature range between day and night and between areas of sunlight and shadow also is greater. Temperatures may decrease noticeably in the shadows and shortly after sunset. Surfaces both gain heat rapidly and lose heat rapidly to the thinner atmosphere.

Also, at higher elevations, the insolation received is more intense because of the reduced mass of atmospheric gases. As a result of this intensity, the ultraviolet energy component makes sunburn a distinct hazard.

The snowline seen in mountain areas indicates where winter snowfall exceeds the amount of snow lost through summer melting and evaporation. The snowline's location is a function both of elevation and latitude, which means that glaciers can exist even at equatorial latitudes if the elevation is great enough. In equatorial mountains, the snowline occurs at approximately 5000 m (16,400 ft), and permanent ice fields and glaciers exist on equatorial mountain summits in the Andes and East Africa. With increasing latitude toward the poles, snowlines gradually descend in elevation from 2700 m (8850 ft) in the midlatitudes to lower than 900 m (2950 ft) in southern Greenland.

Two cities in Bolivia illustrate the interaction of the two temperature controls, latitude and elevation. Figure 3.18 displays temperature data for the cities of Concepción and La Paz, which are near the same latitude (about 16° S). Note the elevation, average annual temperature, and

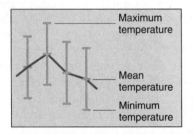

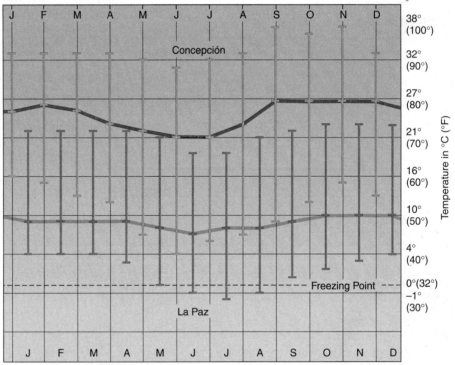

	Station	
	Concepción, Bolivia	**La Paz, Bolivia**
Latitude/longitude	16° 15′ S 62° 03′ W	16° 30′ S 68° 10′ W
Elevation	**490 m (1608 ft)**	**4103 m (13,461 ft)**
Avg. ann. temperature	24°C (75.2°F)	9°C (48.2°F)
Ann. temperature range	5 C° (9 F°)	3 C° (5.4 F°)
Ann. precipitation	121.2 cm (47.7 in.)	55.5 cm (21.9 in.)
Population	10,000	810,300 (Administrative division 1.6 million)

FIGURE 3.18 Combined effects of latitude and altitude.
Comparison of temperature patterns in La Paz (4103 m) and Concepción (490 m), Bolivia.

FIGURE 3.19 High-elevation farming.
People in these high-elevation villages grow potatoes and wheat in view of the permanent ice-covered peaks of the Andes Mountains. The combination of low latitude and high elevation creates consistent but moderate temperatures throughout the year, averaging about 9°C (48°F). [Photo by Mireille Vautier/Woodfin Camp & Associates.]

in this chapter. In general, clouds lower daily maximum temperatures by reflecting insolation as a result of their high albedo values, and they raise nighttime minimum temperatures by acting as insulation and radiating long-wave energy, preventing rapid energy loss. Clouds also reduce latitudinal and seasonal temperature differences.

Clouds are the most variable factor influencing Earth's radiation budget, making them the subject of much investigation and simulation in computer models of atmospheric behavior. The International Satellite Cloud Climatology Project, part of the World Climate Research Programme, is presently in the midst of such research. The Clouds and the Earth Radiant Energy System (CERES) sensors aboard the *TRMM* and *Terra* satellites are assessing cloud effects on longwave, shortwave, and net radiation patterns as never before possible. (For more about clouds, see the ISCCP and WCRP home pages at http://isccp.giss.nasa.gov/ and http://wcrp.wmo.int/wcrp-index.html.)

precipitation for each location. The hot, humid climate of Concepción at its much lower elevation stands in marked contrast to the cool, dry climate of highland La Paz.

The moderate temperature and moisture conditions characteristic of La Paz lead to the formation of more fertile soils than those found in the warmer, wetter climate of Concepción. People living around La Paz actually grow wheat, barley, and potatoes—crops characteristically grown in the cooler midlatitudes—despite the fact that La Paz is 4103 m (13,461 ft) above sea level (Figure 3.19). (For comparison, the summit of Pikes Peak in Colorado is at 4301 m or 14,111 ft, and Mount Rainier, Washington, is at 4392 m or 14,410 ft.)

Cloud Cover Orbiting satellites reveal that approximately 50% of Earth is cloud covered at any given moment. Clouds moderate temperature, and their effect varies with cloud type, height, and density, discussed earlier

Land-Water Heating Differences

The irregular arrangement of landmasses and water bodies on Earth contributes to the overall pattern of temperature. The physical nature of the substances themselves—rock and soil versus water—is the reason for these **land-water heating differences**. More moderate temperature patterns are associated with water bodies, compared with more extreme temperatures inland. Figure 3.20 visually summarizes the following discussion.

Evaporation More of the energy arriving at the ocean's surface is expended for *evaporation* than is expended over a comparable area of land. An estimated 86% of all

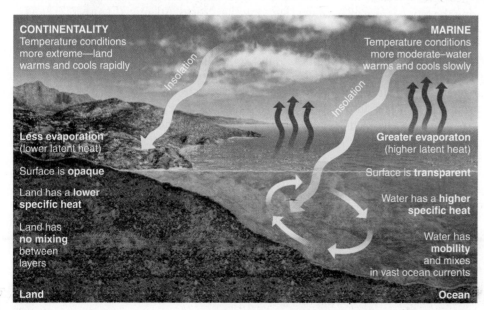

CONTINENTALITY
Temperature conditions more extreme—land warms and cools rapidly

Insolation

Less evaporation (lower latent heat)

Surface is **opaque**

Land has a **lower specific heat**

Land has **no mixing** between layers

Land

MARINE
Temperature conditions more moderate—water warms and cools slowly

Insolation

Greater evaporaton (higher latent heat)

Surface is **transparent**

Water has a **higher specific heat**

Water has **mobility** and mixes in vast ocean currents

Ocean

FIGURE 3.20 Land-water heating differences.
The differential heating of land and water produces contrasting marine (more moderate) and continental (more extreme) temperature regimes.

evaporation on Earth is from the oceans. When water evaporates and thus changes to water vapor, *heat energy is absorbed in the process and is stored in the water vapor*. This stored heat energy is *latent heat*, discussed in Chapter 5. You experience this evaporative heat loss (cooling) by wetting the back of your hand and then blowing on the moist skin. Sensible heat energy is drawn from your skin to supply some of the energy for the water's evaporation, and you feel the cooling. Similarly, as surface water evaporates, substantial energy is absorbed, resulting in a lowering of nearby air temperatures. The land, containing far less water, experiences less evaporation and therefore its temperature is moderated less by evaporative cooling.

Transparency The transmission of light obviously differs between soil and water; solid ground is opaque, whereas water is *transparent*. Consequently, light striking a soil surface does not penetrate but is absorbed, heating the ground surface. That heat is accumulated during times of exposure and is rapidly lost at night or in shadows. Maximum and minimum temperatures generally are experienced right at ground level. Below the surface, even at shallow depths, temperatures remain about the same throughout the day.

In contrast, when light reaches a body of water, it penetrates the surface because of water's **transparency**, transmitting light to an average depth of 60 m (200 ft) in the ocean. This illuminated zone is known as the *photic layer* and reaches a depth in some ocean waters of 300 m (1000 ft). This characteristic of water results in the distribution of available heat energy over a much greater depth and volume, forming a larger heat reservoir than that made up of the surface layers of the land.

Specific Heat When equal volumes of water and land are compared, water requires far more heat to raise its temperature than does land. In other words, *water can absorb more heat than can soil or rock*, and therefore water is said to have a higher **specific heat**. A given volume of water represents a more substantial heat reservoir than an equal volume of land, so that changing the temperature of the oceanic heat reservoir is a slower process than changing the temperature of land.

Movement Land is a rigid, solid material, whereas water is a fluid and is capable of movement. Differing temperatures and currents result in a mixing of cooler and warmer waters, and that *mixing spreads the available heat over an even greater volume* than if the water were still. Surface water and deeper waters mix, redistributing heat energy. Both ocean and land surfaces radiate heat at night, but land loses its heat energy more rapidly than does the moving mass of the oceanic heat reservoir.

Ocean Currents and Sea-Surface Temperatures Although ocean currents are discussed in greater detail in the next chapter, the influence of currents on temperature and weather requires mention here. In an air mass, water-vapor content is affected by ocean temperatures, because

FIGURE 3.21 The Gulf Stream.
Satellite image of the warm Gulf Stream—the streamlike flow in red, orange, and yellow colors—as it flows northward along the North American East Coast. Instruments sensitive to longwave radiation produced this remote-sensing image. Temperature differences are distinguished by computer-enhanced coloration: reds/oranges = 25°–29°C (76°–84°F), yellows/greens = 17°–24°C (63°–75°F); blues = 10°–16°C (50°–61°F); and purples = 2°–9°C (36°–48°F). [Imagery by Rosenstiel School of Marine and Atmospheric Science, University of Miami.]

warm water tends to energize overlying air through high evaporation rates and transfers of latent heat. (See http://podaac.jpl.nasa.gov/ for more on sea-surface temperatures and physical oceanography.)

As a specific example, the **Gulf Stream** moves northward off the east coast of North America, carrying warm water far into the North Atlantic (Figure 3.21). As a result, the southern third of Iceland experiences much milder temperatures than would be expected for a latitude of 65° N, just south of the Arctic Circle (66.5°). In Reykjavík, on the southwestern coast of Iceland, monthly temperatures average above freezing during all months of the year. The Gulf Stream affects Scandinavia and northwestern Europe in the same manner.

In the western Pacific Ocean, the Kuroshio, or Japan, Current, similar to the Gulf Stream, functions much the same in its warming effect on Japan and the Aleutians. In contrast, along midlatitude west coasts, cool ocean currents moderate air temperatures.

Remote sensing from satellites is providing sea-surface temperature (SST) data that correlate well with actual sea-surface measurements. This correlation permits a thorough global assessment of SSTs in programs such as Tropical Ocean Global Atmosphere (TOGA) and Coupled Ocean-Atmosphere Response Experiment (COARE). (See http://lwf.ncdc.noaa.gov/oa/coare/index.html.) Figure 3.22 displays SST data for February and July from satellites in the NOAA/NASA Pathfinder AVHRR data set, aboard *NOAA*-7, -9, -11, and -14. The Western Pacific Warm Pool is the red and maroon area in the southwestern Pacific Ocean (north of New Guinea) with temperatures above 30°C (86°F). This region has the highest average ocean temperatures in the world. Note the seasonal change in ocean temperatures; for example, compare the waters around Australia on the two images. Worldwide average SSTs increased from 1982 through 2008 to record levels for the past several hundred years of ship and coastal records. In the record hurricane season of 2005, storms fed on warm Gulf of Mexico waters that exceeded 33.3°C (92°F).

Summary of Marine Effects versus Continental Effects
Figure 3.20 summarizes the operation of all the land-water temperature controls: evaporation, transparency, specific heat, movement, and ocean currents and sea-surface temperatures. The term **marine effect**, or *maritime*, is used to describe locations that exhibit the moderating influences of the ocean, usually along coastlines or on islands. **Continental effect**, or a condition of *continentality*, refers to the condition of areas that are less affected by the sea and therefore have a greater range between maximum and minimum temperatures diurnally and seasonally.

The cities of San Francisco, California, and Wichita, Kansas, provide us with a comparison of marine and continental conditions (Figure 3.23). Both cities are at approximately 37° 40′ N latitude and are at 5 m (16.4 ft) and 403 m (1321 ft) in elevation, respectively. Summer fog plays a role in delaying until September the warmest summer month in San Francisco. In 100 years of weather records there, summer maximum temperatures have risen above 32.2°C (90°F) only a few days a year and rarely do winter minimums drop below freezing.

In contrast, Wichita is susceptible to freezes from late October to mid-April, with daily variations slightly increased by its elevation, experiencing −30°C (22°F) as a record low. January 2006 was 13 C° (23.4 F°) above normal when it hit an average of 6.2°C (43.2°F) for the month. West of Wichita, winters increase in severity with

(a) February

(b) July

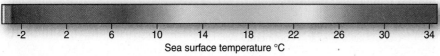

-2 2 6 10 14 18 22 26 30 34
Sea surface temperature °C

Global Sea-surface Temperatures
SATELLITE

FIGURE 3.22 Sea-surface temperatures.
Average annual sea-surface temperatures for (a) February and (b) July 1999 from satellites in the NOAA/NASA Pathfinder AVHRR data set. [Satellite image data from the NASA Physical Oceanography Distributed Active Archive Center at the Jet Propulsion Laboratory, California Institute of Technology.]

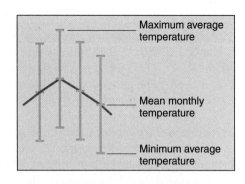

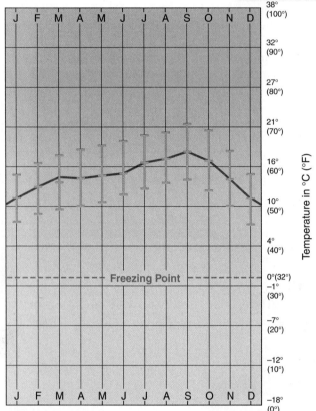

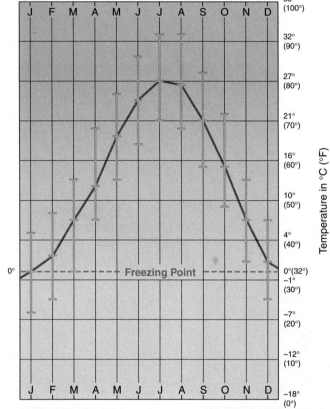

Station: San Francisco, California
Lat/long: 37° 37′ N 122° 23′ W
Avg. ann. temp.: 14.6°C (58.3°F)
Total ann. precip.:
    56.6 cm (22.3 in.)

Elevation: 5 m (16.4 ft)
Population: 777,000
Ann. temp. range:
    11.4 C° (20.5 F°)

Station: Wichita, Kansas
Lat/long: 37° 39′ N 97° 25′ W
Avg. ann. temp.: 13.7°C (56.6°F)
Total ann. precip.:
    72.2 cm (28.4 in.)

Elevation: 402.6 m (1321 ft)
Population: 327,000
Ann. temp. range:
    27 C° (48.6 F°)

FIGURE 3.23 Marine and continental cities—United States.
Comparison of temperatures in coastal San Francisco, California, and continental Wichita, Kansas. [Photos by Bobbé Christopherson.]

increasing distance from the moderating influences of invading air masses from the Gulf of Mexico. Wichita's temperature reaches 32.2°C (90°F) or higher more than 65 days each year, with 46°C (114°F) as a record high. In 1980, 1990, 2000, 2005, and 2006 temperatures exceeded 38°C (100°F) for two to three weeks in a row—23 nonconsecutive days reached over 100°C in 2003.

Earth's Temperature Patterns

Earth's temperature patterns result from the several controlling factors just discussed. Now let's look at their combined effect, portrayed by the temperature maps in this section. These maps show worldwide mean monthly air temperatures for January (Figure 3.24) and July (Figure 3.27). Along with Figure 3.29, which shows annual temperature ranges (January compared to July), these maps are useful in identifying areas that experience the greatest annual extremes. We use maps for January and July instead of the solstice months (December and June) because, as explained earlier, a lag occurs between insolation received and maximum or minimum temperatures experienced.

The lines on temperature maps are known as *isotherms*. An **isotherm** is an isoline—a line along which there is a constant value—that connects points of equal temperature and portrays the temperature pattern, just as a contour line on a topographic map illustrates points of equal elevation. Isotherms help with the spatial analysis of temperatures.

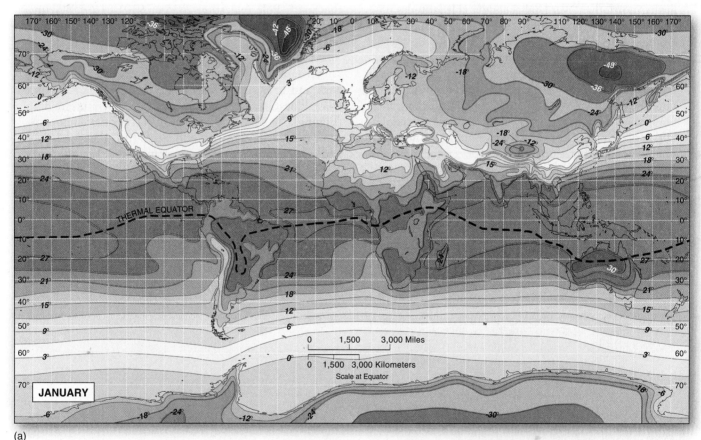

(a)

FIGURE 3.24 Global temperature for January.

Temperatures are in Celsius (convertible to Fahrenheit by means of the scale) as taken from separate air temperature databases for ocean and land. Note the inset map of North America and the equatorward-trending isotherms in the interior. Compare to Figure 3.27. [Adapted and redrawn from National Climatic Data Center, *Monthly Climatic Data for the World* 47, no. 1, January 1994. Prepared in cooperation with the WMO and NOAA.]

Global Surface Temperatures,
Land and Ocean
SATELLITE

January Temperature Map

Figure 3.24 is a map of mean January temperatures. In the Southern Hemisphere, the higher Sun altitude means longer days and summer weather conditions, whereas in the Northern Hemisphere, January marks winter's shortened daylength and lower Sun angles. Isotherms indicate the general decrease in insolation and net radiation with distance from the equator.

The **thermal equator**, a line connecting all points of highest mean temperature (the dashed-red line on the map, roughly 27°C, 80°F), trends southward into the interior of South America and Africa, indicating higher temperatures over landmasses. In the Northern Hemisphere, isotherms trend equatorward as cold air chills the continental interiors. The oceans, on the other hand, appear more moderate, with warmer conditions extending farther north than over land at comparable latitudes.

As an example, in Figure 3.24a follow along 50° N latitude (the 50th parallel) and compare isotherms: 2° to 4°C in the North Pacific and 4° to 10°C in the North Atlantic, as contrasted to −18°C in the interior of North America and −24° to −30°C in central Asia. Also, note the orientation of isotherms over areas where there are mountain ranges and how they illustrate the cooling effects of altitude. Check the South American Andes as an example of the effects of altitude.

The January temperatures for the North Polar and South Polar regions (Figure 3.25a and b) are featured on two maps. Remember: The North Polar region is an ocean surrounded by land, in contrast to the South Polar region, which is the Antarctic continent surrounded by ocean. On the northern map (a), the island of Greenland has a summit elevation of 3240 m (10,630 ft) on Earth's second-largest ice sheet, the highest elevation north of the Arctic Circle. Two-thirds of the island is north of the Arctic Circle—its north shore is only 800 km (500 mi) from the North Pole. This combination of high latitude and interior high elevation on the ice sheet produces cold midwinter temperatures.

In Antarctica (b), "summer" is underway in December. This is Earth's coldest and highest average elevation landmass. January average temperatures for the three scientific bases noted on the map are: McMurdo Station, at the coast on Ross Island, −3°C; the Admundsen–Scott Station at the South Pole (elevation 2835 m, 9301 ft), −28°C; and the Russian Vostok Station at its more continental location (elevation 3420 m, 11,220 ft), −32°C (respectively, 26.6°F, −18.4°F, and −25.6°F).

Trondheim, Norway, is near the latitude of Verkhoyansk and at a similar elevation. However, Trondheim's coastal location moderates its annual temperature regime (Figure 3.26a): January minimum and maximum temperatures range between −17° and +8°C (+1.4° and +46°F), and the minimum/maximum range for July is from

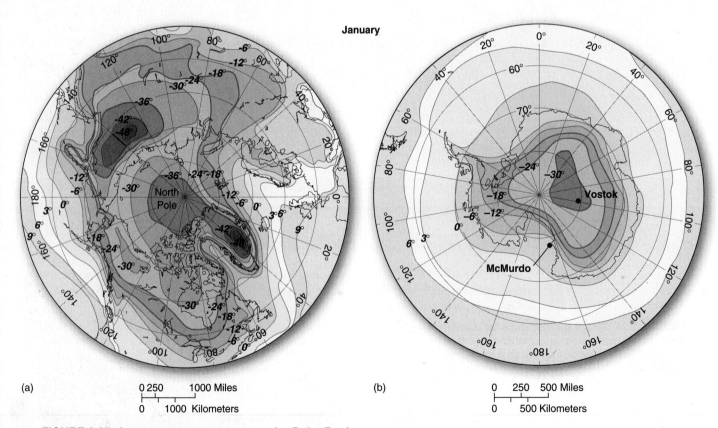

January

(a)

0 250   1000 Miles

0   1000 Kilometers

(b)

0   250   500 Miles

0   500 Kilometers

FIGURE 3.25 January mean temperatures for Polar Regions.
January temperatures in (a) North Polar region and (b) South Polar region. Use temperature conversions in Figure 3.24. Note that each map is at a different scale. [Author-prepared maps using same sources as Figures 3.24 and 3.27.]

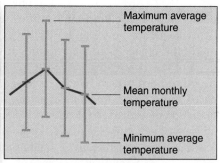

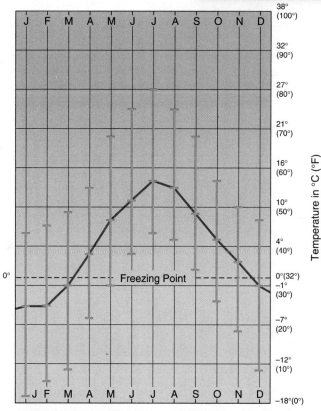

Station: Trondheim, Norway
Lat/long: 63° 25′ N 10° 27′ E
Avg. ann. temp.: 5°C (41°F)
Total ann. precip.:
    85.7 cm (33.7 in.)

Elevation:115 m (377.3 ft)
Population: 139,000
Ann. temp. range:
    17 C° (30.6 F°)

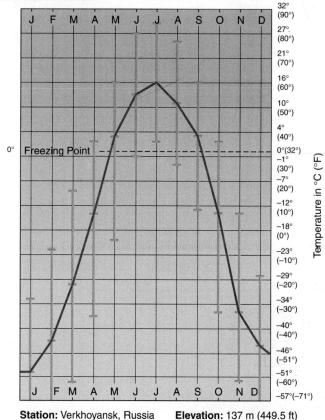

Station: Verkhoyansk, Russia
Lat/long: 67° 35′ N 135° 23′ E
Avg. ann. temp.: −15°C (5°F)
Total ann. precip.:
    15.5 cm (6.1 in.)

Elevation: 137 m (449.5 ft)
Population: 1400
Ann. temp. range:
    63 C° (113.4 F°)

(a)

(b)

FIGURE 3.26 Marine and continental cities—Eurasia.
Comparison of temperatures in coastal Trondheim, Norway (a) and continental Siberian Russia (b).
Note that the freezing levels on the two graphs are positioned differently to accommodate the
contrasting data. [Photos by (a) Norman Benton/Peter Arnold, Inc.; (b) TASS/Sovfoto/Eastfoto.]

+5° to +27°C (+41° to +81°F). The most extreme minimum and maximum temperatures ever recorded in Trondheim are −30° and +35°C (−22° and +95°F)—quite a difference from the extremes at Verkhoyansk.

For the coldest area on these maps outside of Antarctica we go to northeastern Siberia in Russia. The cold experienced there relates to consistent clear, dry, calm air; small insolation input; and an inland location far from any moderating maritime effects. Both Verkhoyansk and Oymyakon, Russia, have experienced a minimum temperature of −68°C (−90°F) and a daily average of −50.5°C (−58.9°F) for January (Figure 3.26b). Verkhoyansk experiences at least 7 months of temperatures below freezing, including at least 4 months below −34°C (−30°F)! People do live and work in Verkhoyansk, which has a population of 1400. In contrast, this same region experiences maximum temperatures of +37°C (+98°F) in July.

July Temperature Map

Average July temperatures are presented in Figure 3.27. The longer days of summer and higher Sun altitude now are in the Northern Hemisphere. Winter occupies the Southern Hemisphere, although it is milder there because large continental landmasses are absent and oceans and seas dominate. The thermal equator shifts northward with the high summer Sun and reaches the Persian Gulf–Pakistan–Iran area. The Persian Gulf is the site of the highest recorded SST of 36°C (96°F), difficult to imagine for a sea or ocean body.

During July in the Northern Hemisphere, isotherms trend poleward over land as higher temperatures dominate continental interiors. July temperatures in Verkhoyansk average more than 13°C (56°F), which represents a 63 C° (113 F°) seasonal variation between winter and

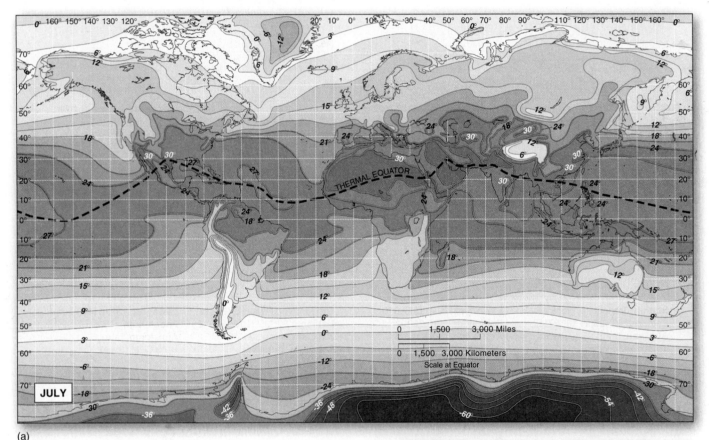

(a)

FIGURE 3.27 Global temperatures for July.
Temperatures are in Celsius (convertible to Fahrenheit by means of the scale) as taken from separate air temperature databases for ocean and land. Note the inset map of North America and the poleward-trending isotherms in the interior. Compare to Figure 3.24. [Adapted from National Climatic Data Center, *Monthly Climatic Data for the World* 47, no. 7, July 1994. Prepared in cooperation with the WMO and NOAA.]

Global Surface Temperatures, Land and Ocean

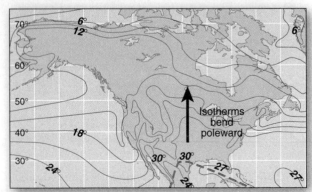

summer averages. The Verkhoyansk region of Siberia is probably Earth's most dramatic example of continental effects on temperature.

The hottest places on Earth occur in Northern Hemisphere deserts during July. These deserts are areas of clear skies and strong surface heating, with virtually no surface water and few plants. Locations such as the Sonoran Desert of North America and the Sahara of Africa are prime examples. Africa has recorded shade temperatures in excess of 58°C (136°F), a record set on September 13, 1922 at Al 'Azī-zī-yah, Libya (32° 32' N; 112 m or 367 ft elevation).

The highest maximum and annual average temperatures in North America occurred in Death Valley, California, where the Greenland Ranch Station (37° N; −54.3 m or −178 ft below sea level) reached 57°C (134°F) in 1913. During June 1994 temperatures hit a high of 53.3°C (128°F) in Death Valley. Such arid and hot lands are discussed further in Chapter 12.

The temperature maps for the North Polar and South Polar regions (Figure 3.28a and b) are featured on two maps. July is "summer" in the Arctic Ocean (a) and brings a thinning of permanent (multiyear) and seasonal sea ice. Look at the Midnight Sun photo that opens Chapter 2 to see what continuous daylight is like. Open cracks and leads stretched all the way to the North Pole over the past several years and all-time lows in the extent of sea-ice are occurring.

July is a time of 24-hour-long nights in Antarctica. The lowest natural temperature reported on Earth occurred in 1983, at the Russian research base at Vostok, Antarctica (78° 27' S, elevation 3420 m or 11,220 ft)—a frigid −89.2°C (−128.6°F). For comparison, such a temperature is 11 C° (19.8 F°) colder than dry ice (solid carbon dioxide)! The average July temperature around Vostok is −68°C (−90.4°F). For average temperature comparisons, the Admundsen–Scott Station at the South Pole experiences −60°C (−76°F) and McMurdo Station has a −26°C (−14.8°F) reading (Figure 3.28b). Note that the coldest temperatures in Antarctica are usually in August, not July, just before the equinox sunrise in September at the end of the long polar night. News Report 3.1 presents an overview of temperature trends in the polar regions.

Annual Range of Temperatures

The pattern of marine effects versus continental effects emerges dramatically when the range in averages is mapped for a full year (Figure 3.29). As you might expect, the largest temperature ranges occur in subpolar locations in North America and Asia, where average ranges of 64 C° (115 F°) are recorded. The Southern Hemisphere, in contrast, has little seasonal variation in mean temperatures, owing to the vast expanses of water that moderate temperature extremes and the lack of large landmasses.

For example, in January (see Figure 3.24) Australia is dominated by isotherms of 20° to 30°C (68° to 86°F),

July

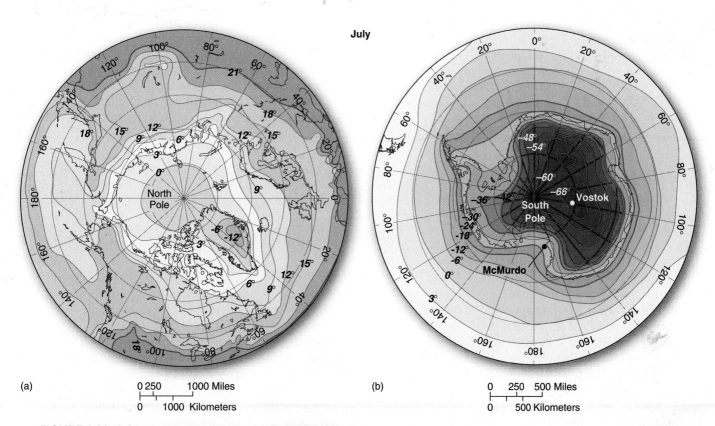

(a) 0 250   1000 Miles
0   1000 Kilometers

(b) 0   250   500 Miles
0   500 Kilometers

FIGURE 3.28 July mean temperatures for Polar Regions.
July temperatures in (a) North Polar region and (b) South Polar region. Use temperature conversions in Figure 3.27. Note that each map is at a different scale. [Author-prepared maps using same sources as Figures 3.24 and 3.27.]

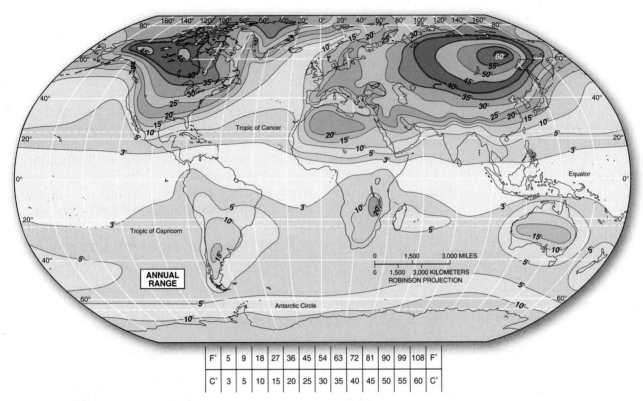

F°	5	9	18	27	36	45	54	63	72	81	90	99	108	F°
C°	3	5	10	15	20	25	30	35	40	45	50	55	60	C°

FIGURE 3.29 Global annual temperature ranges.
A generalized portrait of the annual range of global temperatures in Celsius degrees, C°, with conversions to Fahrenheit degrees, F°, shown on scale. The mapped data show the difference between January and July temperature means.

whereas in July (see Figure 3.27) Australia experiences isotherms of 10° to 20°C (50° to 68°F). Generally speaking, Southern Hemisphere patterns are more marine, whereas Northern Hemisphere patterns are more continental. The Northern Hemisphere, with greater land area overall, registers a slightly higher average surface temperature than does the Southern Hemisphere.

Imagine living in some of these regions and the degree to which you would need to make adaptations for your personal comfort. See Focus Study 3.2 for more on air temperature and the human body. The summer heatwave deaths during the last 10 years in the United States, Canada, and Europe remain powerful reminders of the impact of air temperatures on our lives.

NEWS REPORT 3.1

Overview of Trends in the Polar Regions

Climate models used by scientists to predict and direct measurements show that the polar regions are experiencing some dramatic global climate-change impacts. These changes lead some to say that the high latitudes are providing the global community an early warning signal. The question is, how do we read changing temperature signals, and what are the spatial implications for the rest of the planet?

Arctic
Many proxy and direct measurements confirm that the twentieth century was a time of increasing temperatures at both middle and low latitudes and in the polar regions, especially during the last half-century. Daily maximum and minimum temperatures show there is warming in the Arctic at a rate of about 5 C° (9 F°) per century. Since 1978, warming increased to a rate of 1.2 C° (2.2 F°) per decade,

which means the last 20 years warmed at nearly 7 times the rate of the last 100 years.

In response to rising air and water temperatures, Arctic sea ice is on a significant decline. The sea ice is melting at a rate of about 9% per decade since 1978, especially in the eastern and central Arctic Ocean, and losses are accelerating. Atmospheric pressure patterns cause sea ice to shift as well. There have been record losses since 2000, as sea ice reached its smallest spatial distribution in the satellite record. Figure 3.1.1 is a composite satellite image for 1979 and 2007 comparing sea-ice coverage in the Arctic. September 2007 marked the lowest sea-ice extent on record during the era of satellite surveillance.

The sea ice thins as warmer air melts the surface and warmer waters attack the base of the ice. The summer melt period is increasing as spring arrives earlier and fall is delayed. Since 1970, about half of the Arctic sea ice has disappeared. More open water means darker surfaces that absorb more insolation that leads to more warming. These failing ice conditions threaten polar bears and other Arctic animals and ecosystems—one forecast states that two-thirds of the remaining 23,000 polar bears will be gone before midcentury.

Chapters 1 and 14 discuss the 400% increase in meltponds on the surface of the Greenland ice sheet, ice shelves, and glaciers. This increase represents a significant positive feedback mechanism. In 2008, cracks and wastage occurred along the Petermann Glacier in northern Greenland and the Markham and Ward Hunt ice shelves on Ellsmere Island—masses of ice that had been stable for more than 4500 years. Ward Hunt began breaking up in 2003; by July 2008, eight other ice shelves in the region were disintegrating. In response to the increased discharge of freshwater in the Arctic, the northern oceans are freshening, that is, actually decreasing in salinity. There is valid concern that these meltwaters riding atop saline water can disrupt important circulation systems in the northern oceans.

In Alaska, temperatures rose over the past 30 years at a rate of 3 C° (5 F°) per decade. All 11 mountain ranges experienced glacial ice losses. Glacial thinning reached 1.8 m per year by the late 1990s across the state. In response to this unloading of ice mass, Earth's crust is rebounding upward. A News Report in Chapter 14 explains

1979

2007

FIGURE 3.1.1 **Loss of Arctic sea ice.**
A composite of satellite data (DMSP, Special Sensor Microwave Imager) and surface data from 1979 and 2007 shows remarkable loss of Arctic sea ice. The record low sea-ice extent in 2007 was 24% below the previous low record set in 2005. Each composite is a portrait of minimum sea-ice concentrations during summer months. Not only did the atmosphere experience increased warmth but the Arctic Ocean warmed as well.
[Images courtesy of Scientific Visualization Studio, GSFC/NASA.]

the negative net mass balance of 67 Alaskan glaciers.

All of these changes are affecting living systems as specific high-latitude ecosystems shift across the landscape poleward at 6.1 km (3.8 mi) per decade and spring arrives earlier at a pace of 2.3 days per decade. Changes in tundra soils and the increased depth of active-layer melting in permafrost indicate regional warming, as discussed in Chapter 14. These thawing lands provide less nutrition to caribou and other grazing animals, worsen the threat of

forest fires, and are an additional source of carbon dioxide to the atmosphere.

Antarctica

In Antarctica and the Southern Ocean, the temperature records of the permanently occupied research stations over the last 50 years demonstrate an overall warming. This record includes periods of slight cooling within this time frame. The present warming rate is less than that of the Arctic, at about a 0.9° to 1.2 C° increase per century.

(continued)

News Report 3.1 *(continued)*

However, along the Antarctic Peninsula, the location of numerous ice-shelf collapses and disintegrations, the rate of warming is more than twice this rate, as the January 0° isotherm continues to shift southward. In 2008, the Wilkins Ice Shelf collapsed, at 71° S, the farthest south to fail so far along the peninsula. The exposed land is open to exotic plants and animals that move into these altered ecosystems. In terms of sea ice, the Antarctic trend is not as pronounced as that of the Arctic, with sea-ice extent reduced by 0.4°−1.8° of latitude over the last century. Continued changes will affect marine life and birds.

For the latest science on both polar regions, please see: "Polar Regions (Arctic and Antarctic)," in *Climate Change 2007: Climate Change Impacts, Adaptation, and Vulnerability,*" Working Group II Contribution to the Fourth Assessment Report of the Intergovernmental Panel on Climate—the web site is http://www.ipcc.ch/.

FOCUS STUDY 3.2

Air Temperature and the Human Body

We humans are capable of sensing small changes in temperature in the environment around us. Our perception of temperature is described by the terms *apparent temperature* or *sensible tempera-*

*The traditional value for "normal" body temperature, 37°C (98.6°F), was set in 1868 using old methods of measurement. According to Dr. Philip Mackowiak of the University of Maryland School of Medicine, a more accurate modern assessment places normal at 36.8°C (98.2°F), with a range of 2.7 C° (4.8 F°) for the human population (*Journal of the American Medical Association*, September 23–30, 1992).

ture. This perception of temperature varies among individuals and cultures. Through complex mechanisms, our bodies maintain an average internal temperature ranging within a degree of 36.8°C (98.2°F), slightly lower in the morning or in cold weather and slightly higher at emotional times or during exercise and work.*

Taken together, the water vapor content of air, wind speed, and air temperature affect each individual's sense of comfort. High temperatures, high humidity, and low winds produce the most heat discomfort, whereas low humidity and strong winds enhance cooling rates. Although modern heating and cooling systems, if available and affordable, can reduce the impact of uncomfortable temperatures indoors, the danger to human life from excessive heat or cold persists outdoors. When changes occur in the surrounding air, the human body reacts in various ways to maintain its core temperature and to protect the brain at all cost.

The *wind-chill index* is important to those who experience winters with

FIGURE 3.2.1

Wind-Chill Temperature Index. WCT Index factor for various temperatures and wind speeds. A new version of the wind-chill chart—effective November 2001. (In English units, the new formula is: Wind chill $(F°) = 35.74 + 0.6125T - 35.75(V^{0.16}) + 0.4275T(V^{0.16})$; where T = air temperature in °F, V = wind speed in mph. In Canada, heat loss is expressed in watts per square meter.) [Adapted from the National Weather Service and Meteorological Services of Canada, version 11/01/01.]

Wind speed, kmph (mph)	Calm	4° (40°)	−1° (30°)	−7° (20°)	−12° (10°)	−18° (0°)	−23° (−10°)	−29° (−20°)	−34° (−30°)	−40° (−40°)
8 (5)		2° (36°)	−4° (25°)	−11° (13°)	−17° (1°)	−24° (−11°)	−30° (−22°)	−37° (−34°)	−43° (−46°)	−49° (−57°)
16 (10)		1° (34°)	−6° (21°)	−13° (9°)	−20° (−4°)	−27° (−16°)	−33° (−28°)	−41° (−41°)	−47° (−53°)	−54° (−66°)
24 (15)		0° (32°)	−7° (19°)	−14° (6°)	−22° (−7°)	−28° (−19°)	−36° (−32°)	−43° (−45°)	−50° (−58°)	−57° (−71°)
32 (20)		−1° (30°)	−8° (17°)	−16° (4°)	−23° (−9°)	−30° (−22°)	−37° (−35°)	−44° (−48°)	−52° (−61°)	−59° (−74°)
40 (25)		−2° (29°)	−9° (16°)	−16° (3°)	−24° (−11°)	−31° (−24°)	−38° (−37°)	−46° (−51°)	−53° (−64°)	−61° (−78°)
48 (30)		−2° (28°)	−9° (15°)	−17° (−1°)	−24° (−12°)	−32° (−26°)	−39° (−39°)	−47° (−53°)	−55° (−67°)	−62° (−80°)
56 (35)		−2° (28°)	−10° (14°)	−18° (0°)	−26° (−14°)	−33° (−27°)	−41° (−41°)	−48° (−55°)	−56° (−69°)	−63° (−82°)
64 (40)		−3° (27°)	−11° (13°)	−18° (−1°)	−26° (−15°)	−34° (−29°)	−42° (−43°)	−49° (−57°)	−57° (−71°)	−64° (−84°)
72 (45)		−3° (26°)	−11° (12°)	−19° (−2°)	−27° (−16°)	−34° (−30°)	−42° (−44°)	−50° (−58°)	−58° (−72°)	−66° (−86°)
80 (50)		−3° (26°)	−11° (12°)	−19° (−3°)	−27° (−17°)	−35° (−31°)	−43° (−45°)	−51° (−60°)	−59° (−74°)	−67° (−88°)

Actual Air Temperature in °C (°F)

Frostbite times: 30 min.   ■ 10 min.   ■ 5 min.

freezing temperatures. The wind-chill factor indicates the enhanced rate at which body heat is lost to the air. As wind speeds increase, heat loss from the skin increases. A formula for calculating these relationships was developed in 1945, but this older index tended to overestimate heat loss from skin. The National Weather Service (NWS, http://www.nws.noaa.gov/om/windchill/) and the Meteorological Services of Canada (MSC, http://www.msc.ec.gc.ca/education/windchill/index_e.cfm) revised the wind-chill formula. Figure 3.2.1 is the new version that went into service in 2001. The new Wind Chill Temperature (WCT) Index improves the accuracy of heat loss calculations.

For example, using the new chart, if the air temperature is −7°C (20°F) and the wind is blowing at 32 kmph (20 mph), skin temperatures will be at −16°C (4°F). The lower wind-chill values present a serious freezing hazard to exposed flesh. Imagine the wind chill experienced by a downhill ski racer going 130 kmph (80 mph)—frostbite is a definite possibility during a 2-minute run. The wind chill does omit consideration of sunlight intensity, a person's physical activity, and the use of protective clothing, such as a windbreaker, that prevents wind access to your skin.

The *heat index* (HI) indicates the human body's reaction to air temperature and water vapor. The water vapor in air is expressed as relative humidity, a concept presented in Chapter 5. For now, we simply can assume that the amount of water vapor in the air affects the evaporation rate of perspiration on the skin because the more water vapor in the air (the higher the humidity), the less water from perspiration the air can absorb through evaporation, thus reducing natural evaporative cooling. The heat index indicates how the air feels to an average person—in other words, its apparent temperature.

Figure 3.2.2 is an abbreviated version of the heat index used by the National Weather Service (NWS) and now included in its daily weather summaries during appropriate months. The table beneath the graph describes

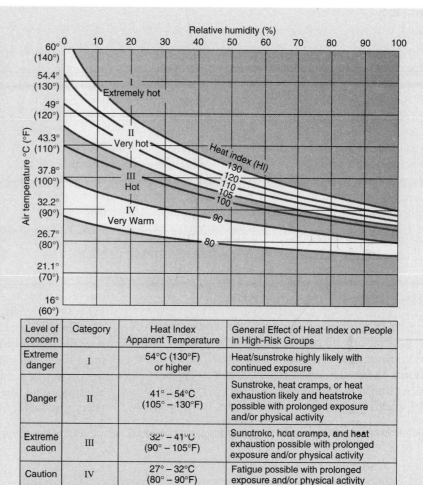

Level of concern	Category	Heat Index Apparent Temperature	General Effect of Heat Index on People in High-Risk Groups
Extreme danger	I	54°C (130°F) or higher	Heat/sunstroke highly likely with continued exposure
Danger	II	41° – 54°C (105° – 130°F)	Sunstroke, heat cramps, or heat exhaustion likely and heatstroke possible with prolonged exposure and/or physical activity
Extreme caution	III	32° – 41°C (90° – 105°F)	Sunstroke, heat cramps, and heat exhaustion possible with prolonged exposure and/or physical activity
Caution	IV	27° – 32°C (80° – 90°F)	Fatigue possible with prolonged exposure and/or physical activity

FIGURE 3.2.2 Heat index.
Heat-index graph for various temperatures and relative humidity levels. [Courtesy of National Weather Service.]

the effects of heat-index categories on higher-risk groups. A combination of high temperature and high humidity can severely reduce the body's natural ability to regulate internal temperature (see http://www.nws.noaa.gov/om/heat/index.shtml); a heat-index calculator is at http://www.crh.noaa.gov/jkl/?n=heat_index_calculator.

Summer Heat-Index Deaths
For nearly a week during July 1995, Chicago's heat-index values went to category I in dwellings that lacked air conditioning. Chicago had never experienced a 48-hour period during which temperatures did not go below 31.5°C (89°F)! With high pressure (hot, stable air) dominating from the Midwest to the Atlantic and moist air from the Gulf of Mexico, all the ingredients for disaster were present. Afternoon temperatures were above

32°C (90°F) for a week in Chicago, 38°C (100°F) in South Dakota, and 40°C (104°F) in Toledo, Ohio, and New York City; the temperature hit 43°C (109°F) in Omaha, Nebraska. Cities in New England that had reached 37.7°C (100°F) only twice in their history exceeded that for 4 or 5 days during July 1995!

This combination of high temperatures and water vapor produced stifling conditions. The sick and elderly, particularly, were affected. On July 13 in Chicago, some apartments without air conditioning exceeded indoor heat-index temperatures of 54°C (130°F) when the official temperature at Midway Airport reached a record 41°C (106°F). The death toll in Chicago was especially severe: 700 people died. Nearly 1000 people died overall in the Midwest and East from these conditions.

(continued)

Focus Study 3.2 (*continued*)

Europe was paralyzed in the summer of 2003 with high heat-index conditions when temperatures topped 40°C (104°F) in June, July, and August. Official estimates across the continent placed the death count at more than 20,000, but this triggered a great undercount controversy, as governments were ill prepared to adequately deal with such a catastrophe. France was hardest hit with more than 14,000 deaths, some

11,000 in the first two weeks of August alone. Although the heat wave began August 1, a full emergency response was not begun until August 13—the day temperatures began to fall.

The rush for medical help under such conditions in the Midwest, Europe, and elsewhere quickly swamps hospitals, which must turn away patients. An added problem in France was that many doctors, nurses, and fellow citizens were

out of town, as this was the beginning of the typical holiday season.

In 2005, for four days in the aftermath of Hurricane Katrina, some 100,000 stranded and homeless New Orleans and Gulf Coast residents suffered from heat-index values exceeding 38°C (100°F), made worse by a lack of drinking water and medical aid—no one will ever know how many perished simply from the heat.

The Urban Environment

For most of you reading this book, the temperatures you feel each day are affected by an urban landscape. Urban microclimates generally differ from those of nearby nonurban areas. In fact, the surface energy characteristics of urban areas are similar to desert locations. Because more than 50% of the world's population now live in cities, urban microclimatology and other specific environmental effects related to cities are important topics for physical geographers. This

statistic is made more meaningful when you consider that in 1950, 30% lived in cities worldwide.

The physical characteristics of urbanized regions produce the **urban heat island**; that is, on average both maximum and minimum temperatures are higher in urban areas than in nearby rural settings. Table 3.1 lists five urban characteristics and the resulting temperature and moisture effects produced. Figure 3.30 illustrates these traits. Every major city produces its own **dust dome** of airborne pollution, which can be blown from the city in elongated plumes;

Table 3.1 Urban Physical Characteristics and Conditions

Urban Characteristics	Results and Conditions
Urban surfaces typically are metal, glass, asphalt, concrete, or stone, and their energy characteristics respond differently than do natural surfaces.	Albedos of urban surfaces are lower, leading to higher net radiation values. Urban surfaces expend more energy as sensible heat than do nonurban areas (70% of the net radiation to H). Surfaces conduct up to 3 times more energy than wet, sandy soil and thus are warmer. During the day and evening, temperatures above urban surfaces are higher than those above natural areas.
Irregular geometric shapes in a city affect radiation patterns and winds.	Incoming insolation is caught in mazelike reflection and radiation "canyons." Delayed energy is conducted into surface materials, thus increasing temperatures. Buildings interrupt wind flows, diminishing heat loss through advective (horizontal) movement. Maximum heat island effects occur on calm, clear days and nights.
Human activity alters the heat characteristics of cities.	In summer, urban electricity production and use of fossil fuels release energy equivalent to 25%–50% of insolation. In winter, urban-generated sensible heat averages 250% greater than arriving insolation, reducing winter heating requirements.
Many urban surfaces are sealed (built on and paved), so water cannot reach the soil.	Central business district surfaces average 50% sealed and suburbs average 20% sealed, producing more water runoff. Urban areas respond as a desert landscape: A storm may cause a flash flood over the hard, sparsely vegetated surfaces, to be followed by dry conditions a few hours later.
Air pollution, including gases and aerosols, greater in urban areas than in comparable natural settings, increases possible convection and precipitation.	Pollution increases the atmosphere's reflectivity above a city, reducing insolation and absorbing infrared radiation, reradiating infrared downward. Increased particulates in pollution are condensation nuclei for water vapor, increasing cloud formation and precipitation. Urban-stimulated increases in precipitation may occur downwind from cities.

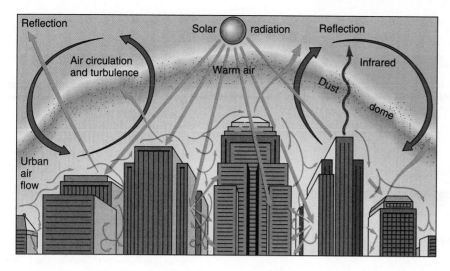

FIGURE 3.30 The urban environment. Insolation, wind movements, and dust dome in city environments.

as noted in the table, such domes affect urban energy budgets. Table 3.2 compares climatic factors of rural and urban environments. The worldwide trend toward greater urbanization is placing more and more people on urban heat islands.

Figure 3.31a illustrates a generalized cross section of a typical urban heat island, showing increasing temperatures toward the downtown central business district. Note that temperatures drop over areas of trees and parks. Sensible heat is lessened because of the latent heat of evaporation and plant effects. Urban forests are important factors in cooling cities. NASA measurements in a mall parking lot

found temperatures of 48°C (118°F); however, a small planter with trees in the same lot was significantly cooler at 32°C (90°F)—16 C° lower in temperature!

NASA launched its Urban Heat Island Pilot Project (UHIPP) in 1997, through its Global Hydrology and Climate Center, to better understand the role of cities in climate. Thermal infrared measurements were made from NASA's research jet over Atlanta, Georgia; Sacramento, California; Baton Rouge, Louisiana; and Salt Lake City, Utah. Images were produced and urban heat island profiles mapped. Teachers and students on the ground assisted by making temperature measurements at the same time as the

Table 3.2 Average Differences in Climatic Elements Between Urban and Rural Environments

Element	Urban Compared with Rural Environs	Element	Urban Compared with Rural Environs
Contaminants		**Precipitation (cont.)**	
Condensation nuclei	10 times more	Snowfall, downwind (lee) of city	10% more
Particulates	10 times more		
Gaseous admixtures	5–25 times more	Thunderstorms	10%–15% more
Radiation		**Temperature**	
Total on horizontal surface	0%–20% less	Annual mean	0.5–3 C° (0.9–5.4 F°) more
Ultraviolet, winter	30% less	Winter minima (average)	1.0–2 C° (1.8–3.6 F°) more
Ultraviolet, summer	5% less	Summer maxima	1.0–3 C° (1.8–3.0 F°) more
Sunshine duration	5%–15% less	Heating degree days	10% less
Cloudiness		**Relative Humidity**	
Clouds	5%–10% more	Annual mean	6% less
Fog, winter	100% more	Winter	2% less
Fog, summer	30% more	Summer	8% less
Precipitation		**Wind Speed**	
Amounts	5%–15% more	Annual mean	20%–30% less
Days with <5 mm (0.2 in.)	10% more	Extreme gusts	10%–20% less
Snowfall, inner city	5%–10% less	Calm	5%–20% more

Source: H. E. Landsberg, *The Urban Climate*, International Geophysics Series, vol. 28 (1981), p. 258. Reprinted by permission from Academic Press.

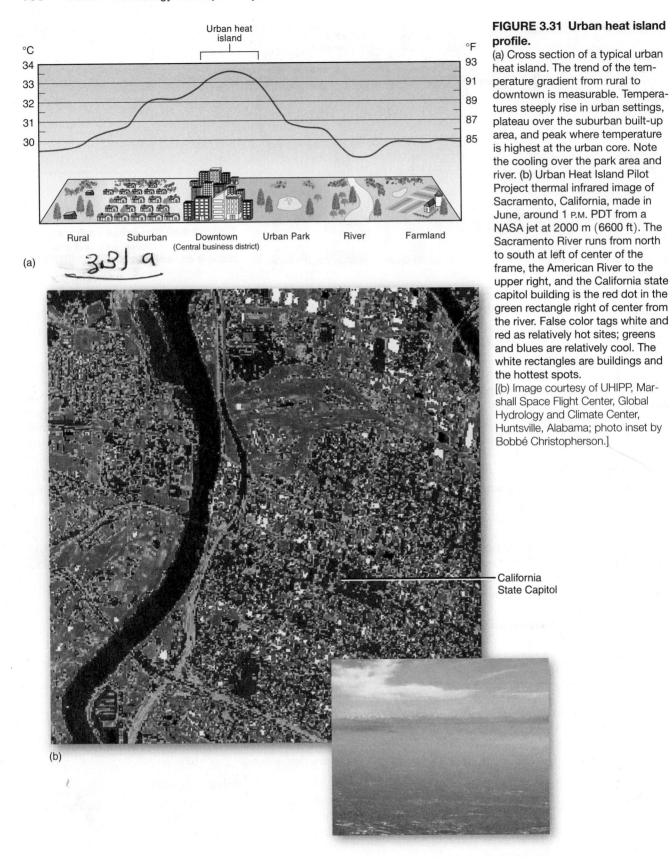

FIGURE 3.31 Urban heat island profile.
(a) Cross section of a typical urban heat island. The trend of the temperature gradient from rural to downtown is measurable. Temperatures steeply rise in urban settings, plateau over the suburban built-up area, and peak where temperature is highest at the urban core. Note the cooling over the park area and river. (b) Urban Heat Island Pilot Project thermal infrared image of Sacramento, California, made in June, around 1 P.M. PDT from a NASA jet at 2000 m (6600 ft). The Sacramento River runs from north to south at left of center of the frame, the American River to the upper right, and the California state capitol building is the red dot in the green rectangle right of center from the river. False color tags white and red as relatively hot sites; greens and blues are relatively cool. The white rectangles are buildings and the hottest spots.
[(b) Image courtesy of UHIPP, Marshall Space Flight Center, Global Hydrology and Climate Center, Huntsville, Alabama; photo inset by Bobbé Christopherson.]

flights. Instruments carried by balloons tracked vertical temperature, relative humidity, and air pressure profiles. (See www.ghcc.msfc.nasa.gov/urban/.)

Figure 3.31b is a thermal infrared image of a portion of Sacramento, California (taken midday). As you look across a small portion of this Sacramento landscape, what strategies do you think would reduce the urban heat island effect? Possibilities include lighter surfaces for buildings, streets, and parking lots; more reflective roofs; and more trees, parks, and open space.

Summary and Review—Atmospheric Energy and Global Temperatures

■ *Identify* the pathways of solar energy through the troposphere to Earth's surface: transmission, scattering, diffuse radiation, refraction, albedo (reflectivity), conduction, convection, and advection.

Earth's biosphere is powered by radiant energy from the Sun that cascades through complex circuits to the surface. Our budget of atmospheric energy comprises shortwave radiation inputs (ultraviolet light, visible light, and near-infrared wavelengths) and longwave radiation outputs (thermal infrared).

Transmission refers to the passage of shortwave and longwave energy through either the atmosphere or water. Insolation encounters an increasing density of atmospheric gases that absorb and reemit it without altering its wavelengths. This **scattering** represents 7% of Earth's reflectivity, or albedo. Some incoming insolation is diffused by clouds and atmosphere and is transmitted to Earth as **diffuse radiation**, the downward component of scattered light. The speed of insolation entering the atmosphere changes as it passes from one medium to another; the change of speed causes a bending action called **refraction**.

A portion of arriving energy bounces directly back into space without being converted into heat or performing any work. **Albedo** is the reflective quality (intrinsic brightness) of a surface. This returned energy is **reflection**. The amount reflected is an important control over the amount of insolation that is available for absorption by a surface. We state albedo as the percentage of insolation that is reflected. Earth and its atmosphere reflect 31% of all insolation when averaged over a year.

An increase in albedo and reflection of shortwave radiation caused by clouds is described by the term **cloud-albedo forcing**. Also, clouds can act as insulation, thus trapping longwave radiation and raising minimum temperatures. An increase in greenhouse warming caused by clouds is described by the term **cloud-greenhouse forcing**. **Global dimming** describes the decline in sunlight reaching Earth's surface owing to pollution, aerosols, and clouds and is perhaps masking the actual degree of global warming that is occurring.

Absorption is the assimilation of radiation by molecules of a substance and its conversion from one form to another—for example, visible light to infrared radiation. **Conduction** is the molecule-to-molecule transfer of energy as it diffuses through a substance. Energy also is transferred by gases and liquids through **convection** (when the physical mixing involves a strong vertical motion) or **advection** (when the dominant motion is horizontal). In the atmosphere or bodies of water, warmer portions tend to rise (they are less dense) and cooler portions tend to sink (they are more dense), establishing patterns of convection.

transmission (p. 76)
scattering (p. 77)
diffuse radiation (p. 77)
refraction (p. 77)
albedo (p. 78)
reflection (p. 78)
cloud-albedo forcing (p. 79)
cloud-greenhouse forcing (p. 79)
global dimming (p. 79)
absorption (p. 79)
conduction (p. 79)
convection (p. 79)
advection (p. 79)

1. Diagram a simple energy balance for the troposphere. Label each shortwave and longwave component and the directional aspects of related flows.
2. Define refraction. How is it related to daylength? To a rainbow? To the beautiful colors of a sunset?
3. List several types of surfaces and their albedo values. Explain the differences among these surfaces. What determines the reflectivity of a surface?
4. Define the concepts transmission, absorption, diffuse radiation, conduction, and convection.

■ *Describe* the greenhouse effect and the patterns of global net radiation and surface energy balances.

The Earth–atmosphere energy system naturally balances itself in a steady-state equilibrium. It does so through energy transfers that are *nonradiative* (convection, conduction, and the latent heat of evaporation) and *radiative* (by longwave radiation between the surface, the atmosphere, and space).

Some longwave radiation is absorbed by carbon dioxide, water vapor, methane, CFCs (chlorofluorocarbons), and other gases in the lower atmosphere and is then radiated to Earth, thus delaying energy loss to space. This process is the **greenhouse effect**. In the atmosphere, longwave radiation is not actually trapped, as it would be in a greenhouse, but its passage to space is delayed (heat energy is detained in the atmosphere) as it is absorbed and radiated. **Jet contrails**, or *condensation trails*, are produced by aircraft exhaust, particulates, and water vapor and can form high cirrus clouds, sometimes called *false cirrus clouds*.

Between the tropics, high insolation angle and consistent daylength cause more energy to be gained than lost (there are energy surpluses). In the polar regions, an extremely low insolation angle, highly reflective surfaces, and up to 6 months of no insolation annually cause more energy to be lost (there are energy deficits). This imbalance of net radiation from tropical surpluses to the polar deficits drives a vast global circulation of both energy and mass.

Surface energy balances are used to summarize the energy expenditure for any location. Surface energy measurements are used as an analytical tool of **microclimatology**. Adding and subtracting the energy flow at the surface lets us calculate **net radiation (NET R)**, or the balance of all radiation at Earth's surface—shortwave (SW) and longwave (LW).

The greatest insolation input occurs at the time of the summer solstice in each hemisphere. Air temperature responds to seasons and variations in insolation input. Within a 24-hour day, air temperature peaks between 3:00 and 4:00 P.M. and dips to its lowest point at or slightly after sunrise.

Air temperature lags behind each day's peak insolation. The warmest time of day occurs not at the moment of maximum insolation but at that moment when a maximum of insolation is absorbed.

greenhouse effect (p. 80)
jet contrails (p. 80)
microclimatology (p. 84)
net radiation (NET R) (p. 84)

5. What are the similarities and differences between an actual greenhouse and the gaseous atmospheric greenhouse? Why is Earth's greenhouse changing?

6. Generalize the pattern of global net radiation. How might this pattern drive the atmospheric weather machine? (See Figure 3.11.)

7. In terms of surface energy balance, explain the term *net radiation* (NET R).

8. What are the expenditure pathways for surface NET R? What kind of work is accomplished?

9. What is the role played by the latent heat of evaporation in surface energy budgets?

10. Compare the daily surface energy patterns of El Mirage, California, and Pitt Meadows, British Columbia. Explain the differences.

11. Why is there a temperature lag between the highest Sun altitude and the warmest time of day? Relate your answer to insolation and temperature patterns during the day.

■ *Review* the temperature concepts and temperature controls that produce global temperature patterns.

Temperature is a measure of the average *kinetic energy* (motion) of individual molecules in matter. Principal controls and influences upon temperature patterns include *latitude* (the distance north or south of the equator), *altitude* and *elevation*, *cloud cover* (reflects, absorbs, and radiates energy), and *land-water heating differences*. The physical nature of land (rock and soil) and water (oceans, seas, and lakes) is the reason for **land-water heating differences**, the fact that land heats and cools faster than water. Moderate temperature patterns are associated with water bodies, and extreme temperatures occur inland. The five controls that differ between land and water surfaces are: evaporation, transparency, specific heat, movement, and ocean currents and sea-surface temperatures.

Light penetrates water because of its **transparency**. Water is clear, and light transmits to an average depth of 60 m (200 ft) in the ocean. This penetration distributes available heat energy over a much greater volume than could occur on opaque land; thus a larger energy reservoir is formed. When equal volumes of water and land are compared, water requires far more energy to increase its temperature than does land. In other words, water can hold more energy than can soil or rock, so water has a higher **specific heat**, the heat capacity of a substance, averaging about four times that of soil.

Ocean currents affect temperature. An example of the effect of ocean currents is the **Gulf Stream**, which moves northward off the east coast of North America, carrying warm water far into the North Atlantic. As a result, the southern third of Iceland experiences much milder temperatures than would be expected for a latitude of 65° N, just below the Arctic Circle (66.5°).

Marine effect, or *maritime*, describes locations that exhibit the moderating influences of the ocean, usually along coastlines or on islands. **Continental effect**, or a condition of continentality, refers to areas that are less affected by the sea and therefore have a greater range between maximum and minimum temperatures diurnally and yearly.

temperature (p. 88)
land-water heating differences (p. 91)
transparency (p. 92)
specific heat (p. 92)
Gulf Stream (p. 92)
marine effect (p. 93)
continental effect (p. 93)

12. Explain the effect of altitude and elevation on air temperature. Why is air at higher altitudes and elevation lower in temperature? Why does it feel cooler standing in shadows at

higher altitudes than at lower altitudes? Briefly, what is the difference between the terms *altitude* and *elevation*?

13. What noticeable effect does air density have on the absorption and radiation of energy? What role does altitude play in that process?

14. How is it possible to grow moderate-climate-type crops such as wheat, barley, and potatoes at an elevation of 4103 m (13,460 ft) near La Paz, Bolivia, so near the equator?

15. Describe the effect of cloud cover with regard to Earth's temperature patterns—review cloud-albedo forcing and cloud-greenhouse forcing.

■ *Interpret* the pattern of Earth's temperatures for January and July and annual temperature ranges.

Maps for January and July instead of the solstice months of December and June are used for temperature comparison because of the natural lag that occurs between insolation received and maximum or minimum temperatures experienced. Each line on these temperature maps is an **isotherm**, an isoline that connects points of equal temperature. Isotherms portray temperature patterns.

Isotherms generally are zonal, trending east-west, parallel to the equator. They mark the general decrease in insolation and net radiation with distance from the equator. The **thermal equator** (isoline connecting all points of highest mean temperature) trends southward in January and shifts northward with the high summer Sun in July. In January it extends farther south into the interior of South America and Africa, indicating higher temperatures over landmasses.

In the Northern Hemisphere in January, isotherms shift equatorward as cold air chills the continental interiors. The coldest area on the map is in Russia, specifically northeastern Siberia. The intense cold experienced there results from winter conditions of consistent clear, dry, calm air; small insolation input; and an inland location far from any moderating maritime effects.

isotherm (p. 95)
thermal equator (p. 96)

16. What is the thermal equator? Describe its location in January and in July. Explain why it shifts position annually.

17. Observe trends in the pattern of isotherms over North America and compare the January average temperature map with the July map. Compare the north polar maps and south polar temperature maps for January and July. What patterns do you observe? Why do the isotherms shift?

18. Describe and explain the extreme temperature range experienced in north-central Siberia between January and July.

19. Where are the hottest places on Earth? Are they near the equator or elsewhere? Explain. Where is the coldest place on Earth?

■ *Contrast* wind chill and heat index, and *determine* human response to these apparent temperature effects.

The *wind-chill factor* indicates the enhanced rate at which body heat is lost to the air. As wind speeds increase, heat loss from the skin increases. The *heat index* (HI) indicates the human body's reaction to air temperature and water vapor. The level of humidity in the air affects our natural ability to cool through evaporation from skin.

20. What is the wind-chill temperature on a day with an air temperature of −12°C (10°F) and a wind speed of 32 kmph (20 mph)?

21. On a day when the temperature reaches 37.8°C (100°F), how does a relative humidity reading of 50% affect apparent temperature?

■ *Portray* typical urban heat island conditions and *contrast* the microclimatology of urban areas with that of surrounding rural environments.

A growing percentage of Earth's people live in cities and experience their unique set of altered microclimatic effects: increased conduction, lower albedos, higher NET R values, increased water runoff, complex radiation and reflection patterns, anthropogenic heating, and the gases, dusts, and aerosols of urban pollution. Urban surfaces of metal, glass, asphalt, concrete, and stone conduct up to three times more energy than wet, sandy soil and thus are warmed as described by the term **urban heat island**. Air pollution, including gases and aerosols, is greater in urban areas than in rural ones. Every major city produces its own **dust dome** of airborne pollution.

urban heat island (p. 104)
dust dome (p. 104)

22. What is the basis for the urban heat island concept? Describe the climatic effects attributable to urban as compared with nonurban environments, or park space, or water bodies. What did NASA determine from the UHIPP overflight of Sacramento (Figure 3.31)?

23. Have you experienced any condition graphed in Focus Study 3.2, or Figures 3.2.1 or 3.2.2? Explain.

24. Assess the potential for solar energy applications in our society. What are some negatives? What are some positives?

NetWork and Critical Thinking Tools

A. Given what you now know about reflection, albedo, absorption, and net radiation expenditures, assess your wardrobe (fabrics and colors), house or apartment (color of walls and roof, orientation relative to the Sun), automobile (color, use of sun shades), bicycle seat (color), and other aspects of your environment to complete a personal energy quiz. What grade do you give yourself? Next summer, be cool!

B. With each temperature map (Figures 3.24 and 3.27), begin by finding your own city or town and noting the temperatures indicated by the isotherms for January, July, and the annual temperature range (Figure 3.29). Record the information from these maps in your notebook. As you work through the different maps throughout this text, note atmospheric pressure and winds, annual precipitation, climate type, landforms, soil orders, vegetation, and terrestrial biomes. By the end of the course you will have a complete physical geography profile for your regional environment.

C. Examine the following photograph. In Antarctica, on Petermann Island off the Graham Land Coast, we noticed this piece of kelp (a seaweed) dropped by a passing bird. The kelp was resting about 10 cm (4 in.) deep in the snow, in a hole about the shape of the kelp. In your opinion, what factors operated to make this scene? Now, expand your conclusion to the issue of mining in Antarctica. There are coal deposits and other minerals in Antarctica. The international Antarctic Treaty (1959, additional measures in 1972, 1982, 1991) blocks mining exploitation, although reconsideration is possible after 2048. Given your analysis of the kelp in the snow, information on surface energy budgets in this chapter, the dust and particulate output of mining, and the importance of a stable sea level, construct a case opposed to such mining activities based on these factors.

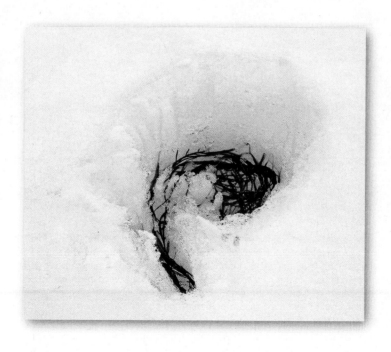

Atmospheric circulation in action—an aerial and ground view of a portion of the Spearville Wind Energy Facility, 27 km (17 mi) east of Dodge City, Kansas. Here, 67 turbines generate 100.5 MW of electricity. The towers are 80 m (262 ft) high, and each turbine blade is 37 m (121 ft) long. Only 13 kmph (8 mph) minimum winds are needed to turn the blades. Turbines are housed in the *nacelle* on the tower tops. Global wind-power capacity is increasing at nearly 30% per year, making it one of the world's fastest-growing sources of energy. Such exploitation of atmospheric circulation could avoid a significant portion of the carbon dioxide emission forcing global warming and unwanted climate change. *[Photos by Bobbé Christopherson.]*

4

Atmospheric and Oceanic Circulations

KEY LEARNING CONCEPTS

After reading the chapter, you should be able to:

▪ **Define** the concept of air pressure and **portray** the pattern of global pressure systems on isobaric maps.

▪ **Define** wind and **describe** how wind is measured, how wind direction is determined, and how winds are named.

▪ **Explain** the four driving forces within the atmosphere—gravity, pressure gradient force, Coriolis force, and friction force—and **describe** the primary high- and low-pressure areas and principal winds.

▪ **Describe** upper-air circulation and its support role for surface systems and **define** the jet streams.

▪ **Overview** several multiyear oscillations of air temperature, air pressure, and circulation in the Arctic, Atlantic, and Pacific Oceans.

▪ **Explain** several types of local winds: land-sea breezes, mountain-valley breezes, katabatic winds, and the regional monsoons.

▪ **Describe** the basic pattern of Earth's major surface and deep thermohaline ocean currents.

STUDENT LEARNING CENTER TOOLS

The *Elemental Geosystems* Student Learning Center provides on-line resources for this chapter at **www.mygeoscienceplace.com**. Some highlights:

▶ Seven animations assist you with wind patterns, Coriolis force, air pressure patterns, global wind circulation, anticyclones and cyclones, jet streams and Rossby waves, and ocean circulation.

▶ Narrated satellite loops display sea-level pressure and global images.

▶ Read a Career Link about a GIS/GPS/environmental geospatial specialist.

▶ More than 9 additional URLs along with Quick Links to all the URLs in the chapter.

Early in April 1815, on the island Sumbawa in present-day Indonesia, the volcano Tambora erupted violently, releasing an estimated 150 km³ (36 mi³) of material, 25 times the volume produced by the 1980 Mount St. Helens eruption in Washington state. Some materials from Tambora—the ash and acid mists—were carried worldwide by global atmospheric circulation, creating a stratospheric dust veil. The result was both a higher atmospheric albedo and absorption of energy by the debris injected into the stratosphere. Beautiful, colorful sunrises and sunsets resulting from the spreading dust were seen worldwide for several years after the eruption. Scientists in 1815 lacked the remote-sensing capability of satellite technology and had no way of knowing the global impact of Tambora's eruption.

After 635 years of dormancy, Mount Pinatubo in the Philippines erupted in 1991 (Figure 4.1a). This event had tremendous atmospheric impact, lofting 15–20 million tons of ash, dust, and sulfur dioxide (SO₂) into the atmosphere. As the sulfur dioxide rose into the stratosphere, it quickly formed sulfuric acid (H_2SO_4) aerosols, which concentrated at 16−25 km (10−15.5 mi) altitude. This debris increased atmospheric albedo about 1.5%, giving scientists an estimate of the aerosol volume generated by the eruption. Technology permits a depth of analysis unknown in the past—satellites now track the atmospheric effects from dust storms, forest fires, industrial haze, warfare, and the dispersal of volcanic explosions, among many things.

Figure 4.1b–e show combined *NOAA-11* images made at 2-week intervals that clearly track the reflected solar radiation from these aerosols, carried by the atmospheric circulation. The earliest image (b), near the time of the eruption, shows winds blowing dust westward from Africa, smoke from the Kuwaiti oil well fires set during the Persian Gulf War, and some haze off the East Coast of the United States. Also visible is the Pinatubo aerosol layer beginning to emerge north of Indonesia. Atmospheric winds spread the eruption debris worldwide in just 60 days. (Figure 1.7 shows this spread and its global spatial effects.)

In this chapter: We begin with a discussion of wind essentials, consisting of air pressure and its measurement and a description of wind. The driving forces that produce surface winds are pressure gradient, Coriolis, and friction. We examine the circulation of Earth's atmosphere and the patterns of global winds, including principal pressure systems and winds. We also consider Earth's wind-driven oceanic currents. The chapter explains multiyear oscillations in atmospheric flows, local winds, and the powerful monsoons. We see that producing electricity using wind-driven turbines offers great promise as an energy source. The energy driving all this movement comes from one source: the Sun.

Wind Essentials

Earth's atmospheric circulation is an important transfer mechanism for both energy and mass. In the process, the imbalance between equatorial energy surpluses and polar energy deficits is partly resolved, Earth's weather patterns are generated, and ocean currents are produced. Human-caused pollution also is spread by this circulation, far from its points of origin. The fluid movement of the atmosphere links together all humanity more than any other natural or cultural factor. Our atmosphere makes all the world a spatially linked society—one person's or country's atmospheric exhalation is another's intake.

Atmospheric circulation is generally categorized at three levels: *primary circulation* (general), *secondary circulation* of migratory high-pressure and low-pressure systems, and *tertiary circulation*, which includes local winds and temporal weather patterns. Winds that move principally north or south along meridians are known as *meridional flows*. Winds moving east or west along parallels of latitude are *zonal flows*.

Air Pressure and Its Measurement

Air pressure, its measurement and expression, is key to understanding wind. The molecules that constitute air create **air pressure** through their motion, size, and number—the factors that determine the temperature and density of the air. Air pressure, then, is a product of the temperature and density of a mass of air. Pressure is exerted on all surfaces in contact with the air.

In A.D. 1643, a pupil of Galileo's named Evangelista Torricelli was working on a mine drainage problem. This led him to discover a method for measuring air pressure (Figure 4.2a). He knew that pumps in the mine were able to draw water upward only about 10 m (33 ft) but did not know why. Careful observation led him to discover that this problem was not caused by the pumps but by the atmosphere itself. Torricelli noted that the water level in the vertical pipe fluctuated from day to day. He figured out that air pressure varied over time and with changing weather conditions. A pump of the type shown in the figure works by creating a vacuum above the water, which allows the pressure of the atmosphere that is pushing down on water in the mine to force water up the pipe.

To simulate the mine-pump situation, Torricelli, at Galileo's suggestion, devised an instrument. Instead of water, he used a much denser fluid, mercury (Hg), so that he could use a glass laboratory tube only 1 m high. He sealed the glass tube at one end, filled it with mercury, and then inverted it into a dish of mercury (Figure 4.2b). Torricelli established that the average height of the column of mercury in the tube was 760 mm (29.92 in.). He concluded that the mass of surrounding air was exerting pressure on the mercury in the dish that counterbalanced the column of mercury.

Thus, the 10-m (33-ft) limit to which the mine pumps could draw water was controlled by the mass of the atmosphere. Using similar instruments today, *normal sea-level pressure* is expressed as 1013.2 millibars of mercury (a way of expressing force per square meter of surface area). In Canada and other countries, normal air pressure is expressed as 101.32 kPa (*kilopascal*; 1 kPa = 10 mb).

Any instrument that measures air pressure is a *barometer*. Torricelli developed a **mercury barometer**. A

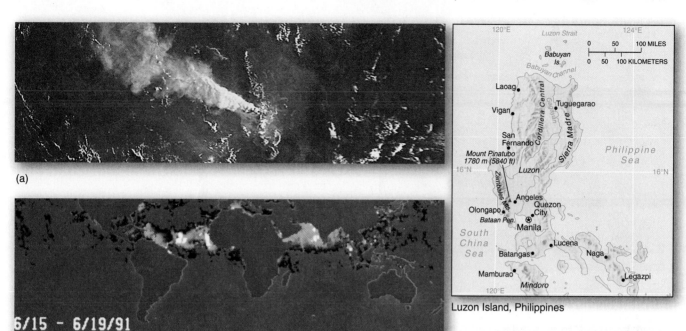

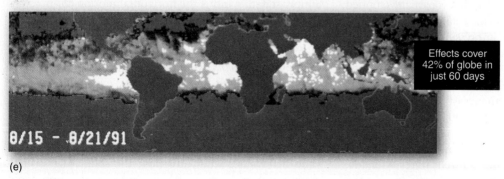

Luzon Island, Philippines

FIGURE 4.1 Volcanic debris spreads worldwide.
These false-color images show aerosols from Mount Pinatubo, smoke from fires, and dust storms, all swept about the globe by the general atmospheric circulation. Image (a) is a satellite image of the volcano as it erupted. The dramatic increase in aerosols from mid-June 1991 (b) to mid-August (e) spanned the area from 20° S to 30° N, covering about 42% of the globe. (The false color denotes the atmosphere's *aerosol optical thickness*, or AOT. Lowest AOT values are dull yellow, medium values are bright yellow, and highest values are white.) [(a) AVHRR satellite eruption image courtesy of EROS Data Center; (b–e) AVHRR satellite images of aerosols courtesy of Dr. Larry L. Stowe, NESDIS, NOAA. Used by permission.]

Effects cover 42% of globe in just 60 days

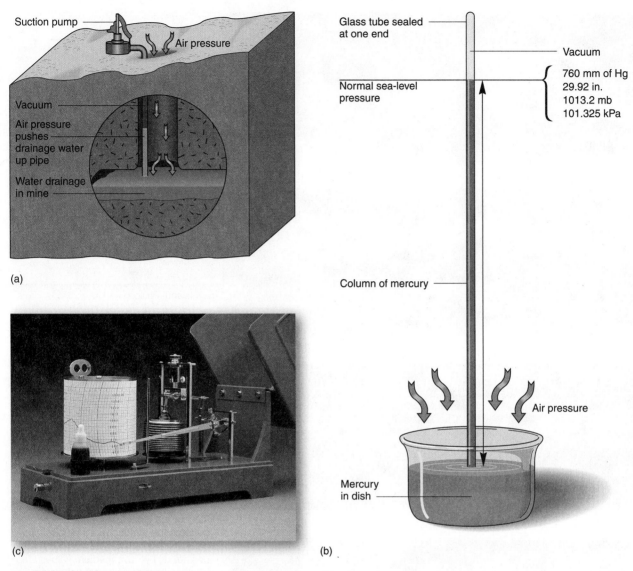

FIGURE 4.2 Developing a barometer.
(a) Evangelista Torricelli developed the barometer to measure air pressure as a by-product of trying to solve a mine drainage problem. Instruments used to measure atmospheric pressure include an idealized model of a mercury barometer (b) and an aneroid barometer (c). Have you used a barometer? If so, what type was it? [(c) Photo courtesy of Qualimetrics, Inc., Sacramento, Calif.]

more compact design that works without a meter-long tube of mercury is an **aneroid barometer** (Figure 4.2c). Inside this type of barometer, a small chamber is partially emptied of air, sealed, and connected to a mechanism that is sensitive to changes in the chamber. As air pressure varies, the mechanism responds to the difference and moves the needle on the dial.

Figure 4.3 illustrates comparative scales in millibars and inches used to express air pressure and its relative force. Strong high pressure to deep low pressure can range from 1050 to 980 mb (31.00 to 29.00 in.). The figure also indicates the extreme highest and lowest pressures ever recorded in the United States, Canada, and on Earth. Hurricane Wilma from 2005 is now the record holder. For comparison, note Hurricanes Gilbert (1988) and Rita and Katrina (2005), and their lowest central pressures.

Wind: Description and Measurement

Simply stated, **wind** is the horizontal motion of air across Earth's surface. Turbulence, wind updrafts and downdrafts, adds a vertical component to this definition. *Differences in air pressure (density) between one location and another produce wind.* Wind's two principal properties are speed and direction, and there are instruments to measure each. An **anemometer** measures wind speed, expressed in kilometers per hour (kmph), miles per hour (mph), meters per second (mps), or knots. (A *knot* is a nautical mile per hour, equivalent to 1.85 kmph, or 1.15 mph.) A **wind vane** determines wind direction; the standard measurement is taken 10 m (33 ft) above the ground to avoid local effects of topography (Figure 4.4).

Winds are named for the direction *from which they originate.* For example, a wind from the west is a *westerly*

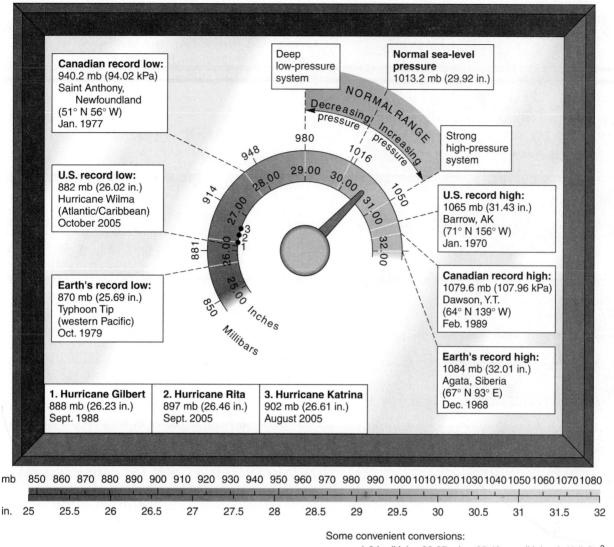

Canadian record low:
940.2 mb (94.02 kPa)
Saint Anthony,
    Newfoundland
(51° N 56° W)
Jan. 1977

U.S. record low:
882 mb (26.02 in.)
Hurricane Wilma
(Atlantic/Caribbean)
October 2005

Earth's record low:
870 mb (25.69 in.)
Typhoon Tip
(western Pacific)
Oct. 1979

Deep
low-pressure
system

**Normal sea-level
pressure**
1013.2 mb (29.92 in.)

Strong
high-pressure
system

U.S. record high:
1065 mb (31.43 in.)
Barrow, AK
(71° N 156° W)
Jan. 1970

Canadian record high:
1079.6 mb (107.96 kPa)
Dawson, Y.T.
(64° N 139° W)
Feb. 1989

Earth's record high:
1084 mb (32.01 in.)
Agata, Siberia
(67° N 93° E)
Dec. 1968

1. Hurricane Gilbert
888 mb (26.23 in.)
Sept. 1988

2. Hurricane Rita
897 mb (26.46 in.)
Sept. 2005

3. Hurricane Katrina
902 mb (26.61 in.)
August 2005

mb	850	860	870	880	890	900	910	920	930	940	950	960	970	980	990	1000	1010	1020	1030	1040	1050	1060	1070	1080

in.	25	25.5	26	26.5	27	27.5	28	28.5	29	29.5	30	30.5	31	31.5	32

Some convenient conversions:
1.0 in. (Hg) = 33.87 mb = 25.40 mm (Hg) = 0.49 lb/in.^{2}
1.0 mb = 0.0295 in. (Hg) = 0.75 mm (Hg) = 0.0145 lb/in.^{2}

FIGURE 4.3 Air pressure readings and conversions.
Scales for expressing barometric air pressure in millibars and inches, with average air pressure
values and recorded pressure extremes. Canadian values include kilopascal equivalents
(10 mb = 1 kPa). The intensity of the 2005 hurricane season is evident; note the positions of
Katrina, Rita, and Wilma on the pressure dial.

wind (it blows eastward); a wind out of the south is a *southerly* wind (it blows northward). Figure 4.5 illustrates a simple wind compass, naming the 16 principal wind directions.

The traditional *Beaufort wind scale* is a descriptive scale useful in visually estimating wind speed (Table 4.1). It is named for Admiral Beaufort of the British Navy, who introduced his wind scale in 1806. In 1926, G. C. Simpson expanded Beaufort's scale to include wind speeds on land. The National Weather Service (old Weather Bureau) standardized the scale in 1955 (see http://www.hpc.ncep.noaa. gov/html/beaufort.shtml). The scale is still referenced on ocean charts, enabling estimation of wind speed without instruments. And although most ships use sophisticated equipment to perform such measurements, you find the Beaufort scale posted on the bridge of many ships. As you walk across campus use the "observed effects" column of the Beaufort Wind Scale to estimate any winds that may be occurring.

Global Winds

The primary circulation of winds across Earth fascinated travelers, sailors, and scientists for centuries, although only in the modern era is a true picture emerging of the pattern of global winds. A remarkable NASA *Seasat* satellite image combining 150,000 measurements made during a single day portrays surface winds (Figure 4.6, p. 118). Wind drives waves on the ocean surface that combine to form patterns of currents. *Seasat* used radar to measure the motion and direction of waves from their backscatter to the satellite, producing this portrait of surface wind fields over the Pacific Ocean.

FIGURE 4.4 Wind vane and anemometer.
Instruments used to measure wind direction (wind vane, flat blade) and wind speed (anemometer, three cups). [Photo by Belfort Instruments.]

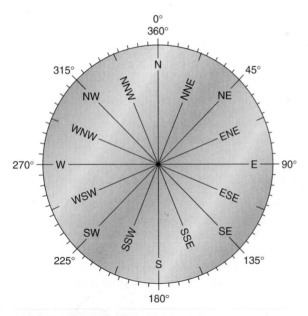

FIGURE 4.5 A wind compass.
Sixteen wind directions identified. Winds are named for the direction from which they originate. For example, a wind from the west is a westerly wind.

The patterns in the figure are the result of specific forces at work in the atmosphere: *pressure gradient force*, *Coriolis force*, *friction force*, and *gravity*. As we progress through this chapter, you may want to refer to this *Seasat* image to identify the winds, eddies, and vortices it portrays.

Driving Forces Within the Atmosphere

Four forces determine both speed and direction of winds:

- Earth's *gravitational force* on the atmosphere is practically uniform, equally compressing the atmosphere near the ground worldwide. Density decreases as altitude increases.
- The **pressure gradient force** drives air from areas of higher barometric pressure to areas of lower barometric pressure, causing winds.
- The **Coriolis force**, a deflective force, appears to deflect wind from a straight path in relation to Earth's rotating surface—to the right in the Northern Hemisphere and to the left in the Southern Hemisphere.
- The **friction force** drags on the wind as it moves across surfaces; it decreases with height above the surface.

All four of these forces operate on moving air and water and affect the circulation patterns of global winds. The following sections describe the actions of the pressure gradient, Coriolis, and friction forces.

Pressure Gradient Force

High- and low-pressure areas that exist in the atmosphere result principally from unequal heating at Earth's surface and from certain dynamic forces in the atmosphere. A pressure gradient is the difference in atmospheric pressure between areas of higher pressure (more dense air) and lower pressure (less dense air). An **isobar** is an *isoline* (a line along which there is a constant value) plotted on a weather map to connect points of equal pressure. The distance between isobars indicates the degree of pressure difference. Isobars on a weather map are used to make a spatial analysis of pressure patterns (Figure 4.7, p. 119).

Just as closer contour lines on a topographic map mark a steeper slope, so do closer isobars denote steepness in the pressure gradient. In Figure 4.7a, note the spacing of the isobars (green lines). A steep gradient causes faster air movement from a high-pressure area to a low-pressure area. Isobars spaced at greater distances from one another mark a more gradual pressure gradient, one that creates a slower air flow. Along a horizontal surface, the pressure gradient force acts at right angles to the isobars. Note the location of steep and gradual pressure

Table 4.1 Beaufort Wind Scale

Wind Speed					Beaufort Wind Scale	
kmph	mph	knots	Beaufort Number	Wind Description	Observed Effects at Sea	Observed Effects on Land
<1	<1	<1	0	Calm	Glassy calm, like a mirror	Calm, no movement of leaves
1–5	1–3	1–3	1	Light air	Small ripples; wavelet scales; no foam on crests	Slight leaf movement; smoke drifts; wind vanes still
6–11	4–7	4–6	2	Light breeze	Small wavelets; glassy look to crests, which do not break	Leaves rustling; wind felt; wind vanes moving
12–19	8–12	7–10	3	Gentle breeze	Large wavelets; dispersed whitecaps as crests break	Leaves and twigs in motion; small flags and banners extended
20–29	13–18	11–16	4	Moderate breeze	Small, longer waves; numerous whitecaps	Small branches moving; raising dust, paper, litter, and dry leaves
30–38	19–24	17–21	5	Fresh breeze	Moderate, pronounced waves; many whitecaps; some spray	Small trees and branches swaying; wavelets forming on inland waterways
39–49	25–31	22–27	6	Strong breeze	Large waves, white foam crests everywhere; some spray	Large branches swaying; overhead wires whistling; difficult to control an umbrella
50–61	32–38	28–33	7	Moderate (near) gale	Sea mounding up; foam and sea spray blown in streaks in the direction of the wind	Entire trees moving; difficult to walk into wind
62–74	39–46	34–40	8	Fresh gale (or gale)	Moderately high waves of greater length; breaking crests forming sea spray; well-marked foam streaks	Small branches breaking; difficult to walk; moving automobiles drifting and veering
75–87	47–54	41–47	9	Strong gale	High waves; wave crests tumbling and the sea beginning to roll; visibility reduced by blowing spray	Roof shingles blown away; slight damage to structures; broken branches littering the ground
88–101	55–63	48–55	10	Whole gale (or storm)	Very high waves and heavy, rolling seas; white appearance to foam-covered sea; overhanging waves; visibility reduced	Uprooted and broken trees; structural damage; considerable destruction; seldom occurring
102–116	64–73	56–63	11	Storm (or violent storm)	White foam covering a breaking sea of exceptionally high waves; small- and medium-sized ships lost from view in wave troughs; wave crests frothy	Widespread damage to structures and trees, a rare occurrence
>117	>74	>64	12–17	Hurricane	Driving foam and spray filling the air; white sea; visibility poor to nonexistent	Severe to catastrophic damage; devastation to affected society

gradients and their relationship to wind intensity on the map in Figure 4.7b.

Figure 4.8 (p. 120) illustrates the combined effect of the forces that direct wind. The pressure gradient force acting alone is shown in Figure 4.8a from two perspectives—top view and side view. Descending air in the high-pressure area creates a field of subsiding, or sinking, air, which then moves out of the high-pressure area in a flow described as diverging. High-pressure areas feature these *descending, diverging* air flows. Conversely, air moving into a low-pressure area converges; thus, low-pressure areas feature *converging, ascending* air flows.

Coriolis Force

You might expect surface winds to move in a straight line from areas of higher pressure to areas of lower pressure, but they do not. The reason for this is the *Coriolis force*,* which deflects from a straight path any object that flies or

*Note that the text uses the word *force*. The label is appropriate because, as the physicist Sir Isaac Newton (1643–1727) stated, when something is accelerating over a space, a force is in operation. This apparent force (in classical mechanics, an inertial force) acts as an effect on moving objects. It is named for Gaspard Coriolis, a French mathematician and researcher of applied mechanics, who first described the phenomenon in 1831.

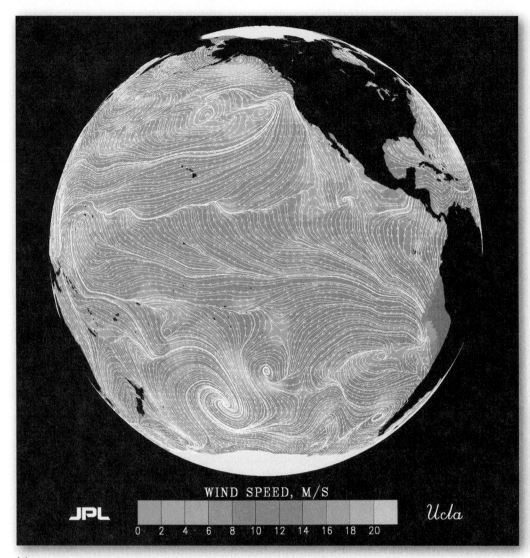

(a)   (b)

FIGURE 4.6 Wind portrait of the Pacific Ocean.
(a) Surface wind measured by radar scatterometer aboard the *Seasat* satellite on September 14, 1978. Scientists analyzed 150,000 measurements to produce this image. Colors correlate to wind speeds (see scale), and the white arrows note wind direction. Can you identify the pattern of trade winds, westerlies, high-pressure cells, and low-pressure cells from the cloud patterns on the image (b)? [*Seasat* Image courtesy of Dr. Peter Woiceshyn, Jet Propulsion Laboratory, Pasadena; satellite image inset from Cornell University.]

flows across Earth's surface, be it wind, an airplane, or ocean currents. The deflection produced by the Coriolis force is a product of Earth's rotation. Because Earth rotates eastward, objects that move in an absolute straight line over a distance (such as winds and ocean currents) appear to curve to the right in the Northern Hemisphere and to the left in the Southern Hemisphere.

Earth's rotational speed varies with latitude, increasing from 0 kmph at the poles to 1675 kmph (1041 mph) at the equator (noted on Figure 4.9b). This means that materials at different latitudes possess different amounts of angular momentum. The Coriolis force is zero along the equator, increasing to half the maximum deflection at 30° N and S latitudes, and to maximum deflection flowing away from the poles (Figure 4.9a).

The key to understanding Coriolis force is one's viewpoint. From the viewpoint of an airplane that is passing over Earth's surface, the surface can be seen to rotate slowly below. But, looking from Earth's surface at the airplane, the surface seems stationary, and the airplane appears to curve off course. The airplane does not actually deviate from a straight path, but it *appears* to do so because

we are standing on Earth's rotating surface beneath the airplane. Because of this apparent deflection, the airplane must make constant corrections in flight path to maintain its "straight" heading relative to a rotating Earth.

As an example of the effect of this force, see Figure 4.9b. A pilot leaves the North Pole and flies due south toward Quito, Ecuador. If Earth were not rotating, the aircraft would simply travel along a meridian of longitude and arrive at Quito. But Earth is rotating eastward beneath the aircraft's flight path. If the pilot does not allow for this rotation, the plane will reach the equator over the ocean along an apparently curved path far to the west of the intended destination. Pilots must correct for this Coriolis deflection in their navigational calculations.

This effect occurs regardless of the direction of the moving object. A flight from California to New York is shown in Figure 4.9c. The Coriolis deflection occurs because, as the airplane flew to New York, Earth continued to rotate eastward, so the destination moved farther to the east. Unless the pilot corrected for Earth's rotational motion, the flight would end up somewhere in North Carolina.

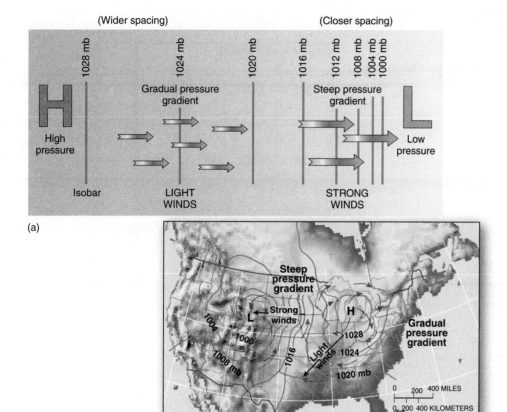

(a)

(b)

FIGURE 4.7 Pressure gradient determines wind speed.
(a) Pressure gradient portrait.
(b) On a weather map the closer spacing of isobars represents a steeper pressure gradient, which produces strong winds; wider spacing of isobars denotes a gradual pressure gradient, which leads to light winds.

We work through a simple explanation here focused on the observed phenomena. If you choose, you can use physics and identify the variables that interplay to produce these observed results. Mathematical formulas explain how the conservation of angular and linear momentum operates to produce the deflection as the distance to Earth's axis changes with latitudinal location, as well as the interplay of the opposing gravitational and centrifugal forces.

How does the Coriolis force affect wind aloft? As air rises from the surface through the lowest levels of the atmosphere, leaving the drag of surface friction behind, its speed increases, thus increasing the Coriolis force, spiraling the winds to the right in the Northern Hemisphere or to the left in the Southern Hemisphere. In the upper troposphere, the Coriolis force just balances the pressure gradient force. Consequently, the winds between high-pressure and low-pressure areas aloft flow parallel to the isobars.

Figure 4.8b illustrates the combined effect of the pressure gradient force and the Coriolis force on air currents aloft, producing **geostrophic winds**. Geostrophic winds are characteristic of upper tropospheric circulation. The air does not flow directly from high to low but around the pressure areas instead, remaining parallel to the isobars and producing the characteristic pattern shown on the upper-air weather map just ahead in Figure 4.15. Note the inset illustration showing the effects of the pressure gradient and Coriolis forces that produce a geostrophic flow of air.

Friction Force

The effect of surface friction extends to a height of about 500 m (1650 ft) and varies with surface texture, wind speed, time of day and year, and atmospheric conditions. Generally, rougher surfaces produce more friction. Figure 4.8c combines all three forces, adding the effect of friction to the Coriolis and pressure gradient forces on wind movements.

Near the surface, friction disrupts the equilibrium between the pressure gradient and Coriolis forces established in geostrophic wind flows—note the inset illustration in Figure 4.8c. Because surface friction decreases wind speed, it reduces the effect of the Coriolis force and causes winds in the Northern Hemisphere to move across isobars at an angle, spiraling out from a high-pressure area clockwise to form an **anticyclone** and spiraling into a low-pressure area counterclockwise to form a **cyclone**. In the Southern Hemisphere these circulation patterns are reversed, flowing out of high-pressure cells counterclockwise and into low-pressure cells clockwise.

Atmospheric Patterns of Motion

With these forces and motions in mind, we are ready to build a general model of total atmospheric circulation and better understand the earlier *Seasat* image (see Figure 4.6). If Earth did not rotate, then the warmer, less dense air of the equatorial regions would rise, creating low pressure at

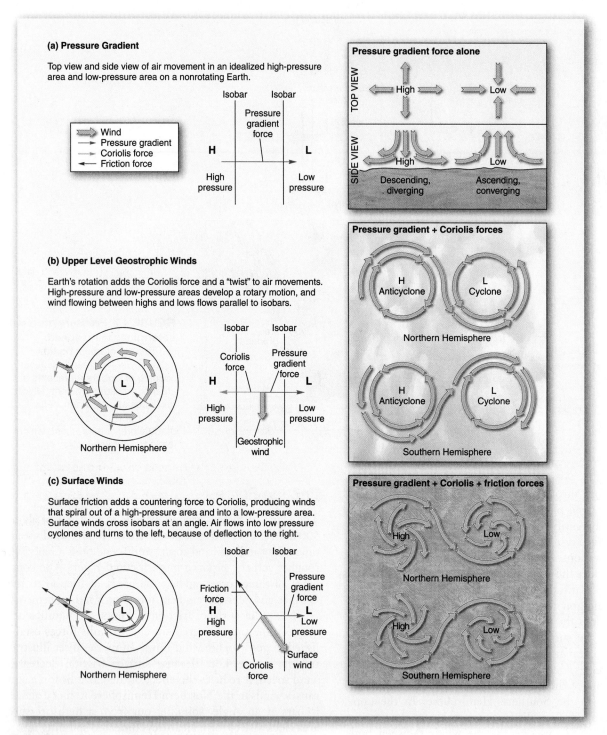

(a) Pressure Gradient

Top view and side view of air movement in an idealized high-pressure area and low-pressure area on a nonrotating Earth.

Wind
Pressure gradient
Coriolis force
Friction force

(b) Upper Level Geostrophic Winds

Earth's rotation adds the Coriolis force and a "twist" to air movements. High-pressure and low-pressure areas develop a rotary motion, and wind flowing between highs and lows flows parallel to isobars.

Northern Hemisphere

(c) Surface Winds

Surface friction adds a countering force to Coriolis, producing winds that spiral out of a high-pressure area and into a low-pressure area. Surface winds cross isobars at an angle. Air flows into low pressure cyclones and turns to the left, because of deflection to the right.

Northern Hemisphere

FIGURE 4.8 Three physical forces that produce winds.
(a) The pressure gradient force. (b) The Coriolis force added to the effects of the pressure gradient force produces a geostrophic wind flow in the upper atmosphere. And (c) the friction force considered together with the other forces produces characteristic surface winds. Note the reverse circulation pattern in the Southern Hemisphere. The three inset diagrams show the interaction of forces that form prevailing wind flow and geostrophic and surface winds. (The gravitational force is assumed.)

ANIMATION

Wind Pattern Development

the surface, and the colder and denser air of the poles would sink, creating high pressure at the surface. The net effect would be a meridional flow of winds established by this simple pressure gradient.

However, in reality, Earth does rotate, producing a more complex system for the transfer of energy and air masses from equatorial surpluses to polar deficits—one with waves, streams, and eddies on a planetary scale. And

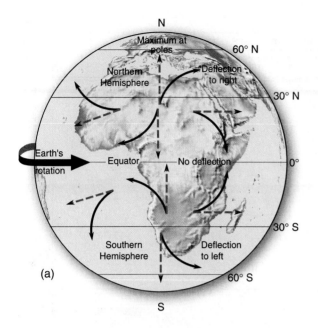

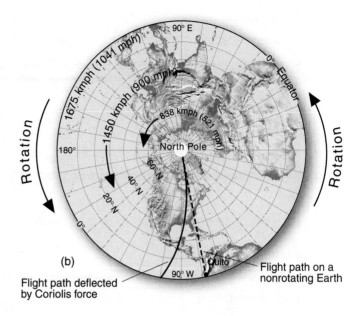

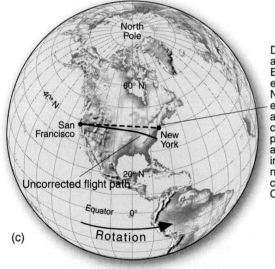

During the airplane flight, Earth rotated eastward, placing New York farther east than it was at takeoff—this change in position produces the apparent deflection in flight path. The navigator must correct for this Coriolis deflection.

Coriolis Force

FIGURE 4.9 The Coriolis force—an apparent deflection.
(a) Apparent deflection to the right of a straight line in the Northern Hemisphere and apparent deflection to the left in the Southern Hemisphere. (b) Coriolis deflection of a flight path between the North Pole and Quito, Ecuador, which is on the equator. (c) The deflection of a flight path between San Francisco and New York. Deflection from a straight path occurs regardless of the direction of movement.

instead of this hypothetical meridional flow, the flow is predominantly zonal (latitudinal) at the surface and aloft: westerly winds (eastward) in the middle and high latitudes and easterly winds (westward) in the equatorial latitudes of both hemispheres.

Primary High-Pressure and Low-Pressure Areas

The following discussion of Earth's pressure and wind patterns refers often to Figure 4.10, which shows January and July isobaric maps of average surface barometric pressure. The primary high- and low-pressure areas on Earth appear on these isobaric maps as interrupted cells or uneven belts of similar pressure that stretch across the face of the planet. Between these areas flow the primary winds that have been noted in adventure tales and myth throughout human experience.

Secondary highs and lows, from a few hundred to a few thousand kilometers in diameter and hundreds to

thousands of meters high, are formed within the primary pressure areas. These pressure systems seasonally migrate to produce changing weather and climate patterns in the regions over which they pass.

Four identifiable pressure areas cover the Northern Hemisphere; a similar set is found in the Southern Hemisphere. In each hemisphere, two of the pressure areas are specifically stimulated by *thermal* (temperature) factors: the **equatorial low-pressure trough** and the weak **polar high-pressure cells,** both north and south. The other two areas are formed by *dynamic* factors: the **subtropical high-pressure cells** and the **subpolar low-pressure cells.** Table 4.2 summarizes the characteristics of these pressure areas, which we now examine.

Equatorial Low-Pressure Trough: Clouds and Rain
The equatorial low-pressure trough is an elongated, narrow band of low pressure (converging, ascending airflow), cloudiness, and precipitation that nearly girdles Earth, following an undulating linear axis. Constant high Sun

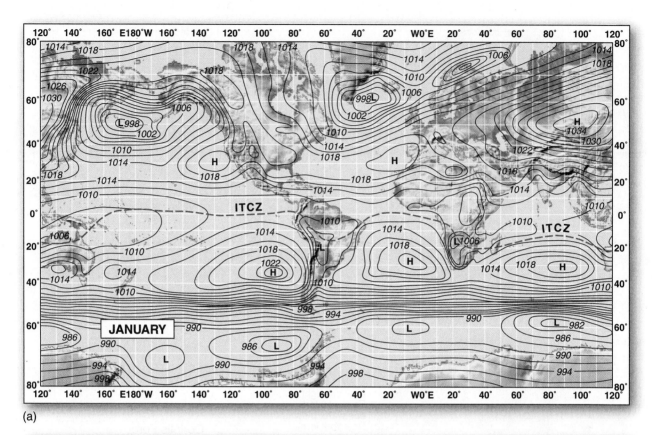

(a)

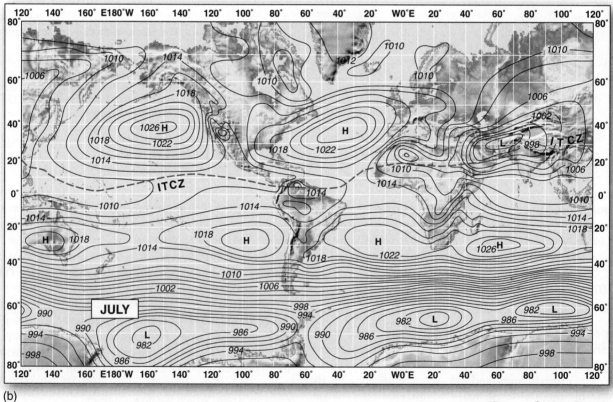

(b)

FIGURE 4.10 Global barometric pressures.

Average surface barometric pressure (millibars) for January and July. The dashed red line marks the intertropical convergence zone (ITCZ). [Adapted from National Climatic Data Center, *Monthly Climatic Data for the World* 46, no. 1, January and July 1993. Prepared in cooperation with the WMO and NOAA.]

ANIMATION Global Patterns of Pressure

SATELLITE Global Sea-level Pressure

Table 4.2 Four Hemispheric Pressure Areas

Name	Cause	Location	Air Temperature/Moisture
Polar high-pressure cells	Thermal	90° N 90° S	Cold/dry
Subpolar low-pressure cells	Dynamic	60° N 60° S	Cool/wet
Subtropical high-pressure cells	Dynamic	20° to 35° N and S	Hot/dry
Equatorial low-pressure trough	Thermal	10° N to 10° S	Warm/wet

altitude and consistent daylength make large amounts of energy available in this region throughout the year. The warming creates lighter, less dense, ascending air, with winds converging all along the extent of the trough. This converging air is extremely moist and full of latent heat energy. Vertical cloud columns frequently reach the tropopause, and precipitation is heavy throughout this zone. The combination of heating and convergence forces air aloft and forms the **intertropical convergence zone (ITCZ)**.

Figure 4.11 features a satellite image showing the equatorial low-pressure trough of the ITCZ and associated precipitation. A broken band of precipitation reveals this low-pressure system and its position slightly north of the equator in July. Note the position of the ITCZ on Fig. 4.10b and compare this with the precipitation pattern captured by the TRMM (Tropical Rainfall Measuring Mission) sensors on several satellites.

During summer, a marked wet season accompanies the shifting ITCZ over various regions. The maps in Figure 4.10 show the location of the ITCZ in January and July. In January, the zone crosses northern Australia and dips southward in eastern Africa and South America, whereas by July the zone has shifted northward in the Americas and as far north as Pakistan and southern Asia.

Figure 4.12 shows two views of this circulation and resulting *Hadley cells*, named for the eighteenth-century English scientist who described the *trade winds*. This equatorial low-pressure system of converging, ascending air generates these cells. The winds converging on the equatorial low-pressure trough are known as the *northeast trade winds* (in the Northern Hemisphere) and *southeast trade winds* (in the Southern Hemisphere), or generally as **trade winds**.

The trade winds pick up large quantities of moisture as they return through the Hadley circulation cell for another cycle of uplift and condensation. Within the ITCZ, winds are calm or mildly variable because of the even pressure gradient and the strong vertical ascent of air. Sixteenth-century sailors complained about getting caught here as being in the *doldrums*.

The rising air from the equatorial low-pressure area spirals upward into a geostrophic flow to the north and south. These upper-air winds turn eastward, flowing from west to east, beginning at about 20° N and S, forming descending anticyclonic flows above the subtropics. In each hemisphere, this circulation pattern appears most vertically symmetrical near the equinoxes of March and September.

Subtropical High-Pressure Cells: Hot, Dry Desert Air

Between 20° and 35° latitude in both hemispheres, a broad high-pressure zone of hot, dry air is evident across the globe (see Figures 4.10 and 4.12). The dynamic cause of these subtropical anticyclones is too complex to detail here, but generally they form as air in the Hadley cell descends in these latitudes, constantly supported (pushed down) by a dynamic mechanism in the upper-air circulation. As air above the subtropics is mechanically pushed downward, it is heated as it compresses on its descent to the surface, as illustrated in Figure 4.12. Warmer air has a greater capacity to absorb water vapor than does cooler air, making this descending warm air relatively dry. Heavy precipitation along the equatorial circulation also removes moisture from the air.

Surface air diverging from the subtropical high-pressure cells generates Earth's principal surface winds: the westerlies and the trade winds. The **westerlies** are the dominant surface winds from the subtropics to high latitudes. They diminish somewhat in summer and are stronger in winter.

If you examine the maps in Figure 4.10, you will find several high-pressure areas. In the Northern Hemisphere,

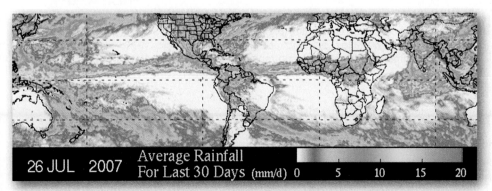

FIGURE 4.11 July precipitation accumulation portrays equatorial and subtropical circulation patterns. An interrupted band of average precipitation in millimeters per day for 30 days (June 27–July 26, 2007) along the intertropical convergence zone (ITCZ). Note in the subtropics to the north and south of the ITCZ several high-pressure systems and clear skies (lacking coloration). [Image from *TRMM* Microwave Imager on several geosynchronous satellites, courtesy of GSFC/NASA.]

26 JUL 2007 Average Rainfall For Last 30 Days (mm/d) 0 5 10 15 20

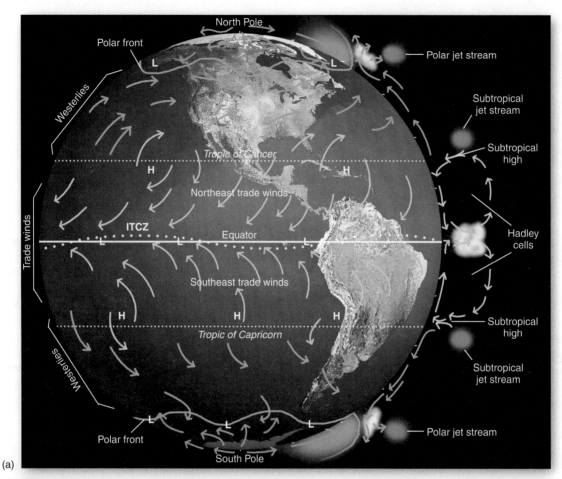

(a)

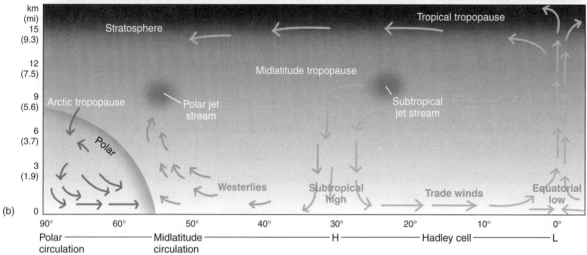

(b)

FIGURE 4.12 General atmospheric circulation model.
Two views of the general atmospheric circulation: (a) general circulation schematic; (b) equator-to-pole cross section for the Northern Hemisphere. Both views show Hadley cells, subtropical highs, the polar front, the subpolar low-pressure cells, and approximate locations of the subtropical and polar jet streams.

ANIMATION

Global Wind
Circulation,
Hadley Cells

SATELLITE

Global Infrared

the Atlantic subtropical high-pressure cell is called the *Bermuda high* (in the western Atlantic) or the *Azores high* (when it migrates to the eastern Atlantic). The Sargasso Sea area under this subtropical high features clear, warm waters and large quantities of *Sargassum* (a seaweed), which gives

the area its name. The *Pacific high*, or *Hawaiian high*, dominates the Pacific in July, retreating southward in January. In the Southern Hemisphere, three large high-pressure centers dominate the Pacific, Atlantic, and Indian Oceans and tend to move along parallels in shifting zonal positions.

The entire high-pressure system migrates with the summer high Sun, fluctuating about 5°–10° in latitude. The eastern (right-hand) sides of these anticyclonic systems are drier and more stable and feature cooler ocean currents than do the western sides. Thus, subtropical and midlatitude west coasts are influenced by the drier eastern side of these systems and therefore experience dry-summer conditions. In fact, Earth's major deserts generally occur within the subtropical belt and extend to the west coast of each continent except Antarctica. Sailing crews lamented the hot, dry, stable air and calms in the high-pressure cells in the era of sailing ships, calling them the *horse latitudes*. The *Oxford English Dictionary* is "uncertain" of this negative nickname's actual origin.

The western (left-hand) sides of subtropical high-pressure cells tend to be moist and unstable, overlying warm ocean currents, as characterized in Figure 4.13. This condition causes warm, moist conditions in Hawai'i, Japan, southeastern China, and the southeastern United States.

Subpolar Low-Pressure Cells: Cool and Moist Air

The January map in Figure 4.10 shows two low-pressure cyclonic cells over the oceans around 60° N latitude. The North Pacific *Aleutian low* is near the Aleutian Islands; the North Atlantic *Icelandic low* is near Iceland. Both cells are dominant in winter and weaken significantly or disappear altogether in the summer with the strengthening of high-pressure systems in the subtropics. The area of contrast between cold and warm air masses forms a contact zone

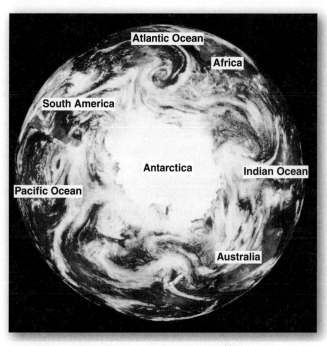

FIGURE 4.14 Clouds portray subpolar and polar circulation patterns.
Centered on Antarctica, this satellite image shows a series of subpolar low-pressure cyclones in the Southern Hemisphere. Antarctica is fully illuminated by a midsummer, December Sun. This natural-color image was made by the *Galileo* spacecraft during its 1990 flyby of Earth. [Image from the late Dr. W. Reid Thompson. Used by permission.]

known as the **polar front**, which encircles Earth at this latitude, concentrated in these low-pressure areas.

Figure 4.12a and b illustrate this confrontation between warmer, moist air from the westerlies and colder, drier air from the polar and Arctic regions. The heavier cold air displaces warmer air upward, forcing cooling and condensation in the lifted air. Low-pressure cyclonic storms migrate out of the Aleutian and Icelandic frontal areas and may produce precipitation in North America and Europe, respectively.

In the Southern Hemisphere, a noncontinuous belt of subpolar cyclonic pressure systems surrounds Antarctica. The spiraling cloud patterns produced by these cyclonic systems are visible on the satellite image in Figure 4.14. Severe cyclonic storms can cross Antarctica, producing strong winds and new snowfall, though small in amount.

Polar High-Pressure Cells: Frigid, Dry Deserts

The weakness of the polar high-pressure cell at each pole may argue against their inclusion among pressure areas on Earth. The polar atmospheric mass is small, receiving little energy to put it into motion. The variable cold, dry winds moving away from the polar region are anticyclonic, descending and diverging clockwise in the Northern Hemisphere (counterclockwise in the Southern Hemisphere) and forming the weak, variable winds of the **polar easterlies**.

Of the two polar regions, the **Antarctic high** is significant in terms of both strength and persistence. Less pronounced is the polar high-pressure cell in the Arctic. When

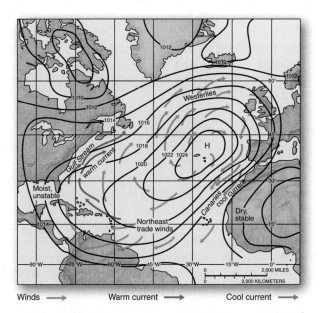

Winds ⟶   Warm current ⟶   Cool current ⟶

FIGURE 4.13 Subtropical high-pressure system in the Atlantic.
Characteristic circulation and climate conditions related to Atlantic subtropical high-pressure anticyclone in the Northern Hemisphere. Note the clockwise wind pattern and different ocean temperatures off each coast—cool currents and deserts in Africa, warm currents and humid conditions in the southeastern United States.

it does form, it tends to locate over the colder northern continental areas in winter (Canadian and Siberian highs) rather than directly over the relatively warmer Arctic Ocean.

Upper Atmospheric Circulation

Circulation in the middle and upper troposphere is an important component of the atmosphere's general circulation. Just as sea level is a reference datum for evaluating air pressure at the surface, we use a pressure level such as 500 mb as a **constant isobaric surface** for a pressure-reference datum in the upper atmosphere.

On upper-air pressure maps, we plot the height above sea level at which an air pressure of 500 mb occurs. The isobaric chart and illustration for an April day in Figure 4.15a and b illustrate such an undulating isobaric surface, upon which all points have the same pressure. In contrast, on surface weather maps we plot different pressures at the fixed elevation of sea level—a *constant height surface*.

We use this 500-mb level to analyze upper-air winds and the possible support they provide for surface weather conditions. Similar to surface maps, closer spacing of the isobars indicates faster winds; wider spacing indicates slower winds. On this constant isobaric pressure surface, altitude variations from the reference datum are called *ridges* for high pressure (with isobars on the map bending poleward) and *troughs* for low pressure (with isobars on the map bending equatorward). Looking at the figure, can you identify such ridges and troughs in the isobaric surface?

The pattern of ridges and troughs in the upper-air wind flow is important in sustaining surface cyclonic (low-pressure) and anticylonic (high-pressure) circulation. Near ridges in the isobaric surface, winds slow and converge (pile up), whereas winds near the area of maximum wind speeds along the troughs in the isobaric surface accelerate and diverge (spread out). Note the wind-speed indicators and labels in Figure 4.15a near the ridge (over Alberta, Saskatchewan, Montana, and Wyoming); now compare these with the wind-speed indicators around the trough (over Kentucky, West Virginia, the New England states, and the Maritimes). Also, note the wind relationships off the Pacific Coast.

Now look at Figure 4.15c. Divergence in the upper-air flow is important to cyclonic circulation at the surface because it creates an outflow of air aloft that stimulates an inflow of air into the low-pressure cyclone (such as what happens when you open a chimney damper to create an upward draft). Convergence aloft, on the other hand, drives descending airflows and divergent winds at the surface, moving out from high-pressure anticyclones.

Rossby Waves Within the westerly flow of geostrophic winds are great waving undulations called **Rossby waves**, named for meteorologist C. G. Rossby, who first described them mathematically. The polar front is the contact between colder air to the north and warmer air to the south (Figure 4.16). The Rossby waves bring tongues of cold air southward, with warmer tropical air moving northward. The development of Rossby waves in the upper-air circulation is

shown in the figure. As these disturbances mature, distinct cyclonic circulations form, with warmer air and colder air mixing along distinct fronts. These wave-and-eddy formations and other upper-air flows support the development of cyclonic storm systems at the surface and develop in relation to jet-stream flows. An experiment in 1998 inadvertently provided confirming information on the flow of the jet stream; see News Report 4.1 for details.

Jet Streams The most prominent movement in these upper-level westerly wind flows is the **jet stream**, an irregular, concentrated band of wind occurring at several different locations. Rather flattened in vertical cross section, the jet streams normally are 160−480 km (100−300 mi) wide by 900−2150 m (3000−7000 ft) thick, with core speeds that can exceed 300 kmph (190 mph).

The *polar jet stream* is located at the tropopause along the polar front, at altitudes between 7600 and 10,700 m (24,900−35,100 ft), meandering between 30° and 70° N latitude. Shifting of daily and seasonal locations supports surface weather patterns. The polar jet stream can migrate as far south as Texas, steering colder air masses into North America and influencing surface storm paths traveling eastward. In the summer, the polar jet stream exerts less influence on storms by staying far poleward. Figure 4.17 (p. 130) shows a stylized view of a polar jet stream and two jet streams over North America. See News Report 4.2 about the jet stream and its impact on airline schedules.

In subtropical latitudes, near the boundary between tropical and midlatitude air, another jet stream flows near the tropopause. This *subtropical jet stream* ranges from 9100 to 13,700 m (29,850−45,000 ft) in altitude, and although it is generally weaker, it can reach greater speeds than those of the polar jet stream. The subtropical jet stream meanders from 20° to 50° latitude and may occur over North America simultaneously with the polar jet stream. The cross section of the troposphere in Figure 4.12b depicts the general location of these two jet streams.

Multiyear Oscillations in Global Circulation

Several system fluctuations that occur in multiyear or shorter periods are important in the global circulation picture, yet an understanding of them is just emerging. The most famous of these is the El Niño–Southern Oscillation (ENSO) phenomenon discussed later in Chapter 7. Multiyear oscillations affect temperatures, air masses, and air pressure patterns and thus affect global winds and climates. Here we briefly overview three of these hemisphere-scale oscillations.

North Atlantic Oscillation A north–south fluctuation of atmospheric variability marks the *North Atlantic Oscillation* (NAO) as pressure differences between the Icelandic low and the Azores high in the Atlantic alternate in strength. The *NAO Index* is in its *positive phase* when the Icelandic low-pressure system is lower in pressure than normal and the Azores (west of Portugal) high-pressure

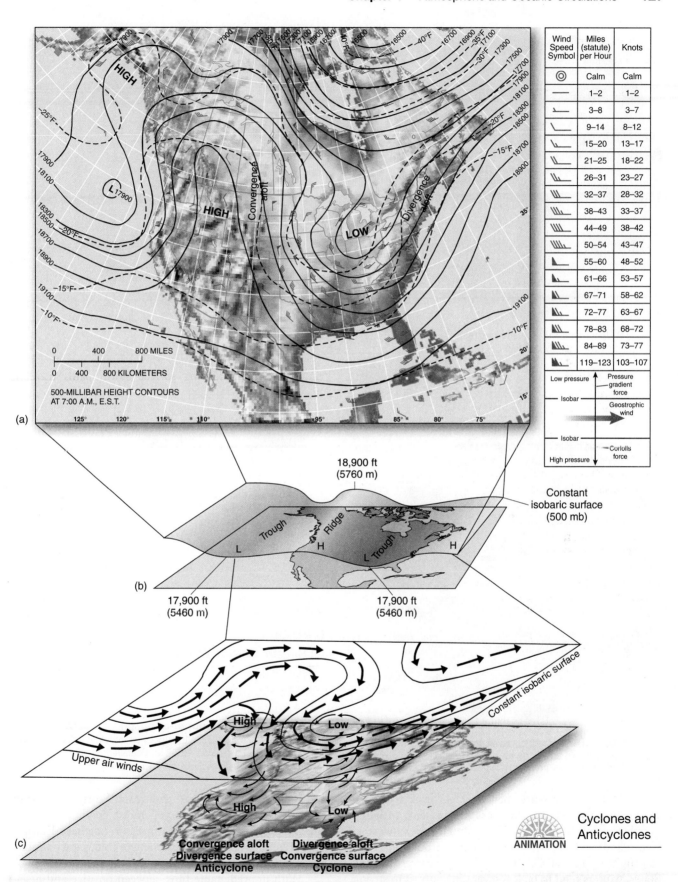

Wind Speed Symbol	Miles (statute) per Hour	Knots
◎	Calm	Calm
—	1–2	1–2
∟	3–8	3–7
∟	9–14	8–12
∟	15–20	13–17
∟	21–25	18–22
∟	26–31	23–27
∟	32–37	28–32
∟	38–43	33–37
∟	44–49	38–42
∟	50–54	43–47
∟	55–60	48–52
∟	61–66	53–57
∟	67–71	58–62
∟	72–77	63–67
∟	78–83	68–72
∟	84–89	73–77
∟	119–123	103–107

FIGURE 4.15 Analysis of a constant isobaric surface for an April day.
(a) Contours show elevation (in feet) at which 500-mb pressure occurs—a constant isobaric surface. The pattern of contours reveals a 500-mb isobaric surface and geostrophic wind patterns in the troposphere ranging from 16,500 to 19,100 ft elevation. (b) Note on the map and in the sketch beneath the chart the "ridge" of high pressure over the Intermountain West, at 5760 m altitude, and the "trough" of low pressure over the Great Lakes region and off the Pacific Coast, also at 5460 m altitude. (c) Note areas of convergence aloft (corresponding to surface divergence) and divergence aloft (corresponding to surface convergence)—upper-atmosphere conditions at the 500-mb level support surface cyclones and anticyclones. [Data adapted for map in (a) from the National Weather Service, NOAA.]

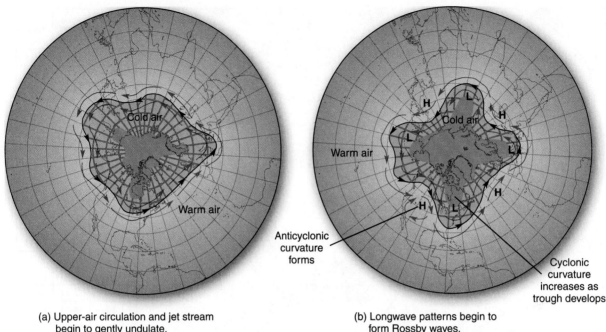

(a) Upper-air circulation and jet stream begin to gently undulate.

(b) Longwave patterns begin to form Rossby waves.

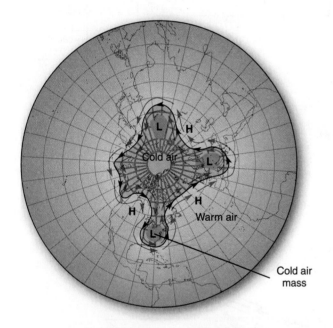

(c) Strong development of waves produces cells of cold and warm air—high-pressure ridges and low-pressure troughs.

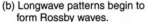

Warm air · Cool air

Polar stereographic projections azimuthal and conformal

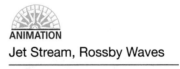

ANIMATION

Jet Stream, Rossby Waves

FIGURE 4.16 Rossby upper-atmosphere waves.
Development of longwaves in the upper-air circulation first described by C. G. Rossby in 1938 and detailed by J. Namias in 1952. [Adapted from J. Namias, NOAA.]

cell is higher in pressure than normal. Strong westerly winds and the jet stream cross the eastern Atlantic. Northerly winds blow in the Labrador Sea (west of Greenland) and southerly winds move through the Norwegian Sea (east of Greenland). In the eastern United States, winters tend to be less severe in contrast to strong, warm, wet storms hitting northern Europe; however, the Mediterranean region is dry.

In its *negative phase* the NAO features a weaker pressure gradient than normal between Iceland and the Azores and reduced westerlies and jet stream. Storm tracks shift southward in Europe, bringing moist conditions to the Mediterranean and cold, dry winters to northern Europe. The eastern United States experiences cold, snowy winters as Arctic air masses plunge to lower latitudes.

With unpredictable flips between positive and negative phases, sometimes changing from week to week, a trend to a more positive phase than negative has emerged since 1960. Since 1980, the NAO has been predominantly positive. (See http://www.ldeo.columbia.edu/NAO/.)

MANTRA's Big Balloon—Blowing in the Wind

A 25-story balloon named MANTRA (Middle Atmosphere Nitrogen Trend Assessment) launched from Vanscoy, Saskatchewan, August 24, 1998. The unpiloted mission was intended to measure concentrations of human-made chemicals and ozone in the stratosphere using a sophisticated instrument array.

After its 3:25 A.M. liftoff, the balloon rose to a planned 38-km altitude and collected valuable data throughout the day. Upon mission completion, explosive charges were designed to release the instrument payload and a parachute for a safe landing in nearby Manitoba. However, a malfunction left the payload attached to the balloon, adrift in the stratosphere in the tow of upper wind streams. An unexpected voyage was underway.

The balloon's path traced upper-atmospheric wind circulation and the flowing polar jet stream across Manitoba, northern Ontario and Québec, Labrador, the tip of Newfoundland, and out over the Atlantic (Figure 4.1.1). Realizing the potential threat to commercial airplanes, Canadian military jets were dispatched to shoot MANTRA down. Thousands of rounds of ammunition were fired into the balloon, to no avail because MANTRA was a zero-pressure balloon.

The helium was not under pressure in the plastic, so the holes created by the ammunition merely produced slow leaks. American and British military radar tracked MANTRA's contin-

FIGURE 4.1.1 **Flight path of MANTRA 1998's big balloon.** Westerly winds carried the balloon east from Vanscoy, Saskatchewan, over Manitoba, and on to eastern Canada. The path turned northeastward and passed between Iceland and Greenland. It then veered northward toward the Svalbard Islands and then south to Russia. Finally, the balloon headed west to land in Finland. [Reference: Minister of Public Works and Government Services Canada; Natural Resources Canada, Geological Survey of Canada.]

uing voyage. Imagine this balloon adrift in the Rossby waves depicted in Figure 4.16.

The upper winds drove MANTRA northeastward over the Denmark Straits between Greenland and Iceland, across the Norwegian Svalbard Islands, poleward of 80° N, and over the Russian Franz Josef Land and Novaya Zemlya islands groups. Then winds took the balloon south over the Russian coast and westward

across Finland. The depleted balloon began its descent September 2, after a 9-day odyssey, eventually settling to Earth in Mariehamn, a city on Aland Island, in the Gulf of Bothnia. Finnish meteorologists gathered the payload and shipped it back to Canada. MANTRA's 1998 unscheduled voyage to Europe dramatically demonstrated the course of the polar jet stream within Earth's upper-atmospheric circulation.

Jet Streams Affect Flight Times

As airplane travel increased during World War II, flight crews reported strong headwinds on routes to the west. In some cases, airplanes leaving San Francisco for the Pacific were turned back by opposing wind in the upper troposphere. These were the jet streams.

Next time you plan a flight, note that airline schedules reflect the presence of these upper-level westerly

winds, for they allot shorter flight times from west to east and longer flight times from east to west. We recently flew round-trip between Sacramento, California, and Boston, Massachusetts, and experienced a difference of 1 hour and 15 minutes between shorter eastbound (tailwind) and longer westbound (headwind) flights—a difference caused by a particularly strong jet stream.

Also important to both military and civilian aircraft is the effect jet streams have on fuel consumption and the existence of air turbulence. Seasonal adjustments are necessary because the jet streams in both hemispheres tend to weaken during each hemisphere's summer and strengthen during winter as the streams shift closer to the equator.

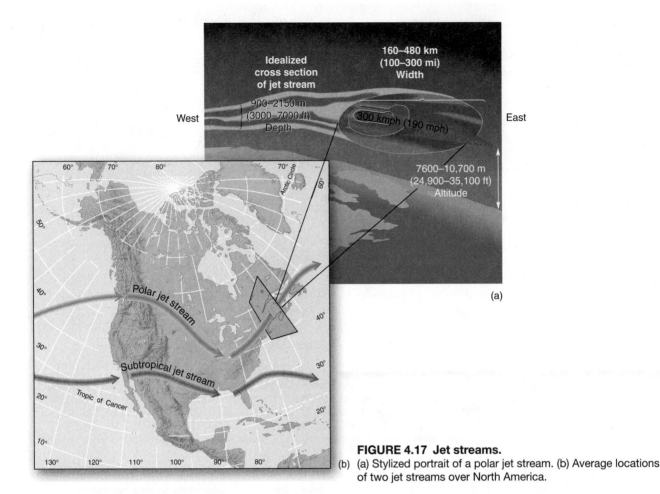

FIGURE 4.17 Jet streams.
(a) Stylized portrait of a polar jet stream. (b) Average locations of two jet streams over North America.

Arctic Oscillation Variable fluctuations between middle- and high-latitude air mass conditions over the Northern Hemisphere produce the *Arctic Oscillation* (AO). The AO is associated with the NAO especially in winter, and the two phases of the *AO Index* correlate to the two-phased *NAO Index*. In the *warm phase* of the AO Index (positive NAO), the pressure gradient is affected by lower pressure than normal over the North Pole region and relatively higher pressures at lower latitudes. This sets up stronger westerly winds and warmer Atlantic water currents into the Arctic Ocean. Cold air masses do not migrate as far south in winter, whereas winters are colder than normal in Greenland.

In the *cold phase* of the AO Index (negative NAO), the pattern is reversed, with higher-than-normal pressure over the North Pole region and relatively lower pressure in the central Atlantic. Here the air masses of winter bring cold conditions to the eastern United States, northern Europe, and Asia, and sea ice in the Arctic Ocean becomes a bit thicker. Greenland is warmer than normal.

Since the 1980s, the NAO and AO indices are in a strong positive, warm phase. In response, the ice drift in the eastern Arctic Ocean is counterclockwise and air pollution from Europe and Russia moves across the region to Canada. Since the expected variability in these oscillations is essentially random, the emergence of a strong positive, warm phase is challenging to understand. Scientists are examining correlations with changes in sea-

surface temperatures to understand what is happening and perhaps develop a forecast capability. For instance, researchers have discovered linkages between the behavior of the NAO and the increasingly warm Indian Ocean.

Since ocean temperatures are warming, given the present trends of freshening in the northern ocean (reduced salinity) and increasing salinity in the tropical ocean, there is concern about a further strengthening of these oscillations and what effects this might have on weather and climate. Such is the excitement of physical geography, as global connections are found linking diverse physical systems (see **http://nsidc.org/arcticmet/patterns/arctic_oscillation.html**).

Pacific Decadal Oscillation Across the Pacific Ocean the *Pacific Decadal Oscillation* (PDO) is longer-lived, at 20 to 30 years in duration, than the 2- to 12-year variation in the ENSO. The PDO term came into use in 1996. Involved are two regions of sea-surface temperatures and related air pressure: the northern and tropical western Pacific (region #1) and the area of the eastern tropical Pacific, along the West Coast (region #2).

Between 1947 and 1977, higher-than-normal temperatures dominated region #1 and lower temperatures were in region #2; this is the PDO *negative phase* (or cool phase). A switch to a *positive phase* (or warm phase) in the PDO ran from 1977 to the 1990s, when lower temperatures were in region #1 and higher-than-normal temperatures dominated region #2, coinciding with a time of more intense ENSO events. In 1999, a switch to a *negative phase* began. Unfortunately for the US Southwest, this PDO negative phase can mean a decade or more of drier conditions for the already drought-plagued region.

Causes of the PDO and its cyclic variability over time are unknown. Scientists monitor conditions in the Pacific and look for patterns. A better understanding of the PDO will help scientists predict ENSO events as well as regional drought cycles. (See http://topex-www.jpl.nasa.gov/science/pdo.html.)

Local Winds

Several winds that form in response to local terrain deserve mention in our discussion of atmospheric circulation. These local effects can, of course, be overwhelmed by weather systems passing through an area.

Land-sea breezes occur on most coastlines (Figure 4.18). The different heating characteristics of land and water surfaces create these winds (Chapter 3). During the day, land heats faster and becomes warmer than the water offshore. Because warm air is less dense, it rises and triggers an onshore flow of cooler marine air, usually stronger in the afternoon. At night, inland areas cool (radiate heat) faster than offshore waters. As a result, the cooler air over the land subsides and flows offshore over the warmer water, where the air lifts. This night pattern reverses the process that developed during the day.

Well inland from the Pacific Ocean—160 km (100 mi)—Sacramento, California, demonstrates the sea-breeze effect. The city is at 39° N and 5 m (17 ft) elevation. The average July maximum (day) and minimum (night) temperatures are 34°C and 14°C (93°F and 57°F). The evening cooling by the natural flow of marine air establishes a monthly mean of only 24°C (75°F), despite high daytime temperatures.

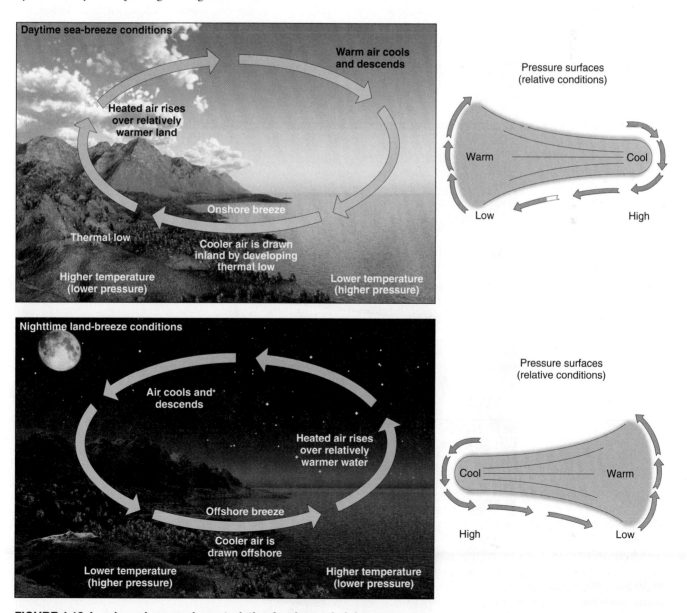

FIGURE 4.18 Land-sea breeze characteristics for day and night.

A somewhat similar exchange creates **mountain-valley breezes**. Mountain air cools rapidly at night, and valley air heats rapidly during the day (Figure 4.19). Thus, warm air rises upslope during the day, particularly in the afternoon; at night, cooler air subsides downslope into the valleys. Chapter 5 discusses another type of downslope wind.

Katabatic winds, or gravity drainage winds, are significant on a larger regional scale than mountain-valley breezes, under certain conditions. These winds usually are stronger than mountain-valley winds. An elevated plateau or highland is essential for katabatic wind, where layers of air at the surface cool, become denser, and flow downslope. The ferocious winds that can blow off the ice sheets of Antarctica and Greenland are katabatic in nature.

Worldwide, a variety of terrains produce such winds and bear many local names. The *mistral* of the Rhône Valley in southern France can cause frost damage to vineyards as the cold north winds move over the region to the Gulf of Lions and the Mediterranean Sea. The frequently stronger *bora*, driven by the cold air of winter high-pressure systems inland, flows across the Adriatic Coast to the west and south. In Alaska such winds are called the *taku*.

Santa Ana winds are another local wind type, generated by high pressure over the Great Basin of the western United States. A strong, dry wind is produced that flows out across the desert to southern California coastal areas. The air is heated by compression as it flows from higher to lower elevations, and with increasing speed it moves through constricting valleys to the southwest. These winds irritate with their dust, dryness, and heat. Powerful Santa Ana winds were in force in southern California during the record wildfire seasons during 2007 and 2008, blowing smoke far into the Pacific (see Figure 2.24a).

Wind represents a substantial source of renewable energy. Focus Study 4.1 briefly explores the rapid growth and potential for development of wind resources.

Monsoonal Winds Regional wind systems that seasonally change direction are important in some areas. Examples occur in the tropics over Southeast Asia, Indonesia,

(text continued on page 135)

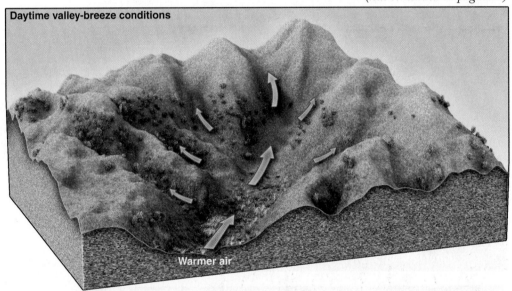

Daytime valley-breeze conditions

Warmer air

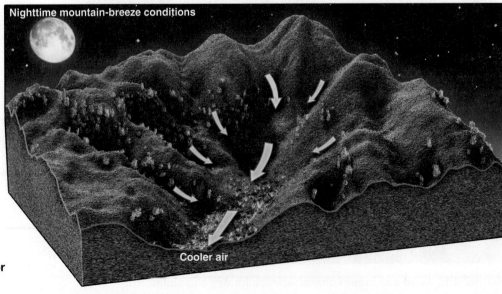

Nighttime mountain-breeze conditions

Cooler air

FIGURE 4.19 Patterns of mountain-valley breeze for day and night.

FOCUS STUDY 4.1

Wind Power: An Energy Resource for the Present and Future

The principles of wind power are ancient, but the technology is modern and the benefits are worth pursuing. In more-developed countries, energy sources are dominated by the use of nonrenewable fuels—coal, gas, oil, nuclear—and centralized energy production. In less-developed countries, however, many rely on renewable energy—small hydroelectric plants, wind-power systems, diminishing wood supplies, and solar energy—for cooking, heating, and pumping water. These resources are considered renewable because they are not depleted in the span of a human lifetime.

The Nature of Wind Energy

Power generation from wind is site-specific, requiring regions with adequate winds. Electricity is generated at wind farms, by groups of wind machines, or in individual installations. Figure 4.1.1 and the chapter-opening photos show such wind farms in California, Kansas, and Scotland. Several basic settings favor wind resources: (1) along coastlines influenced by trade winds and westerly winds; (2) where mountain passes constrict air flow and interior valleys develop thermal low-pressure areas, thus drawing air across the landscape; or (3) where localized winds occur, such as an expanse of relatively flat prairies or in areas with katabatic or monsoonal winds. Cash-short developing countries are generally located in areas blessed by such steady winds, such as the trade winds across the tropics. Where wind is reliable less than 25%–30% of the time, only small-scale uses are economically feasible.

The potential of wind power in the United States is enormous. For example, the winds of North and South Dakota and Texas could meet all US electrical needs; such investments would benefit rural regions of the country. In the California Coast Ranges, because of land-sea breezes

FIGURE 4.1.1 Installed wind-generating capacity.
(a) Wind farm in the Cajon Pass of southern California, along Interstate 15. (b) The Gray County Wind Farm, 40 km (25 mi) west of Dodge City, near Montezuma, Kansas, generates 112 MW of electricity. All the towers and roads combined occupy only 6 acres of land, allowing the balance of 12,000 acres to stay in agricultural production. These towers and turbines are smaller than the ones in the chapter-opening photo at 65 m (214 ft) tall. (c) Burger Hill Wind Farm at Birsay Moors, in the Orkney Islands, Scotland, includes the largest single wind turbine in the UK, at 2 MW size. [All photos by Bobbé Christopherson.]

(continued)

Focus Study 4.1 *(continued)*

between the Pacific and Central Valley, peak winds blow from April to October, matching peak electrical demands.

On the eastern shore of Lake Erie sits a closed Bethlehem Steel mill, an environmental clean-up mess, and eight 2.5-MW (megawatt) wind-power turbines (Figure 4.1.2). This redeveloping *brownfield* site is supplying the equivalent of 75% of the electricity for Lackawanna, New York. Twelve turbines will soon be added, making this a 50-MW electrical generation facility, the largest urban installation in the country. Plans for a surrounding park and other development are in the works. This former steel town is using wind power to lift itself out of an economic depression, calling itself a "Green city." Similar brownfield wind farms are in operation in Pennsylvania.

Wind-power development is enhanced by the income it brings. Farmers in Iowa and Minnesota receive $2000 in annual electrical production income from a leased turbine and $20,000 a year from an owned turbine—requiring only one-quarter acre to site the wind machine. The corn harvested annually from an acre is worth approximately $100. Farmers are exploring this income source with assistance from the American Corn Growers Association (ACGA) and are forming wind-energy cooperatives. The limitation seems to be a shortage of transmission-line capacity from rural sites to cities (see the ACGA report "2001 ACGA Corn Producers Survey Wind Energy & Climate Change" at http://www.acga.org/). The Midwest is on the brink of an economic boom if this wind-energy potential is developed and transmission-line capacity installed.

Wind-Power Status

Wind-generated energy resources are the fastest-growing energy technology, in terms of new capacity—capacity has risen worldwide more than 700% since 1998. Total world capacity exceeded 75,000 MW (75,000 million watts) by

FIGURE 4.1.2 Wind farms in Lackawanna, a new "Green" city.
Known as "The Steel Wind," eight turbines deliver 20 MW of wind power in Lackawanna, New York, on the eastern shore of Lake Erie, south of Buffalo. The former industrial site is to expand to 50 MW generated by 20 2.5-MW turbines. [Photo by Bobbé Christopherson.]

the end of 2006—growing at more than 30% for each of these past 3 years. Globally, the wind-power industry employs more than 100,000 people, with global investments passing the $18-billion USD mark in 2006. Average-size wind farms exceed 100 MW.

Most of the jobs and about 80% of the industry are focused in Europe. Germany now produces almost 7% of its electricity from wind power, and in three of its states more than 30%. The European Wind Energy Association announced installed capacity exceeding 48,000 MW at the end of 2006, more than the worldwide total in 2004; see http://www.ewea.org/. Fifty percent of this capacity is in two countries, Germany and Spain, followed by Denmark, France, Portugal, and the UK. The European Union has a goal of 20% of all energy from renewable sources by 2020.

The United States is third behind Germany and Spain. In 2006, the United States had 11,600 MW installed capacity, adding 5244 MW through 2007, a 45% increase in a single year—16,844 MW total now operating in 34 states. The American Wind Energy Association (http://www.awea.org) reports the impact of this US installed capacity:

- Protects consumers from volatile fossil-fuel prices for fuels used to generate electricity.
- Improved air quality, reduced damage from oil spills and coal

mining, and improved balance of trade.
- Reduces global warming emissions, avoiding some 28 million tons of carbon dioxide emitted annually, or the equivalent of taking almost 5 million cars off the highways.
- Conserves water since wind power does not require cooling water. Steam-electric power use about 48% of water withdrawals in the United States.

AWEA reports the five leading states for installed wind-power capacity are: Texas (4356 MW), California (2439 MW), Minnesota (1299 MW), Iowa (1273 MW), and Washington (1163 MW). In the West Texas region, there is an economic boom underway linked to the wind-power revolution. This flow of capital into America's rural regions is significant because of the generally poor economic conditions present and the depopulation trend. The wind industry's labor intensity and diffuse tax and profit advantages benefit more people. In Nolan County, Texas, as of mid-2008, 11% of the labor force were employed manufacturing, transporting, maintaining, and repairing wind turbines.

To put numbers in meaningful perspective, every 10,000 MW of wind-generation capacity reduces carbon dioxide emissions by 33 million metric tons if it replaces coal, or 21 million metric tons if it replaces mixed fossil fuels. As an example, if countries rally and create a proposed $600-billion-dollar industry by installing 1,250,000 MW of wind capacity by 2020, that would supply 12% of global electrical need. Note that this assumes a 65% increase in electricity demand by that date. If realized, this would avoid substantial carbon dioxide emissions. *Wind Force 12* is the name of this EWEA plan.

The Benefits of the Wind Resource

Because of the economic benefits from using wind resources, especially where peak winds are in concert with peak electrical demand for air-cooling, space heating, or agricultural water pumping, economic reality should override further delaying actions.

In Europe, smaller turbines produce at 4 cents per kWh. Two Stanford University engineers calculated that the contracted price for wind power of 3 cents/kWh in the United States is cheaper than coal, when considering all costs. Some coal-cost estimates exclude consideration of the health impacts of coal-mine dust, acid deposition, smog, and global warming. (See M. Z. Jacobsen and G. M. Masters, "Exploiting Wind vs. Coal," *Science* 293 (August 24, 2001): 1438. Also see American Wind Energy Association, "Wind Power Outlook," annual publication.)

Long-term costs favor wind-energy deployment. With *all* costs considered, wind energy is cost-competitive and actually cheaper than oil, coal, natural gas, and nuclear power. Relative to concerns about wind turbines hurting birds, a survey studied human-caused bird mortality and found wind turbines are lowest when compared to communication towers, pesticides, high-tension lines, domestic cats, and collisions with buildings and windows.

Whether or not governments and transnational energy corporations support full-scale implementation of renewable resources such as wind, the energy realities of the near future leave no alternative, and global political realities affirm this urgency. By the middle of this century, wind-generated electricity could be routine, along with other renewable energy sources, conservation, and energy efficiency.

India, northern Australia, equatorial Africa, and southern Arizona. These winds involve an annual cycle of returning precipitation with the summer Sun and are named after the Arabic word for season, *mausim*, or **monsoon**.

The monsoons of southern and eastern Asia (Figure 4.20) are driven by the location and size of the Asian landmass and its proximity to the Indian Ocean. Also important to the generation of monsoonal flows are wind and pressure patterns in the upper-air circulation.

The extreme temperature range from summer to winter over the Asian landmass is due to its continentality, reflecting its isolation from the modifying effects of the ocean. An intense high-pressure anticyclone dominates the continental landmass in winter (Figure 4.10a), whereas the equatorial low-pressure trough dominates the central area of the Indian Ocean. Resultant cold, dry winds blow from the Asian interior over the Himalayas, downslope, and across India, producing average temperatures between 15° and 20°C (60° and 68°F) at lower elevations. These dry winds desiccate (dehydrate) the landscape and then give way to hot weather from March through May.

During the June–September wet period, the subsolar point shifts northward to the Tropic of Cancer, near the mouths of the Indus and Ganges Rivers. Note the precipitation distribution on the inset climograph for Nagpur, India, in Figure 4.20. The intertropical convergence zone (ITCZ) shifts northward over southern Asia, and the Asian continental interior develops a thermal low pressure, associated with high average temperatures. Meanwhile, the Indian Ocean, with a surface temperature of 30°C (86°F), is under the influence of subtropical high pressure. As a result, hot, dry subtropical air sweeps over the warm ocean, producing extremely high evaporation rates (Figure 4.20b).

By the time this mass of air reaches India and the ITCZ, it is laden with moisture in thunderous, dark clouds. The warmth of the land lends additional lifting to the incoming air, as do the Himalayas, forcing the air mass to higher altitudes. These conditions produce the wet monsoon of India, where world-record rainfalls have occurred. When the monsoonal rains arrive in June, they are welcome relief from the dust, heat, and parched land of Asia's springtime.

The fact that the monsoons of southern Asia involve vast global pressure systems leads us to ask whether future climate change might affect present monsoonal patterns. Researchers are conducting a field experiment in the Indian Ocean, the Indian Ocean Experiment (INDOEX), to examine this question (http://www-indoex.ucsd.edu/ProjDescription.html). Concern increases following events such as on July 27, 2005, when a monsoonal deluge drowned Mumbai (Bombay)—in only a few hours, 94.2 cm (37.1 in.) of rain fell, causing widespread flooding. Again in August 2007 and 2008, extensive flooding from intense monsoons struck across India repeatedly, causing widespread loss of life.

Oceanic Currents

Earth's atmospheric and oceanic circulations are closely interrelated. The driving force for ocean currents is the frictional drag of the winds. Also important in shaping these currents is the interplay of the Coriolis force, density differences caused by temperature and salinity of the water, the configuration of the continents and ocean floor, and astronomical forces (the tides).

Surface Currents

The general patterns of major ocean currents are shown in Figure 4.21. Coriolis force deflects ocean currents because they flow over distance and through time. However, their pattern of deflection is not as tightly circular as that of the

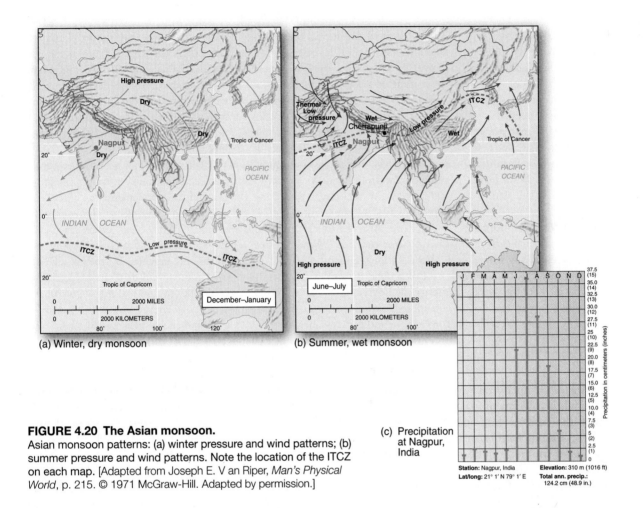

FIGURE 4.20 The Asian monsoon.
Asian monsoon patterns: (a) winter pressure and wind patterns; (b) summer pressure and wind patterns. Note the location of the ITCZ on each map. [Adapted from Joseph E. Van Riper, *Man's Physical World*, p. 215. © 1971 McGraw-Hill. Adapted by permission.]

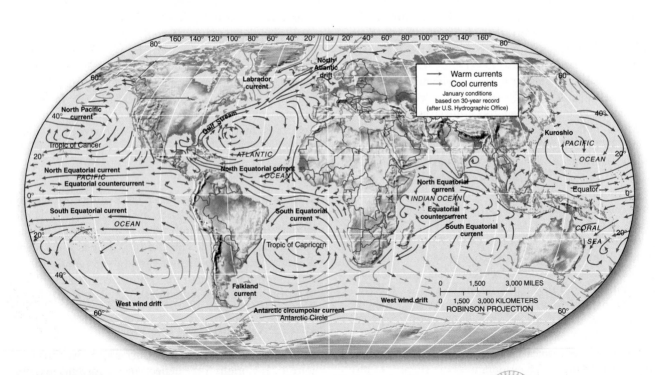

FIGURE 4.21 Major ocean currents.
[After the US Navy Hydrographic Office.]

ANIMATION   Ocean Circulation

atmosphere. Compare this map with that of Earth's pressure and wind systems (Figure 4.10), and you can see that ocean currents are driven by the circulation around subtropical high-pressure cells in both hemispheres. (Remember: In the Northern Hemisphere, winds and ocean currents move *clockwise* about high-pressure cells—note the currents in the North Pacific and North Atlantic on the map. In the Southern Hemisphere, the *counterclockwise* circulation about a high is evident on the map.) News Report 4.3 dramatically portrays this clockwise flow around the Pacific Ocean basin in the Northern Hemisphere. These circulation systems are known as *gyres*.

Along the full extent of ocean areas adjoining the equator, trade winds drive the oceans westward in a concentrated channel. These currents are kept near the equator by a Coriolis force influence that weakens near the equator. As the surface current approaches the western margins of the oceans, the water actually piles up an average of 15 cm (6 in.). This phenomenon is the **western intensification**. From this western edge, ocean

NEWS REPORT 4.3

A Message in a Bottle and Rubber Duckies

To give you an idea of the dynamic circulation of the ocean, consider the following little drama. A 9-year-old child at Dana Point, California (33.5° N), a small seaside community south of Los Angeles, placed a letter in a glass juice bottle in July 1992 and tossed it into the waves. Thoughts of distant lands and fabled characters filled the child's imagination as the bottle disappeared. The vast circulation around the Pacific high, a clockwise-circulating gyre, took command (Figure 4.3.1).

Three years passed before ocean currents carried the bottle to the coral reefs and white sands of Mogmog, a small island in Micronesia (7° N). A 7-year-old child there found a pen pal from afar and immediately mailed a photo and card to the sender of the message. Imagine the journey of that note from California—traveling through storms and calms, clear moonlit nights and typhoons—as it floated on ancient currents as the Spanish galleons once had.

In January 1994, a powerful storm ravaged a large container ship from Hong Kong loaded with toys and other goods. One of the containers on board split apart in the wind off the coast of Japan, dumping nearly 30,000 rubber ducks, turtles, and frogs into the North Pacific. Westerly winds and the North Pacific current swept this floating cargo across the ocean to the coasts of Alaska, Canada, Oregon, and California. Other toys, still adrift, went through the Bering Sea and into the Arctic Ocean (this route is indicated by the dashed line in Figure 4.3.1). These toys will eventually end up in the Atlantic Ocean as they drift around the Arctic Ocean frozen in pack ice.

The high-floating message bottle and rubber ducks offered a much better opportunity to study winds than did an earlier spill of 60,000 athletic shoes near Japan. The low-floating shoes were tracked across the Pacific Ocean until they landed in the Pacific Northwest. (By the way, the shoes that did not make landfall headed, or footed, back around the Pacific gyre into the tropics and westward, back toward Japan!) Scientists are using incidents such as these to learn more about wind systems and ocean currents.

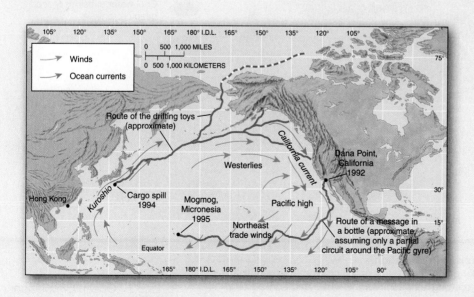

FIGURE 4.3.1 Pacific Ocean currents transport artifacts. The approximate route of a message in a bottle from Dana Point, California, to Mogmog, Micronesia, assumes a partial circuit around the Pacific gyre. Given the 3-year travel time, it is possible that the message circumnavigated the Pacific Ocean more than once. The rubber duckies voyaged eastward across the Pacific and beyond.

water then spills northward and southward in strong currents.

In the Northern Hemisphere, the *Gulf Stream* and the *Kuroshio* move forcefully northward with their speed and depth increased by the constriction of the area they occupy. The warm, deep-blue water of the ribbonlike Gulf Stream usually is 50–80 km (30–50 mi) wide and 1.5–2.0 km (0.9–1.2 mi) deep, moving at 3–10 kmph (1.8–6.2 mph). In 24 hours, ocean water can move 70–240 km (40–150 mi) in the Gulf Stream, although a complete circuit around an entire gyre may take a year.

Upwelling and Downwelling Flows Where surface water is swept away from a coast, either by surface divergence (induced by the Coriolis force) or by offshore winds, an **upwelling current** occurs. This cool water generally is nutrient-rich and rises from great depths to replace the vacating water. Such cold upwelling currents exist off the Pacific coasts of North and South America and the subtropical and midlatitude west coast of Africa. These areas are some of Earth's prime fishing regions.

In other regions where there is an accumulation of water—such as the western end of an equatorial current, or the Labrador Sea, or along the margins of Antarctica—the excess water gravitates downward in a **downwelling current**. These are the deep currents that flow vertically as well as along the ocean floor and travel the full extent of the ocean basins, carrying heat energy and salinity.

Thermohaline Circulation—The Deep Currents
Differences in temperatures and salinity produce density differences important to the flow of deep currents. This is Earth's **thermohaline circulation**, and it is different from wind-driven surface currents. Traveling at slower speeds

than surface currents, the thermohaline circulation hauls larger volumes of water. Figure 13.2 illustrates the ocean's physical structure and profiles of temperature, salinity, and dissolved gases; note the temperature and salinity differences with depth in this illustration.

To picture such a deep current, imagine a continuous channel of water beginning with cold water downwelling in the North Atlantic and upwelling in the Indian Ocean and North Pacific (Figure 4.22). Here it warms and then is carried in surface currents back to the North Atlantic. A complete circuit of the current may require 1000 years from downwelling in the Labrador Sea off Greenland to its reemergence in the southern Indian Ocean and to its return. Even deeper Antarctic bottom water flows northward in the Atlantic Basin beneath these currents.

These current systems appear to play a profound role in global climate; in turn, global warming has the potential to disrupt the downwelling in the North Atlantic and the thermohaline circulation. If this sinking of surface waters slowed or stopped, the ocean's ability to redistribute absorbed heat energy would be disrupted and surface changes of heat energy would alter existing climate patterns.

There is a freshening of ocean surface waters in both polar regions, contrasted with large increases in salinity in the surface waters at lower latitudes. Increased rates of glacier, sea-ice, and ice-sheet melting produce these fresh, lower-density surface waters that ride on top of the denser saline water. The concern is that such changes in ocean temperature and salinity could dampen the rate of deep-water downwelling in the North Atlantic. Such slowing could send Europe into a lower-than-normal temperature regime.

This is a vital research frontier because Earth's hydrologic cycle is still little understood. Research groups such as

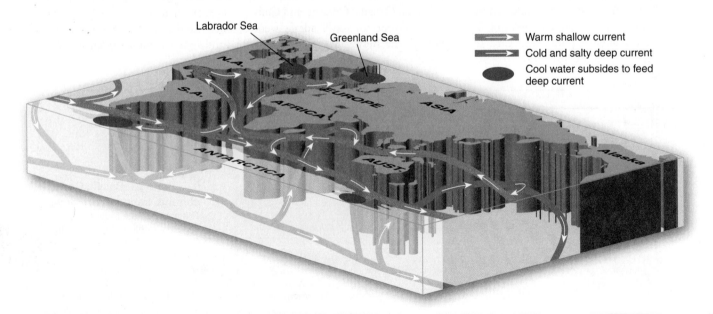

FIGURE 4.22 Deep-ocean thermohaline circulation.
Scientists are deciphering centuries-long deep circulation in the oceans. This global circulation mimics a vast conveyor belt of water drawing heat from some regions and transporting it for release in others.

　Ocean Circulation

Scripps Institution of Oceanography (http://sio.ucsd.edu/), Woods Hole Oceanographic Institution (http://www.whoi.edu/), Institute for the Study of Earth, Oceans, and Space (http://www.eos.sr.unh.edu/), Potsdam Institute for Climate Impact Research (http://www.pik-potsdam.de/), and Bedford Institute of Oceanography (http://www.bio.gc.ca/welcome-e.html), to name only a few of many, are places to periodically check for updated information.

Summary and Review— Atmospheric and Oceanic Circulations

■ *Define* the concept of air pressure and *portray* the pattern of global pressure systems on isobaric maps.

Volcanic eruptions such as those of Tambora in 1815 and Mount Pinatubo in 1991 dramatically demonstrate the power of global winds to disperse aerosols and pollution worldwide in a matter of weeks. Atmospheric circulation facilitates important transfers of energy and mass on Earth, thus maintaining Earth's natural balances. Earth's atmospheric and oceanic circulations represent a vast heat engine powered by the Sun.

The weight (created by motion, size, and number of molecules) of the atmosphere is **air pressure**, which exerts an average force of approximately 1 kg/cm^2 (14.7 lb/in.^{2}). Air pressure is measured with a **mercury barometer** at the surface (mercury in a tube—closed at one end and open at the other, with the open end placed in a vessel of mercury—that changes level in response to pressure changes) or an **aneroid barometer** (a closed cell, partially evacuated of air, that detects changes in pressure).

air pressure (p. 112)
mercury barometer (p. 112)
aneroid barometer (p. 114)

1. How does air exert pressure? Describe the basic instrument used to measure air pressure. Compare the operation of the two different types of instruments discussed.
2. What is normal sea-level pressure in millimeters? Millibars? Inches? Kilopascals?
3. What is a possible explanation for the beautiful sunrises and sunsets during the summer of 1992 in North America? Relate your answer to global circulation.
4. Explain this statement: "The atmosphere links together all humanity, making all the world a spatially linked society." Illustrate your answer with some examples.

■ *Define* wind and *describe* how wind is measured, how wind direction is determined, and how winds are named.

Wind is the horizontal movement of air across Earth's surface. Turbulence adds wind updrafts and downdrafts and a vertical component to this definition. Its speed is measured with an **anemometer** (a device with cups that are pushed by the wind) and its direction with a **wind vane** (a flat blade or surface that is directed by the wind).

wind (p. 114)
anemometer (p. 114)
wind vane (p. 114)

5. Define wind. How is it measured? How is its direction determined?
6. Distinguish among primary, secondary, and tertiary general classifications of global atmospheric circulation.
7. How was the image in Figure 4.6a produced? What does it demonstrate about winds in the Pacific?

■ *Explain* the four driving forces within the atmosphere— gravity, pressure gradient force, Coriolis force, and friction force—and *describe* the primary high- and low-pressure areas and principal winds.

Earth's gravitational force on the atmosphere operates uniformly worldwide. Winds are directed and driven by the **pressure gradient force** (air moves from areas of high pressure to areas of low pressure); the **Coriolis force** (an apparent deflection in the path of winds or ocean currents caused by the rotation of Earth, deflecting objects to the right in the Northern Hemisphere and to the left in the Southern Hemisphere); and **friction force** (Earth's varied surfaces exert a drag on wind movements in opposition to the pressure gradient). Maps portray air pressure patterns using the **isobar**—an isoline that connects points of equal pressure. A combination of the pressure gradient and Coriolis forces alone produces **geostrophic winds**, which move parallel to isobars, characteristic of winds above the surface frictional layer.

Winds descend and diverge, spiraling outward to form an **anticyclone** (clockwise in the Northern Hemisphere), and they converge and ascend, spiraling upward to form a **cyclone** (counterclockwise in the Northern Hemisphere). The pattern of high and low pressures on Earth in generalized belts in each hemisphere produces the distribution of specific wind systems. These primary pressure regions are the **equatorial low-pressure trough**, the weak **polar high-pressure cells** (at both the North and South poles), and the **subtropical high-pressure cells** and **subpolar low-pressure cells**.

All along the equator winds converge into the equatorial low, creating the **intertropical convergence zone (ITCZ)**. Air rises along the equator and descends in the subtropics in each hemisphere. The winds returning to the ITCZ from the northeast in the Northern Hemisphere and from the southeast in the Southern Hemisphere produce the **trade winds**.

Winds flowing out of the subtropics to higher latitudes produce the **westerlies** in either hemisphere. The subtropical high-pressure cells on Earth, generally between 20° and 35° in either hemisphere, are variously named the *Bermuda high*, *Azores high*, and *Pacific high*.

Along the polar front a series of low-pressure cells, the *Aleutian low* and *Icelandic low*, dominate the North Pacific and Atlantic, respectively. This region of contrast between colder air toward the poles and warmer air equatorward is the **polar front**. The weak and variable **polar easterlies** diverge from the polar high-pressure cells, particularly the **Antarctic high**.

pressure gradient force (p. 116)
Coriolis force (p. 116)
friction force (p. 116)
isobar (p. 116)
geostrophic winds (p. 119)
anticyclone (p. 119)
cyclone (p. 119)
equatorial low-pressure trough (p. 121)
polar high-pressure cells (p. 121)

subtropical high-pressure cells (p. 121)
subpolar low-pressure cells (p. 121)
intertropical convergence zone (ITCZ) (p. 123)
trade winds (p. 123)
westerlies (p. 123)
polar front (p. 125)
polar easterlies (p. 125)
Antarctic high (p. 125)

8. What does an isobaric map of surface air pressure portray? Contrast pressures over North America for January and July.

9. Describe the effect of the Coriolis force. Explain how it apparently deflects atmospheric and oceanic circulations.

10. What are geostrophic winds, and where are they encountered in the atmosphere?

11. Describe the horizontal and vertical air motions in a high-pressure anticyclone and in a low-pressure cyclone.

12. Construct a simple diagram of Earth's general circulation, including the four principal pressure belts or zones and the three principal wind systems.

13. How is the intertropical convergence zone (ITCZ) related to the equatorial low-pressure trough? How might it appear on a satellite image?

14. Characterize the belt of subtropical high pressure on Earth. Name the specific cells. Describe the generation of westerlies and trade winds. Discuss sailing conditions.

15. What is the relation among the Aleutian low, the Icelandic low, and migratory low-pressure cyclonic storms in North America? In Europe?

■ *Describe* upper-air circulation and its support role for surface systems and *define* the jet streams.

On upper-air pressure maps a pressure level, such as 500 mb, is a **constant isobaric surface** used as a pressure-reference datum of the height above sea level at which an air pressure of 500 mb occurs. Vast, flowing, longwave undulations in upper-air westerlies form wave motions called **Rossby waves**. The prominent movements in upper-level, westerly winds are streams of high-speed winds called the **jet streams**. Depending on their latitudinal position in either hemisphere, they are termed the *polar jet stream* or the *subtropical jet stream*.

constant isobaric surface (p. 126)
Rossby waves (p. 126)
jet streams (p. 126)

16. What is the relation between wind speed and the spacing of isobars?

17. How are undulations in the upper-air circulation (ridges and troughs) related to surface pressure systems? To divergence aloft and surface lows? To convergence aloft and surface highs? (See Figures 4.12 and 4.15.)

18. Relate the jet-stream phenomenon to general upper-air circulation. How is the presence of this circulation related to airline schedules from New York to San Francisco and the return trip to New York?

■ *Overview* several multiyear oscillations of air temperature, air pressure, and circulation in the Arctic, Atlantic, and Pacific oceans.

Several system fluctuations that occur in multiyear or shorter periods are important in the global circulation picture. The most famous of these is the El Niño–Southern Oscillation

(ENSO) phenomenon. A north–south fluctuation of atmospheric variability marks the *North Atlantic Oscillation (NAO)* as pressure differences between the Icelandic low and the Azores high in the Atlantic alternate in strength. The *Atlantic Oscillation (AO)* is the variable fluctuation between middle- and high-latitude air mass conditions over the Northern Hemisphere. The AO is associated with the NAO especially in winter, and the two phases of the *AO Index* correlate to the two-phased *NAO Index*.

Across the Pacific Ocean, the *Pacific Decadal Oscillation (PDO)* involves variability between two regions of sea-surface temperatures and related air pressure: the northern and tropical western Pacific (region #1) and the area of the eastern tropical Pacific, along the West Coast (region #2). The PDO switches between positive and negative phases in 20- to 30-year cycles.

19. What phases are identified for the AO and NAO indices? What winter weather conditions generally affect the eastern United States during each phase?

20. What is the apparent relation between the PDO and the strength of El Niño events? Between PDO phases and the intensity of drought in the southwestern United States?

■ *Explain* several types of local winds: land-sea breezes, mountain-valley breezes, katabatic winds, and the regional monsoons.

Different heating characteristics of land and water surfaces create **land-sea breezes**. **Mountain-valley breezes** are caused by temperature differences during the day and evening between valleys and mountain summits. **Katabatic winds**, or gravity drainage winds, are of larger regional scale and are usually stronger than mountain-valley breezes, under certain conditions. An elevated plateau or highland is essential for katabatic wind, where layers of air at the surface cool, become denser, and flow downslope.

Intense, seasonally shifting wind systems occur in the tropics over Southeast Asia, Indonesia, India, northern Australia, equatorial Africa, and southern Arizona. These winds involve an annual cycle of returning precipitation with the summer Sun and are named after the Arabic word for season, *mausim*, or **monsoon**. The monsoons of southern and eastern Asia are driven by the location and size of the Asian landmass and its proximity to the Indian Ocean.

land-sea breezes (p. 131)
mountain-valley breezes (p. 132)
katabatic winds (p. 132)
monsoon (p. 135)

21. People living along coastlines generally experience variations in winds from day to night. Explain the factors that produce these changing wind patterns.

22. The arrangement of mountains and nearby valleys produces local wind patterns. Explain the day and night winds that might develop.

23. Describe the seasonal pressure patterns that produce the Asian monsoonal wind and precipitation patterns. Contrast January and July conditions.

■ *Describe* the basic pattern of Earth's major surface and deep thermohaline ocean currents.

Ocean currents are primarily caused by the frictional drag of wind and occur worldwide at varying intensities, temperatures, and speeds, both along the surface and at great depths in the oceanic basins. The circulation around subtropical high-pressure cells in both hemispheres is notable on the ocean circulation map—these *gyres* usually are offset toward the western side of each ocean basin.

As the surface current approaches the western margins of the oceans driven by the trade winds, the water actually piles up an average of 15 cm (6 in.). This phenomenon is the **western intensification.** The trade winds converge along the ITCZ and push enormous quantities of water westward, generating the Gulf Stream and Kuroshio currents. Where surface water is swept away from a coast, either by surface divergence (induced by the Coriolis force) or by offshore winds, an **upwelling current** occurs. This cool water generally is nutrient-rich and rises from great depths to replace the vacating water. In other portions of the sea where there is an accumulation of water, the excess water gravitates downward in a **downwelling current**. These currents generate important mixing actions that flow along the ocean floor and travel the full extent of the ocean basins, carrying heat energy and salinity.

Differences in temperatures and salinity produce density differences important to the flow of deep, sometimes vertical, currents; this is Earth's **thermohaline circulation**. Traveling at slower speeds than wind-driven surface currents, the thermohaline circulation hauls larger volumes of water. There is scientific concern that increased surface temperatures in the ocean and atmosphere, coupled with climate-related changes in salinity, can alter the rate of thermohaline circulation in the oceans and further affect climate.

western intensification (p. 137)
upwelling current (p. 138)
downwelling current (p. 138)
thermohaline circulation (p. 138)

24. What is the relationship between global atmospheric circulation and ocean currents? Relate oceanic gyres to patterns of subtropical high pressure.

25. Define the western intensification. How is it related to the Gulf Stream and Kuroshio currents?

26. Where on Earth are upwelling currents experienced? What is the nature of these currents?

27. What is meant by deep-ocean thermohaline circulation? At what rates do these currents flow? How might this circulation be related to the Gulf Stream in the western Atlantic Ocean?

28. Relative to Question 27, what relation do these deep currents have to global warming and possible climate change?

NetWork and Critical Thinking Tools

A. Find an anemometer on your campus. Where is it located? Describe the installation after a visual inspection. Is it part of a complete weather station? If so, who is operating the station? What were the record wind speeds ever reported on your campus?

B. Go to http://www.awea.org/ and the American Wind Energy Association. Sample the materials presented as you assess Focus Study 4.1 and the potential for wind-generated electricity. What are your thoughts concerning this resource: its potential, reasons for delays, and any necessities you perceive? What countries seem to be leading the way? Locate the nearest wind farm.

C. On the Student Learning Center under "Destinations" and the "Current Wind Speed" site maintained by Ohio State University: Sample surface winds, upper air winds, satellite, and radar sections. What do you find? Check both "stills" and "loops" as you review the various products available. Do you think you might bookmark this URL for your weather information?

Part II   Water, Weather, and Climate Systems

Jan Mayen Island (71° N), in the North Atlantic between the Greenland and Norwegian seas, interacts with the passage of a low-pressure system, producing incredible cloud forms. The Beerenberg volcano at 2277-m (7470-ft) elevation, on the northern part of the island, disrupts winds and clouds, sometimes producing downstream eddies (Karman vortices) swirling hundreds of kilometers across the Atlantic. Topics of water, weather, and climate in such high latitudes were part of the scientific focus during the International Polar Year that began March 2007 and extended through two summer seasons to March 2009. *[Photo by Bobbé Christopherson.]*

Earth is the water planet among planets in the Solar System. In PART II, we analyze the origin, quantity, distribution, and remarkable qualities of water. Water has a leading role in the vast drama played out daily on Earth's stage. In this ongoing play, the actors are large air masses that come into conflict, moving and shifting to dominate different regions. Chapter 5 examines weather and its causes. Topics include the daily weather map and analysis of violent phenomena in thunderstorms, tornadoes, and hurricanes, and the recent trends for each of these.

We examine water resources—the specifics of the hydrologic cycle and how water circulates over Earth. We look at the water-balance concept as a useful way to understand water-resource relationships, whether global, regional, or local. Important water resources include rivers, lakes, groundwater, and oceans. As an ultimate output of these systems, we see the spatial implications over time of the energy–atmosphere and water–weather systems that generate Earth's climate patterns. And we analyze the human impact on the climate system. There is a scientific consensus that human activity is the primary driver of present-day climate change—global warming.

Cumulonimbus clouds towering to more than 9000 m (29,500 ft) in altitude, indicating strong convection in the atmosphere. In 2008, Hurricane Gustav (lower-left) made a September 1 landfall just west of New Orleans along Bayou Lafourche, an area described in News Report 11.3 in this book. Damage and flooding was widespread. In 2008, Hurricane Ike (center) smashed Galveston and Bolivar Islands on September 13 in southeastern Texas, erasing entire towns along the barrier islands (see Figure 13.14). Hurricane Katrina in 2005 (lower right) devastated Louisiana, here on August 28 approaching the coast at category 5 strength, sustained winds of 298 kmph (185 mph), with gusts to 315 kmph. The Gulf of Mexico surface temperatures are 32°−33°C (90°−92°F). Chapter 5 discusses water and explains the processes that produce such cloud formations, the daily weather we experience, dramatic violent weather, and the recent increasing intensity of storms, such as in these three hurricanes that caused more than $145 billion USD in damage. *[Photo by Bobbé Christopherson; satellite images* Terra *MODIS sensor, NASA/GSFC.]*

Hurricane Katrina 2005

Atmospheric Water and Weather

KEY LEARNING CONCEPTS

After reading the chapter, you should be able to:

- **Describe** the origin of Earth's waters, the present distribution, and water's unique heat properties.

- **Define** humidity, relative humidity, and dew-point temperature and **illustrate** stable and unstable atmospheric conditions.

- **Explain** cloud formation and **identify** major cloud classes and types, including fog.

- **Describe** air masses that affect North America and **identify** four types of atmospheric lifting mechanisms.

- **List** the measurable elements of weather and **describe** the life cycle of a midlatitude cyclone.

- **Analyze** various types of violent weather—thunderstorms, tornadoes, and tropical cyclones, including recent trends.

STUDENT LEARNING CENTER TOOLS

The *Elemental Geosystems* Student Learning Center provides on-line resources by clicking on the book cover at **www.mygeoscienceplace.com**. Some highlights:

▶ Do "Data Analysis" of dew-point temperatures reading a satellite map.

▶ Read two Career Links describing a consulting meteorologist with *The Weather Network* in Canada and a research meteorologist who works at the Earth Systems Research Laboratory in Boulder, Colorado.

▶ View the 2005 hurricane season and sea-surface temperatures, see *27 Storms, Arlene to Zeta Movie*.

▶ Seven animations assist you with Earth's water and hydrologic cycle, water phase changes, an interactive lab on atmospheric stability, cold and warm fronts, midlatitude cyclones, and tornado and hurricane wind functions.

▶ More than 34 additional URLs along with Quick Links to all the URLs in the chapter.

Walden is blue at one time and green at another, even from the same point of view. Lying between the earth and the heavens, it partakes of the color of both.... A lake is the landscape's most beautiful and expressive feature. It is earth's eye; looking into which the beholder measures the depth of his own nature.... Sky water. It needs no fence. Nations come and go without defiling it. It is a mirror which no stone can crack.... Nature continually repairs ... a mirror in which all impurity presented to it sinks, swept and dusted by the sun's hazy brush. A field of water ... is continually receiving new life and motion from above. It is intermediate in its nature between land and sky.*

T hus, did Thoreau speak of the water so dear to him—Walden Pond in Massachusetts, where he lived along the shore.

Water is critical to our daily lives and is an extraordinary compound in nature. Water covers 71% of Earth (by area) and in the Solar System occurs in such significant quantities only on our planet. At Mars in 2004, three orbiting spacecraft and two robot landers imaged and detected polar water ice, erosional features formed by flowing water early in Martian history, and evidence of subsurface water. The *Phoenix* lander in 2008 confirmed the presence of water in the Martian polar region, although the bulk of Martian water was lost early in its past.

Pure water is naturally colorless, odorless, and tasteless and weighs 1 gram (g) per cubic centimeter or 1 kg per liter (62.3 lb/ft^3, or 8.337 lb/gal). Water constitutes nearly 70% of our bodies by weight and is the major ingredient in plants, animals, and our food. A human can survive 50 to 60 days without food, but only 2 or 3 days without water. The water we use must be adequate in quantity as well as quality for its many tasks—everything from personal hygiene to vast national water projects. Water indeed occupies that place between land and sky, mediating energy and shaping both the lithosphere and atmosphere, as Thoreau revealed.

In this chapter: We examine water on Earth and the dynamics of atmospheric moisture and stability—the essentials of weather. The key questions answered include: What were the origins of Earth's water? How much is there? Where is it located? What are its unique heat properties and how is it critical in powering Earth's weather systems?

Condensation of water vapor and atmospheric conditions of stability and instability are key to cloud formation. We follow huge air masses across North America, observe powerful lifting mechanisms in the atmosphere, examine migrating cyclonic systems with attendant cold and warm fronts, and conclude with a tour of violent weather. Coverage includes dramatic recent trends in tornadoes and tropical cyclones—Katrina, Wilma, Rita, Fay, Gustav, and Ike, among others. The spatial implications of weather and its relationship to human activities strongly link meteorology to physical geography and this chapter.

Water on Earth: Location and Properties

Water on Earth formed within the planet, reaching Earth's surface in an ongoing **outgassing** process, by which water and water vapor emerge from deep within the crust—25 km (15.5 mi) or deeper below Earth's surface. Figure 5.1 presents two such areas, among many sites worldwide, in New Zealand and Iceland. The Haukadalur Geysir geothermal area is in southeast Iceland, about 110 km from its capital Reykjavík (Figure 5.1b). Written accounts of the Geysir activity date back to A.D. 1294, and it is the origin of the term *geyser* we use today.

In the early atmosphere, massive quantities of outgassed water vapor condensed and fell back to Earth in torrential rains. The lowest places across the face of Earth began to fill with water: first ponds, then lakes and seas, and eventually ocean-sized bodies of water. Massive flows of water washed over the landscape, carrying both dissolved and undissolved elements to the early seas and oceans.

Worldwide Equilibrium

Today, water is the most common compound on the surface of Earth, having achieved the present volume of 1,359,208,000 km^3, or 326,074,000 mi^3, approximately 2 billion years ago. This quantity has remained relatively constant, even though water is continuously being lost from the system, escaping to space or breaking down and forming new compounds with other elements. Lost water is replaced by pristine water not previously at the surface, which emerges from within Earth. The net result of these inputs and outputs to water quantity is a steady-state equilibrium in Earth's hydrosphere.

Despite this overall net balance in quantity, worldwide changes in sea level do occur; this change is **eustasy**. Eustatic sea-level changes are specifically related to changes in volume of water in the oceans. Some of these changes are explained by the amount of water stored on Earth as ice. As more water is bound up in glaciers (in mountains worldwide) and in ice sheets (Greenland and Antarctica), sea level lowers, whereas a warmer era reduces the quantity of water stored as ice and raises sea level.

Some 18,000 years ago, during the most recent ice age, sea level was 100 m (330 ft) less than today; 40,000 years ago it was 150 m (490 ft) lower. Over the past 100 years, glacial ice has retreated, causing mean sea level to rise worldwide at a rate not previously seen as higher temperatures melt more ice. *Glacio-eustatic* effects on sea level are

(a)

(b)

FIGURE 5.1 Water outgassing from the crust.
Outgassing of water from Earth's crust in geothermal areas: (a) near Wairakei on the North Island of New Zealand; and (b) the Smidur boiling spring in the Geysir area in Haukadalur, Iceland.
[Photos by (a) Bill Bachman/Photo Researchers, Inc.; (b) Bobbé Christopherson.]

discussed further in Chapter 13. Chapter 7 discusses global climate change, ice melt, and resulting sea-level rise.

Distribution of Earth's Water

If you briefly examine a globe, it is obvious that most of Earth's continental land is in the Northern Hemisphere, whereas water dominates the Southern Hemisphere. In fact, from certain perspectives Earth appears to have a distinct oceanic hemisphere and a distinct land hemisphere (Figure 5.2). An illustration of the present location of all of Earth's liquid and frozen water—whether fresh or saline, surface or underground—is in Figure 5.3. The oceans contain 97.22% of all water. Only 2.78% of all of Earth's water is freshwater, or nonoceanic (Figure 5.3a).

Of all the freshwater on Earth, about 22% exists as groundwater and soil moisture (Figure 5.3b). On the surface, the greatest single repository of freshwater is ice. Ice sheets and glaciers account for the majority of all freshwater on Earth. The remaining freshwater, although very familiar to us in lakes, rivers, and streams, represents but a small quantity—less than 1% (Figure 5.3c).

Think for a moment of the weather occurring in the atmosphere worldwide and the many flowing rivers and streams. Combined they amount to only 0.033% of all freshwater! This small amount is dynamic, however. A water molecule traveling along atmospheric and surface-water paths moves through the entire ocean–atmosphere–precipitation–runoff cycle in less than 2 weeks, whereas a water molecule located in deep-ocean circulation, ground-water, or a glacier moves slowly, taking thousands of years to migrate through the system.

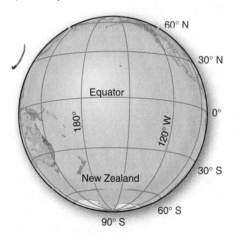

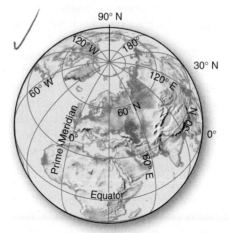

FIGURE 5.2 Land and water hemispheres.
Two perspectives that divide Earth's surface into an ocean hemisphere and a land hemisphere.

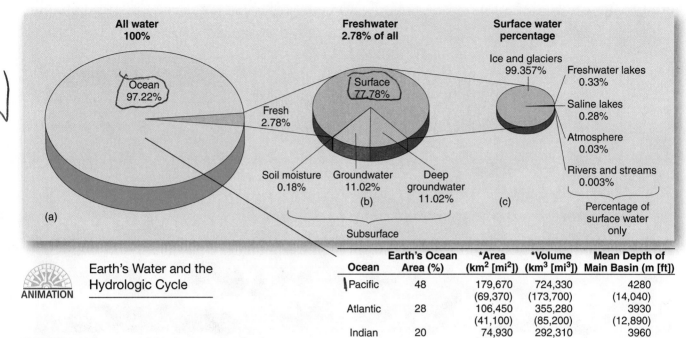

Ocean	Earth's Ocean Area (%)	*Area (km² [mi²])	*Volume (km³ [mi³])	Mean Depth of Main Basin (m [ft])
Pacific	48	179,670 (69,370)	724,330 (173,700)	4280 (14,040)
Atlantic	28	106,450 (41,100)	355,280 (85,200)	3930 (12,890)
Indian	20	74,930 (28,930)	292,310 (70,100)	3960 (12,900)
Arctic	4	14,090 (5440)	17,100 (4100)	1205 (3950)

ANIMATION Earth's Water and the Hydrologic Cycle

FIGURE 5.3 Ocean and freshwater distribution on Earth.
(a) All of Earth's water; (b) freshwater, surface and subsurface; (c) distribution of surface freshwater alone.

*Data in thousands (000): includes all marginal seas.

Unique Properties of Water

Earth's distance from the Sun places it within a remarkable temperate zone compared to the other planets. This temperate location allows all states of water—ice, water, and water vapor—to occur naturally on Earth. Two atoms of hydrogen and one of oxygen, which readily bond, compose a water molecule. The resulting molecule exhibits a unique *stability*, is a *versatile solvent*, and possesses *extraordinary heat characteristics*. As the most common compound on the surface of Earth, water exhibits uncommon properties.

Because of the nature of the hydrogen–oxygen bond, the hydrogen side of the water molecule has a positive charge and the oxygen side a negative charge (see Figure 5.4, upper right). This *polarity* of the water molecule explains why water "acts wet" and dissolves so many other molecules and elements. Thus, because of its solvent activity, pure water is rare in nature.

Water molecules are attracted to each other because of their polarity; the positive (hydrogen) side of one water molecule is attracted to the negative (oxygen) side of another. (Note this bonding pattern in the molecular illustrations in Figure 5.4b and c.) The bonding between water molecules is *hydrogen bonding*. The effects of hydrogen bonding in water are observable in everyday life. Hydrogen bonding creates *surface tension* and is the cause of capillarity, which you observe when you use a paper towel. The towel draws water upward through its fibers because each molecule is pulling on its neighbor. Capillary action is an important component of soil moisture processes, discussed in Chapters 6 and 15. It is important to note that without hydrogen bonding, water would be a gas at normal surface temperatures.

Heat Properties

For water to change from one state to another (solid, liquid, gas), heat energy must be added to it or released from it. To cause a change of state, the amount of heat energy must be sufficient to affect the hydrogen bonds between the molecules. This relationship between water and heat energy is an important driving force in producing weather. In fact, the heat exchanged in the phase changes of water provides over 30% of the energy powering the general circulation of the atmosphere. Figure 5.4 presents the terms used to describe each **phase change**.

Melting and *freezing* describe the phase change between solid and liquid. The terms *condensation* and *evaporation*, or *vaporization* at boiling temperature, apply to the change between liquid and vapor. The term **sublimation** refers to the direct change of water vapor to ice or ice to water vapor; sometimes the term *deposition* is used if water vapor attaches itself directly to an ice crystal. The deposition of water vapor as ice may form frost on surfaces.

Ice, the Solid Phase As water cools, it behaves like most compounds and contracts in volume, reaching its greatest density at 4°C (39°F). But below that temperature, water behaves very differently from other compounds and begins to expand as more hydrogen bonds form among the slower-moving molecules, creating the hexagonal structures shown in Figure 5.4c. This expansion continues to a temperature of −29°C (−20°F) with up to a 9% increase in volume possible.

As shown in Figure 5.5a, the rigid internal structure of ice dictates the six-sided appearance of all ice crystals, which can loosely combine to form snowflakes. This

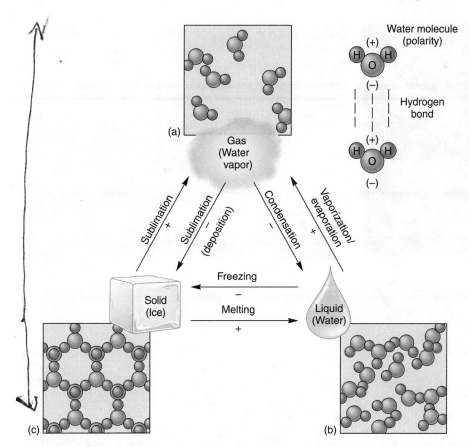

FIGURE 5.4 Three states of water and water's phase changes.
The three physical states of water: (a) water vapor, (b) water, and (c) ice. Note the molecular arrangement in each state and the terms that describe the changes from one phase to another as energy is either absorbed or released. Also note how the polarity of water molecules bonds them to one another, loosely in the liquid state and firmly in the solid state. The plus and minus symbols on the phase changes denote whether heat energy is absorbed (+) or liberated (−) during the phase change.

ANIMATION

Water Phase Changes

six-sided preference applies to ice crystals of all shapes: plates, columns, needles, and dendrites (branching or treelike forms)—a unique interaction of randomness and the determinism of physical principles. (For more on ice crystals and snowflakes, see http://www.its.caltech.edu/~atomic/snowcrystals/ or see http://emu.arsusda.gov/snowsite/default.html.)

In nature, this density varies slightly given the age of the ice and air content. Therefore icebergs float with slightly varying displacements that average about 1/7 (14%) of their mass exposed and 6/7 (86%) in a submerged portion hidden beneath the ocean's surface (Figure 5.5d). With underwater portions melting faster than the portions above water, icebergs are inherently unstable and will overturn. You see in Figure 5.5b ruffled and fluted ice that was previously underwater. News Report 5.1 (p. 152) discusses the power exerted by the volume increase of frozen water.

Water, the Liquid Phase As a liquid, water assumes the shape of its container and is a noncompressible fluid, quite different from solid, rigid ice. For ice to melt, heat energy must increase the motion of the water molecules to break some of the hydrogen bonds. Despite the fact that there is no change in sensible temperature between ice at 0°C and water at 0°C, 80 calories* of heat must be

added for the phase change of 1 g of ice to 1 g of water (Figure 5.6, upper left). This heat is **latent heat** because it is stored within the water and is liberated with a phase reversal when a gram of water freezes. These 80 calories are the *latent heat of melting* and *of freezing*.

To raise the temperature of 1 g of water from freezing at 0°C (32°F) to boiling at 100°C (212°F), we must add 100 additional calories, gaining an increase of 1 C° (1.8 F°) for each calorie added (Figure 5.6, top center).

Water Vapor, the Gas Phase Water vapor is an invisible and compressible gas in which each molecule moves independently (Figure 5.4a). To accomplish the phase change from liquid to vapor at boiling temperature, under normal sea-level pressure, 540 calories must be added to 1 g of boiling water (Figure 5.6, upper right). Those calories are the **latent heat of vaporization**. When water vapor condenses to a liquid, each gram gives up its hidden 540 calories as the **latent heat of condensation**.

In summary, taking 1 g of ice at 0°C and changing its phase to water, then to water vapor at 100°C—from a solid, to a liquid, to a gas—*absorbs* 720 calories (80 cal + 100 cal + 540 cal). Reversing the process, or changing the phase of 1 g of water vapor at 100°C to water, then to ice at 0°C, *liberates* 720 calories into the surrounding environment. Your Student Learning Center web site that supports this book has an excellent animation illustrating these concepts.

Heat Properties of Water in Nature In a lake, stream, or in soil water, at 20°C (68°F), every gram of water that

*Remember from Chapter 2: A calorie is the amount of energy required to raise the temperature of 1 g of water (at 15°C) 1°C and is equal to 4.184 joules.

(a)

(b)

(c)

(d)

FIGURE 5.5 The uniqueness of ice forms.
(a) Computer-enhanced photo reveals ice-crystal patterns, which are dictated by the internal structure between water molecules. This open structure also explains the lower density and why ice floats in water. (b) Icebergs along the Greenland coast that overturned, exposing underwater portions that were sculpted by currents and melt channels. (c) Ice needles form in the freezing process of a meltwater stream. (d) A bergy bit (small iceberg) off the coast of Antarctica illustrates the density–buoyancy state of floating ice. [(a) Photo enhancement © Scott Camazine/Photo Researchers, Inc., after W. A. Bentley; (b), (c), and (d) photos by Bobbé Christopherson.]

breaks away from the surface through evaporation must absorb from the environment approximately 585 calories as the **latent heat of evaporation** (see the natural scene in Figure 5.6). This is slightly more energy than would be required if the water were boiling (540 cal). You can feel this absorption of latent heat as evaporative cooling on your skin when it is wet. This latent heat exchange is the dominant cooling process in Earth's energy budget.

The process reverses when air that contains water vapor is cooled. The vapor eventually condenses back into the liquid state, forming moisture droplets and thus liberating 585 calories as the *latent heat of condensation* for every gram of water. Because of these unique heat properties, water is a major contributor of energy to the atmosphere. Let's now take a look at water vapor in the atmosphere.

Humidity

Humidity refers to water vapor in the air. The capacity of air for water vapor is primarily a function of temperature—the temperatures of both the air and the water vapor, which are usually the same. There are several ways of expressing humidity, as discussed in this section.

We are all aware of humidity in the air, for its relationship to air temperature determines our sense of comfort. North Americans spend billions of dollars a year to adjust humidity, either with air-conditioning (extracting water vapor and cooling) or with air humidifying (adding water vapor). To determine the energy available to empower weather, it is essential to know the water vapor content of air.

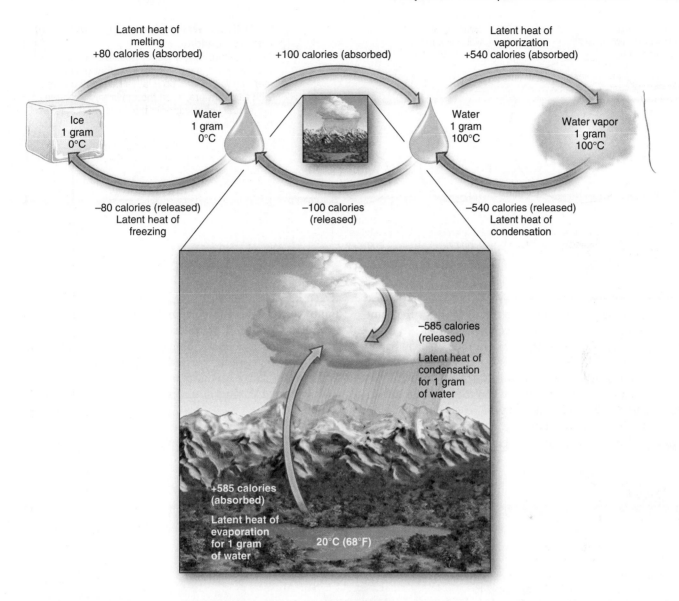

FIGURE 5.6 Water's heat energy characteristics.
The phase changes of water absorb or release a lot of latent heat energy. To transform
1 g of ice at 0°C to 1 g of water vapor at 100°C requires 720 calories: 80 + 100 + 540. The
landscape illustrates phase changes between water (lake at 20°C) and water vapor under typical
conditions in the environment.

ANIMATION
Water Phase Changes

Relative Humidity

Next to air temperature and barometric pressure, the most common piece of information in local weather broadcasts is relative humidity. **Relative humidity** is a ratio (expressed as a percentage) of the amount of water vapor that is actually in the air compared to the maximum water vapor possible in the air at a given temperature.

Warmer air increases the evaporation rate from water surfaces, whereas cooler air tends to increase the condensation rate of water vapor to water surfaces. Because there is a maximum amount of water vapor that can exist in a volume of air at a given temperature, the rates of evaporation and condensation can reach equilibrium at some point; the air is then saturated with humidity. Relative

humidity varies because of evaporation, condensation, or temperature changes. All three affect both the moisture in the air (the numerator) and the water vapor possible (the denominator) in the air. The formula to calculate the relative humidity ratio and express it as a percentage is simply:

$$\text{Relative humidity} = \frac{\text{Actual water vapor in the air}}{\begin{array}{c}\text{Maximum water vapor possible}\\\text{in the air at the temperature}\end{array}} \times 100$$

In Figure 5.7 at 5 P.M. the evaporation rate exceeds condensation at the higher temperatures of this time of day and relative humidity is at 20%. At 11 A.M. the evaporation rate still exceeds condensation, though not as much

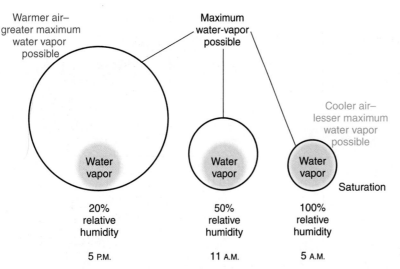

Warmer air–
greater maximum
water vapor
possible

Maximum
water-vapor
possible

Cooler air–
lesser maximum
water vapor
possible

Water
vapor

Water
vapor

Water
vapor

Saturation

20%
relative
humidity

50%
relative
humidity

100%
relative
humidity

5 P.M.

11 A.M.

5 A.M.

FIGURE 5.7 Water vapor, temperature, and relative humidity.
The maximum water vapor possible in warm air is greater (net evaporation) than that of cold air (net condensation), so relative humidity changes with temperature, even though in this example the actual water vapor in the air stays the same during the day.

since daytime temperatures are not as high, so the same volume of water vapor now occupies 50% of the maximum possible capacity. At 5 A.M., in the cooler morning air, saturation equilibrium exists and any further cooling or addition of water vapor produces net condensation.

As stated, air is **saturated** when the rate of evaporation and the rate of condensation reach equilibrium—that is, 100% relative humidity. In saturated air, the net transfer of water molecules between a moist surface and the air reaches equilibrium. Saturation indicates that any further addition of water vapor or any decrease in temperature that reduces the evaporation rate will result in active condensation (clouds, fog, or precipitation). Therefore, relative humidity indicates the nearness of air to a saturated condition and when active condensation will begin.

The temperature at which a given mass of air becomes saturated is the **dew-point temperature**. In

NEWS REPORT 5.1

Breaking Roads and Pipes and Sinking Ships

Road crews are out in the summer in many parts of the country repairing streets and freeways. Winter damage to highways often is the result of the phase change from water to ice. Rainwater seeps into roadway cracks and then expands as it freezes, thus breaking up the pavement. Perhaps you have noticed that bridges are subjected to the greatest roadbed damage. The reason is that cold air circulating beneath a bridge produces more freeze-thaw cycles on the bridge than in the roadbed on rock and soil.

The expansion of freezing water exerts a tremendous force—enough to crack plumbing or an automobile engine block (Figure 5.1.1). People living in very cold climates use antifreeze and engine heaters to avoid such damage. Wrapping water pipes with insulation is a common winter task in many places.

Historically, this physical property of water was put to useful work in

quarrying rock for building materials. Holes were drilled and filled with water before winter, so that when cold weather arrived the water would freeze and expand, cracking the rock into manageable shapes.

Floating ice poses a major shipping hazard in higher latitudes. Since an iceberg sits with approximately 0.86 the density of water, approximately 6/7

FIGURE 5.1.1 Stronger than cast iron.
Expansion in volume as water freezes and ice forms is undeniable, as this busted pipe demonstrates. People living in cold climates must take precautions to avoid such damage. [Photo by Steven K, Huhtala.]

of its mass is below water level. The irregular edges of submarine ice can impact the side of a passing ship. Hitting an iceberg buckled plates and substandard rivets along the side of the RMS *Titanic* on its maiden voyage in 1912, causing its sinking and perhaps a momentary lowering of society's faith in technology.

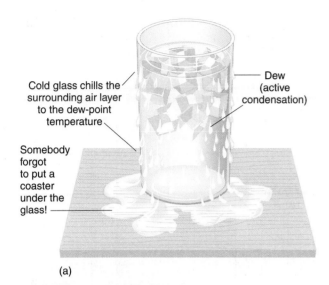

Cold glass chills the surrounding air layer to the dew-point temperature

Dew (active condensation)

Somebody forgot to put a coaster under the glass!

(a)

(b)

FIGURE 5.8 Dew-point temperature examples.
(a) The low temperature of the glass chills the surrounding air layer to the dew point and saturation. Thus, water vapor condenses out of the air and onto the glass as dew. (b) Cold air above the rain-soaked rocks is at the dew point and is saturated. Condensation shrouds the rock in a changing veil of clouds. [Photo by author.]

other words, *air is saturated when the dew-point temperature and the air temperature are the same.* When temperatures are below freezing, the term *frost point* is sometimes used. A cold drink in a glass provides a common example of these conditions (Figure 5.8a). The water droplets that form on the outside of the glass condense from the air because the air layer next to the glass is chilled to below its dew-point temperature and thus is saturated.

Satellites now routinely sense water-vapor content of the lower atmosphere. Water vapor absorbs infrared wavelengths, making it possible to distinguish areas of relatively high water-vapor content from areas of low water-vapor content using infrared sensors. Figure 5.9 portrays the Western Hemisphere and water-vapor content, recorded by satellite. This knowledge is important to forecasting because it shows the energy available (latent heat) and precipitation potential of those systems.

FIGURE 5.9 Image of water vapor in the atmosphere.
Water-vapor content over the full Western Hemisphere seen in a *GOES* infrared image. Note subpolar low-pressure circulation. [*GOES* image courtesy of NESDIS Satellite Services Division, NOAA.]

During a typical day, air temperature and relative humidity relate inversely—as temperature rises, relative humidity falls (Figure 5.10). Relative humidity is highest at dawn, when air temperature is lowest. If you park outdoors,

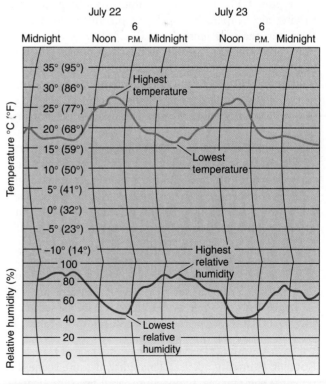

FIGURE 5.10 Typical daily temperature and relative humidity.
Typical daily variations demonstrate temperature and relative humidity relations—higher relative humidity in the A.M., lower relative humidity in the P.M.

you know about the wetness of the dew that condenses on the car overnight. Relative humidity is lowest in the late afternoon, when higher temperatures increase the rate of evaporation. As shown in Figure 5.7, the actual water vapor present in the air may remain the same throughout the day, as relative humidity varies.

Expressions of Humidity

Humidity and relative humidity are expressed in several ways, each with its own utility and application. Two measures involve vapor pressure and specific humidity.

Vapor Pressure   One way of describing humidity is related to air pressure. As free water molecules evaporate from a surface into the atmosphere, they become water vapor, one of the gases in air. That portion of total air pressure that is made up of water vapor molecules is the **vapor pressure**. Like air pressure, it is expressed in millibars (mb).

Water-vapor molecules continue to evaporate from a moist surface, slowly diffusing into the air, until the increasing vapor pressure in the air causes some molecules to return to the surface. Saturation is reached when the movement of water molecules between surface and air is in equilibrium. Air that contains as much water vapor as possible at a given temperature is at the *saturation vapor pressure*, which indicates the maximum pressure that water-vapor molecules can exert. Any increase or decrease in temperature causes the saturation vapor pressure to change.

Figure 5.11 graphs the saturation vapor pressure at varying air temperatures. The graph illustrates that for every temperature increase of 10 C° (18 F°), the saturation water-vapor pressure in air nearly doubles. This relation explains why warm tropical air over the ocean can hold so much water vapor, thus providing great latent heat to power tropical storms. It also explains why cold air is "dry" and why cold air toward the poles does not produce a lot of precipitation (it contains too little water vapor, even though it is near the dew-point temperature).

As the graph shows, air at 20°C (68°F) has a saturation vapor pressure of 24 mb; that is, the air is saturated if the water-vapor portion of the air pressure is at 24 mb. Thus, if the water vapor actually present is exerting a vapor pressure of only 12 mb in 20°C air, the relative humidity is 50% (12 mb ÷ 24 mb = 0.50 × 100 = 50%).

The inset in Figure 5.11 compares saturation vapor pressure over water and over ice surfaces at subfreezing temperatures. You can see that saturation vapor pressure is greater above a water surface than over an ice surface—that is, it takes more water vapor molecules to saturate air above water than it does above ice.

Specific Humidity   A useful humidity measure is one that remains constant as temperature and pressure change. **Specific humidity** refers to the mass of water vapor (in grams) per mass of air (in kilograms) at any specified temperature. Because it is measured in mass, specific humidity is not affected by changes in temperature or pressure, such as

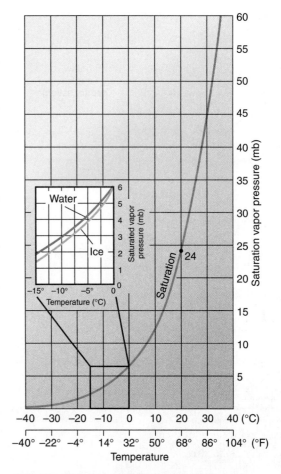

FIGURE 5.11 Saturation vapor pressure.
Saturation vapor pressure of air at various temperatures—the maximum water vapor possible as measured by the pressure it exerts. As the graph shows, air at 20°C (68°F) has a saturation vapor pressure of 24 mb; that is, the air is saturated if the water-vapor portion of the air pressure is at 24 mb.

when an air parcel rises to higher elevations. Specific humidity stays constant despite volume changes.*

The maximum mass of water vapor possible in a kilogram of air at any specified temperature is the *maximum specific humidity*, plotted in Figure 5.12. The graph shows that a kilogram of air could hold a maximum specific humidity of 47 g of water vapor at 40°C (104°F), 15 g at 20°C (68°F), and about 4 g at 0°C (32°F). Therefore, if a kilogram of air at 40°C has a specific humidity of 12 g, its relative humidity is 25.5% (12 g ÷ 47 g = 0.255 × 100 = 25.5%). Specific humidity is useful in describing the moisture content of large air masses that are interacting in a weather system and provides necessary information for weather forecasting.

Instruments for Measurement   Relative humidity is measured with various instruments. The *hair hygrometer* is based on the fact that human hair changes as much as 4% in length between 0% and 100% relative humidity. The

---

*Another similar measure used in relative humidity that approximates specific humidity is the mixing ratio, that is, the ratio of the mass of water vapor (grams) per mass of dry air (kilograms), as in g/kg.

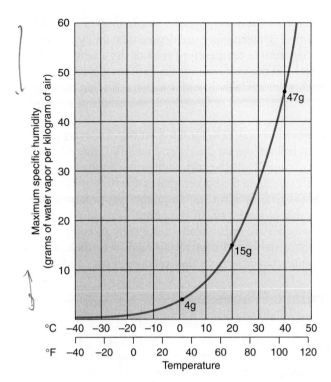

FIGURE 5.12 Maximum specific humidity.
Maximum specific humidity for a mass of air at various temperatures—the maximum water vapor possible, expressed in mass of water vapor per unit mass of air (g/kg). Note the 47 g of water vapor at 40°C (104°F), 15 g at 20°C (68°F), and about 4 g at 0°C (32°F), keyed to the text discussion.

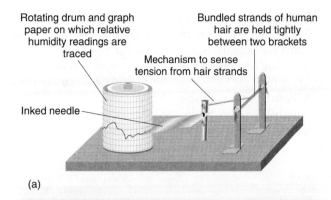

(a)

(b)

FIGURE 5.13 Instruments that measure relative humidity.
(a) The principle of a hair hygrometer. (b) Sling psychrometer with wet and dry bulbs. [Photo by Bobbé Christopherson.]

instrument connects a standardized bundle of human hair through a mechanism to a gauge. As the hair absorbs or loses water in the air, its change in length indicates relative humidity (Figure 5.13a).

Another instrument used to measure relative humidity is a *sling psychrometer*. This device has two thermometers mounted side by side on a metal holder. One is the *dry-bulb thermometer*; it simply records the ambient (surrounding) air temperature. The other thermometer is the *wet-bulb thermometer*. This bulb is covered by a cloth wick, which is moistened. The psychrometer is then spun by its handle (Figure 5.13b). (Other versions use a motor to spin the thermometers, or they are set up with a fan.) After a minute or two, the temperature of each bulb is compared on a relative humidity (psychrometric) chart, from which relative humidity can be determined.

The rate at which water evaporates from the wick depends on the relative saturation of the surrounding air. If the air is dry, water evaporates quickly, absorbing the latent heat of evaporation from the wet-bulb thermometer, causing the temperature to drop (wet-bulb depression). In an area of high humidity, much less water evaporates from the wick, resulting in a smaller temperature difference between the two thermometers.

Now that you know something about atmospheric moisture, dew point, and relative humidity, let's examine the concept of stability in the atmosphere.

Atmospheric Stability

Meteorologists use the term *parcel* to describe a body of air that has specific temperature and humidity characteristics. Think of an air parcel as a volume of air, perhaps 300 m (1000 ft) in diameter or more. Differences in temperature create changes in density within the parcel. Warm air produces a lower density in a given volume of air; cold air produces a higher density. Two opposing forces work on a parcel of air: an upward *buoyancy force* and a downward *gravitational force*. A parcel of lower density will rise (is more buoyant); a rising parcel expands as external pressure decreases. A parcel of higher density will descend (is less buoyant); a falling parcel compresses as external pressure increases. Figure 5.14a shows air parcels and illustrates these relationships.

Stability refers to the tendency of an air parcel, with its water-vapor cargo, either to remain in place or to

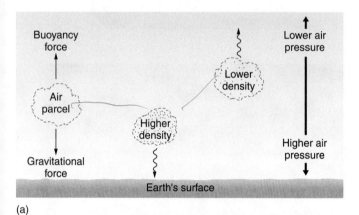

(a)

ANIMATION

Atmospheric Stability

(b)

FIGURE 5.14 Forces acting on an air parcel and balloon launches.
(a) Buoyancy and gravitational forces work on an air parcel. Different densities produce rising or falling parcels in response to imbalance in these forces. (b) Hot-air balloons launched in southern Utah illustrate the principles of stability. As the temperature inside a balloon increases, the air in the balloon becomes less dense than the surrounding air and the buoyancy force causes the balloon to rise, acting like a warm-air parcel. [Photos by Steven K. Huhtala.]

change vertical position by ascending (rising) or descending (falling). Determining the degree of stability involves measuring simple temperature relationships between the air parcel and the surrounding air. Such temperature measurements are made daily with balloon soundings (instrument packages called *radiosondes*) at thousands of weather stations.

To visualize this, imagine a hot-air balloon launch. The air-filled balloon sits on the ground with the same air temperature inside as in the surrounding environment, like a stable air parcel. As the burner ignites, the balloon fills with hot (less-dense) air and rises buoyantly (Figure 5.14b), like an unstable parcel of air. In the photo, why do you think these launches are taking place in the early morning? The concepts of stable and unstable air lead us to the specific temperature characteristics that produce these conditions.

Adiabatic Processes

The **normal lapse rate**, as introduced in Chapter 2, is the average decrease in temperature with increasing altitude, a value of 6.4 C° per 1000 m (3.5 F° per 1000 ft). This average for temperature change is for still, calm air, but it can differ greatly under varying weather conditions, and so the actual lapse rate at a particular place and time is labeled the **environmental lapse rate**.

An ascending parcel of air tends to cool by expansion, responding to the reduced pressure at higher altitudes (Figure 5.15a). In contrast, *descending air tends to heat by compression* (Figure 5.15b). Both ascending and descending temperature changes are assumed to occur *without any heat exchange between the surrounding environment and the vertically moving parcel of air*. The warming and cooling rates for a parcel of expanding or compressing air are termed **adiabatic**. (*Diabatic* means occurring with an exchange of heat; *adiabatic* means occurring *without* a loss or gain of heat.) Adiabatic temperatures are measured with one of two specific rates, depending on moisture conditions in the parcel: *dry* adiabatic rate (DAR) and *moist* adiabatic rate (MAR).

Dry Adiabatic Rate   The **dry adiabatic rate (DAR)** is the rate at which "dry" air cools by expansion (if ascending) or heats by compression (if descending). "Dry" air is less than saturated, with a relative humidity less than 100%. The DAR is 10 C° per 1000 m (5.5 F° per 1000 ft), as illustrated in Figure 5.15. For example, in Figure 5.15a, consider an unsaturated parcel of air at the surface, whose temperature measures 27°C (81°F). It rises, expands, and cools at the DAR as it lifts from the ground to 2500 m (approximately 8000 ft). The temperature of the parcel at 2500-m altitude is 2°C (36°F), having cooled by expansion as it rose:

(10 C°/1000 m) × 2500 m = 25 C° of total cooling
(5.5 F°/1000 ft) × 8000 ft = 44 F° of total cooling

Subtracting the 25 C° (44 F°) of cooling from the starting temperature of 27°C (81°F) gives the temperature at

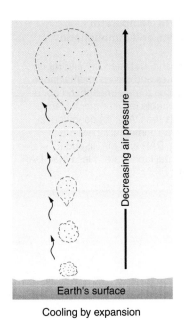

(a) Air parcel *cools* adiabatically at the DAR

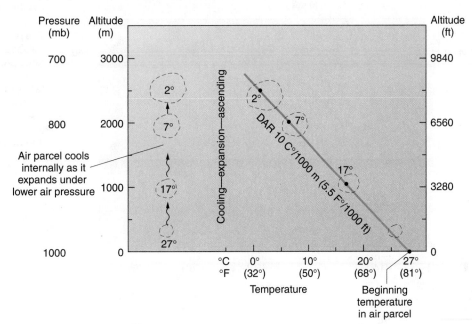

Air parcel cools internally as it expands under lower air pressure

Cooling by expansion

(b) Air parcel *heats* adiabatically at the DAR

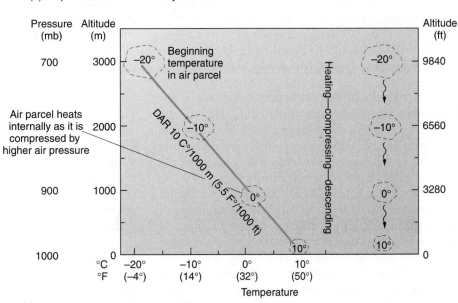

Air parcel heats internally as it is compressed by higher air pressure

Heating by compression

FIGURE 5.15 Vertically moving air experiences temperature changes—adiabatic cooling and heating.
Vertically moving air parcels expand when they rise (because air pressure is less with increasing altitude) and are compressed when they descend. (a) A rising air parcel that is less than saturated cools adiabatically at the dry adiabatic rate (DAR). (b) A descending air parcel that is less than saturated heats adiabatically by compression at the DAR.

ANIMATION
Atmospheric Stability

2500 m of 2°C (36°F). (Note that the parcel cooled adiabatically, without a loss of heat to the environment.)

In Figure 5.15b, assume that an unsaturated air parcel with a temperature of −20°C at 3000 m (−4°F at 9800 ft) descends to the surface, heating adiabatically. Using the dry adiabatic lapse rate, we can determine the temperature of the air parcel when it arrives at the surface:

(10 C°/1000 m) × 3000 m = 30 C° of total warming
(5.5 F°/1000 ft) × 9800 ft = 54 F° of total warming

Adding the 30 C° of adiabatic warming to the starting temperature of −20°C gives a temperature in the air parcel at the surface of 10°C (50°F).

Moist Adiabatic Rate The **moist adiabatic rate (MAR)** is the average rate at which ascending air that is moist (saturated) cools by expansion. The *average* MAR is 6 C° per 1000 m (3.3 F° per 1000 ft), or roughly 4 C° (2 F°) less than the DAR. However, the MAR varies with moisture content and temperature and can range from 4 C° to

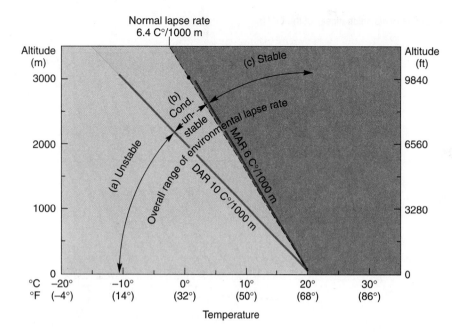

Normal lapse rate
6.4 C°/1000 m

FIGURE 5.16 Temperature relationships and atmospheric stability.
The relationship between dry and moist adiabatic rates and environmental lapse rates produces three atmospheric conditions: (a) unstable (environmental lapse rate exceeds the DAR), (b) conditionally unstable (environmental lapse rate is between the DAR and MAR), and (c) stable (environmental lapse rate is less than DAR and MAR).

10 C° per 1000 m (2 F° to 5.5 F° per 1000 ft). The reason is that in a saturated air parcel, latent heat of condensation is liberated as sensible heat, which reduces the adiabatic rate of cooling. The release of latent heat may vary, which affects the MAR. The MAR is much lower than the DAR in warm air, whereas the two rates are more similar in cold air.

Stable and Unstable Atmospheric Conditions Now we bring this discussion together to determine atmospheric stability. The relationship among the DAR, MAR, and the environmental (actual) lapse rate is complex and determines the stability of the atmosphere over an area. You can see the range of possible relationships in Figure 5.16.

Let's apply this concept to a specific situation. Figure 5.17 illustrates two of these temperature relationships in the atmosphere that lead to different conditions: unstable and stable. For the sake of illustration, both examples begin with a parcel of air at the surface at 25°C (77°F). In each example, examine the temperature relationships between the parcel of air and the surrounding environment. Assume that a lifting mechanism is present (we examine lifting mechanisms shortly).

Under the unstable conditions in Figure 5.17a, the air parcel continues to rise through the atmosphere because it is warmer (and therefore less dense and more buoyant) than the surrounding environment. The air surrounding the lifting parcel is cooler by an environmental lapse rate of 12 C° per 1000 m (6.6 F° per 1000 ft). By 1000 m (3280 ft), the parcel has adiabatically cooled by expansion to 15°C (59°F), but the surrounding air is at 13°C (55.4°F). The temperature in the parcel is 2 C° (3.6 F°) warmer than the surrounding air, and therefore less dense, so it continues to lift, cool, and approach saturation (100% relative humidity and cloud formation).

The stable conditions in Figure 5.17b result from an environmental lapse rate of 5 C° per 1000 m (3 F° per 1000 ft), which is less than both the DAR and the MAR.

Thus, the parcel of air is forced to settle back to its original position because it is cooler (higher density, less buoyant) than the surrounding environment. The parcel lacks a physical motive for liftoff.

With these stability relationships in mind, let's look at the most visible expressions of stability in the atmosphere and indicators of weather: clouds and fog.

Clouds and Fog

Clouds are beautiful indicators of atmospheric stability, moisture content, and weather conditions. A **cloud** is an aggregation, or grouping, of moisture droplets and ice crystals that are suspended in air and are great enough in volume and density to be visible to the human eye. Although cloud types are too numerous to fully describe in this text, their classification scheme is simple and is summarized here.

Clouds are not initially composed of raindrops. Instead, they are made up of a multitude of **moisture droplets**, each individually invisible to the human eye without magnification. An average raindrop, at 2000 μm diameter (0.2 cm, or 0.078 in.), is made up of a million or more of these moisture droplets. Scientists determined how these droplets form during condensation and what processes cause them to coalesce into raindrops.

A parcel of air may rise to an altitude where it becomes saturated—that is, the air cools to the dew-point temperature and 100% relative humidity. Further cooling of the air parcel due to lifting produces active condensation of water vapor to water. This condensation requires microscopic particles—**condensation nuclei**—which are always present in the atmosphere. In continental air masses, which average 10 billion nuclei per cubic meter, condensation nuclei are typically ordinary dust, volcanic and forest-fire soot and ash, and particles from fuel combustion. Given the air composition over cities, great

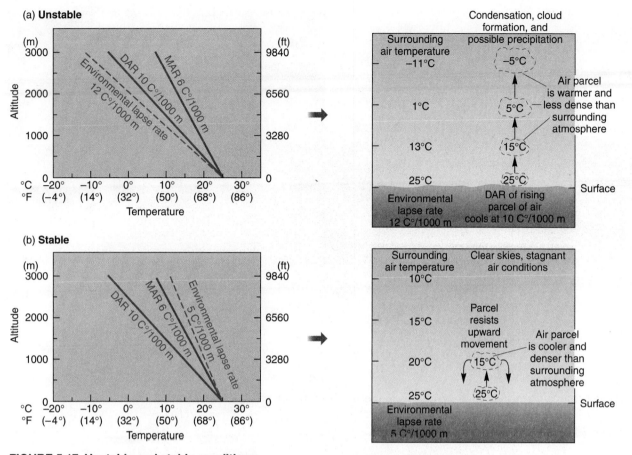

FIGURE 5.17 Unstable and stable conditions.
Specific examples of (a) unstable and (b) stable conditions in the lower atmosphere. Note the environmental lapse rate in each graph: higher than the DAR in (a), lower than the MAR in (b).

concentrations of such nuclei are available. In maritime air masses, which average 1 billion nuclei per cubic meter, the nuclei are formed from a high concentration of sea salts derived from ocean sprays. Salts are particularly attracted to moisture and thus are called *hygroscopic nuclei*.

Given the preconditions of saturated air, availability of condensation nuclei, and the presence of cooling (lifting) mechanisms in the atmosphere, active condensation occurs. Let us look at the clouds that result from these processes.

Cloud Types and Identification

In 1803, an English biologist named Luke Howard, in his article "On the Modification of Clouds," established a classification system and coined Latin names for clouds that we still use. About Howard's accomplishment his biographer stated,

Clouds were no longer exempt from human comprehension, and Howard, in contributing both a system of analysis and a full Latin nomenclature covering their families and genera, had contributed more than anyone to easing the path of understanding. . . . But the naming of clouds was a different kind of gesture for the hand of classification to have made. Here was the naming not of a solid, stable thing but of a series of self-canceling evanescences [disappearing entities]. Here was the naming of a fugitive presence that hastened to its onward dissolution. Here was the naming of clouds.*

Altitude and *shape* are key to cloud classification. Clouds come in three basic forms—flat, puffy, and wispy—which occur in 4 primary altitude classes and 10 basic types. Clouds that develop horizontally—flat and layered—are *stratiform* clouds. Those that develop vertically—puffy and globular—are *cumuliform*. Wispy clouds, usually quite high in altitude and composed of ice crystals, are *cirroform*. These three basic forms occur in 4 altitudinal classes: low, middle, high, and those that vertically develop through these altitude classes. Figure 5.18 illustrates the general appearance of the basic classes and types of clouds and includes representative photographs.

Low clouds, ranging from the surface up to 2000 m (6500 ft) in the middle latitudes, are simply called

*R. Hamblyn, *The Invention of Clouds, How an Amateur Meteorologist Forged the Language of the Skies* (New York: Farrar, Straus, and Giroux, 2001), pp. 165, 171.

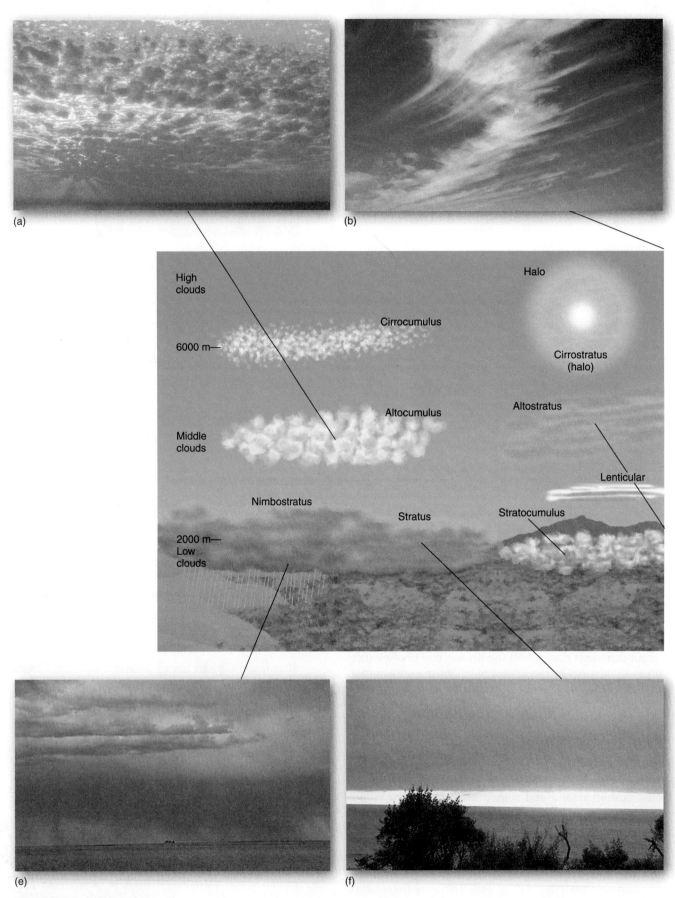

FIGURE 5.18 Principal clouds.
Principal cloud types, classified by form (cirroform, stratiform, and cumuliform) and altitude (low, middle, high, and vertically developed across altitude): (a) altocumulus, (b) cirrus, (c) cirrostratus, (d) cumulonimbus, (e) nimbostratus, (f) stratus, (g) altostratus, and (h) cumulus. [Photos c, e, and f by author; photos a, b, d, g, and h by Bobbé Christopherson.]

(c)

(d)

Cirrus

Cirrostratus

Anvil-shaped head

Clouds with
vertical
development

Cumulonimbus

Cumulus
(fair weather)

(g)

(h)

stratus or *cumulus* (Latin for "layer" and "heap," respectively). **Stratus** clouds appear dull, gray, and featureless. When they yield precipitation, they are **nimbostratus** (*nimbo-* denotes precipitation or rain-bearing), and their showers typically fall as drizzling rain (Figure 5.18e).

   Cumulus clouds appear bright and puffy, like cotton balls (Figure 5.18h). When they do not cover the sky, they float by in infinitely varied shapes. Vertically developed cumulus clouds extend beyond low altitudes into middle and high altitudes and are illustrated to the far right in Figure 5.18.

   Sometimes near the end of the day, lumpy, grayish, low-level **stratocumulus** clouds may fill the sky in patches. Near sunset, these spreading puffy stratiform remnants may catch and filter the Sun's rays, sometimes indicating clearing weather.

   Stratus and cumulus middle-level clouds are denoted by the prefix *alto-*, as in *altostratus*. They are made of water droplets and, when cold enough, can be mixed with ice crystals. **Altocumulus** clouds, in particular, represent a broad category that occurs in many different styles: patchy rows, wave patterns, a "mackerel sky," or lens-shaped (lenticular) clouds (Figure 5.18a).

   Clouds occurring above 6000 m (20,000 ft) are composed principally of ice crystals in thin concentrations. These wispy filaments, usually white except when colored by sunrise or sunset, are **cirrus** clouds (Latin for "curl of hair"), sometimes dubbed mare's tails. Cirrus clouds look as if an artist took a brush and placed delicate feathery strokes high in the sky (Figure 5.18b). Often, cirrus clouds are associated with an oncoming storm, especially if they thicken and lower in elevation. The prefix *cirro-*, as in *cirrostratus* and *cirrocumulus*, indicates other high clouds that form a thin veil or puffy appearance, respectively.

   A cumulus cloud can develop into a towering giant called **cumulonimbus** (again, *-nimbus* denotes precipitation). Such clouds are called thunderheads because of their shape and associated lightning and thunder (Figure 5.19). Note the surface wind gusts, updrafts and downdrafts, heavy rain, and the presence of ice crystals at the top of the rising cloud column. High-altitude winds may shear the top of the cloud into the characteristic anvil shape of the mature thunderhead.

Fog

By international agreement, **fog** is a cloud layer on the ground, restricting visibility to less than 1 km (3300 ft). The presence of fog tells us that the air temperature and the dew-point temperature at ground level are nearly identical, producing saturated conditions. Generally, an inversion layer caps fog, with as much as 22 C° (40 F°) difference in air temperature between the ground under the fog and the clear, sunny skies above. Two forms of fog related to cooling are advection fog and radiation fog. News Report 5.2 describes fog as a water resource for some communities.

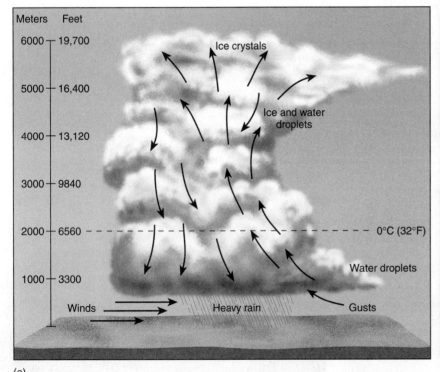

(a)

(b)

FIGURE 5.19 Cumulonimbus thunderhead.
(a) Structure and form of a cumulonimbus cloud. Violent updrafts and downdrafts mark the circulation within the cloud. Blustery wind gusts occur along the ground. (b) Space Shuttle astronauts capture a dramatic cumulonimbus thunderhead as it moves over Galveston Bay, Texas—the same area devastated by Hurricane Ike in 2008.

NEWS REPORT 5.2

Harvesting Fog

Desert organisms have adapted remarkably to the presence of coastal fog along western coastlines in subtropical latitudes. For example, sand beetles in the Namib Desert in extreme southwestern Africa harvest water from the fog. They hold up their wings so condensation collects and runs down to their mouths. For centuries, coastal villages in the deserts of Oman collected water drips from trees that came from coastal fogs.

In the Atacama Desert of Chile and Peru, residents stretch large nets to intercept the fog; moisture condenses on the netting and drips into trays, flowing through pipes to a 100,000-liter (26,000-gallon) reservoir. Large sheets of plastic mesh along a ridge of the El Tofo mountains harvest water from advection fog (Figure 5.2.1). Chungungo, Chile, receives 10,000 liters (2600 gallons) of water from 80 fog-harvesting collectors in a project developed by Canadian (International Development Research

FIGURE 5.2.1 **Fog harvesting.**
In the mountains inland from Chungungo, Chile, polypropylene mesh stretched between two posts captures advection fog for local drinking water supplies. [Photo by Robert S. Schemenauer.]

Center) and Chilean interests and made operational in 1993. At least 30 countries across the globe experience conditions suitable for this water resource technology. (See http://www.idrc.ca/en/ev-5077-201-1-DO_TOPIC.html and http://www.oas.org/dsd/publications/unit/oea59e/ch33.htm.)

Advection Fog   As the name implies, **advection fog** forms when air in one place migrates to another place where saturated conditions exist. When warm, moist air overlays cooler ocean currents, lake surfaces, or snow masses, the layer of air directly above the surface is chilled to the dew point. Off all subtropical west coasts in the world, summer fog forms in this manner (Figure 5.20).

Another type of fog forms when cold air flows over the warm water of a lake, ocean surface, or even a swimming pool. An **evaporation fog**, or steam fog, may form as the water molecules evaporate from the water surface into the cold overlying air, effectively humidifying the air. When visible at sea, the term *sea smoke* is applied to this shipping hazard. Figure 5.21 shows an evaporation fog over a lake at sunrise.

A type of advection fog, involving the movement of air, forms when moist air is forced to higher elevations along a hill or mountain. This upslope lifting leads to cooling by expansion as the air rises. The resulting **upslope fog** forms a stratus cloud at the level of saturation. Along the Appalachians and the eastern slopes of the Rockies, such fog is common in winter and spring. Another fog associated with topography is **valley fog** (Figure 5.22). Because cool air is denser, it settles in low-lying areas, producing a fog in the chilled, saturated layer near the ground.

FIGURE 5.20 **Advection fog.**
San Francisco's Golden Gate Bridge shrouded by an advection fog characteristic of summer conditions along a western coast. [Photo by author.]

FIGURE 5.21 Evaporation fog at sunrise.
Evaporation fog (sea smoke) at dawn on a cold morning at
Donner Lake, California. Later that morning as temperatures
rose, what do you think happened to the evaporation fog?
[Photo by Bobbé Christopherson.]

Radiation Fog   When radiative cooling of a surface chills
the air layer directly above that surface to the dew-point
temperature, creating saturated conditions, **radiation fog**
occurs (Figure 5.23). This fog occurs especially on clear
nights over moist ground; it does not occur over water,
because water does not cool appreciably overnight. Slight
movements of air deliver even more moisture to the cooled
area for more fog formation of greater depth.

In Figure 5.24, can you tell what two kinds of fog are
pictured? The river water is warmer than the cold overlying air, producing an evaporation fog, especially beyond
the bend in the river. The moist farmlands have radiatively cooled overnight, chilling the air along the surface
to the dew point and active condensation. You see wisps of
radiation fog as light air movements flow from right to
left in the photo.

FIGURE 5.23 Radiation fog.
Satellite image of a radiation fog in Central California, November
20, 2002. This fog is locally known as a tule fog (pronounced
"toolee") because of its association with the tule (bulrush)
plants that line the low-elevation islands and marshes of
the Sacramento River and San Joaquin River delta regions.
[*Terra* image courtesy of MODIS Rapid Response Team,
GSFC/NASA.]

Specific conditions of humidity, stability, and cloud coverage occur in regional, homogenous masses of air. These
air masses interact to produce weather patterns—our next
topic.

Air Masses

Each area of Earth's surface imparts its varying characteristics to the air it touches. Such a distinctive body of
air is an **air mass**, and it initially reflects the characteristics of its *source region*. For example, weather reporters

FIGURE 5.22 Valley fog.
Cold air settles in the valleys of the Appalachian Mountains,
chilling the air to the dew point and forming a valley fog.
[Photo by author.]

FIGURE 5.24 Two kinds of fog.
The cold air above the river creates evaporation fog along
portions of the channel, whereas radiation fog is forming over
the farmland. [Photo by Bobbé Christopherson.]

speak of "a cold Canadian air mass" or "moist tropical air mass."

The longer an air mass remains stationary over a region, the more definite its physical attributes become. Within each air mass there is a homogeneity of temperature and humidity that sometimes extends through the lower half of the troposphere. Air masses generally are classified according to the moisture and temperature characteristics of their source regions:

1. *Moisture*—designated **m** for maritime (wetter) and **c** for continental (drier).
2. *Temperature* (latitude)—designated **A** (arctic), **P** (polar), **T** (tropical), **E** (equatorial), and **AA** (antarctic).

The principal air masses that affect North America in winter and summer are mapped and described in Figure 5.25.

Continental polar (cP) air masses form only in the Northern Hemisphere and are most developed in winter when they dominate cold weather conditions. An area covered by cP air in winter experiences cold, stable air, clear skies, high pressure, and anticyclonic wind flow. The Southern Hemisphere lacks the necessary continental masses (continentality) at high latitudes to produce continental polar characteristics.

Maritime polar (mP) air masses in the Northern Hemisphere are northwest and northeast of the North American continent, and within them cool, moist, unstable conditions prevail throughout the year. The Aleutian and Icelandic subpolar low-pressure cells reside within these mP air masses, especially in their well-developed winter state.

Two maritime tropical (mT) air masses—the *mT Gulf/Atlantic* and the *mT Pacific*—influence North America. The humidity experienced in the East and Midwest is created by the mT *Gulf/Atlantic* air mass, which is particularly unstable and active from late spring to early fall. In contrast, the mT *Pacific* is stable to conditionally unstable and generally much lower in moisture content and available energy.

Please review Figure 4.13 and the discussion of subtropical high-pressure cells off the coast of North America—the moist, unstable conditions on the western edge of the Atlantic (east coast) and the drier, stable conditions on the eastern edge of the Atlantic (west coast of Europe). These conditions, coupled, respectively, with warmer and cooler ocean currents, produce the characteristics of each air mass source region.

Air Mass Modification

As air masses migrate from their source regions, their temperature and moisture characteristics modify and slowly take on the characteristics of the land over which they pass. For example, an mT Gulf/Atlantic air mass may carry humidity to Chicago and on to Winnipeg but gradually will lose its initial characteristics of high humidity and warmth with each day's passage northward.

Similarly, below-freezing temperatures occasionally reach into southern Texas and Florida, brought by an invading winter cP air mass from the north. However, that air mass warms from the −50°C (−58°F) of its source region in central Canada. In winter, as a cP air mass moves southward over warmer land, it moderates, warming especially after it leaves areas covered by snow.

Modification of cP air as it moves south and east produces snowbelts that lie to the east of each of the Great Lakes. As below-freezing cP air passes over the warmer Great Lakes, it absorbs heat energy and moisture from the lake surfaces and becomes *humidified*. This enhancement produces heavy lake-effect snowfall downwind into Ontario, Québec, Michigan, Pennsylvania, and New York—some areas receiving in excess of 250 cm (100 in.) in average snowfall a year (Figure 5.26).

A couple of such storms in western New York serve as examples. In October 2006, a lake-effect storm dubbed the "surprise storm" dropped 0.6 m (2 ft) of wet snow.

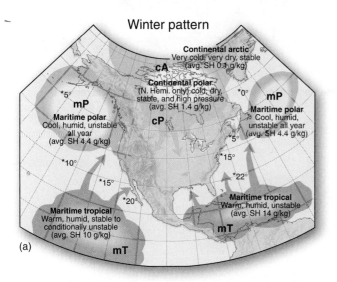

Winter pattern

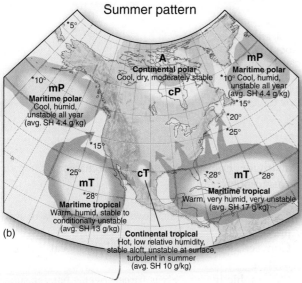

Summer pattern

FIGURE 5.25 Principal air masses.
Air masses that influence North America and their characteristics in (a) winter and (b) summer. [*Temperatures shown are sea-surface temperatures (SSTs) in °C.]

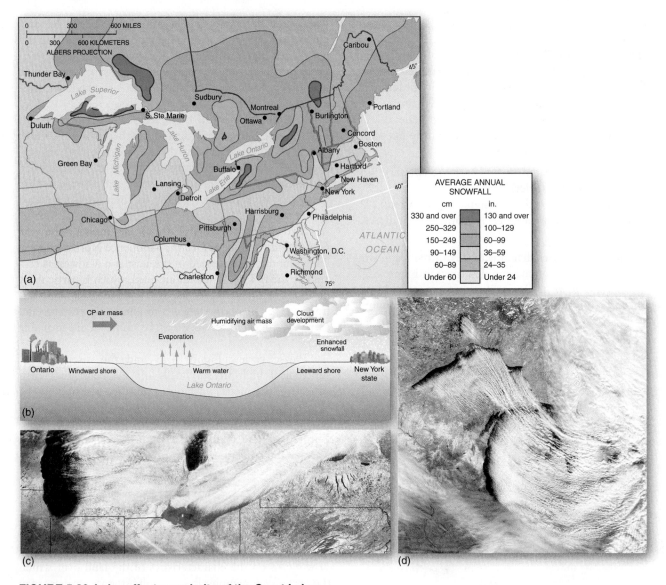

FIGURE 5.26 Lake-effect snowbelts of the Great Lakes.
(a) Locally, heavy areas of snowfall are associated with the lee side of each of the Great Lakes. In winter, cold cP and cA air masses pass from colder land surfaces across the relatively warmer water of these lakes. (b) The air masses are warmed and humidified (water vapor is added) from the lake water. The humid, now unstable air yields heavy snowfall as it moves onshore and becomes chilled. The strongest effect is generally limited to about 50 km (30 mi) inland up to 100 km (60 mi). (c) and (d) Satellite images show the lake-effect weather. [(a) Snowfall data from the *Climatic Atlas of the United States* (Washington, DC: Department of Commerce, NOAA, 1983), p. 53; (c) *Terra* image, December 7, 2002, courtesy of the MODIS Rapid Response Team; and (d) *OrbView-2* image, Dec. 5, 2000, courtesy *SeaWiFS* Project, GeoEye; both NASA/GSFC.]

This was the earliest date for such a storm in 137 years of records and remarkable because the ratio of snow to water content was 6:1. An "avalanche of a storm" hit in February 2007 when 2 m (6.5 ft) of snow fell in 1 day and over 3.7 m (12 ft) over 7 days in the same area.

Atmospheric Lifting Mechanisms

For air masses to cool adiabatically (by expansion) and to reach the dew-point temperature and saturate, condense, form clouds, and perhaps precipitate, they must lift and rise in altitude. Four principal lifting mechanisms operate in the atmosphere:

• Convergent lifting—air flows toward an area of low pressure
• Convectional lifting—stimulated by local surface heating
• Orographic lifting—air is forced over a barrier such as a mountain range
• Frontal lifting—along the leading edges of contrasting air masses

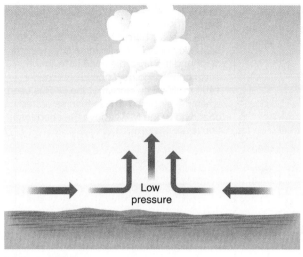

(a) Convergent

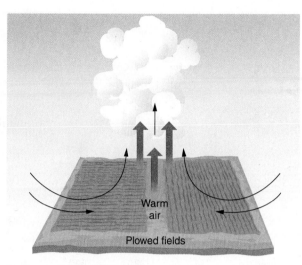

(b) Convectional (local heating)

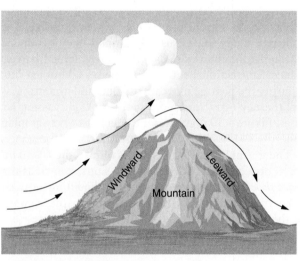

(c) Orographic (barrier)

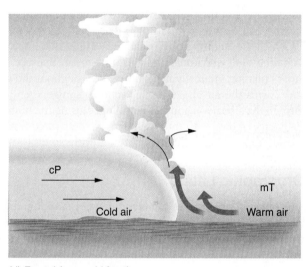

(d) Frontal (e.g., cold front)

FIGURE 5.27 Four atmospheric lifting mechanisms.
(a) Convergent lifting. (b) Convectional lifting. (c) Orographic lifting. (d) Frontal lifting.

Descriptions of all four mechanisms follow and are shown in Figure 5.27.

Convergent Lifting

Air flowing from different directions into the same low-pressure area is converging, displacing air upward in **convergent lifting**. All along the equatorial region, the southeast and northeast trade winds converge, forming the intertropical convergence zone (ITCZ) and areas of extensive uplift, towering cumulonimbus cloud development, and high average annual precipitation (see Figures 4.12 and 4.11).

Convectional Lifting

When an air mass passes from a maritime source region to a warmer continental region, heating from the warmer land causes lifting in the air mass. Other sources of surface

heating might include an urbanized area (heat island) or an area of dark soil in a plowed field—the warmer surfaces produce **convectional lifting**. If conditions are *unstable*, initial lifting is sustained and clouds develop.

Figure 5.28 illustrates convectional lifting stimulated by local heating, with unstable conditions present in the atmosphere. The rising parcel of air continues its ascent because it is warmer (less dense) than the surrounding environment.

In the figure, the MAR is used above the *lifting conden-sation level*, which is visible along the cloud base, where the rising air mass becomes saturated. Buoyancy is added at this level through the release of latent heat by condensa-tion. Continued lifting above this altitude produces cool-ing of the air temperature at the MAR and the dew-point temperature.

Florida's precipitation generally illustrates both these lifting mechanisms: convergence and convection. Heating of the land produces convergence of onshore winds from

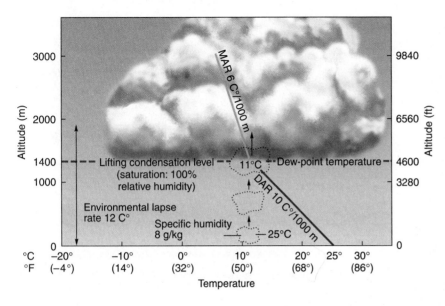

FIGURE 5.28 Local heating and convection.
Given specific humidity, temperature, and unstable conditions within the lifting parcel of air and the unstable conditions in the environment, the dew point is reached at 1400 m (4600 ft). Note that the DAR is used when the air parcel is less than saturated, changing to the MAR above the condensation level.

the Atlantic and the Gulf of Mexico. As an example of local heating and convectional lifting, Figure 5.29 depicts a day on which the landmass of Florida was warmer than the surrounding Gulf of Mexico and Atlantic Ocean. Because the Sun's radiation gradually heats the land throughout the day and warms the air above it, convectional showers tend to form in the afternoon and early evening, causing the highest frequency of days with thunderstorms in the United States. Florida appears highlighted and painted with clouds.

Orographic Lifting

The physical presence of a mountain acts as a topographic barrier to migrating air masses. **Orographic lifting** occurs when air rises upslope as it pushes against a mountain. It

cools adiabatically. Stable air forced upward in this manner may produce stratiform clouds, whereas unstable or conditionally unstable air usually forms a line of cumulus and cumulonimbus clouds. An orographic barrier enhances convectional activity and causes additional lifting during the passage of weather fronts and cyclonic systems, thereby extracting more moisture from passing air masses. Figure 5.30 illustrates the operation of orographic lifting.

The wetter intercepting slope is termed the *windward slope*, as opposed to the drier far-side slope, known as the *leeward slope*. Moisture condenses from the lifting air mass on the windward side of the mountain; on the leeward side the descending air mass is heated by compression, causing evaporation of any remaining water in the air.

SATELLITE

Convectional Heating and Tornado in Florida

FIGURE 5.29 Convectional activity over the Florida peninsula.
Cumulus clouds cover Florida with several cells developing into cumulonimbus thunderheads. Warm, moist air from the Gulf of Mexico and the Atlantic is lifted by local heating as it passes over the land. The remnants of Tropical Storm Edouard are slightly to the east. [*Terra* image from 9/3/2002, courtesy of the MODIS Land Rapid Response Team, NASA/GSFC.]

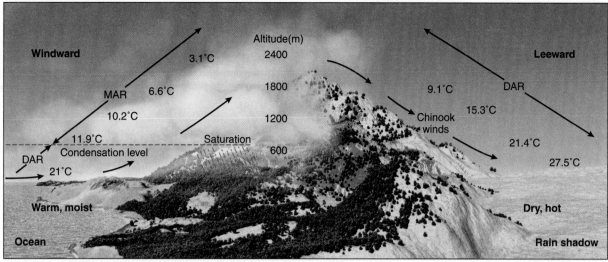

FIGURE 5.30 Orographic precipitation.
(a) Orographic barrier and precipitation patterns—unstable conditions assumed. Warm, moist air is forced upward against a mountain range, producing adiabatic cooling and eventual saturation. On the leeward slopes, compressional heating takes place as the air travels downward, producing relatively hot, dry air in the rain shadow of the mountain. (b) A leeward rain shadow is in stark contrast to the clouds of the windward side. Dust is stirred up by downslope winds. (c) Vegetation and landscapes in this true-color MODIS image from *Terra* demonstrate these precipitation patterns in western (wetter) and central and eastern (drier) Washington state. [Photo (b) by author; (c) *Terra* image courtesy of MODIS Land Response Team, NASA/GSFC.]

Thus, air can begin its ascent up a mountain warm and moist but finish its descent on the leeward slope hot and dry. In North America, *chinook winds* (called *föhn* or *foehn* winds in Europe) are the warm, downslope air flows characteristic of the leeward side of mountains. Such winds can bring a 20 C° (36 F°) jump in temperature and greatly reduced relative humidity.

The term **rain shadow** is applied to dry regions leeward of mountains (Figure 5.30a, b). East of the Cascade Range, Sierra Nevada, and Rocky Mountains, such rain-shadow patterns predominate. In fact, the precipitation pattern of windward and leeward slopes persists worldwide, as confirmed by the precipitation maps for North America in Chapter 6 and for the world in Chapter 7.

The world's greatest average annual precipitation occurs in the United States on the windward slopes of Mount Wai'ale'ale, on the island of Kaua'i, Hawai'i, which rises 1569 m (5147 ft) above sea level. Rainfall averages 1234 cm (486 in., or 40.5 ft) a year (records for 1941–2000). In contrast, the rain-shadow side of Kaua'i receives only 50 cm (20 in.) of rain annually. If no islands existed at this location, this portion of the Pacific Ocean would receive only an average 63.5 cm (25 in.) of precipitation a year.

Frontal Lifting (Warm and Cold Fronts)

The leading edge of an advancing air mass is a *front*. Vilhelm Bjerknes (1862–1951) first applied the term while working with a team of meteorologists in Norway during World War I, because it seemed to them that migrating air mass "armies" were doing battle along their fronts. A front is a place of atmospheric discontinuity—a narrow zone that is the contact between two air masses that differ in temperature, pressure, humidity, wind direction and speed, and cloud development. The leading edge of a cold air mass is a **cold front**, and the leading edge of a warm air mass is a **warm front**.

Cold Front   On weather maps, such as the one in Figure 5.31, a cold front is identified by a line marked with triangular spikes pointing in the direction of frontal movement along an advancing cP air mass. The steep face of the cold air mass suggests its ground-hugging nature, caused by its density and uniform physical character (Figure 5.31a).

Warmer, moist air in advance of the cold front is lifted upward abruptly and is subjected to the same adiabatic rates and factors of stability/instability that pertain to all lifting

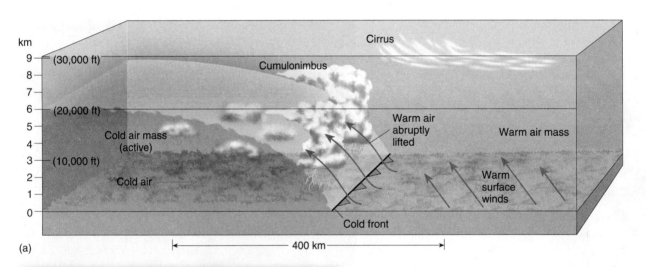

(a)

ANIMATION
Cold and Warm Fronts

(b)

FIGURE 5.31 A typical cold front.
(a) Warm, moist air is forced to lift abruptly by denser, advancing cold air. As the air lifts, it cools by expansion at the DAR, cooling to the dew-point temperature as it rises to a level of active condensation and cloud formation. Cumulonimbus clouds may produce large raindrops, heavy showers, lightning and thunder, and hail. (b) A squall line along a cold front in the Gulf of Mexico is marked by a sharp line of cumulonimbus clouds rising to 16,800 m (55,120 ft). [Space Shuttle photo from NASA.]

air parcels. A day or two ahead of the cold front's passage, high cirrus clouds appear, telling observers that a lifting mechanism is on the way. The cold front's advance is marked by a wind shift, temperature drop, and lowering barometric pressure. Air pressure reaches a local low as the line of most intense lifting passes, usually just ahead of the front itself. Clouds may build up along the cold front into characteristic cumulonimbus types and may appear as an advancing wall of clouds. Precipitation usually is hard, containing large droplets, and can be accompanied by hail, lightning, and thunder. The aftermath usually brings northerly winds in the Northern Hemisphere (southerly winds in the Southern Hemisphere), lower temperatures, increasing air pressure from the invading cooler, denser air, and broken cloud cover.

A fast-advancing cold front can cause violent lifting and create a zone slightly ahead of the front called a **squall line**. Along a squall line such as the one in the Gulf of Mexico shown in Figure 5.31b, wind patterns are turbulent and wildly changing, and precipitation is intense. The well-defined line of clouds in the photo rises abruptly, with new thunderstorms forming along the distinct front. Tornadoes also may develop along such a squall line.

Warm Front A warm front appears on weather maps as a line marked with semicircles facing the direction of frontal movement (see Figure 5.32). The leading edge of an advancing warm air mass is unable to displace cooler, passive air, which is denser. Instead, the warm air tends to push the cooler, underlying air into a characteristic wedge shape, with the warmer air sliding up over the cooler air. Thus, in the cooler-air region, a temperature inversion is present, sometimes causing poor air drainage and stagnation.

Figure 5.32 illustrates a typical warm front in which mT air is gently lifted, leading to stratiform cloud development and characteristic nimbostratus clouds and drizzly precipitation. High cirrus and cirrostratus clouds announce the advancing warm-front system. Closer to the front, the clouds lower and thicken to altostratus and then to stratus within several hundred kilometers of the front.

Midlatitude Cyclonic Systems

Weather is the short-term condition of the atmosphere, as compared with *climate*, which reflects long-term atmospheric conditions and extremes. Weather is, at the same time, both a "snapshot" of atmospheric conditions and a technical status report of the Earth–atmosphere, heat-energy budget. **Meteorology** is the scientific study of the atmosphere (*meteor* means "heavenly" or "of the atmosphere"). Embodied within this science is a study of the atmosphere's physical characteristics and motions; related chemical, physical, and geological processes; the complex linkages of atmospheric systems; and weather forecasting.

We turn to local media for the day's weather report from the National Weather Service (http://www.nws.noaa.gov) in the United States or the Weather Office of the Canadian Meteorological Centre (http://weatheroffice.ec.gc.ca) to see current satellite images and to hear weather analysis and tomorrow's forecast. Internationally, the World Meteorological Organization coordinates weather information (see http://www.wmo.ch/). Many weather sources and related topics are found in this chapter in the *Elemental Geosystems* Student Learning Center.

The conflict between contrasting air masses along fronts leads to conditions appropriate for the development of a **midlatitude cyclone**, or **wave cyclone**, a migrating center of low pressure, with converging, ascending air, spiraling inward counterclockwise in the Northern Hemisphere (or converging clockwise in the Southern Hemisphere). The pressure gradient force, the Coriolis force, and surface friction generate this cyclonic motion (see Chapter 4). Wave cyclones form a dominant type of weather pattern in the middle and higher latitudes of both the Northern and Southern Hemispheres and act as a catalyst for air mass

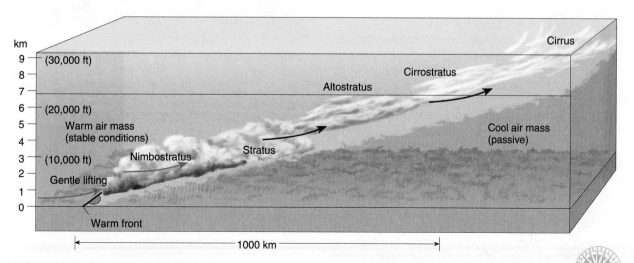

FIGURE 5.32 A typical warm front.
Note the sequence of cloud development as the warm front approaches. Warm air slides upward over a wedge of cooler, passive air near the ground. Gentle lifting of the warm, moist air produces stratus and nimbostratus clouds and drizzly rain showers, in contrast to the more dramatic precipitation associated with the passage of a cold front.

ANIMATION
Cold and Warm Fronts

conflict. Such a *midlatitude cyclone*, or extratropical cyclone, can be born along the polar front, particularly in the region of the Icelandic and Aleutian subpolar low-pressure cells in the Northern Hemisphere (see Figures 4.10 and 4.12).

Life Cycle of a Midlatitude Cyclone

Figure 5.33 shows the birth, maturity, and death of a typical midlatitude cyclone in four stages along with an idealized weather map. Weather conditions for selected

stations are noted using standard weather station symbols. On the average, a midlatitude cyclone takes 3–10 days to progress through these stages and cross the North American continent.

Along the polar front, cold and warm air masses converge and conflict. The polar front forms a discontinuity of temperature, moisture, and winds that establishes potentially unstable conditions. However, for a wave cyclone to form along the polar front, a point of surface air *convergence* must be matched by a compensating area of

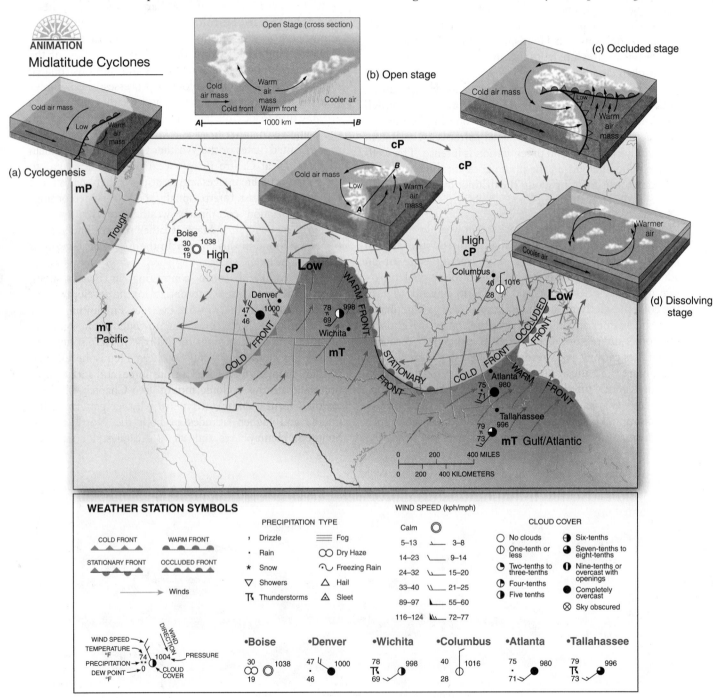

FIGURE 5.33 Idealized stages of a midlatitude wave cyclone.
(a) Cyclogenesis is noted where surface convergence and lifting begin. (b) The open stage. (c) The occluded stage. (d) The dissolving stage is reached at the end of the storm track as the cyclone spins down, no longer energized by the latent heat from condensing moisture. Standard weather symbols are in the inset box. After studying the text and the map, can you describe conditions in Boise, Denver, Wichita, Columbus, Atlanta, and Tallahassee depicted on this weather map?

air *divergence* aloft. Even a slight disturbance along the polar front, perhaps a small change in the path of the jet stream, can initiate the converging, ascending motion of a surface low-pressure system (Figure 5.33a).

Storm Tracks Cyclonic storms—1600 km (1000 mi) wide—and their attendant air masses move across the continent along *storm tracks*, which shift latitudinally with the Sun, crossing North America farther to the north in summer and farther to the south in winter. Development and strengthening of a midlatitude wave cyclone is known as *cyclogenesis*. A map of actual storm tracks for March 1991 in Figure 5.34b demonstrates several areas of cyclogenesis: the northwest over the Pacific Ocean, the Gulf of Mexico (Gulf Lows), the eastern seaboard (Nor'easter, Hatteras Low), and the Arctic. Cyclonic circulation also frequently develops on the lee side of mountain ranges, as along the Rockies from Alberta (Alberta Clipper) south to Colorado (Colorado Low). By crossing the mountains,

such systems gain access to the moisture-laden, energy-rich mT air masses from the Gulf of Mexico and are strengthened.

Open Stage As the midlatitude cyclone matures, the counterclockwise flow (in the Northern Hemisphere) draws the cold air mass from the north and west and the warm air mass from the south. A characteristic open stage forms, as shown in Figure 5.33b and in the enlarged cross section (line A–B).

Such an open stage of a midlatitude cyclone occurred April 20, 2000, with low pressure centered over the Iowa-Missouri border. Figure 5.35 shows you a portion of the daily weather map for that day and an image from *SeaWiFS* sensors aboard satellite *OrbView-2*. Note the cloud pattern stretching along the cold front and the swirling clouds around the low. Compare the cold front and warm front and overall patterns with Figure 5.33.

Occluded Stage Because a cP air mass is more homogeneous in temperature and pressure than an mT air mass, the cold front leads a denser, more unified mass and therefore moves faster than the warm front. Cold fronts can average 40 kmph (25 mph), whereas warm fronts average 16–24 kmph (10–15 mph). Thus, a cold front may overrun the cyclonic warm front, producing an **occluded front** within the overall cyclonic circulation (Figure 5.33c). Precipitation may be moderate to heavy initially and then decrease as the warmer air wedge is lifted higher by the advancing cold air mass.

The final dissolving stage of the midlatitude cyclone occurs when its lifting mechanism is completely cut off from the warm air mass, which was its source of energy and moisture (Figure 5.33d). Remnants of the cyclonic system then dissipate in the atmosphere.

To improve data gathering and weather forecasting, a massive modernization is underway in the agencies responsible. News Report 5.3 (p. 177) overviews aspects of the science of weather forecasting.

Violent Weather

Weather provides a continuous reminder of the flow of energy across the latitudes that at times can set into motion destructive, violent weather conditions. In dollar value, weather-related damage is increasing each year as population increases in areas prone to violent weather and as climate change intensifies weather anomalies. We focus on thunderstorms, tornadoes, and hurricanes.

Weather is often in the news, as weather-related destruction has risen more than 500% over the past two decades. Weather-related losses exceeded $10 billion per year through the 1990s, eclipsing the previous average of less than $2 billion a year. Each year of the twenty-first century has far surpassed the 1990s. Hurricane Katrina and the 2005 season reached $154 billion USD in damages and Hurricanes Gustav and Ike (2008) nearing

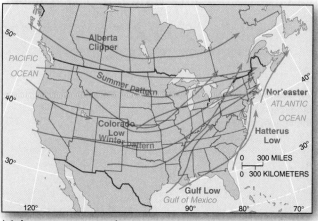

(a) Average storm tracks

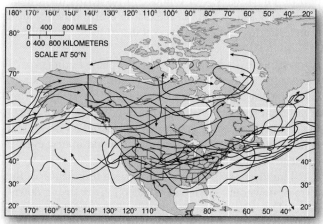

(b) Actual storm tracks in March 1991

FIGURE 5.34 Typical and actual storm tracks.
(a) Cyclonic storm tracks over North America vary seasonally. The tracks indicate several locations of cyclogenesis; note regional names. (b) Actual cyclonic tracks during March 1991 over North America. [(b) From *Storm Data*, vol. 33, no. 3 (March 1991), Asheville, NC: NOAA/NESDIS, National Climatic Data Center.]

(a)   3:00 P.M. EST

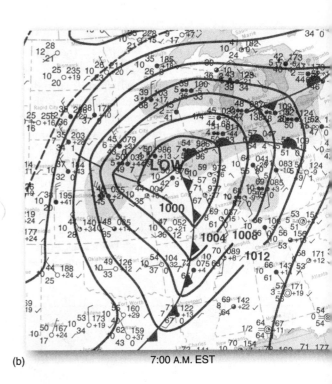

(b)   7:00 A.M. EST

FIGURE 5.35 Open stage of a midlatitude cyclone.
(a) *SeaWiFS* image of a cyclonic system over the Midwest. The cloud patterns are areas of precipitation; clear skies are behind the cold front as cP air mass covers the landscape. (b) A segment from the April 20, 2000, daily weather map, about 8 hours earlier than the image, showing the low-pressure system centered on 997.6 mb (29.45 in.). Counterclockwise winds circulate around the low. [(a) *SeaWiFS* image used with permission of GeoEye/NASA. All rights reserved. (b) Segment of *Daily Weather Map* courtesy of NWS, NOAA.]

$50 billion USD in damages! Government research and monitoring of violent weather is centered at NOAA's National Severe Storms Laboratory and Storm Prediction Center—see http://www.nssl.noaa.gov, http://www.spc.noaa.gov/, and http://www.esrl.noaa.gov/; consult these sites for each of the topics that follow.

Thunderstorms

Tremendous energy is liberated by the condensation of large quantities of water vapor. This process locally heats the air, causing violent updrafts and downdrafts as rising parcels of air pull surrounding air into the column and as the frictional drag of raindrops pulls air toward the ground. Giant cumulonimbus clouds can create dramatic weather moments—squall lines of heavy precipitation, lightning, thunder, hail, blustery winds, and tornadoes. Thunderstorms may develop within an air mass, in a line along a front (particularly a cold front), or where mountain slopes cause orographic lifting.

Thousands of thunderstorms occur on Earth at any given moment. Equatorial regions and the ITCZ experience many of them, as exemplified by the city of Kampala in Uganda, East Africa (north of Lake Victoria). This city sits virtually on the equator and annually averages 242 days of thunderstorms—a record. Figure 5.36 shows the distribution of annual days with thunderstorms across the United States and Canada. You can see that in North America most thunderstorms occur in areas dominated by mT air masses.

Atmospheric Turbulence Most airplane flights experience at least some turbulence—the aircraft at high speed encountering air of different densities or air layers moving at different speeds and directions. This is a natural state of the atmosphere and passengers are asked to keep seat belts fastened when in their seats to avoid injury, even if the seat-belt light is turned off.

Thunderstorms can produce severe turbulence in the form of downbursts, which are exceptionally strong downdrafts. Downbursts are classified by size: a *macroburst* is at least 4.0 km (2.5 mi) wide and in excess of 210 kmph (130 mph), whereas a *microburst* is smaller in size and speed. A microburst causes rapid changes in wind speed and direction, and it produces the dreaded *wind shear* that can bring down aircraft. Such turbulence events are short-lived and hard to detect, although the Forecast Systems Laboratory, among others, is making progress in developing forecasting methods.

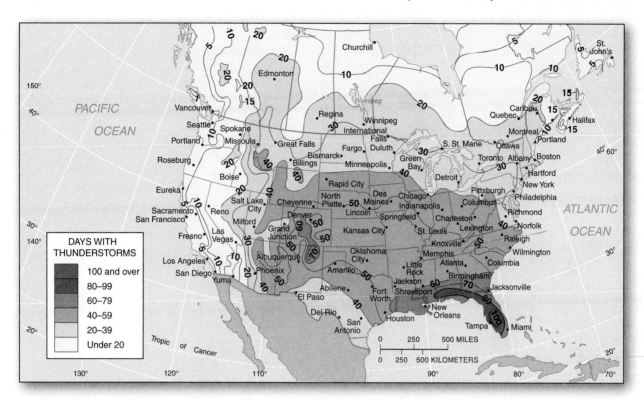

FIGURE 5.36 Thunderstorm occurrence.
Average annual number of days with thunderstorms. [Adapted from the National Weather Service and Map Series 3, *Climatic Atlas of Canada*, Atmospheric Environment Service.]

Lightning and Thunder An estimated 8 million lightning strikes occur each day on Earth. **Lightning** refers to flashes of light caused by enormous electrical discharges—tens to hundreds of millions of volts—that briefly superheat the air to temperatures of 15,000° to 30,000°C (27,000°–54,000°F). The violent expansion of this abruptly heated air sends shock waves through the atmosphere—the sonic bangs known as **thunder**. The greater the distance a lightning stroke travels, the longer the thunder echoes. Lightning at great distance from the observer may not be accompanied by thunder; this is *heat lightning*.

Lightning poses a hazard to aircraft and to people, animals, trees, and structures. Individuals should acknowledge that certain precautions are mandatory when the threat of a lightning discharge is near. Nearly 200 deaths and thousands of injuries occur each year in the United States and Canada; therefore, people should heed severe storm warnings that caution them to remain indoors.

NASA's Lightning Imaging Sensor (LIS) aboard the *Tropical Rainfall Measuring Mission* (*TRMM*) satellite monitors lightning. The LIS can image lightning strikes day or night, within clouds or between cloud and ground. The sensor's data show that about 90% of all strikes occur over land in response to increased convection over relatively warmer continental surfaces, with expected seasonal shifts keyed to the high Sun, as shown in Figure 5.37a and b. (See http://thunder.msfc.nasa.gov/lis/.)

Hail Hail is common in the United States and Canada, although somewhat infrequent at any given location, hitting a specific location perhaps every 1 or 2 years in the highest-frequency areas. Annual hail damage in the United States tops $750 million. **Hail** generally forms within a cumulonimbus cloud when a raindrop circulates repeatedly above and below the freezing level in the cloud, adding layers of ice until its weight can no longer be supported by the circulation in the cloud. Pea-sized and marble-sized hail are common, although hail the size of golf balls and baseballs also is reported at least once or twice a year somewhere in North America. For larger hail to form, the frozen pellets must stay aloft for longer periods. The pattern of hail occurrence across the United States and Canada is similar to that of thunderstorms shown in Figure 5.36, dropping to zero approximately by the 60th parallel.

Derechos

Although tornadoes and hurricanes grab headlines, straight-line winds associated with thunderstorms and bands of showers cause significant damage and crop losses. These strong linear winds in excess of 26 m/s (58 mph) are known as **derechos**, or in Canada sometimes called "plow winds." A derecho in 1998 in eastern Wisconsin exceeded 57 m/s (128 mph). Researchers identified 377 events between 1986 and 2003, an average of about 21 per year. The name, coined by physicist G. Hinrichs in 1888, derives from a Spanish word meaning "direct" or "straight ahead." Wind outbursts from convective storms can produce groups of downburst clusters from a thunderstorm system. These derechos tend to blast in linear paths fanning out along curved-wind fronts over a wide swath of land.

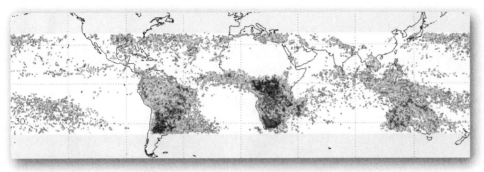

(a)   Winter (Dec. 1999, Jan. and Feb. 2000)

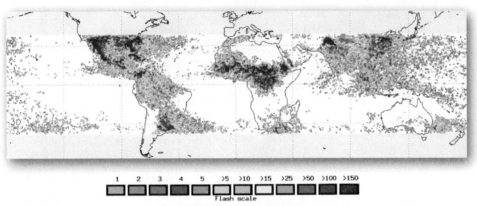

| 1 | 2 | 3 | 4 | 5 | >5 | >10 | >15 | >25 | >50 | >100 | >150 |
Flash scale

(b)   Summer (June, July, August 2000)

(c)

FIGURE 5.37 Seasonal images show global lightning.
A composite of 3 months' data derived from NASA's Lightning
Imaging Sensor (LIS) records all lightning strikes between
35° N and 35° S latitudes during (a) winter (December 1999–
February 2000) and (b) summer (June–August 2000). (c) Multiple
cloud-to-ground lightning strikes captured in a time-lapse photo.
[Images (a) and (b) courtesy of NASA's Global Hydrology and Climate Center, Marshall Spaceflight Center, Huntsville, Alabama;
(c) photo from C. Clark, NOAA Photo Library, NSSL.]

Derechos pose distinct hazards to summer outdoor
activities by overturning boats, hurling flying objects, and
causing broken trees and limbs. Their highest frequency
(about 70%) is from May to August in the region stretching
from Iowa, across Illinois, and into the Ohio River Valley in
the upper Midwest. By September and on through to April,
areas of activity, although less than in summer, migrate
southward to eastern Texas through Alabama. Derecho
conditions are similar to the strong migrating cyclonic systems and rotating thunderstorms that give birth to tornadoes, although they can occur without any tornado activity.

For more information about named derechos and maps of
derecho occurrences by year, see **http://www.spc.noaa
.gov/misc/AbtDerechos/derechofacts.htm.**

Tornadoes

The updrafts associated with a cold-front squall line and
cumulonimbus cloud development appear on satellite
images as pulsing bubbles of clouds. According to one
hypothesis, a spinning, cyclonic, rising column of mid-
troposphere-level air forms a **mesocyclone**. A mesocyclone

NEWS REPORT 5.3

Weather Forecasting

As instruments and software and our knowledge of the interactions that produce weather improve, so too will the accuracy of weather forecasts. *Synoptic analysis* is the evaluation of weather data collected at a selected time. Building a database of wind, pressure, temperature, and moisture conditions is key to *numerical* (computer-based) *weather prediction* and the development of weather-forecasting models. Development of numerical models is a great challenge because the atmosphere operates as a nonlinear (irrational) system, tending toward chaotic behavior. Slight variations in input data or slight changes in the basic assumptions of the model's behavior can produce widely varying forecasts.

Weather data necessary for the preparation of a synoptic map and forecast include:

- Barometric pressure (sea level and altimeter setting)
- Pressure tendency (steady, rising, falling)
- Surface air temperature
- Dew-point temperature
- Wind speed, direction, and character (gusts, squalls)
- Type and movement of clouds
- Current weather
- State of the sky (current sky conditions)
- Visibility, vision obstruction (fog, haze)
- Precipitation since last observation

New developments in supercomputing, an Earth-bound instrument network of 883 Automated Surface Observing System (ASOS) arrays, orbiting observation systems, and Doppler radar installations are rapidly advancing the science of the atmosphere. By 2007, 158 WSR-88D (*Weather Surveillance Radar*) Doppler radar systems as part of the NEXRAD (Next Generation Weather Radar) program were operational through the National Weather Service (125), in conjunction with the

FIGURE 5.3.1 **Weather installation.**
Doppler radar installation at the Indianapolis International Airport operated by the National Weather Service. The radar antenna is sheltered within the dome structure. [Photo by Bobbé Christopherson.]

Federal Aviation Administration (12) and the Department of Defense (21) (Figure 5.3.1). In Canada, the National Radar Project will have 30 CWSR-98 radars in service.

ASOS sensor instrument arrays (AWOS in Canada) are essential to our surface weather-observing network (Figure 5.3.2a). The ASOS installations include: rain gauge (tipping bucket), temperature/dew-point sensor, barometer, present weather identifier, wind speed indicator, direction sensor, cloud height indicator, freezing rain sensor, thunderstorm sensor, and visibility sensor, among other items.

NOAA operates the Earth Systems Research Laboratory (formerly Forecast Systems Lab) in Boulder, Colorado, that is developing new forecasting tools (see http://www.esrl.noaa. gov). A few innovations include: wind profilers using radar to profile winds from the surface to high altitudes; a High Performance Computing System (hundreds of CPUs networked to act as

one, Figure 5.3.2e) to model software for 3-D weather models and other computations; improved international cooperation and weather data dissemination; and standard Advanced Weather Interactive Processing System (AWIPS, Figure 5.3.2c, d). The network of AWIPS stations (including 13 River Forecast Centers) eventually will increase to 148 stations.

The cost of weather-related destruction has risen more than 500% over the past two decades, from an average of $2 billion to $10 billion annually. Floods, droughts, ice storms, tropical cyclones and hurricanes, tornadoes, coastal storm surges, and heat waves add to the need to improve our understanding of weather. A 2005 UN Environment Programme report estimated weather-related damage losses annually (drought, floods, hail, tornadoes, derechos, tropical systems, storm surge, blizzard and ice storms, and wildfires) will exceed $1 trillion by A.D. 2040 (adjusted to current dollars).

(continued)

FIGURE 5.3.2 ASOS weather instruments and AWIPS workstation display.
(a) Automated Surface Observing System (ASOS) installation, one of 883 ASOS stations in use for data gathering as part of the US primary surface weather observing network. (b) Earth Systems Research Lab scientists develop three-dimensional modeling programs to portray weather and simulate future weather for more accurate forecasting. On the image note the gray area that depicts significant cloud water content and the red layer within the cloud where aircraft icing is forecast. The black wind barbs off the two vertical axes denote wind profiles through the atmosphere. The view is from the southeast. (c) Powerful computers are needed to process the data and produce these three-dimensional virtual images on the Advanced Weather Interactive Processing System (AWIPS), employing screens with multiple frames. Of the many items that can be displayed, here we see a simulated three-dimensional pressure analysis, water vapor and visible satellite image—including lightning strikes—and humidity. (d) An AWIPS workstation. (e) A portion of the High Performance Computing System at the ESRL.
[Photos (a) and (e) by Bobbé Christopherson; (b–d) images courtesy of the ESRL, NOAA, Boulder, Colorado.]

can range up to 10 km (6 mi) in diameter and rotate vertically within a *supercell cloud* (the parent cloud) to a height of thousands of meters (Figure 5.38a). A well-developed mesocyclone most certainly will produce heavy rain, large hail, blustery winds, and lightning; some mature mesocyclones will generate tornado activity.

As more moisture-laden air is drawn up into the circulation of a mesocyclone, more energy is liberated, and the rotation of air becomes more rapid. The narrower the mesocyclone, the faster the spin of converging parcels of air being sucked into the rotation. The swirl of the mesocyclone itself is visible, as are smaller, dark gray **funnel clouds** that pulse from the bottom side of the parent cloud. The potential of this stage of development is the lowering of a funnel cloud to Earth—a **tornado**. See Figure 5.38 for an example of a mesocyclone, wall cloud, and tornado.

A tornado can range from a few meters to a few hundred meters in diameter and can last anywhere from a few moments to tens of minutes, although they are increasing in average speed and duration over the last two decades. Greensburg, Kansas, was essentially erased from the map by an EF-5 tornado in May 2007. The statistics are unprecedented: Winds exceeded 330 kmph (205 mph) in the 2.7-km (1.7-mi) wide funnel that was on the ground along a 35-km (22-mi) track. See the Chapter 1 opening collage of photos and Figure 5.38d. Greensburg is attempting to rebuild following "Green" standards for energy conservation and efficiency.

When tornado circulation occurs over water, a **waterspout** forms, and surface water is drawn up into the funnel some 3–5m (10–16 ft). The rest of the waterspout funnel is made visible by the rapid condensation of water vapor.

Tornado Measurement and Science Pressures inside a tornado usually are about 10% less than those in the surrounding air. The inrushing convergence created by such a horizontal pressure gradient causes high wind speeds. The late Theodore Fujita, a noted meteorologist from the University of Chicago, designed the Fujita Scale (1971), which classifies tornadoes according to wind speed as indicated by related property damage.

A refinement in the scale, adopted in February 2007, led to the *Enhanced Fujita Scale*, or *EF Scale* (Table 5.1; http://www.tornadoproject.com/fscale/fscale.htm). The revision filled the need to better assess damage, correlate wind speed to damage caused, and account for structural construction quality. To assist with wind estimates, the EF Scale contains 28 Damage Indicators (DI) relating to types of structures and vegetation affected along with Degrees of Damage (DOD) ratings, both of which you can view at the URL listed below the table.

North America experiences more tornadoes than anywhere on Earth because its latitudinal position and topography permit contrasting air masses to confront one another. Tornadoes have struck all 50 states and all the Canadian provinces and territories. May and June are the peak months, as you see in Figure 5.39, based on 51 years of records. Note the comparison with statistics for the 2003–2005 period and the annual trend for tornado occurrence on the graph. The Storm Prediction Center in Kansas City, Missouri, provides short-term forecasting for thunderstorms and tornadoes to the public and to NWS field offices. Warning times of from 12 to 30 minutes are possible with current technology (see the Storm Prediction Center at http://www.spc.noaa.gov/).

Canada experiences an average of 80 observed tornadoes per year, although unpopulated rural areas go unreported. University of Leeds researchers report that in the United Kingdom observers report 60 to 80 a year. Other continents report a small number of tornadoes each year. In the United States, 51,414 tornadoes were recorded in the 59 years from 1950 through September 2008, causing almost 5000 deaths, or about 85 deaths per year. Of note is the fact that these tornadoes resulted in more than 80,000 injuries and property damage of over $27 billion for this period of record. The yearly average of $500 million in damage is rising each year.

The annual average number of tornadoes before 1990 was 787. Interestingly, since 1990 the average per year has risen to over 1000. Since 1998 the annual average again rose to 1270 tornadoes; 2004 experienced 1819 tornadoes; 2005 had 1264; 2006 had 1105; 2007 had 1200; and preliminary estimates for 2008 report more than 2000 tornadoes.

Table 5.1 The Enhanced Fujita Scale	
EF-number	**3-Second Gust Wind Speed; Damage Specs**
EF-0 Gale	105–137 kmph; 65–85 mph: *light damage*: break branches, damage chimneys.
EF-1 Weak	138–177 kmph; 86–110 mph: *moderate damage*: beginning of hurricane wind-speed designation, roof coverings peel off, mobile homes pushed off foundations.
EF-2 Strong	178–217 kmph; 111–135 mph: *considerable damage*: roofs torn off frame houses, large trees uprooted or snapped, box cars pushed over, small missiles generated.
EF-3 Severe	218–266 kmph; 136–165 mph: *severe damage*: roofs torn off well-constructed houses, trains overturned, trees uprooted, cars thrown.
EF-4 Devastating	267–322 kmph; 166–200 mph: *devastating damage*: well-built houses leveled, cars thrown, large missiles generated.
EF-5 Incredible	More than 322 kmph (200 mph): houses disintegrated, car-sized object thrown more than 100 m

See the EF-Scale pdf file at http://www.wind.ttu.edu/EFScale.pdf.

FIGURE 5.38 Twister!—tornado terror strikes.
(a) A mesocyclone forms as a rotating updraft within the thunderstorm. If a tornado forms, it will descend from the cloud base and the lower portion of the mesocyclone. (b) A supercell tornado descends from the cloud base near Spearman, Texas. Strong hail is falling to the left of the tornado. (c) A tornado emerges from the base of a wall cloud near Oakfield, Wisconsin. (d) An EF-5 tornado destroyed 95% of Greensburg, Kansas, in 2007. The empty lots and debris remain after a year as seen in this 2008 photo. [Photos by (b) Howard Bluestein, all rights reserved; (c) Don Lloyd/AP/Wide World Photos; (d) Bobbé Christopherson.]

Importantly, the number of EF-4 and EF-5 tornadoes is on the increase, reaching more than a dozen each season.

Reasons for the annual increase in tornado frequency in North America range from global climate change and more intense thunderstorm activity to better reporting through Doppler radar and a larger population with videotape recorders. Researchers are pursuing many questions with an enthusiasm for the science and an awe of the subject, as summarized by Howard Bluestein, foremost tornado researcher and meteorologist:

The source of rotation in some tornadoes appears to be preexisting vortices near the ground above where convective storms are growing; in others, the source seems to be linked to the mesocyclone. Does the mesocyclone itself descend to the ground and intensify to become the tornado, or does it trigger events near the ground that create tornadoes from other sources of rotation? Can it do both? We hope the answers to these questions are soon forthcoming. Our quest for discovery has not taken away from the respect we have

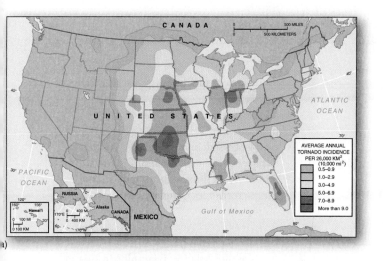

(a)

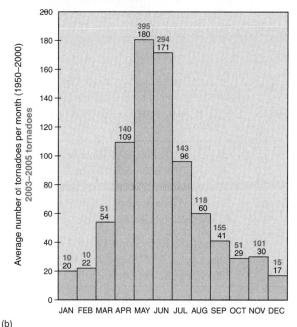

(b)

FIGURE 5.39 Tornado occurrence in the United States.
(a) Average number of tornadoes per 26,000 km² (10,000 mi²)
for 1950 to 2003. (b) Average number of tornadoes per month
1950–2000. Remarkably, the 2003–2005 average tornadoes by
month (in red numbers) show significant increase in numbers.
[Data courtesy of the Storm Prediction Center, National Weather
Service, Kansas City, Missouri.]

for the awesome power the tornado harbors, nor the
thrill for viewing the violent motions in the tornado or
the beauty of the storm. We eagerly await the next act
in the atmospheric play starring the tornado.*

Tropical Cyclones

A powerful manifestation of Earth's energy and moisture
system is the **tropical cyclone**, which originates entirely
within tropical air masses. The tropics extend from the
Tropic of Cancer at 23.5° N to the Tropic of Capricorn at
23.5° S, including the equatorial zone between 10° N and

*H. Bluestein, *Tornado Alley, Monster Storms of the Great Plains* (New
York: Oxford University Press, 1999), p. 162.

10° S. Approximately 80 tropical cyclones occur annually
worldwide. Table 5.2 presents the criteria for classification of tropical cyclones based on wind speed.

Meteorologists now think that the cyclonic motion
begins with slow-moving easterly waves of low pressure in
the trade-wind belt of the tropics, such as the Caribbean
area. It is along the eastern (leeward) side of these migrating troughs of low pressure, a place of convergence and
rainfall, where tropical cyclones form. Surface air flow
spins in toward the low-pressure area (convergence),
ascends, and flows outward aloft. This important divergence aloft acts as a chimney, pulling more moisture-laden
air into the developing system, fueled by an abundant supply of water vapor and thus the necessary latent heat to fuel
these storms. For these processes, sea-surface temperatures must exceed approximately 26°C (79°F).

The map in Figure 5.40 shows areas in which tropical
cyclones form and some of the characteristic storm tracks
traveled and the range of months during which tropical
cyclones are most likely to appear. They tend to occur
when the equatorial low-pressure trough (ITCZ) is farthest
from the equator, during the months following the summer
solstice in each hemisphere. For example, storms that strike
the southeastern United States do so mostly between
August and October, following the June 21 solstice. About
10% of all tropical disturbances have the right ingredients
to intensify into a full-fledged **hurricane** or **typhoon** (see
Table 5.2). For the North Atlantic region from 1871 to
2008 (September), 1205 tropical cyclones (storms and hurricanes) developed. The peak day in this Atlantic region
(assuming a 9-day moving average calculation) is September 10, with 64 cyclones for the 137-year period.

In the Southern Hemisphere, no hurricane was ever
observed turning from the equator into the south Atlantic
(note on the map). However, in March 2004 Hurricane
Catarina made landfall in Brazil south of the resorts
at Laguna, state of Santa Catarina. NASA's *Terra* satellite
in Figure 5.40b clearly shows an organized pattern around
a central eye and rain bands characteristic of a hurricane.
Compare the circulation patterns between Hurricane
Gilbert (see Figure 5.41) and Hurricane Catarina (5.40b);
what do you see? Note the Coriolis force in action,
counterclockwise in the Northern Hemisphere and clockwise in the Southern Hemisphere.

If this Brazilian hurricane wasn't strange enough, in
October 2005, the remnants of Tropical Storm Vince became
the first-ever Atlantic tropical cyclone to strike Spain (5.40c)!
Then in 2006, the first tropical storm to become a category 5
super typhoon developed in the Central Pacific, some
1285 km (800 mi) south of Hawai'i. Super Typhoon Ioke
continued for almost three weeks, with final remnants causing large-scale erosion along Alaskan shores. And in 2007,
Super Cyclone Gonu became the strongest tropical cyclone
on record occurring in the Arabian Sea, eventually hitting
Oman and the Arabian Peninsula. These four storms apparently are signaling changes underway in tropical meteorology
relative to the increased occurrence and intensity of cyclonic
systems, as related to higher oceanic and atmospheric

Table 5.2 Tropical Cyclone Classification		
Designation	**Winds**	**Features**
Tropical disturbance	Variable, low	Definite area of surface low pressure; patches of clouds
Tropical depression	Up to 34 knots 63 kmph (39 mph)	Gale force, organizing circulation; light to moderate rain
Tropical storm	35–63 knots 63–118 kmph (39–73 mph)	Closed isobars; definite circular organization; heavy rain; assigned a name
Hurricane (Atlantic and E. Pacific) Typhoon (W. Pacific) Cyclone (Indian Ocean, Australia)	Greater than 65 knots 119 kmph (74 mph)	Circular, closed isobars; heavy rain, storm surges; tornadoes in right-front quadrant

Saffir–Simpson Hurricane Damage Potential Scale		
Category	**Wind Speed and Central Pressure (mb)**	**Notable Atlantic Examples (rating at landfall, No. and Cent. Americas)**
1	65–82 knots 119–154 kmph (74–95 mph) > 980 mb	
2	83–95 knots 155–178 kmph (96–110 mph) 965–979 mb	1954 Hazel; 1999 Floyd; 2003 Isabel (was a cat. 5), Juan; 2004 Francis; 2008 Dolly
3	96–113 knots 179–210 kmph (111–130 mph) 945–964 mb	1985 Elena; 1991 Bob; 1995 Roxanne, Marilyn; 1998 Bonnie; 2003 Kate; 2004 Ivan (was a cat. 5), Jeanne; 2005 Dennis; Rita and Wilma (were cat. 5); 2007 Henrietta; 2008 Gustav, Ike (were cat. 4)
4	114–135 knots 211–250 kmph (131–155 mph) 920–944 mb	1979 Frederic; 1985 Gloria; 1995 Felix, Luis, Opal; 1998 Georges; 2004 Charley; 2005 Emily, Katrina (were cat. 5)
5	> 135 knots > 250 kmph (> 155 mph) > 920 mb	1935 No. 2; 1938 No. 4; 1960 Donna; 1961 Carla; 1969 Camille; 1971 Edith; 1977 Anita; 1979 David; 1980 Allen; 1988 Gilbert, Mitch; 1989 Hugo; 1992 Andrew; 2007 Dean, Felix

temperatures. (See the National Hurricane Center at http://www.nhc.noaa.gov/ or the Joint Typhoon Warning Center at http://www.npmoc.navy.mil/jtwc.html.)

Hurricanes and Typhoons Tropical cyclones are potentially the most destructive storms experienced by humans, costing an average of several hundred million dollars in property damage and thousands of lives each year. The tropical cyclone that struck Bangladesh in 1970 killed an estimated 300,000 people, and the one in 1991 claimed over 200,000. In the United States, death tolls are much lower, but still significant. The Galveston, Texas, hurricane of 1900 killed 6000; Hurricane Katrina and engineering failures killed 1836 people in Louisiana, Mississippi, and Alabama in 2005; Hurricane Audry (1957), 400; Hurricane Gilbert (1988), 318; Hurricane Camille (1969), 256; Hurricane Agnes (1972), 117. Hurricane Mitch (October 26–November 4, 1998) was the deadliest Atlantic hurricane in two centuries, killing more than 12,000 people in Central America. In 2007 for the first time in history in a single Atlantic season, two category 5 hurricanes simultaneously made landfall, Hurricane Dean (Yucatán) and Hurricane Felix (Honduras).

In 2003, Hurricane Isabel struck the Outer Banks and Cape Hatteras of North Carolina in September, causing extensive flooding and wind damage to these fragile barrier islands as well as 36 deaths and about $2 billion in damages. Hurricane Charley brought category 4 winds to Florida's Gulf Coast in 2004, causing approximately $12 billion in damage and taking about two dozen lives. The 2005 Atlantic hurricane season broke several records, including the most named tropical storms in a single year (featured in Focus Study 5.1). The 2008 season was notable for sixteen named storms, with seven landfalls featuring five major storms, including Hurricanes Gustav hitting Louisiana and Ike hitting Galveston, the southeastern Texas coast, and southern Louisiana.

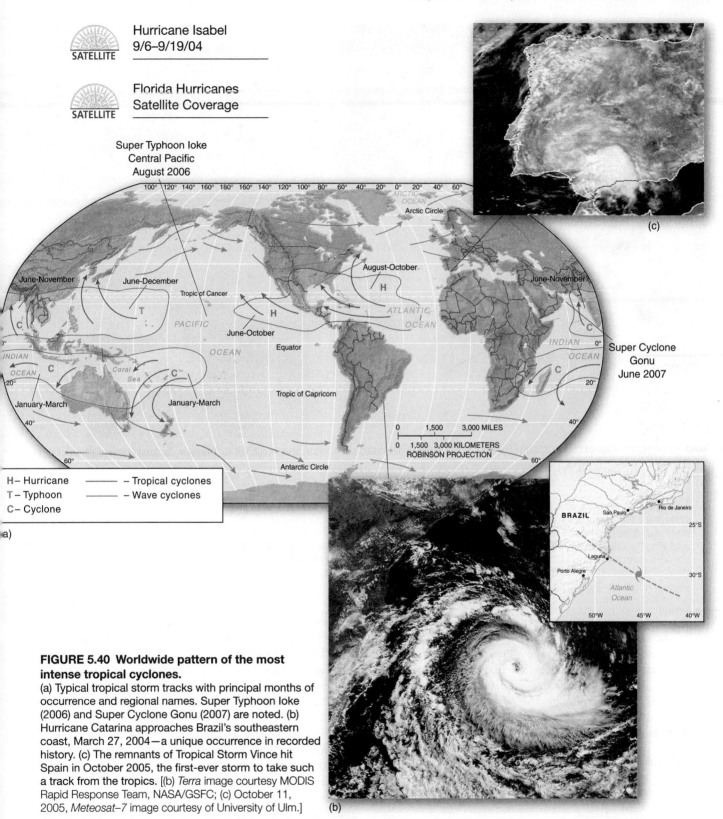

SATELLITE

Hurricane Isabel
9/6–9/19/04

SATELLITE

Florida Hurricanes
Satellite Coverage

Super Typhoon Ioke
Central Pacific
August 2006

Super Cyclone
Gonu
June 2007

H– Hurricane ——— – Tropical cyclones
T– Typhoon ——— – Wave cyclones
C– Cyclone

(a)

(c)

(b)

FIGURE 5.40 Worldwide pattern of the most intense tropical cyclones.
(a) Typical tropical storm tracks with principal months of occurrence and regional names. Super Typhoon Ioke (2006) and Super Cyclone Gonu (2007) are noted. (b) Hurricane Catarina approaches Brazil's southeastern coast, March 27, 2004—a unique occurrence in recorded history. (c) The remnants of Tropical Storm Vince hit Spain in October 2005, the first-ever storm to take such a track from the tropics. [(b) *Terra* image courtesy MODIS Rapid Response Team, NASA/GSFC; (c) October 11, 2005, *Meteosat–7* image courtesy of University of Ulm.]

Warmer Atlantic waters along the 20th parallel between 20° W and 90° W characterize this more intense cycle, an impact of global warming. In 2005, Hurricanes Katrina and Rita passed over Gulf waters that were the warmest since 1890. After 2005, the second-greatest year for the number of Atlantic hurricanes was 1933 (21), followed by 1995 (19), since record keeping began in 1870.

New research appearing in *Nature* in September 2008 confirms earlier studies of the link between higher air temperatures and higher sea-surface temperatures and the increasing intensity of tropical storms:

Atlantic tropical cyclones are getting stronger on average, with a 30-year trend that has been related

(a)

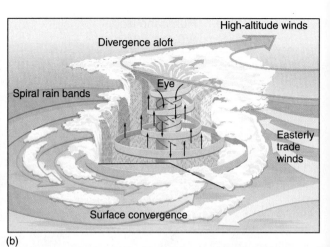

(b)

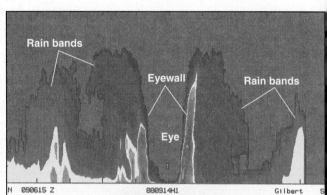

(c)

FIGURE 5.41 Profile of a hurricane.
Category 5 Hurricane Gilbert, September 13, 1988: (a) *GOES-7* satellite image; (b) a stylized
portrait of a mature hurricane, drawn from an oblique view (cutaway shows the eye, rain bands,
and wind flow patterns); (c) side-view radar image (Side-Looking Airborne Radar—SLAR; dBZ in
the legend refers to the decibel value used to rate the radar response) taken from an aircraft flying
through the central portion of the storm. Rain bands of greater cloud density are false-colored in
yellows and reds. [(a) and (c) National Hurricane Center, Miami, and NOAA.]

to an increase in ocean temperatures. . . . The results presented here are conclusive in showing significant increasing trends in the satellite-derived lifetime-maximum wind speeds of the strongest tropical cyclones globally, and are quantitatively consistent with the heat-engine theory of cyclone intensity. Thus, as seas warm, the ocean has more energy that can be converted to tropical cyclone wind.*

Physical Structure Fully organized tropical cyclones possess an intriguing physical appearance. They range in diameter from a compact 160 km (100 mi), to 960 km (600 mi), to some western Pacific typhoons that reach 1300–1600 km (800–1000 mi). Vertically, these storms dominate the full height of the troposphere. The inward-spiraling clouds form dense *rain bands*, with a central area designated the *eye*, around which a thunderstorm cloud called the *eyewall* swirls, producing the area of most intense precipitation. The *eye* remains an enigma, for in the midst

*J. B. Elsner, J. P. Kossin, and T. H. Jagger, "The increasing intensity of the strongest tropical cyclones," *Nature* 455 (September 4, 2008): 92–95.

of devastating winds and torrential rains the eye has quiet, warm air with even a glimpse of blue sky or stars possible.

Hurricane Gilbert (September 1988) illustrates the general characteristics of a hurricane. For a Western Hemisphere hurricane, Gilbert achieved the record size (1600 km or 1000 mi in diameter) and the second-lowest barometric pressure (888 mb, or 26.22 in. of mercury), after Wilma's 882 mb (26.02 in.) record. In addition, it sustained winds of 298 kmph (185 mph), with peak winds exceeding 320 kmph (200 mph). Figure 5.41 portrays Gilbert from overhead and in cross section. From these images you can see the vast counterclockwise circulation, the central eye, eyewall, rain bands, and overall hurricane appearance.

Such a storm travels at 16–40 kmph (10–25 mph). When it moves ashore, or makes *landfall*, winds push *storm surges* of seawater inland. The strongest winds of a tropical cyclone usually are recorded in its right-front quadrant (relative to the storm's directional path), where dozens of fully developed tornadoes may be at the time of landfall. For example, Hurricane Camille in 1969 had up to 100 tornadoes

(*text continued on page 187*)

FOCUS STUDY 5.1

Atlantic Hurricanes and the 2005 Season

Forecasters at the National Hurricane Center (NHC) in Miami were busy throughout the 1995–2008 hurricane seasons. This was the most active 14-year period in the history of the NHC, with 195 named tropical storms, including 107 hurricanes (48 intense, meaning category 3 or higher). This was a record level of activity and individual storm intensity despite the reduced number of tropical cyclones during the 1997 El Niño season.

Statistically, the damage caused by tropical cyclones is increasing substantially as more and more development occurs along susceptible coastlines, whereas loss of life is decreasing in most parts of the world owing to better forecasting of these storms. The journal *Science* offered this "Perspective":

> The risk of human losses is likely to remain low, however, because of a well-established warning and rescue system and ongoing improvements in hurricane prediction. A main concern is the risks of high damage costs (up to $100 billion in a single event) because of ongoing population increases in coastal areas and increasing investment in buildings and extensive infrastructure in general.*

As Atlantic hurricane seasons go, 1995 was the third most active since record keeping began in 1870. Of 19 tropical storms, 11 became hurricanes, and 5 of these achieved category 3 status or higher. Until 2005, only 1933 with 21 tropical storms and 1969 with 12 hurricanes exceeded these two totals. A map of the 1995 season and information on each storm is on our *Elemental Geosystems* Student Learning Center web site under this chapter.

The 2005 Atlantic hurricane season broke several records, including the most named tropical storms in a single year, totaling 27 (the average is

*L. Bengtsson, "Hurricane threats," *Science* 293 (July 20, 2001): 441.

10); most hurricanes, 15 (the average is 5); and 7 of these were rated intense, or greater than category 3 (the average is 2). The 2005 season was the only time three category 5 storms were in the Gulf of Mexico (Katrina, Rita, and Wilma), and the first time that three category 3 or more hurricanes made US landfall in subsequent years (2004 and 2005). The 2005 season also recorded the greatest damage total in 1 year, at nearly $150 billion USD. Hurricane Wilma is now the most intense storm (lowest central pressure) in recorded history for the Atlantic. Figure 5.1.1 presents a map and a detailed table of all the storms in this remarkable 2005 season of tropical cyclones and hurricanes, with five of the storm tracks highlighted.

Each storm went through the stages of tropical depression, tropical storm, hurricane, and extratropical depression, based on wind speed (depression, storm, hurricane) and location of the low-pressure center (tropical or extratropical). Top wind speed and lowest central pressure are noted in the table for each hurricane.

The Gulf Coast and Atlantic Seaboard sustained hits and near misses in 2005 from Hurricanes Cindy, Dennis, Katrina, Ophelia, Rita, Wilma, and Gamma (Figure 5.1.1); and, when added to Charley, Francis, Ivan, and Jeanne in 2004 and Dolly, Fay, Gustav, and Ike in 2008, the United States sustained more than $255 billion USD in damages and more than 2700 lives lost.

Given what we know, how do we cope with the televised images of the Gulf Coast tragedy, stranded victims without supplies, or Katrina's power as sampled in the damage photos in Figure 5.1.2? In the aftermath of Hurricane Ike in 2008, government delays held up water, ice, and food supplies for the 2 million evacuees until 4 or 5 days after landfall, thus repeating past mistakes. As we learn through hearings

and investigations of the poststorm failure to respond and assist these regions and as we see continuing mistakes in planning and engineering, the case for an integrated scientific approach is clearly made. We discuss the engineering/planning, floodplain management, levee considerations, and other mitigation strategy failures in Chapter 11.

Prediction and the Future

The Tropical Prediction Center is at the National Hurricane Center (NHC) in Miami, Florida. Prediction forecasts are posted on the NHC web site (http://www.nhc.noaa.gov/). The center's analysis of weather records for the period 1900–2006 disclosed a significant causal relationship between Atlantic tropical storms, including hurricanes that make landfall along the US Gulf and East Coasts, and several meteorological variables, such as sea-surface temperatures, the presence of upper tropospheric cross winds that cause shearing of vertical tropical storm circulation, and conditions in the Pacific Ocean, among other factors.

Research established a new "total dissipation index" to rate the potential destructiveness of tropical cyclones, integrated over their lifetime. The research relates the marked increase in tropical cyclone intensity. MIT scientist Kerry Emanuel defined an "index of potential destructiveness" based on a hurricane's total dissipation of power over its lifetime, and stated:

> ... this index has increased markedly since the mid-1970s. This trend is due to both longer storm lifetimes and greater storm intensities ... highly correlated with tropical sea surface temperature, reflecting well-documented climate signals, including multidecadal oscillations ... and global warming ... the near doubling of power dissipation over the

(continued)

Focus Study 5.1 *(continued)*

FIGURE 5.1.1 2005 Atlantic hurricane season breaks records.

(a) The remarkable tropical cyclones of 2005 set several records. Each named storm is identified. The inset table gives you all the data. (b) NOAA aerial photo shows the power of the wind and storm surge as entire Mississippi Gulf Coast neighborhoods are obliterated, such as here in Long Beach, with the debris carried a kilometer from shore by the storm surge.

[(a) Data for map from NOAA and the National Hurricane Center; photo (b) courtesy of NOAA aerial photography program.]

 SATELLITE   27 Storms, Arlene to Zeta

T-Tropical Storm, or H (cat. no.)-Hurricane (affected US)	2005 Dates (inclusive)	Top average wind speed (mph/kmph)	Lowest Central Pressure (mb)	Days as: H or T	
				H	T
T-Arlene (US)	6/8 - 6/12	70/113	989	–	3
T-Bret	6/28 - 6/30	40/64	1004	–	1
H1-Cindy (US)	7/03 - 7/06	75/121	991	2	3
H4-Dennis (US)	7/05 - 7/11	150/241	930	4	3
H4-Emily	7/11 - 7/21	155/249	930	7	4
T-Franklin	7/21 - 7/29	70/113	997	–	8
T-Gert	7/23 - 7/25	45/72	1005	–	2
T-Harvey	8/02 - 8/08	65/105	994	–	6
H2-Irene	8/04 - 8/18	100/161	975	3	8
T-Jose	8/22 - 8/23	50/80	1001	–	1
H5-Katrina (US)	8/23 - 8/30	175/282	902	4	3
T-Lee	8/28 - 9/02	40/64	1007	–	1
H3-Maria	9/01 - 9/10	115/185	960	5	3
H1-Nate	9/05 - 9/10	90/145	979	3	4
H1-Ophelia (US)	9/06 - 9/18	85/137	976	7	9
H1-Philippe	9/17 - 9/24	80/129	985	2	5
H5-Rita (US)	9/18 - 9/25	175/282	987	5	4
H1-Stan	10/01 - 10/05	80/129	979	1	3
T-Tammy (US)	10/05 - 10/06	50/80	1001	–	2
H1-Vince	10/09 - 10/11	75/121	987	2	2
H5-Wilma (US)	10/15 - 10/25	175/282	882*	7	2
T-Alpha	10/22 - 10/24	50/80	998	–	2
H3-Beta	10/27 - 10/31	115/185	960	2	4
T-Gamma	11/14 - 11/21	45/72	1004	–	3
T-Delta	11/23 - 11/28	70/113	980	–	6
H1-Epsilon	11/29 - 12/08	80/129	987	3	7
T-Zeta	12/30 - 1/5/06	65/105	992	–	7

* Record low pressure for US.

(b)

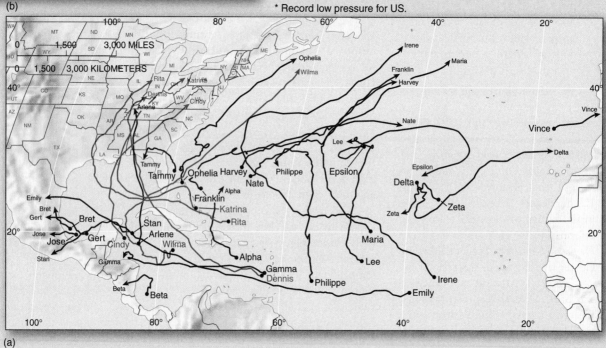
(a)

FIGURE 5.1.2 Hurricane Katrina devastation.
(a) A 91-m (300-ft) break in the London Avenue Canal levee that leads from Lake Pontchartrain partially into New Orleans; inset photo shows the dried, toxic "muck" that covered the region after flood waters retreated (upper-left). (b) Former New Orleans neighborhood next to the 17th-Street Canal where a 153-m (500-ft) section of levee failed; inset photo shows the changing water marks on abandoned homes left by floodwaters in the Lakeview area. The red "X" on houses tells recovery forces the date, search group involved, part of house searched, and victims found (upper-right). (c) Houses missing from their foundations, debris, and broken trees and lives in Long Beach, Mississippi (lower-left). (d) Remnants of the US 90 bridge and causeway out of Bay St. Louis, Mississippi (lower-right). Sadly, after years, the tragedy continues for the region as a failure of government institutions.

ANIMATION
Hurricane Wind Patterns

period of record should be a matter of some concern.[†]

Dr. Emanuel and others cited in this text suggest that future warming

[†]K. Emanuel, "Increasing destructiveness of tropical cyclones over the past 30 years," *Nature* 436 (August 4, 2005): 686–688.

with climate change, when considered alongside increased coastal population settlement, will produce unwanted, yet substantial, hurricane-related damage losses. Recent major hurricanes experienced rapid central pressure drops as they passed over record sea-surface temperatures in the Gulf of Mexico and

the Caribbean. Wilma experienced a 100-mb central pressure drop in a little over 24 hours, October 19, 2005. The record sea-surface temperature produces warmer water to depth; the natural mixing that such storms trigger only brings up more heat to fuel the circulation. The story is not over.

embedded in that quadrant, striking without warning amidst howling winds and pouring wind-driven rain. In 2005, Hurricane Rita spawned 85 tornadoes and Katrina 42.

Typhoons form in the vast region of the Pacific, where the annual frequency of tropical cyclones is greatest. The magnitude and intensity of typhoons generally exceed those

of hurricanes. In fact, the lowest sea-level pressure recorded on Earth was 870 mb (25.69 in.) in the center of Typhoon Tip in October 1979, 837 km (520 mi) northwest of Guam (17° N 138° E). The *Geostationary Meteorological Satellite* (*GMS*), operated by the Japan Weather Association, provides images for tracking these western Pacific storms.

A Final Word The tragedy from recent storms—Andrew in Florida, Mitch in Central America, the three 1999 North Carolina hurricanes, the four 2004 Florida hurricanes, or Arlene, Cindy, Dennis, Katrina, Ophelia, Rita, and Wilma's combined devastation in 2005, and Hurricanes Gustav and Ike in 2008—is the loss of lives, homes, businesses, livestock, and even entire towns. These regions take decades to recover long after media attention fades.

An important lesson is that storms damage human structures more than they damage natural systems such as the Florida Everglades or the salt marshes and wetlands of southern Louisiana. Remember this perspective: *Urbanization, agriculture, pollution, and water diversion pose a greater ongoing threat to the Everglades, coastal wetlands, and coastal barrier islands than do hurricanes.* Couple this reality with increasing storm intensity as stated in a study published in *Science* and you see the problems ahead for the coastal regions.

> ... the trend of increasing numbers of category 4 and 5 hurricanes for the period 1970–2004 is directly linked to the trend in sea-surface temperature.... Higher SST [sea-surface temperatures] was the only statistically significant controlling variable related to the upward trend in global hurricane strength since 1970.*

*C. D. Hoyos, P. A. Agudelo, P. J. Webster, and J. A. Curry, "Deconvolution of factors contributing to the increase in global hurricane intensity," *Science* 312 (5770, April 7, 2006): 94–97.

Ironically, poor perception of this increasing hazard encouraged weak zoning and rapid development of vulnerable coastal lowlands. New buildings, apartments, and government offices are opening right next to still visible rubble and bare foundation pads. Unfortunately, public or private decision makers do not seem to practice thoughtful hazard planning, whether along coastal lowlands, river floodplains, or earthquake fault zones. The consequence is that our entire society bears the financial cost of planning failure from all natural hazards, whether in Florida, North Carolina, California, or the Midwest, not to mention the physical, emotional, and economic hardship of the event shouldered by the direct victims. This recurrent, yet avoidable, cycle—construction, devastation, reconstruction, devastation—was, ironically, reinforced in *Fortune* magazine almost 40 years ago after the destruction caused by Hurricane Camille:

> Before long the beachfront is expected to bristle with new motels, apartments, houses, condominiums, and office buildings. Gulf Coast businessmen, incurably optimistic, doubt there will ever be another hurricane like Camille, and even if there is, they vow, the Gulf Coast will rebuild bigger and better after that.*

Fortune, October 1969, p. 62.

Summary and Review—Atmospheric Water and Weather

■ *Describe* the origin of Earth's waters, the present distribution, and water's unique heat properties.

The next time it rains where you live, pause and reflect on the journey each of those water molecules has made. Water molecules came from within Earth over a period of billions of years, in a process called **outgassing**. Thus began endless cycling of water through the hydrologic system of evaporation-condensation-precipitation. Water covers about 71% of Earth; less than 3% is nonoceanic freshwater—most of it frozen.

The present volume of water on Earth is estimated at 1.36 billion km³ (326 million mi³), an amount achieved roughly 2 billion years ago. This overall steady-state equilibrium might seem in conflict with the many changes in sea level that have occurred over Earth's history, but it is not. Worldwide changes in sea level are called **eustasy** and are related to the change in volume of water in the oceans. Some of these changes are explained by the amount of water stored in glaciers and ice sheets. At present, sea level is rising because of increases in the temperature of the oceans and the record melting of glacial ice.

Part of Earth's uniqueness is that its water exists naturally in all three states—solid, liquid, and gas—owing to Earth's temperate position relative to the Sun. A change from one state to another is called a **phase change**. The change from solid to vapor is called **sublimation**; from liquid to solid, freezing; from solid to liquid, melting; from vapor to liquid, condensation; and from liquid to vapor, vaporization, or evaporation.

The heat energy required for water to change phase is termed **latent heat**, because, once absorbed, it is hidden within the structure of the water, ice, or water vapor. For 1 g of water to become 1 g of water vapor at boiling requires the addition of 540 calories, the **latent heat of vaporization**. When this 1 g of water vapor condenses, the same amount of heat energy is liberated, as 540 calories of the **latent heat of condensation**. The *latent heat of sublimation* is the energy exchanged in the phase change from ice to vapor and vapor to ice. At less than boiling, such as in a natural setting, the **latent heat of evaporation** describes the energy absorbed during phase change. The tremendous amount of latent heat energy involved in the phase changes among the three states of water powers the weather we experience.

outgassing (p. 146)
eustasy (p. 146)
phase change (p. 148)
sublimation (p. 148)
latent heat (p. 149)
latent heat of vaporization (p. 149)
latent heat of condensation (p. 149)
latent heat of evaporation (p. 150)

1. Approximately when did Earth's water reach its present volume?

2. If the quantity of water on Earth has been quite constant in volume for at least 2 billion years, how can sea level have fluctuated? Explain.

3. Describe the locations of Earth's water, both oceanic and fresh. What is the largest repository of freshwater at this time? In what ways is this distribution of water significant to modern society?

4. Why might you describe Earth as the water planet? Explain.

5. Describe the three states of matter as they apply to ice, water, and water vapor.

6. What happens to the physical structure of water as it cools below 4°C (39°F)? What are some visible indications of these physical changes?

7. What is latent heat? How is it involved in the phase changes of water?

8. Take 1 g of water at 0°C and follow it through to 1 g of water vapor at 100°C, describing what happens along the way. What amounts of energy are involved in the changes that take place?

■ *Define* humidity, relative humidity, and dew-point temperature and *illustrate* stable and unstable atmospheric conditions.

The amount of water vapor in the atmosphere is **humidity**. The ability of air to hold water vapor is principally a function of the temperature of the air and of the water vapor (usually the same). **Relative humidity** is a percentage expression of the humidity content of the air compared with the capacity of the air to absorb water vapor at a given temperature—content compared to capacity. Air is said to be **saturated**, or filled to capacity, if it contains all the water vapor it can absorb at a given temperature (100% relative humidity). The temperature at which air achieves saturation is the **dew-point temperature**.

Among the various ways to express humidity and relative humidity are vapor pressure and specific humidity. **Vapor pressure** is that portion of the atmospheric pressure that is produced by the presence of water vapor. A comparison of vapor pressure with the saturation vapor pressure at any moment produces a relative humidity percentage. **Specific humidity** is the mass of water vapor (in grams) per mass of air (in kilograms) at any specified temperature. A comparison of specific humidity with the maximum specific humidity produces a relative humidity percentage.

Meteorologists use the term *parcel* to describe a body of air that has specific temperature and humidity characteristics. Warm air has a lower density in a given volume of air; cold air has a higher density. **Stability** refers to the tendency of an air parcel, with its water-vapor cargo, either to remain in place or to change vertical position by ascending (rising) or descending (falling). An air parcel is *stable* if it resists displacement upward or, when disturbed, it tends to return to its starting place. An air parcel is *unstable* if it continues to rise until it reaches an altitude where the surrounding air has a density (air temperature) similar to its own.

An ascending (rising) parcel of air cools by expansion, responding to the reduced air pressure at higher altitudes. A descending (falling) parcel heats by compression. The **normal lapse rate**, as introduced in Chapter 2, is the average decrease in temperature with increasing altitude, a value of 6.4 C°/1000 m (3.5 F°/1000 ft). This rate of temperature change is for still, calm air. It can vary greatly under different weather conditions. Consequently, the actual lapse rate at a particular place and time is labeled the **environmental lapse rate**.

The warming and cooling rates for a parcel of expanding or compressing air are termed **adiabatic**. The **dry adiabatic rate (DAR)** is the rate at which "dry" air cools by expansion (if ascending) or heats by compression (if descending). The term *dry* is used when air is less than saturated (relative humidity less than 100%). The DAR is 10 C°/1000 m (5.5 F°/1000 ft). The **moist adiabatic rate (MAR)** is the average rate at which ascending air that is moist (saturated) cools by expansion or descending air warms by compression. The average MAR is 6 C°/1000 m (3.3 F°/1000 ft). This is roughly 4 C° (2 F°) less than the dry rate. The MAR, however, varies with moisture content and temperature and can range from 4 C° to 10 C°/1000 m (2 F° to 5.5 F°/1000 ft).

A simple comparison of the DAR and MAR in a vertically moving parcel of air with that of the environmental lapse rate in the surrounding air determines the atmosphere's stability, that is, whether it is unstable (lifting of air parcels continues) or stable (air parcels resist vertical displacement).

humidity (p. 151)
relative humidity (p. 151)
saturated (p. 152)
dew-point temperature (p. 152)
vapor pressure (p. 154)
specific humidity (p. 154)
stability (p. 156)
normal lapse rate (p. 156)
environmental lapse rate (p. 156)
adiabatic (p. 156)
dry adiabatic rate (DAR) (p. 156)
moist adiabatic rate (MAR) (p. 157)

9. What is humidity? How is it related to the energy present in the atmosphere? To our personal comfort and how we perceive apparent temperatures?

10. Define relative humidity. What does the concept represent? What is meant by the terms *saturation* and *dew-point temperature*?

11. Using different measures of humidity in the air as given in the chapter, calculate relative humidity values (vapor pressure/saturation vapor pressure; specific humidity/maximum specific humidity).

12. How do the two instruments described in this chapter measure relative humidity?

13. Differentiate between stability and instability relative to a parcel of air rising vertically in the atmosphere.

14. What are the forces acting on a vertically moving parcel of air? How are they affected by the density of the air parcel?

15. How do the adiabatic rates of heating or cooling in a vertically displaced air parcel differ from the normal lapse rate and environmental lapse rate?

16. What would atmospheric temperature and moisture conditions be on a day when the weather is unstable? When it is stable? Relate in your answer what you would experience if you were outside watching.

■ *Explain* cloud formation and *identify* major cloud classes and types, including fog.

A **cloud** is an aggregation of tiny moisture droplets and ice crystals suspended in the air. Clouds are reminders of the powerful heat-exchange system in the environment. **Moisture droplets** in a cloud form when saturated air and the presence of **condensation nuclei** lead to *condensation*.

Low clouds, ranging from the surface up to 2000 m (6500 ft), are **stratus** (flat clouds, in layers) or **cumulus** (puffy clouds, in heaps). Sometimes near the end of the day, lumpy, grayish, low-level clouds, **stratocumulus**, may fill the sky in patches. When stratus clouds yield precipitation, they are **nimbostratus**. Middle-level clouds are denoted by the prefix *alto-*. Altocumulus clouds, in particular, represent a broad category that occurs in many different styles. Clouds at high altitude, principally composed of ice crystals, are **cirrus**. A cumulus cloud can develop into a towering giant called **cumulonimbus** (*-nimbus* denotes precipitation). Such clouds are called *thunderheads* because of their shape and their associated lightning, thunder, surface wind gusts, updrafts and downdrafts, heavy rain, and hail. **Fog** is a cloud that occurs at ground level. **Advection fog** forms when air in one place migrates to another place where conditions exist that can cause saturation—for example, when warm, moist air moves over cooler ocean currents. Another type of advection fog forms when cold air flows over the warm water of a lake, ocean surface, or swimming pool. This **evaporation fog**, or *steam fog*, may form as the water molecules evaporate from the water surface into the cold overlying air. **Upslope fog** is produced when moist air is forced to higher elevations along a hill or mountain. Another fog caused by topography is **valley fog**, formed because cool, denser air settles in low-lying areas, producing fog in the chilled, saturated layer near the ground. **Radiation fog** is caused by radiative cooling of a surface that chills the air layer directly above the surface to the dew-point temperature, creating saturated conditions and fog.

cloud (p. 158)
moisture droplet (p. 158)
condensation nuclei (p. 158)
stratus (p. 162)
cumulus (p. 162)
stratocumulus (p. 162)
nimbostratus (p. 162)
altocumulus (p. 162)
cirrus (p. 162)
cumulonimbus (p. 162)
fog (p. 162)
advection fog (p. 163)
evaporation fog (p. 163)
upslope fog (p. 163)
valley fog (p. 163)
radiation fog (p. 164)

17. Specifically, what is a cloud? Describe the droplets that form a cloud.

18. Explain the condensation process; what are the two essential requirements?

19. What are the basic forms of clouds? Using Figure 5.18, describe how the basic cloud forms vary with altitude.

20. Explain how clouds might be used as indicators of the conditions of the atmosphere and of expected weather.

21. What type of cloud is fog? List and define the principal types of fog.

■ *Describe* air masses that affect North America and *identify* four types of atmospheric lifting mechanisms.

Specific conditions of humidity, stability, and cloud coverage occur in a regional, homogenous **air mass**. The longer an air mass

remains stationary over a region, the more definite its physical attributes become. Air masses are categorized by their *moisture content*—**m** for maritime (wetter) and **c** for continental (drier)—and their *temperature* (a function of latitude)—designated **A** (arctic), **P** (polar), **T** (tropical), **E** (equatorial), and **AA** (antarctic).

Air masses can be lifted by **convergent lifting** (air flows conflict, forcing some of the air to lift), **convectional lifting** (air passing over warm surfaces gains buoyancy), **orographic lifting** (passage over a topographic barrier), and *frontal lifting*. In North America, *chinook winds* (called *föhn* or *foehn* winds in Europe) are the warm, downslope air flows characteristic of the leeward side of mountains. Orographic lifting creates wetter windward slopes and drier leeward slopes situated in the **rain shadow** of the mountain.

Conflicting air masses at a front produce either a **cold front** (and sometimes a zone of strong wind and rain) and possibly a zone right along or slightly ahead of the front, called a **squall line**, with turbulent and wildly changing wind patterns and intense precipitation, or a **warm front**, where cooler, passive air is pressed into a ground-hugging wedge, as warmer air rides gently over it, producing stratus clouds and the possibility of drizzly, steady precipitation.

air mass (p. 164)
convergent lifting (p. 167)
convectional lifting (p. 167)
orographic lifting (p. 168)
rain shadow (p. 169)
cold front (p. 170)
squall line (p. 171)
warm front (p. 170)

22. How does a source region influence the type of air mass that forms over it? Give specific examples of each basic classification.

23. Of all the air masses, which are of greatest significance to the United States and Canada? What happens to them as they migrate to locations different from their source regions? Give an example of air mass modification.

24. Explain why it is necessary for an air mass to rise if there is to be precipitation.

25. What are the four principal lifting mechanisms that cause air masses to ascend, cool, condense, form clouds, and perhaps produce precipitation? Briefly describe each.

26. Differentiate between the structure of a cold front and a warm front.

■ *List* the measurable elements of weather and *describe* the life cycle of a midlatitude cyclone.

Weather is the short-term condition of the atmosphere; **meteorology** is the scientific study of the atmosphere. Analyzing and understanding patterns of wind, air pressure, temperature, and moisture conditions portrayed on daily weather maps is key to numerical (computer-based) weather forecasting.

A **midlatitude cyclone**, or **wave cyclone**, is a vast low-pressure system that migrates across the continent, pulling air masses into conflict along fronts. *Cyclogenesis*, the birth of the low-pressure circulation, can occur off the west coast of North America, along the polar front, along the lee slopes of the Rockies, in the Gulf of Mexico, and along the east coast. An **occluded front** is produced when a cold front overtakes a warm front in the maturing cyclone. These systems are guided by the jet streams of the upper troposphere along seasonally shifting *storm tracks*.

weather (p. 171)
meteorology (p. 171)
midlatitude cyclone (p. 171)
wave cyclone (p. 171)
occluded front (p. 173)

27. Differentiate between a cold front and a warm front as types of frontal lifting and what you would experience with each one.

28. How does a midlatitude cyclone act as a catalyst for conflict between air masses?

29. Diagram a midlatitude cyclonic storm during its open stage. Label each of the components in your illustration and add arrows to indicate wind patterns in the system.

30. What is your principal source of weather data, information, and forecasts? Where does your source obtain its data? Have you used the Internet and World Wide Web to obtain weather information? In what ways will you personally apply this knowledge in the future? What benefits do you see?

■ *Analyze* various types of violent weather—thunderstorms, tornadoes, and tropical cyclones, including recent trends.

The violent power of some weather phenomena poses a hazard to society. Thunderstorms produce **lightning** (electrical discharges in the atmosphere), **thunder** (sonic bangs produced by the rapid expansion of air after intense heating by lightning), and **hail** (ice pellets formed within cumulonimbus clouds).

Strong linear winds in excess of 26 m/s (58 mph), known as **derechos**, are associated with thunderstorms and bands of showers crossing a region. Straight-line winds can cause significant damage and crop losses. A spinning, cyclonic column rising to midtroposphere level—a **mesocyclone**—is sometimes visible as the swirling mass of a cumulonimbus cloud. Dark gray **funnel clouds** may pulse from the bottom side of the parent cloud. A **tornado** is formed when the funnel connects with Earth's surface. A **waterspout** forms when a tornado circulation occurs over water.

Within tropical air masses, large low-pressure centers can form along easterly wave troughs. Under the right conditions, a tropical cyclone is produced. Depending on wind speeds and central pressure, a **tropical cyclone** can become a **hurricane**, **typhoon**, or *cyclone* when winds exceed 65 knots (119 kmph, 74 mph). As forecasting and the public's perception of weather-related hazards have improved, loss of life has decreased, although property damage continues to increase.

lightning (p. 175)
thunder (p. 175)
hail (p. 175)
derechos (p. 175)
mesocyclone (p. 176)
funnel clouds (p. 179)
tornado (p. 179)
waterspout (p. 179)
tropical cyclone (p. 181)
hurricane (p. 181)
typhoon (p. 181)

31. What constitutes a thunderstorm? What type of cloud is involved? What type of air mass would you expect in an area of thunderstorms in North America?

32. Lightning and thunder are powerful phenomena in nature. Briefly describe how they develop.

33. Describe the formation process of a mesocyclone. How is this development associated with that of a tornado?

34. Evaluate the pattern of tornado activity in the United States. What generalizations can you make about the distribution and timing of tornadoes? Are there any trends in frequency or intensity?

35. What are the different classifications for tropical cyclones? List the various names used worldwide for hurricanes.

36. What forecast factors did scientists use to predict the 1995 and other Atlantic hurricane seasons? Are there any trends in activity since 1995? How did the 2005 Atlantic tropical cyclone season compare? Please explain.

37. Relative to improving weather forecasting, what are some of the technological innovations discussed in this chapter?

NetWork and Critical Thinking Tools

A. Examine the iceberg photographs in Figure 5.5b and 5.5d. Determine what caused the "shoreline" watermark above the ocean surface around the iceberg. Why do you think the iceberg appears to be riding higher in the water than several weeks earlier? In rough terms, how much ice (as a percentage) do you think is beneath the surface in comparison to the amount you see above sea level? Explain why this physical trait can be a hazard to shipping.

B. Using Figure 5.18, begin to observe clouds on a regular basis. See if you can relate the cloud type to particular weather conditions at the time of observation. You may want to keep a log in your notebook throughout the course. Seeing clouds in this way will help you to understand weather. Also, use "Understanding Clouds and Fog" on the Student Learning Center "Destination" link.

C. Relative to coastal devastation from tropical weather, the following line appears in the text: "This recurrent, yet avoidable cycle—construction, devastation, reconstruction, devastation...." This describes the ever-increasing dollar losses to property from tropical storms and hurricanes at a time when improved forecasts have resulted in a significant reduction in loss of life. In your opinion, what is the solution to halting these increasing losses, this cycle of destruction? How would you implement your plan? (In Chapter 11, Figures 11.22 and 11.28 highlight the engineering failures. In Chapter 13, these issues are discussed further in Focus Study 13.1, along with the 2005 Hurricane Katrina and 2008 Hurricane Ike flooding in Figures 13.15 and 13.16.)

D. Under the "Destinations" section in the Student Learning Center sample the many sites listed and sample the information available about the 2005 and 2008 hurricane seasons in the Atlantic; summarize in your notebook.

1983
July

2007
September

Water Resources

KEY LEARNING CONCEPTS

After reading the chapter, you should be able to:

- *Illustrate* the hydrologic cycle with a simple sketch and *label* it with definitions for each water pathway.

- *Relate* the importance of the water-budget concept to an understanding of the hydrologic cycle, water resources, and soil moisture for a specific location.

- *Construct* the water-balance equation as a way of accounting for the expenditures of water supply and *define* each of the components in the equation and its specific operation.

- *Describe* the nature of groundwater and *define* the elements of the groundwater environment.

- *Identify* critical aspects of freshwater supplies for the future and *cite* specific issues related to sectors of use, regions and countries, and potential remedies for any shortfalls.

STUDENT LEARNING CENTER TOOLS

The *Elemental Geosystems* Student Learning Center provides on-line resources for this chapter by clicking on the book cover at **www.mygeoscienceplace.com**. Some highlights:

▶ "Data Analysis" allows you to work with actual water balance data for analysis.

▶ Three animations of the hydrologic cycle, groundwater and the water table, and wells.

▶ Satellite maps and narration illustrate water balance components.

▶ More than 15 additional URLs along with Quick Links to all the URLs in the chapter.

The western United States is experiencing a decade-long drought. Hoover Dam on the Arizona–Nevada border backs up Lake Mead on the Colorado River. The maximum Lake Mead reservoir elevation of 375 m (1229 ft) occurred in July 1983 (upper photo). The rarely used emergency spillways (foreground) operated to prevent the lake from topping the dam. Twenty-four years later, in September 2007, Lake Mead was at 338 m (1111 ft), or about 37 m (121.4 ft) lower than 1983. By August 2008, lake levels dropped to 336.8 m (1105 ft)—a 38.2-m drop (125.3 ft). The Colorado River annual flow is running far below average. *[Photos by (1983) author; (2007) Bobbé Christopherson.]*

The physical reality of life is defined by water. It is the essence of our existence and is therefore the most critical resource supplied by Earth systems—the output of the water–weather system. Our bodies are composed of about 70% water by weight, as are plants and animals. We use water to cook, bathe, wash clothing, and dilute our wastes. We water small gardens and vast agricultural tracts. Most industrial processes are impossible without water.

Fortunately, water is a renewable resource, constantly cycling through the environment, endlessly renewed. Even so, some 1.1 billion people lack safe drinking water. People in 80 countries face impending water shortages, either in quantity or quality, or both. Approximately 2.4 billion people lack adequate sanitary facilities—80% of these in Africa, 13% in Asia. This translates to approximately 2 million deaths a year due to lack of water and 5 million deaths a year from waterborne infections and disease. An investment in safe drinking water, sanitation, and hygiene could decrease these numbers.

During the first half of this century, water availability per person will drop by 74%, as population increases and adequate water decreases. Author Peter Gleick's *The World's Water*, a biennial report first published in 1998 (Washington, DC: Island Press; http://www.worldwater.org/), summarized:

The overall economic and social benefits of meeting basic water requirements far outweigh any reasonable assessment of the costs of providing those needs.... Water-related diseases cost society on the order of $125 billion per year.... Yet the cost of providing new infrastructure needs for all major urban water sectors has been estimated at around $25 to $50 billion per year. While these costs are far below the costs of failing to meet these needs, they are two to three times the average rate of spending for water during the 1980s and 1990s.

Gordon Young, coordinator of the World Water Assessment Program, UNESCO (http://www.unesco.org/water/wwap/), called for an integrated approach:

As we enter the twenty-first century a global water crisis is threatening the security, stability, and environmental sustainability of all nations, particularly those in the developing world. Millions die each year from water-related diseases, while water pollution and ecosystem destruction grow.... Current thinking accepts that the management of water resources must be undertaken using an integrated approach.

Thus the role for geographic science with its integrative, spatial approach is front and center in analyzing a systematic, comprehensive global portrait of the changing status of water resources.

In this chapter: The hydrologic cycle and global water balance give us a model for understanding the global plumbing system. Water spends time in the ocean, in the air, on the surface, and underground as groundwater. Water availability to plants from precipitation and from the soil is critical to water-resource issues.

We look at water resources using a water-budget approach—similar in many ways to a money budget—in which we examine water "receipts" and "expenses" at specific locations. Precipitation provides the principal receipt of moisture, whereas evaporation and plant transpiration are the principal expenditures. This budget approach can be applied at any scale, from a small garden, to a farm, to a regional landscape.

Water is not always naturally available where and when we want it. Consequently, we rearrange surface-water resources to suit our needs. We drill wells, build cisterns and reservoirs, and dam and divert streams to redirect water either spatially (geographically, from one area to another) or temporally (over time, from one part of the calendar to another). All of this activity constitutes water-resource management.

This chapter concludes by considering the quantity and quality of the water we withdraw and consume for irrigation, industrial, and municipal uses—our specific water supply. Adequate water supplies in terms of quantity and quality loom as *the resource issue* for many parts of the world in this century. Ismail Serageldin, chair of the World Water Commission, is quoted saying bluntly on water issues: "... the wars of the twenty-first century will be fought over water.... Water is the most critical issue facing human development."*

The Hydrologic Cycle

Vast currents of water, water vapor, ice, and energy are flowing about us continuously in an elaborate, open, global plumbing system. Together they form the **hydrologic cycle**, which has operated for billions of years, from the lower atmosphere down to several kilometers beneath Earth's surface. The cycle involves the circulation and transformation of water throughout Earth's atmosphere, hydrosphere, cryosphere, lithosphere, and biosphere. (See NASA's Global Hydrology and Climate Center, http://weather.msfc.nasa.gov/surface_hydrology/.)

A Hydrologic Cycle Model

Figure 6.1 is a simplified model of this complex system. Let's use the ocean as a starting point for our discussion. More than 97% of Earth's water is in the ocean, where most evaporation and precipitation occur. We can trace 86% of all evaporation to the ocean. The other 14% evaporates from the land, including water moving from the soil into plant roots and passing through their leaves (a process called *transpiration*, described later in this chapter).

*M. De Villiers, *Water, The Fate of Our Most Precious Resource* (New York: Houghton Mifflin Co., 2000), pp. 13–14.

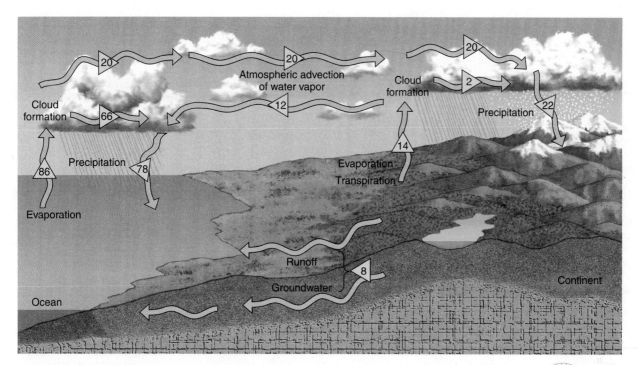

FIGURE 6.1 The hydrologic cycle model.
The hydrologic cycle model with percentages and directional arrows denoting flow paths. Global average values are shown as percentages. Note that all evaporation (86% + 14% = 100%) equals all precipitation (78% + 22% = 100%), when all of Earth is considered, depending on climate. Locally, various parts of the cycle will vary, creating imbalances—surpluses in one region and deficits in another.

ANIMATION
Earth's Water and the Hydrologic Cycle

In the figure, you can see that of the ocean's evaporated 86%, 66% combines with 12% advected from the land to produce 78% of all precipitation that falls back into the ocean. The remaining 20% of moisture evaporated from the ocean, plus 2% of land-derived moisture, produces the 22% of all precipitation that falls over land. Clearly, the bulk of continental precipitation comes from the oceanic portion of the cycle.

Surface Water

Precipitation that reaches Earth's surface follows two basic pathways—overland flow and movement into the ground. Water soaks in through **infiltration**, or penetration of the soil surface. It then permeates soil or rock through the vertical movement of **percolation**.

The atmospheric advection of water vapor from sea to land and land to sea at the top of Figure 6.1 appears to be unbalanced: 20% moving inland but only 12% moving out to sea. However, this exchange is balanced by 8% as runoff that flows from land to sea. Most of this runoff—about 95%—comes from surface waters washing across land as *overland flow* and *streamflow*. Only 5% of this runoff is attributable to slow-moving subsurface groundwater. These percentages indicate that the small amount of water in rivers and streams is very dynamic, whereas the large quantity of subsurface groundwater is sluggish in comparison and comprises only a small portion of runoff.

The residence time of water in any part of the hydrologic cycle determines its relative importance in affecting Earth's climates. The short time spent by water in transit through the atmosphere (10-day average) is reflected in temporary fluctuations in regional weather patterns. Long residence times, such as the 3000–10,000 years in deep-ocean circulations, groundwater aquifers, and glacial ice, act to moderate temperatures and climates. These slower parts of the cycle work as a "system memory"; heat is stored and released to buffer change.

To observe and describe the hydrologic cycle and its related energy budgets, scientists established in 1988 the Global Energy and Water Cycle Experiment (GEWEX); this is part of the World Climate Research Program. As part of this effort, the GEWEX Americas Prediction Project (GAPP), a multiscale, hydrometeorological investigation of the hydrologic cycle, is underway (see http://www.ogp.noaa.gov/mpe/gapp/). The goal is to correlate land surface, general circulation, regional climate, and hydrologic models to better predict seasonal and annual water-resource changes. Now that you are acquainted with the hydrologic cycle, let's examine the concept of the soil-water budget as a method of assessing water resources.

Soil-Water Budget Concept

A **soil-water budget** can be established for any area of Earth's surface—a continent, country, region, or field—by calculating the total precipitation input and the total water output. The water budget is a portrait of the hydrologic

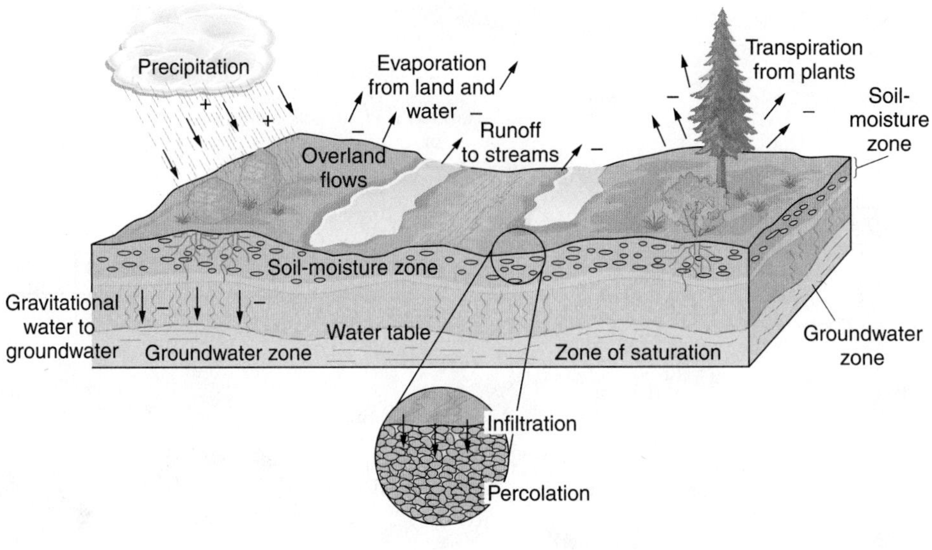

FIGURE 6.2 The soil-moisture environment.
Precipitation supplies the soil-moisture environment. The principal pathways for water include interception by plants; throughfall to the ground; collection on the surface, forming overland flow to streams; transpiration and evaporation from plants; evaporation from land and water; and gravitational water moving to subsurface groundwater.

cycle at a specific site or area for any period of time, including estimation of streamflow. C. W. Thornthwaite (1899–1963), a geographer pioneering in applied water-resource analysis, worked with others to develop a water-balance methodology. They related water-balance concepts to geographical problems, especially to accurately determining irrigation quantity and timing.

The Soil-Water Balance Equation

To understand Thornthwaite's water-balance methodology and "accounting" or "bookkeeping" procedures, we must first understand some terms and concepts. Figure 6.2 illustrates the essential aspects of a soil-water budget. Precipitation (mostly rain and snow) provides the moisture input. The object, as with a money budget, is to account for the ways in which this supply is distributed: actual water taken by evaporation and plant transpiration, extra water that exits in streams and subsurface groundwater, and recharge or utilization of soil-moisture storage.

Figure 6.3 organizes the water-balance components into an equation. As in all equations, the two sides must balance; that is, the precipitation receipt (left side) must be fully accounted for by expenditures (right side). Follow this water-balance equation as you read the following paragraphs. To help you learn these concepts, we use letter abbreviations (such as PRECIP) for the components.

Precipitation (PRECIP) Input   The moisture supply to Earth's surface is **precipitation** (PRECIP, or P). It arrives in several different forms depending upon temperature and moisture supply. Table 6.1 summarizes the different types of precipitation. As you read through this table, recall which types of precipitation you have experienced—rain, sleet, snow, and hail being most common.

Precipitation is measured with the **rain gauge**, which receives precipitation through an open top so that water can be measured by depth, weight, or volume. Wind can

cause an undercatch because the drops or snowflakes are not falling vertically. For example, a wind of 37 kmph (23 mph) produces an undercatch as great as 40%, meaning that an actual 1-in. rainfall might gauge at only 0.6 in. The windshields you see above the gauge's opening reduce this error by catching raindrops that arrive at an angle (Figure 6.4).

According to the World Meteorological Organization (http://www.wmo.ch/), more than 40,000 weather monitoring stations are operating worldwide, with over 100,000 places measuring precipitation. Precipitation patterns over the United States and Canada are shown in Figure 6.5. A map displaying the pattern of world precipitation is included in the next chapter as Figure 7.2.

Actual Evapotranspiration (ACTET)   The net movement of water molecules away from a wet surface into air that is less than saturated is **evaporation**. **Transpiration** in plants is the outward movement of water through small openings (stomata) in the underside of leaves. Both evaporation and transpiration are water-balance outputs, and their rates directly respond to temperature and humidity. In addition, transpiration rates are partially controlled by the plants themselves as they conserve or release water by means of control cells around their stomata. Transpired quantities can be significant: On a hot day, a single tree may transpire hundreds of liters of water; a forest, millions of liters. Evaporation and transpiration are combined into one term—**evapotranspiration** (14% of evaporation and transpiration occurs from land and plants).

Potential Evapotranspiration (POTET, or PE)   The water that would evaporate and transpire under optimum moisture conditions (adequate precipitation and full soil-moisture capacity) is **potential evapotranspiration**. Filling a bowl with water and letting it evaporate illustrates this concept: When the bowl becomes dry, is there still an evaporation demand? A demand is present, of course,

Water balance equation:

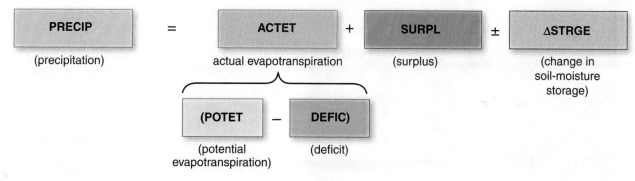

Explanation:

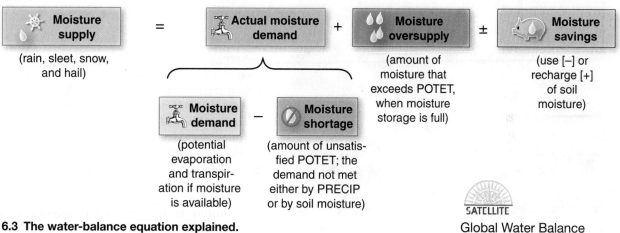

FIGURE 6.3 The water-balance equation explained.
The outputs (components to the right of the equal sign) are an accounting of expenditures from the precipitation moisture-supply input (to the left).

Global Water Balance Components

Table 6.1	Types of Precipitation	
Type	**Characteristic**	**Typical Amounts**
Dew	Condensation on surfaces, especially vegetation; hoar frost when frozen, formed by direct deposition	0.1 to 1.0 mm per night, small input to supply, may be locally significant
Fog-drip	Deposited on vegetation and other surfaces from fog; rime when frozen	Can be 0.1 mm per night; a small input to supply; may be significant in foggy climates
Drizzle (mist)	Numerous, uniform water drops	Light 0.3 mm/hr; moderate 0.3−0.5 mm/hr; heavy >0.5 mm/hr; if over 1 mm/hr, it's rain
Hail	Roughly spherical, irregular lumps of ice >5 mm in diameter with a layered structure of opaque and clear ice in cross section; produced in convective clouds, almost always cumulonimbus	Highly variable, can cover the ground; hailstones can be more than 15 cm in diameter
Rain	Water drops >0.5 mm in diameter; widely scattered drops might be smaller	Light <0.25 cm/hr; moderate 0.26−0.76 cm/hr; heavy >0.76 cm/hr
Snowflakes	Cluster of ice crystals, or a single crystal, that fall from a cloud; an aggregate can exceed 6 cm in diameter	Highly variable
Snow Grains (granular snow)	Quite small, flat, opaque grains of ice; solid equivalent of drizzle	Highly variable
Snow Pellets (graupel or soft hail)	Opaque, approximately round pellets of ice, 2 to 5 mm in diameter falling in showers; often before or with snow	Highly variable
Ice pellets	Clear ice encasing a snowflake or snow; transparent pellets of ice <5 mm in diameter	Highly variable
Sleet (US, UK)	Mixture of rain and partially melted snow; *freezing precipitation* in Canada; can form layer on below-freezing ground	Highly variable

FIGURE 6.4 A rain gauge.
A standard rain gauge is cylindrical. A funnel guides water into a bucket that rests on an electronic weighing device. The gauge is designed to minimize evaporation, which causes low readings. The wind shield around the top of the gauge minimizes the undercatch produced by wind. [Photo by Bobbé Christopherson.]

protect against measurement errors created by wind, which accelerates evaporation.

A **lysimeter** is a relatively elaborate device for measuring POTET (Figure 6.7). It is a buried tank, open at the surface and approximately a cubic meter in size. Using an actual portion of a field, a lysimeter isolates soil, subsoil, and plants so that the moisture moving through the sampled area can be measured. Lysimeters and evaporation pans are somewhat limited in availability across North America but provide a database with which to develop ways of calculating POTET. This meteorological database means that methods of estimating POTET can be easily implemented for regional applications because of the ready availability of precipitation and temperature data. (See the Agricultural Research Service at http://www.ars.usda.gov/.)

Deficit POTET can be satisfied by PRECIP or by moisture derived from the soil, or it can be met artificially through irrigation. If these sources are inadequate, the location experiences a moisture shortage. The unsatisfied moisture demand is recorded as a **deficit** (DEFIC). By subtracting this deficit from the POTET demand, we determine the **actual evapotranspiration**, or ACTET, that takes place. Under ideal conditions, the potential and actual evapotranspiration rates are about the same, so that plants do not experience a water shortage or prolonged deficits that would lead to drought.

Surplus If POTET is satisfied and soil moisture is full, then additional water input becomes **surplus** (SURPL), or water oversupply, as shown in the equation in Figure 6.3. This excess water may sit on the surface or flow through the soil to groundwater storage. Surplus water that flows across the surface toward stream channels is termed *overland flow*. Such flow combines with other precipitation and subsurface flows in river channels to make up the **total runoff** from the area—in essence the *surface-water resource*. Because streamflow, or runoff, is generated mostly from surplus water, the water-balance approach is a useful tool for indirectly estimating streamflow.

Soil-Moisture Storage A "savings account" of water that can receive deposits and allow withdrawals as conditions change in the water balance is **soil-moisture storage**. Soil-moisture storage (ΔSTRGE) refers to the amount of water that is stored in the soil and is accessible to plant roots. (Soil moisture is discussed in more detail in Chapter 15.)

Moisture retained in the soil comprises two classes of water, only one of which is accessible to plants. The inaccessible class is *hygroscopic water*, the thin molecular layer that is tightly bound to each soil particle by the natural hydrogen bonding of water molecules (Figure 6.8, left). Such water exists even in the desert but is not available to meet moisture demands. Soil is said to be at the **wilting point** when all that is left in the soil is unextractable water; the plants wilt and eventually die after a prolonged period of such moisture stress. Unfortunately, most of us have watched a houseplant, or dorm-room plant, go through this process.

even when the bowl is dry. Therefore, the amount of water that would evaporate and/or transpire if water always was available is the POTET, or the amount that would evaporate from the bowl if it was constantly supplied with water.

Figure 6.6 presents POTET values derived by an estimation method that Thornthwaite developed for the United States and Canada. Please compare this POTET (demand) map with Figure 6.5, the PRECIP (supply) map for the same region. The relationship between PRECIP supply and POTET demand determines the remaining components of the water-balance equation. From the two maps, can you identify regions where PRECIP is higher than POTET? (Eastern United States.) Where POTET is higher than PRECIP? (Southwestern United States.) Where you live, is the water demand usually met or exceeded by the precipitation supply? Or does your area experience a natural water shortage? How might you find out?

Determining POTET The easiest method for measuring POTET employs an **evaporation pan**, or *evaporimeter*. As evaporation occurs, water in measured amounts is automatically replaced in the pan. Mesh screens over the pan

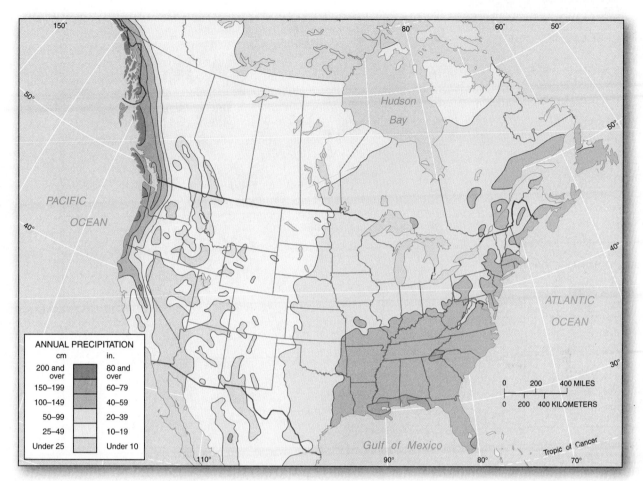

FIGURE 6.5 Precipitation (water supply, PRECIP) in North America.
Annual precipitation (water supply) in the United States and Canada. Compare to Figure 6.6.
[Adapted from US National Weather Service, US Department of Agriculture, and Environment Canada.]

SATELLITE
Global Water Balance
Components

The soil moisture that is generally accessible to plant roots is **capillary water**, held in the soil by surface tension and hydrogen bonding between the water and the soil (Figure 6.8, middle). Most capillary water that remains in the soil is **available water** in soil-moisture storage and is removable to meet moisture demands through the action of plant roots and surface evaporation. After water drains from the larger pore spaces, the available water remaining for plants is termed **field capacity**, or soil-storage capacity. Field capacity is specific to each soil type and is an amount that can be determined by soil surveys. Generally, maintaining soil moisture close to field capacity is essential to plant growing success.

When soil becomes saturated after rainfall or snowmelt, some soil-water surplus becomes **gravitational water**. This excess water percolates downward under the influence of gravity from the shallower capillary zone to the deeper groundwater zone (Figure 6.8, right).

Figure 6.9 shows the relation of soil texture to soil-moisture content. Different plant species send roots to different depths and therefore reach different amounts of soil moisture. A soil blend that maximizes available water is best for plants. On the basis of Figure 6.9, can you determine the soil texture with the greatest quantity of available water?

Plant roots exert a tensional force on soil moisture to absorb it. As water transpires through the leaves, a pressure gradient is established throughout the plant, thus "pulling" water into the plant roots. When soil moisture is at field capacity, plants obtain water with less effort. As the soil water is reduced by **soil-moisture utilization**, the plants must exert greater effort to extract the same amount of moisture. As a result, even though a small amount of water may remain in the soil, plants may be unable to exert enough pressure to utilize it. The unsatisfied demand is a deficit. Rainwater, snowmelt, and irrigation water provide **soil-moisture recharge**. The texture and structure of the soil dictate the rate of soil-moisture recharge. Remember that soil-moisture storage is a water savings account with withdrawals (utilization) and deposits (recharge).

Drought

Drought might seem a simple thing to define: less precipitation and a drier place. However, there is no precise definition of drought. The key is to know the water-resource demand as it relates to the lack of water. There must be a consideration of not just precipitation, temperature, and

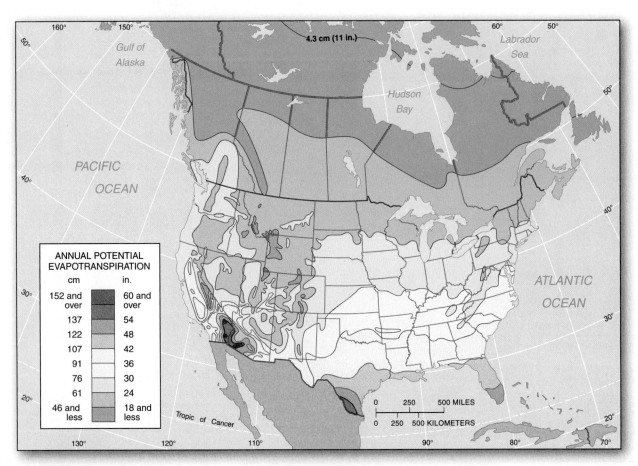

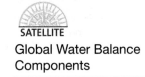

FIGURE 6.6 Potential evapotranspiration (water demand, POTET) for the United States and Canada.
Compare to Figure 6.5. [From Charles W. Thornthwaite, "An approach toward a rational classification of climate," *Geographical Review* 38 (1948): 64. Adapted by permission from the American Geographical Society. Canadian data adapted from Marie Sanderson, "The climates of Canada according to the new Thornthwaite classification," *Scientific Agriculture* 28 (1948): 501–516.]

SATELLITE

Global Water Balance Components

FIGURE 6.7 A weighing lysimeter for measuring evaporation and transpiration.
Various pathways of water that falls on the soil above the lysimeter are traced: Some water remains as soil moisture, some is incorporated into plant tissues, some drains from the bottom of the lysimeter, and the remainder is credited to evapotranspiration. Given natural conditions, the lysimeter measures actual evapotranspiration. [Adapted from illustration courtesy of Lloyd Owens, Agricultural Research Service, USDA, Coshocton, Ohio.]

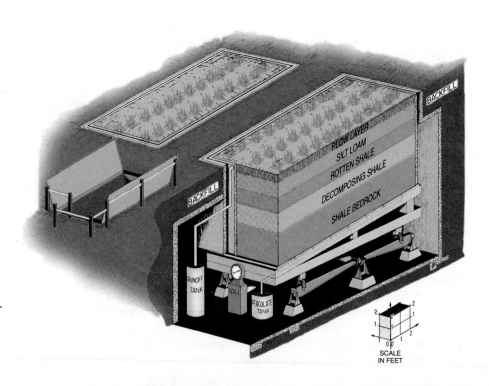

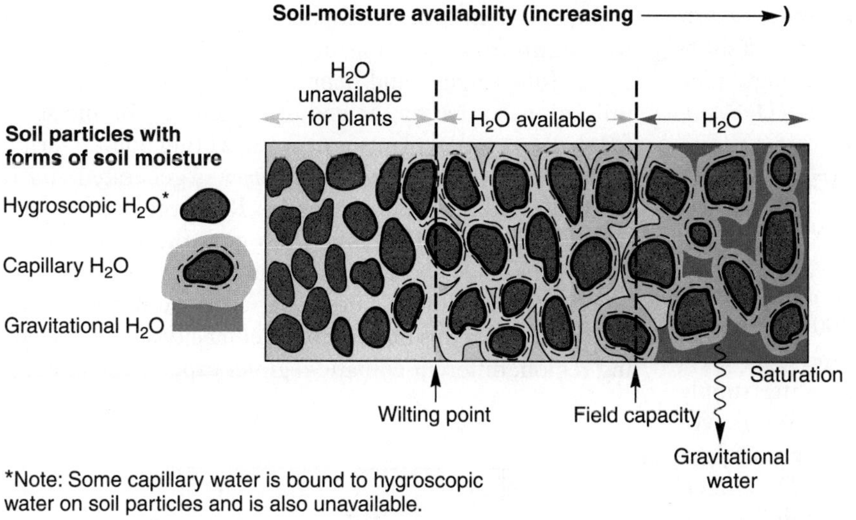

Soil-moisture availability (increasing ———————➤)

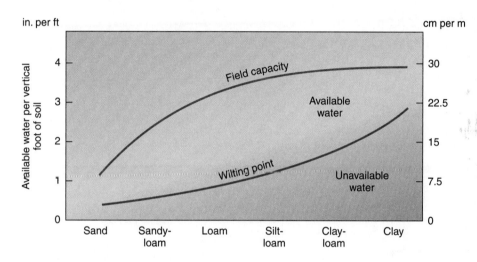

Soil particles with forms of soil moisture

Hygroscopic H₂O*

Capillary H₂O

Gravitational H₂O

H_2O unavailable for plants ← → H₂O available ← → H₂O →

Saturation

Wilting point   Field capacity

Gravitational water

*Note: Some capillary water is bound to hygroscopic water on soil particles and is also unavailable.

SATELLITE

Global Water Balance Components

FIGURE 6.8 Types of soil moisture.
Hygroscopic and gravitational water are unavailable to plants; only capillary water is available. [After Donald Steila, *The Geography of Soils*, © 1976, p. 45. Reprinted by permission of Prentice Hall, Inc., Englewood Cliffs, NJ.]

in. per ft

cm per m

Available water per vertical foot of soil

Field capacity

Available water

Wilting point

Unavailable water

Sand   Sandy-loam   Loam   Silt-loam   Clay-loam   Clay

FIGURE 6.9 Soil-moisture availability.
The relation between soil-moisture availability and soil texture determines the area between the two curves that show field capacity and wilting point. A loam soil (one-third each of sand, silt, and clay) has roughly the most available water per vertical foot of soil exposed to plant roots. [After US Department of Agriculture, *1955 Yearbook of Agriculture-Water*, p. 120.]

soil-moisture content, but whether crops are in the growing season, or if low precipitation occurs during winter and the season for snow pack accumulation, or of the spatial relation of precipitation patterns to the climate of a region. We define four possible types of drought.

A *meteorological drought* relates to dry conditions of lower precipitation and higher temperatures and reductions in soil moisture. An *agricultural drought* occurs when changes in soil moisture and weather elements affect crop yields. Losses can be significant; droughts across the United States and Canada produce damages in the billions each year, yet receive little news coverage. When reservoir levels drop, streamflow decreases, and groundwater mining increases, we have a *hydrologic drought*. A more comprehensive measure considers loss of life, water rationing, wildfire events, and other impacts as they become widespread in *socioeconomic drought*.

The National Drought Mitigation Center, University of Nebraska–Lincoln (NDMC, http://drought.unl.edu/) publishes a weekly "Drought Monitor" map, released each Thursday, which presents a spatial portrait of drought conditions. Intensity is divided into five categories and color coded on each map. Droughts of various intensities are occurring on every continent through 2008. Australia

was in the seventh year of its worst drought in 100 years. Regarding the US Southwest drought, underway since 2000, scientists reported in May 2007 this remarkable situation linked directly to global climate change:

> The projected future climate of intensified aridity in the Southwest is caused by ... a poleward expansion of the subtropical dry zones. The drying of subtropical land areas ... imminent or already underway is unlike any climate state we have seen in the instrumental record. It is also distinct from the multi-decadal megadroughts that afflicted the American Southwest during Medieval times. ... The most severe future droughts will still occur during persistent La Niña events, but they will be worse than any since the Medieval period, because the La Niña conditions will be perturbing a base state that is drier than any state experienced recently.*

The Scripps Institution of Oceanography forecasted in a September 2008 report a 50% probability that Lake Mead, on the Colorado River behind Hoover Dam, could be

*Richard Seager et al., "Model projections of an imminent transition to a more arid climate in southwestern North America," *Science* 316, no. 5828 (May 25, 2007): 1184.

completely dry in 12 years and without enough water to turn turbines in 5 years. Such is the state of this drought and shifting of global pressure systems with climate change. (See **http://scrippsnews.ucsd.edu/Releases/?releaseID=876.**)

Three Examples of Water Balances

Using all these concepts, we can graph the water-balance components for several cities. We look at Kingsport, Tennessee; Ottawa, Ontario; and Phoenix, Arizona. Kingsport is at 36.6° N, 82.5° W, with an elevation of 390 m (1280 ft). Its water-balance data appear on the graph in Figure 6.10.

Figure 6.10 shows Kingsport's average water supply (PRECIP) and water demand (POTET) for a year. We can see that Kingsport experiences a net supply from January through May and from October through December. Soil-moisture estimates are at field capacity from January through May, with the excess supply providing a surplus each month. However, the warm days and months from June through September result in net demands for water.

In June, the *net moisture demand* begins, most of which is withdrawn from soil-moisture storage for shallow-rooted crops. July, August, and September feature soil-moisture utilization, so that by the end of September soil moisture for shallow-rooted plants is near the wilting point. During October, however, *net moisture supply* is generated and is credited as soil-moisture recharge. Recharge occurs again in November and December, bringing the soil moisture back up to field capacity by the end of the year.

Obviously, not all cities experience the surplus moisture patterns of this humid-continental city and surrounding region; different climatic regimes experience different

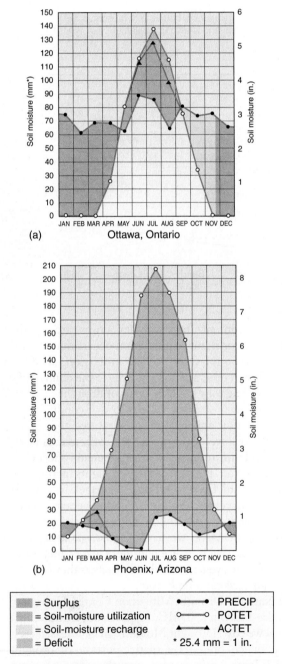

(a) Ottawa, Ontario

(b) Phoenix, Arizona

	= Surplus		●—● PRECIP
	= Soil-moisture utilization		○—○ POTET
	= Soil-moisture recharge		▲—▲ ACTET
	= Deficit		* 25.4 mm = 1 in.

FIGURE 6.11 Sample water budgets for two cities.
Sample water-balance regimes for (a) Ottawa, Ontario, Canada, and (b) Phoenix, Arizona.

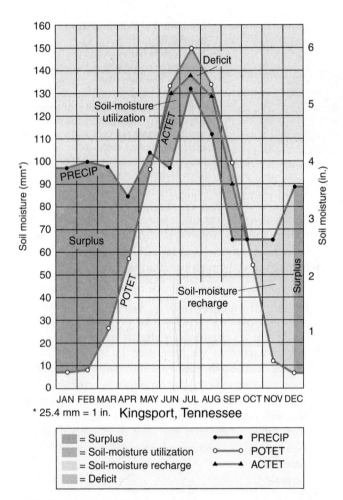

* 25.4 mm = 1 in. Kingsport, Tennessee

	= Surplus		●—● PRECIP
	= Soil-moisture utilization		○—○ POTET
	= Soil-moisture recharge		▲—▲ ACTET
	= Deficit		

FIGURE 6.10 Sample water budget.
Annual average water-balance components graphed for Kingsport, Tennessee. The comparison of plots for precipitation inputs and potential evapotranspiration outputs determines the condition of the soil-moisture environment. A typical pattern of spring surplus, summer soil-moisture utilization, a small summer deficit, autumn soil-moisture recharge, and ending surplus highlights the year.

interactions among their water-balance components. Also, medium- and deep-rooted plants produce different results. Figure 6.11 presents water-balance graphs for the moist and cooler climate of Ottawa, Ontario, Canada, and the drier and warmer climate of Phoenix, Arizona.

An interesting application of the water-balance approach to water resources is analyzing an event, such as the Dust Bowl, an urban flood, or a hurricane. Focus Study 6.1 presents Hurricane Camille (1969) and its generally positive effects on regional water resources, in contrast to its coastal damage and flooding.

Water Budget and Water Resources

Water distribution is uneven over space and time. Because we require a steady supply, we build large-scale management projects intended to redistribute the water resource either geographically, by moving water from one place to another, or over time, by storing water from time of receipt until it is needed. Substantial human effort has been expended to override natural water-balance regimes and recurring drought. The analysis of stream contribution to water resources is an important by-product of using the soil-water budget approach.

FOCUS STUDY 6.1

Hurricane Camille, 1969: Water-Balance Analysis Points to Moisture Benefits

Hurricane Camille was one of the most devastating hurricanes of the twentieth century. Hurricanes Ivan in 2004, Rita, Katrina, and Wilma in 2005, Humberto in 2007, and Dolly, Gustav, and Ike in 2008 brought back memories of Camille to Gulf Coast residents. Ironically, Camille was significant not only for the disaster it brought (256 dead, $1.5 billion damage) but also for the drought it ended.

Figure 6.1.1 shows Camille inland from the Gulf Coast north of Biloxi, Mississippi. Hurricane-force winds sharply diminished after the hurricane made landfall, leaving a vast rainstorm that traveled from the Gulf Coast through Mississippi, western Tennessee, Kentucky, and into central Virginia (Figure 6.1.2). Severe flooding drowned the Gulf Coast near landfall and the James River basin of Virginia, where torrential rains produced record floods. But Camille actually had beneficial aspects; it ended a year-long drought along major portions of its storm track.

Figure 6.1.2a maps the precipitation from Camille (moisture supply) and portrays the storm's track from landfall in Mississippi to the coast of Virginia and Delaware. By comparing actual water budgets along Camille's track with water budgets for the same three days with Camille's rainfall totals removed artificially, I analyzed the moisture impact of the storm on the region's water budget. Figure 6.1.2b

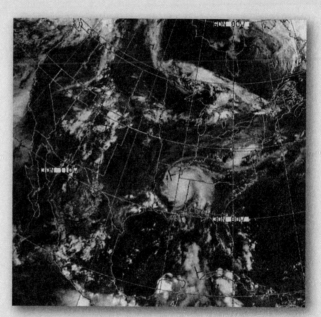

FIGURE 6.1.1 *ESSA-9* satellite image of Hurricane Camille.
Hurricane Camille made landfall on the night of August 17, 1969, and continued inland on August 18. The storm is progressing northward through Mississippi in the image. [ESSA-9 satellite image from NOAA.]

maps the moisture shortages that were avoided because of Camille's rains—termed "deficit abatement." Think of this as drought that did not continue because Camille's rains occurred. Over vast portions of the affected area, Camille reduced dry-soil conditions, restored pastures, and filled low reservoirs.

Thus, Camille's monetary benefits inland outweighed its damage by an estimated 2-to-1 ratio. (Of course, the tragic loss of life does not fit into a financial equation.) In 2007,

Hurricane Humberto, a category 1 storm, brought some needed relief across the Southeast once it moved inland. Hurricanes should be viewed as normal and natural meteorological events that not only have terrible destructive potential, mainly to coastal lowlands, but also contribute to the precipitation regimes of the southern and eastern United States. According to weather records, about one-third of all hurricanes making landfall in the United States provide beneficial precipitation to local water budgets.

(continued)

Focus Study 6.1 *(continued)*

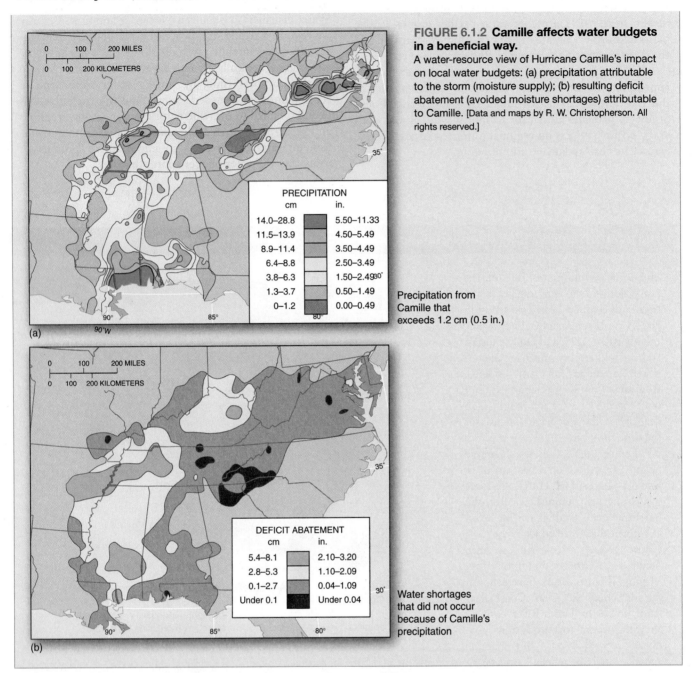

FIGURE 6.1.2 Camille affects water budgets in a beneficial way.
A water-resource view of Hurricane Camille's impact on local water budgets: (a) precipitation attributable to the storm (moisture supply); (b) resulting deficit abatement (avoided moisture shortages) attributable to Camille. [Data and maps by R. W. Christopherson. All rights reserved.]

PRECIPITATION

cm	in.
14.0–28.8	5.50–11.33
11.5–13.9	4.50–5.49
8.9–11.4	3.50–4.49
6.4–8.8	2.50–3.49
3.8–6.3	1.50–2.49
1.3–3.7	0.50–1.49
0–1.2	0.00–0.49

Precipitation from Camille that exceeds 1.2 cm (0.5 in.)

(a)

DEFICIT ABATEMENT

cm	in.
5.4–8.1	2.10–3.20
2.8–5.3	1.10–2.09
0.1–2.7	0.04–1.09
Under 0.1	Under 0.04

Water shortages that did not occur because of Camille's precipitation

(b)

Streams may be perennial (constantly flowing; *perennial* is Latin for "through the year") or intermittent. In either case, the total runoff that moves through them comes from surplus surface-water runoff, subsurface throughflow, and groundwater. Figure 6.12 maps annual global river runoff for the world. Highest runoff amounts are along the equator within the tropics, reflecting the continual rainfall along the intertropical convergence zone (ITCZ). Southeast Asia also experiences high runoff, as do northwest coastal mountains in the Northern Hemisphere. In countries having great seasonal fluctuations in runoff, groundwater becomes an important reserve. Regions of lower runoff coincide with Earth's subtropical

deserts, leeward rain-shadow areas, and continental interiors, particularly in Asia.

Here are a few examples of water redistribution projects. The new Three Gorges Dam controls the treacherous Yangtze River in China (Figure 6.13). At a length of 2.3 km (1.5 mi) and a height of 185 m (607 ft), it is the largest dam and related construction in the world in overall size. Forced relocation of entire cities and more than 1 million people made room for the 600-km- (370-mi-) long reservoir. Controversies surrounded the immense scale of environmental, historical, and cultural losses. Flood control, water resources for redistribution, and power production are considered gains. Eventual electrical production

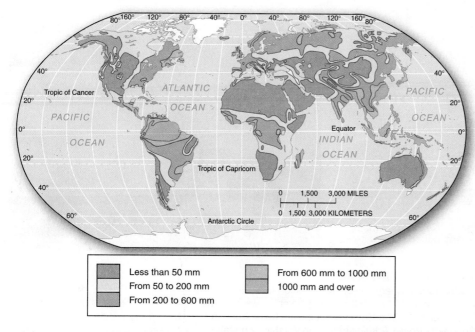

SATELLITE

Global Water Balance
Components

FIGURE 6.12 Annual global river runoff.
Distribution of runoff closely correlates with climatic region, as expected, but poorly correlates with human population distribution and density. [Data from Institute of Geography, Russian Academy of Sciences, Moscow, as compiled and presented by the World Resources Institute.]

Map legend:
- Less than 50 mm
- From 50 to 200 mm
- From 200 to 600 mm
- From 600 mm to 1000 mm
- 1000 mm and over

FIGURE 6.13 Major dam in China.
Satellite views in 2000 (construction) and 2006 (completion) of the $25-billion Three Gorges Dam on the Yangtze River, flowing from upper left to right. For scale, the main portion of the dam is 2.3 km in length. This is the world's largest individual construction project. The channel to the north of the dam is a system of locks for shipping. [Images courtesy of ASTER Science Team, Jesse Allen Earth Observatory, NASA/GSFC/MITI/ERSDAC/JAROS.]

capacity will reach 22,000 MW when the project is built out in 2010.

In Canada, the Nelson River, Manitoba, project; the Churchill Falls, Newfoundland and Labrador, project; and the proposed $12 billion Gull Island Dam and related hydroelectric projects in Québec are significant. The Gull Island project, as originally proposed, would be second only to the Three Gorges project with regard to the size of proposed construction.

In the United States, several large water-management projects already operate: the Bonneville Power Authority in Washington state, the Tennessee Valley Authority in the Southeast, the California Water Project, and the Central Arizona Project (Figure 6.14). Literally, the California Water Project rearranges the water budget in the state, holding back winter runoff for release in summer and pumping water from the north to the south. Completed in 1971, this 1207-km- (750-mi-) long "river" flows from the Sacramento River delta to the Los Angeles region, servicing irrigated agriculture in the San Joaquin Valley along the way. The best sites for multipurpose hydroelectric projects are already taken, so battles among conflicting interests and analyses of negative environmental impacts invariably accompany new project proposals. Chapter 11 discusses river management through such construction in further detail.

Groundwater Resources

Groundwater is a part of the hydrologic cycle that lies beneath the surface but is tied to surface supplies. Groundwater is the largest potential source of freshwater in the hydrologic cycle—larger than all surface lakes and streams combined. Between Earth's land surface and a depth of 4 km (13,100 ft) worldwide, some 22% of Earth's freshwater resides. Despite this volume and its obvious importance, groundwater is widely abused. In many areas it is polluted or

(a)

(b)

FIGURE 6.14 Two major water projects in the United States.
(a) The California Aqueduct transports water from northern to southern California, as part of the California Water Project. Completed in 1971, this 1207-km (750-mi) long "river" flows from the Sacramento River delta to the Los Angeles region. (b) The Grand Coulee Dam holds back the Columbia River, as part of the federal Bonneville Power Administration, in Washington state. Comprising 31 hydroelectric dams and a transmission grid, BPA began in 1937. [Photos by (a) Bobbé Christopherson; (b) author.]

consumed in quantities beyond natural recharge rates. Remember: *Groundwater is not an independent source of water, for it is tied to surface supplies for recharge.*

About 50% of the US population derives a portion of its freshwater from groundwater sources. In some states, such as Nebraska, groundwater supplies 85% of water needs and as high as 100% in rural areas. In Canada, about 6 million people (two-thirds of them live in rural areas) rely on groundwater for domestic needs— 1.5 billion m³/year (53 billion ft³/year). Between 1950 and 2000, annual groundwater withdrawal in these countries increased more than 150%. Figure 6.15 shows potential groundwater resources in the United States and Canada.

Groundwater Profile and Movement

Figure 6.16 illustrates the following discussion. Groundwater is fed by surplus water, which percolates downward from the zone of capillary water as gravitational water. This excess surface water moves through the **zone of aeration**, where soil and rock are less than saturated (an unsaturated zone) and some pore spaces contain air. Eventually, water reaches an area of collected subsurface water known as the **zone of saturation**, where the pores contain only water. Like a sponge made of sand, gravel, and rock, this zone stores water within its structure, filling all available pores and voids.

The **porosity** of this layer is dependent on the arrangement, size, and shape of individual component particles; the cement between them; and their degree of compaction. Subsurface structures are referred to as **permeable** or **impermeable**, depending on whether they permit or obstruct water flows.

An **aquifer** is a rock layer that is permeable to groundwater flow in usable amounts. A water-bearing layer that is not confined by an overlying impermeable layer is an **unconfined aquifer**. An **aquiclude** is a body of rock that does not conduct water in usable amounts. The upper limit of the water that collects in the zone of saturation is the **water table**; it is the contact surface between the zones of saturation and aeration. (See these components on the left section in Figure 6.16.)

The slope of the water table, which generally follows the contours of the land surface, controls groundwater movement. Water wells must penetrate the water table to optimize their potential flow. Where the water table intersects the surface, it creates springs or feeds lakes or riverbeds (Figure 6.17). Ultimately, groundwater may enter stream channels and flow as surface water (near the center in Figure 6.16). During dry periods the water table can act to sustain river flows.

Figure 6.18 illustrates the relation between the water table and surface streams in two different climatic settings. In humid climates, where the water table generally supplies a continuous base flow to a stream and is higher than the stream channel, the stream is *effluent* because it receives the water flowing out (effluent) from the surrounding ground. The Mississippi River and countless other streams are examples. In drier climates, water from *influent* streams flows into the adjacent ground, sustaining deep-rooted vegetation along the stream. The Colorado River and the Rio Grande of the American West are examples of influent streams.

The Arkansas River no longer flows across western and central Kansas. As you see in Focus Study 6.2, "The High Plains Aquifer Overdraft," groundwater mining is dominant. When the bottom of the streambed is no longer in contact with the declining water table, stream flow seeps into the aquifer (Figure 6.18c). These influent conditions produce dry river, stream, and creek beds across the region.

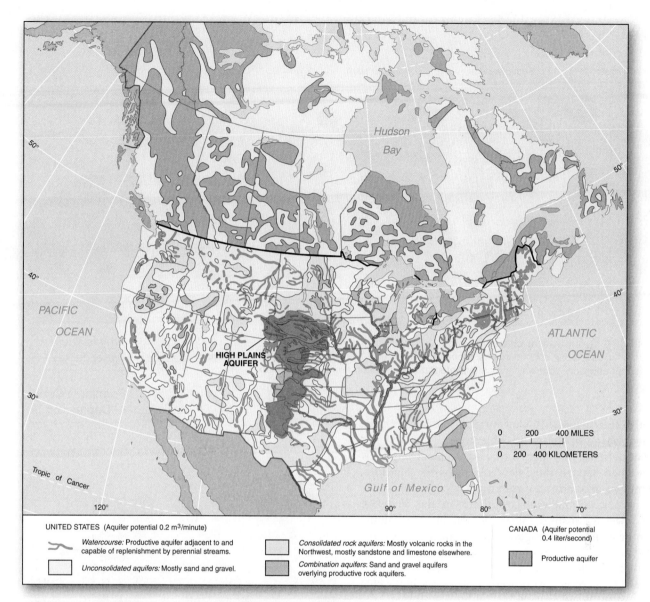

FIGURE 6.15 Groundwater resource potential for the United States and Canada.
Highlighted areas of the United States are underlain by productive aquifers capable of yielding
freshwater to wells at 0.2 m³/min or more (for Canada, 0.4 L/s). [Courtesy of Water Resources
Council for the United States and the *Inquiry on Federal Water Policy* for Canada.]

A confined aquifer differs from an unconfined aquifer in
several important ways. A **confined aquifer** is bounded
above and below by impermeable layers of rock or sediment
(Figure 6.16, right), whereas an *unconfined aquifer* is not. In
addition, the two aquifers differ in **aquifer recharge area**,
which is the surface area where water enters an aquifer to
recharge it. For an unconfined aquifer, that area generally
extends above the entire aquifer to the ground surface. But in
the case of a confined aquifer, the recharge area is far more
restricted (area to the right of the industrial pollution site in
Figure 6.16). Once these recharge areas are identified and
mapped, the appropriate government body should zone them
to prohibit pollution discharges, septic and sewage system
installations, or hazardous-material dumping. Of course, an
informed population would insist on such action. Right?

Aquifers also differ in water pressure. A well drilled
into an unconfined aquifer (Figure 6.16, left) must be
pumped to make the water level rise above the water
table. In contrast, the water in a confined aquifer is under
the pressure of its own weight, creating a pressure level
called the *potentiometric* surface, which actually can be
above ground level (to the right in the figure). Under this
condition, **artesian water**, or groundwater confined
under pressure, may rise up in wells and even flow out at
the surface if the head of the well is below the potentiometric surface. In other wells, however, pressure may be
inadequate, and the artesian water must be pumped the
remaining distance to the surface.

Groundwater Utilization

As water is pumped from a well, the adjacent water table
within an unconfined aquifer will experience a
drawdown, or become lower, if the rate of pumping

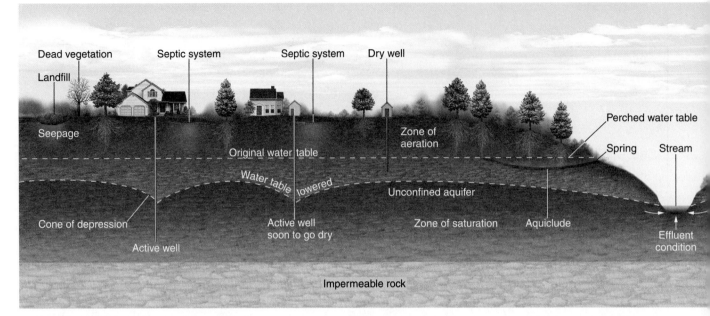

FIGURE 6.16 Groundwater characteristics.
Subsurface groundwater characteristics, processes, water flows, and human interactions. Begin at the far left and work your way across this integrative illustration as you read the text.

ANIMATION · Why Is There a Water Table?

ANIMATION · Forming a Cone of Depression

exceeds the horizontal flow of water in the aquifer. The resultant shape of the water table around the well is a **cone of depression** (Figure 6.16, left). Aquifers frequently are pumped beyond their flow and recharge capacities, resulting in **groundwater mining**. Large tracts experience chronic groundwater overdrafts in the Midwest, West, lower Mississippi Valley, Florida, and the Palouse region of eastern Washington state. In many places, the water table or artesian water level has declined more than 12 m (40 ft). In the United States, groundwater mining is of special concern in the great High Plains Aquifer, which is the topic of Focus Study 6.2.

About half of India's irrigation water and half of industrial and urban water needs are met by the groundwater reserve. In rural areas, groundwater supplies 80% of domestic water from some 3 million hand-pump bore holes. And, in approximately 20% of India's agricultural districts, groundwater mining through more than 17 million wells (motorized dug wells and tube wells) is beyond recharge rates.

> Like any renewable resource, groundwater can be tapped indefinitely as long as the rate of extraction does not exceed the rate of replenishment. But just like a bank account, a groundwater reserve will dwindle if withdrawals exceed deposits. Few governments have established and enforced rules and regulations to insure that groundwater sources are exploited at a sustainable rate. No government has yet adequately tackled the issue of groundwater depletion, but it is at least getting more attention.*

*L. Brown et al. and the Worldwatch Institute, *Vital Signs, The Environmental Trends That Are Shaping Our Future* (New York: W.W. Norton & Co., 2000), pp. 122, 123.

FIGURE 6.17 An active spring.
Springs provide evidence of groundwater emerging at the surface, here flowing into Crystal Lake, Salt River Range, Wyoming. [Photo by Bobbé Christopherson.]

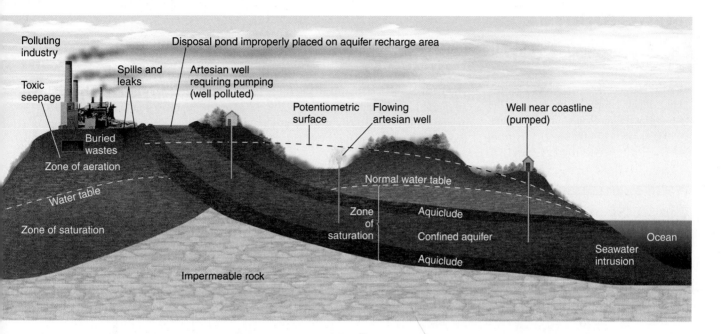

Groundwater depletion and water problems are severe in the Middle East, as detailed in News Report 6.1. The groundwater resource beneath Saudi Arabia accumulated over thousands of years, but the increasing withdrawals are not being naturally recharged to any appreciable degree at present due to the desert climate. Some researchers have suggested that groundwater in the region will be unavailable in less than a decade, although worsening water-quality problems will no doubt arise before this date.

Collapsing Aquifers A possible effect of water removal from an aquifer is that the aquifer, which is a layer of rock or sediment, will lose its internal support. Water in the pore spaces between rock grains is not compressible, so it adds structural strength to the rock. If the water is removed through overpumping, air infiltrates the pores. Air is readily compressible, and the tremendous weight of overlying rock may crush the aquifer. On the surface, the visible result may be land subsidence, cracked house foundations, and changes in drainage.

Houston, Texas, provides an example. The removal of groundwater and crude oil caused land within an 80-km (50-mi) radius of Houston to subside more than 3 m (10 ft) over the years. Yet another example is the Fresno area of California's San Joaquin Valley. After years of intensive pumping of groundwater for irrigation, land levels have dropped almost 10 m (33 ft) because of a combination of water removal and soil compaction from agricultural activity. Surface mining ("strip mining") for coal in eastern Texas, northeastern Louisiana, Wyoming, Arizona, and other locales also destroys aquifers. Aquifer collapse is an

(text continued on page 212)

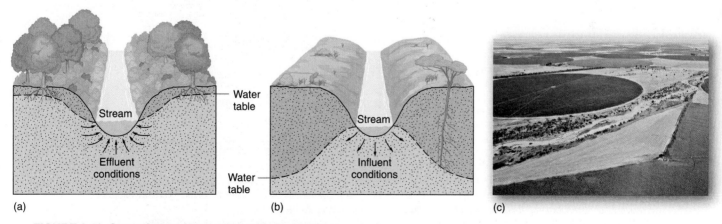

FIGURE 6.18 Groundwater interaction with streamflow.
(a) An effluent stream base flow is partially supplied by a high water table, characteristic of humid regions. (b) An influent stream supplies a lower water table, characteristic of drier regions. The water table may drop below the stream channel, effectively drying it out. (c) The former Arkansas River is dry as water flows into subsurface groundwater aquifers. The river channel appears to be used as a dirt-bike course in the photo. [Photo by Bobbé Christopherson.]

FOCUS STUDY 6.2

High Plains Aquifer Overdraft

Earth's largest known aquifer is the High Plains aquifer. It lies beneath the American High Plains, an eight-state, 450,000-km² (174,000-mi²) area from southern South Dakota to Texas (Figure 6.2.1a). Precipitation over the region varies from 30 cm in the southwest to 60 cm in the northeast (12 to 24 in.). For several hundred thousand years, the aquifer's sand and gravel were charged with meltwaters from retreating glaciers.

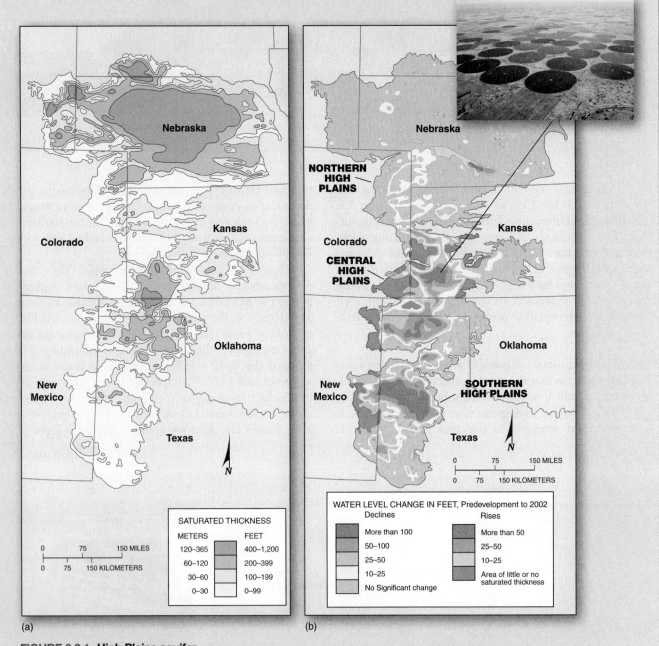

(a)

(b)

FIGURE 6.2.1 High Plains aquifer.

(a) The average saturated thickness of the High Plains aquifer, the largest known aquifer in the world. (b) Using 1950 to represent "predevelopment," this map of changes in groundwater levels was derived from sampling some 7000 wells. The color scale in the legend indicates widespread declines and a few areas of rises in water levels. Inset photo shows dry stream bed (lower right) resulting from lower water tables in the heart of the declining groundwater. [(a) After D. E. Kromm and S. E. White, "Interstate groundwater management preference differences: The High Plains region," *Journal of Geography* 86, no. 1 (January–February 1987): 5; (b) map adapted from "Water-level changes in the High Plains aquifer, predevelopment to 2002, 1980 to 2002, and 2001 to 2002," by V. L. McGuire, *USGS Fact Sheet* 2004-3026, Fig. 1. Inset photo by Bobbé Christopherson]

For the past 100 years, however, High Plains groundwater was heavily mined; mining intensified after World War II with the introduction of central-pivot irrigation devices (Figure 6.2.2). These large, circular devices provide vital water to wheat, sorghums, cotton, corn, and about 40% of the grain fed to cattle in the United States.

The High Plains aquifer irrigates about one-fifth of all US cropland: 120,000 wells provide water for 5.7 million hectares (14 million acres). This withdrawal is down from the peak of 170,000 wells in 1978. In 1980, water was being pumped from the aquifer at the rate of 26 billion cubic meters (21 million acre-feet) a year, an increase of more than 300% since 1950. By 1995, withdrawals had decreased 10% due to declining well yields, increasing pumping costs, and implementation of some water efficiency and conservation techniques.

During the past five decades, the water table in the aquifer has dropped more than 30 m (100 ft), and throughout the 1980s it averaged a 2-m (6-ft) drop each year. The USGS estimates that recovery of the High Plains aquifer (those portions that have not been crushed or subsided) would take at least 1000 years if groundwater mining stopped today.

Figure 6.2.1b maps changes in water levels from 1980 to 1995. Declining water levels are most severe in northern Texas, where the saturated thickness of the aquifer is least, through the Oklahoma panhandle and into Kansas. Rising water levels are noted in portions of south-central Nebraska and a portion of Texas owing to recharge from surface irrigation, a period of above-normal precipitation years, and downward percolation from canals and reservoirs.

The USGS began monitoring this groundwater mining in 1988. A six-year High Plains Regional Ground Water Study began in 1999 and concluded in 2005. Beginning with the central High Plains (northern Texas, Oklahoma panhandle, and Kansas), the study covered the entire

(a)

(b)

FIGURE 6.2.2 Central-pivot irrigation.
(a) Myriad central-pivot irrigation systems water crops in central Kansas. The High Plains aquifer is at depths of more than 80 m (260 ft) with a water table dropping more than 30 m (100 ft) since 1950 in this part of Kansas. (b) From ground level, corn requires from 10 to 20 revolutions of the sprinkler arm, depending on the weather, applying about 3 cm (1.81 in.) of water per revolution. [Photos by Bobbé Christopherson.]

region. The goal is an overall assessment of the resource, including groundwater quality and contaminant sources. (See http://webserver.cr.usgs .gov/nawqa/hpgw/HPGW_home .html.)

Obviously, billions of dollars of agricultural activity cannot be abruptly halted, but neither can profligate water mining continue. This issue raises tough questions: How best to manage cropland? Can extensive irrigation continue? Can the region continue to meet the demand to produce commodities for export—principally grain and packaged beef? Should we continue high-volume farming of certain crops that

are in chronic oversupply? Should we rethink federal policy on crop subsidies and price supports? What would be the impact on farmers and rural communities of any changes to the system?

Present irrigation practices, if continued, will destroy about half of the High Plains aquifer resource (and two-thirds of the Texas portion) by the year 2020. Add to this the approximate 10% loss of soil moisture due to climatic warming, as forecast by computer models for this region by 2050, and we have a portrait of a major regional water problem in this century and a spatial challenge for society.

NEWS REPORT 6.1

Middle East Water Crisis: Running on Empty

The Persian Gulf states soon may run out of freshwater. Their vast groundwater resource is overpumped to such an extent that salty seawater is encroaching into aquifers tens of kilometers inland. In this decade, groundwater on the Arabian Peninsula may become undrinkable. Imagine having the hottest issue in the Middle East become water, not oil!

Remedies for groundwater overuse are neither easy nor cheap. In the Persian Gulf area, additional freshwater is obtained by desalination of seawater, using desalination plants along the coasts. These processing plants remove salt from seawater by distillation and evaporation processes. In fact, approximately 60% of the world's 4000 desalination plants are presently operating in Saudi Arabia and other Persian Gulf states (see Figure 6.21a).

One water pipeline plan would bring water overland from Turkey in the Middle East through two branches. The Gulf water line would run southeast through Jordan and Saudi Arabia, with extensions into Kuwait, Abu Dhabi, and Oman. The other branch would run south through Syria, Jordan, and Saudi Arabia to the cities of Makkah (Mecca) and Jeddah. This pipeline would import 6 million cubic meters (1.68 billion gallons) of water a day over some 1500 km (930 mi), the distance from New York City to St. Louis. (See the links listed under Middle East Water Information Network at http://www.columbia.edu/cu/lweb/indiv/mideast/cuvlm/water.html.)

Other remedies are possible. Traditional agricultural practices could be modernized to use less water. Urban water use could be made more efficient

to reduce groundwater demand. Aquifers and rivers could be shared, although at present no negotiated accords exist for this purpose in the Middle East. Conflicts at various scales pose a grave threat to water pipelines and desalination facilities throughout the region.

> There are severe water shortages in the Middle East.... The resources of the Nile, the Tigris-Euphrates, and the Jordan are overextended owing both to natural causes and to those deriving from human behavior.... In addition to a severe shortage in the quantity of water, there is a growing concern over water quality.*

*N. Kliot, *Water Resources and Conflict in the Middle East* (London: Routledge, 1994), p. 1.

increasing problem in West Virginia, Kentucky, and Virginia, where low-sulfur coal is mined to meet current air pollution requirements for coal-burning electric power plants.

Unfortunately, collapsed aquifers may not be rechargeable, even if surplus gravitational water becomes available, because pore spaces may be permanently collapsed. The water-well fields that serve the Tampa Bay–St. Petersburg, Florida, area are a case in point. Pond and lake levels, swamps, and wetlands in the area are declining and land surfaces subsiding as the groundwater drawdown for export to the cities increases.

An additional problem arises when aquifers are overpumped near the ocean or coastal seas. Along a coastline, fresh groundwater and salty seawater establish a natural interface, or *contact surface*, with the less-dense freshwater flowing on top of the saline groundwater. But excessive withdrawal of freshwater can cause this interface to migrate inland. As a result, wells may become contaminated with saltwater, and the aquifer may become useless as a freshwater source. Figure 6.16 illustrates this seawater intrusion on the far right margin.

Pollution of the Groundwater Resource

When surface water is polluted, groundwater also becomes contaminated because it is recharged from surface-water supplies. Groundwater migrates slowly compared to surface water. Surface water flows rapidly and flushes pol-

lution downstream, but sluggish groundwater, once contaminated, remains polluted virtually forever.

Pollution can enter groundwater from many sources: industrial injection wells (wastes pumped into the ground), septic tank outflows, seepage from hazardous-waste disposal sites, industrial toxic waste, agricultural residues (pesticides, herbicides, fertilizers), and urban solid-waste landfills. As an example, leakage from some 10,000 underground gasoline storage tanks at US gasoline stations is suspected. Some cancer-causing additives to gasoline, such as the oxygenate MTBE (methyl tertiary butyl ether), are contaminating thousands of local water supplies.

For management purposes, about 35% of pollution is *point source* (such as a gasoline tank or septic tank); 65% is *nonpoint source* (from a broad area, such as an agricultural field or urban runoff). Regardless of the spatial nature of the source, pollution can spread over a great distance, as illustrated in Figure 6.16. Since the nature of aquifers makes them inaccessible, the extent of groundwater pollution is underestimated.

In the face of government inaction, and even attempts to reverse some protection laws, serious groundwater contamination continues nationwide, adding validity to a quote from 25 years ago:

> One characteristic is the practical irreversibility of groundwater pollution, causing the cost of clean-up to be prohibitively high.... It is a questionable ethical practice

to impose the potential risks associated with groundwater contamination on future generations when steps can be taken today to prevent further contamination.*

Our Water Supply

Human thirst for adequate water supplies, both in quantity and quality, will be a major issue in this century. Internationally, increases in per capita water use are double the rate of population growth. Since we are so dependent on water, it seems that humans should cluster where good water is plentiful. But accessible water supplies are not well correlated with population distribution or the regions where population growth is greatest.

Table 6.2 shows estimated world water supplies, world population in 2008, estimated population figures for 2025, a population-change forecast between 2008 and 2050 at present growth rates, and a comparison of population and global runoff percentages for each continent. Also note estimates for urban and rural access to improved sanitation for each region. The table gives an idea of the unevenness of

*J. Tripp, "Groundwater Protection Strategies," in *Groundwater Pollution, Environmental and Legal Problems* (Washington, DC: American Association for the Advancement of Science, 1984), p. 137. Reprinted by permission.

Earth's water supply, which is tied to climatic variability, and water demand, which is tied to level of development, affluence, and per capita consumption.

Water is critical for survival. Just to produce our food requires enormous quantities of water for growing, cleaning, processing, and waste disposal. See News Report 6.2 for personal water use and water measurements. We sometimes take for granted the quality and quantity of our water supply because it always seems to be there when we need it.

In northern China, 550 million people living in approximately 500 cities lack adequate water supplies. For comparison, note that the 1990 floods cost China $10 billion, whereas the costs of water shortages to their economy are running at more than $35 billion a year. The Yellow River, a water resource for many Chinese, runs dry every year, and in 1997 it failed to reach the sea for almost 230 days. An analyst for the World Bank stated that China's water shortages pose a more serious threat than floods during this century, yet in news coverage we hear only of floods when they happen.

In Africa, 56 countries draw from a varied water-resource base; these countries share more than 50 river and lake watersheds. Increasing population growth and percentage of urban dwellers cause water demands to rise and lead to an overexploitation of the resource. Twelve

Table 6.2 Estimate of Available Global Water Supply Compared by Region and Population

Region (2008 population in millions)	Land Area in Thousands of km²	(mi²)	Share of Mean Annual Discharge in km³/yr	(BGD)	Global Stream Runoff (%)	Projected Global Population, 2025 (%)	Projected Population, 2025 (millions)	Projected Population Change (+) 2008–2050 (%)	Population with Access to Improved Sanitation (%) for 2006 Urban/Rural
Africa (967)	30,600	(11,800)	4,220	(3,060)	10.8	17.0	1,358	100	62/30
Asia (4052)	44,600	(17,200)	13,200	(9,540)	34.0	60.0	4,793	34	74/31
Australia–Oceania (35)	8,420	(3,250)	1,960	(1,420)	5.1	0.53	42	40	98/58
Europe (736)	9,770	(3,770)	3,150	(2,280)	8.1	9.0	726	−7	100/100
North America (446) (Canada, Mexico, US)	22,100	(8,510)	5,960	(4,310)	15.3	6.5	517	38	100/100 Mexico: 90/39
Central and South America (469)	17,800	(6,880)	10,400	(7,510)	26.7	7.0	563	40	84/43
Global (6705) (excluding Antarctica)	134,000	(51,600)	38,900	(28,100)	—	100.00	8,000	39	More developed: 100/92 Less developed: 73/31

Note: BGD, billion gallons per day. Population data from *2008 World Population Data Sheet* (Washington, DC: Population Reference Bureau, 2008). Numbers are rounded.

NEWS REPORT 6.2

Personal Water Use and Water Measurement

Each of us indirectly accounts for a much greater water demand than we directly consume, because we all consume food and other products produced with water. For the United States, dividing the total water withdrawal (408 billion gallons a day, BGD, in 2000, the latest year data are available) by a population of 285 million people (2001) yields a per capita direct and indirect use of 5419 liters (1431 gallons) of water per day! As population and per capita affluence increase, greater demands are placed on the water-resource base, which is essentially fixed. (This calculation is approximate since water withdrawal data are unavailable for years since 2000.)

Simply providing the variety of food we enjoy requires voluminous water. For example: 77 g (2.7 oz) of broccoli requires 42 l (11 gal) of water to grow and process; producing 250 ml (8 oz) of milk requires 182 l (48 gal) of water; producing

28 g (1 oz) of cheese requires 2121 (56 gal); producing 1 egg requires 238 l (63 gal); and a 113-g (4-oz) beef patty requires 2,314 l (616 gal).

And then there are the toilets, the majority of which still flush approximately 4 gallons of water. Imagine the spatial complexity of servicing the desert city of Las Vegas, with 135,000 hotel rooms times the number of toilet flushes per day in those rooms; of providing all support services for 38 million visitors per year, plus 38 golf courses. One hotel confirms that it washes 14,000 pillowcases a day. In addition, there is the usage of a resident population of 1.5 million people—together, both tourists' and residents' water demand exceeds Nevada's annual allotment from the depleted Colorado River.

Clearly, the average American lifestyle depends on enormous quantities of water and thus is vulnerable to shortfalls or quality problems. Demand-side water conservation and

efficiency remain as potentially our best water resource.

How is all this water measured? In most of the United States, hydrologists measure streamflow in cubic feet per second (ft^3/s); Canadians use cubic meters per second (m^3/s). In the eastern United States, or for large-scale assessments, water managers use millions of gallons a day (MGD), billions of gallons a day (BGD), or billions of liters per day (BLD).

In the western United States, where irrigated agriculture is so important, the measure frequently used is acre-feet per year. One acre-foot is an acre of water, 1 foot deep, equivalent to 325,872 gallons (43,560 ft^3, or 1234 m^3, or 1,233,429 liters). An acre is an area that is about 208 feet on a side, or 0.4047 hectares. For global measurements, 1 km^3 = 1 billion cubic meters = 810 million acre-feet; 1000 m^3 = 264,200 gallons = 0.81 acre-feet. For smaller measures, 1 m^3 = 1000 liters = 264.2 gallons.

African countries experience water stress (less than 1700 m^3 per person/year) and 14 have water scarcity (1000 m^3 per person/year).

Water resources are different from other resources in that there is *no alternative substance to water*. Water-resource stress related to quantity and/or quality shortfalls will dominate future political agendas. Water shortages increase the probability for international conflict, endanger public health, reduce agricultural productivity, and damage life-supporting ecological systems. For the twenty-first century, a transition to a new approach is required, away from large, centralized structures toward decentralized, community-based strategies, focused on demand management and efficient technologies.

Water Supply in the United States

The US water supply derives from surface and groundwater sources that are fed by an average daily precipitation of 4200 BGD (billion gallons a day). That sum is based on an average annual precipitation value of 76.2 cm (30 in.) divided evenly among the 48 contiguous states (excluding Alaska and Hawai'i).

The 4200 billion gallons of average daily precipitation mentioned is unevenly distributed across the country and unevenly distributed throughout the year. For example, New England's water supply is so abundant that only about 1% of available water is consumed each year. (The same is true in Canada, where the resource greatly exceeds that in the United States.) But in the dry Colorado River Basin mentioned earlier, the discharge is completely consumed. In fact, by treaty and compact agreements, the Colorado actually is budgeted beyond its average flow! This paradox results from political misunderstanding of the water resource.

Figure 6.19 shows how the US supply of 4200 BGD is distributed in the 48 contiguous states. Viewed daily, the national water budget has two general outputs: 71% actual evapotranspiration (ACTET) and 29% surplus (SURPL). The 71% actual evapotranspiration involves 2970 BGD of the daily supply. It passes through nonirrigated land, including farm crops and pasture, forest and browse (young leaves, twigs, and shoots), and nonagricultural vegetation. Eventually, it returns to the atmosphere to continue its journey through the hydrologic cycle. The remaining 29% surplus is what we directly use.

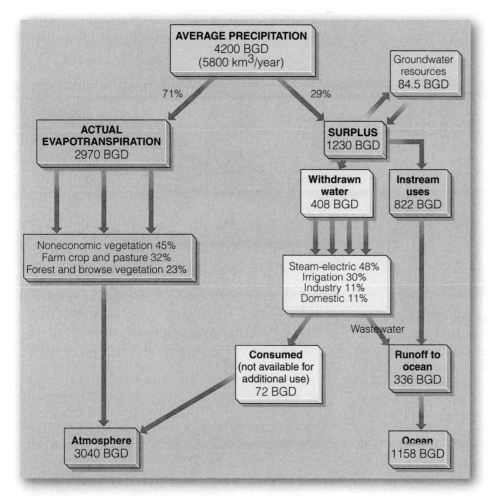

FIGURE 6.19 US water budget.
Daily water budget for the contiguous 48 states in billions of gallons a day (BGD). As of 2000, approximately 33% of available surplus was withdrawn for agricultural (irrigation-livestock), municipal (domestic-commercial), and industrial (industry-mining-thermoelectric) use. [Data from Susan Hutson, Nancy Barber, Joan Kenny, Kristin Linsey, Deborah Lumia, and Molly Maupin, *Estimated Use of Water in the United States in 2000* (Denver: USGS Circular 1268, 2000)—the latest year for which data are available.]

Instream, Nonconsumptive, and Consumptive Uses

The surplus 1230 BGD is runoff, available for withdrawal, consumption, and various instream uses.

- *Instream* uses are those that use streamwater in place: navigation, wildlife and ecosystem preservation, waste dilution and removal, hydroelectric power production, fishing resources, and recreation.
- **Withdrawal**, or *nonconsumptive*, uses remove water from the supply, use it, and then return it to the same supply. Industry, agriculture, municipalities, and steam-electric power generation use nonconsumptive water. A portion of water withdrawn is consumed.
- **Consumptive** uses remove water from a stream but do not return it, so it is not available for a second or subsequent use. Some consumptive examples include water that evaporates or is vaporized in steam-electric plants.

When water is returned to the system, water quality usually is altered—water is contaminated chemically with pollutants or waste or thermally with heat energy. In Figure 6.19, this wastewater portion of the budget is returned to runoff and eventually the ocean. The water resource can be extended through recovery, treatment, and reuse of wastewater.

Contaminated or not, returned water becomes a part of all water systems downstream. A good example is New

Orleans, the last city to withdraw municipal water from the Mississippi River. New Orleans receives diluted and mixed contaminants added throughout the entire Missouri-Ohio-Mississippi River drainage. Abnormally high cancer rates among citizens living along the Mississippi River between Baton Rouge and New Orleans have led to the ominous label "Cancer Alley" for the region. The nutrient discharge from the wastes and fertilizers in the river creates the "Dead Zone" in the Gulf of Mexico, discussed in Chapter 16, Figure 16.10.

The estimated US withdrawal of water for 2000 was 408 BGD, almost three times the usage rate in 1940, but a decrease of 10% since the peak year of 1980 and similar to 1990. Total per capita use dropped from 1340 gal/day in 1990 to 1280 gal/day in 2000. A shift in emphasis from supply-side to demand-side planning, pricing, and management produced these efficiencies. The four main uses of withdrawn water in the United States in 2000 were steam-electric power (48%), industry-mining (11%), irrigation-livestock (30%), and domestic-commercial (11%). In contrast, Canada uses only 12% of its withdrawn water for irrigation and 70% for industry. Figure 6.20 compares regions by their use of withdrawn water during 1998. It graphically illustrates the differences in water usage between more-developed and less-developed parts of the world. (For studies of water use in the United States, see http://water.usgs.gov/public/watuse/.)

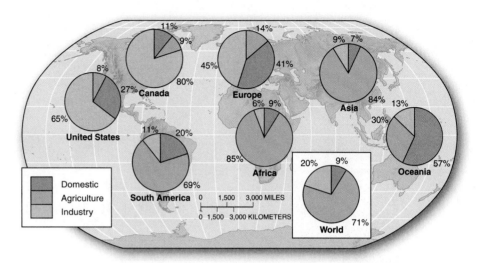

FIGURE 6.20 Water withdrawal by sector.
Compare industrial water use among the geographic areas, as well as agricultural and municipal uses. [After World Resources Institute, *World Resources 2000–2001*, Data Table FW.1, pp. 276–77.]

Desalination

Desalination of seawater to augment diminishing groundwater supplies is becoming increasingly important as a freshwater source. Plant construction is expected to increase 140% between 2007 and 2017, amounting to $60 billion in investments; somewhere between 1500 and 2000 plants are now in operation worldwide. Australia has $4.9 billion in construction underway. In the Middle East, desalination is presently in widespread use. Desalination is an alternative to further groundwater mining and saltwater intrusion under Saudi Arabia (Figure 6.21a). Also, along the coast of southern California and in Florida desalination is slowly increasing in use.

Tampa–St. Petersburg, Florida, took a giant step forward in 2003 with the opening of its own Tampa Bay desalination plant producing 25 million gallons a day (MGD), making it the largest in the United States. The seawater intake is from the Big Bend Power Station's cooling water discharge. Drinking water is produced through a series of filters to remove algae and particles and a reverse osmosis process where the water is forced through semipermeable membranes. The plant is to expand to 35 MGD, or about 10% of the regional water supply in the near future (Figure 6.21b).

Future Water Considerations

Budgeting precipitation input makes apparent the limits of water resources. How can we satisfy the growing demand for water? Water supplies available per person decline as population increases; thus, world population growth since 1970 has reduced per capita water supplies by a third. Also, pollution limits the water-resource base, so that even before *quantity* constraints are felt, *quality* problems may limit the health and growth of a region. A composite index comparing available water resources to current use patterns, supplies, and national income reveals regional scarcities in Africa, the Middle East, Asia, Peru, and Mexico (Figure 6.22).

Working Group II, *Climate Change Impacts, Adaptation, and Vulnerability*, in their "Summary for Policy Makers" (Intergovernmental Panel on Climate Change, *Fourth Assessment Report*, April 2007), concluded that global climate change and greenhouse warming will cause added water problems:

By mid-century, annual river runoff and water availability are projected to increase 10–40% at high latitudes

(a)

(b)

FIGURE 6.21 Water desalination.
(a) Freshwater is supplied to Saudi Arabia from the Jabal water desalination plant along the Red Sea. Saudi Arabia obtains a significant amount of its water from desalination of seawater.
(b) The Tampa Bay desalination plant provides 25 MGD to the region's water resources through a filtration and reverse osmosis process. [Photos by (a) Liaison Agency; (b) Tampa Bay Water.]

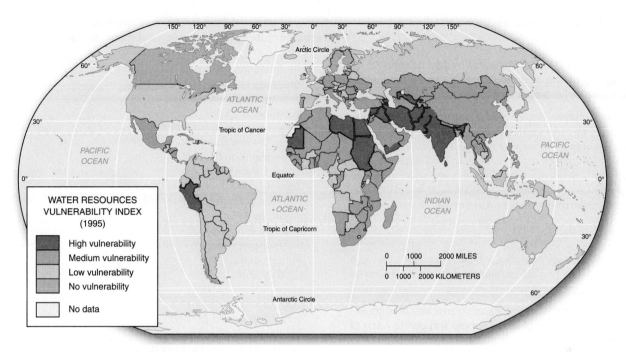

FIGURE 6.22 Global water scarcity.
Water-resource vulnerability, a composite index. [Compiled by The World Resources Institute from
data gathered by the Stockholm Environment Institute, in *Comprehensive Assessment of the Freshwater Resources of the World*, appearing in *World Resources 1998–1999* (New York: The World
Resources Institute, 1998), p. 223.]

and in some wet tropical areas, and decrease by
10–30% over some dry regions at mid-latitude and in
dry tropics, some of which are presently water stressed
areas.... Drought-affected areas will likely increase in
extent.... In the course of the century, water supplies
stored in glaciers and snow cover are projected to
decline, reducing water availability in regions supplied by
meltwater from mountain ranges, where more than one-
sixth of the world population currently lives.

The IPCC concluded that given higher temperatures, water
demand increases to meet potential evapotranspiration
needs and algal blooms flourish in the warmer runoff waters.
Those areas that experience increased intensity of heavy
precipitation events will experience surface and groundwater contamination episodes. Also, where groundwater mining occurs to meet rising demands, the drawdown of the
water table and saltwater intrusion near coastlines will
worsen. An important consideration to remember is this:

Unlike other important commodities such as oil, copper, or wheat, freshwater has no substitutes for most of
its uses. It is also impractical to transport the large
quantities of water needed in agriculture and industry
more than several hundred kilometers. Freshwater is
now scarce in many regions of the world, resulting in
severe ecological degradation, limits on agriculture
and industrial production, threats to human health,
and increased potential for international conflict.*

Clearly, cooperation is needed, yet we continue
toward a water crisis without a concept of a *world water
economy* as a frame of reference. The geospatial question
is: When will more international coordination begin, and
which country or group of countries will lead the way to
sustain future water resources? The soil-water-budget
approach detailed in this chapter is a place to start.

*S. L. Postal et al., "Human appropriation of renewable fresh water,"
Science 271, no. 5250 (February 9, 1996): 785.

Summary and Review—Water Resources

■ *Illustrate* the hydrologic cycle with a simple sketch and
label it with definitions for each water pathway.

The flow of water links the atmosphere, ocean, and land through
energy and matter exchanges. The **hydrologic cycle** is a model
of Earth's water system, which has operated for billions of years
from the lower atmosphere to several kilometers beneath Earth's
surface. Water soaks into the subsurface through **infiltration**, or
penetration of the soil surface. It further permeates soil or rock
through vertical movement called **percolation**.

hydrologic cycle (p. 194)
infiltration (p. 195)
percolation (p. 195)

1. Sketch and explain a simplified model of the complex flows
of water on Earth—the hydrologic cycle.
2. What are the possible routes that a raindrop may take on its
way to and into the soil surface?
3. Compare precipitation and evaporation volumes from
the ocean with those over the land. Describe advection

flows of moisture and countering surface and subsurface runoff.

■ *Relate* the importance of the water-budget concept to an understanding of the hydrologic cycle, water resources, and soil moisture for a specific location.

A **soil-water budget** can be established for any area, from region to field, of Earth's surface by measuring the precipitation input and the output of various water demands in the area considered. Water surpluses supply streams and soil-moisture storage needs and satisfy potential evapotranspiration demands. Similar to financial accounting, only through budgeting can water resources be truly understood and managed.

soil-water budget (p. 195)

4. How might an understanding of the hydrologic cycle in a particular locale or a soil-moisture budget of a site assist you in assessing water resources? Give some specific examples.

■ *Construct* the water-balance equation as a way of accounting for the expenditures of water supply and *define* each of the components in the equation and its specific operation.

The moisture supply to Earth's surface is **precipitation** (PRE-CIP), arriving as rain, sleet, snow, and hail. Precipitation is measured with the **rain gauge**. **Evaporation** is the net movement of free water molecules away from a wet surface into air. **Transpiration** is the movement of water through plants and back into the atmosphere; it is a cooling mechanism for plants. Evaporation and transpiration are combined into one term—**evapotranspiration**. The ultimate demand for moisture is **potential evapotranspiration** (POTET), the amount of water that *would* evaporate and transpire under optimum moisture conditions (adequate precipitation and adequate soil moisture). Evapotranspiration is measured with an **evaporation pan** (*evaporimeter*) or the more elaborate **lysimeter**.

Unsatisfied POTET is a **deficit** (DEFIC). By subtracting DEFIC from POTET, we determine **actual evapotranspiration**, or ACTET. Ideally, POTET and ACTET are about the same, so that plants have sufficient water. If POTET is satisfied and the soil is full of moisture, then additional water input becomes **surplus** (SURPL), which may puddle on the surface, flow across the surface toward stream channels, or percolate through the soil underground. The overland flow includes precipitation and groundwater flows into river channels to make up the **total runoff** from the area.

A "savings account" of water that receives deposits and provides withdrawals as water-balance conditions change is the **soil-moisture storage** (Δ STRGE). This is the volume of water stored in the soil that is accessible to plant roots. In soil, *hygroscopic water* is inaccessible because it is a molecule-thin layer that is tightly bound to each soil particle by hydrogen bonding. As available water is utilized, soil reaches the **wilting point** (all that remains is unextractable water). **Capillary water** is generally accessible to plant roots because it is held in the soil by surface tension and hydrogen bonding between water and soil. Almost all capillary water that remains in the soil is **available water** in soil-moisture storage. After water drains from the larger pore spaces, the available water remaining for plants is termed **field capacity**, or storage capacity. When soil is saturated after a precipitation event, surplus water in the soil becomes **gravitational water** and percolates to groundwater. As **soil-moisture utilization** removes soil water, the plants work

harder to extract the same amount of moisture. **Soil-moisture recharge** is the rate at which needed moisture enters the soil.

precipitation (p. 196)
rain gauge (p. 196)
evaporation (p. 196)
transpiration (p. 196)
evapotranspiration (p. 196)
potential evapotranspiration (p. 196)
evaporation pan (p. 198)
lysimeter (p. 198)
deficit (p. 198)
actual evapotranspiration (p. 198)
surplus (p. 198)
total runoff (p. 198)
soil-moisture storage (p. 198)
wilting point (p. 198)
capillary water (p. 199)
available water (p. 199)
field capacity (p. 199)
gravitational water (p. 199)
soil-moisture utilization (p. 199)
soil-moisture recharge (p. 199)

5. Discuss the meaning of this statement: "The soil-water budget is an assessment of the hydrologic cycle at a specific site."

6. What are the components of the water-balance equation? Construct the equation and place each term's definition below its abbreviation in the equation.

7. Compare the annual water-balance data for Kingsport, Tennessee; Ottawa, Ontario; and Phoenix, Arizona. What differences do you see in the components? What similarities? What does this tell you about local water resources in each area?

8. Explain how to derive actual evapotranspiration (ACTET) in the water-balance equation.

9. What is potential evapotranspiration (POTET)? How do we go about estimating this potential rate? What factors did Thornthwaite use to determine this value?

10. Explain the operation of soil-moisture storage, soil-moisture utilization, and soil-moisture recharge. Include discussion of the field capacity, capillary water, and wilting point concepts.

11. In the case of silt-loam soil from Figure 6.9, what is, roughly, the available water capacity? How is this value derived?

12. For water resources, what does it mean to move water geographically or across the calendar?

13. In terms of water balance and water management, explain the logic behind the California Water Project.

■ *Describe* the nature of groundwater and *define* the elements of the groundwater environment.

Groundwater is a part of the hydrologic cycle, but it lies beneath the surface beyond the soil-moisture root zone. Groundwater replenishment is tied to surface surpluses because groundwater does not exist independently. Excess surface water moves through the **zone of aeration**, where soil and rock are less than saturated. Eventually, the water reaches the **zone of saturation**, where the pores are completely filled with water.

The texture and the structure of the soil dictate available pore spaces, or **porosity**. **Permeable** describes the degree to which water can flow through soil; **impermeable** describes the condition where flow is restricted. Permeability depends on particle sizes and the shape and packing of soil grains. An **aquifer** is a rock layer that is permeable to groundwater flow in usable amounts. An **unconfined aquifer** has a permeable layer on top and an impermeable one beneath. The upper limit of the water that collects in the zone of saturation is the **water table**; it is the contact surface between the zones of saturation and aeration. An **aquiclude** (aquitard) is a body of rock that does not conduct water in usable amounts.

A **confined aquifer** is bounded above and below by impermeable layers of rock or sediment. The **aquifer recharge area** extends over an entire unconfined aquifer. Water in a confined aquifer is under the pressure of its own weight, creating a pressure level to which the water can rise on its own, called the *potentiometric surface*, which can be above ground level. Groundwater confined under pressure is **artesian water**; it may rise up in wells and even flow out at the surface without pumping, if the head of the well is below the potentiometric surface.

As water is pumped from a well, the surrounding water table within an unconfined aquifer will experience a **drawdown**, or become lower, if the rate of pumping exceeds the horizontal flow of water in the aquifer around the well. This excessive pumping causes a **cone of depression**. Aquifers frequently are pumped beyond their flow and recharge capacities, a condition known as **groundwater mining**.

groundwater (p. 205)
zone of aeration (p. 206)
zone of saturation (p. 206)
porosity (p. 206)
permeable (p. 206)
impermeable (p. 206)
aquifer (p. 206)
unconfined aquifer (p. 206)
water table (p. 206)
aquiclude (p. 206)
confined aquifer (p. 207)
aquifer recharge area (p. 207)
artesian water (p. 207)
drawdown (p. 207)
cone of depression (p. 208)
groundwater mining (p. 208)

14. Are groundwater resources independent of surface supplies, or are the two interrelated? Explain your answer.

15. Make a simple sketch of the subsurface environment, labeling zones of aeration and saturation and the water table in an unconfined aquifer. Then add a confined aquifer to the sketch.

16. At what point does groundwater utilization become groundwater mining? Use the High Plains aquifer example to explain your answer.

17. What is the nature of groundwater pollution? Can contaminated groundwater be cleaned up easily? Explain.

■ *Identify* critical aspects of freshwater supplies for the future and *cite* specific issues related to sectors of use, regions and countries, and potential remedies for any shortfalls.

Nonconsumptive uses, or water **withdrawal**, remove water from the supply, use it, and then return it to the stream. **Consumptive uses** remove water from a stream but do not return it, so the water is not available for a second or subsequent use. Americans in the 48 contiguous states withdraw approximately one-third of the available surplus runoff for irrigation, industry, and municipal uses. Water-resource planning regionally and globally, using water-budget principles, is essential if Earth's societies are to have enough water of adequate quality. **Desalination** as a water resource involves the removal of organics, debris, and salinity from seawater through distillation or reverse osmosis. This processing yields potable water for domestic uses.

withdrawal (p. 215)
consumptive uses (p. 215)
desalination (p. 216)

18. Describe the principal pathways involved in the water budget of the contiguous 48 states. What is the difference between withdrawal and consumptive use of water resources? Compare these with instream uses.

19. Characterize each of the sectors withdrawing water: irrigation, industry, and municipalities. What are the present usage trends in more-developed and less-developed nations?

20. Briefly assess the status of world water resources. What challenges are there in meeting the future needs of an expanding population and growing economies? How do you think thinking and planning for the world water economy might begin?

21. How does acid deposition (acid rain) tie together the issues of air pollution and energy production and consumption from Chapter 2 and Focus Study 2.2 and water-resource quality in this chapter?

NetWork and Critical Thinking Tools

A. Select your campus, yard, or perhaps a houseplant and apply soil-water budget concepts. What is the supply of water? Estimate the ultimate water supply (PRECIP) and ultimate demand (POTET) for the area selected. Estimate water needs and how they vary with the season as components of the water-budget change.

If you are maintaining a house- or dorm-room plant, this can be an interesting process. You meter out water to satisfy the water demands created by room temperature and the placement of the plant. You might find that work with the soil (tilling) in the container improves capillary water function and soil moisture storage. You are doing the water budget!

B. You have, no doubt, consumed several glasses of water since you woke this morning. Where did this water originate? Obtain the following information: the name of the water supplier or agency, whether the supplier is using surface or groundwater to meet demands, and how the water is metered and billed. If your state or province requires water-quality reporting, obtain a copy of the analysis of your tap water. Lastly, what is your subjective assessment of this water today: taste, smell, hardness, clarity? Compare these perceptions with others in your class.

What about the water on your campus? If the campus has its own wells, how is the quality tested? Who on campus is in charge of supervising these wells?

Palm–oak forest hammock ("shady place") features a mixed hardwood forest. There are bromeliads (epiphytes) growing on the live oak trees and a few on the cabbage, or sabal, palms. This palm is Florida's state tree. A hammock is slightly higher ground, surrounded by wet prairies, mixed swamp, and cypress forest marshes. Such "islands" of vegetation result from having few fires, protected as they are by surrounding marshes. These two photos are in Myakka River State Park, Sarasota County, about 15 miles inland from the central Gulf Coast. *[Overhead and forest-floor photos by Bobbé Christopherson.]*

Chapter

7

Climate Systems and Climate Change

KEY LEARNING CONCEPTS

After reading the chapter, you should be able to:

- **Define** climate and climatology and **explain** the difference between climate and weather.

- **Review** the role of temperature, precipitation, air pressure, air mass patterns, and sea-surface temperatures used to establish climatic regions.

- **Review** the development of climate classification systems and **compare** genetic and empirical systems as ways of classifying climate.

- **Describe** the principal climate classification categories other than deserts and **locate** these regions on a world map.

- **Explain** the precipitation and moisture efficiency criteria used to determine the arid and semiarid climates and **locate** them on a world map.

- **Outline** future climate patterns from forecasts presented and **explain** the causes and potential consequences of climate change.

STUDENT LEARNING CENTER TOOLS

The *Elemental Geosystems* Student Learning Center provides on-line resources for this chapter by clicking on the book cover at **www.mygeoscienceplace.com**. Some highlights:

▶ A 14-month diary logged by staff stationed at the South Pole base, under "Case Studies."

▶ Interactive animation of El Niño/La Niña circulation patterns.

▶ Opportunity to analyze global precipitation data from satellite imagery and an animated sequence of patterns of precipitation.

▶ View high latitude video clips and photo galleries of East Greenland and high latitude animals.

▶ View a dramatic surface temperature analysis movie showing the years from 1891 to 2006—a warming trend is apparent in the movie animation.

▶ Sixteen additional URLs along with Quick Links to all the URLs in the chapter.

Earth experiences an almost infinite variety of *weather*—the condition of the atmosphere at any given place and time. This variability, when considered along with the average conditions at a place over time, constitutes **climate**. Climate includes not just the averages, but the extremes experienced in a place. Today, climatologists know that intriguing global linkages exist in the Earth-atmosphere-ocean system.

Climatologists, among other scientists, are analyzing global climate change—record-breaking global average temperatures, glacial ice melt, drying soil-moisture conditions, changing crop yields, spreading of infectious disease to new locales, changing distributions of plants and animals, declining coral reef health and fisheries, and the thawing of high-latitude lands and seas. Climatologists are concerned about observed changes in the global climate, as these are occurring at a pace not evidenced in the records of the past millennia. Climate and natural vegetation shifts during the next 50 years could exceed the total of all changes since the peak of the last ice-age episode, some 18,000 years ago. In Chapter 1, *Elemental Geosystems* began with these words from scientist Jack Williams:

> By the end of the 21st century, large portions of the Earth's surface may experience climates not found at present, and some 20th-century climates may disappear.... Novel climates are projected to develop primarily in the tropics and subtropics.... Disappearing climates increase the likelihood of species extinctions and community disruption for species endemic to particular climatic regimes, with the largest impacts projected for poleward and tropical montane regions.*

In this chapter: Climates are so diverse that no two places on Earth's surface experience exactly the same climatic conditions; in fact, Earth is a vast collection of microclimates. However, broad similarities among local climates permit their grouping into climatic regions.

Early climatologists faced the challenge of what to use as a basis for climate classification. Many of the physical elements of the environment, studied in the first six chapters of this text, link together to explain climates. *Elemental Geosystems* uses a simplified classification system to discuss climates based on physical factors that help uncover the "why" question—why climates are in certain locations. Though imperfect, this method is easily understood and is based on a widely used classification system devised by climatologist Wladimir Köppen (pronounced KUR-pen). For reference, Appendix B details the Köppen climate classification system.

Climatologists use powerful computer models to simulate complex changing interactions in the atmosphere, hydrosphere, lithosphere, and biosphere. This chapter concludes with a discussion of climate change and its vital implications for our global society. The climatic regions we study in this chapter will migrate during this century.

*Jack Williams et al., "Projected distributions of novel and disappearing climates by A.D. 2100," *Proceedings of the National Academy of Sciences* (April 3, 2007): 5739.

Earth's Climate System and Its Classification

Climatology, the study of climate and its variability, analyzes long-term weather patterns over time or space and the controls that produce Earth's diverse climatic conditions. Note that climate cannot actually be observed and really does not exist at any particular moment. Climate is therefore a conceptual statistical construction from measured weather elements.

One type of climatic analysis locates areas of similar weather statistics and groups them into **climatic regions** that contain characteristic weather patterns. Weather measurements gathered at different locations within a region are plotted on maps and compared to identify climate types. Expected weather and climatic patterns can be disrupted, as is the case with the recurring El Niño/La Niña phenomenon, discussed in Focus Study 7.1.

Climates may be humid with distinct seasons, dry with consistent warmth, or moist and cool—almost any combination is possible. There are places where rainfall exceeds 20 cm (8 in.) each month and where average monthly temperatures remain above 27°C (80°F) throughout the year. Other places may go without rain for 10 years at a time. A climate may have temperatures that average above freezing every month yet still threaten severe frost problems for agriculture. Students reading this text in Singapore experience precipitation every month ranging from 13.1 to 30.6 cm (5.1 to 12.0 in.), or 228.1 cm (89.8 in.) during an average year, whereas students at the university in Karachi, Pakistan, measure only 20.4 cm (8 in.) of rain over an entire year.

Climates greatly influence *ecosystems*, the natural, self-regulating communities formed by plants and animals within their nonliving environment. On land, the basic climatic regions determine to a large extent the location of the world's major ecosystems. These regions, called *biomes*, include forest, grassland, savanna, tundra, and desert. Plant, soil, and animal communities are associated with these biomes. Because climate cycles through periodic change and is never really stable, ecosystems should be thought of as being in a constant state of adaptation and response.

The present global climatic warming trend is producing changes in plant and animal distributions. Figure 7.1 presents a schematic view of Earth's climate system, showing both internal and external processes and linkages that influence climate and these distributions.

Climate Components: Insolation, Temperature, Air Pressure, Air Masses, and Precipitation

Before we look at climatic regions, let's briefly review key concepts from the first six chapters. Uneven insolation over Earth's surface, varying with latitude, is

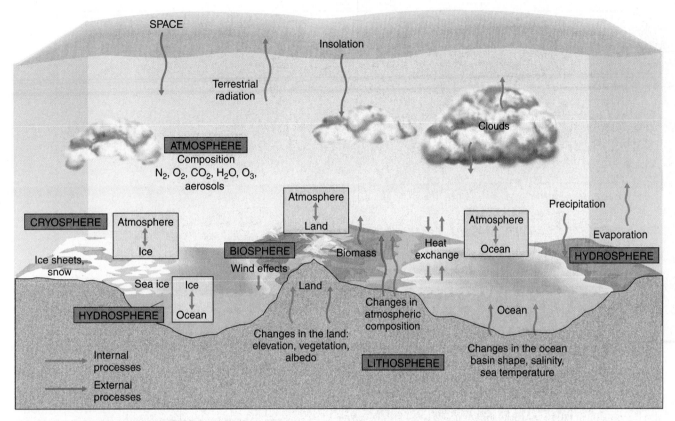

FIGURE 7.1 A schematic of Earth's climate system.
Internal processes that influence climate involve the atmosphere, hydrosphere (streams and ocean), cryosphere (polar ice masses and glaciers), biosphere, and lithosphere (land)—all energized by insolation. All systems interact to produce climatic patterns. *External processes* affect this climatic balance and force climate change. Imagine being hired to write a computer program to model all these linkages and interactions! [After J. Houghton, *The Global Climate* (Cambridge: Cambridge University Press, 1984), and the Global Atmospheric Research Program.]

the energy input for the climate system. Daylength and temperature patterns vary daily and seasonally (Chapter 2). The principal controls of temperature are latitude, altitude/elevation, land–water heating differences, and the amount and duration of cloud cover. The pattern of world temperatures and annual temperature ranges is discussed in Chapter 3 and portrayed in Figures 3.24, 3.27, and 3.29.

The hydrologic cycle provides the basic means of transferring energy and mass through Earth's climate system. Figure 7.2 shows the worldwide distribution of precipitation (rain, sleet, snow, and hail). You can identify several patterns, such as the way precipitation decreases from western Europe inland toward the heart of Asia (continentality). However, high precipitation values dominate in southern Asia, where the landscape receives moisture-laden monsoonal winds from the Indian Ocean in the summer. Southern South America reflects a pattern of wet western (windward) slopes and dry eastern (leeward) slopes (orographic effects). See Figure 6.5 for the distribution of average annual precipitation in North America and an orographic and air mass pattern.

Most of Earth's deserts and regions of permanent drought are located in lands dominated by subtropical

high-pressure cells, with bordering lands transitioning to grasslands and to forests as precipitation increases. The most consistently wet places on Earth straddle the equator in the Amazon region of South America, the Congo (Zaire) region of Africa, and the Indonesian and Southeast Asian area, all of which are influenced by equatorial low pressure and the intertropical convergence zone (Figures 4.10, 4.12).

Simply comparing the two principal climatic components—temperature and precipitation—reveals important relationships (Figure 7.3). These patterns of temperature and precipitation, along with other weather factors, provide the basis for climate classification.

Classification of Climatic Regions

The ancient Greeks simplified their view of world climates into three zones: the "torrid zone" referred to warmer areas south of the Mediterranean, the "frigid zone" was to the north, and the area where they lived was labeled the "temperate zone," which they considered the optimum climate. Through exploration and satellite observations, we know that Earth's myriad climatic variations are more complex than the ancients imagined.

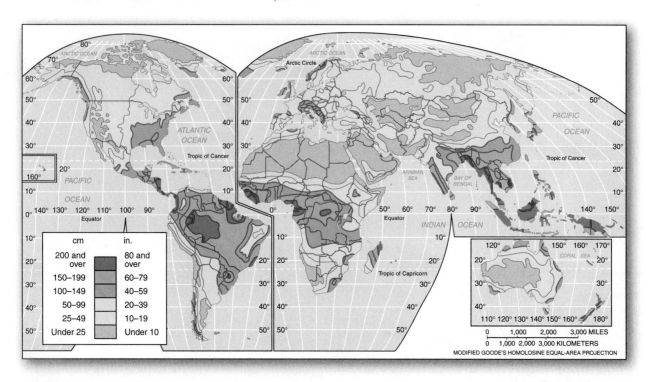

FIGURE 7.2 Worldwide average annual precipitation.
The causes that produce these patterns should be recognizable to you: temperature and pressure patterns; air mass types; convergent, convectional, orographic, and frontal lifting mechanisms; and the general energy availability that decreases toward the poles.

ANIMATION Global Patterns of Precipitation

Classification is the process of ordering or grouping related data or phenomena into classes. Such generalizations are important organizational tools in science and are especially useful for the spatial analysis of climatic regions. A climate classification based on *causative* factors—for example, the interaction of air masses—is a **genetic classification**. This approach explores the "why" question as to the mix of climatic ingredients in certain locations. A climate classification based on *statistics* or other data is an **empirical classification**.

Climate classifications based on temperature and precipitation data are examples of empirical classifications, determined by measurement of observed effects. One empirical classification system, published by C. W. Thornthwaite, identified moisture regions using aspects of the water-balance approach (discussed in Chapter 6) and vegetation types. Another empirical classification system is the Köppen classification system.

Wladimir Köppen (1846–1940), a German climatologist and botanist, designed the widely recognized Köppen classification system. In 1928, after years of research and development, his first wall map showing world climates, coauthored with his student Rudolph Geiger, was widely adopted. Köppen continued to refine this system until his death. In Appendix B you will find a description of his system and the detailed

(text continued on page 227)

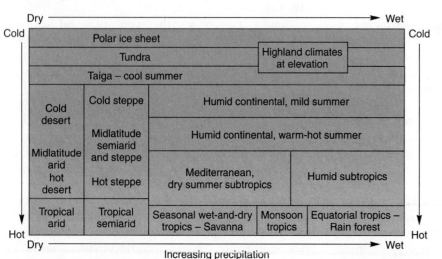

FIGURE 7.3 Climatic relationships.
Temperature and precipitation schematic showing climatic relationships. Based on your general knowledge of where your college is located, can you identify approximately its place on the schematic? How about your birthplace?

FOCUS STUDY 7.1

The El Niño Phenomenon—Record Intensity, Global Linkages

Climate is the consistent behavior of weather over time, but average weather conditions also include extremes that depart from normal. The El Niño–Southern Oscillation (ENSO) in the Pacific Ocean forces the greatest inter-annual variability of temperature and precipitation on a global scale. The two strongest ENSO events in 120 years hit in 1997–1998 and 1982–1983. The spring wildflower bloom in Death Valley in 1998 provides visible evidence of the resultant heavy rains (Figure 7.1.1). Peruvians coined the name El Niño ("the boy child") because these episodes seem to occur around the tra-ditional December celebration time of Christ's birth. Actually, El Niños can occur as early as spring and summer and persist through the year.

Revisit Figure 6.20 and see that the northward-flowing Peru current dominates the region off South America's west coast. These cold waters move toward the equator and join the westward movement of the south equatorial current. The Peru current is part of the normal coun-terclockwise circulation of winds and surface ocean currents around the subtropical high-pressure cell domi-nating the eastern Pacific in the Southern Hemisphere. As a result, a location such as Guayaquil, Ecuador, normally receives 91.4 cm (36 in.) of precipitation each year under domi-nant high pressure, whereas islands in the Indonesian archipelago receive more than 254 cm (100 in.) under dominant low pressure. This normal alignment of pressure is shown in Figure 7.1.2a.

What Is ENSO?

Occasionally, for unexplained reasons, pressure patterns and surface ocean temperatures shift from their usual locations. Higher pressure than nor-mal develops over the western Pacific and lower pressure develops over the eastern Pacific. Trade winds normally moving from east to west weaken and can be reduced or even replaced by an eastward (west-to-east) flow. The shifting of atmospheric pressure and wind patterns across the Pacific is the *Southern Oscillation*. Chapter 4 dis-cussed the Pacific Decadal Oscillation (PDO) and its interrelation with ENSO.

Sea-surface temperatures increase, sometimes more than 8 C° (14 F°) above normal in the central and eastern Pacific during an ENSO, replacing the normally cold, upwelling, nutrient-rich water along Peru's coastline. Such ocean-surface warming, the "warm pool," may extend to the International Date Line. This surface pool of warm water is known as El Niño. Thus, the designation ENSO is derived—El Niño–Southern Oscillation. This con-dition is shown in Figure 7.1.2b in illustration and satellite image.

The thermocline (boundary of colder, deep-ocean water) lowers in depth in the eastern Pacific Ocean. The change in wind direction and warmer surface water slows the nor-mal upwelling currents that control nutrient availability. This loss of nutrients affects the phytoplankton and food chain, depriving many fish, marine mammals, and predator birds of nourishment.

Scientists at the National Oceano-graphic and Atmospheric Administra-tion (NOAA) speculate that ENSO events occurred more than a dozen times since the fourteenth century.

(a)

(b)

FIGURE 7.1.1 El Niño's impact on the desert.
Death Valley, southeastern California, (a) in full spring bloom following record rains triggered by the 1997–1998 El Niño and (b) the same scene in spring 2002 in its stark desert grandeur. A dramatic effect caused by changes in the distant tropics of the Pacific Ocean. [Photos by Bobbé Christopherson.]

(continued)

Focus Study 7.1 *(continued)*

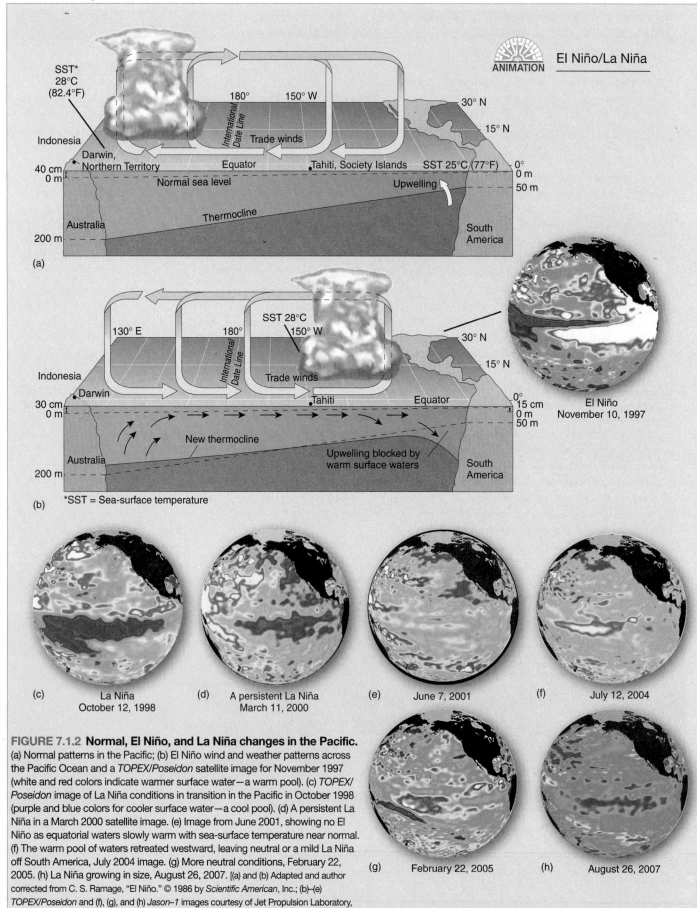

El Niño/La Niña

FIGURE 7.1.2 Normal, El Niño, and La Niña changes in the Pacific.
(a) Normal patterns in the Pacific; (b) El Niño wind and weather patterns across the Pacific Ocean and a *TOPEX/Poseidon* satellite image for November 1997 (white and red colors indicate warmer surface water—a warm pool). (c) *TOPEX/Poseidon* image of La Niña conditions in transition in the Pacific in October 1998 (purple and blue colors for cooler surface water—a cool pool). (d) A persistent La Niña in a March 2000 satellite image. (e) Image from June 2001, showing no El Niño as equatorial waters slowly warm with sea-surface temperature near normal. (f) The warm pool of waters retreated westward, leaving neutral or a mild La Niña off South America, July 2004 image. (g) More neutral conditions, February 22, 2005. (h) La Niña growing in size, August 26, 2007. [(a) and (b) Adapted and author corrected from C. S. Ramage, "El Niño." © 1986 by *Scientific American*, Inc.; (b)–(e) *TOPEX/Poseidon* and (f), (g), and (h) *Jason–1* images courtesy of Jet Propulsion Laboratory, NASA.]

More recently, there was an ENSO in 1982–1983 (second-strongest event), 1986–1987, 1991–1993 (one of the longest), and the most intense episode in 1997–1998 that disrupted global weather. The PDO shifted into its negative phase in 1999, which seems to dampen ENSO cycles and bring drier conditions to the American West.

The expected interval for recurrence is 3 to 5 years, but the interval may range from 2 to 12 years. The frequency and intensity of ENSO events increased through the twentieth century, a topic of much research by scientists to see if there is a relation to global climate change. Recent studies suggest ENSO might be more responsive to global change than previously thought.

La Niña—El Niño's Cousin

When surface waters in the central and eastern Pacific cool to below normal by 0.4 C° (0.7 F°) or more, the condition is dubbed La Niña, Spanish for "the girl." This is a weaker condition and less consistent than El Niño. There is no correlation between the strength or weakness of each. For instance, following the record 1997–1998 ENSO event, the subsequent La Niña was not as strong as predicted and shared the Pacific with lingering warm water.

Between 1900 and 1998, there were 13 La Niñas of note, the latest in 1988, 1995, late 1998 to 2000 (Figure 7.1.2c and d), and 2007 (Figure 7.1.2h).

According to National Center for Atmospheric Research (NCAR) scientist Kevin Trenberth, El Niños occurred 31% and La Niñas 23% of the time since 1950, with the remaining 46% of the time the Pacific being in a more neutral condition. Don't look for symmetry and opposite effects between the two events, for there is great variability possible, except perhaps in Indonesia where remarkable drought (El Niño) and heavy rain (La Niña) correlations seem strong.

Global Effects Related to ENSO and La Niña

Effects related to ENSO and La Niña occur worldwide: droughts in South Africa, southern India, Australia, and the Philippines; strong hurricanes in the Pacific, including Tahiti and French Polynesia; and flooding in the southwestern United States and mountain states, Bolivia, Cuba, Ecuador, and Peru. In India, every drought for more than 400 years seems linked to ENSO events. The Atlantic hurricane season weakens during El Niño years and strengthens during La Niñas.

Precipitation in the southwestern United States is greater in El Niño than La Niña years. The Pacific Northwest is wetter with La Niña than El Niño. El Niño–enhanced rains in 1998 produced the wildflower bloom in Death Valley pictured in Figure 7.1.1. The Colorado River flooding shown in the Chapter 6

opening photo and discussed in Chapter 12's Focus Study 12.1 was in part attributable to the 1982–1983 El Niño. Yet, as conditions vary, other El Niños have produced drought in the very regions that flooded during a previous episode.

Since the 1982–1983 event, and with the development of remote-sensing satellites and computing capability, scientists now are able to identify the complex global interconnections among surface temperatures, pressure patterns in the Pacific, occurrences of drought in some places, excessive rainfall in others, and the disruption of fisheries and wildlife.

Discovery of these truly Earth-wide relations and spatial impacts is at the heart of physical geography. The climate of one location is related to climates elsewhere, although it should be no surprise that Earth operates as a vast integrated system. "It is fascinating that what happens in one area can affect the whole world. As to why this happens, that's the question of the century. Scientists are trying to make order out of chaos," says NOAA scientist Alan Strong. (For ENSO monitoring and forecasts, see the Climate Prediction Center at http://www.ncep .noaa.gov/, the Jet Propulsion Laboratory at http://topex-www.jpl.nasa. gov/mission/jason-1.html/ or http:// sealevel.jpl.nasa.gov/science/jason1, or NOAA's El Niño Theme Page at http://www.pmel.noaa.gov/toga-tao/ el-nino/nino-home.html.)

---

criteria he used to distinguish climatic regions and their boundaries.

As we devise spatial categories and boundaries, we must remember that boundaries really are transition zones of gradual change. The trends and overall patterns of boundary lines are more important than their precise placement, especially with the small scales generally used for world maps. Figure 7.4 portrays six basic climate categories and their regional types that provide a structure for our discussion.

- Tropical (equatorial and tropical latitudes)
    rain forest (rainy all year)
    monsoon (6 to 12 months rainy)
    savanna (less than 6 months rainy)

- Mesothermal (midlatitudes, mild winter)
    humid subtropical (hot summers)
    marine west coast (warm to cool summers)
    Mediterranean (dry summers)

- Microthermal (mid- and high latitude, cold winter)
    humid continental (hot to warm summers)
    subarctic (cool summers to very cold winters)

- Polar (high latitudes and polar regions)
    tundra (high latitude or high elevation)
    ice caps and ice sheets (perpetually frozen)
    polar marine

- Highland (compared to lowlands at the same latitude, highlands have lower temperatures—recall the normal lapse rate)

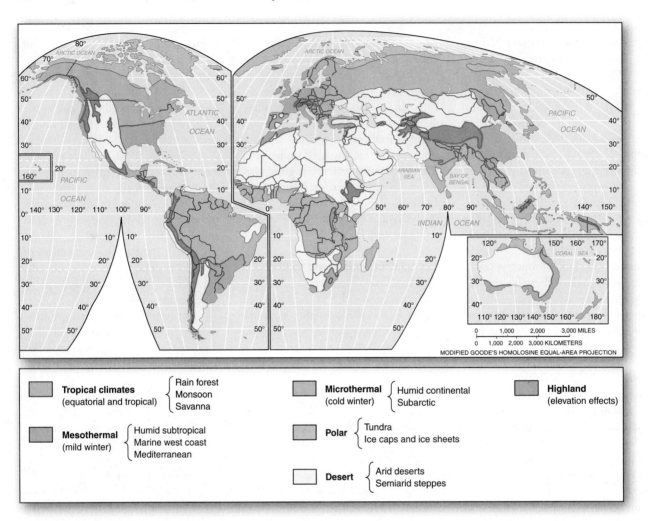

Tropical climates (equatorial and tropical) { Rain forest / Monsoon / Savanna }

Mesothermal (mild winter) { Humid subtropical / Marine west coast / Mediterranean }

Microthermal (cold winter) { Humid continental / Subarctic }

Polar { Tundra / Ice caps and ice sheets }

Desert { Arid deserts / Semiarid steppes }

Highland (elevation effects)

FIGURE 7.4 Climate regions generalized.
Six general climate categories. Relative to the question asked about your campus and birthplace in the caption to Figure 7.3, locate these two places on this map.

ANIMATION

Global Climate Maps, World Map References

Only one climate category is based on moisture efficiency as well as temperature:

- Desert (permanent moisture deficits)
  - arid deserts (tropical, subtropical hot and midlatitude cold)
  - semiarid steppes (tropical, subtropical hot and midlatitude cold)

Here we use a system that is somewhat of a middle ground between genetic and empirical. We focus on temperature and precipitation measurements, and for the desert areas, moisture efficiency. We give a description and some ideas why the climate regions are found where they are on a world climate map.

Global Climate Patterns Adding detail to Figure 7.4, we develop the world climate map presented in Figure 7.5, featuring causal elements that produce these climatic patterns. The following sections describe specific climates, organized around each of the main climate categories

listed previously. An opening box at the beginning of each climate section gives a simple description of the climate category and causal elements that are in operation. A world map showing distribution and the featured representative cities also is in the introductory box for each climate. In this discussion the names of the climates appear in italics.

Climographs exemplify particular climates for selected cities. A **climograph** is a graph that shows monthly temperature and precipitation, location coordinates, average annual temperature, total annual precipitation, elevation, the local population, annual temperature range, annual hours of sunshine (if available, as an indication of cloudiness), and a location map. Along the top of each climograph are the dominant weather features that are influential in that climate.

Discussions of soils, vegetation, and major terrestrial biomes that fully integrate these global climate patterns are in Chapters 15 and 16. Table 16.2 synthesizes all this information and will enhance your understanding of this chapter, so please place a tab on that page and refer to it as you read.

Tropical Climates (equatorial and tropical latitudes)

Tropical climates, the most extensive, occupy about 36% of Earth's surface, including both ocean and land areas. The tropical climates straddle the equator from about 20° N to 20° S, roughly between the Tropics of Cancer and Capricorn, thus the name. Tropical climates stretch northward to the tip of Florida and south-central Mexico, central India, and Southeast Asia, and southward to northern Australia, Madagascar, central Africa, and southern Brazil. These climates truly are winterless. Important causal elements include:

- Consistent daylength and insolation input, which produce consistently warm temeratures;
- Intertropical convergence zone (ITCZ), which brings rains as it shifts seasonally with the high Sun;
- Warm ocean temperatures and unstable maritime air masses.

Tropical climates have three distinct regimes: *tropical rain forest* (ITCZ present all year), *tropical monsoon* (ITCZ 6 to 12 months), and *tropical savanna* (ITCZ less than 6 months).

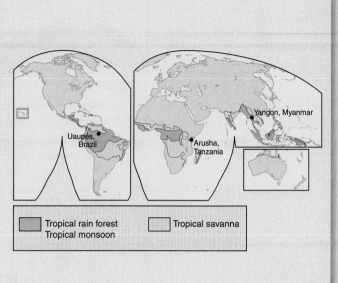

Tropical rain forest
Tropical monsoon

Tropical savanna

Tropical Rain Forest Climates

The *tropical rain forest* climate is constantly moist and warm. Convectional thunderstorms, triggered by local heating and trade-wind convergence, peak each day from midafternoon to late evening inland. These thunderstorms may hit earlier in the day where marine influence is strong. This precipitation follows the migrating intertropical convergence zone (ITCZ, Chapter 4). Its extreme positions for July and January are plotted in Figure 7.5. The ITCZ shifts northward and southward with the summer Sun throughout the year, but it influences *tropical rain forest* regions during all 12 months. Not surprisingly, water surpluses are enormous, creating the world's greatest stream discharges in the Amazon and Congo rivers.

High rainfall sustains lush evergreen broadleaf tree growth, producing Earth's equatorial and tropical rain forests. Their leaf canopy is so dense that little light diffuses to the forest floor, leaving the ground surface dim and sparse in plant cover. Dense surface vegetation occurs along riverbanks, where light is abundant (see Figure 7.6b). Widespread deforestation of Earth's rain forest is detailed in Chapter 16.

High temperature promotes energetic bacterial action in the soil so that organic material is quickly consumed. Heavy precipitation washes away certain minerals and nutrients. The resulting soils are somewhat sterile and can support intensive agriculture only if supplemented by fertilizer.

Uaupés, Brazil (Figure 7.6), is characteristic of *tropical rain forest*. On the climograph you see that the lowest-precipitation month receives nearly 15 cm (6 in.), and the annual temperature range is barely 2 C° (3.6 F°). The ITCZ is present all year. In all such climates, the diurnal (day to night) temperature range exceeds the annual average minimum–maximum (coolest to warmest) range: Day–night

temperatures can range more than 11 C° (20 F°), more than 5 times the annual monthly average range.

The only interruption of *tropical rain forest* climates is in the highlands of the South American Andes and in East Africa (see Figure 7.5). There, higher elevations produce lower temperatures; Mount Kilimanjaro is less than 4° S of the equator, but at 5895 m (19,340 ft) it has permanent glacial ice on its summit (although this ice is disappearing due to higher air temperatures). Such mountainous sites fall within the highland climate category.

Tropical Monsoon Climates

The *tropical monsoon* climates feature a dry season that lasts one or more months. Rainfall brought by the ITCZ affects these areas from 6 to 12 months of the year, with the dry season occurring when the convergence zone is not overhead. Yangon, Myanmar (formerly Rangoon, Burma), is an example of a monsoon climate in the tropics, as illustrated by the climograph and photograph in Figure 7.7. Mountains prevent cold air masses from central Asia getting into Yangon, resulting in its high average annual temperatures.

About 480 km (300 mi) north in another coastal city, Sittwe (Akyab), Myanmar, on the bay of Bengal, annual precipitation rises to 515 cm (203 in.) compared to Yangon's 269 cm (106 in.). Therefore, Yangon is a drier tropical monsoon area than farther north along the coast but still receives more than the 250-cm criteria.

Tropical monsoon climates lie principally along coastal areas within the tropical rain forest climatic realm and experience seasonal variation of winds and precipitation. Evergreen trees grade into thorn forests on the drier margins near the adjoining tropical savannas.

(text continued on page 233)

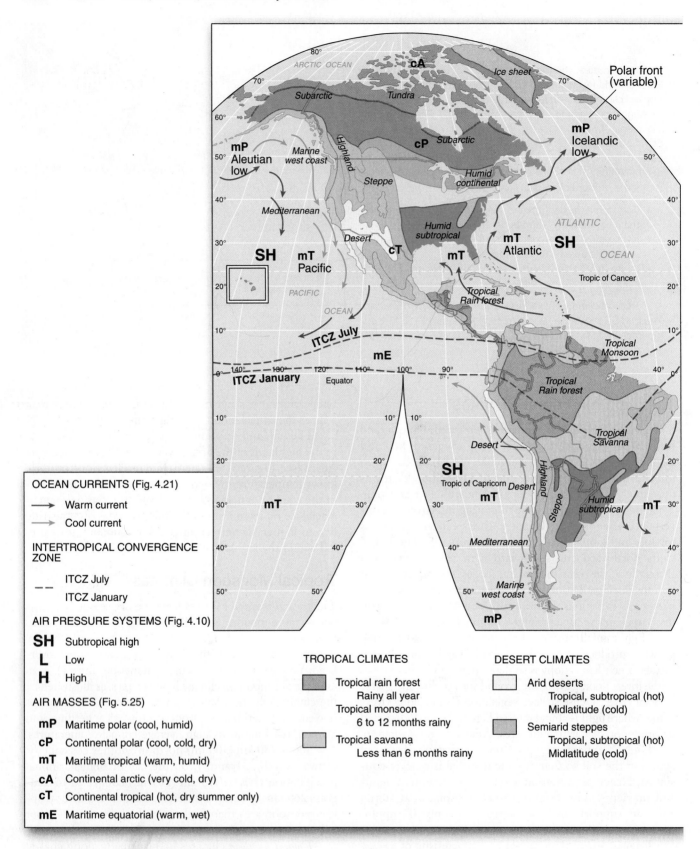

OCEAN CURRENTS (Fig. 4.21)

→ Warm current

→ Cool current

INTERTROPICAL CONVERGENCE
ZONE

-- ITCZ July
   ITCZ January

AIR PRESSURE SYSTEMS (Fig. 4.10)

SH Subtropical high

L Low

H High

AIR MASSES (Fig. 5.25)

mP Maritime polar (cool, humid)

cP Continental polar (cool, cold, dry)

mT Maritime tropical (warm, humid)

cA Continental arctic (very cold, dry)

cT Continental tropical (hot, dry summer only)

mE Maritime equatorial (warm, wet)

TROPICAL CLIMATES

■ Tropical rain forest
    Rainy all year
    Tropical monsoon
    6 to 12 months rainy

▨ Tropical savanna
    Less than 6 months rainy

DESERT CLIMATES

□ Arid deserts
    Tropical, subtropical (hot)
    Midlatitude (cold)

▨ Semiarid steppes
    Tropical, subtropical (hot)
    Midlatitude (cold)

FIGURE 7.5 World climate classifications.
Annotated on this map are selected air masses, near-shore ocean currents, pressure systems,
and the January and July locations of the ITCZ. Use the colors in the legend to locate various
climate types; some labels of the climate names appear in italics on the map to guide you.

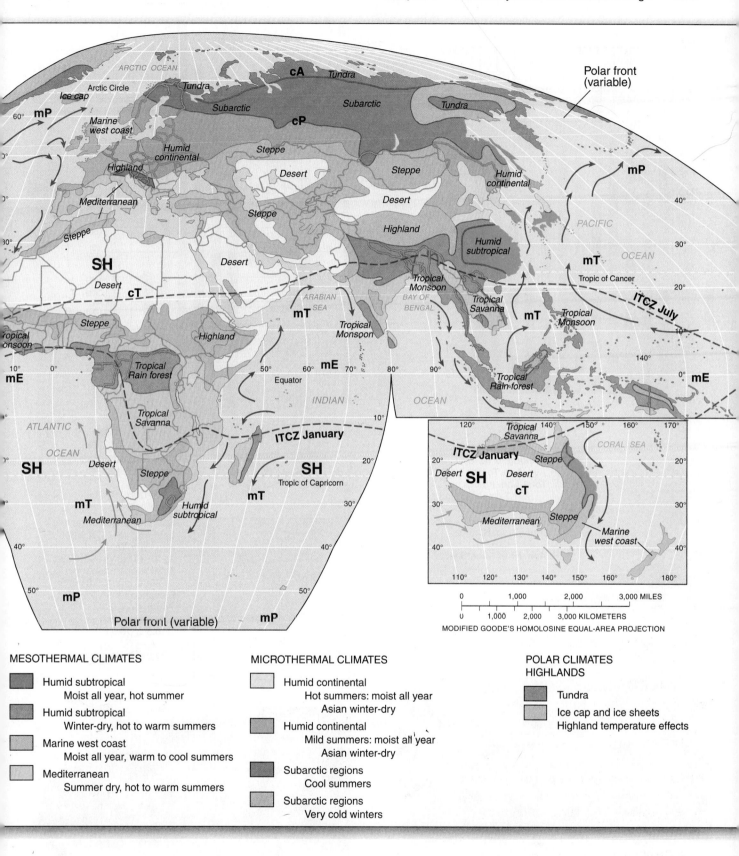

Polar front (variable)

ITCZ July

ITCZ January

Polar front (variable)

MODIFIED GOODE'S HOMOLOSINE EQUAL-AREA PROJECTION

MESOTHERMAL CLIMATES

- Humid subtropical
  Moist all year, hot summer
- Humid subtropical
  Winter-dry, hot to warm summers
- Marine west coast
  Moist all year, warm to cool summers
- Mediterranean
  Summer dry, hot to warm summers

MICROTHERMAL CLIMATES

- Humid continental
  Hot summers: moist all year
  Asian winter-dry
- Humid continental
  Mild summers: moist all year
  Asian winter-dry
- Subarctic regions
  Cool summers
- Subarctic regions
  Very cold winters

**POLAR CLIMATES
HIGHLANDS**

- Tundra
- Ice cap and ice sheets
  Highland temperature effects

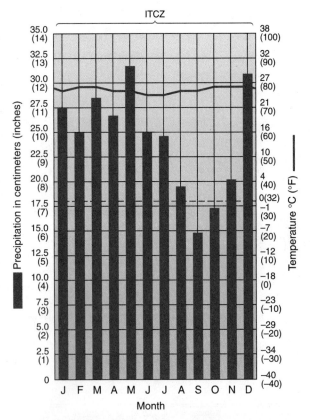

Station: Uaupés, Brazil
Lat/long: 0°08' S 67°05' W
Avg. Ann. Temp.: 25°C (77°F)
Total Ann. Precip.:
    291.7 cm (114.8 in.)
Elevation: 86 m (282.2 ft)
Population: 10,000
Ann. Temp. Range:
    2 C° (3.6 F°)
Ann. Hr of Sunshine: 2018

(a)

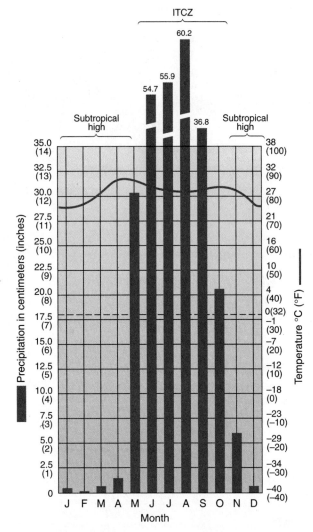

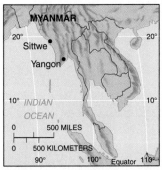

Station: Yangon, Myanmar*
Lat/long: 16°47' N 96°10' E
Avg. Ann. Temp.:
    27.3°C (81.1°F)
Total Ann. Precip.:
    268.8 cm (105.8 in.)
Elevation: 23 m (76 ft)
Population: 6,000,000
Ann. Temp. Range:
    5.5 C° (9.9 F°)

*(Formerly Rangoon, Burma)

(a)

(b)

FIGURE 7.6 Tropical rain forest climate.
(a) Climograph for Uaupés, Brazil (*tropical rain forest*). (b) The rain forest near Uaupés along a tributary of the Rio Negro.
[Photo by Will and Deni McIntyre/Photo Researchers, Inc.]

(b)

FIGURE 7.7 Tropical monsoon climate.
(a) Climograph for Yangon, Myanmar (formerly Rangoon, Burma) (*tropical monsoon*); city of Sittwe
also noted on map. (b) The monsoonal forest near Malang, Java, at the Purwodadi Botanical
Gardens. [Photo by Tom McHugh/Photo Researchers, Inc.]

Tropical Savanna Climates

Tropical savanna climates occur poleward of the *tropical rain forest* climates. The ITCZ dominates these climates for 6 months or less of the year as it migrates with the summer Sun. Summers are wetter than winters because convectional rains accompany the shifting ITCZ when it is overhead. This produces a notable dry condition when the ITCZ is farthest away and high pressure dominates. Thus, the natural water demand exceeds the natural water supply in winter, causing soil-moisture shortages.

Temperatures vary more in tropical savanna climates than in tropical rain forest regions. The tropical savanna regime can have two temperature maximums because the Sun's direct rays are overhead twice during the year (before and after the summer solstice in each

hemisphere as the Sun moves between the equator and the tropic). Dominant grasslands with scattered, drought-resistant trees, able to cope with the highly variable precipitation, characterize the *tropical savanna* regions (Figure 7.8b).

Arusha, Tanzania, is a characteristic *tropical savanna* city (Figure 7.8). This city of more than 1 million people is east of the famous Serengeti Plain savanna grassland and Olduvai Gorge, site of human origins, and north of Tarangire National Park. Temperatures are consistent with tropical climates, despite the elevation of the station. Note the marked dryness from June to October, which defines changing dominant pressure systems rather than annual changes in temperature. This region is near the transition to the dryer hot-desert steppe climates to the northeast.

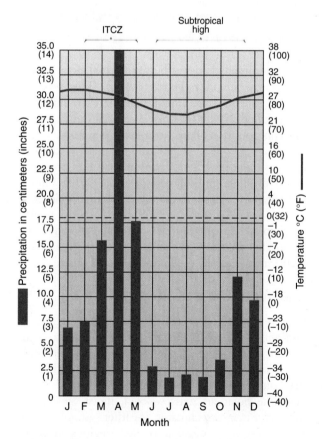

Station: Arusha, Tanzania
Lat/long: 3°24′ S 36°42′ E
Avg. Ann. Temp.:
  26.5°C (79.7°F)
Total Ann. Precip.:
  119 cm (46.9 in.)

Elevation: 1387 m (4550 ft)
Population: 1,368,000
Ann. Temp. Range:
  4.1 C° (7.4 F°)
Ann. Hr of Sunshine: 2600

(a)

(b)

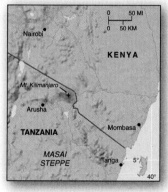

FIGURE 7.8 Tropical savanna climate.
(a) Climograph for Arusha, Tanzania (*tropical savanna*); note the intense dry period. (b) Characteristic landscape in nearby Kenya to the north, with plants and animals adapted to seasonally dry water budgets. [Photo by Stephen J. Krasemann/DRK Photo.]

Mesothermal Climates (midlatitudes, mild winters)

Mesothermal, meaning "middle temperature," describes these warm and temperate climates, where true seasonality begins and seasonal contrasts in vegetation, soil, and human lifestyle adaptations are evident. More than half the world's population— approximately 55%—resides in these mesothermal climates, which occupy the second-largest percentage of Earth's land and sea surface behind the tropical climates, about 27%. Together, the tropical and mesothermal climates dominate over half of Earth's oceans and about one-third of its land area.

The mesothermal climates, and nearby portions of the microthermal (cold winter) climates, are regions of great weather variability, for these are the latitudes of greatest air mass interaction. Causal elements include:

- Shifting air masses of maritime and continental origin are guided by upper-air westerly winds and undulating Rossby waves and jet streams.
- Migrating cyclonic (low-pressure) and anticyclonic (high-pressure) systems bring changeable weather conditions and air mass conflicts.
- Sea-surface temperatures influence air mass strength: cooler along west coasts (weaken) and warmer along east coasts (strengthen).
- Summers transition from hot to warm to cool moving away from the tropics. Climates are humid, except

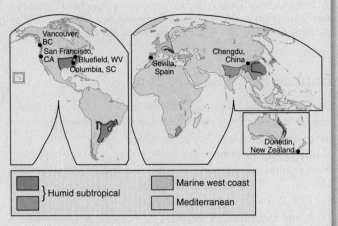

where subtropical high pressure produces dry-summer conditions.

Mesothermal climates have four distinct regimes based on precipitation variability: *humid subtropical hot-summer* (moist all year), *humid subtropical winter-dry* (hot to warm summers, in Asia), *marine west coast* (warm to cool summers, moist all year), and *Mediterranean* (warm to hot summers, dry summers).

Humid Subtropical Hot-Summer Climates

The *humid subtropical hot-summer* climates are either moist all year or have a pronounced winter-dry period, as occurs in eastern and southern Asia. Maritime tropical air masses generated over warm waters off eastern coasts influence *humid subtropical hot-summer* climates during summer. This warm, moist, unstable air produces convectional showers over land. In fall, winter, and spring, maritime tropical and continental polar air masses interact, generating frontal activity and frequent midlatitude cyclonic storms. Overall, precipitation averages 100–200 cm (40–80 in.) a year. In North America, Columbia, South Carolina, is characteristic (Figure 7.9) of this climate regime, with precipitation totals of 126.5 cm (49.8 in.), hot, humid summers, and mild winters; compare to Atlanta's 122 cm (48 in.) annually (Figure 7.9b).

In eastern and southern Asia, winter precipitation is a bit less because of the effects of the Asian monsoon, unlike the precipitation of humid subtropical cities in the United States (Atlanta, Memphis, Norfolk, New Orleans). These Asian *humid subtropical winter-dry* climates relate to the seasonal pulse of the monsoons. They extend poleward from tropical savanna climates and have a summer month that receives 10 times more precipitation than their driest winter month.

A representative station for this climate in Asia is Chengdu, China. Figure 7.10 demonstrates the strong correlation between precipitation and the high-summer Sun (review the monsoonal maps in Figure 4.20). In May 2008, this city and the surrounding region was hit with a magnitude 7.9 earthquake, with many strong aftershocks,

causing widespread damage, killing 85,000 people, and injuring nearly 400,000 (Figure 7.10b). This was the worst earthquake to strike China since the 1976 Tangshan quake that killed 250,000 people, discussed in Chapter 9.

The concentration of people in north-central India, the bulk of China's 1.3 billion people, and the many who live in climatically similar portions of the United States prove the habitability of the humid subtropical climates. The intense summer rains of the Asian monsoon can cause problems, as happened in the 2005 through 2008 seasons (News Report 7.1). Occasional dramatic thunderstorms and tornadoes are notable in the southeastern United States, with tornado strikes currently breaking each previous year's record.

Marine West Coast Climates

Marine west coast climates feature mild winters and cool summers and dominate Europe and other middle- to high-latitude west coasts. In the United States and Canada, these climates, with their cooler summers, contrast with the hot-summer humid climate of the southeastern United States.

Maritime polar air masses—cool, moist, unstable—control *marine west coast* climates. Weather systems formed along the polar front move into these regions throughout the year, which makes weather quite unpredictable. Coastal fog, annually totaling 30 to 60 days, is a part of the moderating marine influence. Frosts are possible and tend to shorten the growing season.

Marine west coast climates are unusually mild for their latitude owing to marine influences. They extend along

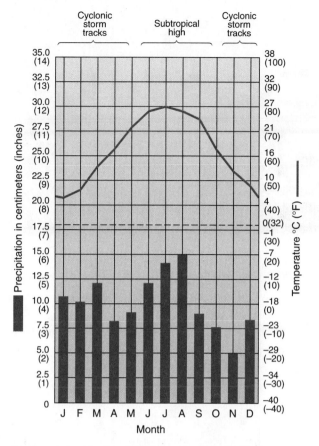

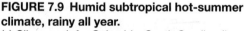

Station: Columbia, South Carolina
Lat/long: 34° N 81′ W
Avg. Ann. Temp.:
  17.3°C (63.1°F)
Total Ann. Precip.:
  126.5 cm (49.8 in.)

Elevation: 96 m (315 ft)
Population: 116,000
Ann. Temp. Range:
  20.7 C° (37.3 F°)
Ann. Hr of Sunshine:
  2800

(a)

(c)

FIGURE 7.9 Humid subtropical hot-summer climate, rainy all year.

(a) Climograph for Columbia, South Carolina (*humid subtropical*). Note the more consistent precipitation pattern, as Columbia receives seasonal cyclonic storm activity and summer convection showers within maritime tropical air. (b) The mixed deciduous and evergreen forest south of Atlanta, Georgia, typical of the humid subtropical southeastern United States. (c) This scene is from the humid subtropical portion of Argentina along the Paraná River, northwest of Buenos Aires. [(b) and (c) Photos by Bobbé Christopherson.]

the coastal margins of the Aleutian Islands in the North Pacific, cover the southern third of Iceland in the North Atlantic and coastal Scandinavia, and dominate the British Isles. It is hard to imagine that at such high-latitude locations average monthly temperatures are above freezing throughout the year.

Unlike *marine west coast* climates in Europe, these climates in Canada, Alaska, Chile, and Australia are backed by mountains and remain restricted to coastal environs. The climograph for Vancouver demonstrates the moderate temperature patterns and the annual temperature range for a *marine west coast* city (Figure 7.11). The climograph for

Dunedin, New Zealand, demonstrates the moderate temperature patterns and the annual temperature range for a *marine west coast* station in the Southern Hemisphere (Figure 7.12).

An interesting anomaly occurs in the eastern United States. In portions of the Appalachian highlands, increased elevation moderates summer temperatures in the surrounding *humid subtropical hot-summer* climate, producing a *marine west coast* cooler summer. The climograph for Bluefield, West Virginia (Figure 7.13), reveals marine west coast temperature and precipitation patterns, despite its location in the east. Vegetation similarities between the Appalachians and the Pacific Northwest are quite noticeable and have

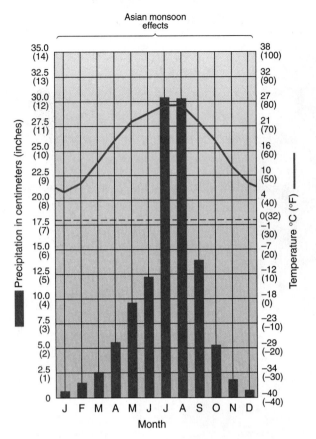

Asian monsoon effects

Station: Chengdu, China
Lat/long: 30°40′ N 104°04′ E
Avg. Ann. Temp.: 17°C (62.6°F)
Total Ann. Precip.:
114.6 cm (45.1 in.)

Elevation: 498 m (1633.9 ft)
Population: 2,500,000
Ann. Temp. Range:
20 C° (36 F°)
Ann. Hr of Sunshine: 1058

(a)

(b)

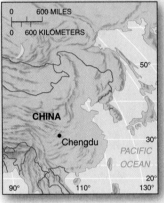

FIGURE 7.10 Humid subtropical winter-dry climate.
(a) Climograph for Chengdu, China. (b) People flee high-rise buildings following a major earthquake in Chengdu. [Photo (b) courtesy of China Daily (chinadaily.com).]

enticed many emigrants from the East to settle in these climatically familiar environments in the Northwest.

Mediterranean Dry-Summer Climates

Across the planet during summer months, shifting subtropical high-pressure cells block moisture-bearing winds from adjacent regions. For example, the continental tropical air mass over the Sahara in Africa shifts northward in summer over the Mediterranean region and blocks maritime air masses and cyclonic systems. The shifting of stable, warm-to-hot, dry air over an area in summer and away from these regions in the winter creates a unique dry-summer and wet-winter pattern. Mediterranean climates

NEWS REPORT 7.1

Record Rains from the Asian Monsoons

The subtropical monsoonal climates hold several precipitation records. Cherrapunji, India, in the Assam Hills south of the Himalayas, is the all-time precipitation record holder for a single year and for every other time interval from 15 days to 2 years. Because of the summer monsoons that pour in from the Indian Ocean and the Bay of Bengal, Cherrapunji has received 930 cm (30.5 ft) of rainfall in one month and 2647 cm (87.8 ft) in one year—

both world records. Because of extensive deforestation, these rains produce tremendous soil erosion. Sediments generated by this erosion are deposited in the Bay of Bengal as fragile, temporary islands that attract settlements. Floodwaters and heavy monsoonal downpours easily overwhelm these fertile islands. Two massive tropical cyclones struck the nearby coast of India in 1999, further worsening the hazardous floodplain conditions.

In 2005, all records were shattered on July 27, 2005, when 94.2 cm (37.1 in.) of rain fell in only a few hours. Think of a yardstick and that a depth of water that great covered the land, all arriving from the sky! During August 2008, record monsoonal floods again struck portions of northeastern India, forcing a million evacuees to flee. A troubling trend in Asian monsoons seems at hand.

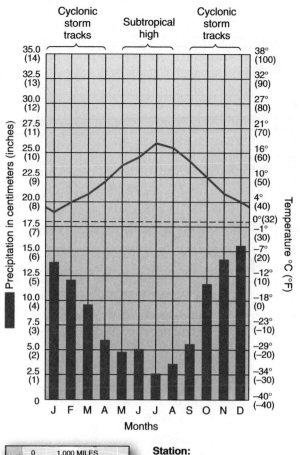

Cyclonic storm tracks | Subtropical high | Cyclonic storm tracks

Station:
Vancouver, British Columbia
Lat/long: 49°11′ N, 123°10′ W
Avg. Ann. Temp.:
  10° C (50° F)
Total Annual Precipitation:
  104.8 cm (41.3 in.)
Elevation: sea level
Population: 1,580,000
Ann. Temp. Range:
  16 C° (28.8 F°)
Ann. Hrs. of Sunshine:
  1723

(a)

(b)

(c)

FIGURE 7.11 A marine west coast climate.
(a) Climograph and locator map for Vancouver, British Columbia, a *marine west coast* climate. (b) The Vancouver waterfront and skyline. (c) Temperate rain forest in the Macmillan Provincial Park, central Vancouver Island, Canada, features a Douglas fir forest, with western red cedar and hemlock. [Photos by (b) author; (c) Bobbé Christopherson.]

experience at least 70% of their annual precipitation during the winter months. This is in contrast to the majority of the world, which experiences summer-maximum precipitation.

Added to the high-pressure influence, cool offshore currents (the California current, Canary current, Peru current, Benguela current, and West Australian current) produce stability in overlying air masses along west coasts, poleward of subtropical high pressure. The world climate map (Figure 7.5) shows these currents and Mediterranean dry-summer regions along the western margins of North America, central Chile, and the southwestern tip of Africa, as well as across southern Australia and the Mediterranean Basin—the climate's namesake region.

Figure 7.14 compares the climographs of the Mediterranean dry-summer cities of San Francisco, California, and Sevilla (Seville), Spain. Coastal maritime effects moderate San Francisco's climate, producing a cool summer. The transition to hotter summers occurs no more than 24–32 km (15–20 mi) inland from San Francisco.

Along these west coasts, the warm, moist air that contacts cool ocean water produces frequent summer fog, discussed in Chapter 5. However, this type of fog is not associated with the enclosed Mediterranean Sea region itself because offshore water temperatures there are higher than water temperatures near other similar climatic regions during the summer months. Instead, the Mediterranean regime features relatively high humidity values.

The *Mediterranean dry-summer* climate brings natural summer water-resource shortages. Winter precipitation recharges soil moisture, but water usage usually exhausts it by late spring. Large-scale agriculture requires irrigation; some subtropical fruits, nuts, and vegetables are uniquely suited to these conditions. Natural vegetation features a hard-leafed, drought-resistant variety known locally as *chaparral* in the western United States. (Chapter 16 discusses other local names for this type of vegetation in other parts of the world.)

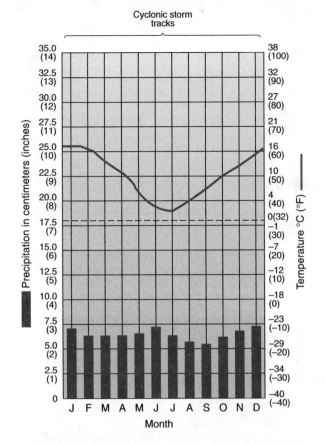

Cyclonic storm
tracks

Station: Dunedin, New Zealand
Lat/long: 45°54′ S 170°31′ E
Avg. Ann. Temp.: 10.2°C (50.3°F)
Total Ann. Precip.:
   78.7 cm (31.0 in.)

Elevation: 1.5 m (5 ft)
Population: 120,000
Ann. Temp. Range:
   14.2 C° (25.5 F°)

(a)

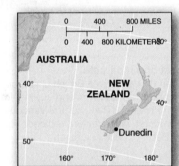

(b)

FIGURE 7.12 A Southern Hemisphere marine west coast climate.
(a) Climograph for Dunedin, New Zealand, demonstrates a *marine west coast* climate with a mild summer. (b) Meadow, forest, and mountains on South Island, New Zealand. [Photo by Brian Enting/ Photo Researchers, Inc.]

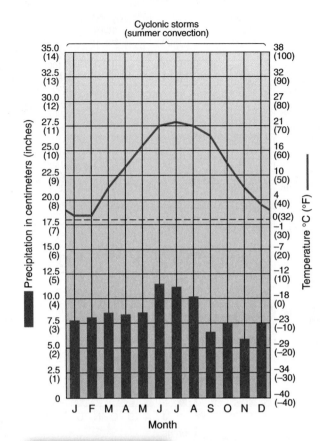

Cyclonic storms
(summer convection)

(a)

Station: Bluefield, West Virginia
Lat/long: 37°16′ N 81°13′ W
Avg. Ann. Temp.:
   12°C (53.6°F)
Total Ann. Precip.:
   101.9 cm (40.1 in.)
Elevation: 780 m (2559 ft)
Population: 11,000
Ann. Temp. Range:
   21 C° (37.8 F°)

(b)

FIGURE 7.13 Marine west coast climate in the Appalachians.
(a) Climograph for Bluefield, West Virginia, exhibits *marine west coast* characteristics. (b) Characteristic mixed forest of Dolly Sods Wilderness in the Appalachian highlands. [Photo by David Muench Photography, Inc.]

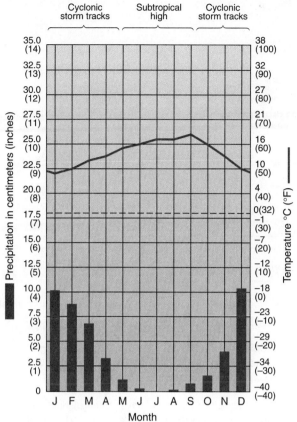

Station: San Francisco, California
Lat/long: 37°37′ N 122°23′ W
Avg. Ann. Temp.:
  14°C (57.2°F)
Total Ann. Precip.:
  47.5 cm (18.7 in.)

Elevation: 5 m (16.4 ft)
Population: 747,000
Ann. Temp. Range:
  9 C° (16.2 F°)
Ann. Hr of Sunshine:
  2975

(a)

Station: Sevilla, Spain
Lat/long: 37°22′ N 6°00′ W
Avg. Ann. Temp.:
  18°C (64.4°F)
Total Ann. Precip.:
  55.9 cm (22 in.)

Elevation: 13 m (42.6 ft)
Population: 1,764,000
Ann. Temp. Range:
  16 C° (28.8 F°)
Ann. Hr of Sunshine:
  2862

(b)

(c)

(d)

FIGURE 7.14 Mediterranean climates, California and Spain.

(a) Climographs for San Francisco (*Mediterranean cool-summer*) and (b) Sevilla, Spain (*Mediterranean hot-summer*). (c) Central California landscape of oak savanna. (d) The countryside around Olvera, Andalusia, Spain. [Photos by (c) Bobbé Christopherson; (d) Kaz Chiba/Gamma-Liaison Agency, Inc.]

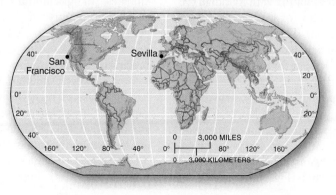

Microthermal Climates (mid- and high-latitude, cold winter)

Humid microthermal climates have a winter season with some summer warmth. Here the term *microthermal* means cool temperate to cold. Approximately 21% of Earth's land surface is influenced by these climates, equaling about 7% of Earth's total surface.

These climates occur poleward of the mesothermal climates and experience great temperature ranges related to continentality and air mass conflicts. Temperatures decrease with increasing latitude and toward the interior of continental landmasses and result in intensely cold winters. In contrast to moist-all-year regions (northern tier across the United States and Canada, eastern Europe through the Ural Mountains) is the winter-dry association with the Asian dry monsoon and cold air masses across the region.

In Figure 7.5, note the absence of microthermal climates in the Southern Hemisphere. Because the Southern Hemisphere lacks substantial landmasses, microthermal climates develop there only in highlands. Important causal elements include:

- Increasing seasonality (daylength and Sun altitude) and greater temperature ranges (daily and annually);
- Upper-air westerly winds and undulating Rossby waves, which bring warmer air northward and colder air southward for cyclonic activity, and convectional thunderstorms from mT air masses in summer;
- Asian winter-dry pattern for the microthermal climates, increasing east of the Ural Mountains to the Pacific Ocean and eastern Asia;

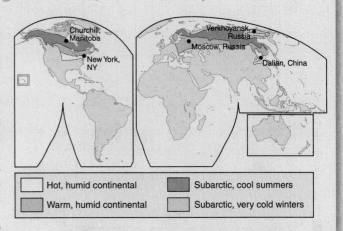

☐ Hot, humid continental	☐ Subarctic, cool summers
☐ Warm, humid continental	☐ Subarctic, very cold winters

- Hot summers cooling northward from the mesothermal climates, and short spring and fall seasons surrounding winters that are cold to very cold;
- Continental interiors serving as source regions for intense continental polar (cP) air masses that dominate winter, blocking cyclonic storms.

Microthermal climates have four distinct regimes based on increasing cold with latitude and precipitation variability: *humid continental hot-summer* (Chicago, New York); *humid continental warm-summer* (Duluth, Toronto, Moscow); *subarctic cool summers* (Churchill, Manitoba); and the formidable extremes of a frigid *subarctic very cold winter* in Verkhoyansk and northern Siberia.

Humid Continental Hot-Summer Climates

Humid continental hot-summer climates are differentiated by their annual precipitation distribution. In the summer, maritime tropical air masses influence both humid continental moist-all-year and winter-dry climates. In North America, frequent stormy weather is possible from conflicting air masses—maritime tropical and continental polar—especially in winter. The climographs for New York City (Figure 7.15a) and Dalian, China (7.15b), illustrate these two humid hot-summer regimes (Figure 7.15).

Originally, forests covered the *humid continental hot-summer* region of the United States west to the Indiana-Illinois border. Beyond that approximate line, tall-grass prairies extended westward to about the 98th meridian (98° W) and the approximate location of the 51-cm (20-in.) *isohyet* (line of equal precipitation). The short-grass prairies extended to the west, where precipitation is less.

Deep sod made farming difficult for the first settlers, as did the climate. However, native grasses soon were replaced with domesticated wheat and barley. Various inventions brought from the East (barbed wire, the self-scouring steel plow, well-drilling techniques, and the railroads) helped open the region further. In the United States today, the *humid continental hot-summer* region is the location of corn, soybean, hog, and cattle production (Figure 7.16a).

The region around Buffalo, New York, is in a climatic transition from milder to hotter summers. As the average July temperature increases further above 22°C (71.6°F), Buffalo's climatic designation will become more like that of New York and Chicago. The surrounding forest in the latter half of this century will shift from the present maple–beech–birch, with some elm and ash, to a dominant oak–hickory–hemlock mix (Figure 7.16b). Such shifting of climatic boundaries and their related ecosystems is occurring worldwide.

Humid Continental Mild-Summer Climates

Soils are thinner and less fertile in the cooler microthermal climates, yet agricultural activity is important and includes dairy cattle, poultry, flax, sunflowers, sugar beets, wheat, and potatoes. Frost-free periods range from fewer than 90 days in the north to as many as 225 days in the south. Overall, precipitation is less than in the hot-summer regions to the south. However, notably heavier snowfall is important to soil-moisture recharge when it melts. Various snow-capturing strategies are used, including fences and tall stubble (plant stalks left standing after harvest) in fields to create snow drifts and thus more moisture retention on the soil.

Characteristic cities are Duluth, Minnesota, and Saint Petersburg, Russia. Figure 7.17 presents a climograph for

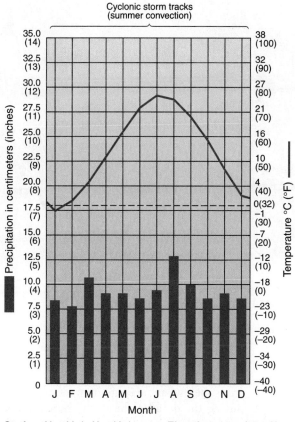

Cyclonic storm tracks
(summer convection)

Station: New York, New York   **Elevation:** 16 m (52.5 ft)
Lat/long: 40°46′ N 74°01′ W   **Population:** 8,092,000
Avg. Ann. Temp.:   **Ann. Temp. Range:**
  13°C (55.4°F)   24 C° (43.2 F°)
Total Ann. Precip.:   **Ann. Hr of Sunshine:**
  112.3 cm (44.2 in.)   2564

(a)

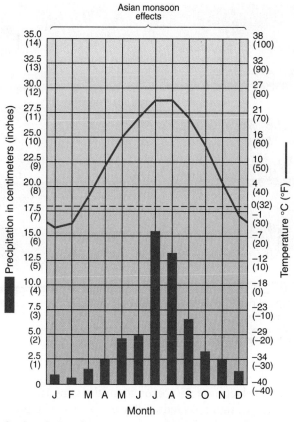

Asian monsoon
effects

Station: Dalian, China   **Elevation:** 96 m (314.9 ft)
Lat/long: 38°54′ N 121°54′ E   **Population:** 5,550,000
Avg. Ann. Temp.:   **Ann. Temp. Range:**
  10°C (50°F)   29 C° (52.2 F°)
Total Ann. Precip.:   **Ann. Hr of Sunshine:**
  57.8 cm (22.8 in.)   2762

(b)

(c)

(d)

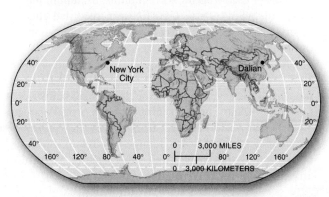

FIGURE 7.15 Humid continental hot-summer climate, New York and China.
Climographs for (a) New York City (*humid continental, moist all year*), and (b) Dalian, China (*humid continental winter-dry*). (c) The "Literary Walk" with a canopy of American elms in New York City's Central Park emerging from winter just as spring and the return of leaves and warmth begin. (d) Dalian, China, cityscape and park in summer. [Photos by (c) by Bobbé Christopherson; (d) courtesy of the Paul Louis Collection.]

(a)   (b)

FIGURE 7.16 Central Indiana and Buffalo, NY, landscapes.
(a) Typical humid continental deciduous forest and ready-to-harvest soybean field in central Indiana near Zelma in the American Midwest. (b) Forest in Buffalo, NY, with its characteristic mix of maple–beech–birch, with some elm and ash, is in transition as regional temperatures rise.
[Photos by Bobbé Christopherson.]

Moscow, which is at 55° N, or about the same latitude as the southern shore of Hudson Bay in Canada. The photos of landscapes near Moscow and Sebago Lake (inland from Portland, Maine) show summer and late winter scenes, respectively.

The dry winter of the vast Asian landmass, specifically Siberia, is exclusively associated with the extremely dry and frigid winter high-pressure anticyclone that dominates. This system produces the dry monsoons of southern and eastern Asia in the winter months, as winds blow out of Siberia toward the Pacific and Indian oceans (see Figure 4.20). The intruding dry cold of continental air is a significant winter feature. A representative *humid continental mild-summer* climate is Vladivostok, Russia, on the Sea of Japan, usually one of only two ice-free ports in that country.

Subarctic Climates

Farther poleward, seasonal change becomes greater. The short growing season is more intense during long summer days. The subarctic climates include vast stretches of Alaska, Canada, and northern Scandinavia with their cool summers, and Siberian Russia with its very cold winters. Discoveries of minerals and petroleum reserves and the Arctic Ocean sea-ice losses have led to new interest in portions of these regions.

Areas that receive 25 cm (10 in.) or more precipitation a year on the northern continental margins and are covered by the so-called snow forests of fir, spruce, larch, and birch are the *boreal forests* of Canada and the *taiga* of Russia. These forests are in transition to the more open northern woodlands and to the tundra region of the far north. Forests thin out to the north when the warmest

month drops below an average temperature of 10°C (50°F). During the decades ahead, the boreal forests are shifting northward into the tundra in response to climate change and higher temperatures (Figure 7.18b).

Soils are thin in these lands once scoured by glaciers. Precipitation and potential evapotranspiration both are low, so soils are generally moist and either partially or totally frozen beneath the surface, a phenomenon known as *permafrost* (discussed in Chapter 14).

The Churchill, Manitoba, climograph (Figure 7.18) shows average monthly temperatures below freezing for 7 months of the year, during which time light snow cover and frozen ground persist. High pressure dominates Churchill during its cold winter—this is the source region for the continental polar air mass. Churchill is representative of the *subarctic* climate, with a cool summer, annual temperature range of 40 C° (72 F°), and low precipitation of 44.3 cm (17.4 in.).

The *subarctic* climates that feature a dry and very cold winter occur only within Russia. The intense cold of Siberia and north-central and eastern Asia is difficult to comprehend, for these areas experience an average temperature lower than freezing for 7 months; minimum temperatures of below –68°C (–90°F) were recorded there, as described in Chapter 3. Yet summer maximum temperatures in these same areas can exceed +37°C (+98°F).

An example of this extreme subarctic climate with very cold winters is Verkhoyansk, Siberia (Figure 7.19). For 4 months of the year, average temperatures fall below –34°C (–30°F). Verkhoyansk has probably the world's greatest annual temperature range from winter to summer: a remarkable 63 C° (113.4 F°) range. Winters feature brittle metals and plastics, triple-thick windowpanes, and temperatures that render straight antifreeze a solid.

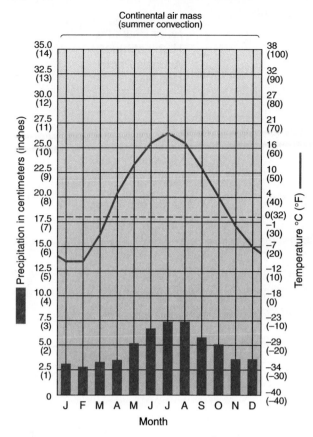

Station: Moscow, Russia
Lat/long: 55°45′ N 37°34′ E
Avg. Ann. Temp.: 4°C (39.2°F)
Total Ann. Precip.:
57.5 cm (22.6 in.)

Elevation: 156 m (511.8 ft)
Population: 11,460,000
Ann. Temp. Range:
29 C° (52.2 F°)
Ann. Hr of Sunshine:
1597

(a)

FIGURE 7.17 Humid continental mild-summer climate.
(a) Climograph for Moscow, Russia. (b) Springtime fields near Saratov, Russia, during the short
summer season. (c) Winter scene of Sebago Lake and forests inland from Portland, Maine.
[Photos by (b) Wolfgang Kaehler/Gamma, Inc.; (c) Bobbé Christopherson.]

Polar and Highland Climates

The polar climates have no true summer as do lower latitudes.
Dominating the South Pole is the Antarctic continent, surrounded by the Southern Ocean, whereas the North Pole region
is the Arctic Ocean, surrounded by the continents of North
America and Eurasia. Poleward of the Arctic and Antarctic circles, daylength increases in summer until daylight becomes
continuous, yet average monthly temperatures never rise above
10°C (50°F). These temperature conditions are intolerant to tree
growth. See the polar region temperature maps for January and
July in Figure 3.25 and Figure 3.28. Principal climatic factors in
these frozen and barren regions are the following:

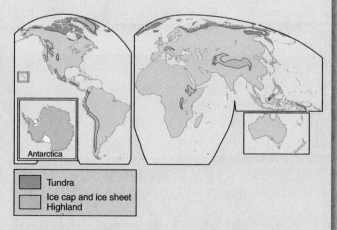

Tundra
Ice cap and ice sheet
Highland

- Extremes of daylength between winter and summer
  determine the amount of insolation received.
- Low Sun altitude even during the long summer days is
  the principal climatic factor.
- Extremely low humidity produces low precipitation
  amounts—Earth's frozen deserts.
- Light-colored surfaces of ice and snow reflect substantial
  energy away from the ground surface, thus reducing net
  radiation.

Polar climates have three regimes: *tundra* (high latitude, or
elevation); *polar marine* (oceanic association, slight moderation

of extreme cold); and *ice caps* and *ice sheets* (perpetually
frozen).

Also in this climate category we include highland climates, for even at low latitudes the effects of elevation can
produce tundra and polar conditions. Glaciers on tropical
mountain summits attest to the cooling effects of elevation.
Highland climates on the map follow the pattern of Earth's
mountain ranges.

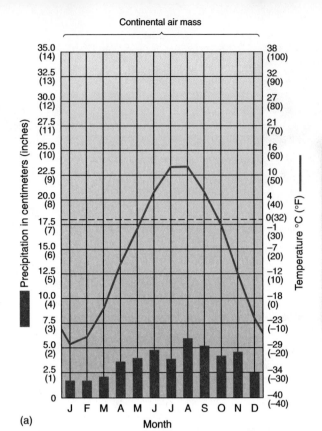

Continental air mass

NOTEBOOK

Polar Bear and
High Latitude
Animals Photo
Galleries

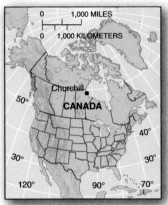

Station: Churchill, Manitoba
Lat/long: 58°45′ N 94°04′ W
Avg. Ann. Temp.: −7°C (19.4°F)
Total Ann. Precip.:
  44.3 cm (17.4 in.)
Elevation: 35 m (114.8 ft)
Population: 1400
Ann. Temp. Range:
  40 C° (72 F°)
Ann. Hr of Sunshine:
  1732

(a)

(b)

(c)

(d)

(e)

FIGURE 7.18 Subarctic cool-summer climate.
(a) Climograph for Churchill, Manitoba (*subarctic, cool summer*). (b) Winter scene on the edge of the boreal forest; young pioneer spruce trees move into the tundra. (c) Lone polar bear hunkered down in a protective dugout next to a frozen pond. (d) Mom and two cubs-of-the-year in the tundra near Hudson Bay, west of Churchill, characteristic willows in the background. (e) November street scene in Churchill. [All photos by Bobbé Christopherson.]

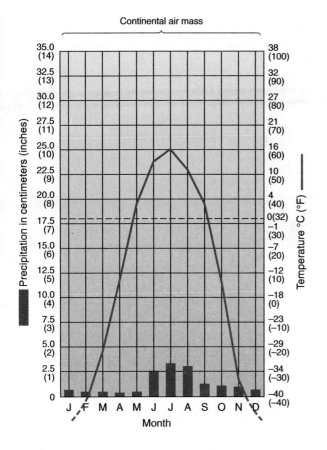

Continental air mass

Station: Verkhoyansk, Russia
Lat/long: 67°35′ N 133°27′ E
Avg. Ann. Temp.: −15°C (5°F)
Total Ann. Precip.:
  15.5 cm (6.1 in.)
Elevation: 137 m (449.5 ft)
Population: 1500
Ann. Temp. Range:
  63 C° (113.4 F°)

(a)

(b)

FIGURE 7.19 Extreme subarctic winter-dry climate.
(a) Climograph for Verkhoyansk, Russia (*subarctic, very cold winter*). (b) Scene in the town of Verkhoyansk during its brief summer. [Photo by Dean Conger/National Geographic Society.]

Tundra Climate

In a *tundra* climate, land is under continuous snow cover for 8–10 months, with the warmest month above 0°C yet never warming above 10°C (50°F). Because of elevation, the summit of Mount Washington in New Hampshire (1914 m, or 6280 ft) statistically qualifies as a highland *tundra* climate on a small scale.

In spring when the snow melts, numerous plants appear—stunted sedges, mosses, flowering plants, and lichens. Some of the little (7.5-cm, 3-in.-tall) willows can exceed 300 years in age. The September photo shows the emerging fall colors of these small plants (Figure 7.20a). Much of the area experiences permafrost and ground ice conditions; these are Earth's periglacial regions (see Chapter 14).

Approximately 410,500 km² (158,475 mi²) of Greenland are ice-free, an area of tundra and rock about the size of California. The rest of Greenland is ice sheet covering 1,756,000 km² (677,900 mi²). Despite the severe climate, a permanent population of 56,500 lives in this island province, legally part of Denmark. There are only a couple of towns along Greenland's east coast. Scoresby Sund, a village on the tundra, has 850 permanent residents (Figure 7.20b).

Global warming is bringing dramatic changes to the tundra and its plants, animals, and permafrost ground conditions. Temperatures as much as 5 C° above average in parts of Canada and Alaska are a regular occurrence, setting many records. As organic peat deposits in the tundra thaw, vast stores of carbon are released to the atmosphere, further adding to the greenhouse gas problem. Temperatures in the Arctic are warming at a rate twice that of the global average increase.

Tundra climates are strictly a Northern Hemisphere occurrence, except for elevated mountain locations in the Southern Hemisphere and a portion of the Antarctic Peninsula, where plants are now growing for the first time.

Ice-Cap and Ice-Sheet Climate

Most of Antarctica and Greenland fall within the *ice-cap* and *ice-sheet* climate, as does the North Pole, with all months averaging below freezing. These regions are dominated by dry, frigid air masses, with vast expanses that never warm above freezing. The area of the North Pole is actually ocean covered by ice, whereas Antarctica is a substantial continental landmass covered by Earth's greatest ice sheet. For comparison, winter minimums at the South Pole (July) frequently drop below the temperature of solid carbon dioxide or "dry ice" (−78°C, or −109°F).

Ice caps are smaller in extent than an ice sheet, roughly less than 50,000 km² (19,300 mi²), yet they completely bury the landscape like an ice sheet. An example is the Vatnajökull Ice Cap in southeastern Iceland.

Antarctica is constantly snow-covered but receives less than 8 cm (3 in.) of precipitation each year. Antarctic ice has accumulated to several kilometers thickness and is the largest repository of freshwater on Earth. Earth's two

(b)

(a)

FIGURE 7.20 Greenland tundra and a small town.
(a) Tundra is marked by an uneven, hummocky surface of mounds resulting from an active layer that freezes and thaws with the seasons, as here in east Greenland (see Chapter 14). Large trees are absent in the tundra; however, a relatively lush vegetation for the harsh conditions includes willow, dwarf birch and shrubs, sedges, moss, lichen, and cotton grass (white tufts). (b) A town in the tundra is Ittoqqortoormiit (Scoresby Sund), Greenland, one of only a few settlements along the entire east coast. In the foreground, sledge (sled) dogs rest to get ready for the winter's work ahead. [Photos by Bobbé Christopherson.]

ice sheets cover the Antarctic continent and most of the island of Greenland. Figure 7.21 shows two scenes of these repositories of multiyear ice.

This ice contains a vast historical record of Earth's atmosphere. Within it, evidence of thousands of past volcanic eruptions from all over the world and ancient combinations of atmospheric gases lie trapped in frozen bubbles. Chapter 14 presents analysis of ice cores taken from Greenland

and the latest one from Antarctica, which pushed the climate record to 800,000 years before the present.

Polar Marine Climate

Polar marine areas are more moderate than other polar climates in winter, with no month averaging below –7°C (20°F), yet they are not as warm as *tundra* climates. Because

(a)

(b)

FIGURE 7.21 Earth's ice sheets—Antarctica and Greenland.
These are Earth's frozen freshwater reservoirs. (a) Sunset in Antarctica on February 25, 2004, across the snow and ice of Hovgaard Island. The waters are Penola Strait, along the Antarctic Peninsula. The temperature was –1°C (30°F). (b) Glaciers move offshore into Davis Strait from West Greenland and the ice sheet beyond. [Photos by Bobbé Christopherson.]

of marine influences, annual temperature ranges are reduced. This climate exists along the Bering Sea, the southern tip of Greenland, northern Iceland, Norway, and in the Southern Hemisphere, generally over oceans between 50° S and 60° S. Macquarie Island at 54° S in the Southern Ocean, south of New Zealand, is polar marine. Precipitation, which frequently falls as sleet (ice pellets), is greater in these regions than in continental polar climates.

Arid and Semiarid Climates (permanent moisture deficits)

The *arid* and *semiarid* regions are where we consider moisture efficiency along with temperature to understand the climate. Dry climates are the world's arid deserts and semiarid regions, with their unique plants, animals, and physical features. The mountains, rock strata, long vistas, and resilient struggle for life are magnified by the dryness.

These dry regions occupy more than 35% of Earth's land area and clearly are the most extensive climate over land. Sparse vegetation leaves the landscape bare; water demand exceeds the precipitation water supply throughout, creating permanent water deficits (water balance is discussed in Chapter 6). The extent of these deficits distinguishes arid deserts from semiarid steppe climatic regions. (See specific annual and daily desert temperature regimes, including the highest recorded temperatures and surface energy budgets, in Chapter 3; desert landscapes in Chapter 12; and desert environments in Chapter 16.)

Important causal elements in these dry lands include:

- Dry, subsiding air in subtropical high-pressure systems dominates.
- Midlatitude deserts and steppes form in the rain shadow of mountains, those regions to the lee of precipitation-intercepting mountains.

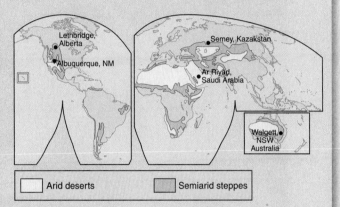

Arid deserts	Semiarid steppes

- Continental interiors, particularly central Asia, are far from moisture-bearing air masses.
- Shifting subtropical high-pressure systems produce semi-arid steppe lands around the periphery of arid deserts.

Dry climates are distributed by latitude and the amount of moisture deficits in four distinct regimes: *arid deserts* (tropical, subtropical, and midlatitude) and *semiarid steppes* (tropical, subtropical, and midlatitude).

Desert Characteristics

Desert vegetation is typically *xerophytic*: drought-resistant, waxy, hard-leafed, or otherwise adapted to aridity and low transpiration loss. Along stream channels, plants called *phreatophytes*, or "water-well plants," have roots that penetrate to great depths for the water they need (Figure 7.22).

The world climate map in Figure 7.5 reveals the pattern of Earth's dry climates, which cover broad regions between 15° and 30° N and S latitudes. In these areas, subtropical high-pressure cells predominate, with subsiding, stable air and low relative humidity. Under generally cloudless skies, these subtropical deserts extend to western continental margins, where cool, stabilizing ocean currents operate offshore and summer advection fog forms. The Atacama Desert of Chile, the Namib Desert of Namibia, the Western Sahara of Morocco, and the Australian Desert each lies adjacent to a coastline.

Extension of dry regions into higher latitudes of North and South America is associated with rain shadows, caused by orographic lifting over western mountains. The isolated interior of Asia, far distant from any moisture-bearing air masses, falls within these *arid* and *semiarid* dry climates.

Major subdivisions include: *deserts* (precipitation supply roughly less than one-half of the natural moisture demand) and *semiarid steppes* (precipitation supply roughly more than one-half of natural moisture demand, but not satisfying it). Important is whether precipitation falls principally in the winter with a dry summer, in the summer with a dry winter, or is evenly distributed. Winter rains are most effective because they fall at a time of lower moisture demand. Relative to temperature, the lower-latitude deserts and steppes tend to be hotter with less seasonal change than the midlatitude deserts and steppes, where mean annual temperatures are below 18°C (64.4°F) and freezing winter temperatures are possible.

Tropical, Subtropical Hot Desert Climates

Tropical, subtropical hot desert climates are Earth's true tropical and subtropical deserts and feature annual average temperatures above 18°C (64.4°F). They generally are concentrated on the western sides of continents, although Egypt, Somalia, and Saudi Arabia also fall within this climate regime. Rainfall is from local summer convectional showers. Some regions receive nearly nothing, whereas others may receive up to 35 cm (14 in.) of precipitation a year. A representative *subtropical hot desert* station is Ar Riyāḍ, (Riyadh), Saudi Arabia (Figure 7.23a and b).

Along the Sahara's southern margin is a drought-tortured region. Human populations suffered great hardship in the last several decades as desert conditions

FIGURE 7.22 Desert landscapes.
Desert plants are particularly well adapted to the harsh environment, typically *xerophytic*: drought-resistant, waxy, hard-leafed, and adapted to aridity and low transpiration loss. (a) Ocotillo (to left) and creosote (to right) in the Anza-Borrego Desert, southern California; such plants are particularly well adapted to the harsh environment. (b) Undependable water flows between the towering sandstone walls surrounding Chinle Wash in Canyon de Chelly, northeastern Arizona. (c) Colorful rock formations and view along a winding desert highway in Valley of Fire State Park, southern Nevada.
[Photos (a) and (c) by Bobbé Christopherson; (b) author.]

gradually expanded into their homelands. The sparse environment sets the stage for a rugged lifestyle and subsistence economies, pictured here near Timbuktu, Mali (Figure 7.23c). Chapter 12 presents the process of desertification (expanding desert conditions that afflict this region). Death Valley, California, features such a hot desert climate with an average annual temperature of 24.4°C (76°F). July and August average temperatures are 46°C and 45°C (115°F, 113°F), respectively. Temperatures over 50°C (122°F) are not uncommon.

During this decade there has been much reporting from Iraq on the war and related political events. What seemed overlooked in many reports was the fact the air temperatures in Baghdad (located on the map in Figure 7.23) were actually hotter than Death Valley during some days in July and August. Soldiers and civilians experience temperatures of 50°C (122°F) and higher in the city. Such readings broke records for Baghdad in 2007 and again in 2008. In January, averages for Death Valley (11°C; 52°F)

and Baghdad (9.4°C; 49°F) are comparable. Death Valley is drier with 5.9 cm of precipitation compared to 14 cm in Baghdad (2.33 in.; 5.5 in.), both low amounts. Baghdad's May to September period is remarkable, with zero precipitation, dominated as it is by an intense subtropical high-pressure system. (Baghdad is 34 m elevation at 33.3°N. Death Valley is at –54 m elevation at 36.5°N.) Keep these hot desert climates in mind as you follow events.

Midlatitude Cold Desert Climates

Midlatitude cold desert climates cover only a small area: the countries bordering southern Russia, the Gobi Desert, and Mongolia in Asia; the central third of Nevada and areas of the American Southwest, particularly at high elevations; and Patagonia in Argentina. Because of lower temperatures and lower natural moisture demand, rainfall also must be low for an area to be a *midlatitude cold desert*. Here we find total annual average rainfall of about 15 cm (6 in.).

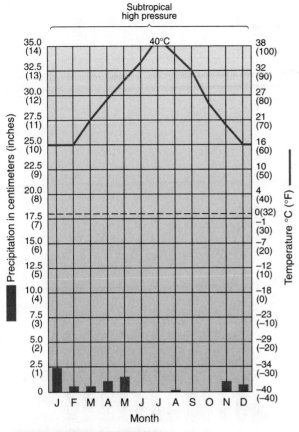

(b)

(c)

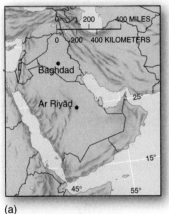

(a)

Station: Ar Riyāḍ (Riyadh),
Saudi Arabia
Lat/long: 24°42′ N 46°43′ E
Avg. Ann. Temp.: 26°C (78.8°F)
Total Ann. Precip.:
  8.2 cm (3.2 in.)
Elevation: 609 m (1998 ft)
Population: 5,024,000
Ann. Temp. Range:
  24 C° (43.2 F°)

FIGURE 7.23 Tropical, Subtropical Hot Desert Climates.
(a) Climograph for Ar Riyāḍ, (Riyadh), Saudi Arabia (*tropical, subtropical hot desert*). (b) The
Arabian desert sand dunes near Ar Riyāḍ. (c) Herders bring a few cattle to market near Timbuktu,
Mali. Precipitation has been below normal in the region since 1966. [Photos by (b) Ray Ellis/Photo
Researchers, Inc.; (c) Betty Press/Woodfin Camp & Associates.]

A representative city is Albuquerque, New Mexico,
with 20.7 cm (8.1 in.) of precipitation and an annual
average temperature of 14°C (57.2°F) (Figure 7.24a).
Across central Nevada stretches a characteristic expanse
of midlatitude cold desert, greatly modified by more
than a century of grazing and other extensive uses
(Figure 7.24b). Comparing the *midlatitude cold desert* of
Albuquerque and the *subtropical hot desert* of Ar Riyāḍ,
Saudi Arabia, reveals an interesting similarity in the
annual temperature range and a distinct difference in
precipitation patterns.

Tropical, Subtropical Hot Steppe Climates

Tropical, subtropical hot steppe climates generally exist around
the periphery of hot deserts, where shifting subtropical high-
pressure cells create a distinct summer-dry and winter-wet
pattern. This climate type is seen around the Sahara's
periphery and in Iran, Afghanistan, and the Turkistan,
Kazakstan, region. Average annual precipitation in *subtropical
hot steppe* areas usually is below 60 cm (23.6 in.). Walgett, in
interior New South Wales, Australia, provides a Southern
Hemisphere example of this climate (Figure 7.25).

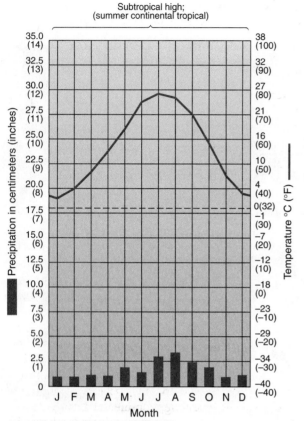

Subtropical high;
(summer continental tropical)

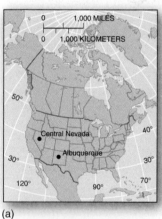

Station: Albuquerque,
New Mexico
Lat/long: 35°03′ N 106°37′ W
Avg. Ann. Temp.: 14°C (57.2°F)
Total Ann. Precip.:
20.7 cm (8.1 in.)
Elevation: 1620 m (5315 ft)
Population: 449,000
Ann. Temp. Range:
24 C° (43.2 F°)
Ann. Hr of Sunshine:
3420

FIGURE 7.24 Midlatitude cold desert climate.
(a) Climograph for Albuquerque, New Mexico (*midlatitude cold desert*). (b) Cold desert landscape of the Basin and Range Province in Nevada, east of Ely along highway US 50. [Photo by Bobbé Christopherson.]

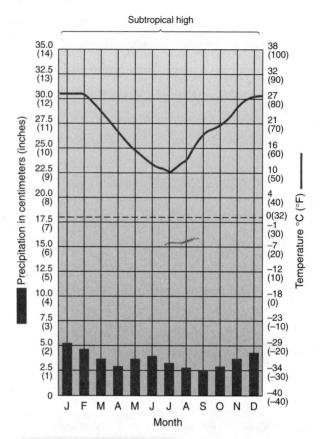

Subtropical high

Station: Walgett, New South
Wales, Australia
Lat/long: 30°0′ S 148°07′ E
Avg. Ann. Temp.: 20°C (68°F)
Total Ann. Precip.:
45.0 cm (17.7 in.)
Elevation: 133 m (436 ft)
Population: 2160
Ann. Temp. Range:
17 C° (31 F°)

FIGURE 7.25 Tropical, subtropical hot steppe climate.
(a) Climograph for Walgett, New South Wales, Australia (*tropical, subtropical hot steppe*). (b) Vast plains characteristic of north-central New South Wales. [Photo by Otto Rogge/Stock Market.]

Midlatitude Cold Steppe Climates

The *midlatitude cold steppe* climates occur poleward of about 30° latitude and the midlatitude desert climates. Such midlatitude steppes are not generally found in the Southern Hemisphere. As with other dry climate regions, rainfall in the steppe climate area is variable and undependable, ranging from 20 to 40 cm (7.9 to 15.7 in.). Not all rainfall is convectional, for cyclonic systems penetrate the continents; however, most storms produce little precipitation. Figure 7.26 presents a comparison between Asian and North American *midlatitude cold steppe* regions; consider Semey (Semipalatinsk) in Kazakstan (greater temperature range, precipitation evenly distributed) and Lethbridge in Alberta (lesser temperature range, summer maximum precipitation).

Global Climate Change

Significant climatic change has occurred on Earth in the past and most certainly will occur in the future. There is nothing society can do about long-term influences that cycle Earth through swings from ice ages to warmer periods. However, our global society must address short-term changes that are influencing global temperatures within the life span of present generations. This is especially true since these changes are due to human activities—an anthropogenic forcing of climate.

Record-high global temperatures have dominated the past two decades—on both land and ocean and during both day and night. The record year for warmth was 2005, with 2007 slightly ahead of 1998 as second warmest, with all the years since 1995 near this record. Air temperatures are the highest since recordings began in earnest more than 140 years ago and higher than at any time in the last 120,000 years, according to the ice-core record. This warming trend is *very likely** due to a buildup of greenhouse gases. Understanding this warming and all its related impacts is an important applied topic of Earth systems science and an opportunity for spatial analysis in physical geography.

Global Warming

Twenty years ago, climatologists Richard Houghton and George Woodwell described the present climatic condition:

> The world is warming. Climatic zones are shifting. Glaciers are melting. Sea level is rising. These are not hypothetical events from a science fiction movie; these changes and others are already taking place, and we expect them to accelerate over the next years as the amounts of carbon dioxide, methane, and

other trace gases accumulating in the atmosphere through human activities increase.[†]

Your author knows this quote well because it has been in every edition of *Elemental Geosystems* since the first edition in 1995. During the intervening years, we watched global temperatures rise and climate-change science mature. There is a strong scientific consensus that global warming is occurring and that human activities are the cause, namely the burning of fossil fuels.

The 2007 Intergovernmental Panel on Climate Change (IPCC) *Fourth Assessment Report* (AR4) concluded:

> Warming of the climate system is unequivocal, as is now evident from observations of increases in global average air and ocean temperatures, widespread melting of snow and ice, and rising global average sea level. . . . Most of the observed increase in globally averaged temperatures since the mid-20th century is *very likely* due to the observed increase in anthropogenic greenhouse gas concentrations. This is an advance since the *Third Assessment Report*'s conclusion that "most of the observed warming over the last 50 years is *likely* to have been due to the increase in greenhouse gas concentrations." Discernible human influences now extend to other aspects of climate, including ocean warming, continental-average temperatures, temperature extremes and wind patterns.[‡]

The IPCC, formed in 1988, is an organization operating under the United Nations Environment Programme (UNEP) and the World Meteorological Organization (WMO) and is the scientific organization coordinating global climate-change research, climate forecasts, and policy formulation. In 2007, the IPCC shared the Nobel Peace Prize with former Vice President Albert Gore for its two decades of work raising understanding and awareness about global climate-change science. The Nobel Committee said that Gore is responsible ". . . for convincing world governments that climate change was real, caused by human activity, and posed a threat to society."

Various organizations and agencies coordinate and conduct global climate-change research. (News Report 7.2 offers an overview and contact information for climate-change science.)

The Arctic Climate Impact Assessment (ACIA) Symposium met in 2004 and released a lengthy scientific report. The ACIA summarized:

> Human activities, primarily the burning of fossil fuels (coal, oil, natural gas), and secondarily the clearing of land, have increased the concentration of carbon dioxide, methane, and other heat-trapping ("greenhouse") gases in the atmosphere . . . this is projected to lead

---

*As a standard scientific reference on climate change, the IPCC uses the following to indicate levels of confidence: *Virtually certain*, > 99% probability of occurrence; *Extremely likely*, > 95%; *Very likely*, > 90%; *Likely*, > 66%; *More likely than not*, > 50%; *Unlikely*, < 33%; *Very unlikely*, < 10%; and *Extremely unlikely*, < 5%.

[†]R. Houghton and G. Woodwell, "Global climate change," *Scientific American* (April 1989): 36.

[‡]IPCC AR4, Working Group I, *Climate Change 2007: The Physical Science Basis* (Geneva, Switzerland: IPCC Secretariat, 2007), pp. 5, 10.

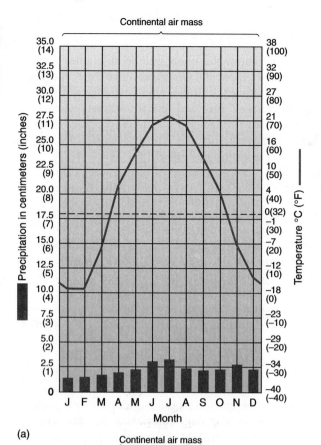

(a)

Station: Semey (Semipalatinsk), Kazakstan
Lat/long: 50°21' N 80°15' E
Avg. Ann. Temp.: 3°C (37.4°F)
Total Ann. Precip.:
26.4 cm (10.4 in.)

Elevation: 206 m (675.9 ft)
Population: 270,500
Ann. Temp. Range:
39 C° (70.2 F°)

(b)

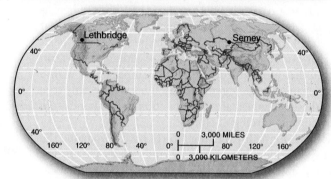

Station: Lethbridge, Alberta
Lat/long: 49°42' N 110°50' W
Avg. Ann. Temp.: 2.9°C (37.3°F)
Total Ann. Precip.:
25.8 cm (10.2 in.)

Elevation: 910 m (2985 ft)
Population: 73,000
Ann. Temp. Range:
24.3 C° (43.7 F°)

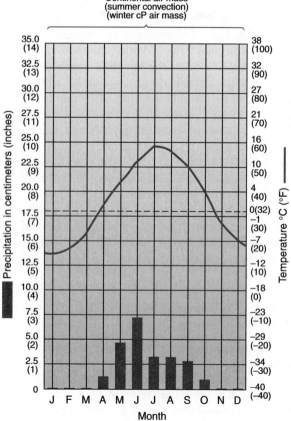

(c)

(d)

FIGURE 7.26 Midlatitude cold steppe climates, Kazakstan and Canada.
(a) Climograph for Semey (Semipalatinsk) in Kazakstan and (b) herders in the region near Semey.
(c) Climograph for Lethbridge in Alberta. (d) Canadian prairies and grain elevators of southern Alberta. [Photos by (b) Sovfoto/Eastfoto; (d) author.]

NEWS REPORT 7.2

Coordinating Global Climate Change Research

A cooperative global network of United Nation members participate in the United Nations Environment Programme (UNEP, http://www .unep.org) and the World Meteorological Organization (WMO, http:// www.wmo.ch/). The World Climate Research Programme (WCRP, http:// wcrp.wmo.int/) and its network under the supervision of the Global Climate Observing System (GCOS, http://www.wmo.ch/web/gcos/gco shome.html) coordinate data gathering and research. Measurements made at the Mauna Loa Observatory are at http://cdiac.esd.ornl.gov/ftp/ maunaloa-co2/maunaloa.co2, the home address being http://www.mlo .noaa.gov/home.html.

The ongoing climate assessment process within the UNEP is conducted by the Intergovernmental Panel on Climate Change (IPCC, http://www.ipcc.ch/), whose three working groups issued completed reports in 1990, 1992 (a supplementary report), 1995, and the *Third Assessment Report* in 2001. The latest is the *Fourth Assessment Report* (AR4), with all three working groups reporting in 2007: Working Group I, "Climate Change 2007: The Physical Basis"; WGII, "Climate Change

Impacts, Adaptation, and Vulnerability"; and WGIII, "Mitigation of Climate Change"; and a final *Synthesis Report*, released at the end of the year. The IPCC AR4 involved 2500 scientific expert reviews, 800 contributing authors, 450 lead authors from 130 countries, and 6 years of work. You can download their "Summary for Policy Makers" for free from the www.ipcc.ch/ web site.

The Arctic Council (at http:// www.arctic-council.org/) initiated the Arctic Climate Impact Assessment (ACIA, http://www.acia.uaf. edu/) in 2000; its three working groups completed the research— Arctic Monitoring and Assessment Program (AMAP, http://www.amap .no/), the Conservation of Flora and Fauna (CAFF, http://www.caff.is/), and the International Arctic Science Committee (IASC, http://www.iasc .no/), comprised of national scientific organizations, including 18 national academies of science.

In the United States, coordination is found at the US Global Change Research Program (http:// www.usgcrp.gov/). An overall source for information is http:// globalchange.gov, which publishes an on-line monthly summary of all

related developments. Also important are programs and services at NASA agencies, such as Dr. James Hansen's work at Goddard Institute for Space Studies (GISS, http:// www.giss.nasa.gov/), Global Hydrology and Climate Center (GHCC, http://www.ghcc.msfc.nasa.gov), and at NOAA agencies at the National Climate Data Center (NCDC, http:// www.ncdc.noaa.gov/) and the National Environmental Satellite, Data, and Information Service (NESDIS, http://www.nesdis.noaa.gov/), among others. The Pew Center on Global Climate Change offers credible analysis and overview and has issued several policy reports at http://www.pewclimate.org/. The multiagency National Ice Center is at http://www.natice.noaa.gov/. Important research is done at the National Center for Atmospheric Research (http://www.ncar.ucar.edu/).

For Canada, information and research is coordinated by Environment Canada (http://www.ec.gc.ca/ climate/). The effect of global warming on permafrost, which involves half of Canadian land area, is found at http://www.socc.ca/. The International Polar Year, 2007–2009, is at http://us-ipy.org.

---

to significant and persistent changes in climate... to wide-ranging consequences including significant impact on coastal communities, animal and plant species, water resources, and human health and well being.*

The American Association for the Advancement of Science (AAAS) reported in "The scientific consensus on climate change" [*Science* 306 (December 3, 2004): 1686] the results of a survey of all 928 climate-change papers published in refereed scientific journals between 1993 and 2003. The papers were divided into six categories and analyzed. As study author Naomi Oreskes concluded, "Remarkably, none of the papers disagreed with the consensus position." There is a consensus that human activi-

ties are heating Earth's surface and lower atmosphere. The author concluded:

> This analysis shows that scientists publishing in the peer-reviewed literature agree with IPCC, the National Academy of Sciences, and public statements of their professional societies. Politicians, economists, journalists, and others may have the impression of confusion, disagreement, or discord among climate scientists, but that impression is incorrect... there is a scientific consensus on the reality of anthropogenic climate change.

In terms of **paleoclimatology**, the science that studies past climates (discussed in Chapter 14), proxy indicators (ice-core data, sediments, coral reefs, ancient pollen, tree-ring density, among others) point to the present time as the warmest in the last 120,000 years, and further, that the increase in temperature during the twentieth century is

*ACIA, *Impacts of a Warming Arctic* (London: Cambridge University Press, 2004), p. 2.

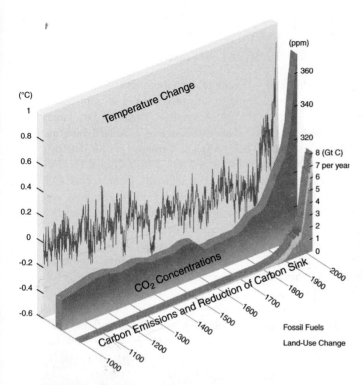

FIGURE 7.27 A thousand years of records and the covariance of carbon dioxide and temperature.
Carbon emissions, CO_2 concentrations, and temperature change correlate over the past 1000 years. This graph was prepared before the record-setting temperatures of 2005. By May 2008, with CO_2 levels at 389 ppm, carbon emissions were approaching 8 Gt annually (gigatons, or 8 billion tons of carbon). Earth systems are in record territory. [Illustration from ACIA, *Impacts of a Warming Arctic* (London: Cambridge University Press, 2004), p. 2. Used by permission of ACIA.]

very likely the largest in any century over the past 1000 years (Figure 7.27). Earth is less than 1 C° (1.8 F°) from equaling the highest average temperature of the past 125,000 years.

The latest Antarctic ice core at the Dome C location provides climatic data going back 800,000 years (described in Chapter 14). The rate of warming in the past 30 years exceeds any comparable period in the entire measured temperature record in Dome C, according to NASA scientists.

Figure 7.28a plots observed annual surface-air temperatures and 5-year mean temperatures from 1880 through 2007: Worldwide, 2005 was the warmest year; 2006 was fifth warmest but warmest ever for North America, with 2007 slightly ahead of 1998 for second warmest in the record.

The map in 7.28b uses the same base period as the graph (1951–1980) to give you an idea of temperature anomalies across the globe in 2007. Note the Arctic region, where new records for land and water temperatures and sea-ice melt were set in 2006 and again broken in 2007; 2008 sea-ice melt was nearly as bad. Anomalies exceeding 2.5 C° (4.5 F°) are portrayed on the map in the Canadian Arctic and portions of Greenland, Siberia, and the Antarctic Peninsula. The fact that 2007–2009 is the International Polar Year underscores the global scientific concerns about the polar climates and lands.

Scientists are working to determine the difference between *forced fluctuations* (human-caused) and *unforced fluctuations* (natural) as a key to predicting future climate trends. Figure 7.29 from the IPCC *Fourth Assessment Report* shows a comparison of climate models of forcing from solar output changes and volcanoes (shaded blue area) with anthropogenic (human) forcing (shaded red area) for global temperature anomalies. The black line on the graph plots observations from 1906 to 2005. Clearly, only those coupled climate models that include anthropogenic factors accurately simulate increasing temperatures. Solar variability and volcanic output alone do not explain the increase.

Because the enhanced greenhouse gases are anthropogenic in origin, various management strategies are possible to reduce human-forced climatic changes. Let's begin by examining the problem at its roots.

Carbon Dioxide and Global Warming   Carbon dioxide and water vapor are the principal radiatively active gases causing Earth's natural greenhouse effect. *Radiatively active* gases include atmospheric gases, such as carbon dioxide (CO_2), methane (CH_4), nitrous oxide (N_2O), chlorofluorocarbons (CFCs), and water vapor, which absorb and radiate longwave energy. Figure 7.30 plots changes in three of these greenhouse gases over the past 10,000 years.

These gases are transparent to light but opaque to the longer wavelengths radiated by Earth. Thus, they transmit light from the Sun to Earth but delay heat-energy loss to space. While detained, this heat energy is absorbed and emitted over and over, warming the lower atmosphere. As concentrations of these greenhouse gases increase, more heat energy remains in the atmosphere and temperatures increase.

The present CO_2 concentration tops anything over the last 800,000 years in the Dome C ice core. In fact, over this ice-core record, changes of as much as 30 ppm in CO_2 took at least 1000 years, yet the concentration has changed 30 ppm in just the last 14 years. With 4.5% of the world's population, the United States produces 24% of global CO_2 emissions. China, with 20% of global population, is responsible for 18% of CO_2 emissions. Clearly, per capita CO_2 emissions in the United States are far above that of an average person in China, respectively 19.2 tons compared to 4.4 tons per person in 2007.

Methane and Global Warming   Another radiatively active gas contributing to the overall greenhouse effect is methane (CH_4), which, at more than 1% per year, is increasing in concentration even faster than carbon dioxide. In ice cores, methane levels never topped 750 ppb in the past 800,000 years, yet in Figure 7.30 we see present levels at 1780 ppb. We are at an atmospheric concentration of methane that is higher than at any time in the past 800 millennia.

Methane is generated by such organic processes as digestion and rotting in the absence of oxygen (anaerobic processes). About 50% of the excess methane comes from bacterial action in the intestinal tracts of livestock and from organic activity in flooded rice fields. Burning of

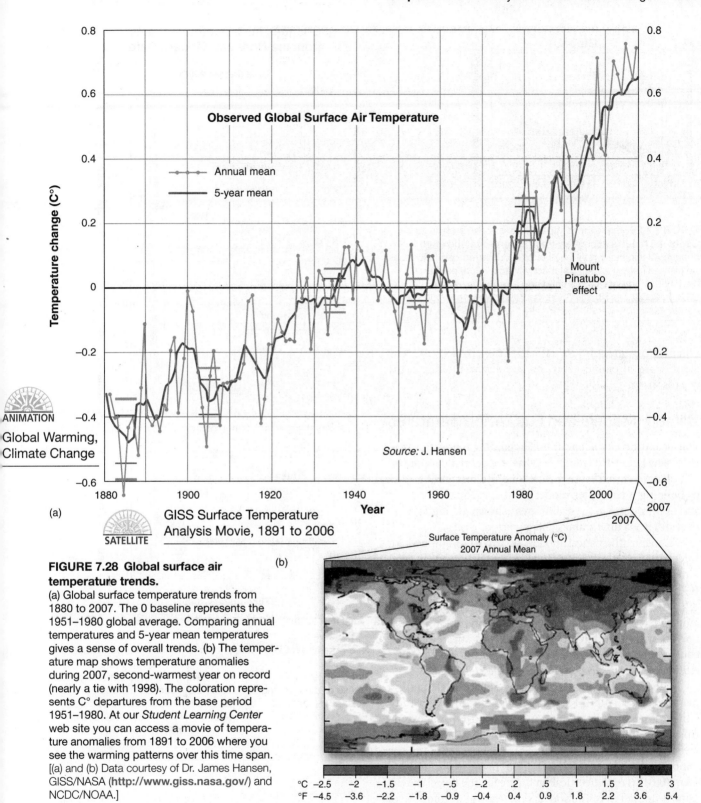

(a)

GISS Surface Temperature
Analysis Movie, 1891 to 2006

ANIMATION
Global Warming,
Climate Change

**FIGURE 7.28 Global surface air
temperature trends.**
(a) Global surface temperature trends from
1880 to 2007. The 0 baseline represents the
1951–1980 global average. Comparing annual
temperatures and 5-year mean temperatures
gives a sense of overall trends. (b) The temperature map shows temperature anomalies
during 2007, second-warmest year on record
(nearly a tie with 1998). The coloration represents C° departures from the base period
1951–1980. At our *Student Learning Center*
web site you can access a movie of temperature anomalies from 1891 to 2006 where you
see the warming patterns over this time span.
[(a) and (b) Data courtesy of Dr. James Hansen,
GISS/NASA (**http://www.giss.nasa.gov/**) and
NCDC/NOAA.]

vegetation causes another 20% of the excess, and bacterial action inside the digestive systems of termite populations also is a significant source. Methane is thought responsible for at least 19% of the total atmospheric warming.

Other Greenhouse Gases   Nitrous oxide (N_2O) is the third most important greenhouse gas that is forced by human activity—up 17% in atmospheric concentration

since 1750, higher than at any time in the past 10,000 years (Figure 7.30). Fertilizer use increases the processes in soil that emit nitrous oxide, although more research is needed to fully understand the relationships. Chloro-fluorocarbons (CFCs) and other halocarbons also con-tribute to global warming. CFCs absorb longwave energy missed by carbon dioxide and water vapor in the lower troposphere. As radiatively active gases, CFCs

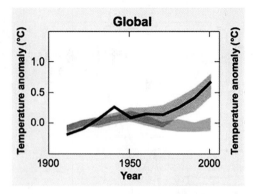

FIGURE 7.29 Explaining global temperature changes.
Computer models accurately track observed temperature
change (black line) when they factor in human-forced influences
on climate (red shading). Solar activity and volcanoes do not
explain the increases (blue shading). [Graph from *Climate
Change 2007: The Physical Science Basis*, Working Group I, IPCC
Fourth Assessment Report , February 2007, Fig. SPM-4, p. 11.]

enhance the greenhouse effect in the troposphere and
are a cause of ozone depletion and slight cooling in the
stratosphere.

Climate Models and Future Temperatures

The scientific challenge in understanding climate change
is to sense climatic trends in what is essentially a nonlinear, chaotic natural system. Imagine the tremendous task
of building a computer model of all climatic components
and programming these linkages (shown in Figure 7.1)
over different time frames and at various scales.

Using mathematical models originally established for
forecasting weather, scientists developed a complex computer climate model known as a **general circulation model
(GCM)**. There are at least a dozen established GCMs now
operating around the world. Submodel programs for the
atmosphere, ocean, land surface, cryosphere, and biosphere
operate within the GCM. The most sophisticated models
couple atmosphere and ocean submodels and are known as
Atmosphere-Ocean General Circulation Models (AOGCMs).

The first step in describing a climate is defining a
manageable portion of Earth's climatic system for study.
Climatologists create dimensional "grid boxes" that extend
from beneath the ocean to the tropopause, in multiple
layers (Figure 7.31). Resolution of these boxes in the
atmosphere is about 250 km (155 mi) in the horizontal and
1 km (0.6 mi) in the vertical; in the ocean the boxes use the
same horizontal resolution and a vertical resolution of
about 200 to 400 m (650–1300 ft). Analysts deal not only
with the climatic components within each grid layer but
also with the interaction among the layers on all sides.

A comparative benchmark among the operational
GCMs is *climatic sensitivity* to doubling of carbon dioxide
levels in the atmosphere. GCMs do not predict specific
temperatures, but they do offer various scenarios of global
warming. GCM-generated maps correlate well with the
observed global warming patterns experienced since 1990.

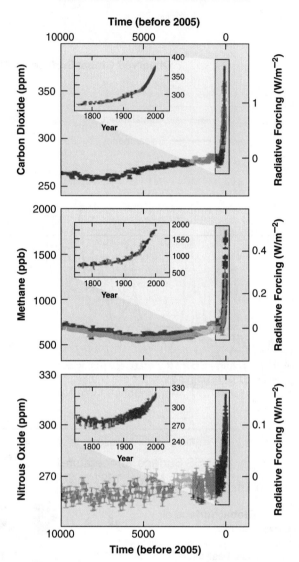

**FIGURE 7.30 Greenhouse gas changes over the last
10,000 years.**
Ice-core and modern data show the trends in carbon dioxide,
methane, and nitrous oxide over the last 10,000 years. It is
immediately obvious that we are living in unique territory on
these graphs. During May 2008, CO_2 reached the 389 ppm
level, CH_4 reached 1780 ppb, and N_2O approached 330 ppb.
[Graphs from *Climate Change 2007: The Physical Science Basis*,
Working Group I, IPCC *Fourth Assessment Report* , February
2007, Fig. SPM-1, p. 3. CO_2 data from **http://cdiac.esd.ornl
.gov/ftp/maunaloa-co2/maunaloa.co2.**]

The 2007 IPCC *Fourth Assessment Report*, using a variety of GCM forecast scenarios, predicted a range of average surface warming for this century. Figure 7.32 illustrates
six of these scenarios, each with its own assumptions of economics, population, degree of global cooperation, and
greenhouse gas emission levels. The choices and policies
we make determine which scenario happens. The orange
line is the simulation experiment where greenhouse gas
emissions are held at 2000 values with no increases. The
gray bars give you the best estimate and likely ranges of

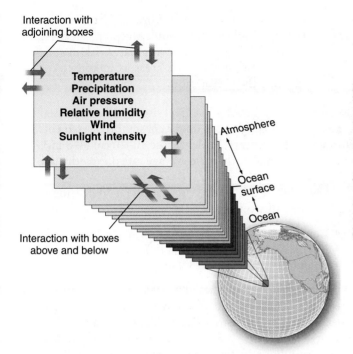

FIGURE 7.31 A general circulation model scheme.
Temperature, precipitation, air pressure, relative humidity, wind, and sunlight intensity are sampled in myriad grid boxes. In the ocean, sampling is limited, but temperature, salinity, and ocean current data are considered. The interactions within a grid layer, and between layers on all six sides, are modeled in a general circulation model program.

outcomes; for example, from a "low forecast" in B1 to a "high forecast" in A1Fl. Even the "B" scenario represents a significant increase in global land and ocean temperatures and will produce consequences. Although regionally variable and subject to revision, the IPCC temperature forecasts for the twenty-first century are:

- High forecast: 6.4 C° (11.5 F°)
- Middle forecast: 1.8 C°–4.0 C° (3.1 F°–7.2 F°)
- Low forecast: 1.1 C° (2.0 F°)

Consequences of Global Warming

The consequences of uncontrolled atmospheric warming are complex. Regional climate responses are expected as temperature, precipitation, soil-moisture, and air mass characteristics change. Although the ability to accurately forecast such regional changes is still evolving, some consequences of warming have been forecasted and in several regions are already underway. The challenge for science is to analyze such effects on a global scale.

The following list is a brief overview from the IPCC *Fourth Assessment Report*, "Summary for Policy Makers," and other sources, that summarizes global impacts emerging from climate change.

- The observed pattern of tropospheric warming and stratospheric cooling is *very likely* due to greenhouse gas increases and stratospheric ozone depletion.
- Widespread changes in extreme temperatures have been observed over the last 50 years. Cold

Multi-model Averages and Assessed Ranges for Surface Warming

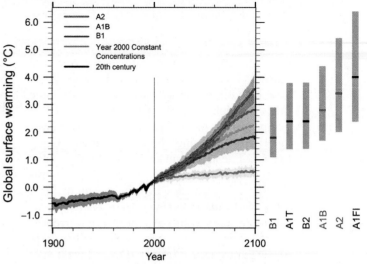

FIGURE 7.32 Model scenarios for surface warming.
Ranging from business-as-usual, lack of international cooperation, and continuation of fossil fuels ("A2Fl") to a low-emission scenario ("B1"), each scenario spells climatic catastrophe to some degree. Holding emissions at the 2000 concentration (orange line plot) results in the least temperature impacts. [Graph from *Climate Change 2007: The Physical Science Basis*, Working Group I, IPCC *Fourth Assessment Report*, February 2007, Fig. SPM-5, p. 14.]

days, cold nights, and frost are less frequent, while hot days, hot nights, and heat waves are more frequent.

- Observations since 1961 show the average global ocean temperature increased to depths of 3000 m and the ocean absorbed more than 80% of climate system heating. Such warming causes thermal expansion of seawater, contributing to sea level rise.
- There is observational evidence of increased intensity of tropical cyclones correlated with increases of tropical sea-surface temperatures. Total "power dissipation" of these storms has doubled since 1970. Worldwide there are more category 3, 4, and 5 tropical storms than in the previous record.
- Mountain glaciers and snow cover declined on average in both hemispheres, contributing to sea-level rise.
- Mount Kilimanjaro in Africa, portions of the South American Andes, and the Himalayas will very likely lose most of their glacial ice within the next two decades, affecting local water resources. Glacial ice continues its retreat in Alaska.
- The average atmospheric water vapor content has increased over land and ocean as well as in the upper troposphere. The increase is consistent with the fact that warmer air can absorb more water vapor.
- Flow speed accelerated for some Greenland and Antarctic outlet glaciers as they drain ice from the interior of the ice sheets. In Greenland this

rate of loss exceeds snowfall accumulation, with losses per year more than doubling between 1996 and 2005 (mass losses of 91 km³ compared to 224 km³).

- Average Arctic temperatures increased at almost twice the global average rate in the past 100 years. Wintertime lower atmosphere temperatures over Antarctica are warming at nearly three times the global average, first reported in 2006.
- Temperatures in the permafrost active layer increased overall since the 1980s in the Arctic up to 3 C°. The maximum area of seasonally frozen ground decreased 7% in the Northern Hemisphere, with a decrease in spring of up to 15%.
- Since 1978, annual average Arctic sea ice extent shrunk in summer by 7.4% per decade, according to satellite measurements. September 2007 Arctic sea ice extent retreated to its lowest area coverage in the record.
- Changes in precipitation and evaporation over the oceans and the melting of ice are freshening mid- and high-latitude oceans and seas, together with increased salinity in low-latitudes. Oceans are acidifying (lower pH) in response to absorption of increasing atmospheric CO_2.
- Mid-latitude westerly winds strengthened in both hemispheres since the 1960s.
- More intense and longer droughts have been observed over wider areas since the 1970s, particularly in the tropics and subtropics. Increased drying linked to higher temperatures and decreased precipitation observed in the Sahel, the Mediterranean, southern Africa, parts of southern Asia, Australia, and the American West.
- Higher spring and summer temperatures and earlier snowmelt are extending the wildfire season and increasing the intensity of wildfires in the western US and elsewhere.
- The frequency of heavy precipitation events increased over most land areas, consistent with warming and observed increases of atmospheric water vapor. Significantly increased precipitation has been observed in eastern parts of North and South America, northern Europe, and northern and central Asia.
- Crop patterns, as well as natural habitats of plants and animals, will shift to maintain preferred temperatures. According to climate models, climatic regions in the midlatitudes could shift poleward by 150 to 550 km (90 to 350 mi) during this century.
- Biosphere models predict that a global average of 30% of the present forest cover (varying regionally from 15% to 65%) will undergo major species redistribution—greatest at high latitudes and least in the tropics.
- Populations previously unaffected by malaria, dengue fever, lymphatic filariasis, and yellow fever (all mosquito vector), schistosomiasis (water snail vector), and sleeping sickness (tsetse fly vector) will be at greater risk in subtropical and midlatitude areas as temperatures increase the vector ranges.

Changes in Sea Level   Sea-level rise must be expressed as a range of values that are under constant reassessment. During the last century, sea level rose 10–20 cm (4–8 in.), a rate 10 times higher than the average rate during the last 3000 years.

The 2007 IPCC forecast scenarios for global mean sea-level rise this century, given regional variations, are:

- Low forecast: 0.18 m (7.1 in.)
- Middle forecast: 0.39 m (15.4 in.)
- High forecast: 0.59 m (23.2 in.)

Unfortunately the new 2006–2008 measurements of Greenland's ice-loss acceleration did not reach the IPCC in time for its report. Scientists are considering at least a 1.2-m (3.94-ft) "high case" for estimates of sea-level rise this century as more realistic given Greenland's present losses coupled with mountain glacial ice losses worldwide. Remember that a 0.3-m rise in sea level would produce a shoreline retreat of 30 m (98 ft) on average. Here is the concern:

> The data now available raise concerns that the climate system, in particular sea level, may be responding more quickly than climate models indicate.... The rate of sea-level rise for the past 20 years is 25% faster than the rate of rise in any 20-year period in the preceding 115 years.... Since 1990 the observed sea-level has been rising faster than the rise projected by models.[*]

These increases would continue beyond 2100 even if greenhouse gas concentrations were stabilized. In Chapter 13, maps present what coastlines will experience in the event of a 1-m rise in sea level. The state of California is using a 1.4-m rise in its planning models.

A quick survey of world coastlines shows that even a moderate rise could bring change of unparalleled proportions. At stake are the river deltas, lowland coastal farming valleys, and low-lying mainland areas, all contending with high water, high tides, and higher storm surges. Particularly tragic social and economic consequences will affect small island states, which are unable to adjust within their present country boundaries—disruption of biological systems, loss of biodiversity, reduction in water resources, and evacuation of residents are among the impacts. There could be both internal and international migration of affected human populations, spread over decades, as people move away from coastal flooding caused by the sea-level rise—a 1-m rise will displace 130 million people. Presently, there is no body of world law that covers "environmental refugees."

---

[*]S. Rahmstorf et al., "Recent climate observations compared to projections," AAAS *Science* 316 (May 4, 2007): 709.

Political Action and "No Regrets"

Reading through all this climate-change science must seem pretty "heavy." Instead, think of this information as empowering and as motivation to take action—personally, locally, regionally, nationally, and globally.

A product of the 1992 Earth Summit in Rio de Janeiro was the United Nations Framework Convention on Climate Change (FCCC). The leading body of the convention is the Conference of the Parties (COP) operated by the countries that ratified the FCCC (186 countries by 2000). Meetings were held in Berlin (*COP-1*, 1995) and Geneva (*COP-2*, 1996). These meetings set the stage for *COP-3* in Kyoto, Japan, in December 1997, where 10,000 participants adopted the *Kyoto Protocol* by consensus. Seventeen national academies of science endorsed the Kyoto Protocol. (For updates on the status of the Kyoto Protocol, see http://unfccc.int/2860.php.) The 2007 gathering, *COP-13*, was in Bali, Indonesia; the 2008 *COP-14* was in Poznan, Poland.

The Kyoto Protocol binds more-developed countries to a collective 5.2% reduction in greenhouse gas emissions as measured at 1990 levels for the period 2008 to 2012. Within this group goal, various countries promised cuts: Canada is to cut 6%, the European Union 8%, and Australia 8%, among many others. The *Group of 77* countries plus China favor a 15% reduction by 2010. With Russian ratification in November 2004, the Kyoto Protocol is now international law, and as of 2008,178 countries are signatories. This is without US participation, although there is a possibility the United States will ratify in 2009.

The Intergovernmental Panel on Climate Change (IPCC) declared that "no regrets" opportunities to reduce carbon dioxide emissions are available in most countries. The IPCC Working Group III defines this as follows:

> No regrets options are by definition greenhouse gas emissions reduction options that have negative net costs, because they generate direct and indirect benefits that are large enough to offset the costs of implementing the options.

Benefits that equal or exceed their cost to society include reduced energy cost, improved air quality and health, reduction in tanker spills and oil imports, and deployment of renewable and sustainable energy sources, among others. This holds true without even considering the benefits of slowing the rate of climate change. For Europe, scientists determined that carbon emissions could be reduced to less than half the 1990 level by 2030, at a negative cost. One key to "no regrets" is the untapped energy-efficiency potential. A California government analysis in 2008 determined that a 30% emissions reduction would produce a net household savings of $500 a year and contribute a $4 billion gain to the state economy per year.

In the United States, five Department of Energy national laboratories (Oak Ridge, Lawrence Berkeley, Pacific Northwest, National Renewable Energy, and Argonne) reported that the United States can meet the Kyoto carbon emission reduction targets with negative overall costs (cash benefit savings) ranging from –$7 to –$34 billion. [For more, see Working Group III, *Climate Change 2001, Mitigation* (London: Cambridge University Press, 2001), pp. 21, 474–76, and 506–507.]

Summary and Review—Climate Systems and Climate Change

■ *Define* climate and climatology and *explain* the difference between climate and weather.

Climate is dynamic, not fixed. **Climate** is a synthesis of weather phenomena at many scales, from planetary to local, in contrast to *weather*, which is the condition of the atmosphere at any given time and place. Earth experiences a wide variety of climatic conditions that can be grouped by general similarities into climatic regions. **Climatology** is the study of climate and attempts to discern similar weather statistics and identify **climatic regions**.

climate (p. 222)
climatology (p. 222)
climatic regions (p. 222)

1. Define climate and compare it with weather. What is climatology?

2. Explain how a climatic region synthesizes climate statistics.

3. How does the El Niño phenomenon produce the largest interannual variability in climate? What are some of the changes and effects that occur worldwide?

■ *Review* the role of temperature, precipitation, air pressure, air mass patterns, and sea-surface temperatures used to establish climatic regions.

Climatic inputs include insolation (pattern of solar energy in the Earth–atmosphere environment); temperature (sensible heat energy content of the air); precipitation (rain, sleet, snow, and hail; the supply of moisture); air pressure (varying patterns of atmospheric density); and air masses (regional-sized homogeneous units of air). Climate is the basic element in ecosystems— self-regulating communities of plants and animals that thrive in specific environments.

4. How do radiation receipts, temperature, air pressure inputs, and precipitation patterns interact to produce climate types? Give an example from a humid environment and one from an arid environment.

5. Evaluate the relationships among a climatic region, ecosystem, and biome.

■ *Review* the development of climate classification systems and *compare* genetic and empirical systems as ways of classifying climate.

Classification is the process of ordering or grouping related data in categories. Relative to climate, a **genetic classification** is based on causative factors, such as the interaction of air masses. An **empirical classification** is one based on statistical data, such as temperature or precipitation. This text analyzes climate using aspects of both approaches, with a map based on climatological elements.

> classification (p. 224)
> genetic classification (p. 224)
> empirical classification (p. 224)

6. What are the differences between a genetic and an empirical classification system?

7. What are some of the climatological elements used in classifying climates?

■ *Describe* the principal climate classification categories other than deserts and *locate* these regions on a world map.

Here we focus on temperature and precipitation measures. Keep in mind these are measurable results produced by interacting elements of weather and climate. These data are plotted on a **climograph** to display the characteristics of the climate.

There are six basic climate categories. Temperature and precipitation considerations form the basis of five climate categories and their regional types:

- Tropical (equatorial and tropical latitudes)
  rain forest (rainy all year)
  monsoon (6 to 12 months rainy)
  savanna (less than 6 months rainy)
- Mesothermal (midlatitudes, mild winter)
  humid subtropical (hot summers)
  marine west coast (warm to cool summers)
  Mediterranean (dry summers)
- Microthermal (middle and high latitude, cold winter)
  humid continental (hot to warm summers)
  subarctic (cool summers to very cold winters)
- Polar (high latitudes and polar regions)
  tundra (high latitude or high elevation)
  ice caps and ice sheets (perpetually frozen)
  polar marine
- Highland (compared to lowlands at the same latitude, highlands have lower temperatures—recall the normal lapse rate)

Only one climate category is based on moisture efficiency as well as temperature:

- Deserts (permanent moisture deficits)
  arid deserts (tropical, subtropical hot and midlatitude cold)
  semiarid steppes (tropical, subtropical hot and midlatitude cold)
  climograph (p. 228)

8. List and discuss each of the principal climate categories. In which one of these general types do you live? Which category is the only type associated with the annual distribution and amount of precipitation?

9. What is a climograph, and how is it used to display climatic information?

10. Which of the major climate types occupies the most land and ocean area on Earth?

11. Characterize the tropical climates in terms of temperature, moisture, and location.

12. Using Africa's tropical climates as an example, characterize the climates produced by the seasonal shifting of the ITCZ with the high Sun.

13. Mesothermal (subtropical and midlatitude, mild winter) climates occupy the second-largest portion of Earth's entire surface. Describe their temperature, moisture, and precipitation characteristics.

14. Explain the distribution of the *humid subtropical hot-summer* and *Mediterranean dry-summer* climates at similar latitudes and the difference in precipitation patterns between the two types. Describe the difference in vegetation associated with these two climate types.

15. Which climates are characteristic of the Asian monsoon region?

16. Explain how a *marine west coast* climate type can occur in the Appalachian region of the eastern United States.

17. What role do offshore ocean currents play in the distribution of the *marine west coast* climates? What type of fog is formed in these regions?

18. Discuss the climatic conditions for the coldest places on Earth outside of the poles.

■ *Explain* the precipitation and moisture efficiency criteria used to determine the arid and semiarid desert climates and *locate* them on a world map.

The arid and semiarid climates are described by precipitation rather than temperature. Dry climates are the world's arid deserts and semiarid regions, with their unique plants, animals, and physical features. The *arid* and *semiarid* climates occupy more than 35% of Earth's land area, clearly the most extensive climate over land.

Major subdivisions are *arid deserts* in tropical and midlatitude areas (precipitation—natural water supply—less than one-half of natural water demand) and *semiarid steppes* in tropical and midlatitude areas (precipitation more than one-half of natural water demand but not satisfying it).

19. In general terms, what are the differences among the four desert classifications? How are moisture and temperature distributions used to differentiate these subtypes?

20. Relative to the distribution of dry climates, describe at least three locations where they occur across the globe and the reasons for their presence in these locations.

■ *Outline* future climate patterns from forecasts presented and *explain* the causes and potential consequences of climate change.

Various activities of present-day society are producing climatic changes, particularly a global warming trend. The highest average annual temperatures experienced since the advent of instrumental measurements have dominated the last 25 years. There is a scientific consensus that global warming is principally caused by anthropogenic impacts on the natural greenhouse effect.

The 2007 *Fourth Assessment Report* from the Intergovernmental Panel on Climate Change affirms this consensus. IPCC has predicted surface-temperature response to a doubling of carbon dioxide ranging from an increase of 1.1 C° (2.0 F°) to 6.4 C° (11.5 F°) between the present and 2100. Natural climatic variability over the span of Earth's history is the subject of **paleoclimatology**. A **general circulation model (GCM)** forecasts climate patterns and is evolving toward greater capability and accuracy than in the past. People and their political institutions can use GCM forecasts to form policies aimed at reducing unwanted climate change.

> paleoclimatology (p. 253)
> general circulation model (GCM) (p. 256)

21. Explain climate forecasts. How do general circulation models (GCMs) produce such forecasts?

22. Differentiate between natural climatic cycles (unforced) and forced climatic change, as described in the chapter.

23. Describe the potential climatic effects of global warming on polar and high-latitude regions. What are the implications of these climatic changes for persons living at lower latitudes?

24. How is climatic change affecting agricultural production? Natural environments? Forests? The possible spread of disease?

25. What are the present actions being taken to delay the effects of global climate change? What is the Kyoto Protocol?

NetWork and Critical Thinking Tools

A. The text asked that you find the climate conditions for your campus and your birthplace and locate these two places on Figures 7.3 and Figure 7.4. Briefly describe your search for the information and the sources you used: library, Internet, teacher, phone calls to state and provincial climatologists. Now, refer to Appendix B to refine your assessment of climate for your two locations. Briefly show how you worked through the Köppen climatic categories to find the climate classification for your two cities.

B. On the *Elemental Geosystems* Student Learning Center web site and in Focus Study 7.1, there are URL references for El

Niño and La Niña. Sample three or four sources and determine the status of the phenomena at the present as you work in this book. How are current conditions different from the record El Niño event in 1997–1998? Or La Niña in 2007? What Internet links were most helpful to you in completing this status report?

C. Many external factors force climate. The chart "Global and annual mean radiative forcing for the year 2000, relative to 1750" is presented here [from *Climate Change 2001, The Scientific Basis* (Washington: Cambridge University Press, 2001), Figure 3, p. 8, and Figure 6.6, p. 392].

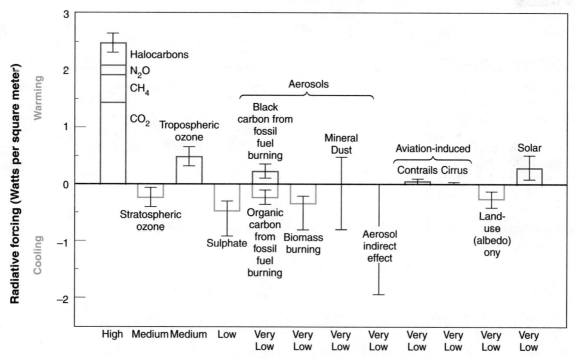

The Global Mean Radiative Forcing of the Climate System for the Year 2000, Relative to 1750

The estimates of radiative forcing in units of watts per square meter are given on the *y*-axis (vertical axis). The level of scientific understanding is noted along the *x*-axis (horizontal axis), arranged from "high" to "very low." Those columns above the "0" value (in red) indicate *positive forcing*, such as the greenhouse gases grouped in the far-left column. Columns that fall below the "0" value (in blue) indicate *negative forcing*, such as the haze from sulfate aerosols, fourth column from the left. The vertical line between the markers on each column is an estimate of the uncertainty range. Where no column appears but

there is instead a line denoting a range, there is no central estimate given present uncertainties, such as for mineral dust.

Assume you are a policymaker with a goal of reducing the rate of global warming, that is, reducing positive radiative forcing of the climate system. What strategies do you suggest to alter the height of the columns and adjust the mix of elements that cause warming? Assign priorities to each suggested strategy to denote most effective to least effective in moderating climate change. Brainstorm and discuss your strategies with others.

Part III | Earth's Changing Landscape Systems

A dynamic landscape, this aerial view looks south along the Green River into Flaming Gorge National Recreation Area in southwestern Wyoming. The Whalen Butte formation just east of the river (near top of photo) rises to 2175 m (7135 ft) elevation. The dominant rock is the Green River formation of Eocene Age (33.7 to 54.8 m.y.a), formed at the end of the Laramide Orogeny episode, out of an ancient lake formation that today is rich in fossils—plants, fish, bird, and dinosaur. The work of fluvial action dissects the land. PART III covers many aspects of this photo. *[Photo by Bobbé Christopherson.]*

arth is a dynamic planet whose surface is actively shaped by physical agents of change. PART III is organized around two broad systems of these agents—endogenic and exogenic. The **endogenic system** (Chapters 8 and 9) encompasses internal processes that produce flows of heat and material from deep below the crust, powered by radioactive decay—*the solid realm* of Earth. Earth's surface responds by moving, warping, and breaking, sometimes in dramatic episodes of earthquakes and volcanic eruptions.

At the same time, the **exogenic system** (Chapters 10–14) involves external processes that set into motion air, water, and ice, all powered by solar energy—*the fluid realm* of Earth's environment. These media are sculpting agents that carve, shape, and reduce the landscape. One such process, weathering, breaks up and dissolves the crust. Erosion picks up these materials; transports them in rivers, winds, coastal waves, and flowing glaciers; and deposits them along the way. Thus, Earth's surface is the interface between two vast open systems: one that builds the landscape and topographic relief and one that tears it down into sedimentary plains.

NOTEBOOK   Earth's Varied Landscapes

Iceland is a basaltic island sitting atop a hot spot on the mid-ocean ridge. South of Reykjavík, lava flows produced a remarkable landscape. Beneath the thin surface crust, lava tubes formed as conduits carried molten lava. Gas-charged flows puffed up and broke through the surface, cracking and swelling the cooled rock to form a lava blister. Lava was active on this landscape as recent as 1000 years ago. *[Photo by Bobbé Christopherson.]*

Chapter

8

The Dynamic Planet

KEY LEARNING CONCEPTS

After reading the chapter, you should be able to:

- *Distinguish* between the endogenic and exogenic systems, *determine* the driving force for each, and *explain* the pace at which these systems operate.

- *Diagram* Earth's interior in cross section and *describe* each distinct layer.

- *Illustrate* the geologic cycle and *relate* the rock cycle and rock types to endogenic and exogenic processes.

- *Describe* Pangaea and its breakup and *relate* several physical proofs that crustal drifting is continuing today.

- *Portray* the pattern of Earth's major plates and *relate* this pattern to the occurrence of earthquakes, volcanic activity, and hot spots.

STUDENT LEARNING CENTER TOOLS

The *Elemental Geosystems* Student Learning Center provides on-line resources for this chapter by clicking on the book cover at **www.mygeoscienceplace.com**. Once you are in our Learning Center, bookmark the URL address. For this chapter find reviewing tools, eBook links, self-tests and quizzes, critical thinking items, figure animations, photo galleries, and satellite loops. Some highlights:

▶ Twelve animations specific to figures and principals in this chapter: relative dating principles and the rock cycle, igneous and metamorphic rock processes and features, and plate tectonics, including sea-floor spreading, transform faults, mantle convection, motions at plate boundaries, the breakup of Pangaea, and hot spot volcanic tracks.

▶ Thinking spatially you work with specific figures from the text.

▶ Thirteen additional URLs along with Quick Links to all the URLs in the chapter.

The twentieth century was a time of great discovery about Earth's internal structure and dynamic crust, yet much remains on the scientific frontier of Earth systems science. Discoveries have revolutionized our understanding of how continents and oceans came to be arranged as they are. A new era of understanding is emerging, combining various sciences within the study of physical geography. One task of physical geography is to explain the *spatial implications* of all this new information.

In this chapter: Earth's interior is organized as a core surrounded by roughly concentric shells of material from the mantle to the crust. The interior is unevenly heated by the radioactive decay of unstable elements. Physical convection flows rise in some regions and sink in others. A rock cycle produces three classes of rocks through igneous, sedimentary, and metamorphic processes. Internal processes coupled with the rock cycle, hydrologic cycle, and tectonic cycle result in a varied crustal surface.

Earth features extensive mountain ranges on both land and the ocean floor, drifting continental and oceanic crust, and frequent earthquakes and volcanic events. All of this movement of material results from endogenic forces within Earth—the subject of this chapter.

The Pace of Change

The **geologic time scale** is a summary timeline of all Earth history. Figure 8.1 reflects currently accepted names of time intervals that encompass Earth's history, from vast *eons* through briefer *eras*, *periods*, and *epochs*. Present thinking places Earth's age at 4.567 billion years (or rounded to 4.6 billion). The Moon is about 30 million years younger, formed when a Mars-sized object struck the early Earth.

The sequence in this scale is based on the relative positions of rock strata above or below each other—*relative time*. An important general principle is that of *superposition*, which states that rock and sediment always are arranged with the youngest beds "superposed" near the top of a rock formation and the oldest at the base, if they have not been disturbed.

The absolute ages on the scale, determined by scientific methods such as dating by radioactive isotopes, help refine the time-scale sequence—*absolute time*. Thus, the geologic time scale is a product of both relative and absolute time measures. See News Report 8.1 for more on this absolute dating method. Also on the geologic time scale, see the labels denoting the six major extinctions of life forms in Earth history. These range from 440 million years ago (m.y.a.) to the present one caused by modern civilization. For more on the geologic time scale, see http://www.ucmp.berkeley.edu/exhibit/histgeoscale.html.

The name Holocene, approximately the last 11,500 years, is given to the youngest epoch in the geologic time scale, characteristic of postglacial conditions since the retreat of the continental glaciers. Because of the impact of human society on planetary systems, discussion is underway about designating a new name, *Anthropocene*, for the human epoch.

However, when would such a time begin? Some think the beginning of the Industrial Revolution in the eighteenth century is the starting point, whereas others think an appropriate time is some 5000 years ago when Asians began flooding rice fields that released methane gas into the atmosphere. Or perhaps 8000 years ago is a better chronology with the spread of domesticated crops, widespread clearing of forests, and the human use of fire to modify environments. Human impacts on the atmosphere began at these roots of civilization. What do you think?

A guiding principle of Earth science is uniformitarianism, first proposed by James Hutton in his *Theory of the Earth* (1795) and later amplified by Charles Lyell in his *Principles of Geology* (1830). **Uniformitarianism** assumes that *the same physical processes active in the environment today have been operating throughout geologic time*. "The present is the key to the past" describes this principle. Evidence from modern scientific exploration and from the landscape record of volcanic eruptions, earthquakes, and Earth's processes support uniformitarianism.

Within the principle of uniformitarianism, catastrophic events such as massive landslides, earthquakes, volcanic episodes, extraterrestrial asteroid impacts, and sometime cyclic episodes of mountain building punctuate geologic time. Vast super floods occurred as ice-blocked lakes tore through their frozen dams in the later days of the last ice age, such as glacial Lake Missoula, which left evidence across a broad region. Within the principle of uniformitarianism, these localized catastrophic events occur as small interruptions in the generally uniform processes that shape the slowly evolving landscape. Here, the *punctuated equilibrium* concept (interruptions in the flow of events, systems jumping to new operation levels) studied in the life sciences and paleontology might apply to aspects of Earth's long developmental history.

We start our journey deep within the planet. A knowledge of Earth's internal structure and energy is key to understanding the surface.

Earth's Structure and Internal Energy

Along with the other planets and the Sun, Earth is thought to have condensed and congealed from a nebula of dust, gas, and icy comets about 4.6 billion years ago (Chapter 2). Until recently, the oldest known surface rocks on Earth were in northwestern Canada. The Acasta Gneiss is a 3.96-billion-year-old formation; nearby the Slave Province rocks date to 4.03 billion years. Rocks from Greenland date back to 3.8 billion years old.

Detrital zircons (particles of preexisting zirconium silica oxides that formed rock) in Western Australia date to between 4.2 and 4.4 billion years old and are possibly the oldest materials in Earth's crust. Recent research in 2008 found mineral grains in the Nuvvuagittuq greenstone belt

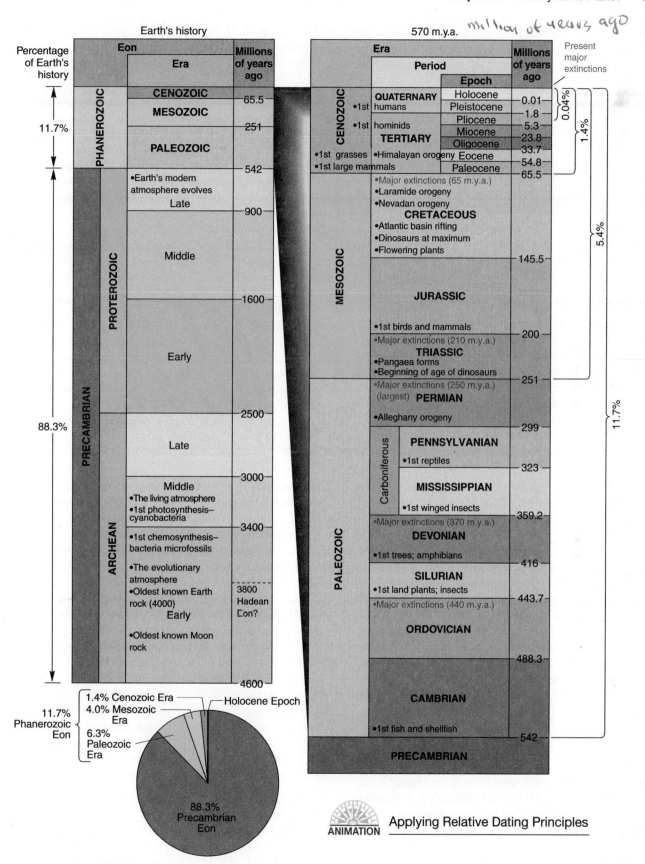

FIGURE 8.1 Geologic time scale.
Both relative and absolute dating methods calibrate the geologic time scale. In the column at the left, note that 88% of geologic time occurred during the Precambrian Eon. Highlights of Earth's history are shown in the figure as bulleted items, including the six major extinctions or depletions of life forms, printed in red (the sixth extinction episode is underway at the present time), and dates appear in m.y.a. [Data and update from Geological Society of America and *Nature* 429 (May 13, 2004): 124–25.]

NEWS REPORT 8.1

Radioactivity: Earth's Time Clock

The age of Earth and the age of the earliest known crustal rock are astounding, for we think in terms of Earth's annual trips around the Sun and the pace of our own lives and birthdays. We need something greater than human time to measure the vastness of geologic time. Nature has provided a way: *radiometric dating*. It is based on the steady decay of certain atoms.

Protons and neutrons reside in an atom's nucleus. Certain forms of atoms (isotopes) have unstable nuclei; that is, the protons and neutrons do not remain together indefinitely. As particles break away and the nucleus disintegrates, radiation is emitted and the atom decays into a different element—this process is *radioactivity*.

Radioactivity provides the steady time clock needed to measure the age of ancient rocks. It works because the decay rates for different isotopes are determined precisely, and they do not vary. These rates are expressed as *half-life*, the time required for one-half of the unstable atoms in a sample to decay. Some examples of unstable elements that become stable elements and their half-lives include uranium-238 to lead-206 (4.5 billion years); thorium-232 to lead-208 (14.1 billion years); potassium-40 to argon-40 (1.3 billion years); and, in organic (living) materials, carbon-14 to nitrogen-14 (5730 years), although with correlation from ice-core data, carbon-dating accuracy is being pushed back to more than 24,000 years.

The presence of these decaying elements and stable end products in sediment or rock allows scientists to read the radiometric "clock." They compare the amount of original isotope in the sample with the amount of decayed end product in the sample. If the two are in a ratio of 1:1 (equal parts), one half-life has passed. Errors can occur if the sample has been disturbed or subjected to natural weathering processes that might alter its radioactivity.

To increase accuracy, investigators may check a sample using more than one radiometric measurement. Calibration of the past 10,000 years is accomplished using tree-ring analysis; the past 45,000 years, using fossils found in lake sediments; and the past 500,000 years and more, using ice-core records from Greenland and Antarctica, among several tools available. Using several radiometric methods, Earth's oldest known rocks are absolutely dated.

of northern Quebec, Canada, that date to 4.28 billion years. These discoveries tell us something significant: Earth was forming continental crust at least 4 billion years ago, during the Archean Eon—the term *protocrust* is applied to this phenomenon. Although not an official designation, note the Hadean Eon in Figure 8.1, a proposal to describe the period before the Archean.

Throughout Earth's formation process, heat was accumulating. As the protoplanet compacted into a smaller volume, energy was transformed into heat through compression. In addition, significant heat was trapped inside Earth from the decay of isotopes (unstable forms) of uranium, thorium, potassium, and other radioactive elements.

Earth in Cross Section

As Earth solidified, heavier elements slowly gravitated toward the center, and lighter elements slowly welled upward to the surface, concentrating in the crust. Consequently, Earth's interior is arranged roughly in concentric layers, each one distinct either in chemical composition or temperature, with heat radiating outward from the center by conduction and then by physical convection in the more plastic levels nearer the surface.

Our knowledge of Earth's *internal differentiation* into these layers is acquired entirely through indirect evidence, because we are unable to drill more than a few kilometers into the crust. There are several physical properties of Earth materials that enable us to approximate the nature of the interior. For example, when an earthquake sends shock waves through the planet, the cooler areas, which generally are more rigid, transmit these **seismic waves** at a higher velocity than do the hotter areas. Density also has an effect on seismic velocities, and fluid or plastic zones simply do not transmit some types of seismic waves; they absorb them. Some seismic waves reflect off interior density changes; others are refracted, or bent, as they travel through Earth. Thus, the distinctive ways in which seismic waves pass through Earth and the time they take to travel between two surface points help seismologists deduce the structure of Earth's interior (Figure 8.2).

Figure 8.3 illustrates the dimensions of Earth's interior in comparison to surface distances in North America to give you a sense of scale. An airplane flying from Anchorage, Alaska, to Fort Lauderdale, Florida, would travel the same distance as that from Earth's center to its surface. Note, in Florida, Earth's thin crust extends only about 30 km inland from the coast.

Earth's Core　A third of Earth's entire mass, but only a sixth of its volume, lies in its dense core. The **core** is differentiated into two regions—*inner core* and *outer core*—divided by a transition zone of several hundred kilometers (Figure 8.2). Estimated core temperatures range from 3000°C (5400°F) to as high as 6650°C (12,000°F). The inner core is thought to be solid iron and well above the melting temperature of iron at the surface. It remains solid because of tremendous pressures. The iron is not pure but probably is combined with silicon, and possibly

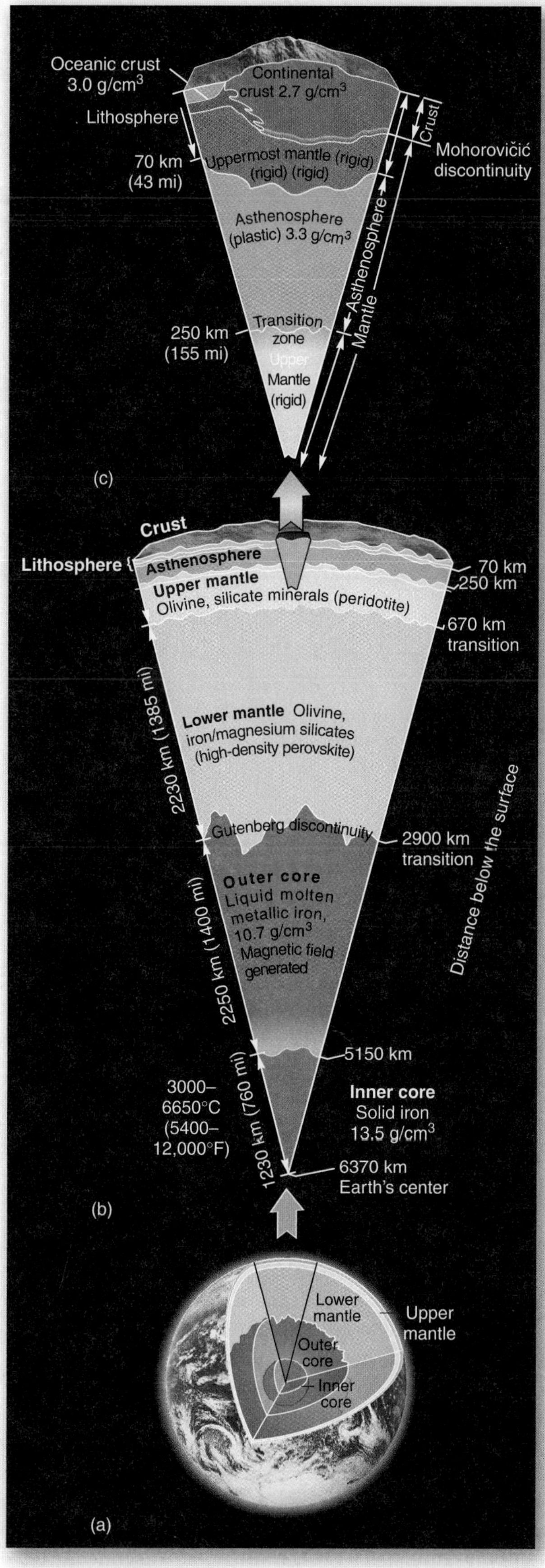

Oceanic crust
3.0 g/cm³

Continental
crust 2.7 g/cm³

Lithosphere

Mohorovičić
discontinuity

70 km
(43 mi)

Uppermost mantle (rigid)
(rigid) (rigid)

Asthenosphere
(plastic) 3.3 g/cm³

250 km
(155 mi)

Transition
zone

Upper
Mantle
(rigid)

(c)

Crust

Lithosphere {

Asthenosphere

Upper mantle
Olivine, silicate minerals (peridotite)

70 km
250 km

670 km
transition

Lower mantle Olivine,
iron/magnesium silicates
(high-density perovskite)

2230 km (1385 mi)

Gutenberg discontinuity

2900 km
transition

Outer core
Liquid molten
metallic iron,
10.7 g/cm³
Magnetic field
generated

2250 km (1400 mi)

Distance below the surface

5150 km

3000–
6650°C
(5400–
12,000°F)

1230 km (760 mi)

Inner core
Solid iron
13.5 g/cm³

6370 km
Earth's center

(b)

Lower
mantle

Upper
mantle

Outer
core

Inner
core

(a)

FIGURE 8.2 Earth in cross section.
(a) Cutaway showing Earth's interior. (b) Earth's interior in cross section, from inner core to the crust. (c) Detail of the structure of the lithosphere and its relation to the asthenosphere. A revised estimate of Earth's mass (or weight) made in 2000 set it at 5.972 sextillion metric tons (5972 followed by 18 zeros). [Photo from NASA.]

oxygen and sulfur. The outer core is molten, metallic iron with lighter densities than the inner core.

Earth's Magnetism The fluid outer core generates at least 90% of Earth's magnetic field and the magnetosphere that surrounds and protects the surface from the solar wind. One hypothesis for Earth's magnetic field describes how Earth's rotation influences circulation patterns in the outer core. A new discovery suggests that Earth's solid inner core rotates slightly faster than the rest of the planet. This circulation generates electric currents, which in turn induce the magnetic field. In 2007, the north magnetic pole (NMP) was near 84.4° N by 121.7° W in the Canadian Arctic, and the south magnetic pole was off the coast of Wilkes Land, East Antarctica. These surface expressions of Earth's magnetic field migrate; for example, the NMP moved 1100 km (685 mi) to its present location during the last century. (See http://gsc.nrcan.gc.ca/geomag/nmp/northpole_e.php.)

An intriguing feature of Earth's magnetic field is that its polarity sometimes fades to zero and then returns to full strength with north and south magnetic poles reversed. The field does not blink on and off but instead oscillates slowly down to low intensity and then rapidly regains its full strength. This **geomagnetic reversal** took place 9 times during the past 4 million years and hundreds of times over Earth's history.

The average period of a magnetic reversal is 500,000 years; occurrences possibly vary from as short as several thousand years to as long as tens of millions of years. Given present rates of magnetic field decay over the last 150 years, we are perhaps 1000 years away from entering the next phase of field changes, although there is no expected pattern to forecast the timing. There appears to be a trend in recent geologic time toward more frequent reversals with shorter intervals. When Earth is without polarity in its magnetic field, a random pattern of magnetism results. During the transition interval of low strength, Earth's surface receives higher levels of cosmic radiation and solar particles, but note that past reversals do not correlate with species extinctions. The evolution of life has weathered many of these transitions.

The reasons for these magnetic reversals are unknown; however, the spatial patterns they create at Earth's surface give us important evidence for understanding the evolution of landmasses and the movement of the continents. When new iron-bearing rocks solidify from molten material at Earth's surface, small magnetic particles in the rocks align according to the orientation of the magnetic poles at that time. This pattern is then locked in place as the rocks cool and harden. All across Earth, rocks of comparable types and ages bear this identical record of magnetic reversals. In addition, these magnetic patterns record the migration of the continents over time in relation to Earth's magnetic poles.

Earth's Mantle A transition zone of several hundred kilometers divides the outer core from the mantle. Scientists at the California Institute of Technology analyzed more than 25,000 earthquakes and determined that this

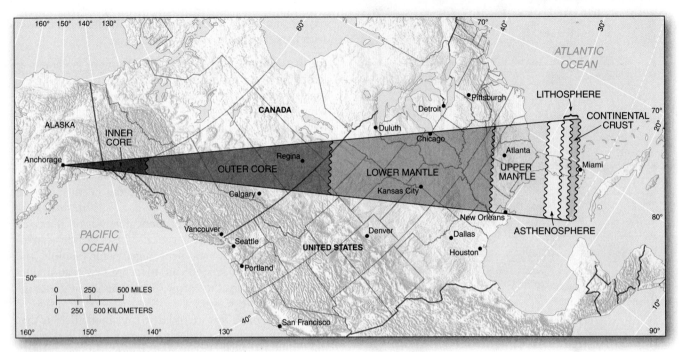

FIGURE 8.3 Distances from core to crust.
The distance from Anchorage, Alaska, to Fort Lauderdale, Florida, is the same distance as that
from Earth's center to its outer crust. Earth's cross section overlays the surface map to illustrate
Earth's radius. The continental crust thickness only extends to the outskirts of Fort Lauderdale
and its suburbs, some 30 km (18.6 mi) from the coast.

transition area is bumpy and uneven, with ragged peak-and-valley-like formations. Some of the motions in the mantle may be created by this rough texture at the *Gutenberg discontinuity* (Figure 8.2b). A *discontinuity* is a place where physical differences occur between adjoining regions in Earth's interior or surface rock, such as between the outer core and lower mantle.

The **mantle** is about 80% of Earth's total volume. It is rich in oxides and silicates of iron and magnesium (FeO, MgO, and SiO_2) and is denser and tightly packed at depth, grading to lesser densities toward the surface. Of the mantle's volume, 50% is in the lower mantle, which is composed of rock that probably has never been at Earth's surface. A broad transition zone of several hundred kilometers separates the upper mantle from the lower mantle. The entire mantle experiences a gradual temperature increase with depth.

Three fairly distinct layers divide the upper mantle. Just below the crust is a high-velocity zone approximately 45–70 km (25–43 mi) thick, so called because seismic waves transmit rapidly through this rigid, cooler layer. This *uppermost mantle*, along with the crust, makes up the *lithosphere*.

Below the lithosphere, from about 70 km down to 250 km (43–155 mi), is the **asthenosphere**, or plastic layer, from the Greek *asthenos*, meaning "weak." It contains pockets of increased heat from radioactive decay and is susceptible to slow convective currents in these hotter (and therefore less dense) materials. Because of this dynamic condition, the asthenosphere is the least rigid region of the mantle. About 10% of the asthenosphere is molten in asymmetrical patterns and hot spots. The

resulting slow movement in this zone disturbs the overlying crust and creates tectonic activity—folding, faulting, and general deformation of surface rocks. Below the asthenosphere resides the third layer, the rest of the *upper mantle*. Here the rocks are solid again, a zone of greater density, high seismic velocity, and increasing pressures.

The depth affected by convection currents is the subject of much scientific research. One body of evidence states that mixing occurs throughout the entire mantle, upwelling from great depths at the core-mantle boundary, sometimes in small blobs of material and other times in "megablobs" of mantle convecting toward and away from the crust. Another view states that mixing in the mantle is layered, segregated above and below the 670-km boundary. Presently, evidence indicates some truth in both positions. As an example, there are hot spots on Earth, such as those under Hawai'i and Iceland, that appear to be at the top of tall plumes of rising mantle rock that are anchored deep in the lower mantle.

Lithosphere and Crust The lithosphere includes the entire **crust**, which represents only 0.01% of Earth's mass, and the uppermost mantle down to a depth of about 70 km (43 mi). The boundary between the crust and the high-velocity portion of the lithospheric upper mantle is another discontinuity called the **Mohorovičić discontinuity**, or **Moho** for short, named for the Yugoslavian seismologist who determined that seismic waves change at this depth, owing to sharp contrasts of materials and densities.

Figure 8.2c illustrates the relationship of the crust to the rest of the lithosphere and the asthenosphere below. Crustal areas below mountain masses extend lower into the asthenosphere, perhaps to 50–60 km (31–37 mi), whereas the crust beneath continental interiors averages about 30 km (19 mi) in thickness, and oceanic crust averages only 5 km (3 mi). Drilling into the mantle through the crust remains an elusive scientific goal; see News Report 8.2 for more on these efforts.

The composition and texture of continental and oceanic crusts are quite different, and this difference is a key to the concept of drifting continents. The oceanic crust is denser than continental crust. In collisions, the denser oceanic material plunges beneath the lighter, more buoyant continental crust.

- *Continental crust* is essentially **granite**; it is crystalline and high in silica, aluminum, potassium, calcium, and sodium. (Sometimes continental crust is called *sial*, shorthand for *si*lica and *al*uminum.) Continental crust is relatively low in density, averaging 2.7 g/cm^3 (or 2700 kg/m^3). Compare this with other densities given in Figure 8.2.

NEWS REPORT 8.2

Drilling the Crust to Record Depths

Scientists wanting to sample mantle material directly have unsuccessfully tried for decades to penetrate Earth's crust to the Moho discontinuity. The longest-lasting deep-drilling attempt is on the northern Kola Peninsula near Zapolyarny, Russia, 250 km north of the Arctic Circle—the *Kola Borehole* (*KSDB*). Twenty years of high-technology drilling (1970–1989) produced a hole 12.262 km deep (7.6 mi, or 40,230 ft), purely for exploration and science. Crystalline rock 1.4 billion years old at 180°C (356°F) was reached. The site has other active boreholes, and the fifth is underway. (See an analysis log at **http://www.icdp-online.de/**.) A record holder for depth of a gas well is in Oklahoma; it was stopped at 9750 m (32,000 ft)

when the drill bit ran into molten sulfur.

Oceanic crust is thinner than continental crust and is the object of several drilling attempts. The Integrated Ocean Drilling Program (IODP), a cooperative effort between the governments of Japan and the United States, replaced the former ODP effort directed by Texas A & M University in 2003. However, the university program is still the Science Operator of the research ship *Joides Resolution* and lead coordinator (Figure 8.2.1a).

Previously, ODP drilled a 2.5-km-deep hole in an oceanic rift near the Galápagos Islands from 1975 to 1993, still the record. An ocean floor of distinct layers of lava and deeper magma chambers was found. The

Moho and Earth's mantle remain untapped. The new IODP plans further drilling at the site. See **http://www-odp.tamu.edu/** or **http://www.oceandrilling.org/** for the IODP home page and information.

The first expeditions of the largest deep-ocean drilling ship, Japan's *Chikyu*, were in 2006 and 2007. The ship is the first capable of high-latitude drilling in the Arctic. The *Chikyu* is "riserless," in that it uses seawater as the primary drilling fluid (8.2.1b). The new IODP ship is able to drill to 7 km, more than 3 times the *JOIDES Resolution* limits. This *JOIDES*, workhorse of the ODP, is being refurbished and converted to upgraded riserless status and reenters active duty in late 2009 for the new IODP program.

(a)

(b)

FIGURE 8.2.1 Ocean drilling ship. (a) A modern ocean-floor drilling ship, the *JOIDES Resolution*, operated by the International Ocean Drilling Program. Since it began operations in 1984, through 2003, *JOIDES* has spent more than 5000 days at sea, traveled more than half a million kilometers, drilled 1555 holes into the seafloor, and recovered hundreds of kilometers of core for analysis. (b) The new *Chikyu* began drilling operations in 2006. [Photos by IODP/Texas A & M University.]

- *Oceanic crust* is **basalt**; it is granular and high in silica, magnesium, and iron. (Sometimes oceanic crust is called *sima*, shorthand for *si*lica and *ma*gnesium.) It is denser than continental crust, averaging 3.0 g/cm^3 (or 3000 kg/m^3).

Buoyancy is the principle that something less dense, such as wood, floats in something denser, such as water. The principles of buoyancy and balance were combined in the 1800s into the important principle of **isostasy**. Isostasy explains certain vertical movements of Earth's crust. Think of Earth's outer crust as floating on the denser layers beneath, much as a boat floats on water. With a greater load (for example, glaciers, sediment, mountains), the crust tends to ride lower in the asthenosphere. Without that load (for example, when a glacier melts), the crust rides higher—an uplift known as an *isostatic rebound*. Thus, the entire crust is in a constant state of compensating adjustment, or isostasy, slowly rising and sinking in response to its own weight as it is pushed, and dragged, and pulled over the asthenosphere (Figure 8.4). For example, with the rapid loss of glacial ice

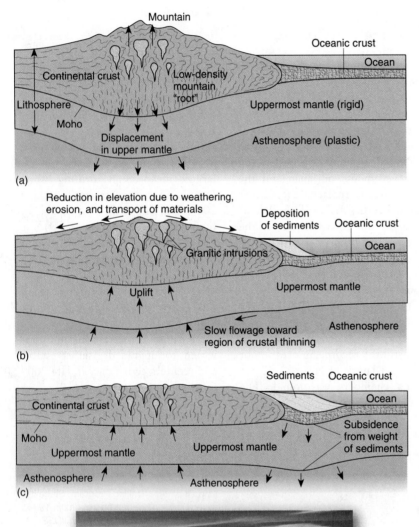

(a)

(b)

(c)

(d)

FIGURE 8.4 Isostatic adjustment of the crust.
Earth's entire crust is in a constant state of compensating adjustment, as suggested by these three sequential stages: (a) The mountain mass slowly sinks, displacing mantle material. (b) Because of the loss of mass through erosion and transportation, the crust isostatically adjusts upward, and sediments accumulate in the ocean. (c) As the continent thins and rises, the heavy sediment load offshore begins to deform the lithosphere beneath the ocean. (d) The melting of ice from the last ice age and losses of overlying sediments are thought to produce an ongoing isostatic uplift of portions of the Sierra Nevada batholith. [Photo by author.]

NEWS REPORT 8.3

Isostatic Rebound in Alaska

The retreat of glacial ice following the last ice-age cycle unloaded much weight off the crust in Alaska, among those regions affected. Researchers from the Geophysical Institute at the University of Alaska–Fairbanks, using an array of global positioning systems (GPS) receivers, measured isostatic rebound of the crust following glacial ice losses. They expected to find a slowed rate of crustal rebound in southeastern Alaska as compared to the more rapid response in the distant past when the ice first retreated.

Instead, they detected some of the most rapid vertical motion on Earth, averaging about 36 mm (1.42 in.) per year. Scientists attribute this isostatic rebound to the loss of glaciers in the region, especially the areas from Yakutat Bay and the Saint Elias Mountains in the north, through Glacier Bay, Juneau, and on south through the inland passage. Only about 1 mm of this uplift is from postglacial response. This rapid rebound is attributable to glacial melt and retreat over the past 150

years and correlates with accelerating ice losses and record temperatures across Alaska (see News Report 14.1) as climate continues its warming shift. For instance, Glacier Bay experienced a retreat of ice of more than 97 km (60 mi) since 1794. The land that was beneath this ice has isostatically rebounded 5.5 m (18 ft). The vertical action appears to be affecting some of the fault systems in the region, although the extent of this and its consequences are unknown at this writing.

across southern Alaska since the mid-1990s, scientists are measuring relatively rapid isostatic uplift of the region—some 36 mm (1.4 in.) a year. This interesting development in Alaska relates this concept of isostatic rebound and climate change, as described in News Report 8.3.

Earth's crust is the outermost irregular shell that resides restlessly on a dynamic and diverse interior. Let us now examine the processes at work on this crust and the variety of rock types that compose the landscape.

The Geologic Cycle

Earth's crust is in an ongoing state of change, being formed, deformed, moved, and broken by physical, chemical, and biological processes. The endogenic (internal) system is at work building landforms while the exogenic (external) system is wearing them down. This vast give-and-take at the Earth–atmosphere interface is the **geologic cycle**. The overall cycle is fueled by Earth's internal heat and by solar energy, influenced by the leveling force of Earth's gravity. Figure 8.5 illustrates the geologic cycle, combining many of the topics discussed in this text.

As you can see in the figure, the geologic cycle is composed of three principal cycles—hydrologic, rock, and tectonic. The *hydrologic cycle*, along the top of Figure 8.5, through erosion, transportation, and deposition, processes Earth materials with the chemical and physical action of water, ice, and wind. The *rock cycle* produces the three basic rock types found in the crust—igneous, metamorphic, and sedimentary. The *tectonic cycle* brings heat energy and new material to the surface and recycles material, creating movement and deformation of the crust.

The Rock Cycle

Only eight natural elements compose 99% of Earth's crust, and of these only two—oxygen and silicon—account for 74.3% (Table 8.1). Oxygen is the most reactive gas in the lower atmosphere, readily combining with other elements. For this reason, the percentage of oxygen is higher in the crust than its concentration in the atmosphere (21%). The internal differentiation process in which less dense elements migrated toward the surface explains the relatively large percentages of lightweight elements such as silicon and aluminum in the crust.

A **mineral** is an element or combination of elements that forms an inorganic natural compound. A mineral can be described with a specific symbol or chemical formula and possesses specific qualities, including a crystalline structure. *Mineralogy* is the study of the composition, properties, and classification of minerals (see http://webmineral.com/).

One of the most widespread mineral families on Earth is the *silicates*, because silicon (Si) readily combines with oxygen and other elements. The common mineral *quartz* is silicon dioxide and has a distinctive six-sided crystal (see photo in Table 8.1). This mineral family also includes feldspar, clay minerals, and numerous gemstones. Silicates comprise roughly 95% of Earth's crust.

Another important mineral family is the *carbonate* group, which features carbon in combination with oxygen and other elements, such as calcium, magnesium, and potassium. An example is the mineral calcite ($CaCO_3$), a form of calcium carbonate. *Oxides* are another group of minerals in which oxygen combines with metallic elements, such as iron to form hematite, Fe_2O_3. Also, there are the *sulfide* and *sulfate* groups in which sulfur compounds combine with metallic elements to form pyrite (FeS_2) and anhydrite ($CaSO_4$), respectively.

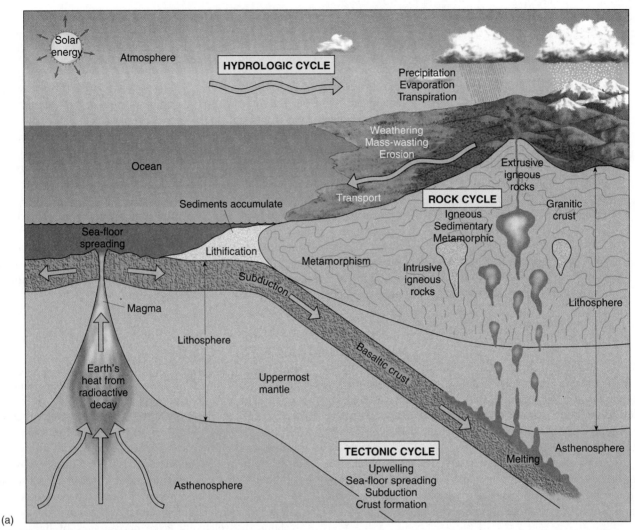

(a)

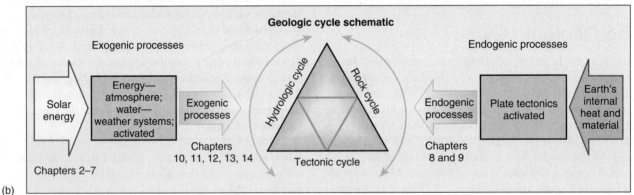

(b)

FIGURE 8.5 The geologic cycle.
(a) The geologic cycle is a model showing the interactive relation among the hydrologic cycle, rock cycle, and tectonic cycle. (b) Earth's surface is where two dynamic systems interact—the endogenic (internal) and exogenic (external).

Convection in a Lava Lamp

Of more than 4200 minerals, about 30 comprise the *rock-forming minerals* most commonly found in nature. A **rock** is an assemblage of minerals bound together (such as granite, a rock containing three minerals), or it may be a mass of a single mineral (such as rock salt). Literally thousands of rocks have been identified, all the result of three rock-forming processes: *igneous* (melted), *sedimentary* (from settling out), and *metamorphic* (altered). Figure 8.6

illustrates the interrelationships among these three processes in the *rock cycle*. The next three sections examine each rock-forming process.

Igneous Processes Rocks that solidify and crystallize from a molten state are **igneous rocks**. They form from **magma**, which is molten rock beneath the surface (hence the name *igneous*, which means "fire-formed" in Latin).

Table 8.1 Common Elements in Earth's Crust

Element	Percentage of Earth's Crust by Weight
Oxygen (O)	46.6
Silicon (Si)	27.7
Aluminum (Al)	8.1
Iron (Fe)	5.0
Calcium (Ca)	3.6
Sodium (Na)	2.8
Potassium (K)	2.6
Magnesium (Mg)	2.1
All others	1.5
Total	100.00

Note. A quartz crystal (SiO_2) consists of Earth's two most abundant elements, silicon (Si) and oxygen (O). Inset photo from *Laboratory Manual in Physical Geology*, 3rd ed., R. M. Busch, ed. © 1993 by Macmillan Publishing Co.

Magma is fluid, highly gaseous, and under tremendous pressure. It is either *intruded* into crustal rocks, known as country rock, or *extruded* onto the surface as **lava**.

The cooling history of the rock—how fast it cooled and how steadily the temperature dropped—determines its texture and degree of crystallization. Textures range from coarse-grained (slower cooling, with more time for larger crystals to form) to fine-grained or glassy (faster cooling). Most rocks in the crust are igneous, although sedimentary rocks, soil, or the ocean frequently cover them. Familiar igneous examples are granite, basalt, and rhyolite. Figure 8.7 illustrates the variety of occurrences of igneous rocks, both on and beneath Earth's surface.

Intrusive igneous rock that cools slowly in the crust forms a **pluton**, a general term for any intrusive igneous rock body regardless of size or shape. It is named after the Roman god of the underworld, Pluto. The largest pluton form is a **batholith** (Figure 8.7), defined as an irregular-shaped mass with a surface exposure greater than 100 km² (39 mi²) that has invaded layers of crustal rocks. Batholiths form the mass of many large mountain ranges, for example, the Sierra Nevada batholith in California, the Idaho batholith, and the Coast Range batholith of Canada and extreme northern Washington state (Figure 8.8a, p. 278).

Smaller plutons include the magma conduits of ancient volcanoes that have cooled and hardened. Those that form parallel to layers of sedimentary rock are *sills*; those that cross layers of the rock they invade are *dikes*. You see these two forms in Figure 8.7 at Petermann Island, Antarctica. Magma also can bulge between rock strata and produce a lens-shaped body called a *laccolith*, a type of sill. In addition, magma conduits themselves may solidify in roughly cylindrical forms that stand starkly above the landscape when finally exposed by weathering and erosion. Shiprock *volcanic neck* in New Mexico is such a feature, rising 518 m (1700 ft) above the surrounding plain, as suggested by the art in Figure 8.7 and shown in the inset photos; note the radiating dikes in the aerial photo. Weathering action of air, water, and ice can expose all of these intrusive forms.

Basalt is the most common fine-grained extrusive igneous rock here on Earth and in the Solar System. It makes up the bulk of the ocean floor, constituting 71% of Earth's surface. Lava flows such as those on the big island of Hawai'i cool to form basalt (Figure 8.8b).

Sedimentary Processes Most sedimentary rocks are derived from existing rocks or from organic materials such as bone and shell. The exogenic processes of weathering and erosion provide the raw materials needed to form sedimentary rocks. Bits and pieces of former rocks—principally quartz, feldspar, and clay minerals—are eroded and then mechanically transported (by water, ice, wind, and gravity) to other sites, where they are deposited. In addition, some minerals go into solution and form sedimentary deposits by precipitating from those solutions; this is an important process in the oceanic environment.

Various sedimentation forms are created in different environments—deserts, glaciers, beaches, the tropics, and so on. Characteristically, sedimentary rocks are laid down by wind, water, and ice in horizontally layered beds. These layered strata are important records of past ages, and **stratigraphy** is the study of their sequence (superposition), thickness, and spatial distribution—clues to the age and origin of the rocks. Figure 8.9a (p. 279) is a sandstone sedimentary rock in a desert landscape. Note the various layers in the formation and how differently they resist erosional processes. The various layers visible in the rock formation tell the story of past floods and wetter times, and past droughts and drier times. Ancient sand dunes are hardened near the top of the rock. Figure 8.9c shows sedimentary rock strata on Mars that imply water deposition of sediments sometime in the distant Martian past. Findings from the *Phoenix* lander in 2008 confirmed clays and limestone that could only be formed in the presence of water.

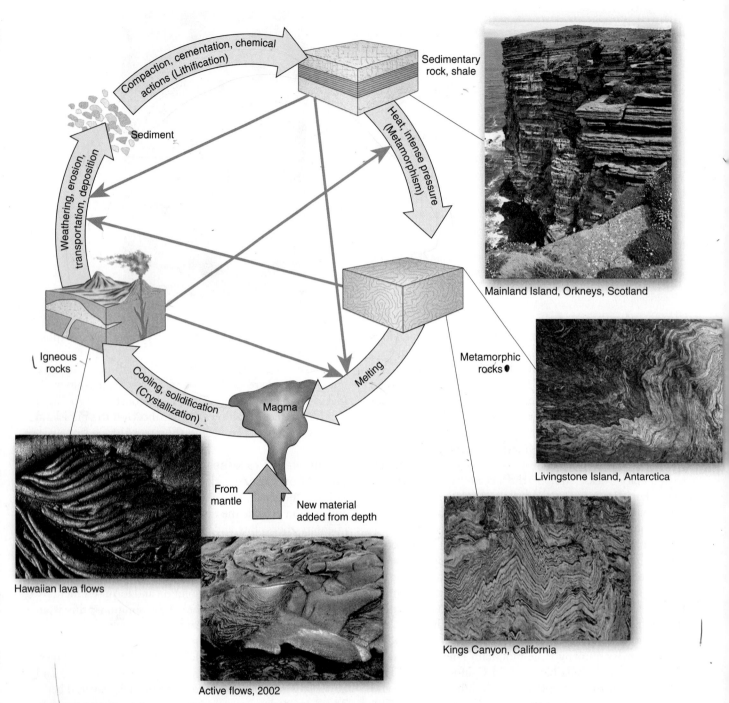

Sedimentary rock, shale

Compaction, cementation, chemical actions (Lithification)

Sediment

Heat, intense pressure (Metamorphism)

Weathering, erosion, transportation, deposition

Mainland Island, Orkneys, Scotland

Metamorphic rocks

Igneous rocks

Cooling, solidification (Crystallization)

Melting

Magma

Livingstone Island, Antarctica

From mantle

New material added from depth

Kings Canyon, California

Hawaiian lava flows

Active flows, 2002

FIGURE 8.6 The rock cycle.
A rock-cycle schematic demonstrates the relation among igneous, sedimentary, and metamorphic processes. The arrows indicate that each rock type can enter the cycle at various points and be transformed into other rock types. Each inset photo example is labeled as to location. [Art adapted by permission from R. M. Busch, ed., *Laboratory Manual in Physical Geology*, 3rd ed., © 1993 by Macmillan Publishing Co. All photos by Bobbé Christopherson.]

 NOTEBOOK   The Rock Cycle

The two primary sources of sedimentary rocks are the mechanically transported bits and pieces of former rock, the *clastic sediments*, and the minerals in solution, the *chemical sediments*. Clastic sedimentary rocks, such as sandstone, are derived from weathered and fragmented rocks that are further worn in transport—everything from boulders to microscopic clay particles. The process

of cementation, compaction, and hardening of sediments into **sedimentary rocks** is **lithification**. Various cements fuse rock particles together; lime ($CaCO_3$, or calcium carbonate) is the most common, followed by iron oxides (Fe_2O_3) and silica (SiO_2). Drying (dehydration), heating, or chemical reactions also unite particles.

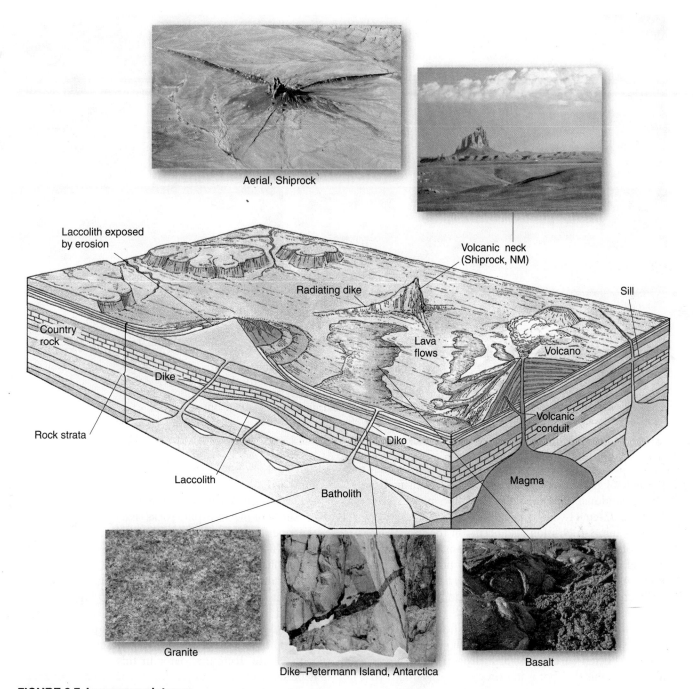

Aerial, Shiprock

Volcanic neck
(Shiprock, NM)

Sill

Laccolith exposed
by erosion

Radiating dike

Country
rock

Lava
flows

Volcano

Dike

Volcanic
conduit

Rock strata

Dike

Laccolith

Batholith

Magma

Granite

Dike–Petermann Island, Antarctica

Basalt

FIGURE 8.7 Igneous rock types.
The variety of occurrences of igneous rocks, both intrusive (below the surface) and extrusive (on
the surface). Photographs show samples of White Mountain Conway granite and an Antarctic dike
(intrusive), Hawaiian basalt (extrusive), and the Shiprock volcanic neck. [Photos: Granite, basalt
and Shiprock aerial photos by Bobbé Christopherson; Shiprock volcanic neck photo by author.]

 ANIMATION

Formation of Intrusive
Igneous Features

Chemical sedimentary rocks are not formed from physical pieces of broken rock, but instead are dissolved, transported in solution, and chemically precipitated out of solution (they are essentially nonclastic). The most common chemical sedimentary rock is **limestone**, which is lithified *calcium carbonate*, ($CaCO_3$). A similar form is *dolomite*, which is lithified *calcium-magnesium carbonate*, $CaMg(CO_3)_2$. About 90% of all limestone comes from organic sources; it is biochemical—derived from shell and

bone produced by biological activity. Once formed, these rocks are vulnerable to chemical weathering, which produces unique landforms, as discussed in the weathering section of Chapter 10 and as shown in Figure 8.9b.

Chemical sediments also form from inorganic sources when water evaporates and leaves behind a residue of salts. These **evaporites** may exist as common salts, such as gypsum or sodium chloride (table salt), to name only two, and often appear as flat, layered deposits across a dry

(a)

(b)

FIGURE 8.8 Intrusive and extrusive rocks.
(a) Exposed granites of the Sierra Nevada batholith. The boulders sitting on the granite are erratics left behind by glacial ice that melted thousands of years ago. (b) Basaltic lava flows on Hawai'i. The glow is from a skylight into an active lava tube where molten lava is visible; the shiny surface is where lava recently flowed out of the skylight. (Do you see a suggestion of a dragon head and glowing red eye in the lava sculpture?) [Photos by Bobbé Christopherson.]

landscape. This process is dramatically demonstrated in the pair of photographs in Figure 8.10, taken in Death Valley National Park one day and one month after a record 2.57-cm (1.01-in.) rainfall.

Chemical deposition also occurs in the water of natural hot springs from chemical reactions between minerals and oxygen; this is a sedimentary process related to *hydrothermal activity*. A deposit of *travertine*, a form of calcium carbonate, at Mammoth Hot Springs in Yellowstone National Park is an example (Figure 8.11a). Hydrothermal activity also occurs on the ocean floor along the rift valleys formed by sea-floor spreading. These "black smokers" belch dark clouds of hydrogen sulfides, minerals, and metals that hot water (in excess of 380°C; 716°F) leached from the basalt. In contact with seawater, metals and minerals precipitate into deposits (Figure 8.11b). Along the Mid-Atlantic Ridge, one region of hydrothermal activity has 30- to 60-m-tall vents of calcium carbonate that are at least 30,000 years old and still active. This area is nicknamed "The Lost City" because of the way the towering chimneys look from the observation submersibles.

Metamorphic Processes   Any rock, either igneous or sedimentary, may be transformed into a **metamorphic rock** by going through profound physical and/or chemical changes under pressure and increased temperature. (The name *metamorphic* comes from the Greek, meaning to "change form.") Metamorphic rocks generally are more compact than the original rock and therefore are

harder and more resistant to weathering and erosion (Figure 8.12a).

The ancient roots of mountains are composed predominantly of metamorphic rocks. Exposed at the bottom of the inner gorge of the Grand Canyon in Arizona, a Precambrian metamorphic rock called the Vishnu Schist is a remnant of such an ancient mountain root (Figure 8.12b).

Several conditions can cause metamorphism, particularly when subsurface rock is subjected to high temperatures and high compressional stresses occurring over millions of years. Igneous rocks are compressed during collisions of portions of Earth's crust (described under the section "Plate Tectonics" later in this chapter), or they may be sheared and stressed along earthquake fault zones. Sometimes they simply are crushed under great weight when a crustal area is thrust beneath other crust.

Another metamorphic condition occurs when sediments collect in broad depressions in Earth's crust and, because of their own weight, create enough pressure in the bottommost layers to transform the sediments into metamorphic rock, termed *regional metamorphism*. Also, molten magma rising within the crust may "cook" adjacent rock, a process called *contact metamorphism*.

Metamorphic rocks may be changed both physically and chemically from the original rocks. If the mineral structure demonstrates a particular alignment after metamorphism, the rock is *foliated*, and some minerals may appear as wavy striations (streaks or lines) in the new rock. In contrast, parent rock with a more homogeneous (evenly mixed) makeup may produce a *nonfoliated rock*.

(a)

(b)

Sandstone

Limestone

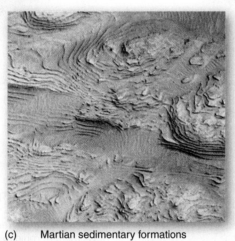

(c)   Martian sedimentary formations

FIGURE 8.9 Sedimentary rock types.
Two types of sedimentary rock. (a) Sandstone formation with sedimentary strata subjected to differential weathering; note weaker underlying siltstone. (b) A limestone landscape, formed by chemical sedimentary processes, in south-central Indiana; inset samples have whole shells and some clasts (parts) cemented together. (c) Ancient sediment deposition through repeated cycles of erosion and deposition is evident in the Martian western Arabia Terra (8° N and 7° W). [(a) Photos by author; (b) photos by Bobbé Christopherson; (c) image from Mars Global Surveyor courtesy of NASA/JPL/ Malin Space Science Systems.]

(a)

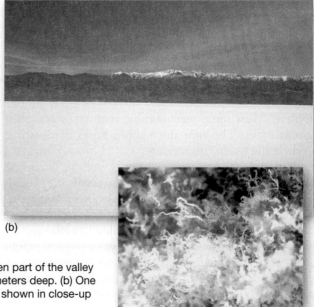

(b)

FIGURE 8.10 Death Valley, wet and dry.
A Death Valley landscape (a) one day after a record rainfall event when part of the valley was covered by several square kilometers of water only a few centimeters deep. (b) One month later the water evaporated, leaving evaporites (borated salts), shown in close-up inset photo, on the same playa (lake bed). [Photos by author.]

(a)

(b)

FIGURE 8.11 Hydrothermal deposits.
(a) Mammoth Hot Springs, Yellowstone National Park, is an example of a hydrothermal deposit principally composed of travertine ($CaCO_3$) deposited as a chemical precipitate from heated spring water as it evaporated. (b) A hydrothermal vent with black smokers along the Mid-Atlantic Ridge. [(a) Photo by author; (b) image courtesy of Scripps Oceanographic Institute.]

(a)

(b)

FIGURE 8.12 Metamorphic rocks.
(a) A metamorphic rock outcrop in Greenland, the Amitsoq Gneiss, at 3.8 billion years old, is the second-oldest rock found on Earth. (b) Precambrian Vishnu Schist composes these rock cliffs in the inner gorge of the Grand Canyon in northern Arizona. Despite the fact that the schist is harder than steel, the Colorado River has cut down into the uplifted Colorado Plateau, exposing and eroding these ancient rocks. [Photos by (a) Kevin Schafer/Peter Arnold, Inc.; (b) author.]

ANIMATION
Foliation (Metamorphic Rock)

Table 8.2 lists some metamorphic rock types and their parent rocks. The four inset photographs demonstrate foliated and nonfoliated textures.

On the Isle of Lewis (Outer Hebrides), northwest of the Scottish coast, people constructed the Standing Stones of Calanais (Callanish) beginning approximately 5000 years ago (Table 8.2, far-right inset). These Neolithic people used the metamorphic Lewisian gneiss. This gneiss rock dates at 3.1 billion years old and is foliated, with the foliations aligned along the vertical rock structure, as arranged by the ancient peoples in their monument. The standing stone in the photo is about 3.5 m (11.5 ft) tall.

You have seen how the rock-forming processes yield the igneous, sedimentary, and metamorphic materials of Earth's crust. Next we look at how vast *tectonic processes* press, push, and drag portions of the crust in large-scale movements, causing the continents to *drift*.

Plate Tectonics

Have you ever looked at a world map and noticed that a few of the continental landmasses appear to have matching shapes like pieces of a jigsaw puzzle—particularly South America and Africa? The incredible reality is that

Table 8.2   Metamorphic Rocks

Parent Rock	Metamorphic Equivalent	Texture
Shale (clay minerals)	Slate	Foliated
Granite, slate, shale	Gneiss	Foliated
Basalt, shale, peridotite	Schist	Foliated
Limestone, dolomite	Marble	Nonfoliated
Sandstone	Quartzite	Nonfoliated

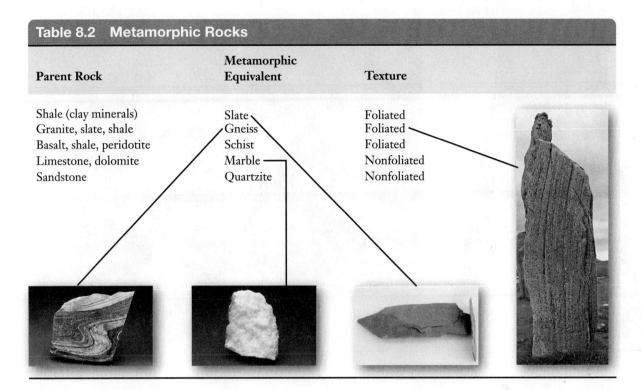

the continental pieces once did fit together! Continental landmasses not only migrated to their present locations but continue to move at speeds up to 6 cm (2.4 in.) per year. We say that the continents are *adrift* because convection currents in the asthenosphere and upper mantle are dragging them around. The key point is that the arrangement of continents and oceans we see today is not permanent but is in a continuing state of change.

A continent such as North America is actually a collage of crustal pieces and fragments that have migrated to form the landscape we know. Through a historical chronology, let us trace the discoveries that produced this major revolution in science of *plate tectonics*. The facts that the continents drift and the young seafloor spreads from rifts are key discoveries in the tectonic cycle that came to light only in the twentieth century.

A Brief History

As early mapping gained accuracy, some observers noticed a symmetry to the coastlines. Abraham Ortelius (1527–1598) described the apparent fit of some continental coastlines in his *Thesaurus Geographicus* (1596). In 1620, the English philosopher Sir Francis Bacon noted gross similarities between the shapes of Africa and South America (although he did not suggest that they had drifted apart). Others wrote—unscientifically—about such apparent relationships, but it was not until much later that a valid explanation was proposed.

In 1912, German geophysicist and meteorologist Alfred Wegener publicly presented in a lecture his idea that Earth's landmasses migrate. His book *Origin of the Continents and Oceans* appeared in 1915. Wegener today is

regarded as the father of this concept, which he called *continental drift*. But scientists at the time, knowing little of Earth's interior structure and bound by an inertial mindset as to how continents and mountains were formed, were unreceptive to Wegener's revolutionary proposal and, in a nonscientific spirit, rejected it outright.

Wegener thought that all landmasses were united in one supercontinent approximately 225 million years ago, during the Triassic period—see Figure 8.15b. (As you read this, it may be helpful to refer to the geologic time scale in Figure 8.1.) This one landmass he called **Pangaea**, for "all Earth." Although his initial model kept the landmasses together far too long, and his idea about the driving mechanism was incorrect, Wegener's overall configuration of Pangaea was right.

To come up with his Pangaea fit, he began studying the research of others, specifically the geologic record, the fossil record, and the climatic record for the continents. He concluded that South America and Africa were related in many complex ways. He also concluded that the large midlatitude coal deposits, which stretch from North America to Europe to China and date to the Permian and Carboniferous periods (245–360 million years ago), existed because these regions once were more equatorial in location and therefore covered by lush vegetation that became coal.

As modern scientific capabilities built the case for continental drift, the 1950s and 1960s saw a revival of interest in Wegener's concepts and, finally, confirmation. Aided by an avalanche of discoveries, the theory today is universally accepted as an accurate model of the way Earth's surface evolves, and virtually all Earth scientists accept the fact that continental masses move about in dramatic ways.

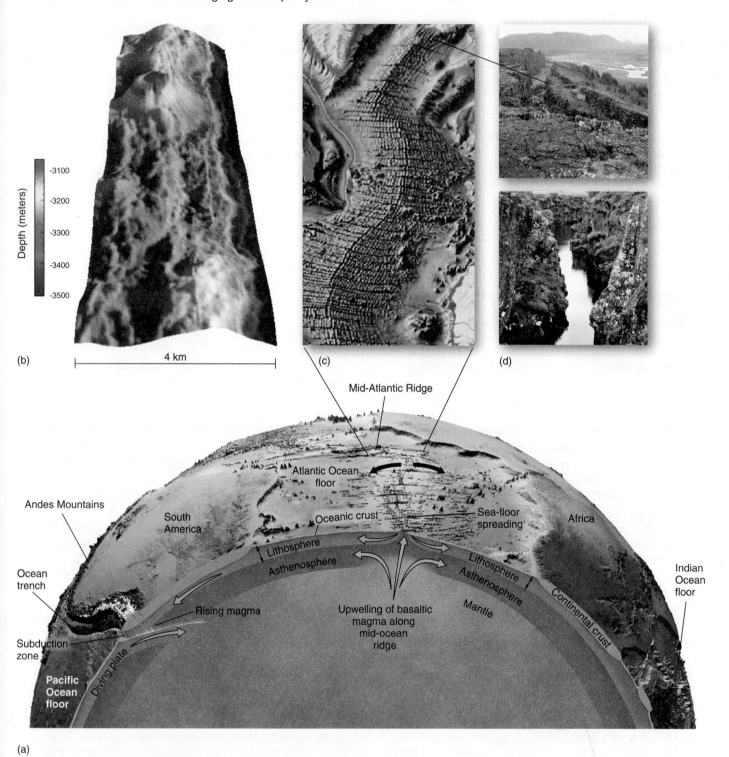

(b)

Depth (meters)

-3100
-3200
-3300
-3400
-3500

4 km

(c)

(d)

Mid-Atlantic Ridge

Atlantic Ocean floor

Andes Mountains

South America

Ocean trench

Subduction zone

Pacific Ocean floor

Oceanic crust

Lithosphere

Asthenosphere

Rising magma

Diving plate

Sea-floor spreading

Upwelling of basaltic magma along mid-ocean ridge

Lithosphere

Asthenosphere

Mantle

Africa

Continental crust

Indian Ocean floor

(a)

FIGURE 8.13 Crustal movements.
(a) Sea-floor spreading, upwelling currents, subduction, and plate movements, shown in cross section. Arrows indicate the direction of the spreading. (b) A 4-km-wide image of the Mid-Atlantic Ridge, showing linear faults, a volcanic crater, a rift valley, and ridges. The image was taken by the *TOBI* (towed ocean-bottom instrument) at approximately 29° N latitude. (c) Detail from Tharp's ocean-floor map. (d) The Mid-Atlantic rift surfaces through Iceland. This lichen- and moss-covered basalt is at Thingvellir in southeastern Iceland, where the North American and European plates of Earth's crust meet. [(b) image courtesy of D. K. Smith, Woods Hole Oceanographic Institute, Woods Hole, Massachusetts. All rights reserved; (c) inset derived from Marie Tharp's *Floor of the Oceans*; and (d) photos by Bobbé Christopherson.]

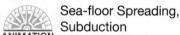

ANIMATION

Sea-floor Spreading, Subduction

Tectonic, from the Greek *tektonikùs*, meaning "building" or "construction," refers to changes in the configuration of Earth's crust as a result of internal forces. **Plate tectonic** processes include upwelling of magma, crustal plate movements, earthquakes, volcanic activity, subduction, warping, folding, and faulting of the crust.

Sea-Floor Spreading and Production of New Crust

The key to establishing the fact of continental drift was a better understanding of the seafloor. The seafloor has a remarkable feature: an interconnected mountain chain (ridge) some 64,000 km (40,000 mi) in extent and averaging more than 1000 km (600 mi) in width. A striking map of this great undersea mountain chain opens Chapter 9. How did this mountain chain get there?

In the early 1960s, geophysicists Harry H. Hess and Robert S. Dietz proposed **sea-floor spreading** as the mechanism that builds this mountain chain and drives continental movement. Hess said that these submarine mountain ranges, the **mid-ocean ridges**, were the direct result of upwelling flows of magma from hot areas in the upper mantle and asthenosphere and perhaps deeper sources. When mantle convection brings magma up to the crust, the crust is fractured and the magma spills out and cools to form new seafloor, building the ridges and spreading laterally. Figure 8.13 shows that the ocean floor is rifted and scarred along mid-ocean ridges; the arrows indicate the direction in which the seafloor spreads. Figure 8.13d shows a portion of the mid-ocean ridge system that surfaces at Thingvellir,

Iceland, forming these rifts—the North American plate along one side and the European plate on the other.

As new crust is generated and the seafloor spreads, the alignment of the magnetic field in force at the time dictates the alignment of magnetic particles in the cooling rock, creating a kind of magnetic tape recording in the seafloor. Each magnetic reversal and reorientation of Earth's polarity is recorded in the oceanic crust. Figure 8.14 illustrates just such a recording, from the Mid-Atlantic Ridge near Iceland. Note the mirror images that develop on either side of the sea-floor rift as a result of these magnetic reversals recorded in the rock.

If you measure the same distance from the rift to either side, you find the same orientation of magnetic content in the rock. Thus, the periodic reversals of Earth's magnetic field have proven a valuable clue to understanding sea-floor spreading. Scientists were provided with a global pattern of magnetic stripes to use in fitting together pieces of Earth's crust.

As the age of the seafloor was determined from these sea-floor recordings of Earth's magnetic-field reversals and other measurements, the complex harmony of the two concepts of continental drift and sea-floor spreading became clearer. The youngest crust anywhere on Earth is at the spreading centers of the mid-ocean ridges. With increasing distance from these centers, Earth's surface gets older. The oldest seafloor, is in the western Pacific near Japan, farthest from a spreading center (see Figure 8.17).

Overall, the seafloor is relatively young at less than 208 m.y.a. The reason is that oceanic crust is being produced along sea-floor spreading centers and consumed

(text continued on page 286)

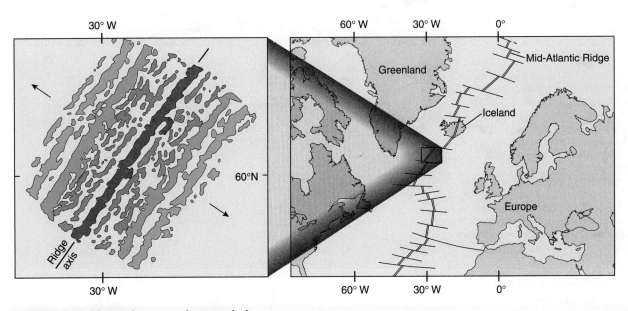

FIGURE 8.14 Magnetic reversals recorded.
Magnetic reversals recorded in the seafloor south of Iceland along the Mid-Atlantic Ridge.
[Magnetic reversals reprinted from *Deep-Sea Research* 13, J. R. Heirtzler, S. Le Pichon, and J. G. Baron. Copyright © 1966, Pergamon Press, p. 247.]

FIGURE 8.15 Continents adrift, from 465 million years ago to the present.
The formation and breakup of Pangaea and the types of motions occurring at plate boundaries. [(a) From R. K. Bambach, "Before Pangaea: The geography of the Paleozoic world," *American Scientist* 68 (1980): 26–38, reprinted by permission; (b) through (e) from Robert S. Dietz and John C. Holden, *Journal of Geophysical Research* 75, no. 26 (September 10, 1970): 4939–56, © American Geophysical Union; (f) remote-sensing image courtesy of S. Tighe, University of Rhode Island, and R. Detrick, Woods Hole Oceanographic Institute, Woods Hole, Massachusetts.]

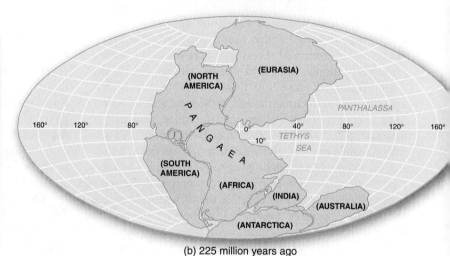

Plate Motions Through Time

ANIMATION

(a) 465 million years ago

Pangaea—all Earth. Panthalassa ("all seas") became the Pacific Ocean, and the Tethys Sea (partly enclosed by the African and the Eurasian plates) became the Mediterranean Sea; trapped portions of former ocean became the present-day Caspian Sea. The Atlantic Ocean did not exist.

Africa shared a common connection with both North and South America. Today, the Appalachian Mountains in the eastern United States and the Lesser-Atlas Mountains of northwestern Africa reflect this common ancestry; they are, in fact, portions of the same mountain range, torn thousands of kilometers apart.

(b) 225 million years ago

New seafloor that formed since the last map is highlighted (gray tint). An active spreading center rifted North America away from landmasses to the east, shaping the coast of Labrador. India was farther along in its journey, with a spreading center to the south and a subduction zone to the north; the leading edge of the India plate was diving beneath Eurasia.

The outlines of South America, Africa, India, Australia, and southern Europe appear within the continent called *Gondwana* in the Southern Hemisphere. North America, Europe, and Asia comprise *Laurasia*, the northern portion.

(c) 135 million years ago

Motions at Plate Boundaries

ANIMATION

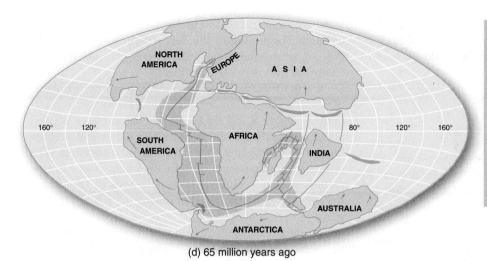

(d) 65 million years ago

Sea-floor spreading along the Mid-Atlantic Ridge grew some 3000 km (almost 1900 mi) in 70 million years. Africa moved northward about 10° in latitude, leaving Madagascar split from the mainland and opening up the Gulf of Aden. The rifting along what would be the Red Sea began. The India plate moved three-fourths of the way to Asia, as Asia continued to rotate clockwise. Of all the major plates, India traveled the farthest— almost 10,000 km.

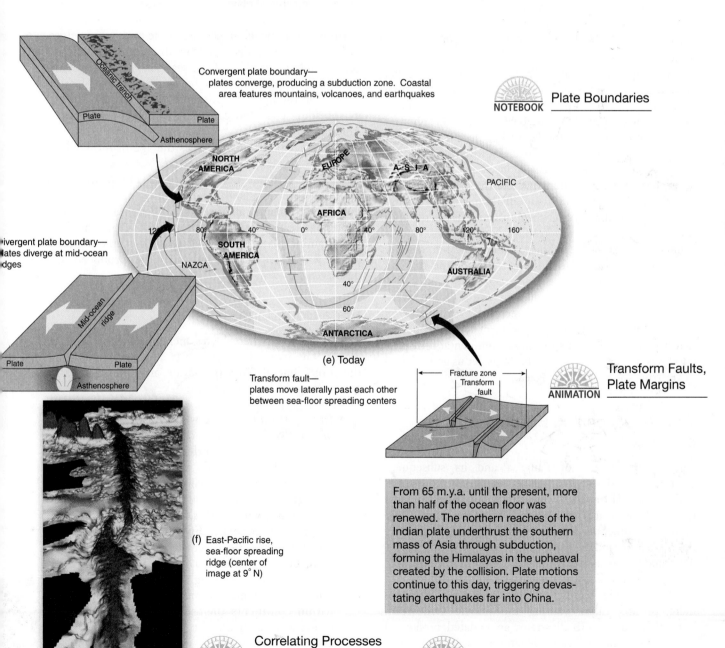

Convergent plate boundary— plates converge, producing a subduction zone. Coastal area features mountains, volcanoes, and earthquakes

NOTEBOOK Plate Boundaries

Divergent plate boundary— plates diverge at mid-ocean ridges

(e) Today

Transform fault— plates move laterally past each other between sea-floor spreading centers

Transform Faults, Plate Margins
ANIMATION

(f) East-Pacific rise, sea-floor spreading ridge (center of image at 9° N)

From 65 m.y.a. until the present, more than half of the ocean floor was renewed. The northern reaches of the Indian plate underthrust the southern mass of Asia through subduction, forming the Himalayas in the upheaval created by the collision. Plate motions continue to this day, triggering devastating earthquakes far into China.

Correlating Processes and Plate Boundaries
ANIMATION

Forming a Divergent Boundary
ANIMATION

along plate margins in areas of crustal subduction. The discovery that the seafloor is so young demolished earlier geologic thinking that the oldest rocks would be found there.

Subduction of Crust

In contrast to the upwelling zones along the mid-ocean ridges are the areas of descending crust elsewhere. We must remember that the basaltic ocean crust has a greater average density, at 3.0 g/cm^3, than continental crust, at 2.7 g/cm^3. As a result, when continental crust and oceanic crust slowly collide, the denser ocean floor dives beneath the lighter continent. This action forms a descending **subduction zone** (see Figure 8.13). The world's **oceanic trenches** are subduction zones and are the deepest features on Earth's surface. The deepest is the Mariana Trench near Guam, which descends to –11,030 m (–36,198 ft) below sea level.

The subducting slab of crust exerts a gravitational pull on the rest of the plate—an important driving force in plate motion. Through shear traction this plate motion can trigger along the base of the lithosphere a flow in the mantle material beneath, where it remelts, is recycled, and eventually migrates back toward the surface through deep fissures and cracks in crustal rock (Figure 8.13a, left). Volcanic mountains like the Andes in South America and the Cascade Range from northern California to the Canadian border form inland of these subduction zones. The fact that spreading ridges and subduction zones are areas of earthquake and volcanic activity provides important clues for the study of the drifting continents.

Sea-floor spreading, subduction, and mantle convection all are phenomena of plate tectonics, which by 1968 had become the all-encompassing term for the concepts that began with Wegener's original continental drift proposal. Using current scientific findings, let us go back and reconstruct the past and Pangaea.

The Formation and Breakup of Pangaea

The supercontinent of Pangaea and its subsequent breakup into today's continents represent only the last 225 million years of Earth's 4.6 billion years, or only the most recent 1/23 of Earth's existence. During the other 22/23 of geologic time, other things were happening, including the formation and breakup of previous supercontinents. The landmasses as we know them were unrecognizable through most of Earth's history.

Figure 8.15a begins with the pre-Pangaea arrangement of 465 m.y.a. (during the middle Ordovician Period). Figure 8.15b illustrates an updated version of Wegener's Pangaea, 225–200 m.y.a. (Triassic–Jurassic Periods). The movement of plates that occurred by 135 m.y.a. (the beginning of the Cretaceous Period) is in

Figure 8.15c. Figure 8.15d presents the arrangement from 65 m.y.a. (beginning of the Tertiary Period). Finally, the present arrangement in modern geologic time (the late Cenozoic Era) is in Figure 8.15e.

Earth's present crust is divided into at least 14 plates, of which about half are major and half are minor, in terms of area (Figure 8.16). Each of these broad plates is composed of literally hundreds of smaller pieces and perhaps dozens of microplates that have migrated together. The arrows in the figure indicate the direction in which each plate is presently moving, and the length of the arrows suggests the rate of movement during the past 20 million years.

Compare Figure 8.16 with the relative-ages map of the ocean floor in Figure 8.17. The complex harmony of the two concepts of plate tectonics and sea-floor spreading thus become clearer. The youngest crust anywhere on Earth is at the spreading centers of the mid-ocean ridges, and with increasing distance from these centers, the crust gets steadily older. The oldest seafloor is in the western Pacific near Japan, dating to the Jurassic Period. Note on the map in the figure the distance between this basin and its spreading center in the South Pacific, west of South America.

The fact that nowhere does the age of the seafloor exceed 208 m.y.a. is remarkable when you remember that Earth's age is about 4.6 billion years. The processes of sea-floor spreading and subduction of crust that produced the formation and breakup of Pangaea explain this young age. Let us take this information about the crust and examine plate boundaries, those dynamic edges of plates where crust and landscapes are produced and destroyed.

Plate Boundaries

The boundaries where plates meet clearly are dynamic places. The block diagram inserts in Figure 8.15e show the three general types of motion and interaction that occur along the boundary areas:

- *Divergent boundaries* (lower left) are characteristic of sea-floor spreading centers, where upwelling material from the mantle forms new seafloor and crustal plates spread apart ("constructional"). These are zones of tension. An example noted in the figure is the divergent boundary along the East Pacific rise, which gives birth to the Nazca plate and the Pacific plate. Figure 8.14 illustrates patterns of magnetic reversals on such a divergent plate boundary south of Iceland along the Mid-Atlantic Ridge. Although most divergent boundaries occur at mid-ocean ridges, there are a few within continents themselves. An example is the Great Rift Valley of East Africa, where continental crust is pulling apart.
- *Convergent boundaries* (upper left) are characteristic of collision zones, where areas of continental

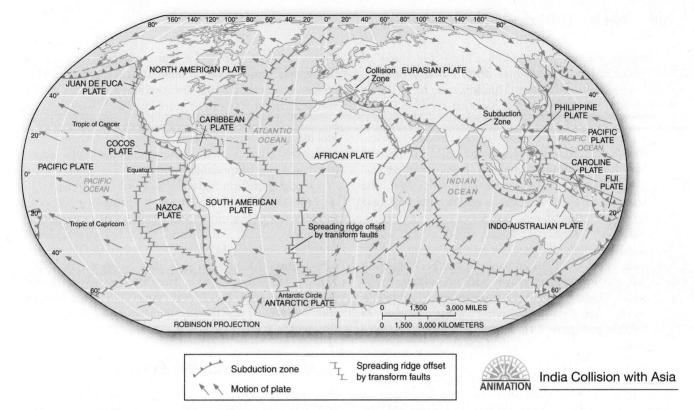

India Collision with Asia

FIGURE 8.16 Earth's major lithospheric plates and their movements.
Each arrow represents 20 million years of movement. The longer arrows indicate that the Pacific and Nazca plates are moving more rapidly than the Atlantic plates. Compare the length of these arrows with those shaded areas on Figure 8.15. [Adapted from US Geodynamics Committee.]

Plate Boundaries

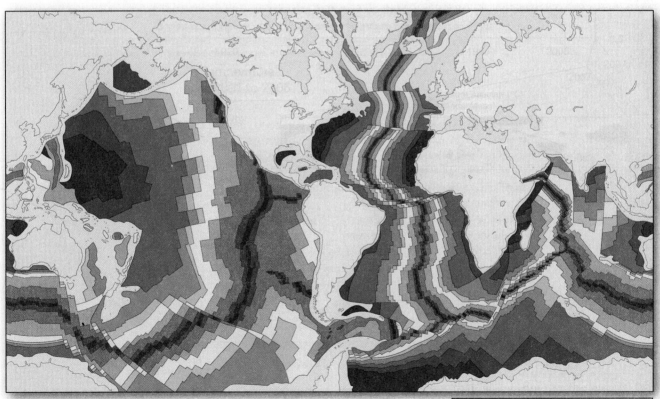

FIGURE 8.17 Relative age of the oceanic crust.
Compare the mid-rust color near the East Pacific rise in the eastern Pacific Ocean with the same mid-rust color along the Mid-Atlantic Ridge. What does the difference in width tell you about the rates of plate motion in the two locations? [Adapted with the permission of W. H. Freeman and Company from *The Bedrock Geology of the World* by R. L. Larson and others, © 1985.]

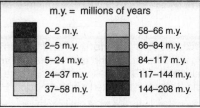

m.y. = millions of years

0–2 m.y.	58–66 m.y.
2–5 m.y.	66–84 m.y.
5–24 m.y.	84–117 m.y.
24–37 m.y.	117–144 m.y.
37–58 m.y.	144–208 m.y.

and/or oceanic crust collide. These are zones of compression and crustal loss ("destructional"). Examples include the subduction zone off the west coast of South and Central America (noted in Figure 8.15e) and the area along the Japan and Aleutian trenches. Along the western edge of South America, the Nazca plate collides with and is subducted beneath the South American plate, creating the Andes Mountains chain and related volcanoes. The collision of India and Asia is another example of a convergent boundary.

- *Transform boundaries* (lower right) occur where plates slide laterally past one another at right angles to a sea-floor spreading center, neither diverging nor converging, and usually with no volcanic eruptions. These are the right-angle fractures stretching across the mid-ocean ridge system worldwide (visible in Figure 8.16 and the Chapter 9 opening illustration).

Related to plate boundaries, another piece of the tectonic puzzle fell into place in 1965 when University of Toronto geophysicist Tuzo Wilson first described the nature of transform boundaries and their relation to earthquake activity. All the spreading centers on Earth's crust

feature these perpendicular scars, which stretch out on both sides. Some *transform faults* are a few hundred kilometers long; others, such as those along the East Pacific rise, stretch out 1000 km or more (over 600 mi). The faults generally are parallel to the direction in which the plate is moving and the fracture zone, which follows these breaks in Earth's crust, is active only along the fault section between ridges of spreading centers.

Along transform faults the motion is one of horizontal displacement—no new crust is formed or old crust subducted. The famous San Andreas fault system in California, where continental crust has overridden a transform system, is related to this type of motion (see Figure 9.12). The fault that triggered the 1995 Kobe earthquake in Japan has this type of horizontal motion, as does the North Anatolian fault system, which triggered the 1999 quake that struck Turkey (see Figure 9.13).

Earthquakes and Volcanoes

Plate boundaries are the primary location of earthquake and volcanic activity, and the correlation of these phenomena is an important aspect of plate tectonics. The next chapter discusses earthquakes and volcanic activity in

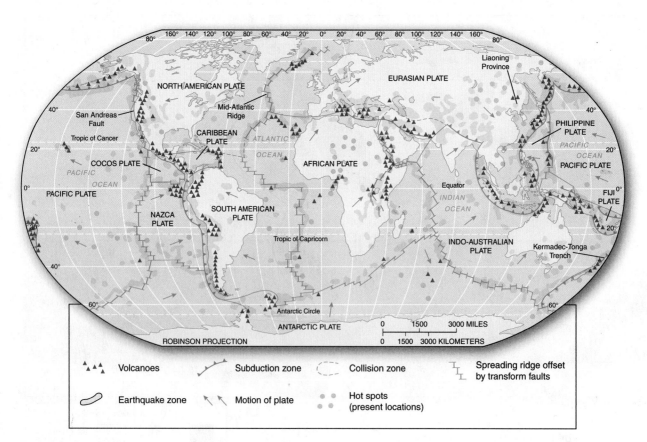

FIGURE 8.18 Earthquake and volcanic activity locations.
Principal hot spots and earthquake and volcanic activity in relation to major tectonic plate boundaries. [Earthquake and volcano data from *Earthquakes*, B. A. Bolt, © 1988 W. H. Freeman and Company; reprinted with permission. Hot spots adapted from USGS data.]

ANIMATION

Correlating Processes and Plate Boundaries

more detail. Figure 8.18 identifies earthquake zones and volcanic sites. The "ring of fire" surrounding the Pacific basin, named for the frequent incidence of volcanoes, is most evident. The subducting edge of the Pacific plate thrusts deep into the crust and upper mantle, producing molten material that makes its way back toward the surface, causing active volcanoes along the Pacific Rim. Similar processes occur at plate boundaries throughout the world.

Hot Spots

A dramatic aspect of Earth's internal dynamics is the estimated 50–100 hot spots across Earth's surface. **Hot spots** are individual sites of upwelling material, noted on Figure 8.18, which arrives at the surface in tall plumes from the mantle, sometimes producing thermal effects in groundwater and the crust. Some of these sites are developed for geothermal power (Focus Study 8.1). Hot

FOCUS STUDY 8.1

Heat from Earth—Geothermal Energy and Power

A tremendous amount of endogenic energy flows from Earth's interior toward the surface. Temperatures at the base of Earth's crust range from 200° to 1000°C (392°–1800°F). Convection and conduction transports this geothermal energy in enormous quantities from the mantle to the crust, yet geothermal power development is limited in extent to certain locations (Figure 8.1.1). Despite such site-specific limitations, applications of geothermal energy for power production are a viable energy source for the present and future.

What Is Geothermal Energy?

Geothermal energy happens when pockets of magma and hot portions of the crust heat groundwater. This energy transmits to the surface by heated water or steam accessed through the drilling of wells. For an effective *underground thermal reservoir* to form, an aquifer strata must have high porosity and high permeability that allows heated water to move freely through connecting pore spaces. (See Figure 6.16 and the accompanying text for illustration and definition of these terms.) An impermeable rock strata above the thermal reservoir aquifer helps preserve the resource and describes a *hydrothermal reservoir*, the most common form of thermal repository. Ideally, this aquifer should heat groundwater from 180° to 350°C

(355°–600°F) and be accessible to drilling within 3 km (1.9 mi) of the surface, although 6–7 km depths are workable (3.7–4.3 mi).

Geothermal energy literally refers to heat from Earth's interior, whereas *geothermal power* relates to specific applied strategies of geothermal electric or geothermal direct applications. *Geothermal electric* uses steam, or hot water that flashes to steam, to drive a turbine-generator. In *geothermal direct*, hot water is used to heat buildings, to cool buildings using a heat-exchange system, to heat greenhouses, to heat soil, for manufacturing processes, for aquaculture, and to heat swimming pools, among many uses.

(a)

(b)

FIGURE 8.1.1 Surface geothermal activity in Iceland.
(a) This is the original Geysir, in Haukadalur, Iceland, from which we derive the name for all other geysers on Earth; historical accounts of Geysir's activity date back to A.D. 1294. (b) The popular Blue Lagoon Spa is filled with hot geothermal waters. The Svartsengi power plant is in the distance in this geothermal field on the Reykjavík Peninsula, Iceland. [Photos by Bobbé Christopherson.]

(continued)

Focus Study 8.1 *(continued)*

For either electrical production or direct use, the end-use facilities must be fairly near the wellhead because of heat losses in piping the hot water or steam over distance. Of course, once electricity is generated, it can feed into the regional power grid for transmission to distant markets.

Geothermal Power Production

The first geothermal electrical generating station was built in Larderello, Italy, in 1904, and it has been in continuous operation since 1913 with an installed capacity of 360 MWe (megawatts electric). Today, geothermal applications are in use in 30 countries, through some 200 plants, producing an output of 8900 MWe and direct heat of 12,000 MWt (megawatts thermal).

In Reykjavík, Iceland, the majority of space heating (87%) and some of the electrical production (18%) is geothermal. The Svartsengi power plant (40 MWe) on the Reykjavík Peninsula is an example of such a modern facility (Figure 8.1.2). In the same area, a large array of wells collects heat from hot water and pumps it to the city for geothermal direct applications.

In the Paris Basin, France, some 25,000 residences are heated using geothermal direct 70°C (158°F) water. In the Philippines, almost 30% of total electrical production is generated with geothermal energy. The top six countries for installed geothermal electrical generation are the United States, Philippines, Italy, Mexico, Indonesia, and Japan. Worldwide operational capacity is 8900 MWe of a potential estimated to be in excess of 80,000 MWe and hundreds of thousands of MWt. The United States, China, and Iceland lead in thermal installations.

The Geysers Geothermal Field (so named despite the lack of any geysers in the area) began production

FIGURE 8.1.2 Svartsengi geothermal power plant.
On the Reykjavík Peninsula, southwest of Iceland's capital, the Svartsengi power plant produces electricity and provides hot water for homes and businesses. The plant's name means "Black Meadow," referring to the extensive lava beds throughout this region of Iceland. Pipelines carry the hot water into towns for radiant space heating. [Photo by Bobbé Christopherson.]

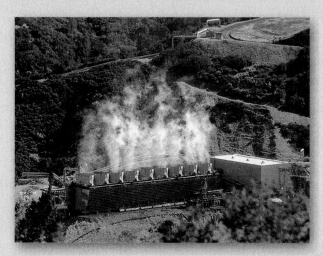

FIGURE 8.1.3 The Geysers Geothermal Field, California.
The Geysers is the largest geothermal electric installation in the world. One of the 21 generating stations is Eagle Rock, Unit 11, with a 73-MWe capacity, in commercial operation since 1975. [Photo courtesy of Calpine, San Jose, California, http://www.geysers.com.]

in 1960, increased to a peak in 1989 at 1967 MWe, and decreased to a present capacity of 1070 MWe (Figure 8.1.3). By the mid-1990s, some 600 wells had been drilled. The average well depth is 2500 m (8200 ft). Declining production relates to the expected problems of decreased yields. The Geysers are operating below the capacity of the field potential.

Some 2800 MWe of geothermal power capacity is operating in California, Hawai'i, Nevada, and Utah, with a new 185-MWe plant proposed in the desert near Brawley, California. Figure 8.1.4 maps low-, medium-, and high-temperature known geothermal resource areas (KGRA) in the United States. About 300 cities are within 8 km (5 mi) of a KGRA. The Department of Energy has identified about 9000 potential development sites. In Canada, a proposed project at Mount Meager, B.C., will produce electricity; geothermal direct heats buildings at Carleton University, Ottawa; and some direct operations are in development in Nova Scotia. For more information, see http://www1 .eere.energy.gov/ geothermal/, http:// geothermal.marin.org/, http://www .geothermal .org/, or http://www .smu.edu/geothermal/.

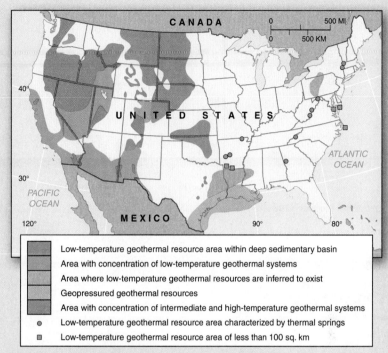

Low-temperature geothermal resource area within deep sedimentary basin
Area with concentration of low-temperature geothermal systems
Area where low-temperature geothermal resources are inferred to exist
Geopressured geothermal resources
Area with concentration of intermediate and high-temperature geothermal systems
Low-temperature geothermal resource area characterized by thermal springs
Low-temperature geothermal resource area of less than 100 sq. km

FIGURE 8.1.4 Known geothermal resource areas.
Geothermal resources in the conterminous United States. [Map courtesy of USGS, from W. A. Duffield and J. H. Sass, *Tapping the Earth's Natural Heat*, USGS Circular 1125, 1994, p. 35.]

spots occur beneath both oceanic and continental crust and are anchored deep in the stiff lower mantle, tending to remain fixed relative to migrating plates, but not always. Thus, the area of a plate that is above a hot spot is locally heated for the brief geologic time it is there (a few hundred thousand or million years).

An example of an isolated hot spot is the one that has formed the Hawaiian-Emperor Islands chain (Figure 8.19). The Pacific plate has moved across this hot, upward-erupting plume for almost 80 million years, with the resulting string of volcanic islands moving northwestward away from the hot spot. Thus, the age of each island or seamount in the chain increases northwestward from the island of Hawai'i, as you can see from the ages indicated in the figure.

The big island of Hawai'i actually took less than 1 million years to build to its present stature. The oldest island in the Hawaiian part of the chain is Kaua'i, at approximately 5 million years old; it is weathered, eroded, and deeply etched with canyon and valley terrain. The youngest island in the chain is still a *seamount*, a submarine mountain that does not reach the surface. It rises 3350 m (11,000 ft) from the ocean floor but is still 975 m (3200 ft) beneath the ocean surface. Even though this new island will not experience the Sun for about 10,000 years, it is already named Lō'ihi.

To the northwest of this active hot spot in Hawai'i, the island of Midway rises as a part of the same system.

From there the Emperor seamounts stretch northwestward until they reach about 40 m.y.a. Their general relationship to the tectonic plates is important to point out here, for at that point, this linear island chain shifts direction northward, evidently reflecting a change in the movement of the Pacific plate at that earlier time. At the the northernmost extreme, the seamounts that formed about 80 million years ago are now approaching the Aleutian Trench, where they eventually will be subducted beneath the Eurasian plate. Recent research suggests that the change in direction of the chain you see in Figure 8.19a could be related to movement of the hot spot itself. Evidently, the mantle flows disrupt the upwelling plume that forms the hot spot.

Iceland is an example of an active hot spot sitting astride a mid-ocean ridge (Figure 8.13d). It also is an example of a segment of mid-ocean ridge rising above sea level. This hot spot has generated enough material to form Iceland and continues to cause volcanic eruptions. As a result, Iceland is still growing in area and volume. The youngest rocks are near the center of Iceland, with rock age increasing toward the eastern and western coasts.

Take a moment to locate these features (mid-ocean ridges, subduction zones, Hawaiian and Icelandic hot spots) on the Chapter 9 opening map illustration. Earth is "The Dynamic Planet" of our chapter title!

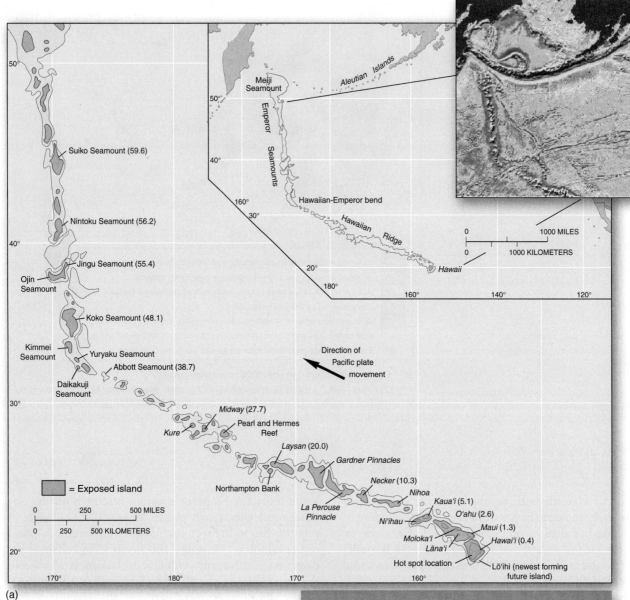

(a)

ANIMATION   Hot Spot Volcano Tracks

FIGURE 8.19 Hot spot tracks across the North Pacific.

Hawai'i and the linear volcanic chain of islands known as the Emperor Seamounts. (a) The islands and seamounts in the chain are progressively younger toward the southeast. Ages, in m.y.a., are shown in parentheses. Note that Midway Island is 27.7 million years old, meaning that the site was over the plume 27.7 m.y.a. (b) Lō'ihi is forming 975 m (3200 ft) beneath the Pacific Ocean; presently an undersea volcano (seamount), it will continue to grow into the next Hawaiian island. The photo is from South Point, Hawai'i; Lō'ihi is east of this headland. [(a) After D. A. Clague, "Petrology and K–Ar (potassium–argon) ages of dredged volcanic rocks from the Western Hawaiian Ridge and the Southern Emperor Seamount Chain," *Geological Society of America Bulletin* 86 (1975): 991; inset from global gravity anomaly map image, Scripps Institution of Oceanography. All rights reserved. (b) Photo by Bobbé Christopherson.]

(b)

Summary and Review—The Dynamic Planet

■ *Distinguish* between the endogenic and exogenic systems, *determine* the driving force for each, and *explain* the pace at which these systems operate.

The Earth–atmosphere interface is where the **endogenic system** (internal), powered by heat energy from within the planet, interacts with the **exogenic system** (external), powered by insolation and influenced by gravity. These systems work together to produce Earth's diverse landscape. The **geologic time scale** is an effective device for organizing the vast span of geologic time. It depicts the sequence of Earth's events (relative time) and the approximate actual dates (absolute time).

The most fundamental principle of Earth science is **uniformitarianism**. Uniformitarianism assumes that the *same physical processes active in the environment today have been operating throughout geologic time.*

endogenic system (p. 263)
exogenic system (p. 263)
geologic time scale (p. 266)
uniformitarianism (p. 266)

1. To what extent is Earth's crust active at this time in its history?
2. Define the endogenic and the exogenic systems. Describe the driving forces that energize these systems.
3. How is the geologic time scale organized? What is the basis for the time scale in relative and absolute terms? What era, period, and epoch are we living in today?
4. Describe uniformitarianism in Earth's development. How can this flow of events and time be interrupted?

■ *Diagram* Earth's interior in cross section and *describe* each distinct layer.

We have learned about Earth's interior from indirect evidence—the way its various layers transmit **seismic waves**. The core is differentiated into an *inner core* and an *outer core*—divided by a transition zone. Earth's magnetic field is generated almost entirely within the outer core. Polarity reversals in Earth's magnetism are recorded in cooling magma that contains iron minerals. The pattern of **geomagnetic reversals** frozen in rock helps scientists piece together the story of Earth's mobile crust.

Outside Earth's core lies the **mantle**, differentiated into lower mantle and upper mantle. It experiences a gradual temperature increase with depth and stiffening due to increased pressures. The upper mantle is divided into three fairly distinct layers. The uppermost mantle, along with the crust, makes up the *lithosphere.* Below the lithosphere is the **asthenosphere**, or *plastic layer*. It contains pockets of increased heat from radioactive decay and is susceptible to slow convective currents in these hotter materials. An important internal boundary between the **crust** and the high-velocity portion of the uppermost mantle is the **Mohorovičić discontinuity**, or **Moho**. *Continental crust* is basically **granite**; it is crystalline and high in silica, aluminum, potassium, calcium, and sodium. *Oceanic crust* is **basalt**; it is granular and high in silica, magnesium, and iron. The principles of buoyancy and balance produce the important principle of **isostasy**. Isostasy explains certain vertical movements of Earth's crust.

seismic waves (p. 268)
core (p. 268)
geomagnetic reversal (p. 269)
mantle (p. 270)
asthenosphere (p. 270)
crust (p. 270)
Mohorovičić discontinuity (Moho) (p. 270)
granite (p. 271)
basalt (p. 272)
isostasy (p. 272)

5. Make a simple sketch of Earth's interior, label each layer, and list the physical characteristics, temperature, composition, and range of size of each on your drawing.
6. What is the present thinking on how Earth generates its magnetic field? Is this field constant, or does it change? Explain the implications of your answer.
7. Describe the asthenosphere. Why is it also known as the plastic layer? What are the consequences of its convection currents?
8. What is a discontinuity? Describe the principal discontinuities within Earth.
9. Define *isostasy* and *isostatic rebound* and explain the crustal equilibrium (balance between buoyancy and gravity) concept. Relative to isostasy, what has been occurring in Alaska during the past decade?
10. Diagram the uppermost mantle and crust. Label the density of the layers in grams per cubic centimeter. What two types of crust were described in the text in terms of rock composition?

■ *Illustrate* the geologic cycle and *relate* the rock cycle and rock types to endogenic and exogenic processes.

The **geologic cycle** is a model of the internal and external interactions that shape the crust. A **mineral** is an inorganic natural compound having a specific chemical formula and possessing a crystalline structure. A **rock** is an assemblage of minerals bound together (such as granite, a rock containing three minerals), or it may be a mass of a single mineral (such as rock salt). Thousands of different rocks have been identified.

The geologic cycle comprises three cycles: the *hydrologic cycle*, the *tectonic cycle*, and the *rock cycle*. The rock cycle describes the three principal rock-forming processes and the rocks they produce. **Igneous rocks** form from **magma**, which is molten rock beneath the surface. Magma is fluid, highly gaseous, and under tremendous pressure. It either *intrudes* into crustal rocks, cools, and hardens, or it *extrudes* onto the surface as **lava**. Intrusive igneous rock that cools slowly in the crust forms a **pluton**. The largest pluton form is a **batholith**.

Stratigraphy is the study of the sequence (superposition), thickness, and spatial distribution of strata that yield clues to the age and origin of the rocks. The cementation, compaction, and hardening of sediments into **sedimentary rocks** is called **lithification**. These layered strata form important records of past ages. Clastic sedimentary rocks are derived from bits and pieces of weathered rocks. Chemical sedimentary rocks are not formed from physical pieces of broken rock, but instead are dissolved minerals, transported in solution, and chemically precipitated out of solution (they are essentially nonclastic). The most common chemical sedimentary rock is **limestone**, which is lithified calcium carbonate, $CaCO_3$.

Chemical sediments also form from inorganic sources when water evaporates and leaves behind a residue of salts. These **evaporites** may exist as common salts, such as gypsum or sodium chloride (table salt), to name only two, and often appear as flat, layered deposits across a dry landscape.

Any rock, either igneous or sedimentary, may be transformed into a **metamorphic rock** by going through profound physical or chemical changes under pressure and increased temperature.

geologic cycle (p. 273)
mineral (p. 273)
rock (p. 274)
igneous rock (p. 274)
magma (p. 274)
lava (p. 275)
pluton (p. 275)
batholith (p. 275)
stratigraphy (p. 275)
sedimentary rock (p. 276)
lithification (p. 276)
limestone (p. 277)
evaporite (p. 277)
metamorphic rock (p. 278)

11. Illustrate the geologic cycle and define each component: rock cycle, tectonic cycle, and hydrologic cycle.

12. What is a mineral? A mineral family? Name the most common minerals on Earth. What is a rock?

13. Describe igneous processes. What is the difference between intrusive and extrusive types of igneous rocks?

14. Briefly explain how the rock formation in Figure 8.9a demonstrates through its layers a record of past climates.

15. Briefly describe sedimentary processes and lithification. Describe the sources and particle sizes of sedimentary rocks.

16. What is metamorphism, and how are metamorphic rocks produced? Name some original parent rocks and metamorphic equivalents.

■ *Describe* Pangaea and its breakup and *relate* several physical proofs that crustal drifting is continuing today.

The present configuration of the ocean basins and continents is the result of *tectonic processes* involving Earth's interior dynamics and crust. Alfred Wegener coined the phrase *continental drift* to describe his idea that the crust moves in response to vast forces within the planet. **Pangaea** was the name he gave to a single assemblage of continental crust some 225 m.y.a., which subsequently broke apart. Earth's crust is fractured into huge slabs or plates, each moving in response to flowing currents in the mantle. The all-encompassing theory of **plate tectonics** includes **sea-floor spreading** along **mid-ocean ridges** and denser oceanic crust diving beneath lighter continental crust along **subduction zones**— places where deep **oceanic trenches** form. Supporting evidence includes: the pattern of magnetic reversals in rocks worldwide, matching animal and plant fossil records in adjoining continents, patterns of relative and absolute dating of crust rock, and deposits of fossil fuels, among many items.

Pangaea (p. 281)
plate tectonics (p. 283)
sea-floor spreading (p. 283)
mid-ocean ridge (p. 283)
subduction zone (p. 286)
oceanic trenches (p. 286)

17. Briefly review the history of the theory of continental drift, sea-floor spreading, and the all-inclusive plate tectonics theory. What was Alfred Wegener's role?

18. Define *upwelling* and describe related features on the ocean floor. Define *subduction* and explain the process.

19. What was Pangaea? What happened to it during the past 225 million years?

20. Characterize the three types of plate boundaries and the actions associated with each type.

■ *Portray* the pattern of Earth's major plates and *relate* this pattern to the occurrence of earthquakes, volcanic activity, and hot spots.

Occurrences of often damaging earthquakes and volcanic eruptions are correlated with plate boundaries. Three types of plate boundaries form: divergent, convergent, and transform. Along the offset portions of mid-ocean ridges, horizontal motions produce *transform faults*.

As many as 50 to 100 hot spots exist across Earth's surface, where tall plumes of magma, anchored in the lower mantle, remain fixed as eruptions penetrate drifting plates. Geothermal energy literally refers to heat from Earth's interior, whereas geothermal power relates to specific applied strategies of geothermal energy.

hot spots (p. 289)
geothermal energy (p. 289)

21. What is the relation between plate boundaries and volcanic and earthquake activity?

22. What is the nature of motion along a transform fault? Name a famous example of such a fault.

23. How is the Hawaiian-Emperor chain of islands and seamounts an example of plate motion and hot-spot activity? Correlate your answer with the ages of the Hawaiian Islands.

NetWork and Critical Thinking Tools

A. Using the maps in this chapter, determine your present location relative to Earth's crustal plates. Now, using Figure 8.15b, approximately determine where your present location was 225 m.y.a.; express it in a rough estimate using the equator and the longitudes noted on the map.

B. Go to the Student Learning Center and open the animation "Plate Motions Through Time." As you follow the drifting continents in the animation, track the location you determined in "A" to its location today. Where will your location be 50 million years in the future on the final map frame? Click "Next" and watch the animation drawn with realistic coloration as you track the location.

C. Relative to the motion of the Pacific plate shown in Figure 8.19 (note the scale in the lower-left corner), the island of Midway formed 27.7 m.y.a. over the hot spot that is active under the southeast coast of the big island of Hawai'i today. Given the scale of the map, roughly

determine the average annual speed of the Pacific plate in centimeters per year for Midway to have traveled this distance.

D. Under the heading "The Pace of Change," the text discusses a possible "Anthropocene" Epoch to describe modern times. The section states, in part:

> The name Holocene, approximately the last 11,500 years, is given to the youngest epoch in the geologic time scale, characteristic of postglacial conditions since the retreat of the continental glaciers. Because of the impact of human society on planetary systems, discussion is underway about designating a new name, *Anthropocene*, for the human epoch.

Take a moment and explore this concept of naming our human epoch in the geologic time scale. Determine some arguments for such a change and opposed to such a renaming of the late Holocene. What do you think?

Floor of the Oceans, by Bruce C. Heezen and Marie Tharp.
The scarred ocean floor is clearly visible: sea-floor spreading centers marked by oceanic ridges that stretch over 64,000 km (40,000 mi), subduction zones indicated by deep oceanic trenches, and transform faults slicing across oceanic ridges.

Follow the East Pacific rise (an ocean ridge and spreading center) northward as it trends beneath the west coast of the North American plate, disappearing under earthquake-prone California. The continents, offshore submerged continental shelves, and the expanse of the sediment-covered abyssal plain are all identifiable in this illustration.

On the floor of the Indian Ocean, you can see the wide track along which the India plate traveled northward to its collision with the Eurasian plate. Vast deposits of sediment cover the Indian Ocean floor, south of the Ganges River to the east of India. Sediments derived from the Himalayan Range blanket the floor of the Bay of Bengal (south of Bangladesh) to a depth of 20 km (12.4 mi). These sediments result from centuries of soil erosion in the land of the monsoons.

Isostasy can be examined in action in central and west-central Greenland, where the weight of the ice sheet depresses portions of the land far below sea level—now visible with the ice artificially removed by the cartographer (see the extreme upper-left corner of the map). In contrast, the region around Canada's Hudson Bay became ice-free about 8000 years ago and has isostatically rebounded 300 m (1100 ft).

In the area of the Hawaiian Islands, the hot-spot track is marked by a chain of islands and seamounts that you can follow along the Pacific plate from Hawai'i to the Aleutians. Subduction zones south and east of Alaska and Japan, and along the western coast of South and Central America, are visible as dark trenches. Examine the Atlantic Ocean floor, following northward to find Iceland's position on the Mid-Atlantic Ridge. [© 1980 by Marie Tharp. Reproduced by permission of Marie Tharp.]

Chapter

9

Tectonics, Earthquakes, and Volcanoes

KEY LEARNING CONCEPTS

After reading the chapter, you should be able to:

■ *Describe* first, second, and third orders of relief and *relate* examples of each from Earth's major topographic regions.

■ *Describe* the several origins of continental crust and *define* displaced terranes.

■ *Explain* compressional processes and folding and *describe* four principal types of faults and their characteristic landforms.

■ *Relate* the three types of plate collisions associated with orogenesis and *identify* specific examples of each.

■ *Explain* the nature of earthquakes, their measurement, and the nature of faulting.

■ *Distinguish* between an effusive and an explosive volcanic eruption and *describe* related landforms, using specific examples.

STUDENT LEARNING CENTER TOOLS

The *Elemental Geosystems* Student Learning Center provides on-line resources for this chapter by clicking on the book cover at **www.mygeoscienceplace.com**. Once you are in our Learning Center, bookmark the URL address. For this chapter find reviewing tools, eBook links, self-tests and quizzes, critical thinking items, figure animations, photo galleries, and satellite loops. Some highlights:

▶ Eleven animations to assist in learning key concepts in the chapter: terrane formation; folding, faulting, and transform faults; how seismographs work; and fault zone mechanics.

▶ An interactive animation showing tectonic settings and types of volcanic activity.

▶ An animation profile of the Mt. St. Helens eruption allowing you to trigger the process and see what occurred during the eruption and debris avalanche.

▶ More than 17 additional URLs along with Quick Links to all the URLs in the chapter.

arth's endogenic systems produce flows of heat and material toward the surface to form crust. Ongoing processes produce continental landscapes and oceanic sea-floor crust, sometimes in dramatic episodes. Earth's physical systems move to the front pages whenever an earthquake strikes or after a few of the 50 yearly volcanic eruptions threaten a city, although most are in remote locations or beneath the ocean surface.

In 1999, more than 30,000 people perished in quakes in Turkey, Greece, Taiwan, and Mexico. In 2003, in one earthquake alone 30,000 were killed in Bam, Iran (Figure 9.1a). More than 83,000 were killed by a magnitude (M) 7.6 earthquake with many major aftershocks in the Kashmir region of Pakistan in October 2005 (Figure 9.1c). Chengdu, China, a city of 2.5 million with a metropolitan region exceeding 9 million people, was hit by a M 7.9 quake

in May 2008 killing more than 50,000 people (Figure 9.1d; and see Figure 7.10b). Yet, in other years, the death toll might fall below 1000.

A magnitude 7.9 earthquake near Denali National Park, Alaska, was notable as the largest on Earth in 2002. The remote location prevented loss of life but dramatic effects took place: cracked roads and bridges, huge boulders and rocks that fell onto glacial surfaces, offsets in glacial crevasses, many avalanches, and 150 km (92 mi) of the Alaska pipeline supports broken (Figure 9.1b). The fault rupture extended eastward from the epicenter for over 340 km (210 mi), with right-lateral horizontal offsets of nearly 9 m (26 ft). This quake triggered far-reaching effects: more than 200 small earthquakes in Yellowstone, seismic crustal waves that were felt in Texas, and altered water tables and well water in Pennsylvania.

(a)　(b)　(c)　(d)

FIGURE 9.1 Earthquakes strike Iran, China, Alaska, and Pakistan.
(a) Captured in 1-m resolution, the mud-brick and straw buildings in a section of Bam, Iran, lay in ruins, leveled by a magnitude 6.9 quake. The buildings look as if they were smudged and blurred by some giant thumb. (b) A magnitude 7.9 earthquake struck central Alaska, 66 km (41 mi) southeast of Denali National Park. A portion of the Trans-Alaska oil pipeline was knocked off its supports. (c) A 7.9 quake and numerous large aftershocks destroyed homes in Muzaffarabad, capital of Pakistan-Kashmir, in October 2005. (d) Two villagers react to the destruction of their apartment building from a M 7.9 quake that killed 87,000 people in Chengdu, China. [Image by (a) *IKONOS* satellite, used by permission of Space Imaging, Inc.; photos (b) by Earthquake Engineering Research Institute; (c) Rong Shoujun, Xinhua/Corbis; (d) Dou Zezhong, Xinhua/Corbis.]

Volcanic eruptions pose threats to life and property to those who live in proximity to the hazard. Earth systems science is providing analysis and warnings to affected populations as never before. For instance, since 1998 scientists have been closely monitoring the growing bulge on the western flanks of the South Sister volcano in central Oregon, although the bulge rate has slowed during the past several years. Now approaching its fourth decade of eruption, longest in history, the outpouring of lava from Kīlauea volcano, Hawai'i, continues to amaze both scientists and tourists. Note the Internet links in this chapter and the wealth of information that is readily available allowing the public to be kept up-to-date on these and other hazards.

In this chapter: We examine processes that construct Earth's surface and create world structural regions. Tectonic processes deform, recycle, and reshape Earth's crust. These processes occur sometimes in dramatic episodes but most often in slow, deliberate motions that build the landscape. Continental crust has been forming throughout most of Earth's 4.6-billion-year existence.

The arrangement of continents and oceans, the origin of mountain ranges, the topography of the land and the seafloor, and the locations of earthquake and volcanic activity are all evidence of our dynamic Earth. Principal seismic and volcanic zones occur along plate boundaries and hotspot locations, thus linking plate tectonics to the devastation of major earthquakes and local threats of major volcanic eruptions and their attendant potential global climatic impact.

We begin our look at tectonics and volcanism on the ocean floor, hidden from direct view. The illustration that opens this chapter is a striking representation of Earth with its blanket of water removed, as revealed to us through decades of direct and indirect observation. Careful examination of this portrait gives us a helpful review of the concepts learned in the previous chapter, laying the foundation for this and subsequent chapters. Be sure to take the quick tour presented with this chapter-opening map. Try to correlate this sea-floor illustration with the maps of crustal plates and plate boundaries shown in Figure 8.16 and the age of the ocean floor in Figure 8.17.

Earth's Surface Relief Features

Relief refers to vertical elevation differences in the landscape. Examples include the low relief of Iowa and Nebraska, medium relief in foothills along mountain ranges, and high relief in the Rockies and Himalayas. The undulating form of Earth's surface is **topography**, portrayed effectively on topographic maps—the lay of the land.

The relief and topography of Earth's crustal landforms have played a vital role in human history: High mountain passes both protected and isolated societies; ridges and valleys have dictated transportation routes; and vast plains encouraged better methods of communication and travel. Earth's topography stimulated human invention and adaptation.

Crustal Orders of Relief

Our understanding of Earth's relief and topography is enhanced by modern computer capabilities and tools such as the global positioning system (GPS) for determining location and elevation. Scientists put elevation data in digital form; the data then are available for computer manipulation and display. The resulting *digital elevation models* (*DEMs*) aid in the scientific analysis of topography, area–altitude distributions, slopes, and local stream-drainage characteristics, among many benefits.

The US Geological Survey has prepared a digitized shaded-relief map of the United States. The entire map is made up of 12 million spot elevations, with less than a kilometer between any two. A web site prepared by the USGS, "A Tapestry of Time and Terrain," uses this data, colored geologic regions, and detailed topography in a useful virtual map: see http://tapestry.usgs.gov.

For convenience, let's group the landscape's topography into three *orders of relief*. Orders of relief classify landscapes by scale, from enormous ocean basins and continents down to local hills and valleys.

First Order of Relief The broadest category of landforms includes huge continental platforms and ocean basins. **Continental landmasses** are the masses of crust that reside above or near sea level, including the undersea continental shelves along the coastlines. The **ocean basins** are entirely below sea level and are first order. Approximately 71% of Earth is covered by water.

Second Order of Relief Continental masses, mountain masses, plains, and lowlands are continental features that are in the second order of relief category. A few examples are the Alps, Rocky Mountains (both Canadian and American), west Siberian lowland, and the Tibetan Plateau. The great rock "shields" that form the heart of each continental mass are of this second order. In the ocean basins, second order of relief includes continental rises, slopes, abyssal plains, mid-ocean ridges, submarine canyons, and subduction trenches—all visible in the sea-floor illustration that opens this chapter.

Third Order of Relief The third and most-detailed order of relief includes individual mountains, cliffs, valleys, hills, and other landforms of smaller scale. These features are identifiable as local landscapes.

Hypsometry Figure 9.2 is a *hypsographic curve*, which shows the distribution of Earth's surface by area and elevation (the Greek *hypsos* means "height"). Relative to Earth's diameter of 12,756 km (7926 mi), the surface generally is of low relief; for comparison, Mount Everest is only 8.8 km (5.5 mi) above sea level, and Mauna Kea in Hawai'i is 10.0 km (6.2 mi) tall measured from the seafloor. Note the GPS measurement for the summit of Mount Everest announced in 1999—8850 m (29,035 ft). See News Report 9.1 for more on this geographic milestone.

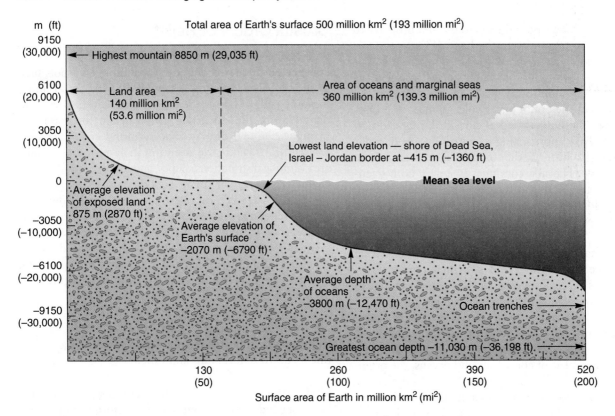

FIGURE 9.2 Earth's hypsometry.
Hypsographic curve of Earth's surface, charting elevation as related to mean sea level. From the highest point above sea level (Mount Everest) to the deepest oceanic trench (Mariana Trench) Earth's overall relief is almost 20 km (12.5 mi). The height given for Mount Everest was announced in 1999.

NEWS REPORT 9.1

Mount Everest at New Heights

The announcement came November 11, 1999, in Washington, DC, at the opening reception of the 87th annual meeting of the American Alpine Club. The press conference was held at National Geographic Society headquarters. Mount Everest had a revised elevation determined by placement of a direct Global Positioning System (GPS) on the mountain's icy summit!

In May 1999, mountaineers Pete Athans and Bill Crouse reached the summit with five Sherpas. The climbers operated Global Positioning System satellite equipment on the top of Mount Everest and determined the precise height of the world's tallest mountain. In Chapter 1, the photograph in News Report 1.2 shows a GPS unit placed by climber Wally Berg

18 m below the summit in 1998. This site is still Earth's highest benchmark, with a permanent metal marker installed in rock. Other GPS instruments in the region permit differential calibration for exact measurement (Figure 9.1.1). This is a continuation of a measurement effort begun in 1995 by the late Bradford Washburn, renowned mountain photographer/explorer, who was honorary director of Boston's Museum of Science.

Washburn announced the new measurement of 8850 m (29,035 ft). Everest's new elevation is close to the previous official measure of 8848 m (29,028 ft) set in 1954 by the Survey of India. Washburn stated an interesting note, "... from these GPS readings it appears that the horizontal position

of Everest seems to be moving steadily and slightly northeastward; between 3 and 6 mm a year [up to 0.25 in. a year]." The mountain range is being driven farther into Asia by plate tectonics—the continuing collision of Indian and Asian landmasses.

Astronaut Thomas D. Jones made the photo of Mount Everest in Figure 9.1.2 during Space Shuttle flight STS-80 in 1996 aboard Space Shuttle *Columbia*. This startling view caught the mountain's triangular East Face bathed in morning light (east is to the lower left, north is to the lower right, south at the upper left; see locator map direction arrow). The North Face is in shadow. Several glaciers are visible flowing outward from the mountains, darkened by rock and debris.

FIGURE 9.1.1 GPS installation measures Mount Everest.
A global positioning system (GPS) installation near Namche Bazar in the Khumbu (Everest) region of Nepal. The station is at latitude 27.8° N, longitude 86.7° E, and 3523 m (11,558 ft). The summit of Mount Everest is 30 km (18.6 mi) distant—visible above the right side of the weather dome. A network of such GPS installations correlated with the unit the climbers took to the summit. [Photo by Charles Corfield, science manager to the 1998 and 1999 expeditions.]

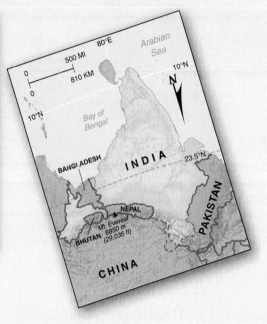

FIGURE 9.1.2 Mount Everest in morning light from orbit aboard Space Shuttle *Columbia*.
Note the orientation of the photo and locator map with north to the lower-right. [Photo courtesy of NASA, made by Dr. Thomas Jones, astronaut, 1996.]

The average elevation of Earth's solid surface is actually under water: –2070 m (–6790 ft) below mean sea level. The average elevation just for exposed land is +875 m (+2870 ft). For the ocean depths, the average elevation is –3800 m (–12,470 ft). From these statistics you can see that, on the average, the oceans are much deeper than continental regions are high. Overall, the underwater ocean basins, ocean floor, and submarine mountain ranges form Earth's largest "landscape."

Earth's Topographic Regions

The three orders of relief can be further generalized into six topographic regions: plains, high tablelands, hills and low tablelands, mountains, widely spaced mountains, and depressions (Figure 9.3). An arbitrary elevation or descriptive limit in common use defines each type of topography (see the legend in Figure 9.3).

Four of the continents possess extensive *plains*, which are identified as areas with local relief of less than 100 m (325 ft) and slope angles of 5° or less. Some plains have high elevations of over 600 m (2000 ft); in the United States the high plains achieve elevations above 1220 m (4000 ft). The Colorado Plateau and Antarctica are notable *high tablelands*, with elevations exceeding 1520 m (5000 ft). *Hills and low tablelands* dominate Africa.

Mountain ranges are characterized by local relief exceeding 600 m (2000 ft) and appear on each continent. Earth's relief and topography are undergoing constant change as a result of processes that form crust.

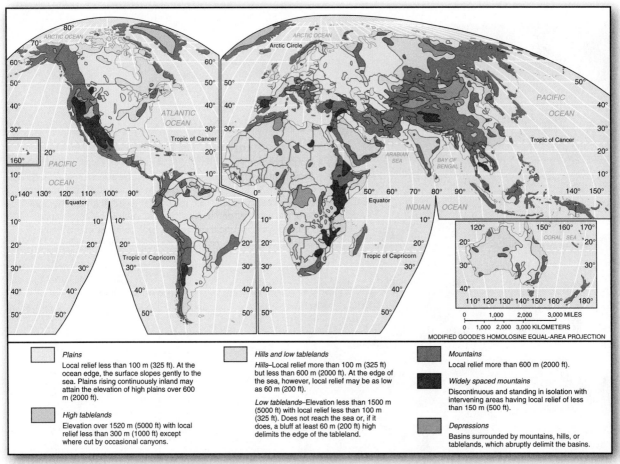

FIGURE 9.3 Earth's topographic regions.
Earth's topography is characterized as plains, high and low tablelands, hills, mountains, and depressions. [After Richard E. Murphy, "Landforms of the World," Map Supplement No. 9, *Annals of the Association of American Geographers* 58, no. 1 (March 1968). Adapted by permission.]

Crustal Formation Processes

How did Earth's continental crust form? What gives rise to the three orders of relief just discussed? Ultimately, the answer is tectonic activity, driven by our planet's internal energy, and the exogenic processes of weathering and erosion, powered by the Sun through the actions of water, air, waves, and ice.

Tectonic activity generally is slow, requiring millions of years. Endogenic (internal) processes result in gradual uplift and new landforms, with major mountain building occurring along plate boundaries. These uplifted crustal regions are varied, but we think of them in three general categories, all discussed in this chapter:

- Residual mountains and stable continental cratons, formed from inactive remnants of ancient tectonic activity;
- Tectonic mountains and landforms, produced by active folding, faulting, and crustal movements;
- Volcanic features, formed by the surface accumulation of molten rock from eruptions of subsurface materials.

Thus, several distinct processes operate in concert to produce the continental crust we see around us.

Continental Shields

All continents have a nucleus of ancient crystalline rock on which the continent "grows" with the addition of crustal fragments and sediments. This nucleus is the *craton*, or heartland region, of the continental crust. Cratons generally have been eroded to a low elevation and relief. Most date to the Precambrian and can exceed 2 billion years of age. The lack of basaltic components in these cratons offers a clue to their stability. The lithosphere in cratonic regions, which consists of crust and lithospheric uppermost mantle, is thicker than the lithosphere beneath younger portions of continents and oceanic crust.

A **continental shield** is a region where a craton is exposed at the surface. Figure 9.4 shows the principal areas of exposed shields and a photo of the Canadian shield. Layers of younger sedimentary rock surround these shields and appear quite stable over time. Examples of such a stable *platform* are the region that stretches from east of the Rockies to the Appalachians and northward

FIGURE 9.4 Continental shields.
(a) Portions of major continental shields that have been exposed by erosion. Adjacent portions of these shields remain covered. (b) Canadian shield landscape in northern Québec, stable for hundreds of millions of years, stripped by past glaciations, and marked by intrusive igneous dikes (magmatic intrusions). Can you see evidence of the directional flow of the continental glaciers? [(a) After Richard E. Murphy, "Landforms of the World," Map Supplement No. 9, *Annals of the Association of American Geographers* 58, no. 1 (March 1968). Adapted by permission; (b) photo by Bobbé Christopherson.]

(b)

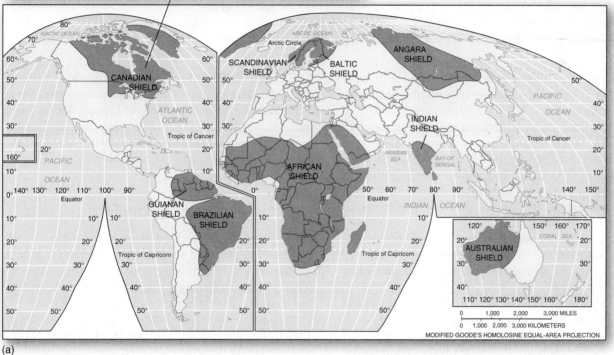

(a)

into central and eastern Canada, a large portion of China, eastern Europe to the Ural Mountains, and across portions of Siberia.

Building Continental Crust and Terranes

Continental crust results from a complex process that involves sea-floor spreading and formation of oceanic crust and its subduction, remelting, and subsequent rise as new magma, as summarized in Figure 8.13.

To understand this process, study Figure 9.5. Begin with the magma that originates in the asthenosphere and wells up along the mid-ocean ridges. Basaltic magma is formed from minerals in the upper mantle that are rich in iron and magnesium. Such magma has less than 50% silica

and has a low-viscosity (thin) texture—it tends to flow. This mafic material rises to erupt at spreading centers and cools to form new basaltic seafloor, which spreads outward to collide with continental crust along its far edges. This denser oceanic crust plunges beneath the lighter continental crust into the mantle, where it remelts. The new magma then rises and cools, forming more continental crust, in the form of intrusive granitic igneous rock.

As the subducting oceanic plate works its way under a continental plate, it takes with it sediment and trapped water, melting and incorporating various elements from the crust into the mixture. As a result, the magma, generally called a *melt*, that migrates upward from a subducted plate contains 50%–75% silica, is high in aluminum, and has a high-viscosity (thick) texture. Note that this melt is

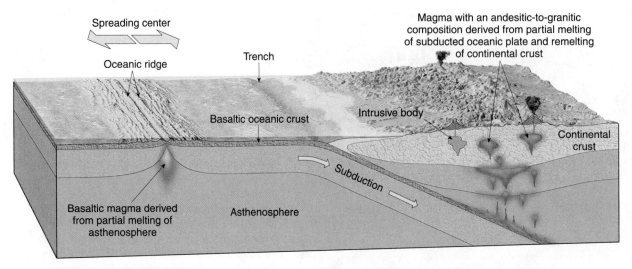

FIGURE 9.5 Crustal formation.
Material from the asthenosphere upwells along sea-floor spreading centers. Basaltic ocean floor is subducted beneath lighter continental crust, where it melts, along with its cargo of sediments, water, and minerals. This melting generates magma, which makes its way up through the crust to form igneous intrusions and extrusive eruptions. [From E. J. Tarbuck and F. K. Lutgens, *Earth, An Introduction to Physical Geology*, 5th ed., art by Dennis Tasa, © Prentice Hall, 1996, Fig. 20.20, p. 502.]

NOTEBOOK   Plate Boundaries

different in composition from the magma that arose directly from the asthenosphere at spreading centers to form new seafloor.

Bodies of the silica-rich magma may reach the surface in explosive volcanic eruptions, or they may stop short and become subsurface intrusive bodies in the crust, cooling slowly to form crystalline plutons such as batholiths (see Figures 8.7 and 8.8). In these processes of crustal formation, you can literally follow the cycling of materials through the tectonic cycle.

A surprising discovery is that each of Earth's major plates actually is a collage of many crustal pieces acquired from a variety of sources. Accretion, or accumulation, occurred as crustal fragments of ocean floor, curved chains (or arcs) of volcanic islands, and other pieces of continental crust were forced against the edges of continental shields and platforms. These migrating crustal pieces, which became attached to the plates, are **terranes** (not to be confused with "terrain," which refers to the topography of a tract of land).

These displaced terranes, sometimes called *microplate* or *exotic terranes*, have histories different from those of the continents that capture them. They are usually framed by fault-zone fractures and differ in rock composition and structure from their new continental homes. As an example, accreted terranes are particularly prevalent in the region surrounding the Pacific. At least 25% of the growth of western North America is attributed to the accretion of terranes since the early Jurassic period (190 million years ago). A good example is the Wrangell Mountains, which lie just east of Prince William Sound and the city of Valdez, Alaska. The *Wrangellia terranes*—a former volcanic island arc and associated marine sediments from near the equator—migrated approximately 10,000 km (6200 mi) to form these mountains (Figure 9.6).

The Appalachian Mountains, extending from Alabama to the Maritime Provinces of Canada, possess bits of land once attached to ancient portions of Europe, Africa, South America, Antarctica, and various oceanic islands. The discovery of terranes, made only in the 1980s, demonstrates one of the ways continents are assembled.

Crustal Deformation Processes

Rocks, whether igneous, sedimentary, or metamorphic, are subject to powerful stress by tectonic forces, gravity, and the weight of overlying rocks. The three types of stress, *tension* (stretching), *compression* (shortening), and *shear* (tearing and twisting), are shown in Figure 9.7. *Strain* is how rocks respond to these stresses, as expressed in *folding* (bending) or *faulting* (breaking). Whether a rock bends or breaks depends on several factors, including its composition and how much pressure is on the rock. An important quality is whether the rock is *brittle* or *ductile*. The patterns created by these processes are evident in the landforms we see today.

Folding and Broad Warping

When layered rock strata are subjected to compressional forces, they become deformed (Figure 9.8). Convergent plate boundaries intensely compress rocks, deforming them in a process known as **folding**, much as layers of thick fabric, stacked flat on a table, would bend and form similar folds if we slowly pushed on opposite ends of the stack (Figure 9.8a).

If we then draw a line down the center axis of a resulting ridge and trough, we are able to see how the names of the folds are assigned. Along the *ridge* of a fold, layers *slope*

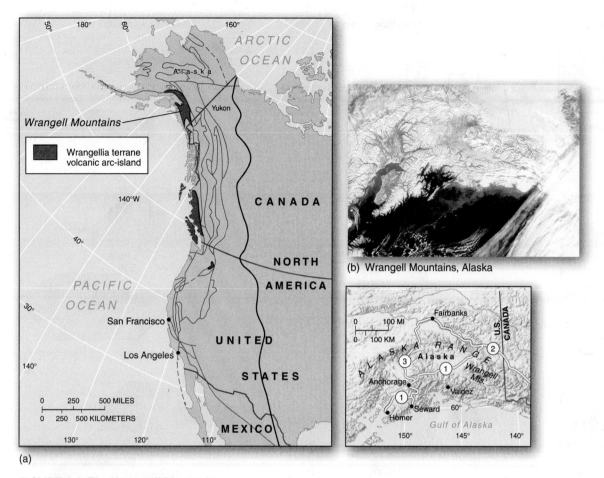

(a)

FIGURE 9.6 The Wrangell Mountains.
(a) Wrangellia terranes, highlighted among the other terranes along the western margin of North America, occur in four segments. (b) Snow-covered Wrangell Mountains north of the Chugach Mountains in this November 7, 2001, image in east-central Alaska and across the Canadian border. Note Mount McKinley (Denali) at 6194 m (20,322 ft) in the upper-left part of the image, north of Cook Inlet, the highest elevation in North America. [(a) Based on data from US Geological Survey; (b) *Terra* image courtesy of MODIS Land Rapid Response Team, NASA/GSFC.]

 Terrane Formation

downward away from the axis, resulting in an **anticline**. In the trough of a fold, however, layers slope *downward toward the axis*, producing a **syncline** (Figure 9.8a, left).

If the axis of either type of fold is not "level" (horizontal, or parallel to Earth's surface), the fold-axis then *plunges*, inclined (dipped down) at an angle. Relative to resource recovery it is important to know the angle of folds to Earth's surface and their location. Petroleum, for example, collects in the upper portions of folds in permeable rock layers such as sandstone.

A residual ridge may form within a syncline due to the weathering response of rock strata of different resistances (a "synclinal ridge," Figure 9.8a, right). An interstate highway roadcut dramatically exposes such a synclinal ridge in western Maryland (Figure 9.8b). Compressional forces often push folds far enough that they actually overturn on their own strata ("overturned anticline," near center of figure). Further stress eventually fractures the rock strata along distinct lines, and some overturned folds are thrust up, causing a considerable shortening of the original strata ("thrust fault"). Areas of

intense stress, compressional folding, and faulting are visible in a roadcut where the San Andreas fault in Southern California passes beneath the Antelope Valley Freeway (Figure 9.8c).

The Canadian Rocky Mountains and the Appalachian Mountains illustrate well the complexity of the resulting folded landscape. Satellites let us view many of these structures from an orbital perspective, as in Figure 9.9 (p. 308). Northwest of the Strait of Hormuz, just north of the Persian Gulf, are the Zagros Mountains of Iran. In the satellite image, anticlines form the parallel ridges and active weathering and erosion processes expose underlying strata.

In addition to the rumpling of rock strata just discussed, Earth's continental crust also is subjected to broad warping actions. These actions produce similar up-and-down bending of strata, but the bends are far greater in extent than those produced by folding. Warping forces include mantle convection, isostatic adjustment such as from the loss of weight from previous ice loads across northern Canada, or swelling from an

Stress	Resulting strain	Surface expressions

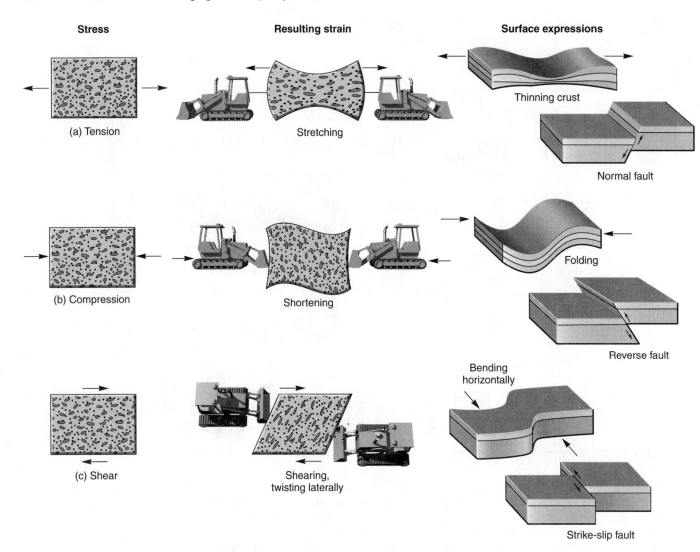

(a) Tension — Stretching — Thinning crust / Normal fault

(b) Compression — Shortening — Folding / Reverse fault

(c) Shear — Shearing, twisting laterally — Bending horizontally / Strike-slip fault

FIGURE 9.7 Three kinds of stress, strain, and resulting surface expressions.
The bulldozers represent the stress, or force, on the rock strata. (a) Tensional stress produces a stretching and thinning of the crust and a *normal fault*. (b) Compression produces shortening and folding and *reverse faulting* of the crust. (c) In a back-and-forth horizontal motion, shear stress produces a bending of the crust and, on breaking, a *strike-slip fault*.

underlying hot spot. Warping features include small, individual, foldlike structures called *basins* and *domes* (Figure 9.10a, b). Or they can range up to regional features the size of the Ozark Mountain complex in Arkansas and Missouri, the Colorado Plateau in the West, the Richat dome in Mauritania, or the Black Hills of South Dakota (Figure 9.10c, d).

Faulting

A freshly poured concrete sidewalk is smooth and strong. But stress the sidewalk by driving heavy equipment over it, and the resulting strain might cause a fracture. Pieces on either side of the fracture may move up, down, or horizontally, depending on the direction of stress. When rock strata are strained beyond their ability to remain a solid unit, they express the strain as a fracture. Displacement of rock formations on either side of the fracture is the process of **faulting**.

Thus, *fault zones* are areas of crustal movement. At the moment of fracture the fault line shifts and a sharp release of energy occurs—an **earthquake** or *quake* (Figure 9.11). The fracture surface along which the two sides of a fault move is the *fault plane*. The tilt and orientation of the fault plane provide the basis for naming the three basic types of faults.

Normal Fault   A **normal fault**, or tension fault, occurs when rocks pull apart. They move vertically along an inclined fault plane so that one "block" of rock ends up lower than the other (Figure 9.11a). The downward-shifting side is the *hanging wall*; it drops relative to the *footwall block*. The exposed fault plane sometimes is visible along the base of faulted mountains, where individual ridges truncate owing to the movements of the fault and appear as triangular facets at the ends of the ridges, as shown in the two photos. A cliff formed by faulting is a *fault scarp*, or *escarpment*.

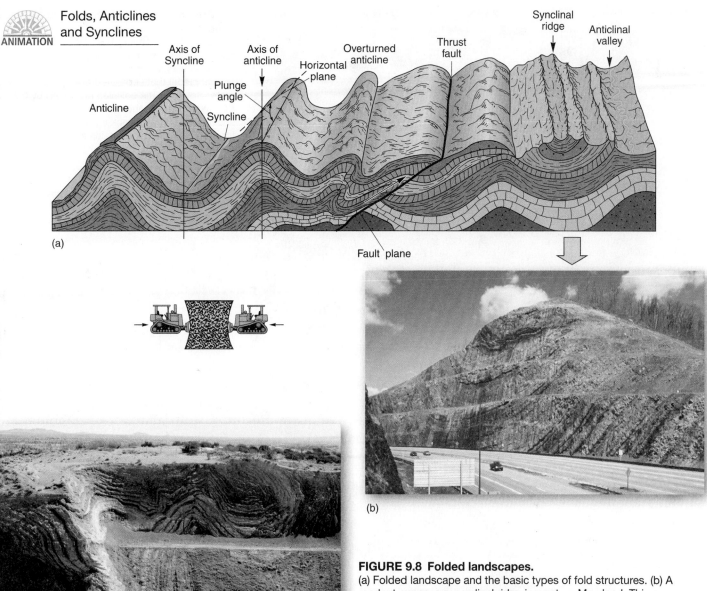

Folds, Anticlines and Synclines

ANIMATION

(a)

(b)

(c)

FIGURE 9.8 Folded landscapes.
(a) Folded landscape and the basic types of fold structures. (b) A roadcut exposes a synclinal ridge in western Maryland. This syncline is a natural outdoor classroom, and the state of Maryland built an interpretive center with a walkway above the highway for observation. (c) Compressional folding and faulting in a roadcut along the San Andreas fault zone in Southern California. [Photos by (b) Mike Boroff/Photri-Microstock; (c) Bobbé Christopherson.]

Reverse (Thrust) Fault Compressional forces associated with converging plates force rocks to move *upward* along the fault plane, producing a **reverse fault**, or compression fault (Figure 9.11b). On the surface it appears similar in form to a normal fault. If the fault plane forms a low angle relative to the horizontal, the fault is a **thrust fault**, or *overthrust fault*, indicating that the overlying block has shifted far over the underlying block (Figure 9.8a, "thrust fault"). Place one hand palm-down on the back of the other and move your hands past each other—this is the motion of a low-angle thrust fault with one side pushing over the other.

In the Alps, several such overthrusts result from compressional forces of the ongoing collision between the African and Eurasian plates. Beneath the Los Angeles

Basin, such overthrust faults produce a high risk of earthquakes and caused many quakes in the twentieth century, including the $30 billion 1994 Northridge earthquake. These blind (unknown until they rupture) thrust faults beneath the Los Angeles region are a major earthquake threat in the future.

Strike-Slip Fault If movement along a fault plane is horizontal, as produced by a *transform fault*, it forms a **strike-slip fault** (Figure 9.11c). The movement is *right-lateral* or *left-lateral*, depending on the motion perceived when you observe movement on one side of the fault relative to the other side.

Although strike-slip faults do not produce cliffs (scarps), they can create linear *rift valleys* and sag ponds, (text continued on page 310)

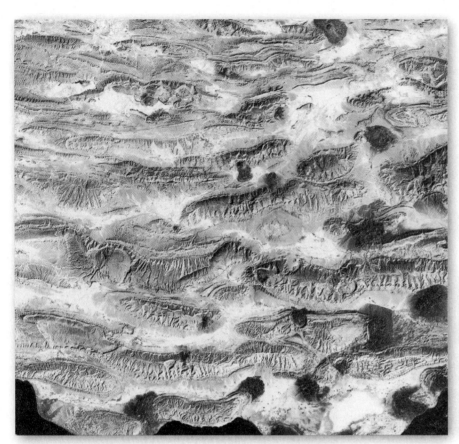

FIGURE 9.9 Folding in the Zagros crush zone, Iran.
Southern Zagros Mountains in the Zagros crush zone between the Arabian and Eurasian plates. This area was a dispersed terrane (migrating crustal piece) that separated from the Eurasian plate. However, the collision produced by the northward push of the Arabian block is now shoving this terrane back into Eurasia and forming the folded mountains shown. [NASA image.]

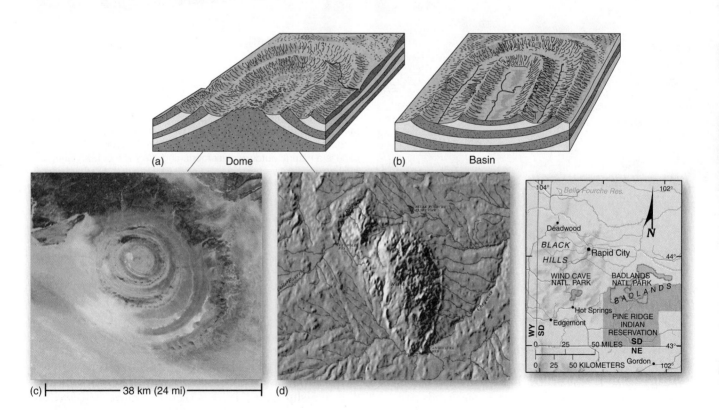

(a) Dome   (b) Basin

(c) ⊢———— 38 km (24 mi) ————⊣   (d)

FIGURE 9.10 Domes and basins.
(a) An upwarped dome. (b) A structural basin. (c) The Richat dome structure of Mauritania. (d) The Black Hills of South Dakota is a dome structure, shown here on a digitized relief map. [(c) Image from *Terra* ASTER sensor, October 7, 2000, courtesy of NASA/GSFC/MITI and US/Japan ASTER Science Team; (d) from USGS digital terrain map I-2206, 1992.]

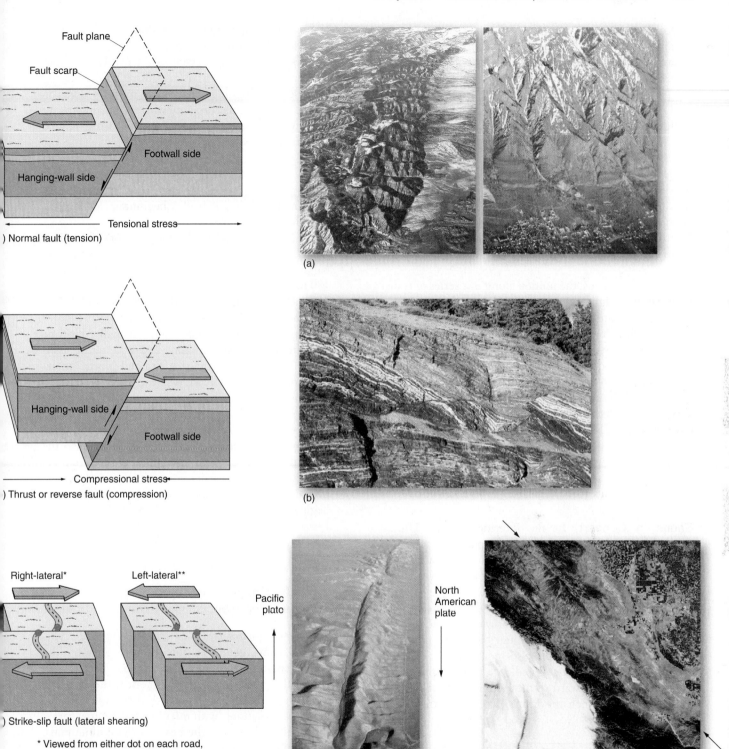

Fault plane

Fault scarp

Hanging-wall side

Footwall side

Tensional stress

) Normal fault (tension)

Hanging-wall side

Footwall side

Compressional stress

) Thrust or reverse fault (compression)

Right-lateral*   Left-lateral**

Pacific plate

North American plate

) Strike-slip fault (lateral shearing)

* Viewed from either dot on each road, movement to opposite side is *to the right*.
** Viewed from either dot on each road, movement to opposite side is *to the left*.

(a)

(b)

(c)

(d)

FIGURE 9.11 Types of faults.
(a) A normal fault produced by tension in the crust, visible along the edge of mountain ranges in Nevada and along the Wasatch Front in Utah. (b) A thrust, or reverse, fault, produced by compression in the crust, visible in these offset strata in coal seams and volcanic ash in British Columbia. (c) A strike-slip fault produced by lateral shearing, clearly seen looking north along the San Andreas fault rift zone on the eastern edge of the Coast Ranges in the California-Pacific plate to the left and North American plate to the right. (d) Can you find the San Andreas fault on the satellite image? Look for a linear rift stretching from southeast to northwest, at the edge of the coastal mountains and San Joaquin Valley. Use the arrows as a guide. [Photos by (a) Bobbé Christopherson; (b) Fletcher and Baylis/Photo Researchers, Inc.; (c) Kevin Schafer/Peter Arnold, Inc.; (d) *Terra* image courtesy of MISR Team, NASA/GSFC/JPL.]

ANIMATION

Fault Types, Transform Faults, Plate Margins

where disruptions of groundwater flows occur, as is the case with the San Andreas fault system of California. Clearly visible in the photograph is the rift where the edges of the North American and Pacific plates are grinding past each other as a result of transform faults associated with a former sea-floor spreading center. The evolution of this fault system is shown in Figure 9.12.

Note in the figure how the East Pacific rise developed as a spreading center with associated transform faults (1) while the North American plate was progressing westward after the breakup of Pangaea. Forces then shifted the transform faults toward a northwest–southeast alignment along a weaving axis (2). Finally, the western margin of North America overrode those shifting transform faults (3). Note the convergence rate is a rapid 4 cm (1.6 in.) per year.

In *relative* terms, the motion along this series of transform faults is right-lateral, whereas in *absolute* terms the North American plate is still moving westward. Consequently, the San Andreas system is a series of faults that are *transform* (associated with a former spreading center), *strike-slip* (horizontal in motion), and *right-lateral* (one side is moving to the right relative to the other side).

The North Anatolian fault system in Turkey is a strike-slip fault and has a right-lateral motion similar to the San Andreas system (Figure 9.13). A progression of earthquakes have hit along the entire extent of the fault system in Turkey since 1939; the latest devastation was in August and November 1999 and again in February 2002. Strike-slip faulting also struck Kobe, Japan (see Figure 9.21).

Faults in Concert In the interior western United States, the *Basin and Range Province* experiences tensional forces caused by uplifting and thinning of the crust, which crack the surface into aligned pairs of normal faults and a distinctive landscape. (Please refer to Chapter 12 and the section on these basin and range desert landscapes.) The term **horst** is applied to upward-faulted blocks; **graben** refers to downward-faulted blocks. Examples of horst-and-graben landscapes also include the Great Rift Valley of East Africa (associated with crustal spreading), which extends northward to the Red Sea, and the Rhine graben through which the Rhine River flows in Europe (Figure 9.14).

Orogenesis (Mountain Building)

Orogenesis literally means the birth of mountains (*oros* comes from the Greek for "mountain"). An *orogeny* is a mountain-building episode that thickens continental crust over millions of years. Earth's major chains of folded and faulted mountains, called *orogens*, correlate remarkably with the plate tectonics model. Orogenesis can occur through large-scale deformation and uplift of the crust. It also may include the capture of migrating exotic terranes, adding by accretion to the continental margins, or the intrusion of granitic magmas to form plutons. Erosion

often exposes these granite masses following uplift. Uplift is the final act of the orogenic cycle.

No orogeny is a simple event; many involve previous developmental stages dating back far into Earth's past, and the processes are ongoing today. Care must be taken in identifying any particular orogeny with the mountains we see today. Major mountain ranges, and the latest related orogenies that caused them, include:

- Rocky Mountains of North America—*Laramide orogeny*, 40–80 million years ago (m.y.a.); third mountain-building episode, beginning 170 m.y.a., including Sevier Orogeny;
- Sierra Nevada and Klamath Mountains of California—*Nevadan orogeny*, with faulting 29–35 m.y.a. and older batholithic intrusion emplacements dating back 80 to 180 million years;
- Appalachian Mountains and the Ridge and Valley Province (nearly parallel ridges and valleys) of the eastern United States and Canadian Maritime Provinces—*Alleghany orogeny*, 250–300 m.y.a., associated with the collision of Africa and North America, preceded by at least two earlier orogenies; in Europe this is contemporary with the *Hercynian orogeny*;
- Alps of Europe—*Alpine orogeny*, 2 to 66 m.y.a., principally in the Tertiary Period and continuing to the present across southern Europe and the Mediterranean, with many earlier episodes (Figure 9.15);
- Himalayas of Asia—*Himalayan orogeny*, 45–54 m.y.a., beginning with the collision of the India plate and Eurasia plate and continuing to the present.

Types of Orogenies

Figure 9.16 (p. 314) illustrates convergent plate collisions associated with orogenesis:

a. *Oceanic plate–continental plate collision orogenesis.* This type of convergence is now occurring along the Pacific coast of the Americas and has formed the Andes, the Sierra of Central America, the Rockies, and other western mountains. We see folded sedimentary formations, with intrusions of magma forming granitic plutons at the heart of these mountains. Their buildup has been augmented by the capturing of displaced terranes, cemented during their collision with the continental mass. Also, note the associated volcanic activity inland from the subduction zone.

b. *Oceanic plate–oceanic plate collision orogenesis.* Such collisions can produce either simple volcanic island arcs or more complex arcs, such as Indonesia and Japan, which include deformation and metamorphism of rocks and granitic intrusions. These processes formed the chains of island arcs and volcanoes that continue from the southwestern Pacific to the western Pacific, the Philippines, the Kurils, and on through portions of the Aleutians.

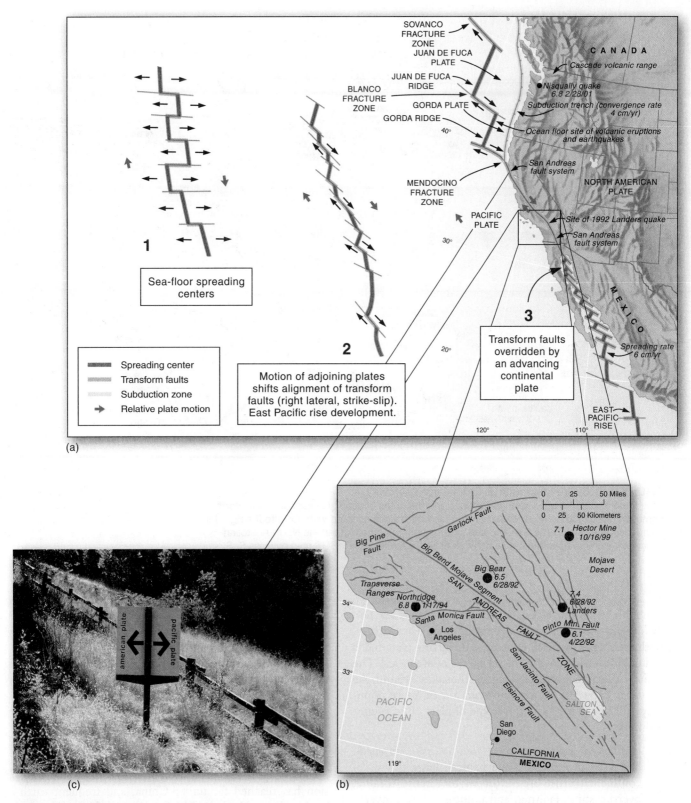

FIGURE 9.12 San Andreas fault formation.
(a) Formation of the San Andreas fault system as a series of transform faults in three successive stages.
Note the epicenter location of the January 2001 magnitude 6.8 Nisqually quake (Washington state),
related to the subduction zone offshore. (b) Enlargement shows the portion of southern California where
the 1992 Landers and Big Bear, the 1994 Northridge (Reseda), and 1999 Hector Mine earthquakes
occurred. (c) Trail sign near the epicenter of the 1906 San Francisco earthquake roughly marks the
Pacific–North American plate boundary zone, Marin County, Point Reyes National Seashore, California.
[(c) Photo by author.]

ANIMATION

Transform Faults,
Plate Margins

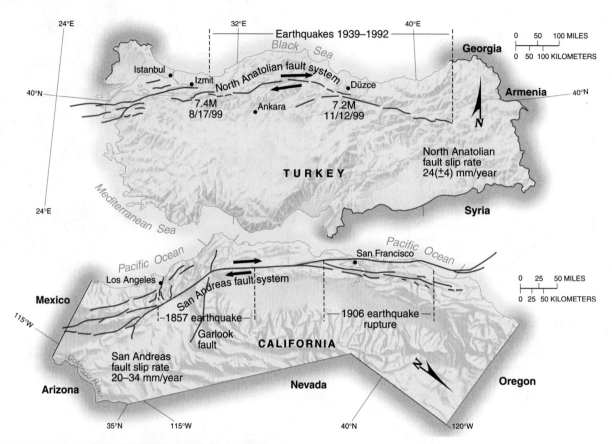

FIGURE 9.13 Strike-slip faults in Turkey and California.
Note these two strike-slip, right-lateral faults and the dates of activity along each—the North
Anatolian (Turkey) and San Andreas (California). California is oriented with north to the right for
comparison purposes. The San Andreas system is more complex than the North Anatolian
because it is involved with a dense network of active faults. [After USGS map by Ross S. Stein.]

Both collision types, (a) oceanic–continental and (b) oceanic–oceanic, are active around the Pacific Rim. Both are thermal in nature, because the diving plate melts and migrates back toward the surface as molten rock. The region of active volcanoes and earthquakes around the Pacific is known as the *circum-Pacific belt* or, more popularly, the *ring of fire*. The third type of plate convergence is different.

  c. *Continental plate–continental plate collision orogenesis.* Here the orogenesis is mechanical; large masses of continental crust are subjected to intense folding, overthrusting, faulting, and uplifting. The converging plates crush and deform both marine sediments and basaltic oceanic crust. The formation of the European Alps is a result of such compression forces and includes considerable crustal shortening, forming great overturned folds called *nappes* (see Figure 9.15).

As mentioned earlier, the collision of India with the Eurasian landmass produced the Himalayan Mountains. That collision is estimated to have shortened the overall continental crust by as much as 1000 km (about 600 mi) and to have produced telescoping sequences of thrust faults at depths of 40 km (25 mi). The resulting Himalayas feature the tallest above-sea-level mountains on Earth, including Mount Everest at 8850 m elevation (29,035 ft), and all 10 of Earth's highest peaks (measured from sea level). Along the 2500-m (1553-mi) fault zone that marks this plunging Indian subcontinent, a 40-km (25-mi) segment snapped in October 2005. A magnitude 7.6 quake hit the Kashmir region of Pakistan, killing more than 83,000 people. The disruption created by this plate collision has reached far under China, and frequent earthquakes there signal the continuation of this rapid-paced collision, as India moves northward at about 6 mm (0.24 in.) a year.

Grand Tetons and the Sierra Nevada The Sierra Nevada of California and the Grand Tetons of Wyoming are in the later stages of mountain-building orogenesis. Each is a *tilted-fault block* mountain range, in which a normal fault on one side of the range produces a tilted

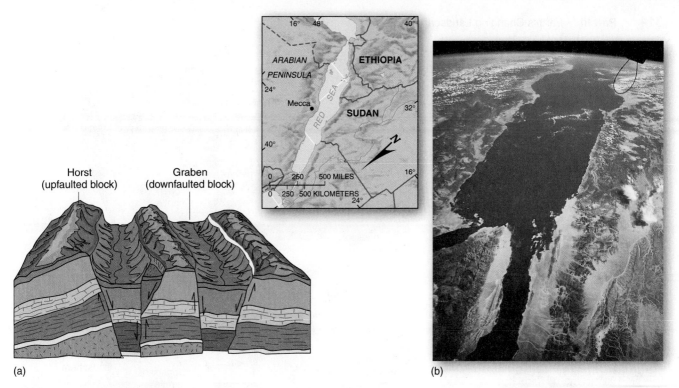

Horst
(upfaulted block)

Graben
(downfaulted block)

(a)

(b)

FIGURE 9.14 Faulted landscapes.
(a) Pairs of faults produce a horst-and-graben landscape characteristic of the Basin and Range
Province in the western United States. (b) The Red Sea occupies a down-dropped block that is
part of the rift system through East Africa. [(b) *Gemini* photo from NASA.]

FIGURE 9.15 European Alps.
Western (France), Central (Italy), and Eastern (Austria) segments comprise
the crescent shape of the Alps. Complex overturned faults and crustal
shortening due to compressional forces occur along convergent plates.
The Alps are some 1200 km (750 mi) in length, occupying 207,000 km^2
(80,000 mi^2). Note the aerosols and other pollutants concentrated south
of the Alps in northern Italy in the image. [*Terra* image courtesy of MODIS
Land Rapid Response Team, NASA/GSFC.]

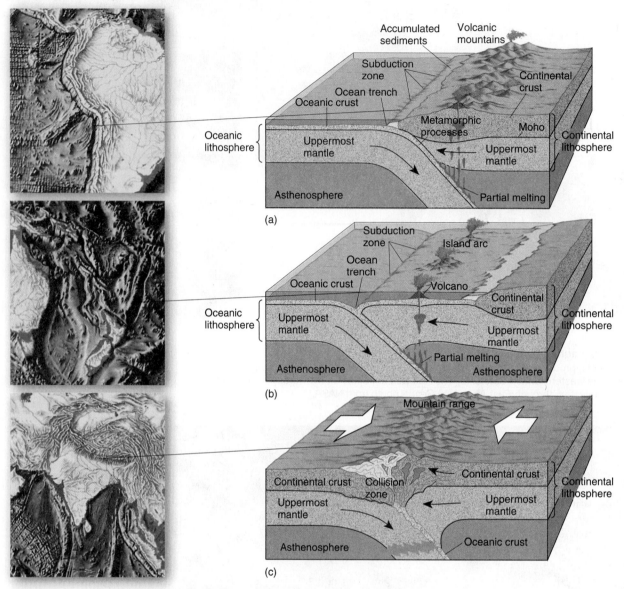

FIGURE 9.16 Three types of plate convergence.
Real-world examples illustrate three types of crustal collisions. (a) Oceanic-continental (example: Nazca plate–South American plate collision and subduction). (b) Oceanic-oceanic (example: New Hebrides Trench near Vanuatu, 16° S 168° E). (c) Continental-continental (example: India plate and Eurasian landmass collision and resulting Himalayan Mountains). Can you identify more of these plate convergence areas on the chapter-opening map? [Inset illustrations derived from *Floor of the Oceans*, 1975, by Bruce C. Heezen and Marie Tharp. © 1980 by Marie Tharp.]

Plate Boundaries

landscape of dramatic relief (Figure 9.17). Slowly cooling magma intruded into those blocks and formed granitic cores of coarsely crystalline rock. After tremendous tectonic uplift and the removal of overlying material through weathering, erosion, and transport, those granitic masses are now exposed in each mountain range. In some areas, the overlying material originally covering these batholiths was more than 7500 m (25,000 ft) thick.

Recent research in the Sierra Nevada disclosed that some of the uplift was isostatic, caused by the erosion of overburden and loss of melting ice mass (that is, loss in overlying weight) following the last ice age some 18,000 years ago. The accumulation of sediments in the adjoining valleys depressed the crust, thus enhancing relief in the landscape.

The Appalachian Mountains The Appalachians are older than the higher mountains of western North America (35–80 m.y.a.). The eroded, fold-and-thrust belt of the eastern United States and southeastern Canada has origins dating to the formation of Pangaea and the collision of Africa and North America (250–300 m.y.a.). As noted, the complexity of the *Alleghany orogeny* derives from at least two earlier orogenic cycles of uplift and the accretion of several captured terranes.

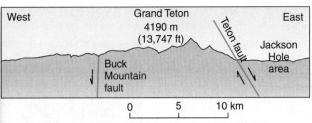

FIGURE 9.17 Tilted-fault block.
The Teton Range in Wyoming is an example of a tilted-fault block, a range of scenic beauty featuring 2130 m (7000 ft) of rugged relief between Jackson Hole and the summits. The Grand Teton is the highest peak in the range at 4190 m (13,747 ft). [Photo by author.]

Structure and composition link the mountain ranges separated by the Atlantic Ocean and the Pangaea collision and separation. In fact, the Lesser (or Anti-) Atlas Mountains of Mauritania and northwestern Africa were connected to the Appalachians at some time in the past, but the mountains embedded in the African plate rafted apart from the Appalachians.

The Appalachian Mountain region comprises several landscape subregions (refer to the locator map in Figure 9.18): the Ridge and Valley Province (elongated sequences of folded sedimentary rock); the Blue Ridge Province (principally of crystalline rock, highest where North Carolina, Virginia, and Tennessee converge); the Piedmont (hilly to gentle terrain along most of the eastern and southern margins of the mountains); the east coastal plain (from gentle hills to flat plains that extend to the coast); and the Canadian maritime provinces.

Figure 9.18 displays the linear folds of the Appalachian system. Note how these dissected ridges are cut through by rivers, forming *water gaps*. These important breaks in the rugged ridges greatly influenced migration, settlement patterns, and the diffusion of cultural traits in the 1700s. The initial flow of people, goods, and ideas was guided by this topography. The Susquehanna River in east-central Pennsylvania is one watercourse that interrupts the ridges and valleys of the Appalachians.

World Structural Regions

Figure 9.19 defines seven essential structural regions, highlights Earth's two major continental mountain systems, and allows interesting comparison with the chapter-opening illustration of Earth's continental crust. Looking at the distribution of world structural regions helps summarize the information presented about the three rock-forming processes (igneous, sedimentary, metamorphic), plate tectonics, landform construction, and orogenesis.

The map reveals two large mountain chains—Earth's *alpine system*. The relatively young mountains along the western margins of the North and South American plates stretch from the tip of Tierra del Fuego to the massive peaks of Alaska, forming the *Cordilleran system*. The

mountains of southern Asia, China, and northern India continue in a belt through the upper Middle East to Europe and the European Alps, constituting the *Eurasian-Himalayan system*.

As you examine the map, identify the continental shields at the heart of each landmass. These areas are surrounded by continental platforms composed of sedimentary deposits. Various mountain chains, rifted regions, and isolated volcanic areas are portrayed. On the continent of Australia, you see older mountain sequences to the east, sedimentary layers covering basement rocks west of these ranges, and portions of the original Gondwana, an ancient landscape in the central and western region. (Remember that Gondwana was a landmass that included Antarctica, Australia, South America, Africa, and the southern portion of India; it broke away from Pangaea some 200 m.y.a.)

Earthquakes

Crustal plates do not move smoothly past one another. Instead, stress builds strain in the rocks along plate boundaries until the sides release and lurch into new positions. Devastating earthquakes can result from these rapid shifts of crust. This reality was brought home to millions of television viewers as they watched the 1989 World Series from Candlestick Park near San Francisco. Less than half an hour before the call to "play ball," a powerful earthquake rocked the region and turned sportscasters into newscasters and sports fans into disaster witnesses.

In 2007, southern Sumatra, Indonesia, was hit with M 8.4 and M 7.9 quakes, near the same plate boundary where there have been four other major quakes since 1998. Along the coast, the Australian Plate moves northeast, subducting beneath the Sunda Plate on which Sumatra rides. The speed of motion is about 6 cm (2.4 in.) a year. This is where the M 9.1 earthquake struck December 26, 2004, that triggered the devastating Indian Ocean tsunami (more on this in Chapter 13). For complete listings of recent earthquakes, go to **http://earthquake.usgs.gov/eqcenter/.**

Area of photo (b)

(a)

(b)

FIGURE 9.18 The Appalachian Mountains.
(a) Appalachian Mountain region of Pennsylvania south through Maryland, Virginia, and West Virginia in a *Terra* MODIS sensor image from October 11, 2000, showing true fall colors. The highly folded nature of this entire region is clearly visible in the satellite image. (b) The folded Little and Line Mountains end at the Susquehanna River, along the left edge of this aerial photo.
[(a) *Terra* image courtesy of MODIS Land Science Team, NASA/GSFC; photo (b) by Bobbé Christopherson.]

Expected Quakes and Those of Deadly Surprise

In the Liaoning Province of northeastern China, ominous indications of tectonic activity began in 1970. Foreboding symptoms included land uplift and tilting, increased numbers of minor tremors, and changes in the region's magnetic field—all of this after almost 120 years of quiet. These precursors of tectonic events continued for almost 5 years before Chinese scientists took the bold step of forecasting an earthquake. Finally, on February 4, 1975, some 3 million people evacuated in what turned out to be a

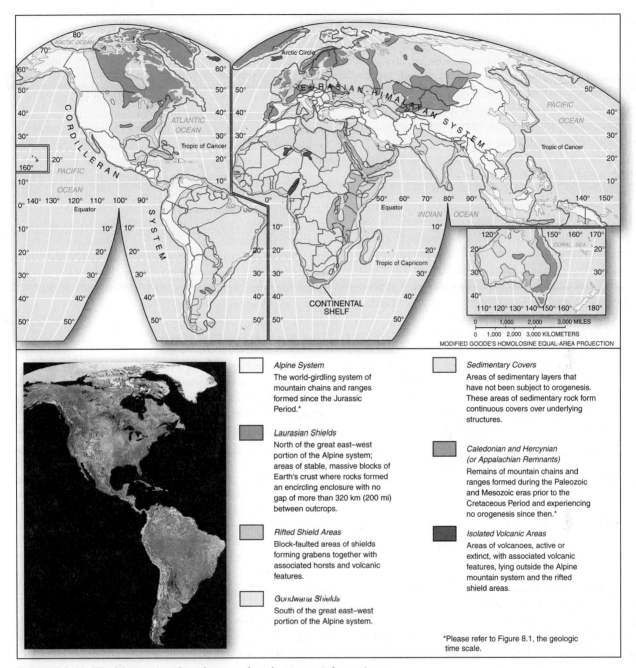

FIGURE 9.19 World structural regions and major mountain systems.
Because each structural region in this figure includes related landforms adjacent to the central feature of the region, some of the regions appear larger than the structures themselves. Structural regions in the Western Hemisphere are visible on this composite *Landsat* image. [After Richard E. Murphy, "Landforms of the World," Map Supplement No. 9, *Annals of the Association of American Geographers* 58, no. 1 (March 1968). Adapted by permission. Image courtesy of EROS Data Center and the National Geographic Society.]

timely manner; the quake struck within the predicted time frame. Ninety percent of the buildings in the city of Haicheng were destroyed, but thousands of lives were saved, and success was proclaimed—an earthquake was forecast and preparatory action taken for the first time in history.

In contrast, only 17 months later, at Tangshan in the northeastern province of Hebei (Hopei), a M 7.4 earthquake occurred on July 28, 1976, without warning, killing a quarter of a million people. (This is the official death toll; other estimates range as high as 650,000 dead.) The earthquake also destroyed 95% of the buildings and was strong enough to throw people against the ceilings of their homes. More recently, in May 2008, the Chengdu quake mentioned in Figure 9.1d hit the Sichuan Province of China at mid-afternoon without warning. How is it possible that these three tectonic events produced such different human consequences—one forecasted and two as surprises?

Earthquake Essentials

Tectonic earthquakes are those quakes associated with faulting. Their vibrations are transmitted as waves of energy throughout Earth's interior and are detected with a **seismograph**, an instrument that records vibrations in the crust—some 4000 seismographs form a global network. Earthquakes are rated on two different scales, an intensity scale and a magnitude scale.

An *intensity scale* is useful in classifying and describing damage to terrain and structures following an earthquake. Earthquake intensity is rated on the arbitrary *Mercalli scale*, a Roman-numeral scale from I to XII representing "barely felt" to "catastrophic total destruction." It was designed in 1902 and modified in 1931 to be more applicable to conditions in North America (Table 9.1).

A system designed by Charles Richter in 1935 estimated earthquake magnitude. In this method, a seismograph located at least 100 km (62 mi) from the origin of the quake (epicenter) records the amplitude of seismic waves. The *Richter scale* is then estimated from the seismogram chart. The scale is open-ended and logarithmic; that is, each whole number on the scale represents a 10-fold increase in the measured *wave amplitude*. Translated into energy, each whole number demonstrates a 31.5-fold increase in the amount of energy released. Thus, a 3.0 on the Richter scale represents 31.5 times more energy than a 2.0 and 992 times more energy than a 1.0.

Today, measuring earthquakes is improved in accuracy and made more quantitative. The original scale did not properly measure or differentiate between quakes of high magnitude. Seismologists wanted to know more about what they call the *seismic moment* to understand a broader range of possible motions during an earthquake.

The **moment-magnitude scale** in use since 1993 is more accurate than Richter's *amplitude-magnitude* scale for large earthquakes. Moment magnitude considers the amount of fault slippage produced by the earthquake, the size of the surface (or subsurface) area that ruptured, and the nature of the materials that faulted, including how resistant they were to failure. A reassessment of past quakes increased the rating of some and decreased others. The 1964 earthquake at Prince William Sound in Alaska had an amplitude magnitude of 8.6, but on the moment magnitude scale its rating increases to a 9.2. The 1906 San Francisco quake goes from an 8.25 down to a 7.7; the Tangshan quake from a 7.6 to a 7.4. Table 9.2 presents a sampling of significant earthquakes.

The National Earthquake Information Center in Golden, Colorado, reports earthquake epicenters. For references, see http://www.neic.cr.usgs.gov/ and the National Geophysical Data Center at http://www.ngdc.noaa.gov/seg/. As part of the "Learning from Earthquakes Project," reconnaissance teams examine the sites of major quakes; see http://www.eqnet.org. In southern California, the USGS is coordinating a Southern California Seismographic Network (SCSN at http://www.trinet.org/scsn/scsn.html) to correlate 350 instruments for immediate earthquake location analysis and disaster coordination.

The Nature of Faulting

The subsurface area along a fault plane where the motion of seismic waves is initiated is the *focus*, or *hypocenter* (see Figure 9.22, p. 322). The area at the surface directly above the focus is the **epicenter**. Shock waves produced by an earthquake radiate outward from both the focus and epicenter areas. An *aftershock* may occur after the main shock, sharing the same general area of the epicenter. A *foreshock* also is possible, preceding the main shock. Usually, the greater the distance from an epicenter, the less severe the shock. At a distance of 40 km (25 mi), only 1/10 of the full effect of an earthquake normally is felt. However, if the ground is unstable, distant effects can be magnified, as they were in Mexico in 1985 and in San Francisco in 1989.

We earlier described specific types of faults and faulting motions. How a fault actually breaks is still under investigation, but the basic process involves the **elastic-rebound theory**. Generally, two sides along a fault appear to be locked by friction, resisting any movement despite the powerful forces acting on adjoining pieces of

Description	Effects on Populated Areas	Moment Magnitude Scale	Modified Mercalli Scale	Number per Year*
Great	Damage nearly total	8.0 and higher	XII	1
Major	Great damage	7–7.9	X–XI	17
Strong	Considerable-to-serious damage to buildings; railroad tracks bent	6–6.9	VIII–IX	134
Moderate	Felt-by-all, with slight building damage	5–5.9	V–VII	1,319
Light	Felt-by-some to felt-by-many	4–4.9	III–IV	13,000 (estimated)
Minor	Slight, some feel it	3–3.9	I–II	130,000 (estimated)
Very minor	Not felt, but recorded	2–2.9	None to I	1,300,000 (estimated)

Table 9.1   Magnitude, Intensity, and Frequency of Earthquakes

*Based on observations since 1990.
Source: USGS, Earthquake Information Center.

ANIMATION   Seismograph, How it Works

Table 9.2 A Sampling of Significant Earthquakes*

Year	Date	Location	Number of Deaths	Mercalli Intensity	Moment Magnitude (Richter)
1556	Jan. 23	Shaanxi Province, China	830,000	?	?
1737	Oct. 11	Calcutta, India	300,000	?	?
1812	Feb. 7	New Madrid, Missouri	Several	XI–XII	?
1857	Jan. 9	Fort Tejon, California	?	X–XI	?
1870	Oct. 21	Montreal to Québec, Canada	?	IX	?
1886	Aug. 31	Charleston, South Carolina	?	IX	6.7
1906	Apr. 18	San Francisco, California	3,000	XI	7.7 (8.25)
1923	Sept. 1	Kwanto, Japan	143,000	XII	7.9 (8.2)
1939	Dec. 27	Erzincan, Turkey	40,000	XII	7.6 (8.0)
1960	May 22	Southern Chile	5,700	XII	9.5 (8.6)
1964	Mar. 28	Southern Alaska	131	X–XII	9.2 (8.6)
1970	May 31	Northern Peru	66,000	X–XII	7.9 (7.8)
1971	Feb. 9	San Fernando, California	65	VII–IX	6.7 (6.5)
1972	Dec. 23	Managua, Nicaragua	5,000	X–XII	6.2 (6.2)
1976	Jul. 28	Tangshan, China	250,000	XI–XII	7.4 (7.6)
1978	Sept. 16	Iran	25,000	X–XII	7.8 (7.7)
1985	Sept. 19	Mexico City, Mexico	7,000	IX–XII	8.1 (8.1)
1988	Dec. 7	Armenia-Turkey border	30,000	XII	6.8 (6.9)
1989	Oct. 17	Loma Prieta (near Santa Cruz, California)	67	VII–IX	7.0 (7.1)
1991	Oct. 20	Uttar Pradesh, India	1,700	IX–XI	6.2 (6.1)
1994	Jan. 17	Northridge (Reseda), California	66	VII–IX	6.8
1995	Jan. 17	Kobe, Japan	5,500	XII	6.9
1996	Feb. 17	Indonesia	110	X	8.1
1997	Feb. 28	Armenia–Azerbaijan	1,100	XII	6.1
1997	May 10	Northern Iran	1,600	XII	7.3
1998	May 30	Afghanistan–Tajikistan	4,000	XII	6.9
1998	Jul. 17	Papua, New Guinea	2,200	X	7.1
1999	Jan. 26	Armenia, Colombia	1,000	VIII–IX	6.0
1999	Aug. 17	Izmit, Turkey	17,100	VIII–XI	7.4
1999	Sept. 7	Athens, Greece	150	VI–VIII	5.9
1999	Sept. 20	Chi-Chi, Taiwan	2,500	VI–X	7.6
1999	Nov. 12	Düzce, Turkey	700	VI–X	7.2
2001	Jan. 26	Gujarat state, India	20,023	X–XII	7.7
2002	Nov. 3	Near Denali National Park, Alaska	1	X	7.9
2003	Dec 26	Bam, Iran	30,000	X–XII	6.9
2004	Dec 26	Sumatra, Indonesia	170,000+**	X–XII	9.3
2005	Oct 8	Kashmir region, Pakistan	83,000+	X–XII	7.6
2008	May 12	Chengdu, Sichuan Province, China	87,000+	X–XII	7.9

*Note that earthquakes are not on an increase; the list merely reflects a greater sample for recent years.
**Estimated deaths in 11 countries from quake and resulting tsunami that traveled across the Indian Ocean.

Seismograph, How it Works

crust. This stress continues to build strain along the fault surfaces, storing elastic energy like a wound-up spring. As the strain builds and exceeds the frictional lock, movement occurs, releasing a burst of mechanical energy, and both sides of the fault return to a condition of less strain. Unbroken portions of the fault adjoining the rupture remain stressed.

Think of the fault plane as a surface with irregularities that act as sticking points, preventing movement, similar to two pieces of wood held together by drops of glue rather than an even coating of glue. Research scientists at the USGS and the University of California identified these small areas of high strain as *asperities*. They are the points that break and release the sides of the fault. If the fracture along the fault line is isolated to a small asperity break, the quake will be small in magnitude. Clearly, as some asperities break (perhaps recorded as small foreshocks), the strain increases on surrounding asperities that remain intact. Thus, small earthquakes in an area may be precursors to a major quake. However, if the break involves the release of

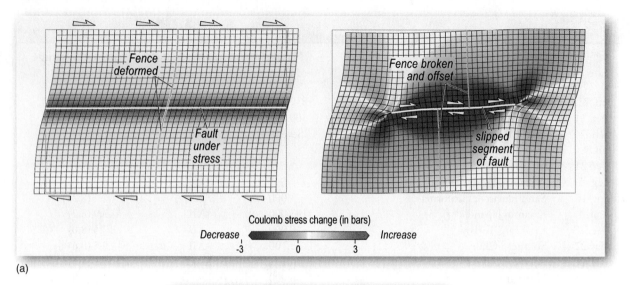

(a)

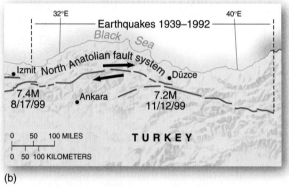

(b)

FIGURE 9.20 Buildup and release of stress and strain along a fault system.
(a) As two sides of this right-lateral, strike-slip fault move, they remain locked by friction, creating
stress. Such stress (a force) produces strain (a deformation) in the rock. Note the fence that crosses
the fault line. This stress continues to build to the break point. (b) In an earthquake, such as the 7.4
magnitude shock that hit Izmit, Turkey, the two sides snap into new positions, strain is released, and
the fence no longer forms a continuous property line. A coulomb is a unit of shear-stress failure
overcoming internal resisting friction and resistance to movement; the color scale ranges from blue
(decreased stress) to red (increased stress). [Illustration (a) courtesy of Serkan Bozkurt, USGS.]

 ANIMATION   Elastic Rebound

strain along several asperities, the quake will be greater in
extent and will involve the shifting of massive amounts of
crust. The latest seismic evidence points to a wavelike pattern, as rupturing spreads along the fault plane, rather than
the entire fault surface giving way at once.

A study of the buildup and release of stress and strain
along the North Anatolian fault system in Turkey (Figure
9.13) illustrates the process (Figure 9.20a). Strain accumulates as the two sides of the fault attempt to move laterally past one another, yet remain locked. Stress forces
continue to build strain until it overwhelms the frictional
sticking points and the two sides snap violently and at
great speed into new positions. An earthquake hits. Note
in Figure 9.20b that strain deformation is relieved in the
area that broke, whereas on either side of the faulted
region the sides remain locked, stresses are stronger, and
resulting strain is increased.

This type of sudden, right-lateral movement rocked
the region around Kobe, Japan, in 1995 with a M 6.9

quake. Early on the morning of January 17, 1995, a year
to the day after the Northridge quake, a section of the
Nojima fault zone, approximately 16 km (10 mi) below
the surface, became the focus of a devastating earthquake
near Kobe. It killed 5500 people, injured 26,000, and
caused over $100 billion in damage. About 200,000 homes
were damaged; some 80,000 buildings collapsed completely; major transportation arteries were severed; and
the busy port was brought to a standstill (Figure 9.21).

The lessons from Kobe are many, especially the
occurrence of liquefaction and the failure of filled land.
Areas of Osaka Bay reclaimed by landfill taken from the
sea liquefied quickly in the shaking. Soil surfaces failed,
collapsing the structures they supported (Figure 12.21b).
Importantly, these human-made landfills and islands are
characteristic of the San Francisco Bay region, where
since the early 1900s about half the original surface area
of the bay has been reclaimed with fill and occupied by
buildings.

(a) (b)

FIGURE 9.21 1995 earthquake in Kobe, Japan.
(a) Catastrophic failure of an elevated freeway; note the failure of the supporting pillars. (b) Ground failure, mainly liquefaction, tilts this building to near collapse; note the space between the building and the ground on the right side. [Photos by Haruyoshi Yamaguchi/Sygma.]

Earthquakes and the San Andreas Fault

In 1906 an earthquake devastated San Francisco, then a city of 400,000 people (several million people live in the metropolitan region today). A M 7.7 (8.25 Richter) tremor rocked the city, felling buildings and initiating a firestorm fed by broken gas pipes. Movement along the San Andreas fault was evident for over 435 km (270 mi). This quake prompted intensive research to discover the nature of faulting. The elastic-rebound theory developed as a result. (Realize that this occurred 6 years before Wegener proposed his continental drift hypothesis.)

The earthquake that disrupted the 1989 World Series, mentioned earlier, involved a portion of the San Andreas fault approximately 16 km (9.9 mi) east of Santa Cruz and 95 km (59 mi) south of San Francisco (Figure 9.22). The quake registered M 7.0, damage totaled $8 billion, 14,000 people were displaced from their homes, 4000 were injured, and 67 killed. A fault ruptured at a focus unusually deep for the San Andreas system—more than 18 km (11.5 mi) below the surface. Unlike previous earthquakes—such as the one in 1906, when the plates shifted a maximum of 6.4 m (21 ft) relative to each other—there was no evidence of a fault plane or rifting at the surface in the Loma Prieta area. Instead, the fault plane suggested in Figure 9.22 shows the two plates moving horizontally approximately 2 m (6 ft) past each other deep below the surface, with the Pacific plate thrusting 1.3 m (4.3 ft) upward. This vertical motion is unusual for the San Andreas fault and perhaps is a clue that this portion of the San Andreas system is more complex than previously thought.

In the San Francisco Bay Area, the present forecast estimates a 62% chance of a M 6.7, or higher, quake striking in the next 30 years (2000–2030) along the San Andreas system of faults. Added to this is a 14% chance of a significant earthquake occurring along some yet unknown fault.

The Southern California Earthquakes

Students at California State University–Northridge, in the San Fernando Valley of southern California, need no reminder of the power of earthquakes. Their campus was near the epicenter of the most devastating earthquake in US history in terms of property damage—$30 billion in destruction across the region. The January 17, 1994, Northridge (Reseda) earthquake, a M 6.8, with over 10,000 aftershocks (through 2000), caused approximately $350 million damage to campus buildings 2 weeks before the beginning of spring classes. Amazingly, the semester began just 3 weeks late in 450 temporary trailers, and graduation was still held in May!

Since the mid-1980s, in an area east of Los Angeles seven earthquakes greater than M 6.0 have struck. After a M 6.1 quake in 1992, a M 7.4 tremor rocked the lightly populated area near Landers, California (see the enlargement map of southern California in Figure 9.12b). Although it was the single largest quake in California in 30 years, the remote location kept injuries and damage slight. Involved in this series were four different faults and some unknown segments. The main faulting caused a displacement of 6.1 m (20 ft). For as yet unexplained reasons, related earthquakes over the next few weeks struck in Mammoth Lakes, about 645 km (400 mi) to the north; at Mount Shasta in northern California; in southern Nevada and Utah; and 1810 km (1125 mi) distant in Yellowstone National Park, Wyoming.

The 1971 San Fernando, 1987 Whittier, 1988 Pasadena, 1991 Sierra Madré, 1994 Northridge (Reseda), and 2008 Chino Hills earthquakes are a few of the many quakes associated with deeply buried thrust faults. Earthquakes roughly originate at foci 18 km (11 mi) deep in the crust. Although not directly aligned with the San Andreas, scientists think that southern California will be affected by more of these thrust-fault actions as strain continues to build along the nearby San Andreas system of faults.

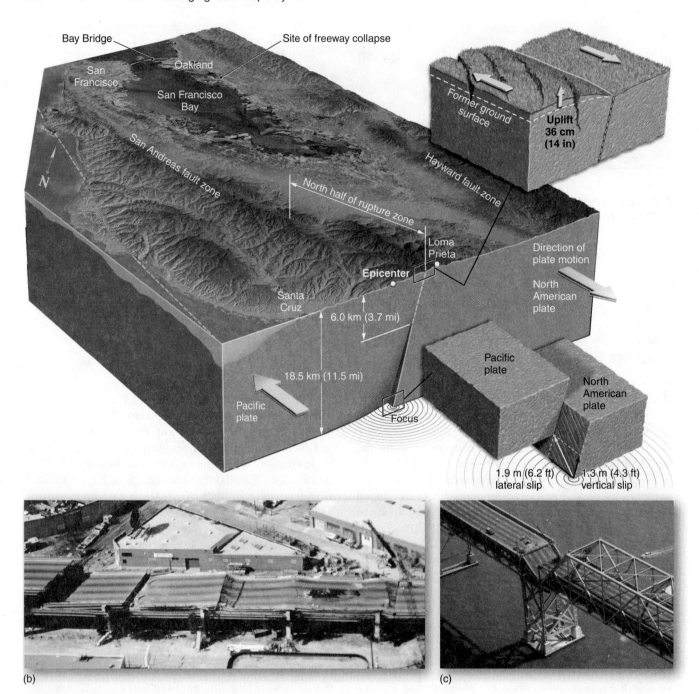

FIGURE 9.22 Anatomy of an earthquake.
(a) The fault-plane solution for the 1989 Loma Prieta, California, earthquake shows the lateral and vertical (thrust) movements occurring at depth. There was no surface expression of this fault plane. (b) In only 15 seconds, more than 2 km (1.2 mi) of the Route 880 Cypress Freeway collapsed. (c) The section failure in the San Francisco–Oakland Bay Bridge actually is the way the bridge handles strain without completely failing. [(a) After P. J. Ward and R. A. Page, *The Loma Prieta Earthquake of October 17, 1989* (Washington, DC: US Geological Survey, November 1989), p. 1; (b) and (c) courtesy of California Department of Transportation.]

 P- and S-Waves, Seismology
ANIMATION

Earthquake Forecasting and Planning

The map in Figure 9.23 plots the epicenters of earthquakes in the United States and southern Canada between 1899 and 1990. These occurrences give some indication of relative risk by region. The challenge is to discover how to predict the *specific time and place* for a quake in the short term.

Actual implementation of an action plan to reduce death, injury, and property damage from earthquakes is difficult. The political environment adds complexity since studies show that an accurate earthquake prediction may threaten a region's economy. Research into the potential socioeconomic impact of earthquake prediction on an

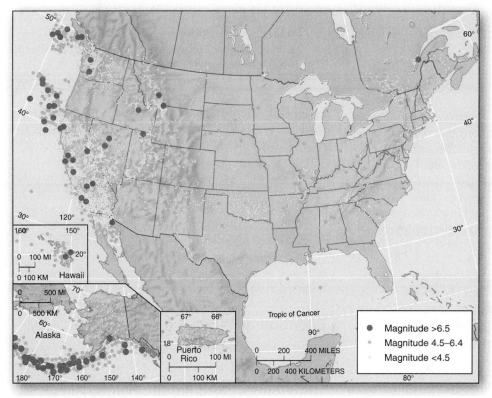

FIGURE 9.23 Seismicity of the United States and southern Canada, 1899–1990. Earthquake occurrences of magnitude 4.5, or greater, indicate areas of greatest seismic risk. Active seismic regions include the West Coast, the Wasatch Front of Utah northward into Canada, the central Mississippi Valley, southern Appalachians, and portions of South Carolina, upstate New York, and Ontario. [From the USGS, National Earthquake Information Center; see http://eqhazmaps.usgs.gov/.]

urban community shows negative economic impacts in the period before the quake hits. Imagine a chamber of commerce, bank, real estate agent, tax assessor, or politician who would welcome an earthquake prediction and such negative publicity for their city. These factors work against effective planning and preparation.

Long-range planning is a complex subject. After the Loma Prieta earthquake in 1989, the National Research Council concluded in a report:

> One of the most jarring lessons from these quakes may be that earthquake professionals have long known many of the things that could have been done to reduce devastation.... The cost-effectiveness of mitigation and the importance of closing the knowledge gap among researchers, building professionals, government officials, and the public are only two of the many lessons from the Loma Prieta that need immediate action.

A valid and applicable generalization is that *humans and their institutions are unable or unwilling to perceive hazards in a familiar environment.* In other words, we tend to feel secure in our homes and communities, even if they are sitting on a quiet fault zone. Such an axiom of human behavior certainly helps explain why large populations continue to live and work in earthquake-prone settings. Similar questions also can be raised about populations in areas vulnerable to floods, droughts, hurricanes and coastal storm surge, and settlements on barrier islands. (See the Natural Hazards Observer at http://www.colorado.edu/hazards/o/ as part of the Natural Hazards Center at the University of Colorado at http://www.colorado.edu/hazards/index.html.)

Volcanism

Volcanic eruptions across the globe remind us of Earth's internal energy. Ongoing eruption activity matches plate tectonic activity, as shown on the map in Figure 8.18. A few recent eruptions include Kīlauea in Hawai'i, Soufriere Hills on Montserrat (Lesser Antilles), Mount Etna (Sicily, Italy), several on the Kamchatka Peninsula (Russia), Rabaul (Papua, New Guinea), Grímsvötn (Iceland), Axial Summit and the Jackson Segment (sea-floor spreading center off the coast of Oregon), Shishaldin (Unimak Island, Alaska; see Figure 9.35), Lascar (Chile), White Island (New Zealand; see Figure 9.30), Nyiragongo (Congo, Africa), and Mayon (Philippines), among many.

Over 1300 identifiable volcanic cones and mountains exist on Earth, although fewer than 600 are active (have had at least one eruption in recorded history). In an average year, about 50 volcanoes erupt worldwide, varying from modest activity to major explosions. (An index to the world's volcanoes is at http://vulcan.wr.usgs.gov/Volcanoes/framework.html, and the National Museum of Natural History Global Volcanism Program lists information about more than 8500 eruptions at http://www.nmnh.si.edu/gvp/. For more volcanic eruption listings, see http://volcano.und.nodak.edu/, or check http://www.geo.mtu.edu/volcanoes/links/observatories.html for links to volcano observatories across the globe.)

Eruptions in remote locations and at depths on the seafloor go largely unnoticed, but the occasional eruption of great magnitude near a population center makes headlines. North America has about 70 volcanoes (mostly inactive) along the western margin of the continent and Hawai'i. Mount St. Helens in Washington state is a famous active example, and over 1 million visitors a year travel to Mount St. Helens Volcanic National Monument to see volcanism for themselves.

Volcanic Features

A **volcano** forms at the end of a central vent or conduit that rises from the asthenosphere through the crust into a volcanic mountain. A **crater**, or circular surface depression, usually forms at the summit. Magma rises and collects in a magma chamber deep below the volcano until conditions are right for an eruption. This subsurface magma carries tremendous heat, and in some areas it boils groundwater, as seen in the thermal features of Yellowstone National Park. The steam given off is a potential energy source if it is accessible.

Lava (molten rock), gases, and **pyroclastics**, or *tephra* (pulverized rock and clastic materials of various sizes ejected violently during an eruption), pass through the vent to the surface and build a volcanic landform. There are two principal forms of flowing basaltic lava, both named in Hawaiian terms (Figure 9.24). **Aa** is a basaltic lava that is rough and jagged with sharp edges. This texture happens because the lava loses trapped gases, flows slowly, and develops a thick skin that cracks into the jagged surface. Another principal lava texture that is more fluid than aa is **pahoehoe**. As this lava flows it forms a thin crust that develops folds and appears "ropy," like coiled, twisted rope. Both forms can come from the same eruption, and sometimes pahoehoe will become aa as the flow progresses. Other types of basaltic magma are described later in this section.

The fact that lava can occur in many different textures and forms accounts for the varied behavior of volcanoes and the different landforms they build. In this section, we look at five volcanic landforms and their origins: cinder cones, calderas, shield volcanoes, plateau basalts, and composite volcanoes.

A **cinder cone** is a small, cone-shaped hill usually less than 450 m (1500 ft) high, with a truncated top formed from cinders that accumulate during moderately explosive eruptions. Cinder cones are made of tephra and scoria

(cindery rock full of air bubbles, or vesicular), as exemplified by Paricutín in southwestern Mexico.

Another distinctive landform is a large, basin-shaped depression, a **caldera** (Spanish for "kettle"). A caldera forms when summit material on a volcanic mountain collapses inward after an eruption or other loss of magma. A caldera may fill with rainwater, as it did in beautiful Crater Lake in southern Oregon. News Report 9.2 discusses the Long Valley Caldera in California.

In the Cape Verde Islands (15° N 24.5° W), Fogo Island has such a caldera in the midst of a volcanic cone, opening to the sea on the eastern side. The best farmland is in the caldera, so the people choose to live and work there. Unfortunately, eruptions began in 1995, destroying farmland and forcing the evacuation of 5000 people. Previously, eruptions occurred between A.D. 1500 and 1750, followed by six more in the next century, each destroying houses and crops.

Location and Types of Volcanic Activity

The location of volcanic mountains on Earth is a function of plate tectonics and hot-spot activity. Volcanic activity occurs in three areas:

1. Along subduction boundaries at continental plate–oceanic plate convergence (such as Mount St. Helens and Kliuchevskoi, Siberia) or oceanic plate–oceanic plate convergence (Philippines and Japan);
2. Along sea-floor spreading centers on the ocean floor (Iceland on the Mid-Atlantic Ridge or off the coast of Oregon and Washington) and areas of rifting on continental plates (the rift zone in east Africa);
3. At hot spots, where individual plumes of magma rise to the crust (such as Hawai'i and Yellowstone National Park).

Figure 9.25 illustrates these three types of volcanic activity, which you can compare to the active volcano sites and plate boundaries shown in Figure 8.18. Consult Figures 9.26 to 9.31, which are arranged around Figure 9.25, as you read this section.

 ANIMATION   Tectonic Settings and Volcanic Activity

(a) Aa   (b) Pahoehoe

FIGURE 9.24 Two types of basaltic lava—Hawaiian examples.
(a) Aa is a rough, sharp-edged lava that is said to get its name from the sounds people make when they take off their shoes and attempt to walk on it. The gold-colored strands are a clump of natural spun glass formed by the lava as it blew out into the air—called Pele's hair. (b) Pahoehoe forms ropy chords in twisted folds. Look at the chapter-opening photo for Chapter 8; can you identify the type of lava flow on the lava blister? [Photos by Bobbé Christopherson.]

NEWS REPORT 9.2

Is the Long Valley Caldera Next?

Trees are dying in the forests on Mammoth Mountain, in a portion of the Long Valley Caldera, near the California–Nevada border. The oval caldera is about 15 by 30 km (9 by 19 mi) long in a north–south direction and sits at 2000 m (6500 ft) elevation, rising to 2600 m (8500 ft) along the western side.

About 1200 tons of carbon dioxide are coming up through the soil in the old caldera each day. Carbon dioxide levels reach 30% to 96% of gases in some soil samples. The source: active and moving magma at a 3-km depth. Elsewhere in the world, this and other gas emissions assist forecasts of potential volcanic activity. At Mammoth Mountain, these gases

signify magmatic activity and may be a portent of things to come (Figure 9.2.1). There are now seven areas of dead trees in the region.

A powerful volcanic eruption 760,000 years ago formed the Long Valley Caldera. This ancient eruption exceeded the volume output of the 1980 Mount St. Helens event by more than 2000 times! Volcanic activity of a lesser extent has occurred in the area over the past several hundred thousand years, most recently between A.D. 1720 and 1850 in the Mono Lake area north of Long Valley.

In the late 1970s and again in 1996, continuing through 1998, swarms of earthquakes shook Long

Valley—thousands of quakes overall, some three dozen greater than M 3.0. The subsurface foci for these quakes are at about 11 km beneath the region. The combination of volcanic gas production and earthquakes is drawing scientific and public attention to this area because of the potential for a future massive eruption somewhere along the Mono-Inyo Craters stretching north of Long Valley. The region is a popular tourist mecca and recreational center, with second-home residential development and a growing year-round population. For the latest information, see http://lvo.wr.usgs .gov/.

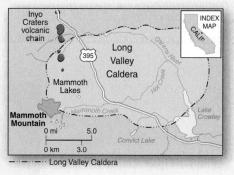

FIGURE 9.2.1 **Trees are dying.** A forest killed by carbon dioxide and a warning sign about the danger. The hazard has killed one person to date. The CO_2 is gassing up through soils from subsurface magma and related volcanic activity. Rates of CO_2 emissions in this forest continue to average 50 to 150 tons per day. [Photo by Bobbé Christopherson.]

The variety of forms among volcanoes makes them hard to classify; most fall in transition between one type and another. Even during a single eruption, a volcano may behave in several different ways. The primary factors in determining an eruption type are (1) the chemistry of its magma, which is related to the magma's source, and (2) its viscosity. Viscosity is magma's thickness (resistance to flow, or degree of fluidity); it ranges from low viscosity (very fluid) to high viscosity (thick and flowing slowly). We consider two types of eruptions—effusive and explosive—and the characteristic landforms they build.

Effusive Eruptions Effusive eruptions are the relatively gentle eruptions that produce enormous volumes of lava annually on the seafloor and in places such as Hawai'i and Iceland. Direct eruptions from the asthenosphere produce a low-viscosity magma that is very fluid and cools to form a dark, basaltic rock (less than 50% silica and rich in iron and magnesium). Gases readily escape from this magma because of its texture, causing an **effusive eruption**

that pours out on the surface, with relatively small explosions and little tephra. However, dramatic fountains of basaltic lava sometimes shoot upward, powered by jets of rapidly expanding gases (Figure 9.28a).

An effusive eruption may come from a single vent or from a flank eruption through side vents in surrounding slopes. If such vents form a linear opening, they are *fissures*; eruptions from fissures sometimes create a dramatic *curtain of fire* (sheets of molten rock spraying into the air). In Iceland, active fissures are spread throughout the plateau landscape. In Hawai'i, rift zones capable of erupting tend to converge on the central crater, or vent. The interior of such a crater is often a sunken caldera, which may fill with low-viscosity magma during an eruption, forming a molten lake that may then overflow lava downslope.

On the island of Hawai'i, the continuing Kīlauea eruption is the longest in recorded history, active since January 3, 1983—the 55th episode of this series began in February 1997. More than 60 separate eruptions occurred at

(text continued on page 328)

FIGURE 9.25 Tectonic settings of volcanic activity.
Magma rises and lava erupts from rifts, through crust above subduction zones, and where thermal plumes at hot spots break through the crust. [Adapted from US Geological Survey, *The Dynamic Planet* (Washington, DC: Government Printing Office, 1989).]

    Forming Types of Volcanoes

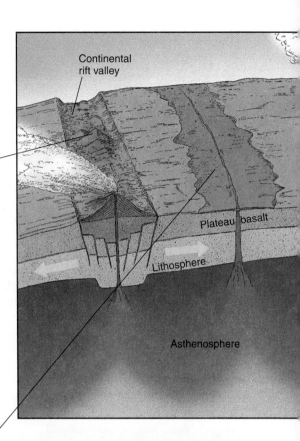

FIGURE 9.26 Rift Valley.
This rift is in Iceland, at the dramatic Gullfoss (waterfalls). Iceland is the only place where the Mid-Atlantic Ridge system is seen above sea level. Iceland is growing as new material increases the size of this island country. [Photo by Bobbé Christopherson.]

FIGURE 9.27 Two expressions of volcanic activity.
In the foreground are plateau (flood) basalts character-istic of the Columbia Plateau in Oregon. Mount Hood, in the background to the west of the plateau, is a com-posite volcano, part of the Cascade Range of active and dormant volcanoes that result from subduction of the Juan de Fuca plate beneath the North American plate. [Photo by author.]

FIGURE 9.28 Volcanic fountaining, boiling surf, and lava falls in Hawai'i
(a) Kilauea erupts a fountain of lava from its East Rift spatter cone on the big island of Hawai'i. (b) Lava flows reach the sea, quenched in the boiling surf. (c) Continuing flows from Earth's most active volcano form spectacular lava falls over the pali in 2002. [(a) Photo by Kepa Maly/US Geological Survey; (b) and (c) photos by Bobbé Christopherson.]

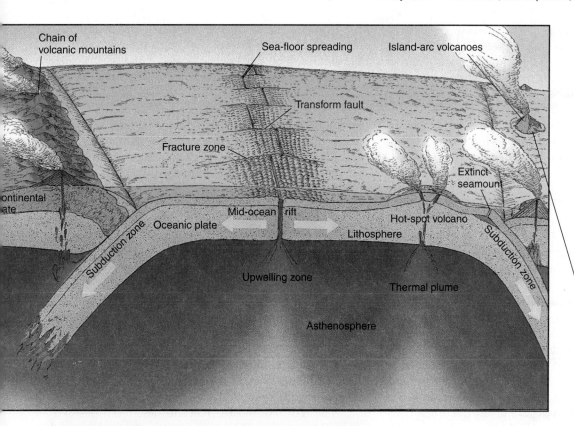

FIGURE 9.31 Extrusive igneous rocks compared.
A dacite volcanic bomb (lighter rock on left) from Mount St. Helens and a basalt lava rock (darker rock on right) from Hawai'i. What does this coloration tell you about the rocks' history and composition? [Photos by Bobbé Christopherson.]

GURE 9.29 Kīlauea landscape.
rial photo of the newest land on the planet produced by the massive
ws of basaltic lavas from the Kīlauea volcano, Hawai'i Volcano National
rk. In the distance, ocean water reluctantly swallows the intense 1200°C
92°F) lava, producing steam and hydrochloric acid mist. [Photo by Bobbé
ristopherson.]

FIGURE 9.30 Volcanic island off the coast of New Zealand.
White Island, New Zealand (37.5° S 177° E), actively erupted without warning through 2001—sometimes gas and steam, other times bursts of pyroclastics and ash. [Photo by Harvey Lloyd/ Stock Market.]

Kīlauea's summit between 1778 (the arrival of Westerners) and 1982. To date, Kīlauea has produced 3.1 km³ (0.7 mi³) of lava, covering 117 km² (45 mi²). Fourteen km (7 mi) of the Chain-of-Craters road are covered up to 35 m (115 ft) deep in lava flows. Needless to say, the NPS Rangers keep their end-of-road checkpoint for tourists movable.

During 1989–1990, lava flows from Kīlauea actually consumed several visitor buildings in the Hawai'i Volcanoes National Park and two housing subdivisions in Kalapana, Kapa'ahu, and Kaimū. Kīlauea has produced more lava than any other vent on Earth in recorded history (Figures 9.28 and 9.29). Beginning on Mothers Day 2002, an aggressive eruption sequence intensified (Figure 9.28c), flowing over Chain of Craters Road and continuing for several years to send streams of lava to the ocean (Figure 9.32). Pu'u O'o is the active crater on Kīlauea and is presently continuing to erupt. In 2007, a portion of Pu'u O'o crater's west side collapsed, filled with lava, and collapsed again (Figure 9.33)—see http://hvo.wr.usgs.gov/.

A typical mountain landform built from effusive eruptions is gently sloped, gradually rising from the surrounding landscape to a summit crater. The shape is similar in outline to a shield of armor lying face up on the ground, and therefore it is a **shield volcano**. The shield shape and size of Mauna Loa in Hawai'i are distinctive when compared with Mount Rainier in Washington, which is a different type of volcano and the largest in the Cascade Range (Figure 9.34). The height of the shield is the result of successive eruptions, flowing one on top of another. Mauna Loa is one of five shield volcanoes that make up the island of Hawai'i, and it took at least 1 million years to accumulate its mass. Mauna Kea is slightly taller, but Mauna Loa is the most massive single mountain on Earth.

In rifting areas, effusive eruptions send material out through elongated fissures, forming extensive sheets on the surface. The Columbia Plateau of the northwestern United States represents the eruption of **plateau basalts**, or *flood basalts* (Figure 9.27). Extensive sheet eruptions also episodically flow from the mid-ocean ridges, forming new seafloor in the process. Worldwide, the volume of material produced in effusive fissure eruptions along spreading centers is far greater than that of the continental eruptions. Iceland continues to form in this effusive manner, with upwelling basaltic flows creating new Icelandic real estate.

Explosive Eruptions Volcanic activity along subduction zones produces the well-known explosive volcanoes. Magma produced by the melting of subducted oceanic plate and other materials is thicker (more viscous) than magma from effusive volcanoes. It is 50%–75% silica and high in aluminum; consequently, it tends to block the magma conduit inside the volcano, trapping and compressing gases, causing pressure and conditions to build for a possible **explosive eruption**. This type of magma forms a lighter rock at the surface, as illustrated in the comparison in Figure 9.31.

The term **composite volcano** describes explosively formed mountains. (They are sometimes called *stratovolcanoes*, but because shield volcanoes also can exhibit a stratified structure, *composite* is the preferred term.) *Composite* volcanoes tend to have steep sides; they are more conical than shield volcanoes. Their alternating layers of lava and ash and cone shape give them their name: *composite cones* (Figure 9.35). If a single summit vent remains stationary, a remarkable symmetry may develop as the mountain grows

FIGURE 9.32 Dramatic Hawaiian landscape where active lava flows into the ocean.
Look closely in this aerial photo for the glowing lava, which is 1200°C (2192°F) as it sluggishly makes its way to the sea, setting the surf to boil. Notice the couch-sized rock blown out as the ocean quenches the lava, as new additions to the Kamoamoa bench form. Even at the safe distance of this aerial photo, the radiated heat from the lava flow is incredible to feel. [Photo by Bobbé Christopherson.]

(a)

(b)

FIGURE 9.33 The active Pu'u O'o Vent, at Kīlauea, Hawai'i.
(a) The cracked and faulted crater of Pu'u O'o Vent, part of the Kīlauea volcano, Hawai'i, in 1999 is a risky environment for volcanic research. (b) Note the changes 8 years later in 2007. Portions of the crater wall collapsed in summer 2007, refilled with lava, and collapsed again. Active lava from the crater continues to make its way to the ocean. [Photo by (a) Bobbé Christopherson; (b) courtesy of USGS Hawaiian Volcano Observatory, 2007).]

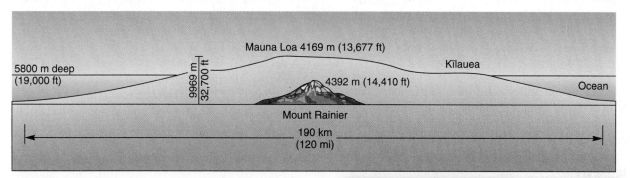

FIGURE 9.34 Shield and composite volcanoes compared.
Comparison of Mauna Loa in Hawai'i (a shield volcano) and Mount Rainier in Washington state (a composite volcano). Their strikingly different profiles signify different tectonic origins. [After USGS, *Eruption of Hawaiian Volcanoes* (Washington, DC: Government Printing Office, 1986). Inset photo of Mauna Loa by Bobbé Christopherson.]

in size, as demonstrated by Mount Orizaba in Mexico, Mount Shishaldin in Alaska, Mount Fuji in Japan, the pre-1980-eruption shape of Mount St. Helens in Washington, and Mount Mayon in the Philippines. The beautifully symmetrical summit of Mount Mayon erupted in 2001 and 2002.

As the magma in a composite volcano forms plugs near the surface, blocked-off passages build tremendous pressure, keeping the trapped gases compressed and liquefied. When the blockage can no longer hold back this pressurized inferno, explosions equivalent to megatons of TNT blast the tops and sides off these mountains. Much less lava is produced than in shield eruptions, but composite volcanoes eject larger amounts of *pyroclastics*, which include volcanic *ash* (<2 mm, or 0.08 in., diameter), dust, cinders, *lapilli* (up to 32 mm, or 1.26 in., diameter), *scoria* (volcanic slag), pumice, and *aerial bombs* (explosively ejected blobs of incandescent lava).

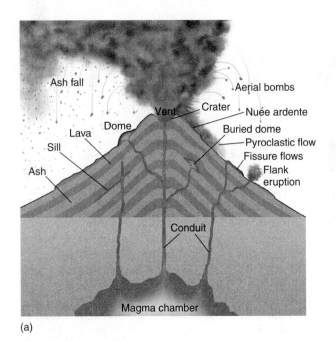

(a)

(b)

FIGURE 9.35 A composite volcano.
(a) A typical composite volcano with its cone-shaped form. (b) Mount Shishaldin, Unimak Island in the eastern Aleutian Islands, Alaska, is a 2857-m (9372-ft) composite volcano. Here shown during a 1995 eruption, it was active through 2001. [(b) Photo courtesy of AeroMap US, Inc., Anchorage, Alaska.]

Unlike the volcanoes in Hawai'i Volcanoes National Park, where tourists gather at observation areas to watch the relatively calm effusive eruptions, composite volcanoes do not invite close inspection and can explode with little warning. Table 9.3 presents a few notable composite volcano eruptions. Focus Study 9.1 details the much-publicized 1980 eruption of Mount St. Helens.

Mount Pinatubo Eruption   In June 1991, after 600 years of dormancy, Mount Pinatubo in the Philippines erupted. The summit of the 1460-m (4795-ft) volcano exploded, devastating many surrounding villages and permanently closing the US Clark Air Force Base (see photo in Figure 1.7). Fortunately, scientists from the US Geological Survey, working with local scientists, accurately predicted the eruption. A resulting timely evacuation of the surrounding countryside saved thousands of lives; however, 800 people were killed.

Although volcanoes are regional events, their spatial implications can be worldwide. The single volcanic eruption of Mount Pinatubo was significant to the global environment and energy budget, as discussed in Chapters 1, 2, 3, 4, and 7 and summarized in Figure 1.7:

- 15–20 million tons of ash and sulfuric acid mist were blasted into the atmosphere, concentrating at 16–25 km (10–15.5 mi) altitude.
- 12 km³ (3.0 mi³) of material was ejected and extruded by the eruption (12 times the volume from Mount St. Helens).
- 60 days after the eruption, about 42% of the globe was affected (from 20° S to 30° N) by the thin, spreading aerosol cloud in the atmosphere.

- Colorful twilight and dawn skies were observed worldwide.
- An increase in atmospheric albedo of 1.5% (4.3 W/m²) occurred.
- An increase in the atmospheric absorption of insolation followed (2.5 W/m²).
- A decrease in net radiation at the surface and a lowering of Northern Hemisphere average temperatures of 0.5 C° (0.9 F°) were measured.
- Atmospheric scientists and volcanologists were able to study the eruption aftermath using satellite-borne orbiting sensors and general circulation model computer simulations.

Volcano Forecasting and Planning

The USGS and the Office of Foreign Disaster Assistance operate the Volcano Disaster Assistance Program (VDAP). For a listing of more than two dozen projects, see **http://vulcan.wr.usgs.gov/Vdap/framework.html**. The need for such a program is evident in that over the past 20 years, volcanic activity has killed 29,000 people, forced more than 800,000 to evacuate their homes, and caused more than $3 billion in damage. The program was established after 23,000 people died in the eruption of Nevado del Ruiz, Colombia, in 1985. The US VDAP is in place to help local scientists with eruption forecasts by setting up mobile volcano-monitoring systems at the most threatened sites.

As an example of the need for accurate forecasting, a volcano watch continues at Soufriere Hills, an erupting composite volcano on the island of Montserrat in the Lesser Antilles (17° N 62° W). The Montserrat Volcano

Table 9.3 Sample of Notable Composite Volcano Eruptions

Date	Location	Number of Deaths	Amount Extruded (mostly pyroclastics) in km³ (mi³)
Prehistoric	Yellowstone, Wyoming	Unknown	2400 (576)
4600 B.C.	Mount Mazama (Crater Lake, Oregon)	Unknown	50–70 (12–17)
1900 B.C.	Mount St. Helens	Unknown	4 (0.95)
A.D. 79	Mount Vesuvius, Italy	20,000	3 (0.7)
1500	Mount St. Helens	Unknown	1 (0.24)
1815	Tambora, Indonesia	66,000	80–100 (19–24)
1883	Krakatau, Indonesia	36,000	18 (4.3)
1902	Mont Pelée, Martinique	29,000	Unknown
1912	Mount Katmai, Alaska	Unknown	12 (2.9)
1943–1952	Paricutín, Mexico	0	1.3 (0.30)
1980	Mount St. Helens	54	4 (0.95)
1985	Nevado del Ruiz, Colombia	23,000	1 (0.24)
1991	Mount Unzen, Japan	10	2 (0.5)
1991	Mount Pinatubo, Philippines	800	12 (3.0)
1992	Mount Spurr/Mount Shishaldin, Alaska	0	1 (0.24)
1993	Galeras Volcano, Colombia	5	1 (0.24)
1993, 2003, 2004	Mount Mayon, Philippines	0	1 (0.24)
2002	Nyiragongo, Congo	70	1 (0.24)
1994–2007	Kliuchevskoi, Russia Soufrière Hills Montserrat, West Indies	0	Ash, steam, pyroclastic flows
2007	Arenal, Costa Rica Karymsky, Kamchatka, Russia	0	Ash, pyroclastics

Observatory reports a dozen evacuations since volcanic activity began in 1995 and strengthened through 1998. Are these precursors to a main event? Or are these just routine rumblings that will grow quiet again? Although the US VDAP and other scientists issue warnings, residents remain skeptical after many false alerts, which only heightens the risk.

In this era of the Internet you can access "volcano cams" positioned around the world to give you 24-hour surveillance of many volcanoes. For an exciting visual adventure, go to the following URL and add it to your bookmarks (note whether it is day or night for the location you are checking): **http://vulcan.wr.usgs.gov/ Photo/volcano_cams.html**.

FOCUS STUDY 9.1

The 1980 Eruption of Mount St. Helens

Probably the most studied and photographed composite (explosive) volcano on Earth is Mount St. Helens, located 70 km (45 mi) northeast of Portland, Oregon, and 130 km (80 mi) south of the Tacoma-Seattle area of Washington. Mount St. Helens is the youngest and most active of the Cascade Range of volcanoes, which form a line from Mount Lassen in California to Mount Meager in British Columbia (Figure 9.1.1a). The Cascade Range is the product of the Juan de Fuca sea-floor spreading center off the coast of northern California,

Oregon, Washington, and British Columbia and the plate subduction that occurs offshore, as identified in the upper right of Figure 9.12.

The mountain had been quiet since 1857. New activity began in March 1980, with a sharp earthquake registering M 4.1. The first eruptive outburst occurred one week later, beginning with a M 4.5 quake and continuing with a thick black plume of ash and the development of a small summit crater. Ten days later, the first volcanic earthquake, a *harmonic tremor*, registered on the many instruments

that were hurriedly placed around the volcano. Harmonic tremors are slow, steady vibrations, unlike the sharp releases of energy associated with earthquakes and faulting. Harmonic tremors told scientists that magma was on the move within the mountain.

Also developing was a massive bulge on the north side of the mountain. This bulge indicated the direction of the magma flow within the volcano. A bulge represents the greatest risk from a composite volcano, for it could signal a potential

(continued)

Focus Study 9.1 (continued)

lateral burst through the bulge and across the landscape.

Early on Sunday, May 18, the area north of the mountain was rocked by a M 5.0 quake, the strongest to date. The mountain, with its distended 245-m (800-ft) bulge, was shaken, but nothing else happened. Then a second quake (M 5.1) hit at 8:32 A.M., loosening the bulge and launching the eruption. David Johnston, a volcanologist with the US Geological Survey, was only 8 km (5 mi) from the mountain, servicing instruments, when he saw the eruption begin. He reportedly radioed headquarters in Vancouver, Washington, saying, "Vancouver, Vancouver, this is it!" He perished in the eruption. For a continuously updated live picture of Mount St. Helens from Johnston's approximate observation point, go to http://vulcan.wr.usgs.gov/Volcanoes/MSH/VolcanoCam/framework.html. The camera is mounted below the roof line at the Johnston Ridge Observatory 8 km (5 mi) from the mountain (Figure 9.1.1b). The observatory is open to the public and is named for this brave scientist.

As the contents of the mountain exploded, a surge of hot gas (about 300°C, or 570°F), a *nuée ardente* (rapidly moving, very hot, steam-filled, explosive ash, pyroclastics, and incandescent gases), moved northward, hugging the ground and traveling at speeds up to 400 kmph (250 mph) for a distance of 28 km (17 mi).

The slumping north face of the mountain produced the greatest landslide in recorded history; about 2.75 km³ (0.67 mi³) of rock, ice, and trapped air, all fluidized with steam, surged at speeds approaching 250 kmph (155 mph). Landslide materials traveled for 21 km (13 mi) into the valley, blanketing the forest, covering a lake, and filling rivers (Figure 9.1.1c). A series of photographs, taken at 10-second intervals from the east looking west, records this sequence (Figure 9.1.2). The eruption continued with intensity for 9 hours, first clearing out old rock from the throat of the volcano and then blasting new material.

Before the eruption, Mount St. Helens was 2950 m (9677 ft) tall; the eruption blew away 418 m (1370 ft). But today, Mount St. Helens is building a lava dome within its crater. As destructive as such eruptions are, they also are constructive, for this is the way in which a volcano eventually builds its height. The thick lava rapidly and repeatedly plugs and breaks in a series of lesser dome eruptions that may continue for several decades. The dome already is over 300 m (1000 ft) high; a new mountain is being born from the eruption of the old. Dome eruptions of varying intensities occurred through 2007. Intensive scientific research and monitoring has paid off, as every eruption since 1980 was successfully forecasted from days to as long as 3 weeks in advance, with the exception of one small eruption in 1984.

An ever-resilient ecology is recovering as plants and animals reclaim the devastated landscape. In Chapter 16, matching photographs from 1983 and 1999 document the slow successional recovery near the volcano (see Figure 16.19). Almost 30 years have passed, yet strong interest in the area continues; scientists and millions of tourists visit the Mount St. Helens Volcanic National Monument every year. For more information, see the Cascades Volcano Observatory at http://vulcan.wr.usgs.gov/ and specifically Mount St. Helens at http://vulcan.wr.usgs.gov/Volcanoes/MSH/framework.html.

 (a)

 (b)

 (c)

FIGURE 9.1.1 Mount St. Helens before and after the eruption.
(a) Mount St. Helens prior to the 1980 eruption. (b) The devastated and scorched land shortly after the eruption in 1980. The scorched earth and tree blowdown area covered some 38,950 hectares (95,000 acres). (c) A landscape in recovery as life moves back in and takes hold to establish new ecosystems in 1999 along the Toutle River and debris flow. [(a) Photo by Pat and Tom Lesson/Photo Researchers; (b) photo by Krafft-Explorer/Photo Researchers, Inc.; (c) photo by Bobbé Christopherson.]

ANIMATION
Debris Avalanche and Eruption of Mt. St. Helens

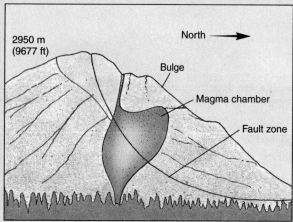

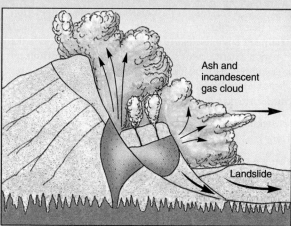

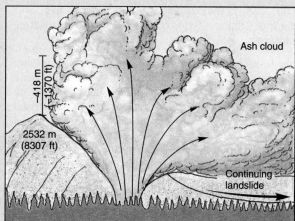

FIGURE 9.1.2 The Mount St. Helens eruption sequence and corresponding schematics.
[Photo sequence by Keith Ronnholm. All rights reserved.]

Summary and Review—Tectonics, Earthquakes, and Volcanoes

■ *Describe* first, second, and third orders of relief and *relate* examples of each from Earth's major topographic regions.

Earth's surface is dramatically shaped by tectonic forces generated within the planet. **Relief** is the vertical elevation difference in a local landscape. The undulating physical surface of Earth, including

relief, is called **topography**. Convenient descriptive categories are termed *orders of relief*. The most general level of landforms includes the **continental landmasses** and **ocean basins**; the finest comprises local hills and valleys.

relief (p. 299)

333

1. How does the map of the ocean floor (chapter-opening illustration) exhibit the principles of plate tectonics? Briefly analyze.
2. What is meant by an "order of relief"? Give an example from each order.
3. Explain the difference between relief and topography.

■ *Describe* the several origins of continental crust and *define* displaced terranes.

The continents are formed as a result of several processes, including upwelling material from below the crust and migrating portions of crust. The source of material generated by plate tectonic processes determines the behavior and the composition of the crust that results.

A continent has a nucleus of ancient crystalline rock called a *craton*. A region where a craton is exposed is termed a **continental shield**. As continental crust forms, it is enlarged through accretion of dispersed **terranes**. An example is the *Wrangellia terrane* of the Pacific Northwest and Alaska.

4. What is a craton? Relate this structure to continental shields and platforms and describe these regions in North America.
5. What is a migrating terrane, and how does it add to the formation of continental masses?
6. Briefly describe the journey and destination of the Wrangellia Terrane.

■ *Explain* compressional processes and folding and *describe* four principal types of faults and their characteristic landforms.

The crust is deformed by folding, broad warping, and faulting, which produce characteristic landforms. Compression causes rocks to deform in a process known as **folding**, during which rock strata bend and may overturn. Along the *ridge* of a fold, layers *slope downward away from the axis*, which is an **anticline**. In the *trough* of a fold, however, layers *slope downward toward the axis*, called a **syncline**.

When rock strata are stressed beyond their ability to remain a solid unit, they express the strain as a fracture. Rocks on either side of the fracture are displaced relative to the other side in a process known as faulting. Thus, *fault zones* are areas where fractures in the rock demonstrate crustal movement. At the moment of fracture, a sharp release of energy, called an **earthquake**, or *quake*, occurs.

When forces pull rocks apart, the tension causes a **normal fault**, sometimes visible on the landscape as a scarp, or escarpment. Compressional forces associated with converging plates force rocks to move upward, producing a **reverse fault**. A low-angle fault plane is referred to as a **thrust fault**. Horizontal movement along a fault plane that produces a linear rift valley is called a **strike-slip fault**. In the US interior West, the Basin and Range Province is an example of aligned pairs of normal faults and a distinctive horst-and-graben landscape. The term **horst** is applied to upward-faulted blocks; **graben** refers to downward-faulted blocks.

7. Diagram a simple folded landscape in cross section and identify the features created by the folded strata.
8. Define the four basic types of faults. How are faults related to earthquakes and seismic activity?
9. How did the Basin and Range Province evolve in the western United States? What other examples exist of this type of landscape?

■ *Relate* the three types of plate collisions associated with orogenesis and *identify* specific examples of each.

Orogenesis is the birth of mountains. An orogeny is a mountain-building episode, occurring over millions of years, that thickens continental crust. It can occur through large-scale deformation and uplift of the crust. It also may include the capture of migrating terranes and cementation of them to the continental margins and the intrusion of granitic magmas to form plutons.

Oceanic plate–continental plate collision orogenesis is now occurring along the Pacific coast of the Americas and has formed the Andes, the Sierra of Central America, the Rockies, and other western mountains. *Oceanic plate–oceanic plate collision* orogenesis produces either simple volcanic island arcs or more complex arcs such as Japan, the Philippines, the Kurils, and portions of the Aleutians. The region around the Pacific contains expressions of each type of collision in the *circum-Pacific belt*, or the *ring of fire*.

Continental plate–continental plate collision orogenesis is quite mechanical; large masses of continental crust, such as the Himalayan Range, are subjected to intense folding, overthrusting, faulting, and uplifting.

10. Define orogenesis. What is meant by the birth of mountain chains?
11. Name some significant orogenies.
12. Identify on a map Earth's two large mountain chains. What processes contributed to their development?
13. How are plate boundaries related to episodes of mountain building? Explain how different types of plate boundaries produce differing orogenic episodes and different landscapes.
14. Relate tectonic processes to the formation of the Appalachians and the Alleghany orogeny.

■ *Explain* the nature of earthquakes, their measurement, and the nature of faulting.

Earthquakes generally occur along plate boundaries; major ones can be disastrous. Earthquakes result from faults, which are under continuing study to determine the nature of faulting, stress and the buildup of strain, irregularities along fault plane surfaces, the way faults rupture, and the relationship among active faults. Earthquake prediction and improved planning are active concerns of *seismology*, the study of earthquake waves and Earth's interior. Seismic motions are measured with a **seismograph**.

Charles Richter developed the *Richter scale*, a measure of earthquake magnitude. A more precise and quantitative scale, the **moment-magnitude scale,** that assesses the *seismic moment* is now used, especially for larger quakes. The area at the surface directly above the focus (where motion initiates seismic waves) is the **epicenter**. The specific mechanics of how a fault breaks are under study, but the **elastic-rebound theory** describes the basic process. In general, two sides along a fault appear to be locked by friction, resisting any movement. This stress continues to build strain along the fault surfaces, storing elastic energy like a wound-up spring. When energy is released abruptly as the rock breaks, both sides of the fault return to a condition of less strain.

seismograph (p. 318)
moment-magnitude scale (p. 318)
epicenter (p. 318)
elastic-rebound theory (p. 318)

15. Describe the differences in human response to the two earthquakes that occurred in China in 1975 and 1976.

16. Compare the North Anatolian, Turkey, and San Andreas, California, fault systems. Discuss their similarities and differences.

17. Differentiate between the Mercalli and moment-magnitude and amplitude scales. How are these used to describe an earthquake? Why has the Richter scale been updated and modified?

18. What is the relationship between an epicenter and the focus of an earthquake? Give examples from the Loma Prieta, California, and Kobe, Japan, earthquakes.

19. What local soil and surface conditions in San Francisco severely magnify the energy felt in earthquakes?

20. How do the elastic-rebound theory and asperities help explain the nature of faulting?

21. Describe the San Andreas fault and its relationship to ancient sea-floor spreading movements along transform faults.

22. How is the seismic gap concept related to expected earthquake occurrences? Are any gaps correlated with earthquake events in the recent past? Explain.

23. What do you see as the biggest barrier to effective earthquake prediction?

■ *Distinguish* between an effusive and an explosive volcanic eruption and *describe* related landforms, using specific examples.

Volcanoes offer direct evidence of the makeup of the asthenosphere and uppermost mantle. A **volcano** forms at the end of a central vent or pipe that rises from the asthenosphere through the crust into a volcanic mountain. A **crater**, or circular surface depression, usually forms at the summit.

Eruptions produce **lava** (molten rock), gases, and **pyroclastics** (pulverized rock and clastic materials ejected violently during an eruption) that pass through the vent to openings and fissures at the surface and build volcanic landforms. Basaltic lava flows occur in two principal textures: **aa**, rough and sharp-edged lava, and **pahoehoe**, smooth, ropy folds of lava. Volcanic activity produces landforms such as a **cinder cone**, a small hill, or a large basin-shaped depression called a **caldera**.

Volcanoes are of two general types, based on the chemistry and the viscosity of the magma involved. **Effusive eruption** produces a **shield volcano** (such as Kīlauea in Hawai'i) and extensive deposits of **plateau basalts**, or flood basalts. Magma of higher viscosity leads to an **explosive eruption** (such as Mount Pinatubo in the Philippines), producing a **composite volcano**. Volcanic activity has produced some destructive moments in history but constantly creates new seafloor, land, and soils.

volcano (p. 324)
crater (p. 324)
lava (p. 324)
pyroclastics (p. 324)
aa (p. 324)
pahoehoe (p. 324)
cinder cone (p. 324)
caldera (p. 324)
effusive eruption (p. 325)
shield volcano (p. 328)
plateau basalts (p. 328)
explosive eruption (p. 328)
composite volcano (p. 328)

24. What is a volcano? In general terms, describe some related features.

25. Where do you expect to find volcanic activity in the world? Why?

26. Compare effusive and explosive eruptions. Why are they different? What distinct landforms are produced by each type? Give examples of each.

27. Describe several recent volcanic eruptions.

NetWork and Critical Thinking Tools

A. Using the topographic regions map in Figure 9.3, assess the region within 100 km (62 mi) and beyond to 1000 km (620 mi) of your campus. Describe the topographic character and the variety of relief within these two regional scales. You may want to consult local maps and atlases in your analysis, and consult Internet sources. Do you perceive that this type of topographic region influences lifestyles? Economic activities? Transportation? History of the region?

B. Determine from the library, Internet, and local agencies the seismic potential, or volcanic potential, if any, for the region in which your campus is located. If a hazard exists, does it influence regional planning? The availability of property insurance? If appropriate, are there existing disaster or emergency plans for your campus? For your community, state, or province? Explain any other significant issues you found in working on this critical thinking challenge.

C. Go to one of the several URLs in this chapter and in the Student Learning Center covering active volcanoes in the world. Select one volcano that is currently erupting and prepare a profile of it, including its physical nature, history, present activity, and overall risk assessment.

Weathered Aztec Sandstone in the Red Rock Canyon National Conservation Area, 24 km (15 mi) west of Las Vegas, Nevada. Joints in the rock provide physical and chemical weathering agents access to work the rock and for opportunistic lichens to harvest water that infrequently courses down the cracks. Desert varnish darkens the sandstone as rare moisture events carry magnesium, iron oxides, and other materials to stain the rock. [Photos by Bobbé Christopherson.]

Weathering, Karst Landscapes, and Mass Movement

KEY LEARNING CONCEPTS

After reading the chapter, you should be able to:

- **Define** the science of geomorphology.

- **Illustrate** the forces at work on materials residing on a slope.

- **Define** weathering and **explain** the importance of the parent rock and joints and fractures in rock.

- **Describe** frost action, crystallization (salt-crystal growth), pressure-release jointing, and the role of freezing water as physical weathering processes.

- **Describe** the susceptibility of different minerals to the chemical weathering processes, including hydration, hydrolysis, oxidation, carbonation, and solution.

- **Review** the processes and features associated with karst topography.

- **Portray** the various types of mass movements and **identify** examples of each in relation to moisture content and speed of movement.

STUDENT LEARNING CENTER TOOLS

The *Elemental Geosystems* Student Learning Center provides on-line resources for this chapter at **www.mygeoscienceplace.com**. Some highlights:

▶ Read Career Links to meet a geographer who conducted the "Cultural Stones of Central Park" field trip (News Report 10.1) and meet a consulting geoscientist who specializes in landslides.

▶ Visit a Case Study on the effects of weathering on sandstone in Petra, Jordan.

▶ Two animations feature physical weathering and the debris avalanche caused by the eruption of Mt. St. Helens.

▶ Notebooks take you to karst areas and the Madison River landslide site.

▶ More than 15 additional URLs along with Quick Links to all the URLs in the chapter.

A benefit you receive from a physical geography course is a new appreciation of the scenery. Whether you go by foot, bicycle, car, train, or plane, travel is an opportunity to experience Earth's varied landscapes and to witness the active processes that produce them. Warning: Seeing Earth's landscapes after taking a physical geography course can make each trip take longer, for there is so much more to see!

In the last two chapters, we discussed the endogenic (internal) processes of our planet and how they produce landforms. However, as the landscape is formed, several exogenic (external) processes simultaneously wear and waste it. This ongoing struggle between tectonics and climate shapes the landscape upon which living systems play out their drama.

Here we begin a five-chapter examination of exogenic processes at work on the landscape. This chapter examines weathering and mass movement of the lithosphere. The next four chapters look at specific exogenic agents and their handiwork—river systems, wind-influenced landscapes,

deserts in water-deficit regions, coastal processes and landforms, and regions worked by ice and glaciers. Whether you enjoy time along a river, love the desert or the waves along a coastline, or live in a place where glaciers once carved the land, you will find something of interest in these chapters.

In this chapter: We look at physical (mechanical) and chemical weathering processes that break up, dissolve, and generally reduce the landscape. Such weathering releases essential minerals from bedrock for soil formation or enrichment. Perhaps you live in a region that has caves and caverns. The dissolving power of water has created enormous underground worlds of mystery and darkness, yet many caves remain undiscovered. In addition, we examine mass-movement processes that continually operate in and upon the landscape. News broadcasts may describe an avalanche in Colombia; mudflows in Los Angeles, China, or Peru; tragic landslides in China, Turkey, Indonesia, or Colombia; massive rockfalls in Yosemite; or an imminent mudflow down the slopes of a volcano in New Zealand or

NEWS REPORT 10.1

Weathering on Bridges in Central Park, NYC

In 2001, geographers Greg Pope of Montclair State University and Patricia Beyer of Bloomsburg University led us on a field trip through Central Park in New York City—called "Cultural Stones of Central Park: Outcrops, Arches, and Statues." They prepared a detailed field guide for our urban expedition.

The popular park has gone through many stages of design and construction over the last 150 years, including a variety of arches and bridges (the terms seem interchangeable here) to accommodate different forms of traffic. Thirty-six bridges remain of the original thirty-nine, built from a great variety of rock. Dr. Pope said, "The stones used by the designers and builders in the art and architecture of the park are truly 'cultural stones.' Sources for all this building stone cover a wide geography, from many locales across the Northeast and into Canada."

All this imported rock material presented a diverse array of surfaces on which physical and chemical weathering processes could work (Figure 10.1.1). With fossil-fuel

FIGURE 10.1.1 Weathered sandstone in Central Park, New York City.
Here on the sides of one of the many arches, constructed in 1859, we see the ravages of both physical and chemical weathering. The patterns etched into this Alberta sandstone, found in New Brunswick, are completely gone in several squares. Encrusted salts, leading to crystallization processes, are visible. [Photo by Bobbé Christopherson.]

consumption and the resulting air pollution, increased acidity in rain and snow hastened rates of weathering. For several bridges weathering processes were worsened by auto traffic and the use of salt on the roads in winter. Some sandstone blocks so

poorly resisted weathering that they were actually replaced by cast concrete. Researchers, along with geography students, study the rates of building-stone weathering around Central Park structures and New York City, and then map the spatial results.

in a southern California suburb. In 2008, Haiti was deluged by three hurricanes that caused massive landslides and mudflows from deforested mountain slopes. We need to understand the processes that caused these events.

Landmass Denudation

Geomorphology is the science of landforms—their origin, evolution, form, and spatial distribution. This discipline is an important aspect of physical geography. **Denudation** is any process that wears away or rearranges landforms. The principal denudation processes affecting surface materials include *weathering, mass movement, erosion, transportation,* and *deposition,* as produced by the agents of moving water, air, waves, ice, and the pull of gravity.

Interactions between the structural elements of the land and denudation processes are complex. They represent a continuing struggle between Earth's internal and external processes, between the resistance of materials and the weathering and erosional processes. The 15-story-tall Delicate Arch in Utah is dramatic evidence of this struggle (Figure 10.1). Differing resistances of the rocks, coupled with variations in the processes at work on the rock, are carving this delicate sculpture—an example of **differential weathering**, where a more resistant cap rock protects supporting strata below.

Dynamic Equilibrium Approach to Landforms

A landscape is an open system, with highly variable inputs of energy and materials. Uplift creates the *potential energy of position* above sea level and therefore a disequilibrium

between relief and energy. The Sun provides radiant energy that is converted into *heat energy*. The hydrologic cycle imparts *kinetic energy* through mechanical motion. *Chemical energy* is made available from the atmosphere and various reactions within the crust.

As physical factors fluctuate in an area, the surface constantly responds in search of equilibrium. Every change produces compensating actions and reactions. The balancing act between tectonic uplift and reduction by weathering and erosion, between the resistance of rocks and the ceaseless attack of weathering and erosion, is the **dynamic equilibrium model**. According to current thinking, landscapes in a dynamic equilibrium show ongoing adaptations to the ever-changing conditions of rock structure, climate, local relief, and elevation.

Endogenic events, such as faulting or a lava flow, or exogenic events, such as a heavy rainfall or a forest fire, may provide new sets of relationships for the landscape. Following such destabilizing events, a landform system arrives at a **geomorphic threshold**—the point at which there is enough energy to overcome the resistance against movement. At this threshold the system breaks through to a new set of equilibrium relationships as the landform and its slopes enter a period of adjustment and realignment. The pattern over time follows a sequence: (1) equilibrium stability (fluctuating around some average), (2) a destabilizing event, (3) a period of adjustment, and (4) development of a new and different condition of equilibrium stability. The disturbed hillslope in Figure 10.2 is in the midst of compensating adjustment. The failure of saturated slopes caused a landslide into the river. As a consequence, the new dam of wasted material threw the stream into a disequilibrium condition between its flow and sediment load.

FIGURE 10.1 Delicate weathering in the desert.
Delicate Arch—a dramatic example of differential weathering in Arches National Park, Utah. The arch stands 15 stories tall! Note the person at the base in the inset photo for a sense of scale. In the distance are the snow-covered Manti LaSal Mountains, extinct volcanoes. [Photos by author.]

FIGURE 10.2 A slope in disequilibrium.
Unstable, saturated soils gave way, leaving a debris dam partially blocking the river. [Photo by author.]

Slow, continuous-change events, such as soil development and erosion, tend to maintain an approximate equilibrium condition. Less frequent, dramatic events such as a major landslide or dam collapse require longer recovery times before a new equilibrium is established. (Figure 1.6 graphically illustrates both the distinction between a steady-state equilibrium and a dynamic equilibrium and the occurrence of a geomorphic threshold.)

Slopes

Material loosened by weathering is susceptible to erosion and transportation. However, if it is to move downslope, the forces of erosion must overcome other forces: friction, inertia (the resistance to movement), and the cohesion of particles to one another (Figure 10.3a). If the angle is steep enough for gravity to overcome frictional forces, or if the impact of raindrops or moving animals or even wind dislodges material, then erosion of particles and transport downslope can occur.

Slopes, or hillslopes, are curved, inclined surfaces that form the boundaries of landforms. Figure 10.3b illustrates essential slope forms that may vary with conditions of rock structure and climate. Slopes generally feature an *upper waxing* slope near the top ("waxing" here means increasing). This convex surface curves downward and grades into the *free face* below. The presence of a free face indicates an outcrop of resistant rock that forms a steep scarp or cliff.

Downslope from the free face is a *debris slope* that receives rock fragments and materials from above. In humid climates, constantly moving water carries away material as it arrives; in arid climates, graded debris slopes persist and accumulate. A debris slope grades into a *waning slope*, a concave surface along the base of the slope that forms a *pediment*, or broad, gently sloping erosional surface. Figure 10.3c provides an example from Wyoming.

Slopes are open systems and seek an *angle of equilibrium* among the forces described here. Conflicting forces work together on slopes to establish an optimum compromise incline that balances these forces. A geomorphic threshold (change point) is reached when any of the conditions in the balance is altered. All the forces on the slope then compensate by adjusting to a new dynamic equilibrium.

The relation between rates of weathering and breakup of slope materials, coupled with the rates of mass movement and material erosion, forms the shape of the slope. A slope is *stable* if its strength exceeds these denudation processes and *unstable* if materials are weaker than these processes. Why are hillslopes shaped in certain ways? How do slope elements evolve? How do hillslopes behave during rapid, moderate, or slow uplift? These are topics of active scientific study and research.

Now, with the concepts of landmass denudation and slope development in mind, let's examine specific processes that operate to reduce landforms.

Weathering Processes

Weathering processes attack rocks at Earth's surface and to some depth below the surface. **Weathering** processes either disintegrate rock into mineral particles or dissolve them in water. Weathering processes are both physical (mechanical) and chemical, and the interplay of the two is complex. There is a synergy, a combined action, between physical and chemical weathering as the suite of processes work the rock.

Weathering does not transport the weathered materials; it simply generates them for transport by the agents of water, wind, waves, and ice—all influenced by gravity. In most areas, the upper surface of bedrock undergoes continual weathering to form broken-up **regolith**. Loose surface material comes from further weathering of regolith and from transported and deposited regolith (Figure 10.4a, b). In some areas regolith may be missing or undeveloped, thus exposing an outcrop of unweathered bedrock.

Bedrock is the *parent rock* from which weathered regolith and soils develop. The bedrock is traceable through similarities in composition. For example, the sand and soils in Figure 10.4c on Earth, and 10.4d on Mars, derive their color and character from the parent rock in the background. Fragmented material, known as **sediment**, combines with weathered rock to form the **parent material** from which soil evolves.

Factors Influencing Weathering Processes

Weathering is greatly influenced by the character of the bedrock: hard or soft, soluble or insoluble, broken or unbroken. *Jointing* in rock is important for weathering processes. **Joints** are fractures or separations in rock that occur without displacement of the sides (as would be the case in faulting). The presence of these usually plane (flat)

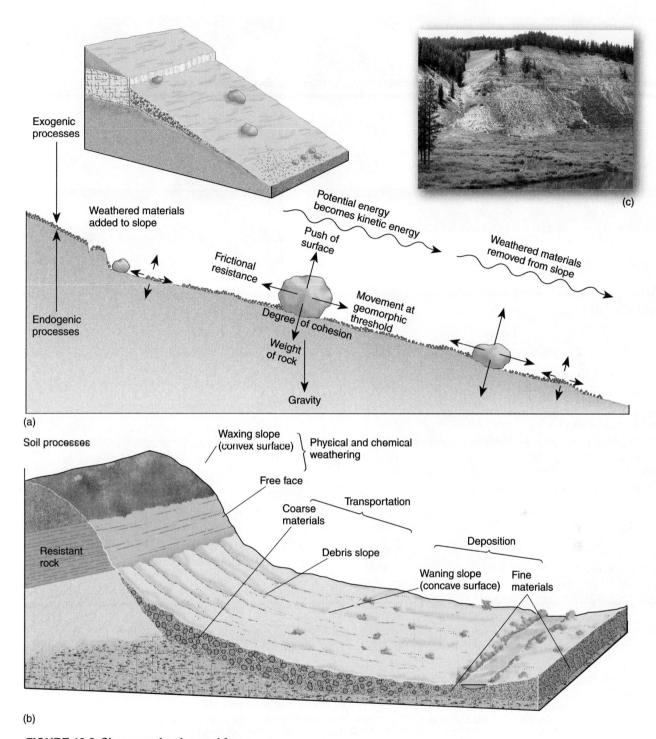

FIGURE 10.3 Slope mechanics and form.
(a) Directional forces (noted by arrows) act on materials along an inclined slope. (b) The principal elements of a slope. (c) A hillslope example; note the role of the rock outcrop as a slope interruption and the rock fragments loosened by frost action. [Photo by author.]

surfaces increases the surface area of rock exposed to both physical and chemical weathering. You see the role joints play in weathering of rock in the chapter-opening photos.

Important controls on weathering rates are climatic elements—precipitation, temperature, and freeze-thaw cycles. Also related to climate and significant weathering is the position of the water table and water movement on and within rock structures and the subsurface environment. Another control over weathering rates is the *geographic orientation* of a slope—whether it faces north, south, east, or west. Orientation controls the slope's exposure to Sun, wind, and precipitation. Slopes facing away from the Sun's rays tend to be cooler, moister, and more vegetated than slopes in direct sunlight. This effect of orientation is especially noticeable in the middle and higher latitudes.

FIGURE 10.4 Regolith, soil, and parent materials.
(a) A cross section of a typical hillside. (b) A cliff exposes hillside components. (c) These reddish-colored dunes in the Navajo Tribal Park, near the Utah-Arizona border, derive their color from the red-sandstone parent materials in the background. (d) This image, from a larger panoramic scene, was made by the Mars Exploration Rover *Spirit* on March 12, 2004. Looking toward the Columbia Hills, you can see weathered rocks and windblown sand in approximate true colors. [Photos (b) and (c) by author; (d) Mars image courtesy of NASA/JPL and Cornell University.]

Vegetation is also a factor in weathering. Although vegetative cover protects rock by shielding it from raindrop impact and provides roots to stabilize soil, it also forms organic acids from the partial decay of organic matter; these acids contribute to chemical weathering. Plant roots can enter crevices and break up a rock, exerting enough pressure to drive rock segments apart, thereby exposing greater surface area to other weathering processes (Figure 10.5a). You may have observed how tree roots heave the sections of a sidewalk or driveway sufficiently to raise and crack the concrete.

The role of the climatic factor warrants further discussion. There is a relation among climate (annual precipitation and temperature), physical weathering, and chemical weathering processes. In general, physical weathering dominates in drier, cooler climates, whereas chemical weathering dominates in wetter, warmer climates. Extreme dryness reduces weathering rates, as is experienced in desert climates (*low-latitude arid desert*). In the hot, wet, tropical and equatorial rain forest climates (*tropical rain forest*), most rocks weather rapidly, and the weathering extends deep below the surface.

The scale at which we analyze weathering processes is important. Research at *microscale* resolution reveals greater complexity in the relation between climate and weathering. At the small scale of actual reaction sites on the rock surface, both physical and chemical weathering processes can occur across varied climate types. Hygroscopic water (a molecule-thin water layer on soil particles) and capillary water (soil water) activate chemical weathering processes, even in the driest landscape. (Review sections in Chapter 6 for these water types.)

Imagine all the factors that influence weathering rates as operating in concert: climatic influence, soil water and groundwater, rock composition and structure (jointing), slope orientation, vegetation, and microscopic boundary-layer conditions at reaction sites. We separate these processes here for convenience of study. Of course, in all of this *time* is the crucial factor, for these processes require long periods of time to operate.

Physical Weathering Processes

When rock is broken and disintegrated without any chemical alteration, the process is **physical weathering**, or *mechanical weathering*. By breaking up rock, physical weathering greatly increases the surface area on which chemical weathering may operate. In the complexity of nature, physical and chemical weathering usually operate together in a combination of processes.

Frost Action   Water expands by as much as 9% of its volume as it freezes (see the section "Ice, the Solid Phase" in Chapter 5). This expansion creates a powerful mechanical force of **frost action,** or freeze-thaw action, that can exceed the tensional strength of rock. Freezing actions are important in humid microthermal climates (*humid continental*) and in subarctic and polar regimes (*subarctic* and *polar* climates).

Repeated cycles of water freezing and thawing break rock segments apart. The work of ice begins in small openings, gradually expanding until rocks are cleaved (split), as demonstrated in Figure 10.5b and c. Figure 10.6 shows this action on blocks of rock, causing *joint-block separation* along existing joints and fractures in the rock. This weathering action, called *frost-wedging*, pushes portions of the rock apart. Cracking and breaking can be in any shape, depending on the rock structure. The softer supporting rock underneath the slabs in the photo already has weathered physically—an example of differential weathering.

Several cultures have used this principle of frost action to quarry rock. Pioneers in the early American West drilled holes in rock, poured water in the holes, and then plugged them. During the cold winter months, expanding ice broke off large blocks along lines determined by

ANIMATION
Physical
Weathering

(a)

(b)

(c)

FIGURE 10.5 Physical weathering examples.
(a) Roots exert a force on the sides of this joint in the rock. (b) Frost action broke this granite rock; ice expansion forced the rock segments apart. (c) The freeze-thaw force of frost action shatters marble, a metamorphic rock. [Photos by (a) author; (b) and (c) Bobbé Christopherson.]

the hole patterns. In spring, the blocks were hauled to cities for use as construction material. Frost action also produces unwanted fractures that damage pavement and split water pipes.

Spring can be a risky time to venture into mountainous terrain. As rising temperatures melt the winter's ice, newly fractured rock pieces fall without warning and may even start rock slides. The falling rock pieces may physically shatter on impact—another form of physical weathering

(a)

(b)

FIGURE 10.6 Physical weathering.
(a) Physical weathering along joints in rock produces discrete blocks in the backcountry of Canyonlands National Park in Utah. (b) Joint-block separation in slate at Alkehornet, Isfjord, on Spitsbergen Island in the Arctic Ocean, where freezing is intense. [Photos by (a) author; (b) Bobbé Christopherson.]

(a)

(b)

FIGURE 10.7 Rockfall.
(a) Shattered rock debris, the product of a large rockfall in Yosemite National Park. (b) A large slab failure produces a rockfall in Zion National Park. The small opening at the cliff base is a window blasted through to the highway tunnel that was excavated inside the cliff. [Photos by author.]

(Figure 10.7a). One such rockfall in Yosemite National Park (1996) involved a crashing drop of 670 m, at 260 kmph (2200 ft, 160 mph), of a 162,000-ton granite slab. The impact shattered the rock, covering 50 acres in powdered rock and felling 500 trees. Another rockfall dropped off Glacier Point in 1999, killing one person. In October 2008, 1380 m³ (1800 yd³) of granite plummeted 610 m (2000 ft) onto a section of Curry Village in Yosemite; a reminder of rockfall realities. Figure 10.7b shows a rockfall of sandstone slabs in Zion National Park.

Salt-Crystal Growth (Salt Weathering) Especially in arid climates, dry weather draws moisture to the surface of rocks. As the water evaporates, crystals develop from dissolved minerals. As this process continues over time and the crystals grow and enlarge, they exert a force great enough to spread apart individual mineral grains and begin breaking up the rock. Such *crystallization*, or *salt-crystal growth*, is a form of physical weathering.

In the Colorado Plateau of the Southwest, deep indentations develop in sandstone cliffs, especially above

impervious layers, where salty water slowly flows out of the rock strata. Crystallization then loosens the sand grains, and subsequent erosion and transportation by water and wind complete the sculpting process. Over 1000 years ago, Native Americans built entire villages in these weathered niches at several locations, including Mesa Verde in Colorado and Arizona's Canyon de Chelly (pronounced "canyon duh shay," Figure 10.8).

Pressure-Release Jointing To review from Chapter 8, rising magma that is deeply buried and subjected to high pressure forms intrusive igneous rocks called plutons, which cool slowly and produce coarse-grained, crystalline granitic rocks (see illustration in Figure 8.8a). As the landscape is subjected to uplift, the weathering regolith that covers such a pluton can be eroded and transported away, eventually exposing the pluton as a mountainous batholith.

As the tremendous weight of the overburden is removed from the granite, the pressure of deep burial is relieved. The batholith responds with a slow but

FIGURE 10.8 Physical weathering in sandstone.
(a) Cliff-dwelling site in Canyon de Chelly, Arizona, used by the Anasazi people until about 900 years ago. The niche in the rock was partially formed by crystallization, which forced apart mineral grains and broke up the sandstone. The dark streaks on the rock are desert varnish, composed of iron oxides with traces of manganese and silica. (b) Water and an impervious sandstone layer helped concentrate weathering processes in the niche. (c) A niche forming in sandstone. [Photos by (a) author; (c) Bobbé Christopherson.]

enormous heave, and in a process known as *pressure-release jointing*, layer after layer of rock peels off in curved slabs or plates, thinner at the top of the rock structure and thicker at the sides. As these slabs weather, they slip off in the process of **sheeting** (Figure 10.9a, d). This *exfoliation process* creates arch-shaped and dome-shaped features on the exposed landscape. An **exfoliation dome** is the largest single weathering feature on Earth (Figure 10.9b, c).

Chemical Weathering Processes

Chemical weathering refers to actual decomposition and decay of the constituent minerals in rock due to chemical alteration of those minerals, always in the presence of water. The chemical breakdown becomes more intense as both temperature and precipitation increase. Although individual minerals vary in susceptibility, no rock-forming minerals are completely unresponsive to chemical weathering. A familiar example of chemical weathering is the

eating away of cathedral façades and the etching of tombstones by acid precipitation.

In Europe, where increasingly acidic rains resulted from the burning of coal, chemical weathering processes are visible on many buildings. For example, Saint Magnus Cathedral, in Kirkwell, Orkney Islands, north of Scotland, was built of red and yellow sandstone, with construction beginning in A.D. 1137—this is 873 years before the copyright date of this book. Almost nine centuries of chemical weathering dissolved cementing materials in the sandstone, breaking down the rock and making many of the sculptured design elements unrecognizable, as if they had "melted" or were "out-of-focus" (Figure 10.10). News Report 10.1 discusses the bridges of Central Park in New York City that display similar weathering.

As an example of the way chemical weathering attacks rock, consider **spheroidal weathering**. The sharp edges and corners of rocks become rounded as the alteration of minerals progresses through the rock. Joints in the rock

(a)

(b)

(c)

(d)

FIGURE 10.9 Exfoliation in granite.
Exfoliation processes loosen slabs of granite, freeing them for further weathering and downslope movement. (a) Great arches form in the White Mountains of New Hampshire. (b) Exfoliated layers of rock are visible in characteristic dome formations in granites. This view is from the east side of Half Dome in Yosemite National Park, California. (c) Half Dome perspective from the west; relief is approximately 1500 m (5000 ft) from the top of the dome to the glaciated valley below. (d) Rock sheeting exposed along Beverly Sund, Nordaustlandet Island in the Arctic Ocean. [Photos (a) and (d) by Bobbé Christopherson; (b) and (c) by author.]

offer more surfaces of opportunity for more weathering. Water penetrates joints and fractures and dissolves the rock's weaker minerals or cementing materials. A boulder can be attacked from all sides, shedding shells of decayed rock, like the layers of an onion. The resulting rounded edges are the basis for the name *spheroidal*. Spheroidal weathering of rock resembles exfoliation, but it does not result from pressure-release jointing. This type of weathering is visible in the photograph of a weathered rock in Figure 10.11; note the spheroidal rock form.

Hydration and Hydrolysis Two processes involving water, one of simple combination with a mineral and the other of chemical reaction with a mineral, work to decompose rock. In this book, these are discussed under chemical weathering because both involve chemical actions.

Hydration, meaning "combination with water," involves little chemical change. Water becomes part of the chemical composition of the mineral (such a hydrate is gypsum, which is hydrous calcium sulfate: $CaSO_4 \cdot 2H_2O$). When some minerals hydrate, they expand, creating a strong mechanical effect that stresses the rock, forcing grains apart.

A cycle of hydration and dehydration can lead to granular disintegration and further susceptibility of the rock to chemical weathering. Hydration works together with carbonation and oxidation to convert feldspar, a common mineral in many rocks, to clay minerals and silica. The hydration process is also at work on the sandstone niches shown in the cliff-dwelling photo.

When minerals chemically react with water, the process is **hydrolysis**. Hydrolysis is a decomposition

FIGURE 10.10 Chemical weathering attacks sandstone cathedral.
The west-front entrance of Saint Magnus Cathedral, in Kirkwell, Scotland, shows the signs of chemical weathering as processes attack the cementing materials in the rock. Construction of this edifice began in A.D. 1137, so the weathering forces have been at work for almost nine centuries. [Photos by Bobbé Christopherson.]

process that breaks down silicate minerals in rocks. Compared with hydration, in which water combines with minerals in the rock, the hydrolysis process involves water in chemical reactions to produce different compounds. Reaction to normal mild acids dissolved in precipitation causes the weathering of feldspar minerals, such as in:

feldspar (K, Al, Si, O) + carbonic acid and water :
residual clays + dissolved minerals + silica

So the by-products of chemical weathering of feldspar in granite include clay (such as kaolinite) and silica. As clay forms from some minerals in the granite, quartz (SiO_2) particles are left behind. The resistant quartz may wash downstream, eventually becoming sand on some distant beach. Clay minerals become a major component in soil and in shale, a common sedimentary rock.

When weaker minerals in rock are changed by hydrolysis, the interlocking crystal network breaks down, so the rock fails and *granular disintegration* takes place. Such disintegration in granite may make the rock appear etched, corroded, and softened (Figure 10.11b).

Oxidation   Another example of chemical weathering occurs when certain metallic elements combine with oxygen to form oxides. This is a chemical weathering process known as **oxidation**. Perhaps the most familiar oxidation form is the "rusting" of iron in rocks or soil that produces a reddish-brown stain of iron oxide (Fe_2O_3). We have all left a tool or nails outside only to find them weeks later, coated with iron oxide. The rusty color is visible on the surfaces of rock and in heavily oxidized soils such as those in the southeastern United States, southwestern deserts, or

(a)

(b)

FIGURE 10.11 Chemical weathering and spheroidal weathering.
(a) Chemical weathering processes act on the joints in granite to dissolve weaker minerals, leading to a rounding of the edges of the cracks. (b) Rounded granite outcrop demonstrates spheroidal weathering and the disintegration of rock. [Photos by Bobbé Christopherson.]

(a)

(b)

FIGURE 10.12 Oxidation processes in rock and soil.
(a) Oxidation of iron minerals produces these brilliant red colors in the sandstone formations of Valley of Fire State Park, Nevada. (b) Ultisols, soils formed under warm, moist conditions, in Sumter County, Georgia, bear the color of iron and aluminum oxides. [Photos by Bobbé Christopherson.]

the tropics (Figure 10.12). Here is a simple oxidation reaction in iron:

$$\text{iron (Fe)} + \text{oxygen (O}_2) : \text{ iron oxide (hematite; Fe}_2\text{O}_3)$$

As iron is removed from the minerals in a rock, the disruption of the crystal structures in the minerals makes the rock more susceptible to further chemical weathering and disintegration.

Dissolution of Carbonates A third form of chemical weathering occurs when a mineral dissolves into *solution*— for example, when sodium chloride (common table salt) dissolves in water. Water is the universal solvent because it is capable of dissolving at least 57 of the natural elements and many of their compounds.

Water vapor readily dissolves carbon dioxide, thereby yielding precipitation containing carbonic acid (H_2CO_3). This acid is strong enough to dissolve many minerals, especially limestone, in the **carbonation** process. *Carbonation* simply means reactions whereby carbon combines with minerals, dissolving them. When rainwater attacks formations of *limestone* (which is calcium carbonate, $CaCO_3$), the rainwater dissolves and washes away constituent minerals, such as in:

$$\text{calcium carbonate} + \text{carbonic acid and water} :$$
$$\text{calcium bicarbonate (Ca(HCO}_3)_2)$$

A walk through an old cemetery shows you the carbonation of marble, a metamorphic form of limestone (Figure 10.13). Limestone and marble, in tombstones or in rock formations, appear pitted and weathered wherever adequate water is available for carbonation. In this era of human-induced increases in acid precipitation, carbonation processes are greatly enhanced because the rain and snow are more acid (see Focus Study 2.2, "Acid Deposition: A Continuing Blight on the Landscape"). You see this

FIGURE 10.13 Dissolution of limestone.
A marble tombstone is chemically weathered beyond recognition in a Scottish churchyard. Marble is a metamorphic form of limestone. Based upon readable dates on surrounding tombstones, this one is about 225-years old. [Photo by Bobbé Christopherson.]

effect on the cathedral stones in Figure 10.10 and the Central Park bridge in News Report 10.1.

The chemical weathering process of carbonation dominates entire landscapes composed of limestone. These are the regions of karst topography, which we look at next.

Karst Topography and Landscapes

Limestone is abundant on Earth, composing many landscapes (Figure 10.14). These areas are susceptible to chemical weathering, which creates a specific landscape of pitted and bumpy surface topography, poor surface drainage, and well-developed channels underground. Remarkable labyrinths of underworld caverns also may develop. These are the hallmarks of **karst topography**, originally named for the Krš Plateau in Slovenia (formerly Yugoslavia), where these processes were first studied. Approximately 15% of Earth's land area has some developed karst, with outstanding examples found in southern China, Japan, Puerto Rico, Cuba, the state of Yucatán in Mexico, Kentucky, Indiana, New Mexico, and Florida. As an example, approximately 38% of Kentucky has sinkholes and related karst features, which are noted on topographic maps.

Formation of Karst

For a limestone landscape to develop into karst topography, there are several necessary conditions:

- The limestone formation must contain 80% or more calcium carbonate for dissolution processes to proceed effectively.

- Complex patterns of joints in the otherwise impermeable limestone are needed for water to form routes to subsurface drainage channels.
- There must be an aerated (containing air) zone between the ground surface and the water table.
- Vegetation cover is required to supply varying amounts of organic acids that enhance the dissolution process.

The role of climate in providing optimum conditions for karst processes remains under debate, although the amount and distribution of rainfall appear important. Karst occurs in arid regions, but it is primarily due to former climatic conditions of greater humidity. Karst is rare in the Arctic and Antarctic because the water, although present, is generally frozen.

As with all weathering processes, time is a factor. Early in the twentieth century, karst landscapes were thought to progress through evolutionary stages of development, as if they were aging. Today, these landscapes are thought to be locally unique, a result of specific conditions, and there is little evidence that different regions evolve in some predictable sequence. Nonetheless, mature karst landscapes do display certain characteristic forms.

Lands Covered with Sinkholes

The weathering of limestone landscapes creates many **sinkholes**, which form in circular depressions. (Traditional

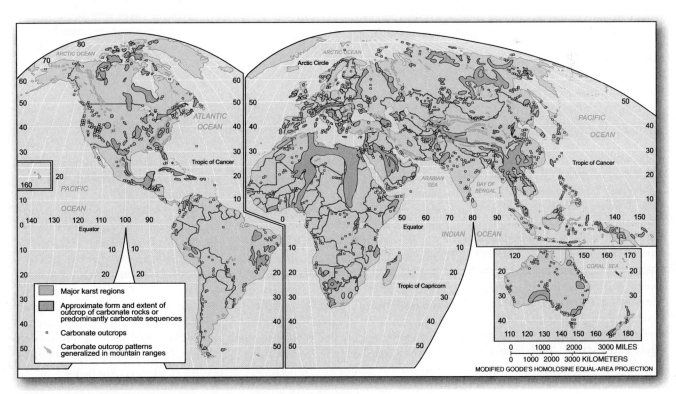

FIGURE 10.14 Global distribution of karst and carbonate rocks.
Major karst regions exist on every continent. The outcrops of carbonate rocks or predominantly carbonate sequences are limestone and dolomite (calcium or magnesium carbonate), but may contain other carbonate rocks. [Map adapted by Pam Schaus, after R. E. Snead, *Atlas of the World Physical Features*, p. 76. © 1972 by John Wiley & Sons; and D. C. Ford and P. Williams, *Karst Geomorphology and Hydrology*, p. 601. © 1989 by Kluwer Academic Publishers. Adapted by permission.]

studies may call a sinkhole a *doline*.) If a solution sinkhole collapses through the roof of an underground cavern, a *collapse sinkhole* forms. A gently rolling limestone plain might be pockmarked by sinkholes with depths of 2 to 100 m (7 to 330 ft) and diameters of 10 to 1000 m (33 to 3300 ft), as shown in Figure 10.15a. Through continuing solution and collapse, sinkholes may coalesce to form a *karst valley*, an elongated depression up to several kilometers long.

The area southwest of Orleans, Indiana, has over 1000 sinkholes in just 2.6 km² (1 mi²). In this area, the Lost River, a "disappearing stream," flows more than 13 km (8 mi) underground before it resurfaces at its Lost River rise (near the Orangeville rise shown in Figure 10.15e); its flow is diverted from the surface through sinkholes and solution channels. Its dry bed can be seen on the lower left of the topographic map in Figure 10.15b.

In Florida, several sinkholes have made news because lowered water tables (lowered by pumping) caused their collapse into underground solution caves, taking with them homes, businesses, and even new cars from an auto dealership. One such sinkhole collapsed in a suburban area in 1981 and others in 1993 and 1998 (Figure 10.16a), an area depicted with the photo and topographic map in Figure 10.16b. The case is obvious in Florida and elsewhere that limestone landscapes are complex and require integrated spatial analysis involving hydrology and geology for all possible development.

A complex landscape in which sinkholes intersect is called a *cockpit karst*. The sinkholes can be symmetrically shaped in certain circumstances; one at Arecibo, Puerto Rico, is perfectly shaped for a radio telescope installation (Figure 10.16c).

Caves and Caverns

Most caves are formed in limestone rock because it is so easily dissolved by carbonation. The largest limestone caverns in the United States are Mammoth Cave in Kentucky, Carlsbad Caverns in New Mexico, and Lehman Cave in Nevada. Limestone formations in which the Carlsbad Caverns formed were themselves deposited 200 million years ago when the region was covered by shallow seas. Regional uplifts associated with the Rockies (the Laramide orogeny 40–80 million years ago) then brought the region above sea level, leading to active cave formation (Figure 10.17).

Caves generally form just beneath the water table, where a later lowering of the water level exposes them to further development. The *dripstones* that form under such conditions are produced as water containing dissolved minerals slowly drips from the cave ceiling. Calcium carbonate precipitates out of the evaporating solution, literally a molecular layer at a time, and accumulates at a point below on the cave floor. Depositional features called *stalactites* grow down from the ceiling and *stalagmites* build up from the floor, sometimes growing until they connect and form a continuous *column*. A dramatic subterranean world is thus created.

Cave science, *speleology*, is an aspect of geomorphology in which amateur cavers make important discoveries about these unique habitats and new life forms (*biospeleology*). More than 90% of known caves are not biologically surveyed, and some 90% of possible caves worldwide still lie undiscovered, making this an important research frontier (see News Report 10.2). For more on caves and related

NEWS REPORT 10.2

Amateurs Make Cave Discoveries

The exploration and scientific study of caves is *speleology*. Physical and biological scientists carry out professional investigations of caves; however, amateur cavers, or "spelunkers," have made many important discoveries. As an example, in the early 1940s George Colglazier, a farmer southwest of Bedford, Indiana, awoke to find his farm pond at the bottom of a deep collapsed sinkhole. This sinkhole is now the entrance to an extensive cave system that includes a subterranean navigable stream.

Cave habitats are unique. They are nearly closed systems—self-contained ecosystems with simple food chains and great stability. In total darkness, bacteria synthesize inorganic elements and produce organic compounds that sustain many types of cave life, including algae, small invertebrates, amphibians, and fish.

In a cave discovered in 1986, near Movile in southeastern Romania, cave-adapted invertebrates were discovered after millions of years of sunless isolation. Thirty-one of these organisms were previously unknown. Without sunlight, the ecosystem in Movile is based on sulfur-metabolizing bacteria that synthesize organic matter using energy from oxidation processes. These chemosynthetic bacteria feed other bacteria and fungi that in turn support cave animals. The sulfur bacteria produce sulfuric acid compounds that may prove to be important in the chemical weathering of some caves.

The mystery, intrigue, and excitement of cave exploration lie in the variety of dark passageways, enormous chambers that narrow to tiny crawl spaces, strange formations, and underwater worlds that can be accessed only by cave diving. Private property owners and amateur adventurers discovered many of the major caves, a fact that keeps this popular science/sport very much alive. (For nearly a thousand worldwide links and information, see the http://www.cbel.com/speleology web site, and for many related links of interest, see http://www.caves.org/.)

FIGURE 10.15 Features of karst topography in Indiana.
(a) Idealized features of karst topography in southern Indiana. (b) Karst topography southwest of the town of Orleans, Indiana. On average, 1022 sinkholes occur per 1 mi^2 in this area. On the map, note the contour lines: Depressions are indicated with small hachures (tick marks) on the downslope side of contour lines. (c) Gently rolling karst landscape and cornfields near Orleans, Indiana. (d) This pond is in a sinkhole depression near Palmyra, Indiana. (e) The Orangeville rise, near Orangeville, Indiana, just north of the Lost River rise. During periods of high rainfall, this rise is almost filled with water. [(a) Adapted from W. D. Thornbury, *Principles of Geomorphology*, illustration by W. J. Wayne, p. 326; © 1954 by John Wiley & Sons; (b) Mitchell, Indiana quadrangle, USGS; photos (c), (d), and (e) by Bobbé Christopherson.]

NOTEBOOK

Karst Farm Park, Indiana

(c)

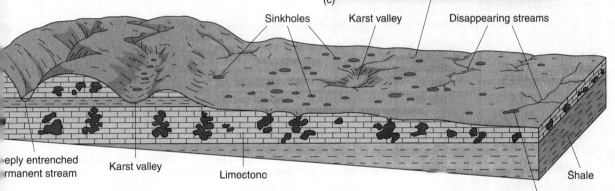

Sinkholes   Karst valley   Disappearing streams

eply entrenched
rmanent stream   Karst valley   Limestone   Shale

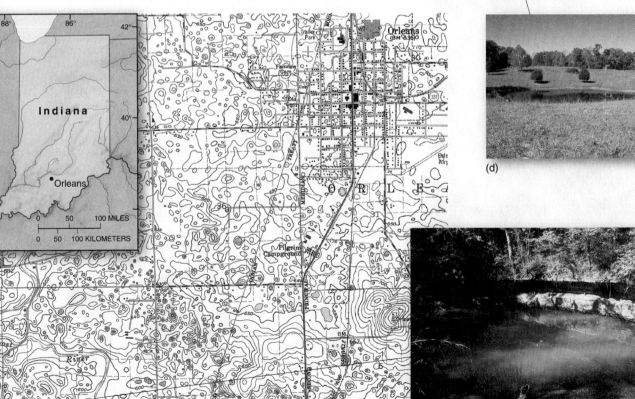

Indiana

Orleans

0   50   100 MILES

0   50   100 KILOMETERS

(d)

(e)

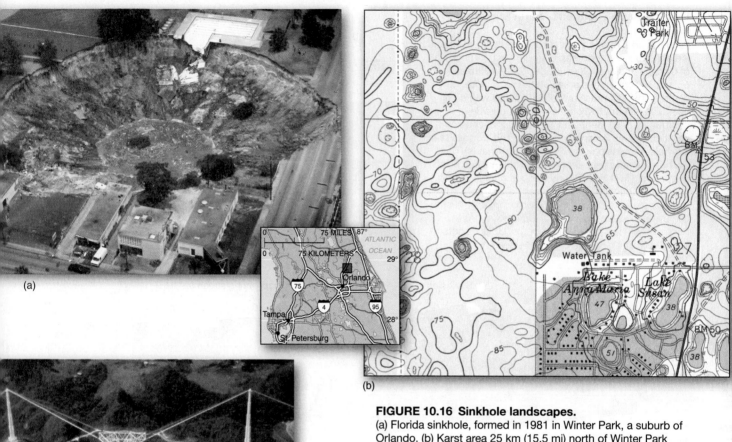

(a)

(b)

(c)

FIGURE 10.16 Sinkhole landscapes.
(a) Florida sinkhole, formed in 1981 in Winter Park, a suburb of Orlando. (b) Karst area 25 km (15.5 mi) north of Winter Park depicted on a topographic map. Note the depressions marked by small hachures (tick marks). (c) Cockpit karst topography near Arecibo, Puerto Rico, provides a natural depression for the dish antenna of a giant radio telescope. The Arecibo Observatory is part of the National Astronomy and Ionosphere Center, which is operated by Cornell University under contract with the National Science Foundation (http://www.naic.edu). [(b) Orange City quadrangle, USGS; photos by (a) Jim Tuten/ Black Star; (c) courtesy of Cornell University.]

formations, see http://www.goodearthgraphics.com/virtcave/virtcave.html.

Mass-Movement Processes

Nevado del Ruiz, northernmost of two-dozen dormant (not extinct, sometimes active) volcanic peaks in the Cordilleran Central of Colombia, had erupted six times during the past 3000 years, killing 1000 people during its last eruption in 1845. On November 13, 1985 at 11 P.M., after a year of earthquakes and harmonic tremors, a growing bulge on its northeast flank, and months of small summit eruptions, Nevado del Ruiz violently erupted in a lateral explosion. The mountain was back in action.

On this night, the familiar pyroclastics, lava, and blast were not the worst problem. The hot eruption quickly melted ice on the mountain's snowy peak, liquefying mud and volcanic ash, sending a hot mudflow downslope. Such a flow is a *lahar*, an Indonesian word referring to mudflows of volcanic origin. This lahar moved rapidly down the Lagunilla River toward the villages below. The wall of mud was at least 40 m (130 ft) high as it approached Armero, a regional center with a population of 25,000. The city slept as the lahar buried its homes: 23,000 people were killed there and in other afflicted river valleys; thousands were injured; 60,000 were left homeless. The volcanic debris flow is now a permanent grave for its victims. Although less devastating to human life, the mudflow generated by the eruption of Mount St. Helens also was a lahar. Not all mass movements are this destructive. Such processes play an important role in the denudation of the landscape.

For more on mass-movement hazards, including landslides, see the web site of the Natural Hazards Center

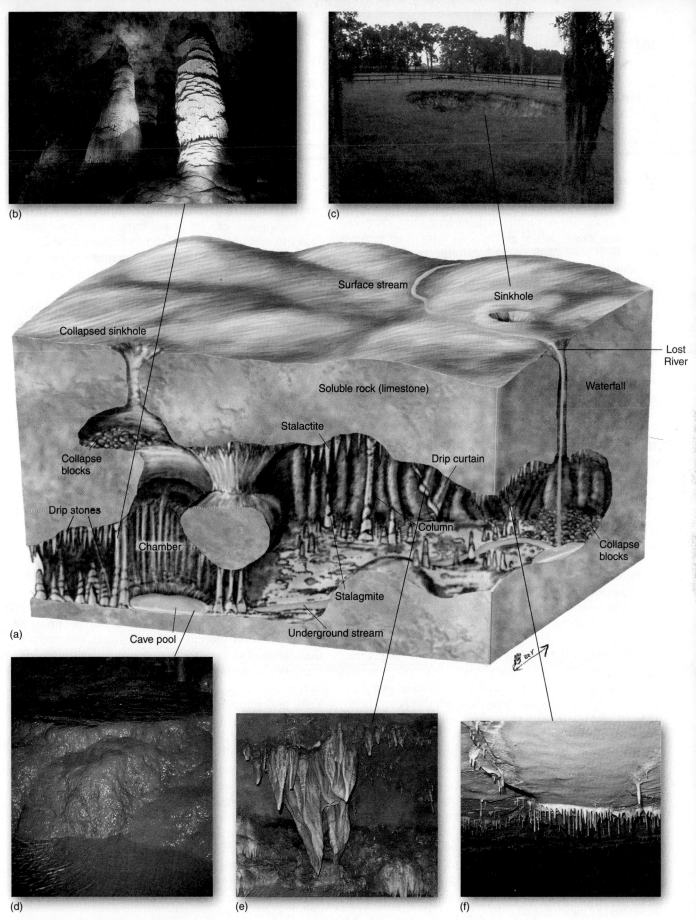

FIGURE 10.17 Cavern features.
(a) An underground cavern and related forms in limestone. (b) A column in Carlsbad Caverns, New Mexico, where a series of underground caverns includes rooms more than 1200 m (4000 ft) long and 190 m (625 ft) wide. Although a national park since 1930, unexplored portions remain. (c) A sinkhole in a Florida pasture. From Marengo Caves, Marengo, Indiana: (d) a flowstone and pool of water; (e) dripstone drapery formations; (f) soda straws hanging from ceiling cracks, forming one molecular layer at a time.
[(b) Photo by author; (c) photo by Thomas M. Scott, Florida Geological Survey; photos (d), (e), (f) by Bobbé Christopherson.]

at the University of Colorado, Boulder, at **http://www. Colorado.edu/hazards/**, or the USGS Geologic Hazards page at **http://landslides.usgs.gov/**. Many significant mass-movement episodes are reported in "Landslides in the News" on the latter web site.

Mass-Movement Mechanics

Physical and chemical weathering processes create an overall weakening of surface rock, making the rock more susceptible to the pull of gravity. The term **mass movement** applies to all downward movements of materials propelled and controlled by gravity, such as the lahar just described. These movements can range from dry to wet, slow to fast, small to large, and from free-falling to gradual or intermittent in motion. The term *mass movement* sometimes is used interchangeably with *mass wasting*, which is the general process involved in mass movements and erosion of the landscape. To combine the concepts, we can say that mass movement of material works to waste slopes and provide raw material for erosion, transportation, and deposition.

The Role of Slopes All mass movements occur on slopes. If we try to pile dry sand on the beach, the grains will flow downward until an equilibrium is achieved. The steepness of the slope that results when loose sand comes to rest depends on the size and texture of the grains; this is the *angle of repose*. This angle represents a balance of driving and resisting forces (gravity and friction) and commonly ranges between 33° and 37° (measured from a horizontal plane).

The *driving force* in mass movements is gravity, working in conjunction with the weight, size, and shape of the surface material, the degree to which the slope is oversteepened, and the amount and form of moisture available—whether frozen or fluid. The greater the slope angle, the more susceptible the surface material is to mass movement when gravity overcomes friction. The *resisting force* is the shearing strength of slope material, that is, its cohesiveness and internal friction, which work against gravity and mass wasting.

Clays, shales, and mudstones are highly susceptible to hydration (physical swelling in response to the presence of water). If such materials underlay rock strata in a slope, the strata will move with less driving-force energy. When clay surfaces are wet, they deform slowly in the direction of movement, and when saturated they form a viscous fluid with little shearing strength (resistance to movement) to hold back the slope. However, if the rock strata are such that material is held back from slipping, then more driving-force energy may be required, such as that generated by an earthquake.

Madison River Canyon Landslide In the Madison River Canyon near West Yellowstone, Montana, on the Wyoming border, a blockade of dolomite (a magnesium-rich

carbonate rock) held back a deeply weathered and *oversteepened slope* (40°–60° slope angle) for untold centuries (white area in Figure 10.18). Then, shortly after midnight on August 17, 1959, an earthquake measuring 7.5 magnitude broke the dolomite structure along the foot of the slope. The break released 32 million m³ (1.13 billion ft³) of mountainside, which moved downslope at 95 kmph (60 mph), causing gale force winds through the canyon. The material continued more than 120 m (390 ft) up the opposite canyon slope, trapping several hundred campers with about 80 m (260 ft) of rock, killing 28 people. These events were a dramatic example of the role of slopes and tectonic forces in creating mass land movements.

The mass of material also effectively dammed the Madison River and created a new lake, dubbed Quake Lake. The landslide debris dam established a new temporary base level for the canyon. A channel was quickly excavated by the US Army Corps of Engineers to prevent a disaster below the dam, for if Quake Lake overflowed the landslide dam, the water would quickly erode a channel and thereby release the entire contents of the new lake onto farmland downstream.

Classes of Mass Movements

In any mass movement, gravity pulls on a mass until the critical shear-failure point is reached—*a geomorphic threshold*. The material then can *fall, slide, flow,* or *creep*—the four classes of mass movement. The Madison River Canyon event was a type of slide, whereas the Nevado del Ruiz lahar mentioned earlier was a flow. Figure 10.19 summarizes the relationship between moisture content and rate of movement in producing the different classes of mass movement. (In contrast to these surface processes, submarine mass movements are disorganized, turbulent flows of mud accumulating as deep-sea depositional fans along the continental slope.)

Falls and Avalanches A **rockfall** is simply a quantity of rock that falls through the air and hits a surface (see the earlier examples in Figure 10.7). During a rockfall, individual pieces fall independently and characteristically form a cone-shaped pile of irregular broken rocks called a *talus slope* at the base of a steep incline, where several *talus cones* coalesce (Figure 10.20).

A **debris avalanche** is a mass of falling and tumbling rock, debris, and soil. It is differentiated from a debris slide or landslide by the tremendous velocity achieved by the onrushing materials. The high speeds often result from ice and water that fluidize the debris. The extreme danger of a debris avalanche is from these tremendous speeds and lack of warning.

In 1962 and again in 1970, debris avalanches roared down the west face of Nevado Huascarán, the highest peak in the Peruvian Andes. The 1962 debris avalanche contained an estimated 13 million m³ (460 million ft³) of

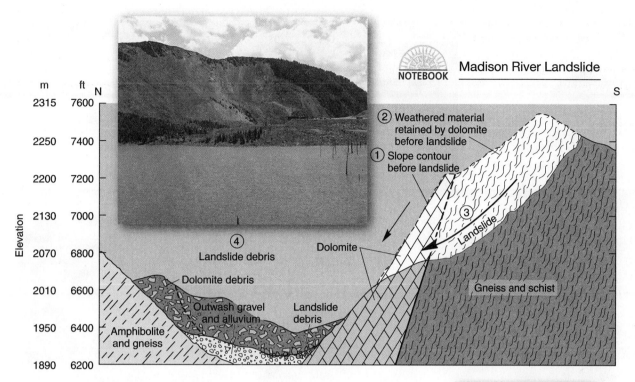

FIGURE 10.18 Madison River landslide.
Cross section showing geologic structure of the Madison River Canyon in Montana, where an earthquake triggered a landslide in 1959. (1) Prequake slope contour. (2) Weathered rock that failed. (3) Direction of landslide. (4) Landslide debris blocking the canyon and damming the Madison River. [After J. B. Hadley, *Landslides and Related Phenomena Accompanying the Hebgen Lake Earthquake of 17 August 1959*, US Geological Survey Professional Paper 435-K, p. 115. Inset photo by Bobbé Christopherson.]

material, burying the city of Ranrahirca and eight other towns, killing 4000 people. An earthquake initiated the 1970 event (Figure 10.21). This time, upward of 100 million m³ (3.53 billion ft³) of debris buried the city of Yungay, where 18,000 people perished. This avalanche attained velocities of 300 kmph (185 mph).

Landslide A sudden rapid movement of a cohesive mass of regolith or bedrock that is not saturated with moisture is a **landslide**—a large amount of material failing simultaneously. Surprise is the danger, for the downward pull of gravity dominates the struggle for equilibrium in an instant. One such surprise event struck near Longarone, Italy, in 1963, and is described in Focus Study 10.1.

To eliminate the surprise element, scientists are using global positioning system (GPS) instruments to monitor landslide movement. With GPS, scientists measure slight land shifts in suspect areas for clues to possible mass wasting. GPS was applied in two cases in Japan and effectively identified prelandslide movements of 2–5 cm per year, providing information to expand the area of hazard concern.

Slides occur in one of two basic forms—translational or rotational (shown in Figure 10.19). *Translational slides* involve movement along a planar (flat) surface roughly parallel to the angle of the slope. The Madison River

Canyon landslide was translational. Flow and creep patterns also are considered translational in nature. *Rotational slides* occur when surface material moves along a concave surface. Frequently, underlying clay presents an impervious surface to percolating water. As a result, water flows along the clay surface, undermining the overlying block. The surface may rotate as a single unit, or it may present a stepped appearance. Continuing rotational mudslides plague La Conchita, California (Figure 10.19a). For thousands of years this area had been unstable; then a slump landslide hit in 1995, burying some homes, and most recently in January 2005, when debris buried 30 homes and took 10 lives in a mudslide episode that followed a period of heavy rains.

In October 2007, mass-movement realities struck upscale La Jolla, California, damaging more than 100 homes along Soledad Mountain Road (Figure 10.22). The rotational slump along a hillside pulled material from beneath the highway, causing a sinkhole-like collapse. The area is inherently unstable, having experienced three other collapse events since 1961, with early warning signs of the 2007 occurrence evident in July. In an area such as this it is difficult to imagine that standard cut-and-fill housing pads were allowed in the planning stage prior to construction, or that homeowners were allowed to irrigate lawns and

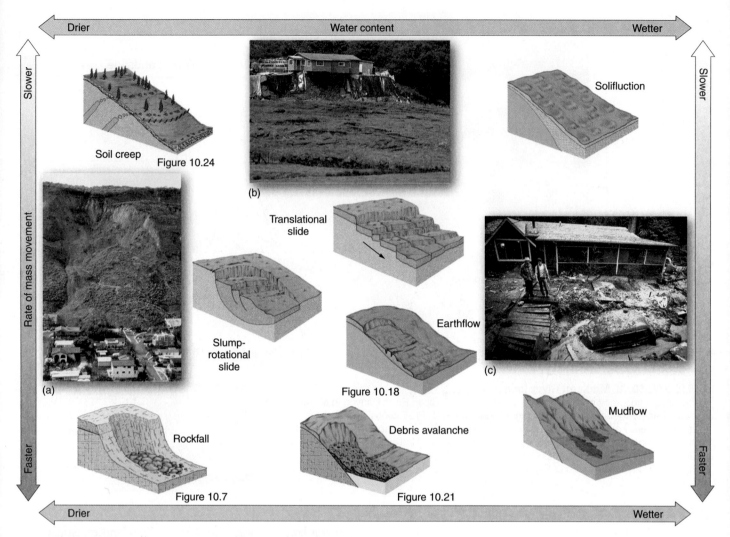

FIGURE 10.19 Mass-movement classes.
Principal types of mass-movement and mass-wasting events. Variations in water content and rates of movement produce a variety of forms. (a) A 1995 rotational slide in La Conchita, California, also site of a 2005 mudslide event; (b) saturated hillsides fail; (c) mudflows 2 m deep, both in Santa Cruz County, California. [Photos by (a) Robert L. Schuster/USGS; (b) Alexander Lowry/ Photo Researchers, Inc.; (c) James A. Sugar.]

 Mass Movements

FIGURE 10.20 Talus slope.
Rockfall and talus deposits at the base of a steep slope along Duve Fjord, Nordaustlandet Island. Can you see the lighter rock strata that are the source for the three talus cones? [Photo by Bobbé Christopherson.]

gardens with no restrictions, feeding drainage water into subsurface strata.

Flow   When the moisture content of moving material is high, the suffix -*flow* is used, as in *earthflow* and *mudflow*, illustrated in Figure 10.19c. An estimated 20,000 people died in Venezuela as the result of massive mudflows in 1999. Heavy rains frequently saturate barren mountain slopes and set them moving, as was the case east of Jackson Hole, Wyoming, in the spring of 1925. A slope above the Gros Ventre River (pronounced "grow vaunt") broke loose and slid downward as a unit. The earthflow occurred because sandstone formations rested on weak shale and siltstone, which became moistened and soft and offered little resistance to the overlying strata (Figure 10.23).

Because of melted snow and rain, the water content of the Gros Ventre landslide was high enough to classify it as an earthflow. About 37 million m^3 (1.3 billion ft^3) of wet

FIGURE 10.22 Landslide under major highway and homes.
More than 100 homes and a portion of Soledad Mountain Road were damaged by this October 2007 rotational slide. The area experiences such failures on a recurring basis. [AP Photo/Chris Park.]

FIGURE 10.21 Debris avalanche, Peru.
A 1970 debris avalanche falls more than 4100 m (2.5 mi) down the west face of Nevado Huascarán, burying the city of Yungay, Peru. The same area was devastated by a similar avalanche in 1962 and by others in pre-Columbian times. A great danger remains for the cities and towns in the valley from possible future mass movements. [Photo by George Plafker.]

soil and rock moved down one side of the canyon and surged 30 m (100 ft) up the other side. The earthflow dammed the river and formed a lake. Two years later, the lake water broke through the temporary dam, transporting a tremendous quantity of debris over the region downstream.

Creep   A persistent mass movement of surface soil is **soil creep**. Expansion of soil moisture as it freezes, or cycles of moistness and dryness, or daily temperature variations, or even grazing by livestock or digging by animals disturb and lift individual soil particles. In the freeze-thaw cycle, freezing soil moisture lifts particles at right angles

ANIMATION   Debris Avalanche and Eruption of Mt. St. Helens

FOCUS STUDY 10.1

Vaiont Reservoir Landslide Disaster

Place: Northeastern Italy, Vaiont Canyon in the Italian Alps, rugged scenery, 680 m (2200 ft) above sea level.

Location: 46.3° N 12.3° E, near the border of the Veneto and Friuli-Venezia Giulia regions.

Situation: Centrally located in the region, an ideal situation for hydroelectric power production.

Site: Steep-sided, narrow, glaciated canyon, opening to populated low-lands to the west; ideal site for a narrow-crested high dam and deep reservoir.

The Plan: Build the second-highest dam in the world, using a new thin-arch design, 262 m (860 ft) high, 190 m (623 ft) in crest length, impounding a reservoir capacity of 150 million m³ (5.3 billion ft³) of water (third-largest reservoir in the world), and generate hydroelectric power.

Geologic Analysis

1. Steep canyon walls composed of interbedded limestone and shale; badly cracked and deformed structures; open fractures in the shale are inclined toward the future reservoir body. The steepness of the canyon walls enhances the strong driving forces (gravitational) at work on rock structures.

2. High potential for bank storage (water absorption by canyon walls)

(continued)

Focus Study 10.1 (*continued*)

into the groundwater system, which will increase water pressure on all rocks in contact with the reservoir.

3. The nature of the shale beds is such that cohesion will be reduced as their clay minerals become saturated.

4. Evidence of ancient rockfalls and landslides on the north side of the canyon.

5. Evidence of creep activity along the south side of the canyon.

Political/Engineering Decision: Begin design and construction of a thin-arch dam at this site.

Events During Construction and Filling: Large volumes of concrete were injected into the bedrock as "dental work" in an attempt to strengthen the fractured rock. During reservoir filling in 1960, 700,000 m³ (2.5 million ft³) of rock and soil slid into the reservoir from the south

side. A slow creep of slope materials along the entire south side began shortly thereafter, increasing to about 1 cm per week by January 1963. Creep rate increased as the level of the reservoir rose. By mid-September 1963, the creep rate exceeded 40 cm (16 in.) a day.

This sequence of events set the stage for one of the worst dam disasters in history. Heavy rains began on September 28, 1963. Alarmed at the runoff from the rain into the reservoir and the increase in creep rate along the south wall, engineers opened the outlet tunnels on October 8 in an attempt to bring down the reservoir level—but it was too late.

The next evening, in only 30 seconds, a landslide of 240 million m³ (8.5 billion ft³) crashed into the reservoir, shaking seismographs across Europe. A 150-m-(500-ft-) thick slab

of mountainside gave way (2 km by 1.6 km in area, 1.2 mi by 1 mi), sending shock waves of wind and water up the canyon 2 km (1.2 mi) and splashing a 100-m (330-ft) wave over the dam. The former reservoir was effectively filled with bedrock, regolith, and soil that almost entirely displaced its water content (Figure 10.1.1). Amazingly, the experimental dam design held.

Downstream, in the unsuspecting town of Longarone, near the mouth of Vaiont Canyon on the Piave River, people heard the distant rumble and were quickly drowned by the 69-m (226-ft) wave of water that came out of the canyon. That night 3000 people perished. As for a lesson or recommendation, we need only to reread the preconstruction geologic analysis that began this focus study. The Italian courts eventually prosecuted the persons responsible.

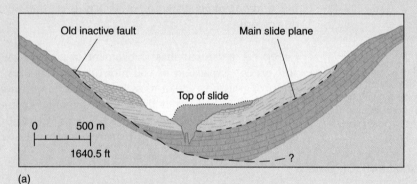

(a)

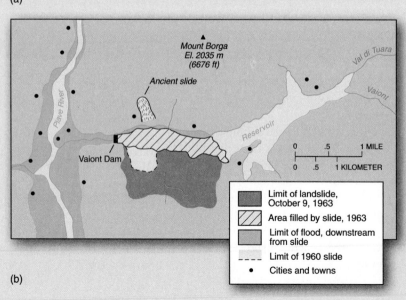

(b)

Limit of landslide, October 9, 1963

Area filled by slide, 1963

Limit of flood, downstream from slide

Limit of 1960 slide

Cities and towns

FIGURE 10.1.1 Vaiont Reservoir disaster.
(a) Cross section of the Vaiont River Valley. (b) Map of the disaster site noted with evidence of previous activity. [After G. A. Kiersch, "The Vaiont Reservoir Disaster," California Division of Mines and Geology, Mineral Information Service, vol. 18, no. 7, pp. 129–138.]

This forested mass is the main portion of the earthflow.

FIGURE 10.23 The Gros Ventre slide near Jackson, Wyoming.
Evidence of the 1925 earthflow is still visible after more than 80 years. [Photo by Steven K. Huhtala.]

to the slope, as shown in Figure 10.24. However, when the ice melts, gravity pulls the particles straight downward. As the process is repeated, the soil cover gradually creeps its way downslope.

The overall wasting of a slope may cover a wide area and may cause fence posts, utility poles, and even trees to lean downslope. Various strategies are used to arrest the mass movement of slope material—grading the terrain,

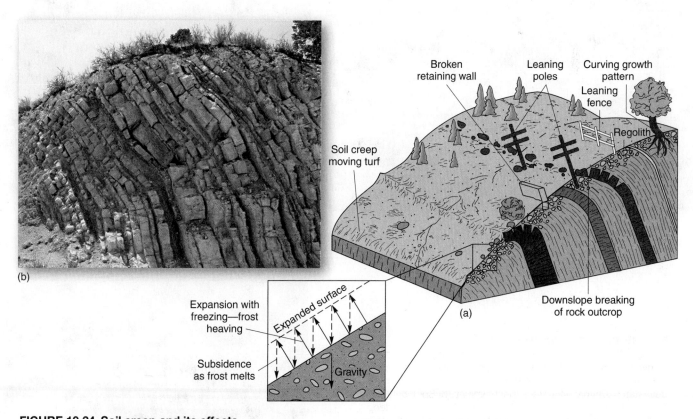

FIGURE 10.24 Soil creep and its effects.
(a) Typical soil-creep features. (b) Note the subsurface effects on rock strata caused by surface creep. [(b) photo by Bobbé Christopherson.]

building terraces and retaining walls, planting ground cover—but the persistence of creep always prevails.

Human-Induced Mass Movements

Every human disturbance of a slope—highway roadcut, surface mining, or building of a shopping mall, housing development, or home—can hasten mass wasting. The newly destabilized and oversteepened surfaces are thrust into a search for a new equilibrium. Imagine the disequilibrium in slope relations created by the Maryland interstate highway roadcut pictured in Figure 9.8b.

Large open-pit surface mines—such as the Bingham Copper Mine west of Salt Lake City, the abandoned Berkeley Pit in Butte, Montana, and numerous large coal surface mines in the US East and West—are examples of human-induced mass movements, generally called **scarification** (Figure 10.25). At the Berkeley Pit, toxic drainage water is threatening the regional aquifers, the Clark Fork River, and the local freshwater supply.

(a)

(b)

(c)

(d)

FIGURE 10.25 Scarification.
(a) Black Mesa, Arizona, strip-mining for coal. (b) Bingham Canyon, Utah, west of Salt Lake City, strip-mining for copper and other minerals. (c) Mountaintop removal coal strip-mining devastates Kayford Mountain, West Virginia. (d) Spoil banks of tailings in a West Virginia coal-mining area. (e) The 55-m-deep Dally slate quarry in Pen Argyl, Pennsylvania. [Photos (a) and (d) by author; (b) and (e) by Bobbé Christopherson; (c) by Dr. Chris Mayda, Eastern Michigan University, all rights reserved.]

Residues of copper, zinc, lead, and arsenic lace one creek, now devoid of life, and are in the soil on Butte Hill, where children play. Wind dispersal is a particular problem with radioactive uranium tailings in the West.

At the Bingham Copper Mine, a mountain has literally been removed since mining began in 1906, forming a pit 4 km wide and 1 km deep (Figure 10.25b), easily the largest human-made excavation on Earth, with ores containing copper, gold, silver, and molybdenum. The disposal of tailings (mined ore of little value) and waste material is a significant problem at surface mines. Such large excavations produce tailing piles that are unstable and susceptible to further weathering, mass wasting, or wind dispersal. The leaching of toxic materials from tailings and waste piles poses an ever-increasing problem to streams, aquifers, and public health across the country.

Where underground mining is common, particularly for coal in the Appalachians, land subsidence and collapse produce further mass movements. Homes, highways, streams, wells, and property values are severely affected. A controversial form of coal mining called *mountaintop removal* is done by removing ridges and summits and dumping the debris into stream valleys. The exposed coal seams are then mined. A change in the Clean Water Act in 2002 removed a 20-year-old, 100-foot restriction to protect streams. Figure 10.25c is an aerial view of mountaintop removal at Kayford Mountain, West Virginia. The 20-story-tall dragline machine is visible center-right. In the background, forest cover and litter are removed using fire, to prepare the site for minimal reclamation, although this is difficult since the landscape is essentially lost. Some 1930 km (1200 mi) of streams are now filled with tailings (Figure 10.25d) and mountaintop scenery is gone.

In eastern Pennsylvania, the Dally slate quarry in Pen Argyl is mining commercial slate deposits formed over millions of years. It is now working a second pit that is 55 m (180 ft) deep (Figure 10.25e). The grain of the metamorphic rock guides removal of slate so as to produce the flattest pieces, which is used for shingles and countertops. The local county turned the first quarry into a landfill—a practical use for the pit.

Scientists can informally quantify the scale of human-induced scarification for comparison with natural denudation processes. Fifteen years ago, R. L. Hooke, Department of Geology and Geophysics, University of Minnesota, used estimates of US excavations for new housing, mineral production (including the three largest—stone, sand and gravel, and coal), and highway construction. He then prorated these quantities of moved earth for all countries; each was rated using its gross domestic product (GDP), energy consumption, and agriculture's effect on river sediment loads. From these, he calculated a global estimate for human earth moving.

Hooke estimated that humans, as a geomorphic agent, annually move 40 to 45 billion tons (40–45 Gt/yr) of the planet's surface (we wonder how much this has increased since 1994). Therefore, humans exceed natural river sediment transfer (14 Gt/yr), movement through stream meandering (39 Gt/yr), haulage by glaciers (4.3 Gt/yr), movement due to wave action and erosion (1.25 Gt/yr), wind transport (1 Gt/yr), sediment movement by continental and oceanic mountain building (34 Gt/yr), or deep-ocean sedimentation (7 Gt/yr). As Hooke stated,

> *Homo sapiens* has become an impressive geomorphic agent. Coupling our earth-moving prowess with our inadvertent adding of sediment load to rivers and the visual impact of our activities on the landscape, one is compelled to acknowledge that, for better or for worse, this biogeomorphic agent may be the premier geomorphic agent of our time.*

*R. L. Hooke, "On the efficacy of humans as geomorphic agents," *GSA Today*, The Geological Society of America, 4, no. 9 (September 1994): 217–226.

Summary and Review—Weathering, Karst Landscapes, and Mass Movement

■ *Define* the science of geomorphology.

Geomorphology is the science that analyzes and describes the origin, evolution, form, and spatial distribution of landforms. The exogenic system, powered by solar energy and gravity, tears down the landscape through processes of landmass **denudation** involving weathering, mass movement, erosion, transportation, and deposition. Different rocks offer differing resistance to these weathering processes and produce a pattern of **differential weathering** on the landscape. This balancing act between tectonic uplift and reduction by weathering and erosion, between the resistance of rocks and the ceaseless attack of weathering and erosion, is summarized in the **dynamic equilibrium model**.

> geomorphology (p. 339)
> denudation (p. 339)
> differential weathering (p. 339)
> dynamic equilibrium model (p. 339)

1. Define geomorphology and describe its relationship to physical geography.
2. Define landmass denudation. What processes are included in the concept?
3. What is the interplay between the resistance of rock structures and weathering variabilities?
4. Explain how Figure 10.1 is an example of differential weathering.
5. What are the principal considerations in the dynamic equilibrium model?

■ *Illustrate* the forces at work on materials residing on a slope.

Slopes are shaped by the relation between rate of weathering and breakup of slope materials and the rate of mass movement and erosion of those materials. A slope is considered stable if it is stronger than these denudation processes; it is unstable if it is weaker. In this struggle against gravity, a slope may reach a **geomorphic threshold**—the point at which there is enough energy to overcome resistance against movement. **Slopes** that form the boundaries of landforms have several general components: *waxing slope*, *free face*, *debris slope*, and *waning slope*. Slopes seek an angle of equilibrium among the operating forces.

> geomorphic threshold (p. 339)
> slopes (p. 340)

6. Describe conditions on a hillslope that is right at the geomorphic threshold. What factors might push the slope beyond this point?

7. Given all the interacting variables, do you think a landscape ever reaches a stable, old-age condition? Explain.

8. What are the general components of an ideal slope?

9. Relative to slopes, what is meant by an "angle of equilibrium"? Can you apply this concept to the photograph in Figure 10.2?

■ *Define* weathering and *explain* the importance of parent rock and joints and fractures in rock.

Weathering processes disintegrate both surface and subsurface rock into mineral particles or dissolve them in water. The upper layers of surface material undergo continual weathering and create broken-up rock called **regolith**. Weathered **bedrock** is the *parent rock* from which regolith forms. The unconsolidated, fragmented material that develops after weathering is called **sediment**, which, along with weathered rock, forms the **parent material** from which soil evolves.

Important in weathering processes are **joints**, the fractures and separations in the rock. Jointing opens up rock surfaces on which weathering processes operate. Factors that influence weathering include: character of the bedrock (hard or soft, soluble or insoluble, broken or unbroken), climatic elements (temperature, precipitation, freeze-thaw cycles), position of the water table, slope orientation, surface vegetation and its subsurface roots, and time. Agents of change include moving air, water, waves, and ice.

> weathering (p. 340)
> regolith (p. 340)
> bedrock (p. 340)
> sediment (p. 340)
> parent material (p. 340)
> joints (p. 340)

10. Describe weathering processes operating on an open expanse of bedrock. How does regolith develop? How is sediment derived?

11. Describe the relationship between climatic conditions and rates of weathering activities. Is the scale at which we consider weathering processes important?

12. What is the relation among parent rock, parent material, regolith, and soil?

13. What role do joints play in the weathering process? Give an example from one of the illustrations in this chapter.

■ *Describe* frost action, crystallization (salt-crystal growth), pressure-release jointing, and the role of freezing water as physical weathering processes.

Physical weathering refers to the breakup of rock into smaller pieces with no alteration of mineral identity. The physical action of water when it freezes (expands) and thaws (contracts) is a powerful agent in shaping the landscape. This **frost action** may break apart any rock. Working in joints, expanded ice can produce *joint-block separation* through the process of *frost-wedging*.

Another process of physical weathering is *crystallization* (salt-crystal growth); as crystals in rock grow and enlarge over time, they force apart mineral grains and break up rock.

As overburden is removed from a granitic batholith, the pressure of deep burial is relieved. The granite slowly responds with *pressure-release jointing*, with layer after layer of rock peeling off in curved slabs or plates. As these slabs weather, they slip off in a process called **sheeting**. This *exfoliation process* creates a dome-shaped feature on the exposed landscape called an **exfoliation dome**.

> physical weathering (p. 343)
> frost action (p. 343)
> sheeting (p. 345)
> exfoliation dome (p. 345)

14. What is physical weathering? Give an example.

15. Why is freezing water such an effective physical weathering agent?

16. What weathering processes produce a granite dome? Describe the sequence of events.

■ *Describe* the susceptibility of different minerals to the chemical weathering processes, including hydration, hydrolysis, oxidation, carbonation, and solution.

Chemical weathering is the chemical decomposition of minerals in rock. **Spheroidal weathering** is chemical weathering that occurs in cracks in the rock. As cementing and binding materials are removed, the rock begins to disintegrate, and sharp edges and corners become rounded.

Hydration occurs when a mineral absorbs water and expands, thus creating a strong mechanical force that stresses rocks. **Hydrolysis** breaks down silicate minerals in rock, as in the chemical weathering of feldspar into clays and silica. Water actively participates in chemical reactions. **Oxidation** is the reaction of oxygen with certain metallic elements, the most familiar example being the rusting of iron, producing iron oxide. *Solution* is chemical weathering. For instance, a mild acid such as carbonic acid in rainwater will cause **carbonation**, wherein carbon combines with certain minerals, such as calcium, magnesium, potassium, and sodium.

> chemical weathering (p. 345)
> spheroidal weathering (p. 345)
> hydration (p. 346)
> hydrolysis (p. 346)
> oxidation (p. 347)
> carbonation (p. 348)

17. What is chemical weathering? Contrast this set of processes to physical weathering.

18. What is meant by the term *spheroidal weathering*? Describe the spheroidal weathering process.

19. What is hydration? What is hydrolysis? Differentiate between these processes. How do they affect rocks?

20. Iron minerals in rock are susceptible to which form of chemical weathering? What characteristic color is associated with this type of weathering?

21. With what kind of minerals do carbon compounds react, and under what circumstances does this reaction occur? What is this weathering process called?

■ *Review* the processes and features associated with karst topography.

Karst topography refers to distinctively pitted and weathered limestone landscapes. Surface circular **sinkholes** form and may combine to form a *karst valley*. A sinkhole may collapse through the roof of an underground cavern, forming a *collapse sinkhole*. The formation of caverns is part of karst processes and groundwater erosion. Limestone caves feature many unique erosional and depositional features, producing a dramatic subterranean world.

> karst topography (p. 349)
> sinkholes (p. 349)

22. Describe the development of limestone topography. What is the name applied to such landscapes? From what area was this name derived?

23. Differentiate among sinkholes, karst valleys, and cockpit karst. Within which form is the radio telescope at Arecibo, Puerto Rico?

24. In general, how would you characterize the region southwest of Orleans, Indiana?

25. What are some of the unique erosional and depositional features you find in a limestone cavern?

■ *Portray* the various types of mass movements and *identify* examples of each in relation to moisture content and speed of movement.

Any movement of a body of material, propelled and controlled by gravity, is called **mass movement**, also called *mass wasting*. The *angle of repose* of loose sediment grains represents a balance of driving and resisting forces on a slope. Mass movement of Earth's surface produces some dramatic incidents, including **rockfall** (a volume of rock that falls), **debris avalanche** (mass of tumbling, falling rock, debris, and soil at high speed), **landslide** (a large amount of material failing simultaneously), *mudflow* (material in motion with a high moisture content), and **soil creep** (a persistent movement of individual soil particles that are lifted by the expansion of soil moisture as it freezes, cycles of wetness and dryness, temperature variations, or the impact of grazing animals). In addition, human mining and construction activities have created massive **scarification** of landscapes.

> mass movement (p. 354)
> rockfall (p. 354)
> debris avalanche (p. 354)
> landslide (p. 355)
> soil creep (p. 357)
> scarification (p. 360)

26. Define the role of slopes in mass movements, using the terms *angle of repose*, *driving force*, *resisting force*, and *geomorphic threshold*.

27. What events occurred in the Madison River Canyon in 1959?

28. What are the classes of mass movement? Describe each briefly and differentiate among these classes.

29. Name and describe the type of mudflow associated with a volcanic eruption.

30. Describe the difference between a landslide and what happened on the slopes of Nevado Huascarán.

31. What is scarification, and why is it considered a type of mass movement? Give several examples of scarification, including a description of mountaintop removal and its impacts. Why are humans a significant geomorphic agent?

NetWork and Critical Thinking Tools

A. Locate a slope, possibly near campus, near your home, or at a local roadcut. Using Figure 10.3a and b, can you identify the forces and forms of a hillslope at your site? How would you go about assessing the stability of the slope? Is there any evidence of the mass wasting of materials, soil creep, or other processes discussed in this chapter?

B. In terms of mass-movement hazards in the region where you attend college, consult several sources to determine if any such hazard exists nearby. How far back in history do you need to track to find the last episode? Under "Destinations" on our Student Learning Center web site go to the USGS Landslides Hazard Program link. Sample recent events over a half year or so. Do you find any mass-movement events similar to hazards you identified in your region?

C. On the Student Learning Center site go to "Critical Thinking." Study the two remote-sensing images. The Himalayan Mountains are the primary source of the sediment transported by the rivers. Imagine yourself as a piece of rock near the top of Mount Everest. Describe a scenario by which you join a collection of sediments in the Bay of Bengal. Discuss the changes in weathering processes acting upon you between the high, cold reaches of the Himalayas and the hot, humid mud flats of the Ganges Delta.

The Niagara River plunges over the Canadian Falls 57 m (187 ft) along the Niagara Escarpment. Discharge averages 600,000 gal/second, or 14.7 million acre-feet a year, flowing between Lake Erie and Lake Ontario on the way to the Atlantic Ocean. More on these falls and how they are migrating upstream in Figure 11.20. *[Photo by Bobbé Christopherson.]*

Chapter 11

River Systems and Landforms

KEY LEARNING CONCEPTS

After reading the chapter, you should be able to:

- **Define** fluvial and **outline** the fluvial processes: erosion, transportation, and deposition.

- **Construct** a basic drainage basin model and **identify** different types of drainage patterns, with examples.

- **Describe** the relation among velocity, depth, width, and discharge and **explain** the various ways that a stream erodes and transports its load.

- **Develop** a model of a meandering stream, including point bar, undercut bank, and cutoff, and **explain** the role of stream gradient in these flow characteristics.

- **Define** a floodplain and **analyze** the behavior of a stream channel during a flood.

- **Differentiate** the several types of river deltas and detail each.

- **Explain** flood probability estimates and **review** strategies for mitigating flood hazards.

STUDENT LEARNING CENTER TOOLS

The *Elemental Geosystems* Student Learning Center provides on-line resources for this chapter by clicking on the book cover at **www.mygeoscienceplace.com**. Some highlights:

- ▶ Three animations assist with stream processes and floodplains, oxbow lake formation, and stream terraces—how they form and evolve.

- ▶ A Notebook takes you to Niagara Falls where you work with the processes active at the falls; see an author video clip made from above and from a boat below the falls.

- ▶ Read a Career Link about a hydrologist/weather forecaster working to coordinate weather events with river response forecasts.

- ▶ Two Case Studies by geographers cover river restoration in the Pacific Northwest and possible sites for dam removals in the US.

- ▶ More than 14 additional URLs along with Quick Links to all the URLs in the chapter.

arth's rivers and waterways form vast arterial networks that both shape and drain the continents, transporting the by-products of weathering, mass movement, and erosion. To call them "Earth's lifeblood" is no exaggeration, inasmuch as rivers redistribute mineral nutrients important for soil formation and plant growth and serve society in many ways.

Rivers not only provide us with essential water supplies, they also receive, dilute, and transport wastes, provide critical cooling water for industry, and form one of the world's most important transportation networks. Rivers have been significant throughout human history. Each devastating flood event demonstrates the interrelationship between rivers and society.

Hydrology is the science of water, its global circulation, distribution, and properties, specifically water at and below Earth's surface. For hydrology links on the web, see http:// www.worldwater.org/links.htm or the Global Hydrology and Climate Center at http://weather.msfc.nasa.gov/ surface_hydrology/; also see the Amazon River at http:// boto.ocean.washington.edu/eos/index.html.

In this chapter: We begin with a look at the largest rivers on Earth. Essential fluvial concepts of base level, drainage basin, drainage density, and patterns follow. With this foundation, we define streamflow discharge, "Q," the essential element in fluvial processes. We discuss factors that affect streamflow characteristics and the work performed by flowing water, including erosion and transport. Stream depositional features such as deltas are illustrated with a detailed look at the Mississippi River delta. We look at the effects of urbanization on hydrology. Human response to floods and floodplain management are important aspects of river management, with details of the recent New Orleans levee and floodwall disaster and tragedy included in the chapter.

Fluvial Processes and Landscapes

At any one moment approximately 1250 km^3 (300 mi^3) of water flows through Earth's waterways. Even though this volume represents only 0.003% of all freshwater, the work performed by this energetic flow makes it the dominant natural agent of landmass denudation. Of the world's rivers, those with the greatest *discharge* (streamflow volume past a point in a given unit of time) are the Amazon (Figure 11.1), the Congo of Africa, the Chàng Jiang (Yangtze) of Asia, and the Orinoco of South America (Table 11.1). In North America, the greatest discharges are from the Mississippi-Missouri-Ohio, Saint Lawrence, and Mackenzie river systems.

Stream-related processes are **fluvial** (from the Latin *fluvius*, meaning "river"). Geographers analyze stream patterns and the fluvial processes that created them. Fluvial systems, like all natural systems, have characteristic processes and produce recognizable landforms. Yet a stream system can behave with randomness and seeming disorder. The term *river* is applied to a trunk, or main stream, or an entire river system. *Stream* is a more general term not necessarily related to size. There is some overlap in usage between the terms *river* and *stream*.

Insolation and gravity are the driving forces of fluvial systems because they power the hydrologic cycle. Individual streams vary greatly, depending on their region's

FIGURE 11.1 Mouth of the Amazon River.
The mouth of the Amazon River, 160 km (100 mi) wide, discharges a fifth of all the freshwater that enters the world's oceans. The Amazon's drainage basin, which is as large as the Australian continent, produces millions of tons of sediment. Large islands of sediment are left where the river's discharge leaves the mouth and flows into the Atlantic Ocean. [*Terra* image courtesy of NASA/GSFC/JPL and the MISR Team.]

Table 11.1 Largest Rivers on Earth Ranked by Discharge Volume

Rank by Volume	Average Discharge at Mouth in Thousands of m³/s (cfs)	River (with Tributaries)	Outflow/Location	Length km (mi)	Rank by Length
1	180 (6350)	Amazon (Ucayali, Tambo, Ene, Apurimac)	Atlantic Ocean/ Amapá-Pará, Brazil	6570 (4080)	2
2	41 (1460)	Congo (Lualaba)	Atlantic Ocean/ Angola, Congo	4630 (2880)	10
3	34 (1201)	Yangtze (Chàng Jiang)	East China Sea/Kiangsu, China	6300 (3915)	3
4	30 (1060)	Orinoco	Atlantic Ocean/Venezuela	2737 (1700)	27
5	21.8 (779)	La Plata estuary (Paraná)	Atlantic Ocean/Argentina	3945 (2450)	16
6	19.6 (699)	Ganges (Brahmaputra)	Bay of Bengal/India	2510 (1560)	23
7	19.4 (692)	Yenisey (Angara, Selenga or Selenge, Ider)	Gulf of Kara Sea/Siberia	5870 (3650)	5
8	18.2 (650)	Mississippi (Missouri, Ohio, Tennessee, Jefferson, Beaverhead, Red Rock)	Gulf of Mexico/Louisiana	6020 (3740)	4
9	16.0 (568)	Lena	Laptev Sea/Siberia	4400 (2730)	11
17	9.7 (348)	St. Lawrence	Gulf of St. Lawrence/ Canada and US	3060 (1900)	21
36	2.83 (100)	Nile (Kagera, Ruvuvu, Luvironza)	Mediterranean Sea/Egypt	6690 (4160)	1

climate, the variety of surface composition and topography over which they flow, the nature of vegetation and plant cover, and the length of time they have existed in a specific setting.

Wind, water, and ice dislodge, dissolve, or remove surface material in the process of **erosion**. Thus, streams produce fluvial erosion, which supplies weathered sediment for **transport** to new locations, where it is laid down in a **deposition** process. A stream is a mixture of water and solids, which are carried in solution, in suspension, and by mechanical transport. **Alluvium** is the general term for the clay, silt, and sand transported by running water.

Base Level of Streams

Base level is a level below which a stream cannot erode its valley. In general, the *ultimate base level* is sea level, the average level between high and low tides. Imagine base level as a surface extending inland from sea level, inclined gently upward under the continents. Ideally, this is the lowest practical level for all denudation processes (Figure 11.2a).

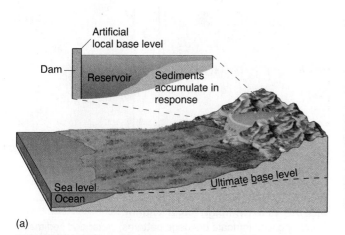

(a)

(b)

FIGURE 11.2 Powell's journey to base level.
(a) The concepts of ultimate base level (sea level) and local base level (natural, such as a lake, or artificial, such as a dam). Note how base level curves gently upward from the sea as it is traced inland; this is the theoretical limit for stream erosion. (b) A local base level was introduced to the landscape above Glen Canyon Dam when it closed off the canyon. Because of the western drought, the level of Lake Powell is lowering, as you can tell by the light-colored rock ringing the lake. [Photo by Bobbé Christopherson.]

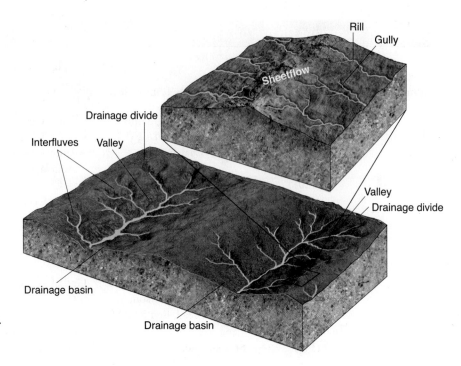

FIGURE 11.3 A drainage basin.
A drainage divide separates a drainage basin and its watershed from other basins. Note the high ground that separates one valley from another—an interfluve.

American geologist and ethnologist John Wesley Powell (1834–1902) put forward the idea in 1875. He was a director of the US Geological Survey, first director of the US Bureau of Ethnology, explorer of the Colorado River, and a pioneer in understanding the western landscape.

Powell also recognized that not every landscape degraded down to sea level, for other base levels seemed to be in operation. A *local base level*, or a temporary one, may control the lower limit of local streams. That local base level might be a river, a lake, a hard and resistant rock structure, or a human-made dam. In arid landscapes, with their intermittent precipitation, valleys, plains, or other low points determine local base level.

Ideally, endogenic processes build and create *initial* landscapes, whereas exogenic processes work toward low relief, an ultimate condition of little change, and the stability of *sequential* landscapes. Various theories were proposed to model these denudation processes and account for the appearance of the landscape at different developmental stages, not only for humid temperate climates but also for coastal, arid desert, and equatorial landscapes.

A *dynamic equilibrium*—a balance among force, form, and process—is the preferred model used by many contemporary geomorphologists. Slope and landform stability are a function of the resistance of rock materials to the attack of denudation processes.

Over time, the work of streams modifies the landscape dramatically. Two processes produce landforms: (1) erosive action of flowing water, and (2) deposition of stream-transported materials. Let us begin our study by examining a basic fluvial unit—the drainage basin.

The Drainage Basin System

Figure 11.3 illustrates drainage basin concepts. Every stream has a **drainage basin**, ranging in size from tiny to vast. Ridges

form *drainage divides* that define every drainage basin. That is, the ridges are the dividing lines that control into which basin runoff water from precipitation drains. Drainage divides define a drainage basin, the catchment (water-receiving) area of the drainage basin that delivers water. In any drainage basin, water initially moves downslope in a thin film of **sheetflow**, or

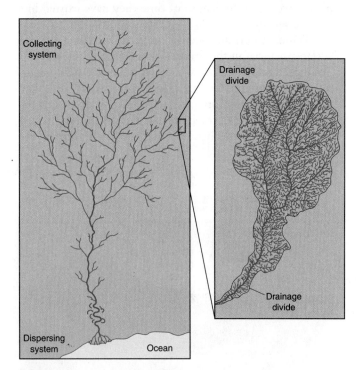

FIGURE 11.4 Drainage basin system.
Beginning with intricate drainage patterns, water and sediment are collected and delivered to the stream tributaries. This fluidized mass transports to the main trunk and the dispersing system at the end of the stream, where it enters another water body.
[Adapted from W. K. Hamblin and E. H. Christiansen, *Earth's Dynamic System*, 10th ed. (Upper Saddle River, NJ: Pearson Prentice Hall, 2004), Figure 12.3, p. 300.]

overland flow. High ground that separates one valley from another and directs sheetflow is an *interfluve*. A small ravine, a *gully*, focuses sheetflow. Surface runoff concentrates in *rills* that are small-scale downhill grooves, which may develop into deeper *channels* then into a stream in the valley.

The drainage basin acts as a collection system of water comprised of many subsystems, from erosion in the headwaters, transportation throughout, to deposition in the lower portions. Figure 11.4 illustrates the operation of the drainage basin system: from the smallest rill and gully, to the main tributaries and stream trunk, to the dispersing lower extremities and mouth, possibly forming a delta or depositing sediments in a water body.

Drainage Divides and Basins   Several high drainage divides, the **continental divides**, cross the United States and Canada. These are extensive mountain and highland regions that separate drainage basins, sending flows to the Pacific, the Gulf of Mexico, the Atlantic, Hudson Bay, or the Arctic Ocean. The principal drainage basins and continental divides in the United States and Canada are on the map in Figure 11.5. These drainage basins

FIGURE 11.5 Drainage basins and continental divides.
Continental divides (blue lines) separate the major drainage basins that empty into the Pacific, Atlantic, and Gulf of Mexico, and to the north through Canada into Hudson Bay and the Arctic Ocean. Subdividing these large-scale basins are major river basins. [After US Geological Survey; *The National Atlas of Canada*, 1985, Energy, Mines, and Resources Canada; and Environment Canada, *Currents of Change—Inquiry on Federal Water Policy—Final Report 1986*.]

provide a spatial framework for water studies and water planning.

A major drainage basin system, such as the one created by the 3.1 million km² (1.2 million mi²) Mississippi-Missouri-Ohio river system, is made up of many smaller drainage basins, which in turn comprise even smaller basins, each divided into specific watersheds. Each drainage basin gathers and delivers its precipitation and sediment to a larger basin, concentrating the volume in the main stream. For example, consider the travels of rainfall in north-central Pennsylvania. This water feeds hundreds of small streams that flow into the Allegheny River (noted on Figure 11.5). The Allegheny River then joins with the Monongahela River at Pittsburgh to form the Ohio River. The Ohio flows southwest and connects with the Mississippi River at Cairo, Illinois, and eventually flows on past New Orleans to the Gulf of Mexico.

Each contributing tributary adds its discharge and sediment load to the larger river. In our example, sediment weathered and eroded in north-central Pennsylvania is transported thousands of kilometers and accumulates as the Mississippi River delta on the floor of the Gulf of Mexico.

Imagine the complexity of an international drainage basin, such as the Danube River in Europe, that flows 2850 km (1770 mi) from western Germany's Black Forest to the Black Sea. The river crosses or forms the borders between nine countries (Figure 11.6). A total area of 817,000 km² (315,000 mi²) falls within the drainage basin, including some 300 tributaries.

The Danube serves many economic functions: commercial transport, municipal water source, agricultural irrigation, fishing, and hydroelectric production. An international struggle is underway to save the river from its burden of industrial and mining wastes, sewage, chemical discharge, agricultural runoff, and drainage from ships. All of this pollution passes through Romania and the deltaic ecosystems in the Black Sea. The river is widely regarded as one of Earth's most polluted.

Political changes in Europe in 1989 allowed the first scientific analysis of the entire Danube system. The United Nations Environment Programme (UNEP) and the European Union, along with other organizations, are dedicated to clearing the Danube, saving the delta, and restoring the environment of this valuable resource; see http://www.icpdr.org/. The river delta is one of the International Biosphere Reserves; see http://www.unesco.org/mab/ and http://www.blacksea-environment.org/.

Drainage Basins as Open Systems   Drainage basins are *open systems* whose inputs include precipitation, the minerals and rocks of the regional geology, and changes of energy with both the uplift and subsidence provided by tectonic activities. System outputs of water and sediment leave through the mouth of the river. Change that occurs in any portion of a drainage basin can affect the entire system, because the stream adjusts to carry the appropriate load of sediment relative to discharge (its volume of flow in a unit of time) and velocity. A stream drainage system exhibits a constant struggle toward an equilibrium among interacting variables of discharge, transported sediment load, channel shape, and channel slope.

An Example: The Delaware River Basin   Let us look at the Delaware River basin, within the Atlantic Ocean drainage region (Figure 11.7). The Delaware River headwaters are in the Catskill Mountains of New York. This basin encompasses 33,060 km² (12,890 mi²) and includes parts of five states in the river's length, 595 km (370 mi) from headwaters to the mouth. The river system ends at Delaware Bay, which eventually enters the Atlantic Ocean. Topography varies from low-relief coastal plains

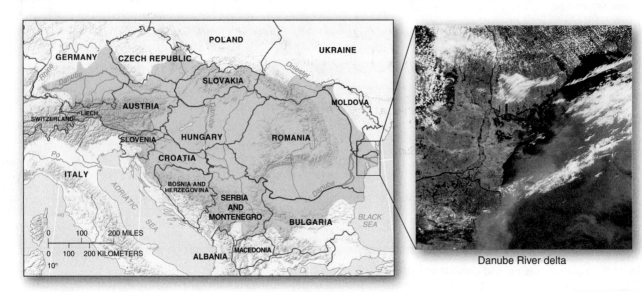

Danube River delta

FIGURE 11.6 An international drainage basin—the Danube River.
The Danube crosses or forms the border of nine countries as it flows across Europe to the Black Sea. The river spews polluted discharge into the Black Sea through its arcuate-form delta. [*Terra* MODIS image, June 15, 2002, courtesy of MODIS Land Rapid Response Team, NASA/GSFC.]

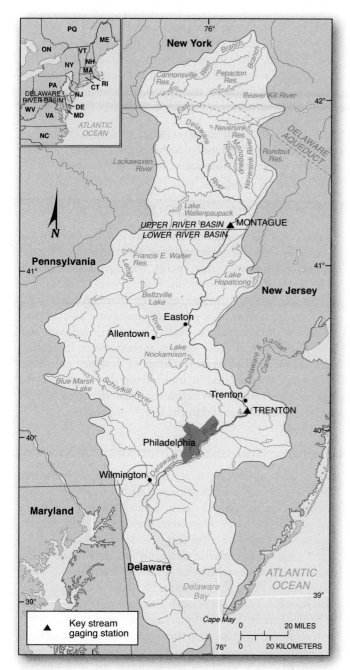

FIGURE 11.7 The Delaware River Basin.
The Delaware River basin was studied by the USGS to assess the potential impact of global warming on a river system, as temperature and precipitation patterns change over the five-state region. [After US Geological Survey, 1986, "Hydrologic events and surface water resources," *National Water Summary 1985*, Water Supply Paper 2300 (Washington, DC: Government Printing Office), p. 30.]

New Jersey). Several reservoirs in the drainage basin allow some control over streamflow and storage for dry periods. The sustainability of this water resource is critical to the entire region.

Drainage Density and Patterns

The **drainage pattern** is an arrangement of channels that is determined by slope, differing rock resistance, climatic and hydrologic variability, relief of the land, and structural controls imposed by the landscape. Consequently, the drainage pattern of any land area on Earth is a remarkable visual summary of every characteristic—geologic and climatic—of that region.

The *Landsat* image in Figure 11.8 is of the Ohio River area near the junction of the West Virginia, Ohio, and Kentucky borders. The high-density drainage pattern and intricate dissection of the land occur because of the region's generally level sandstone, siltstone, and shale strata, which are easily eroded, given the humid mesothermal climate. Fluvial action, along with other denudation processes, is responsible for this dissected topography.

The seven most common drainage patterns encountered in nature are represented in Figure 11.9. The most familiar pattern is *dendritic* (a); this treelike pattern is similar to that of many natural systems, such as capillaries in the human circulatory system or the vein patterns in leaves. Energy expended by this drainage system is efficient because the overall length of the branches is minimized.

The *trellis* drainage pattern (b) is characteristic of dipping or folded topography, which exists in nearly parallel mountains of the Ridge and Valley Province of the East, where drainage patterns are influenced by rock structures of variable resistance and folded strata. Review Figure 9.18, which presents a satellite image of this region showing its distinctive drainage pattern. The principal streams are directed by the parallel folded structures, whereas smaller dendritic streams are at work on nearby slopes, joining the main streams at right angles.

The sketch beside Figure 11.9b suggests that the headward-eroding part of one stream could break through a drainage divide and *capture* the headwaters of another stream in the next valley, and indeed this does happen. The sharp bends in two of the streams in the illustration are *elbows of capture* and are evidence that one stream has breached a drainage divide. This type of capture, or *stream piracy*, can occur in other drainage patterns.

The remaining drainage patterns in Figure 11.9 are responses to other specific structural conditions:

- A *radial* drainage pattern (c) results when streams flow off a central peak or dome, such as occurs on a volcanic mountain.
- *Parallel* drainage (d) is associated with steep slopes.
- A *rectangular* pattern (e) is formed by a faulted and jointed landscape, which directs stream courses in patterns of right-angle turns.

to the Appalachian Mountains in the north. The entire basin lies within a humid, temperate climate and receives an average annual precipitation of 120 cm (47.2 in.).

The Delaware River provides water for an estimated 20 million people, not only within the basin but to cities outside the basin as well. Several major conduits export water from the Delaware River. Note on the map the Delaware Aqueduct to New York City (in the north) and the Delaware & Raritan Canal (near Trenton,

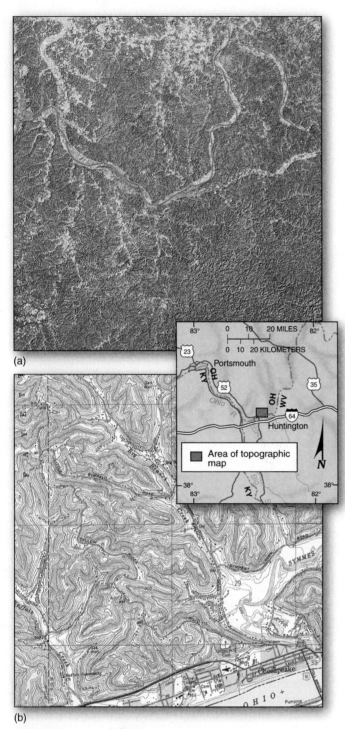

(a)

(b)

FIGURE 11.8 A stream-dissected landscape.
(a) A highly dissected topography (featuring a dendritic drainage pattern, shown in Figure 11.9a) around the junction of West Virginia, Ohio, and Kentucky. All of these borders are formed by rivers. (b) A portion of the USGS topographic map covering the area north of Huntington, West Virginia, reveals the intricate complexity of dissected landscapes. (The image and map are not at the same scale.) [(a) *Landsat* image from NASA; (b) Huntington Quadrangle, USGS.]

- *Annular* patterns (f) are produced by structural domes, with concentric patterns of rock strata guiding stream courses. Figure 9.10c provides an example of annular drainage on a dome structure.

- In areas having disrupted surface patterns, such as the glaciated shield regions of Canada, northern Europe, and some parts of Michigan and other states, a *deranged* pattern (g) is in evidence, with no clear geometry in the drainage and no true stream valley pattern (see Figures 9.4b and 14.13a).

In Figure 11.9h, you see two distinct drainage patterns. Of the seven types illustrated, which two patterns are most like those in the aerial photo made in central Montana?

Streamflow Characteristics

A mass of water positioned above base level in a stream has *potential energy*. As the water flows downstream under the influence of gravity, this energy becomes *kinetic energy*. The rate of this conversion from potential to kinetic energy depends on the steepness of the stream channel.

Stream channels vary in *width* and *depth*. The streams that flow in them vary in *velocity* and in the *sediment load* they carry. All of these factors increase with increasing **discharge**, or the stream's volume of flow per unit time. Discharge is calculated by multiplying the velocity of the stream by its width and depth for a specific cross section of the channel, as stated in the simple expression:

$$Q = wdv$$

where Q = discharge; w = channel width; d = channel depth; and v = stream velocity. As Q increases, some combination of channel width, depth, and stream velocity increases. Discharge is expressed either in cubic meters per second (m³/s) or cubic feet per second (cfs).

Given the interplay of channel width, depth, and stream velocity, the cross section of a stream varies over time, especially during heavy floods. Figure 11.10 shows changes in the San Juan River channel in Utah that occurred during a flood. The increase in discharge increases the velocity and therefore the carrying capacity of the river as the flood progresses. As a result, the river's ability to scour materials from its bed is enhanced. Such scouring represents a powerful clearing action, especially in the excavation of alluvium.

You can see in the figure that the San Juan River's channel was deepest on October 14, when floodwaters were highest (blue outline). Then, as the discharge returned to normal, the kinetic energy of the river reduced, and the bed again filled as new sediments were deposited. The process

FIGURE 11.9 The seven most common drainage patterns.
Each pattern is a visual summary of all the geologic and climatic conditions of its region. Two of these drainage patterns dominate the scene in (h) from central Montana, in response to local relief and rock structure. [(a) through (g) after A. D. Howard, "Drainage analysis in geological interpretation: A summation," *Bulletin of American Association of Petroleum Geologists* 51 (1967): 2248. Adapted by permission; (h) photo by Bobbé Christopherson.]

depicted in Figure 11.10 moved a depth of about 3 m (10 ft) of sediment from this cross section of the stream channel. Three scouring, sediment-redistribution experiments intended to provide new recreational areas and restore wildlife habitats in the Grand Canyon are described in News Report 11.1.

Exotic Streams

Most streams increase discharge downstream because the area being drained increases. The Mississippi River is typical. It starts as many small brooks and grows to a mighty river pouring into the Gulf of Mexico. In contrast, a stream can originate in a humid region and subsequently flow through an arid region. In that case, the *discharge usually decreases with distance*, because of high potential evapotranspiration rates in the arid area. Such a stream is an **exotic stream** (*exotic* means "of foreign origin").

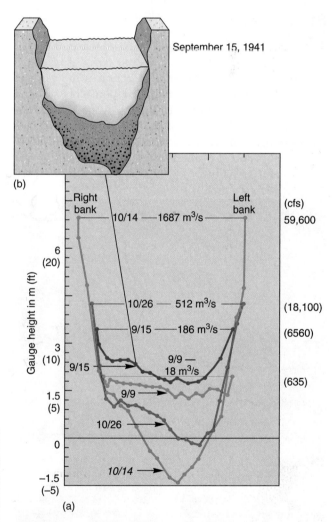

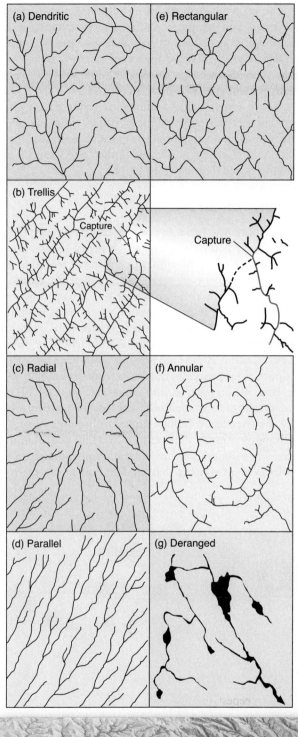

FIGURE 11.10 A flood affects a stream channel.
(a) Stream channel cross sections showing the progress of a 1941 flood on the San Juan River near Bluff, Utah. (b) Detail of stream-channel profile on September 15, 1941.
[After L. Leopold and T. Maddock, Jr., *The Hydraulic Geometry of Stream Channels and Some Physiographic Implications*, US Geological Survey Professional Paper 252 (Washington, DC: Government Printing Office, 1953), p. 32.]

The Nile River exemplifies exotic streams. This great river, Earth's longest, drains much of northeastern Africa. But as it courses through the deserts of Sudan and Egypt, it loses water instead of gaining it, because of evaporation and withdrawal for agriculture. By the time it empties into the Mediterranean Sea, the Nile's flow has dwindled so much that it ranks only 36th in discharge.

The United States has exotic streams too, notably the Colorado River. Its *Q* (discharge) decreases with distance from its source; in fact, the river no longer produces enough natural discharge to reach its mouth in the Gulf of California—only some agricultural runoff remains at its delta. The exotic Colorado River is depleted not only by passage across dry, desert lands but also by upstream removal of water for agriculture and municipal uses. See Focus Study 12.1 in Chapter 12 for a detailed discussion of this river and a satellite image of the river's former mouth. Now, let's take a look at the work streams perform.

Stream Erosion

The erosional work of a stream carves and shapes the landscape through which it flows. Several types of erosional processes are operative. **Hydraulic action** is the work of water moving materials through the exertion of pressure and shearing force. Running water causes hydraulic squeeze-and-release action that loosens and lifts rocks. As this debris moves along, it mechanically erodes the streambed further through **abrasion**, the process whereby rock particles grind and carve the streambed.

The upstream tributaries in a drainage basin usually have small and irregular discharges, with most of the stream energy expended in turbulent eddies. As a result, hydraulic action in these upstream sections is at maximum, even though the coarse-textured load of such a stream is small. The downstream portions of a river, however, move much larger volumes of water past a given point and carry larger suspended loads of sediment. Thus, both volume and velocity are important determinants of the amount of energy expended in the erosion and transportation of sediment (Figure 11.11).

Stream Transport

You may have watched a river or creek after a rainfall, noting the water colored brown by the heavy sediment load being transported. The amount of material available to a stream is dependent on topographic relief, the nature of rock and materials through which the stream flows, climate, vegetation, and the types of processes at work in a drainage basin. *Competence*, which is a stream's ability to move particles of specific size, is a function of stream velocity. *Capacity* is the total possible load that a stream can transport. Four processes transport eroded materials: solution, suspension, traction, and saltation (Figure 11.12).

Solution refers to the **dissolved load** of a stream, especially the chemical solution derived from minerals such as limestone or from soluble salts. The main contributor of material in solution is chemical weathering. Sometimes the undesirable salt content that hinders human use of rivers

(a)

(b)

FIGURE 11.11 Stream velocity and discharge increase together.
(a) A low-volume, low-velocity (but turbulent) mountain stream in the White Mountains, New Hampshire; and, (b) the high-discharge, high-velocity (but laminar, or smooth water flow) portion of the Ocmulgee River near Jacksonville, Georgia. [Photos by Bobbé Christopherson.]

NEWS REPORT 11.1

Scouring the Grand Canyon for New Beaches and Habitats

Glen Canyon Dam sealed the Colorado River gorge north of the Grand Canyon, near Lees Ferry and the Utah-Arizona border (see the map in Focus Study 12.1, Figure 12.1.1). Impoundment of water by the new Lake Powell in many canyons upstream from the dam began in 1963. The lake began collecting tremendous sediment load from the Colorado River, reducing the volume of water in the river below the dam and eliminating the natural seasonal fluctuations in river discharge. The production of hydroelectricity at Glen Canyon further affected the Grand Canyon with highly variable water releases keyed to electrical turbine operations. Water in the canyon rose and fell as much as 4.3 m (14 ft) as the distant lights and air conditioning of Las Vegas and Phoenix went on and off.

All of these changes affected the canyon. Over the years, the beaches were starved for sand, channels filled with sediment, fisheries were disrupted, and backwater channels were depleted of nutrients. In 1996, an unprecedented experiment took place. The Grand Canyon was artificially flooded with a 1260-m³/s (45,000-cfs) release from Glen Canyon (Figure 11.1.1). The flow lasted for 7 days and then was reduced to 224 m³/s (8000 cfs). Flows were decreased gradually to allow the newly formed beaches to drain, in contrast to the rapid opening of the pipes that initiated the flood.

The results were initially thought to be positive: 35% more beach area was created by the scouring of sand and sediments from the channel. Approximately 80% of the aggradation (building up of beaches) took

FIGURE 11.1.1 Glen Canyon Dam floods the Grand Canyon.
Bypass tubes dramatically discharge water from Lake Powell in a 1996 experiment to correct the damaging effects of the dam on the Grand Canyon. Note for scale the workers on the walkway, who are experiencing a deafening roar and vibration. Modified experiments were repeated in 2004 and 2008. [Photo by Tom Smart, *Deseret News.*]

place in the first 40 hours and was completed by the 100-hour mark. Numerous backwater channels were created, flush with fresh nutrients for the humpback chub and other endangered fish species. However, benefits turned out to be limited, and negatives, in the form of existing ecosystem disruption and immediate losses of some sediment deposits, marred the experiment. Present thinking is that the lack of sediment to supply the floodwaters for downstream deposition and beach building was the problem. Some of the initial beach deposits quickly disappeared. For more on this experiment, see http://pubs.usgs.gov/fs/FS-060-99/.

In November 2004 and again in March 2008, hydrologists conducted two additional tests. Scientists timed these artificial floods with natural peak runoff that supplies fresh sediment to the system. Also, instead of 7 days, the artificial flood releases from Glen Canyon lasted 60 hours (2.5 days) during both tests. The release coincided with local storms that fed tributaries, such as the Paria River, with sediment to accomplish the task of rebuilding beaches downstream. Tributaries deliver more than 3 million tons of sediment to the Colorado River. These were attempts at sediment redistribution—analysis of results is continuing. (See Grand Canyon Monitoring and Research Center for progress reports and analysis at http://www.gcmrc.gov/.) Record low river discharge and low reservoir levels are testing the entire Colorado River basin during the current prolonged drought.

comes from dissolved rock formations or from springs in the stream channel; as an example, the San Juan and Little Colorado Rivers that flow into the Colorado River near the Utah-Arizona border add dissolved salts to the system.

The **suspended load** consists of fine-grained, clastic particles (bits and pieces of rock) physically held aloft in the

stream, with the finest particles held in suspension until the stream velocity slows to near zero. Turbulence in the water, with random upward motions, is an important mechanical factor in holding a load of sediment in suspension.

The **bed load** refers to those coarser materials that are dragged along the bed of the stream by **traction** or are

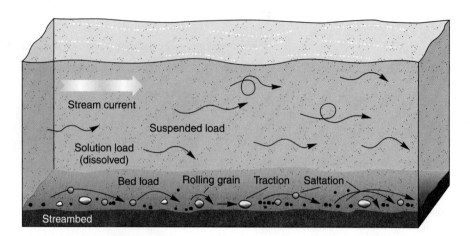

FIGURE 11.12 Fluvial transport.
Fluvial transportation of eroded materials through solution, suspension, traction, and saltation.

rolled and bounced along by **saltation** (from the Latin *saltim*, which means "by leaps or jumps"). With increased kinetic energy, parts of the bed load are rafted up and become suspended load. This fact was demonstrated in the flood-induced channel deepening of the San Juan River shown in Figure 11.10 and the Grand Canyon scouring in News Report 11.1. Saltation is also a process in the transportation of materials by wind (see Chapter 12).

The early explorers who visited the Grand Canyon reported in their journals that they were kept awake at night by the thundering sound of the river and rapids. Imagine the tremendous quantity of material being moved along by the natural Colorado River before any dams were built. Dams and reservoirs now trap sediments that formerly contributed to the river's bed load.

If the load in a stream exceeds its capacity, sediments accumulate as **aggradation** (the opposite of degradation) as the stream channel builds up through deposition. With excess sediment, a stream becomes a maze of interconnected channels that form a **braided stream** pattern. Braiding often occurs when reduced discharge affects a stream's transportation ability, under seasonal conditions, for example, or when a landslide occurs upstream, or where weak banks of sand or gravel exist, producing excess supplies of sediment. Locally, braiding also may result from a new sediment load, which frequently is associated with meltwaters, such as in the glacial materials of rock,

sediment, and glacial flour (ground-up rock), that exceed stream capacity, such as in the 15-km (9.3-mi) stretch of the Brahmaputra River about 35 km (22 mi) south of Lhasa, Tibet, in south-central Asia (Figure 11.13).

Flow and Channel Characteristics

Flow characteristics of a stream are best seen in a cross-section view. The greatest velocities in a stream are near the surface at the center, corresponding with the deepest part of the stream channel (Figure 11.14). Velocities decrease closer to the sides and bottom of the channel because of the frictional drag on the water flow. In a curving stream, the maximum velocity line migrates from side to side along the channel, deflected by the curves.

Where slopes are gradual, stream channels assume a sinuous (snakelike) form weaving across the landscape. This action produces a **meandering stream**, from the Greek *maiandros*, after the ancient Maiandros River in Asia Minor (the present-day Menderes River in Turkey) that had a meandering channel pattern. The outer portion of each meandering curve is subject to the greatest erosive action and can be the site of a steep bank, an **undercut bank**. In contrast, the inner portion of a meander receives sediment fill, forming a **point bar** deposit.

As meanders develop, the scour-and-fill features gradually work their way downstream. As a result, the

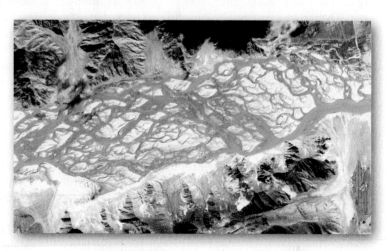

FIGURE 11.13 A braided stream.
A braided Brahmaputra River channel some 35 km (22 mi) south of Lhasa, Tibet, in a narrow valley south of the Tibetan Plateau. These streams reflect excessive sediment load associated with glacial meltwaters filled with fine sediments, or "glacial flour." [Photo by International Space Station astronaut courtesy of Earth Science and Image Analysis Lab, JSC/NASA.]

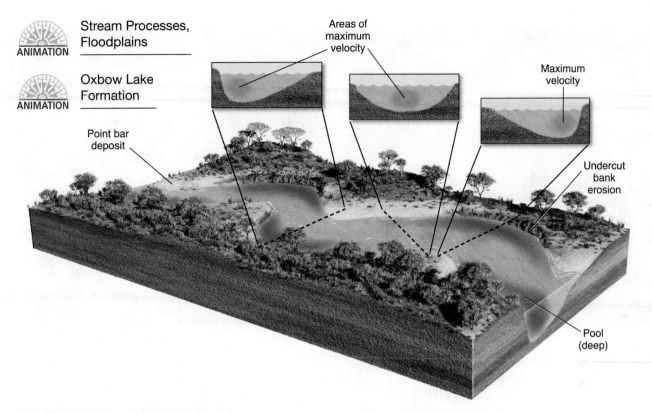

FIGURE 11.14 Meandering stream profile.
Aerial view and cross sections of a meandering stream, showing the location of maximum flow velocity, point-bar deposits, and areas of undercut bank erosion.

landscape near a meandering river bears meander scars of residual deposits from the previous river channels. Former point-bar deposits leave low-lying ridges, creating a bar-and-swale relief (ridges and slight depressions). The photograph of the Itkillik River in Alaska, Figure 11.15a, shows both meanders and meander scars.

Figure 11.15b shows (in four stages) how a stream meander can form a cutoff. The stream erodes its outside bank as the curve migrates downstream (1); the neck of land created by the looping meander (2) eventually erodes through and forms a *cutoff* (3). When the former meander becomes isolated from the rest of the river, the resulting **oxbow lake** (4) may gradually fill with silt or may again become part of the river when it floods. The Mississippi River is many miles shorter today than it was in the 1830s because of artificial cutoffs that were dredged across these necks of land to improve navigation and safety.

Streams often form natural political boundaries, as we saw with the Danube River earlier, but it is easy to see how disagreements might arise when boundaries are based on river channels that change course. For example, the Ohio, Missouri, and Mississippi Rivers can shift their positions quite rapidly during times of flood and, therefore, make confusing the boundaries based on them. Carter Lake, Iowa, provides us with a case in point (Figure 11.16). The Nebraska-Iowa border was originally placed mid-channel in the Missouri River. In 1877, the river cut off the

meander loop around the town of Carter Lake, leaving the town "captured" by Nebraska. The new oxbow lake was called Carter Lake and still is the state boundary.

Boundaries should always be fixed by surveys independent of river locations, because border disputes result when river channels change course. Such surveys have been completed along the Rio Grande near El Paso, Texas, and along the Colorado River between Arizona and California, permanently establishing political boundaries separate from changing river locations.

Stream Gradient

Every stream has a degree of inclination or **gradient**, which is the decline in elevation from its headwaters to its mouth. A stream's gradient generally forms a concave-shaped slope (Figure 11.17). Characteristically, the *longitudinal profile* of a stream (a side view) has a steeper slope upstream and a more gradual slope downstream. This curve assumes its shape for complex reasons related to the stream's ability to do enough work to accomplish the transport of the load it receives.

A **graded stream** condition occurs when the load carried by the stream and the landscape through which it flows become mutually adjusted (balanced); it is in a state of *dynamic equilibrium* among erosion, transported load, deposition, and the stream's capacity. Dynamic equilibrium implies that the stream and landscape work together to maintain

(a)

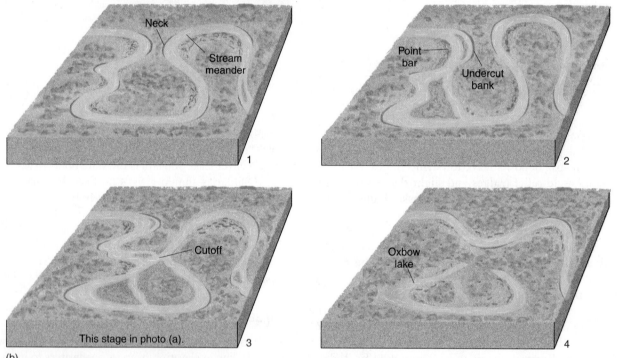

Neck

Stream
meander

Point
bar

Undercut
bank

1

2

Cutoff

Oxbow
lake

This stage in photo (a).

3

4

(b)

FIGURE 11.15 Meandering stream development.
(a) Itkillik River in Alaska. (b) Development of a river meander and oxbow lake simplified in four
stages. [(a) US Geological Survey photo.]

 ANIMATION

Stream Processes,
Floodplains

this balance. Both high-gradient and low-gradient streams can achieve a graded condition. The difference in gradient is the result of variation in each stream's discharge and the nature of the transported load. A stream's profile tells geographers about characteristics of slope, discharge, and load.

A graded stream is one in which, over a period of years, *slope* is delicately adjusted to provide, with available *discharge* and with prevailing *channel characteristics*, just the velocity required for transportation of the load supplied from the drainage basin.*

Attainment of a graded condition does not mean that the stream is at its lowest gradient, but rather that it represents a balance among erosion, transportation, and deposition over time along a specific portion of the stream.

*J. H. Mackin, "Concept of the graded river," *Geological Society of America Bulletin* 59 (1948): 463.

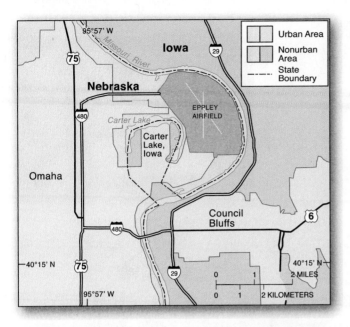

FIGURE 11.16 Rivers make poor political boundaries.
Carter Lake, Iowa, sits within the curve of a former meander that was cut off by the Missouri River. The city and oxbow lake remain part of Iowa even though they are stranded within Nebraska.

One problem with applying the graded stream concept is that an individual stream can have both graded and ungraded portions. It may have graded sections without having an overall graded slope. In fact, variations and interruptions are the rule rather than the exception.

Tectonic uplift of the landscape affects stream gradient and changes relations in base level. Such changes stimulate renewed erosional activity. If tectonic forces slowly lift the landscape, the stream gradient will increase. Imagine this happening to the Colorado River landscape in Figure 11.2b. The region and stream flowing through the landscape become *rejuvenated*. With rejuvenation, river meanders actively return to downcutting and become *entrenched meanders* in the landscape (Figure 11.18).

Nickpoints When the longitudinal profile of a stream shows an abrupt change in gradient, such as at a waterfall or an area of rapids, the point of interruption is a **nickpoint** (also spelled *knickpoint*). At a nickpoint, the conversion of potential energy to kinetic energy is concentrated and works to eliminate the nickpoint. Figure 11.19a shows a stream with two such interruptions. Nickpoints can result from a stream flowing across a zone of hard, resistant rock or from various tectonic uplift episodes, such as might occur along a fault line. Temporary blockage in a channel, caused by a landslide or a logjam, also could be considered a nickpoint; when the logjam breaks, the stream quickly readjusts its channel to its former grade.

One of the more interesting and beautiful gradient breaks is a waterfall. At its edge, a stream becomes free-falling, moving at high velocity and causing increased abrasion on the channel below. This action generally undercuts the waterfall, and eventually the rock ledge at the lip of the fall collapses, causing the waterfall to shift a bit farther upstream. Thus, nickpoints migrate upstream. The height of the waterfall is reduced gradually as debris accumulates at its base.

At Niagara Falls on the Ontario–New York border (see chapter-opening photo), glaciers advanced over the region and then receded. In doing so, they exposed resistant rock strata that are underlain by less-resistant shales. This formation is a *cuesta*, which is a ridge with a cliff on one side and beds gently sloping away on the other side. This Niagara escarpment actually stretches across more than 700 km (435 mi) from east of the falls, northward through Ontario, Canada, the Upper Peninsula of Michigan, curving south through Wisconsin along the western shore of Lake Michigan and the Door Peninsula. As this less-resistant material continues to weather away, the overlying rock strata collapse, allowing Niagara Falls to erode farther upstream toward Lake Erie (Figure 11.20, p. 382).

Niagara Falls is a place where natural processes labor to eliminate a nickpoint, eventually reducing this portion of the river to a series of mere rapids. In fact, the falls have retreated more than 11 km (6.8 mi) from the steep face of the Niagara escarpment (cliff) during the past 12,000 years. In the past, engineers have used control facilities upstream to reduce flows over the American Falls at Niagara for inspection of the cliff to assess the progress of natural processes that are working to eliminate the Niagara Falls nickpoint. (Figure 11.20c); compare this to normal discharge in (d).

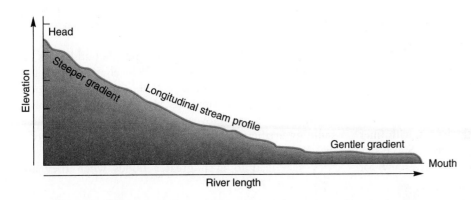

FIGURE 11.17 An ideal longitudinal profile.
Idealized cross section of the longitudinal profile of a stream showing its gradient. Upstream segments have a steeper gradient; downstream the gradient is gentle. The middle and lower portions in the illustration appear graded, or in dynamic equilibrium.

FIGURE 11.18 Entrenched meanders.
The San Juan River (a) and (b) near Mexican Hat, Utah, cuts down into the uplifted Colorado Plateau landscape, producing entrenched meanders, the Goosenecks of the San Juan. Entrenched meanders (c) merge in central Montana. [Photos (a) and (c) by Bobbé Christopherson; (b) by Randall M. Christopherson.]

Stream Deposition

Deposition is the next logical event after weathering, mass movement, erosion, and transportation. In *deposition*, a stream deposits alluvium, or unconsolidated sediments, thereby creating specific depositional landforms, such as floodplains, terraces, and deltas.

Floodplains   The flat, low-lying area along a stream channel that is subjected to recurrent flooding is a **floodplain**. It is formed when the river overflows its channel during times of high flow. Thus, when floods occur, the floodplain is inundated. When the water recedes, alluvial deposits generally mask the underlying rock. Figure 11.21 illustrates a characteristic floodplain, related features in photos, and a representative topographic map of an area near Philip, Mississippi.

On either bank of most streams, **natural levees** develop as by-products of flooding. When flood waters arrive, the river overflows its banks, loses velocity as it spreads out, and drops a portion of its sediment load to form the levees. Larger sand-sized particles drop out first, forming the principal component of the levees, with finer silts and clays deposited farther from the river. Successive floods increase the height of the levees (*levée* is French for

"raising"). The levees may grow in height until the river channel becomes elevated, or *perched*, above the surrounding floodplain.

Notice on Figure 11.21a an area labeled *backswamp* and *yazoo tributary* stream. The natural levees and elevated channel of the river prevent this tributary from joining the main channel, so it flows parallel to the river and through the backswamp area. (The name comes from the Yazoo River in the southern part of the Mississippi floodplain.) On the topographic map (Figure 11.21d), you can see the natural levees represented by several contour lines that run immediately adjacent to the Tallahatchie River. These contour lines (5-ft interval) denote a height of 10–15 ft (3–4.5 m) above the river and the adjoining floodplain. Next time you have an opportunity to see a river and its floodplain, look for levees (they may be low and subtle).

Despite the threat of flooding, people build cities on floodplains because they are nearly level and next to the water. Government assurances of artificial protection from floods or of disaster assistance often encourage people to settle there. Government assistance may include building artificial levees on top of natural levees. Artificial levees do increase the capacity in the channel, but they also lead to even greater floods when they are

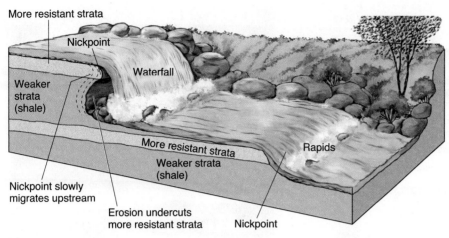

More resistant strata

Nickpoint

Waterfall

Weaker strata (shale)

Nickpoint slowly migrates upstream

Erosion undercuts more resistant strata

More resistant strata

Weaker strata (shale)

Rapids

Nickpoint

(a)

(b)

FIGURE 11.19 Nickpoints interrupt a stream profile.
(a) Longitudinal stream profile showing nickpoints produced by resistant rock strata. Potential energy is converted into kinetic energy and concentrated at the nickpolnt, accelerating erosion, which eventually eliminates the feature. (b) Falls and rapids break the flow of the Big Sioux River in Sioux Falls, South Dakota. [Photo by Bobbé Christopherson.]

topped or when they fail, as happened in New Orleans (Figure 11.22, p. 384).

The catastrophic floods along the Mississippi River and its tributaries in 1993 illustrated the risk of floodplain settlement. Total damage from these floods exceeded $30 billion. Nationally, in 2001, some two-thirds of disaster losses were attributable to floods. Tropical storm Allison left $6 billion in damage in its wandering visit to Texas in June 2001, producing the most flood damage along occupied floodplains. The 2002 floods in Europe produced estimated losses exceeding US$50 billion. Of course, the 2005 levee and floodwall breaks that resulted in catastrophic loss of life and property in New Orleans will forever be in our memories, as floodwaters engulfed 80% of the city. Estimates of losses exceeded $100 billion. In fact, the United States has experienced in just three seasons—2004, 2005, and 2008—more than $255 billion in damage to the Gulf and Atlantic coasts and floods inland from tropical storms and hurricanes.

Perhaps the best use of some floodplains is for crop agriculture, because inundation generally delivers nutrients to the land with each new alluvial deposit. A significant example is the Nile River in Egypt, where annual flooding enriches the soil. However, floodplains that are covered with coarse sediment—sand and gravel—are less suitable for agriculture. Are there river floodplains where you live? If so, how would you assess present land-use patterns, people's perception of hazard, and local planning and zoning?

Stream Terraces   As explained earlier, several factors may rejuvenate stream energy and stream-landscape relationships so that a stream again scours downward with increased erosion. The resulting entrenchment of the river into its own floodplain produces **alluvial terraces** on both sides of the valley, which look like topographic steps above the river. Alluvial terraces generally appear paired at similar elevations on opposite sides of the valley (Figure 11.23, p. 384). If more than one set of paired terraces is present, the valley probably has undergone more than one episode of rejuvenation.

If the terraces on the sides of the valley do not match in elevation, then entrenchment actions must have been continuous as the river meandered from side to side, with each meander cutting a terrace slightly lower in elevation. Thus,

(text continued on page 384)

(a)

(b)

(c)

(d)

FIGURE 11.20 Retreat of Niagara Falls.
(a) Headward retreat of Niagara Falls from the Niagara escarpment. It has taken the falls about 12,000 years to reach this position at a pace of about 1.3 m (4.3 ft) per year. (b) The escarpment seen from below the American Falls. (c) Niagara Falls, with the American Falls portion almost completely shut off by upstream controls for engineering inspection. Horseshoe Falls in the background is still flowing over its 57-m (188-ft) plunge. (d) In contrast to (c), the American Falls in full discharge. [(a) After W. K. Hamblin, *Earth's Dynamic Systems,* 6th ed. (Upper Saddle River, NJ: Pearson Prentice Hall, Inc. © 1992), Figure 12.15, p. 246. Photos: (c) courtesy of the New York Power Authority; (b) and (d) Bobbé Christopherson.]

NOTEBOOK Niagara Falls

NEWS REPORT 11.2

The Nile Delta Is Disappearing

People along the Nile River have depended on its regular flow and annual floods for millennia. Herodotus noted in the fifth century B.C. how the people farmed the fields in the floodplain and delta regions. After harvesting their crops, they retreated from the area to their homes. They would await the annual floods that brought fresh silt and nutrients for next year's planting. This cycle of fertility continued until the completion of the Aswân High Dam in 1964. This structure caused a decrease in the supply of sediment to the delta, and partially as a result the delta coastline continues to

actively recede. Herodotus stated in *The History,* Book Two, "In the part called the Delta, it seems to me that if the Nile no longer floods . . . for all time to come, the Egyptians will suffer."

J. Stanley, an oceanographer at the Smithsonian Institution, proposed an intriguing explanation for the Nile delta's recession, and one that goes beyond the impacts of the Aswân High Dam. Over the centuries, more than 9000 km (5500 mi) of canals were built in the delta to augment the natural distributary system. As the river discharge enters the network of

canals, flow velocity is reduced, stream competence and capacity are lost, and sediment load is deposited far short of where the delta touches the Mediterranean Sea. River flows no longer effectively reach the sea.

The Nile delta is receding from the coast at an alarming 50 to 100 m (165 ft to 330 ft) per year. Seawater is intruding farther inland in both surface water and groundwater. Human action and reaction to this evolving situation will no doubt determine the delta's future. If Herodotus could only see the delta he described as it looks today!

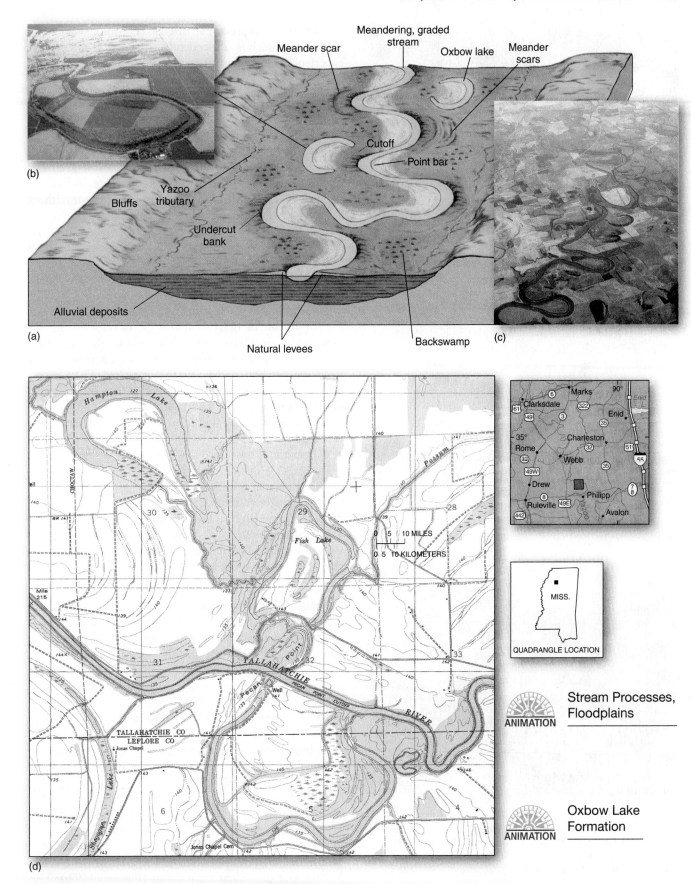

FIGURE 11.21 A floodplain.
(a) Typical floodplain landscape and related landscape features. (b) An oxbow lake. (c) Levees, oxbow lakes, and farmland in a floodplain. (d) A portion of the Philipp, Mississippi, topographic map quadrangle.
[Photos (b) and (c) by Bobbé Christopherson; (d) topographic map from USGS.]

FIGURE 11.22 Levee break and devastation in New Orleans.
August 30, 2005, view across the Inner Harbor Navigation Canal, as broken levees allow floodwaters to drown the city—a dramatic failure of engineering, levee design and construction, and proper planning. [Photo by © Smiley N. Pool/*Dallas Morning News*/Corbis.]

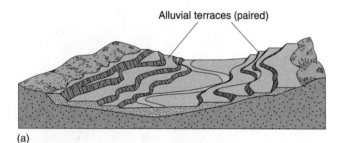

Alluvial terraces (paired)

(a)

FIGURE 11.23 Alluvial stream terraces.
(a) Alluvial terraces are formed as a stream cuts into a valley.
(b) Alluvial terraces along the Rakaia River, in New Zealand.
[(a) After W. M. Davis, *Geographical Essays* (New York: Dover, 1964 [1909]), p. 515; (b) photo by Bill Bachman/Photo Researchers, Inc.]

ANIMATION   Stream Terrace Formation

alluvial terraces represent an original depositional feature, a floodplain, which is subsequently eroded by a stream that experienced changes in stream load and capacity.

River Deltas   The mouth of a river is where it reaches its base level. The river's forward velocity rapidly decelerates as it enters a larger body of water, with the reduced velocity causing its transported load to exceed its capacity. Coarse sediments such as sand and gravel drop out first, with finer clays carried to the extreme end of the deposit. The depositional plain formed at the mouth of a river is a **delta**, named after the Greek letter *delta* (Δ), the triangular shape which Herodotus perceived to be similar to the shape of the Nile River delta. In these ancient times Herodotus already knew the significance of the Nile River delta for food production (see News Report 11.2).

Each flood stage deposits a new layer of alluvium over the surface of the delta so that it grows outward. At the same time, river channels divide into smaller channels known as *distributaries*, which appear as a reverse of the dendritic drainage pattern discussed earlier. The Ganges River delta features an intricate braided pattern of distributaries. Alluvium carried from deforested slopes upstream is the excess sediment that forms the many deltaic islands (Figure 11.24).

The Nile River forms an *arcuate* (arc-shaped) *delta* (Figure 11.25). Also arcuate in form is the Danube River delta in Romania as it enters the Black Sea (see Figure 11.6) and the Ganges and Indus River deltas. In another distinct form, the Seine River in France and the Tiber River in Italy have *estuarian deltas*, or one that is in the process of filling an **estuary**, which is the seaward mouth of a river where the river's freshwater encounters seawater.

Mississippi River Delta   The Mississippi River delta has an interesting history. Over the past 120 million years, the Mississippi has collected sediments throughout its vast basin and deposited them in the Gulf of Mexico. During the past 5000 years, the river has formed a succession of seven distinct deltaic complexes along the Louisiana coast (Figure 11.26a).

Each generalized lobe in the illustration reflects distinct course changes in the Mississippi River, probably where the river broke through its natural levees during episodes of severe flooding, thus changing the configuration of the delta. The seventh and current delta has been building for at least 500 years and is a classic example of a *bird's-foot delta*—a long channel with many distributaries and sediments carried beyond the tip of the delta into the Gulf of Mexico.

The Mississippi River delta clearly is dynamic over time as sediments accumulate on the floor of the Gulf of Mexico and the distributaries shift (Figure 11.26b). The main channel persists because of much effort and expense directed at maintaining the artificial levee system. The 3.25-million-km^2 (1.25-million-mi^2) Mississippi drainage basin produces enough sediment to extend the Louisiana

FIGURE 11.24 The Ganges River enters the Bay of Bengal.
The complex distributary pattern in the "many mouths" of the Ganges River delta in Bangladesh and extreme eastern India imaged from the *Terra* satellite. [*Terra* image courtesy of MODIS Land Team, NASA.]

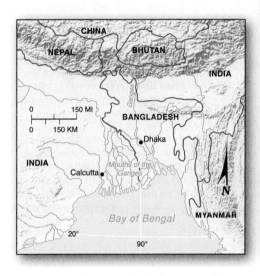

coast 90 m (295 ft) a year—550 million metric tonnes a year—however, several factors are actually causing losses to the delta each year.

Compaction and the tremendous weight of the sediments in the Mississippi River create isostatic adjustments in Earth's crust. These adjustments are causing the entire region of the delta to subside, thereby placing ever-increasing stress on natural and artificial levees and other structures along the lower Mississippi. The many canals and waterways plowed through the delta by the oil and gas industry and shipping interests have worked to deprive the delta of new alluvial building materials. Also, the pumping of tremendous quantities of oil and gas from thousands of onshore and offshore wells is thought to be a cause of regional land deflation and lowering.

The city of New Orleans is almost entirely below river level, with sections of the city below sea level. Severe flooding is a certainty for existing and planned settlements unless further intervention or urban relocation occurs. The building of multiple flood-control structures and extensive reclamation efforts by the US Army Corps of Engineers apparently only delayed the peril, as demonstrated by recent flooding. The fact that four levee breaks during Hurricane Katrina's storm surge, winds, and rainfall in 2005 inundated 80% of the city confirms this concern and issue.

An additional problem for the lower Mississippi Valley is the possibility, in a worst-case flood, that the river could break from its existing channel and seek a new route to the Gulf of Mexico. If you examine the map in Figure 11.26c and look at the sediment plume to the west of the main delta in the satellite image, an obvious alternative to the Mississippi's present channel is the Atchafalaya River (its delta is shown in Figure 11.26e). The Atchafalaya would provide a much shorter route to the Gulf of Mexico, less than one-half the present distance, and it has a steeper gradient than the Mississippi. Presently, this alternative-route distributary carries about 30% of the Mississippi's total discharge. For the Mississippi to bypass New Orleans entirely would be a blessing, for it would remove the flood threat. However, this shift would be a financial disaster, as a major US port would silt in and seawater would intrude into freshwater resources.

At present, artificial barriers block the Atchafalaya from reaching the Mississippi at the point shown; without the floodgates the two rivers do connect. The Old River Control Project (1963) maintains three structures and a lock about 320 km (200 mi) from the Mississippi's mouth to keep these rivers in their channels (Figure 11.26d). Many floods have hit the Mississippi River system in the past, such as in 1993, 2001, 2003 (Figure 11.26g), and 2008, and major storm surges and flooding from tropical storms and hurricanes. Another major flood is only a matter of time, one that might cause the river channel to change back into the Atchafalaya. A case study in News Report 11.3 about

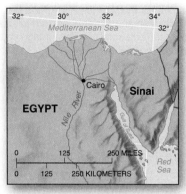

FIGURE 11.25 The Nile River delta—an arcuate delta.
Intensive agricultural activity and small settlements are visible on the delta and along the Nile River floodplain in this true-color image. Cairo is at the apex of the delta. You can see the two main distributaries: Damietta to the east and Rosetta to the west. Because of the disruption of sediment flow into the delta, the Nile delta is receding from the coast at an alarming 50 to 100 m (165 to 330 ft) per year (see News Report 11.2). [January 30, 2001, *Terra* image courtesy of MISR Team, NASA/ GSFC/JPL.]

Bayou Lafourche, a former major distributary of the Mississippi until 1904, highlights many of these issues.

Rivers Without Deltas The Amazon River, whose discharge exceeds 175,000 m³/s (6.2 million cfs) and carries sediments far into the Atlantic, lacks a true delta. Its mouth, 160 km (100 mi) wide, formed an underwater deltaic plain deposited on a sloping continental shelf. As a result, the Amazon's mouth is braided into a broad maze of islands and channels (see satellite image in Figure 11.1). Other rivers lack deltaic formations if they lack significant sediment or if they discharge into strong erosive currents. The Columbia River of the US Northwest is one that lacks a delta because of offshore currents.

Floods and River Management

 Stream Processes, Floodplains
ANIMATION

Throughout history, civilizations have settled on floodplains and deltas, especially since the agricultural revolution of 10,000 years ago and discovery of the fertility of floodplain soils. Early villages generally were built away from the area of flooding, or on stream terraces, because the floodplain was dedicated exclusively to farming. However, as commerce grew, sites near rivers became important for transportation. Port and dock facilities were built, as were river bridges. Also, because water is a basic industrial raw material, used for cooling and for diluting and removing wastes, waterside industrial sites

FIGURE 11.26 The Mississippi River delta.
(a) Evolution of the present delta, from 5000 years ago (1) to the present (7). (b) The bird's-foot delta of the Mississippi River (in 2001) receives a continuous supply of sediments, focused by controlling levees. The delta extends into the Gulf of Mexico, although subsidence of the delta and rising sea level have diminished the overall surface area. (c) Location map of the Old Control Structures and potential capture point (arrow) where the Atchafalaya River may one day divert the present channel. (d) Old River Control Auxiliary Structure, one of the dams to keep the Mississippi in its channel. (e) Where the Atchafalaya River enters the Gulf. (f) The end of the bird's-foot delta stretches far out in the Gulf. (g) Floods occur frequently in the delta as in the spring of 2003. (h) Mississippi delta waterscape; note the house with the first floor raised on stilts. [(a) Adapted from C. R. Kolb and J. R. Van Lopik, "Depositional environments of the Mississippi River deltaic plain," in *Deltas in Their Geologic Framework* (Houston: Houston Geological Society, 1966). (b) *Terra* image, March 5, 2001, courtesy of Liam Gumley, Space Science and Engineering Center, University of Wisconsin, and the MODIS Science Team, NASA. Photos (d) through (h) by Bobbé Christopherson.]

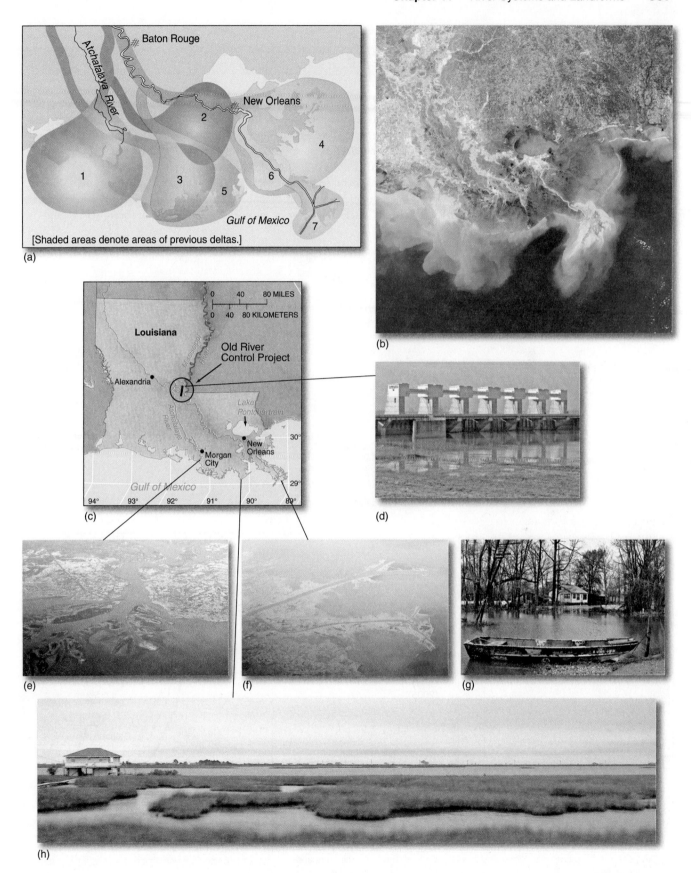

(a) [Shaded areas denote areas of previous deltas.]

(b)

(c)

(d)

(e)

(f)

(g)

(h)

NEWS REPORT 11.3

What Once Was Bayou Lafourche—Analysis of a Photo

Bayou Lafourche (pronounced "La-foosh") was the main distributary of the Mississippi River between the Civil War and 1904, when a dam stopped the flow of this distributary and led to water losses in the bayou and stagnation of the former marshlands. The bayou was closed to navigation by 1930. In 1955, attempts began to reverse the losses to this bayou by restoring some water flow with pumps. Meanwhile, the overall losses to the delta continued.

Bayou is a general term for several water features in the Lower Mississippi River system, including creeks and secondary waterways. These sometimes stagnant, winding watercourses pass through coastal marshlands and swamps and allow tidal waters access to deltaic lowlands.

Human impact on Bayou Lafourche's 175-km (110-mi) stretch of the delta has been severe and is

(a)

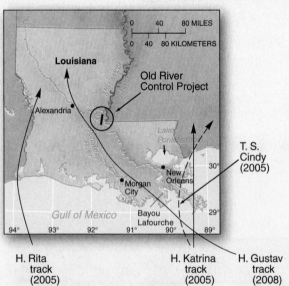

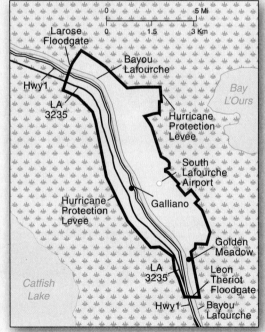

(b)

FIGURE 11.3.1 A hurricane protection levee compound in the Mississippi delta.
(a) Development of settlements, farms, and industry within a 65-km (40-mi) hurricane-protection levee system. Bayou Lafourche is the waterway that runs along the highway, with locks at the north and south ends of the levee system to seal off the interior in times of storm surges. (b) A map of this general region. In 2005, the eye of Hurricanes Katrina and Cindy passed east of the area in the photo, Hurricane Rita to the west; Gustav made landfall nearby in 2008, as noted on the map. [Photo by Bobbé Christopherson.]

mirrored by similar impacts across southern Louisiana. In fact, most of coastal Louisiana, which represents about 40% of the coastal marshes in the United States, is profoundly altered and disrupted by dam and levee construction, flow alterations, oil and gas exploration and pumping, pipelines, and dredging for navigation, logging, and industrial needs. Approximately 65 km² (25 mi²) of these Louisiana coastal lowlands are reclaimed by the Gulf of Mexico in an average year.

Figure 11.3.1 illustrates a portion of Bayou Lafourche in 2003. The towns of Galliano and Golden Meadow and a total population of more than 10,000 people are protected from rising water, tidal fluctuations, and hurricane storm surge within the compound surrounded by the hurricane protection levee system. The 65-km (40-mi) levee system, some 4 m (13 ft) in height and 28 m (91 ft) wide at the base, was begun in the late 1970s. You can see the long-lot land survey system dividing lands at right angles to Bayou Lafourche and Highway 1, with some plots planted in sugar cane. Floodgates at either end of this enclave are closed when the flood threat is imminent. Water tables in the protected area have decreased since the levee enclosure was completed and surface-water flows have been reduced.

With delta subsidence and rising sea levels, increasing levee height is a continuing process. Ongoing construction, known as a "lift" to the levee, is adding another meter of

FIGURE 11.3.2 **Port Fourchon.**
Port Fourchon is the service center for the offshore oil industry, with about 6000 workers using the port as a depot for commuting to offshore drilling rigs. Several recent hurricanes caused damage to the port facilities and offshore oil-drilling platforms.
[Photo by Bobbé Christopherson.]

height in an attempt to maintain 100-year storm-surge protection—a project that is continuing into 2009. The real irony is that the former coastal wetlands were a buffer between the Gulf storm surges and interior lands, yet today because of all the human disruptions there is much open water between Golden Meadow and the former shoreline along the margins of the delta. Disruptions include reduction in sediment supplies to the delta, a rising sea level, and land deflation due to oil and gas removal through thousands of wells.

Only a few marsh mitigation and restoration projects are underway. One is at Port Fourchon at the extreme southern end of the former coast (south of the Figure 11.3.1 photo). This is a major oil and gas industry transshipment port and helicopter base where workers connect for their jobs on drilling rigs in the Gulf of Mexico (Figure 11.3.2).

Hurricanes Katrina, Rita, Cindy, and Gustav picked up huge oil barges, equipment, shrimp boats, other marine craft, and debris and dumped them far inland. Grand Isle, to the east of the

port on the Gulf Coast, experienced major destruction in 2005 and was leveled in 2008. Some 80% of homes, businesses, and fishing camps were destroyed. Many of the homes that are built on stilts were simply gone after these storm events. The hurricane levee compound was surrounded by as much as 2.4 m (8 ft) of water.

Hurricane Gustav in 2008 made landfall just west of the compound near Cocodrie, Louisiana, and Hurricane Ike passed to the south, with huge rainbands that delivered heavy precipitation to southern Louisiana. During both events the parish was evacuated, the levees held, and residents returned to much wind and water damage, but no flooding inside the levee compound.

Meanwhile, Galliano in the levee compound is presently 122 cm (48 in.) above sea level; the Mississippi delta lands are slowly subsiding and sea level is continuing to rise; and the Atlantic tropical cyclone season is in an active cycle of a decade or longer, with storm power enhanced by increasing air and water temperatures. The story continues to unfold.

became desirable. However, all these human activities on vulnerable flood-prone lands require planning, zoning, and restrictions to avoid disaster. Unfortunately, the hazard is not perceived, as a general rule.

Floods and Floodplains

A **flood** is a high-water level that overflows the natural (or artificial) levees along any portion of a stream. Understanding the flood patterns of a drainage basin is as complex as understanding the weather, for floods and weather are equally variable and both include a level of unpredictability.

The abuse and misuse of river floodplains brought catastrophe to North Carolina in 1999. In short succession during September and October, Hurricanes Dennis, Floyd, and Irene delivered several feet of precipitation to the state, each storm falling on already saturated ground. Some 50,000 people were left homeless and at least 50 died; over 4000 homes were lost and an equal amount badly damaged (Figure 11.27a). The dollar estimate for the ongoing disaster through 1999 reached $10 billion. However, the real tragedy will be unfolding for years to come.

Hogs, in factory farms, outnumber humans in North Carolina. More than 10 million hogs, each producing

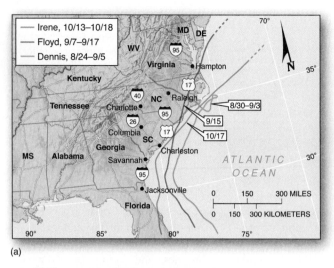

(a)

(b)

FIGURE 11.27 North Carolina 1999 floodplain disaster.
(a) Three hurricanes deluged North Carolina with several feet of rain during September and October 1999, Hurricane Floyd being the worst. (b) Livestock were killed in the floods, which also washed out hundreds of sewage lagoons into wetlands, streams, and the ocean. [(b) Photo by Mel Nathanson, *Raleigh News & Observer*.]

2 tons of waste per year, were located in about 3000 agricultural factories. These generally unregulated operations collect millions of tons of manure into open lagoons, many set on river floodplains. The hurricane downpour flushed out these waste lagoons, spewing hundreds of millions of gallons of untreated sewage into wetlands, streams, and eventually Pamlico Sound and the ocean—a spreading "dead zone." Added to this waste were hundreds of thousands of hog, poultry, and other livestock carcasses, industrial toxins, floodplain junkyard oil, and municipal waste, creating an environmental catastrophe (Figure 11.27b). Final assessment of overall damage will take a decade, if not longer.

Catastrophic floods continue to be a threat, especially in poor nations. Bangladesh is perhaps the most persistent example. Bangladesh is one of the most densely populated countries on Earth, and more than *three-fourths* of its land area is a floodplain. The country's vast alluvial plain sprawls over an area the size of Alabama (130,000 km^2, or 50,000 mi^2). The flooding severity in Bangladesh is a consequence of human economic activities, along with heavy precipitation episodes.

Excessive forest harvesting in the upstream portions of the Ganges–Brahmaputra River watersheds increased runoff. Over time, the increased sediment load carried by the river was deposited in the Bay of Bengal, creating new islands (see Figure 11.24). These islands, barely above sea level, became sites for new farming villages. As a result, many people perished in the 1987, 1988, 1991,1998, and 2007 floods and storm surges. (For information on floods worldwide, see http://www.dartmouth.edu/~floods/Resources.html/ or a daily flood summary at http://www.nws.noaa.gov/oh/hic/current/fln/fln_sum.shtml.)

Floods and floodplains are rated statistically for the expected intervals between floods. A *10-year flood* is

expected to occur once every 10 years (that is, it has a 10% probability of occurring in any one year). Such a frequency labels a floodplain as one of moderate threat. A 50-year or 100-year flood is of greater and perhaps catastrophic consequence, but it is also less likely to occur in a given year. These statistical estimates are probabilities that events will occur randomly during any single year of the specified period. Of course, two decades might pass without a 10-year flood, or 10-year flood volumes could occur 3 years in a row. The record-breaking Mississippi River Valley floods in 1993, or the North Carolina floods of 1999, easily exceeded a 1000-year flood probability of occurrence. Focus Study 11.1 presents a closer look at floodplain strategies.

Engineering Failures in New Orleans, 2005

Under normal conditions the city of New Orleans is almost entirely below river level, with sections of the city below sea level in elevation. Severe flooding is a certainty for existing and planned settlements unless further intervention or urban relocation occurs. The building of multiple flood-control structures and extensive reclamation efforts by the US Army Corps of Engineers apparently have only delayed the peril, as demonstrated by recent flooding. The events of 2005, following the passage of Hurricane Katrina, will remain for many years as one of the government's greatest engineering failures. Four levee breaks and at least four dozen levee breaches (topping) permitted the inundation of a major city, which confirms this conclusion. The polluted water remained for days.

Figure 11.28a and b from *Landsat*–7 are matching images of New Orleans along the shore of Lake Pontchartrain

(text continued on page 394)

FOCUS STUDY 11.1

Floodplain Strategies

Detailed measurements of stream-flows and flood records have been kept rigorously in the United States for only about 100 years, in particular since the 1940s. At any selected location along any given stream, the *probable maximum flood* (PMF) is a hypothetical flood of such a magnitude that there is virtually no possibility it will be exceeded. Because the collection and concentration of rainfall produces floods, hydrologists speak of a corollary, the *probable maximum precipitation* (PMP), for a given drainage basin, which is an amount of rainfall so great that it will never be exceeded.

These parameters are used by hydrologic engineers to establish a *design flood* against which to take protective measures. For urban areas near creeks, planning maps often include survey lines for a 50-year or a 100-year floodplain; such maps have been completed for most US urban areas. The design flood usually is used to enforce planning restrictions and special insurance requirements. Restrictive zoning using these floodplain designations is an effective way of avoiding potential damage. (The floodplain managers' organization at http://www.floods.org/ is a respected source of information.)

Such political action is not generally implemented, and the scenario all too often goes like this: (1) Minimal zoning precautions are not carefully supervised; (2) a flooding disaster occurs; (3) the public is outraged at being caught off-guard; (4) businesses and homeowners are surprisingly resistant to stricter laws and enforcement; (5) eventually another flood refreshes the memory and promotes more knee-jerk planning. As strange as it seems, *there is little indication that our risk perception improves as the risk increases.* In the years to come it will be interesting to see if rebuilding in southern Louisiana and Florida follows this sequence of response.

(a)

(b)

(c)

ANIMATION
Stream Processes, Floodplains

FIGURE 11.1.1 Weir and bypass channels to divert floodwaters.
(a) The Bonnet Carré spillway (a weir), on the lower Mississippi, New Orleans District, is opened in flood events to let water flow to Lake Pontchartrain and bypass New Orleans to the south. (b) The Sacramento weir awaits floodwaters from the Sacramento River, which it will direct through a channel into the Yolo Bypass visible in the distance in (c). When not flooded, the bypass is used for farmland crops. [Photos by Bobbé Christopherson.]

(continued)

Focus Study 11.1 *(continued)*

A Few Planning Strategies

A planning strategy used in some large river systems is to develop artificial floodplains by constructing *bypass channels* to accept seasonal or occasional floods. When not flooded, the bypass channel can serve as farmland, often benefiting from occasional soil-replenishing inundations. When the river reaches flood stage, large gates called weirs are opened, allowing the water to enter the bypass channel. This alternate route relieves the main channel of the burden of carrying the entire discharge.

Along the Lower Mississippi, north of New Orleans, the Bonnet Carré spillway is such a weir, in operation in the photo (Figure 11.1.1a). In high-water flood events, water is let through a floodway into a bypass between a pair of guide levees to relieve the flood discharge flow. The bypass diverts the flood 9.7 km (6 mi) into Lake Pontchartrain, as it has done 8 times since 1937, to protect New Orleans. (For more details, see http://www.mvn.usace.army.mil/pao/bcarre/bcarre.htm.)

In the Sacramento Valley, California, such weirs lead to bypass channels for taking floodwaters past the metropolitan areas. In Figure 11.1.1b we see the Sacramento weir, and in Figure 11.1.1c the bypass is filled with floodwaters. Farther downstream in the Sacramento–San Joaquin delta, artificial levees and pumps attempt to keep farmlands dry that are actually below river level (Figure 11.1.2). To make matters worse, the area in the aerial photo is subsiding due to oxidation of peat in the soils, withdrawal of groundwater, and compaction. One of these delta islands flooded in a levee break in 2004.

Reservoir Considerations

Dams (an artificial structure placed in a river channel) and *reservoirs* (an impoundment of water behind a dam; a human-made lake) are common streamflow control methods within a

FIGURE 11.1.2 A delta island.
Levees surround this delta island and its farms. The central portion of the island is below the river level. Note the former water channel and developing oxbow on the land. [Photo by Bobbé Christopherson.]

(a)

(b)

FIGURE 11.1.3 Reservoir extremes.
Four years separate these comparative photographs of the New Hogan reservoir, central California, during (a) dry and (b) wet weather conditions. During intense storms in 1996, this reservoir came very close to spilling water over the top of the dam. [Photos by author.]

watershed. For conservation purposes, a dam holds back seasonal peak flows for distribution during low-water periods. In this way, streamflows are regulated to assure year-round water supplies. Dams also are constructed for flood control, to hold back excess flows for later release at more moderate discharge levels. Adding hydroelectric power production to these functions of conservation and flood control can define a modern multipurpose reclamation project.

The function of reservoir impoundment is to provide flexible storage capacity within a watershed to regulate river flows, especially in a region with variable precipitation. Figure 11.1.3 shows one reservoir during drought conditions and during a time of wetter weather 4 years later. Unfortunately, the multipurpose benefits of reservoir construction are countered by some negative consequences. The area upstream from a dam becomes permanently drowned. In mountainous regions, this may mean loss of white-water rapids and recreational sections of a river. In agricultural areas, the ironic end result may be that a hectare of farmland is inundated upstream to preserve a hectare of farmland downstream.

Furthermore, dams built in warm and arid climates lose substantial water to evaporation, compared with the free-flowing streams they replace. Reservoirs in the southwestern United States can lose 3–4 meters (10–13 ft) of water a year. Also, sedimentation can reduce the effective capacity of a reservoir and can shorten a dam's life span, as mentioned earlier regarding Glen Canyon Dam. (The World Commission on Dams is at http://www.dams.org/.)

A Final Thought About Floods

The benefit of any levee, bypass, or other project intended to prevent flood destruction is measured in damage avoided and is used to justify the cost of the protection facility. Thus, ever-increasing damage leads to the justification of ever-increasing flood control structures. All such strategies are subjected to cost-benefit analysis, but bias is a serious drawback because such an analysis usually is prepared by an agency or bureau with a vested interest in building more flood control projects.

As suggested in a half-century-old article titled "Settlement Control Beats Flood Control,"* there are other ways to protect populations than with enormous, expensive, sometimes environmentally disruptive projects. Strictly zoning the floodplain is one approach. However, the flat, easily developed floodplains near pleasant rivers are desirable for housing and thus weaken political resolve (Figure 11.1.4). A reasoned zoning strategy would set aside the floodplain for farming or passive recreation, such as a riverine park, golf course, or plant and wildlife sanctuary, or for other uses that are not hurt by natural floods. This study concluded that "urban and industrial losses would be largely obviated [avoided] by set-back levees and zoning and thus cancel the biggest share of the assessed benefits which justify big dams."

(a)

(b)

FIGURE 11.1.4 Living and building in a floodplain.
(a) Near Harpers Ferry, Iowa, homes are located on a forested sandbar in the Mississippi River. (b) New fill is being dumped into the river to extend the land for further development. The river is at a high water level in the photo. [Photos by Bobbé Christopherson.]

*Walter Kollmorgen, *Economic Geography* 29, no. 3 (July 1953): 215.

(a) April 24, 2005

(b) August 30, 2005

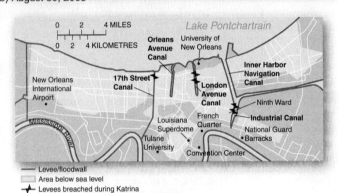

Lake Pontchartrain

0   2   4 MILES
0   2   4 KILOMETRES

Orleans Avenue Canal   University of New Orleans

Inner Harbor Navigation Canal

New Orleans International Airport

17th Street Canal

London Avenue Canal

Ninth Ward

Industrial Canal

Mississippi River

Louisiana Superdome   French Quarter

National Guard Barracks

Tulane University

Convention Center

— Levee/floodwall
▢ Area below sea level
✦ Levees breached during Katrina

FIGURE 11.28 New Orleans before and after disaster.
Landsat–7 images made (a) April 24, 2005, and (b) August 30, 2005, clearly show canal locations and flooded portions of the city; compare features to the locator map. Numerous levee breaks and breaches put 80% of the city underwater for more than a week. [*Landsat*–7, Bands 7, 5, and 3, courtesy of USGS *Landsat* Image Gallery.]

on April 24 and August 30, 2005. The locator map helps you position the levee/floodwalls, areas that are below sea level, and the four major breaches. The extensive floods show up as the darker areas of the city, where some neighborhoods were submerged up to 6.1 m (20 ft). At the time of image (b), New Orleans was approximately 80% under water.

Six investigations by civil engineers, scientists, and political bodies agreed that in concept, design, construction, and maintenance, the "protection" system was flawed. Many of the structures failed before they reached design-failure limits. After all, Hurricane Katrina was diminished in power as it moved onshore. Several of the

17th St. Canal

FIGURE 11.29 Levee breach and break.
The Lakeview neighborhood is underwater (to the right) from the 17th Street Canal levee failure, whereas the Metairie neighborhood is essentially dry (to the left). You see debris caught up north of the bridge where steel beams were placed in an attempt to curb flows. Flotsam (floating debris) is visible in the photo. [Courtesy of NOAA aerial photography program.]

levee floodwalls and pilings were not anchored to the design depth, such as along the 17th Street Canal (Figure 11.29). This inadequate anchoring depth made parts of the system inherently weak. Unexplained are findings in 2007 by civil engineers inspecting levee and floodwall repair and placement that the composition of fill materials is too low in clays and too high in sand—a recipe for future disaster.

Repair and recovery work in New Orleans is not completed, and elements of the completed sections appear inadequate and positioned for repeat disasters, which nearly occurred in 2008 as Hurricane Gustav followed a track just west of New Orleans. Most levees held, some were topped, and much flooding again hit. Gustav made landfall at the southern edge of the hurricane levee enclosure in Bayou Lafourche, featured in News Report 11.3. Hurricane Gustav is regarded as a near-miss from being another Katrina-level disaster for New Orleans. It caused extensive flooding in the parishes west and south of the city, which have no levees, such as in Terrebonne Parish.

As we contemplate all this, we must consider the ongoing rise in sea level coupled with subsidence along this portion of the Gulf Coast. In Chapter 13, a new map portrays a 1-m rise in sea level forecasted for this area. Physical geography has much to study, analyze, and conclude about this enigmatic situation that we will read about throughout our lives.

Streamflow Measurement

To develop the best possible flood management, the behavior of each large watershed and stream is measured and analyzed. Unfortunately, such data often are not available for small basins or for the changing landscapes of urban areas where urban creek flooding is an increasing problem.

The key is to measure *streamflow*—the height and discharge of a stream (Figure 11.30). A *staff gauge*, a pole placed in a stream bank and marked with water heights, is used to measure stream level. With a fully measured cross section, stream level can be used to determine discharge (discharge is equal to width times depth times velocity). A *stilling well* is sited on the stream bank and a gauge is mounted in it to measure stream level (Figure 11.30c). A movable current meter can be used to sample velocity at various locations.

Approximately 11,000 stream gaging stations (hydrologists use this spelling) are used in the United States (an average of over 200 per state). Of these, 7000 are operated by the US Geological Survey and have continuous recorders for stage (level) and discharge (see http://water.usgs.gov/pubs/circ/circ1123/). Many of these stations automatically send telemetry data to satellites, from which information is retransmitted to regional centers (Figure 11.30b). Environment Canada's Water Survey of Canada maintains more than 3000 gaging stations.

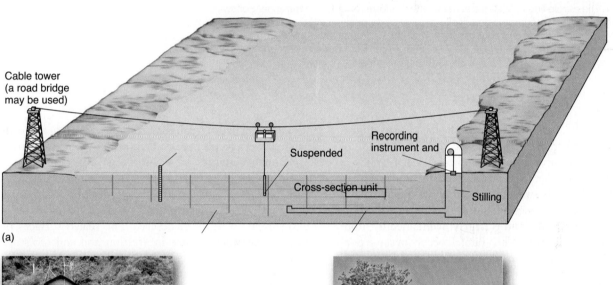

Cable tower (a road bridge may be used)

Suspended

Recording instrument and

Cross-section unit

Stilling

(a)

(b)

(c)

FIGURE 11.30 Streamflow measurement.
(a) A typical streamflow measurement installation: staff gauge, stilling well with recording instrument, and suspended current meter. An automated hydrographic station (b) and stilling well (c) send telemetry to a satellite for collection by the USGS. [Photos (b) courtesy of the California Department of Water Resources; (c) by Bobbé Christopherson.]

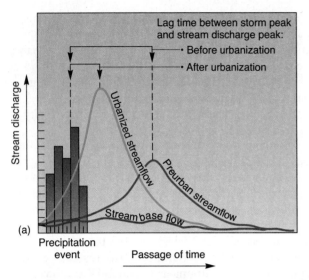

(a)

(b)

FIGURE 11.31 Urban flood profile.
(a) Effect of urbanization on a typical stream hydrograph. The dark blue line indicates normal base flow. The purple line indicates discharge after a storm, prior to urbanization. Following urbanization, stream discharge dramatically increases and peaks earlier, as shown by the light blue line. (b) Severe flooding of an urban area in Linda, California, after a levee break on the Sacramento River in 1986; it flooded again in 1997. [Photo from California Department of Water Resources.]

If we study streamflow measurements, we can understand channel characteristics as conditions vary. A graph of stream discharge over a time period for a specific place is called a **hydrograph**. The hydrograph in Figure 11.31a shows the relation between precipitation input and stream discharge. During dry periods at low-water stages the flow is described as *base flow* and is largely maintained by input from local groundwater. When rainfall occurs in some portion of the watershed, the runoff collects and is concentrated in streams and tributaries. The amount, location, and duration of the rainfall episode determine the *peak flow*. Also important is the nature of the surface in a watershed;

for example, a hydrograph for a specific portion of a stream changes after a forest fire or urbanization of the watershed.

Human activities have enormous impact on water flow in a basin. The effects of urbanization are quite dramatic, both increasing and hastening peak flow and causing damage. In fact, urban areas produce runoff patterns quite similar to those of deserts. The sealed surfaces of the city drastically reduce infiltration and soil moisture recharge, behaving much like the hard, nearly barren surfaces of the desert, producing flash flood events. The flooded region in Figure 11.31b was hit in 1986 and again in 1997.

Summary and Review—River Systems and Landforms

■ *Define fluvial* and *outline* the fluvial processes: erosion, transportation, and deposition.

River systems, fluvial processes and landscapes, floodplains, and river control strategies are important to human populations as demands for limited water resources increase. **Hydrology** is the science of water—its global circulation, distribution, and properties—specifically water at and below Earth's surface. Stream-related processes are **fluvial**. Water dislodges, dissolves, or removes surface material in the process **erosion**. Streams produce *fluvial erosion*, in which weathered sediment is picked up for **transport** and movement to new locations. Sediments are laid down by another process, **deposition**. **Alluvium** is the general term for the clay, silt, and sand deposited by running water.

Base level is a level below which a stream cannot erode its valley further.

> hydrology (p. 366)
> fluvial (p. 366)
> erosion (p. 367)
> transport (p. 367)
> deposition (p. 367)
> alluvium (p. 367)
> base level (p. 367)

1. What role is played by rivers in the hydrologic cycle?
2. What are the five largest rivers on Earth in terms of discharge? Relate these to the weather patterns in each area

and to regional potential evapotranspiration (moisture demand) and precipitation (moisture supply).

3. Define *fluvial*. What is a fluvial process?

4. What is the difference between a local and an ultimate base level?

■ *Construct* a basic drainage basin model and *identify* different types of drainage patterns, with examples.

The basic fluvial system is a **drainage basin**, which is an open system. *Drainage divides* define the drainage basin catchment (water-receiving) area. In any drainage basin, water initially moves downslope in a thin film of **sheetflow**, or *overland flow*. This surface runoff concentrates in *rills*, or small-scale downhill grooves, which may develop into deeper *gullies* and a stream course in a valley. High ground that separates one valley from another and directs sheetflow is termed an *interfluve*. Extensive mountain and highland regions act as **continental divides** that separate major drainage basins. **Drainage pattern** refers to the arrangement of channels in an area as determined by the steepness, variable rock resistance, variable climate, hydrology, relief of the land, and structural controls imposed by the landscape. There are seven basic drainage patterns generally found in nature: dendritic, trellis, radial, parallel, rectangular, annular, and deranged.

drainage basin (p. 368)
sheetflow (p. 369)
continental divides (p. 369)
drainage pattern (p. 371)

5. According to Figure 11.5, in which drainage basin are you located? Where are you in relation to the various continental divides?

6. What is the spatial geomorphic unit of an individual river system? How is it determined on the landscape? Define the several relevant key terms used.

7. On Figure 11.5, follow the Allegheny-Ohio-Mississippi River systems to the Gulf of Mexico, analyze the pattern of tributaries, and describe the channel. What role do continental divides play in this drainage?

8. Describe drainage patterns. Define the various patterns that commonly appear in nature. What drainage patterns exist in your hometown? Where you attend school?

■ *Describe* the relation among velocity, depth, width, and discharge and *explain* the various ways that a stream erodes and transports its load.

Stream channels vary in *width* and *depth*. The streams that flow in them vary in velocity and in the *sediment load* they carry. All of these factors may increase with increasing discharge. **Discharge** is calculated by multiplying the velocity of the stream by its width and depth for a specific cross section of the channel. Most streamflows increase discharge downstream because the area being drained increases. Where a river flows from a humid region into a drier region, stream *discharge usually decreases with distance*, such as in an **exotic stream** (*exotic* means "of foreign origin"), because of high potential evapotranspiration rates in the arid area.

Hydraulic action is the work of *turbulence* in the water. Running water causes hydraulic squeeze-and-release action to loosen and lift rocks and sediment. As this debris moves along, it mechanically erodes the streambed further, through a process of **abrasion**.

Solution refers to the **dissolved load** of a stream, especially the chemical solution derived from minerals such as limestone or dolomite or from soluble salts. The **suspended load** consists of fine-grained, clastic particles held aloft in the stream, with the finest particles held in suspension until the stream velocity slows nearly to zero. **Bed load** refers to coarser materials that are dragged along the streambed by **traction** or are rolled and bounced along by **saltation**. If the load in a stream exceeds its capacity, **aggradation**, or the accumulation of excess sediment, occurs as deposition fills the stream channel. With excess sediment, a stream becomes a maze of interconnected channels that form a **braided stream** pattern.

discharge (p. 372)
exotic stream (p. 373)
hydraulic action (p. 374)
abrasion (p. 374)
dissolved load (p. 374)
suspended load (p. 375)
bed load (p. 375)
traction (p. 375)
saltation (p. 376)
aggradation (p. 376)
braided stream (p. 376)

9. What was the impact of flood discharge on the channel of the San Juan River near Bluff, Utah? Why did these changes take place?

10. How does stream discharge complete its erosive work? What are the processes at work in the channel?

11. Differentiate between stream competence and stream capacity.

12. How does a stream transport its sediment load? What processes are at work?

■ *Develop* a model of a meandering stream, including point bar, undercut bank, and cutoff, and *explain* the role of stream gradient in these flow characteristics.

Where the slope is gradual, stream channels develop a sinuous form called a **meandering stream**. The outer portion of each meandering curve is subject to the fastest water velocity and can be the site of a steep **undercut bank**. On the other hand, the inner portion of a meander experiences the slowest water velocity and forms a **point bar** deposit. When a meander neck is cut off as two undercut banks merge, the meander becomes isolated and forms an **oxbow lake**.

Every stream develops its own **gradient** and establishes a longitudinal profile. A portion of the stream is designated a **graded stream** when that portion's available discharge, channel characteristics, velocity, and the load supplied from the drainage basin are balanced. An interruption in a stream's longitudinal profile is a **nickpoint**. A nickpoint can develop as the stream flows across hard, resistant rock or after tectonic uplift episodes.

meandering stream (p. 376)
undercut bank (p. 376)
point bar (p. 376)

oxbow lake (p. 377)
gradient (p. 377)
graded stream (p. 377)
nickpoint (p. 379)

13. Describe the flow characteristics of a meandering stream. What is the pattern of flow in the channel? What are the erosional and depositional features and the typical landforms created?

14. Explain these statements: (a) All streams have a gradient, but not all streams are graded. (b) Graded streams may have ungraded segments.

15. Why is Niagara Falls an example of a nickpoint? Without human intervention, what do you think would eventually take place at Niagara Falls?

16. Apply these concepts (gradient, graded stream, meandering stream, nickpoint), where appropriate, to a stream in your area. Explain and discuss.

■ *Define* a floodplain and *analyze* the behavior of a stream channel during a flood.

Floodplains have been an important site of human activity throughout history. Rich soils, bathed in fresh nutrients by floodwaters, attract agricultural activity and urbanization. Despite our knowledge of historical devastation by floods, floodplains are settled, raising issues of humans' perception of hazard. The flat, low-lying area along a stream channel that is subjected to recurrent flooding is a **floodplain**. It is formed when the river overflows its channel during times of high flow. On either bank of most streams, **natural levees** develop as by-products of flooding. On the floodplain, backswamps and yazoo tributaries may develop. **Alluvial terraces** are formed by the entrenchment of a river into its own floodplain.

floodplain (p. 380)
natural levees (p. 380)
alluvial terraces (p. 381)

17. Describe the formation of a floodplain. How are natural levees, oxbow lakes, backswamps, and yazoo tributaries produced?

18. How is it possible to travel fewer kilometers on the Mississippi River between St. Louis and New Orleans today than 100 years ago? Explain.

19. Describe any floodplains near where you live or where you go to college. Have you seen any of the floodplain features discussed in this chapter? If so, which ones?

■ *Differentiate* the several types of river deltas and *detail* each.

A depositional plain formed at the mouth of a river is called a **delta**. When the mouth of a river enters the sea and is inundated by seawater in a mix with freshwater it is called an **estuary**.

delta (p. 384)
estuary (p. 384)

20. What is a river delta? What are the various deltaic forms? Give some examples.

21. Based on this chapter, what is your analysis and resulting opinion concerning conditions along the Mississippi River delta as compared to the rest of the century?

22. Describe the Ganges River delta. What factors upstream explain its form and pattern? Assess the consequences of settlement on this delta.

23. What is meant by the statement "the Nile Delta is receding" (caption for Figure 11.25)?

■ *Explain* flood probability estimates and *review* strategies for mitigating flood hazards.

A **flood** occurs when high water overflows the natural or artificial levees of a stream. Both floods and the floodplains they might occupy are rated statistically for the expected time interval between floods. A 10-year flood is the greatest level of flooding that is likely once every 10 years. A graph of stream discharge over time for a specific place is called a **hydrograph**.

Collective efforts by government agencies undertake to reduce flood probability. Such management attempts include the construction of artificial levees, bypasses, straightened channels, diversions, dams, and reservoirs. Society still is learning how to live in a sustainable way with Earth's dynamic river systems.

flood (p. 389)
hydrograph (p. 396)

24. Specifically, what is a flood? How are such flows measured and tracked?

25. Differentiate between a hydrograph from a natural terrain and one from an urbanized area.

26. Based on materials in this chapter and in Chapter 5, assess what occurred in New Orleans. How might New Orleans change later in this century? Explain. In your opinion did Hurricane Katrina cause the disaster or was it a result of engineering failures and hazard-planning shortfalls?

27. What do you see as the major consideration regarding floodplain management? How would you describe the general attitude of society toward natural hazards and disasters?

28. What do you think the author of the article "Settlement Control Beats Flood Control" meant by the title? Explain your answer, using information presented in the chapter.

NetWork and Critical Thinking Tools

A. Determine the name of the river drainage basin within which your campus is located. Where are its headwaters? Where is the river's mouth? If you are in the United States or Canada, use Figure 11.5 to locate the larger drainage basins and divides for your region. Is there any regulatory organization that oversees planning and coordination for this drainage basin? See if there is a topographic map on file in the library, geography department, or at a local outdoor recreation store that covers the portion of the basin near campus. After examining the map, can you discern a prominent drainage pattern for the area (Figure 11.9)?

B. "Destinations" on the Student Learning Center site for this chapter links to the Dartmouth Flood Observatory. Click "River Watch" and navigate to your region, identified in "A." Sample several of the data points; what do you find? Briefly, sample other features on this valuable site.

C. Using your search engine or browser on the Internet, type in "New Orleans flooding, Katrina" or "Gulf Coast damage, Ike." Allow some time to sample through at least 10 of the links. Describe what you found. Did you find photo galleries, engineering reports, news coverage? What is your opinion about this region of the lower Mississippi River, New Orleans, and the Texas Gulf Coast cities and the direction they should take at this point in history: Rebuild things as they were; rethink the hazards and adopt a new strategy of replacing urban with low-density recreation zones; establish possible set-back hazard zoning; walk away except for necessary port functions; continue pumping oil and gas and order corporations to repressurize the subsurface strata to raise ground levels; or take no action since regional subsidence has other causes? What do you think about these complex issues? Remember, as you think this through, sea level is still rising at an unprecedented rate.

Hurricane Katrina, August 28, 2005, 11 A.M., category 5 strength [*Terra* image, NASA/GSFC.]

Three Sisters formation in Monument Valley, Arizona, composed of de Chelly sandstone and organ rock shale below. Attached to one side of Mitchell Mesa, the pinnacles are about 122 m (400 ft) tall and form striking silhouettes at dawn and sunset. The towers are capped by resistant Moenkopi shale that protects the rock below, making such height possible. *[Photo by author.]*

Wind Processes and Desert Landscapes

KEY LEARNING CONCEPTS

After reading the chapter, you should be able to:

- *Characterize* the unique work accomplished by wind and eolian processes.

- *Describe* eolian erosion, including deflation, abrasion, and the resultant landforms.

- *Describe* eolian transportation and *explain* saltation and surface creep.

- *Identify* the major classes of sand dunes and *present* examples within each class.

- *Define* loess deposits and their origins, locations, and landforms.

- *Portray* desert landscapes and *locate* these regions on a world map.

STUDENT LEARNING CENTER TOOLS

The *Elemental Geosystems* Student Learning Center provides on-line resources for this chapter by clicking on the book cover at **www.mygeoscienceplace.com**. Once you are in our Learning Center, bookmark the URL address. For this chapter, find reviewing tools, eBook links, self-tests and quizzes, critical thinking items, figure animations, photo galleries, and satellite loops. Some highlights:

▶ Animations assist you on how wind moves sediment and the formation of cross bedding.

▶ In Critical Thinking, work with satellite images to identify dunes.

▶ Work through the alluvial history of the southern Colorado Plateau.

▶ Interact with figures and maps from the chapter in Thinking Spatially.

▶ More than 11 additional URLs along with Quick Links to all the URLs in the chapter.

Wind is an agent of geomorphic change. Like moving water, moving air (wind) causes erosion, transportation, and deposition of materials. Like moving water, moving air is a fluid, and it behaves similarly, although it has a lower viscosity (it is less dense than water). Although lacking the lifting ability of water, wind processes can modify and move quantities of materials in deserts and along coastlines.

Wind contributes to soil formation in distant places, bringing fine particles from regions where glaciers once were active. Fallow fields (those not planted) give up their soil resource to destructive wind erosion. Scientists are only now getting an accurate picture of the amount of wind-blown dust that fills the atmosphere and crosses the oceans between continents. Chemical fingerprints and satellites trace windblown dust from African soils to South America and from Asian landscapes to Europe. Winds even spread living organisms. One study found related mosses, liverworts, and lichens distributed thousands of miles between islands in the Southern Ocean.

Earth's dry lands stand out in stark contrast on the water planet. In desert environments, an overall lack of moisture and stabilizing vegetation allows wind to create extensive sand seas and dunes of infinite variety. The polar regions are deserts as well, so there should be no surprise that weather monitoring stations in Yuma, Arizona, and in Antarctica report the same amount of annual precipitation. These polar and high-latitude deserts possess unique features related to their cold, dry environment—aspects covered in Chapter 14.

Arid landscapes display distinctive landforms and life forms: "Instead of finding chaos and disorder the observer never fails to be amazed at a simplicity of form, an exactitude of repetition and a geometric order."*

In this chapter: We examine the work of wind, associated erosion, transport, and depositional processes and resulting landforms. Windblown fine particles form vast loess deposits, the basis for rich agricultural soils. We include the discussion of arid lands in this chapter. Most deserts are rocky and covered with desert pavement, whereas other dry landscapes are covered in sand deposits and dunes. In desert environments, landscapes lacking moisture and stabilizing vegetation are set off in sharp, Sun-baked relief. Water is the major erosional agent in the desert, yet water is the limiting resource for human development. We examine the causes and distribution of Earth's desert landscapes. The Colorado River and the Western drought are the subject of an important focus study.

The Work of Wind

The work of wind—erosion, transportation, and deposition—is **eolian** (also spelled *aeolian*, for Aeolus, ruler of the winds in Greek mythology). Ralph Bagnold, a British major stationed in Egypt in 1925, was a pioneer in desert research. An engineering officer, he spent much of

*R. A. Bagnold, *The Physics of Blown Sand and Desert Dunes* (London: Methuen, 1941).

FIGURE 12.1 The work of wind.
Wind-sculpted tree near South Point, Hawai'i. Nearly constant tradewinds keep this tree naturally pruned. [Photo by Bobbé Christopherson.]

his time in the deserts west of the Nile. Bagnold measured, sketched, and developed hypotheses about the wind and desert forms. His often-cited work, *The Physics of Blown Sand and Desert Dunes*, was published in 1941 following the completion in London of wind-tunnel simulations of windy desert conditions.

The ability of wind to move materials is actually small compared with that of other transporting agents such as water and ice, because air is so much less dense than these other media. Yet over time, wind accomplishes enormous work. Bagnold studied the ability of wind to transport sand over the surface of a dune. Wind of 50 kmph (30 mph) can move approximately a half ton of sand per day over a 1-m-wide section of dune! Consistent local wind can prune and shape vegetation (Figure 12.1).

Eolian Erosion

Two principal wind-erosion processes are **deflation**, the removal and lifting of individual loose particles, and **abrasion**, the grinding of rock surfaces with a "sandblasting" action by particles captured in the air. Deflation and abrasion produce a variety of distinctive landforms and landscapes.

Deflation Deflation literally blows away loose or noncohesive sediment and works with rainwater to form a surface resembling a cobblestone street: a **desert pavement** that protects underlying sediment from further deflation and water erosion. Traditionally, deflation was regarded as a key formative process, eroding fine dust, clay, and sand away, leaving behind a concentration of pebbles and gravel as desert pavement (Figure 12.2a).

Another hypothesis that better explains some desert pavement surfaces states that deposition of windblown sediments is the formative agent, not removal. Windblown particles settle between and below coarse rocks

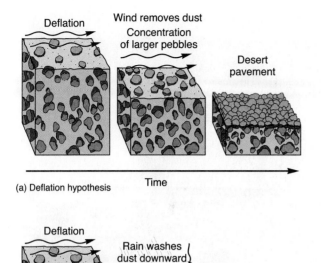

(a) Deflation hypothesis

(b) Sediment-accumulation hypothesis

(c)

FIGURE 12.2 Desert pavement.
(a) One way desert pavement is formed is when larger rocks and fragments are left after deflation and sheetwash (water flowing across the surface). (b) Another way finds wind delivering fine particles that settle and wash downward, gradually displacing gravels upward to form pavement. (c) A typical desert pavement. [Photo by Bobbé Christopherson.]

and pebbles that are gradually displaced upwards. Rainwater is involved as wetting and drying episodes swell and shrink clay-sized particles. The gravel fragments are gradually lifted to surface positions to form the pavement (Figure 12.2b). Desert pavements are so common that many provincial names have been used for them—for example, *gibber plain* in Australia, *gobi* in China, and, in

Africa, *lag gravels* or *serir* (or *reg* desert, if some fine particles remain).

Heavy recreational activity damages fragile desert landscapes. Over 15 million off-road vehicles (ORVs) are in use in the United States. Such vehicles crush plants and animals, disrupt desert pavement, promote deflation losses to wind, and create ruts that easily concentrate sheetwash (water flowing across surfaces) to form gullies.

Military activities can also threaten delicate desert landscapes. A serious environmental impact of the 1991 Persian Gulf War and ongoing Iraq War was the disruption of desert pavement. The resulting loosened sand and silt became available for deflation, plaguing cities and farms with increased dust and sand accumulations (see Figure 12.4b).

Wherever wind encounters loose sediment, deflation also may form basins. **Blowout depressions** range from small indentations of less than a meter up to areas hundreds of meters wide and many meters deep. Chemical weathering, although it operates slowly in the desert, is important in the formation of a blowout, for it removes the cementing materials that give particles their cohesiveness. Large depressions in the Sahara are at least partially formed by deflation. The enormous Munkhafad el Qaṭṭâra (Qaṭṭâra Depression), which covers 18,000 km² (6950 mi²) just inland from the Mediterranean Sea in the Western Desert of Egypt, is now about 130 m (427 ft) below sea level at its lowest point.

Abrasion   Sandblasting is commonly used to clean stone surfaces on buildings or to remove unwanted markings from streets. Abrasion by windblown particles is nature's version of sandblasting and is especially effective at polishing exposed rocks when the abrading particles are hard and angular. People living in desert regions, where frequent sand storms occur, contend with the sandblasting of painted surfaces and etched window glass. Variables that affect the rate of abrasion include the hardness of vulnerable surface rocks and the wind velocity and constancy. Abrasive action is restricted to the area immediately above the ground, usually no more than a meter or two in height, because sand grains are lifted only a short distance.

Rocks exposed to eolian abrasion appear pitted, grooved, or polished and usually are aerodynamically shaped in a specific direction, according to the flow of airborne particles. Rocks that bear such evidence of eolian erosion are **ventifacts**. On a larger scale, deflation and abrasion are capable of streamlining rock structures that are aligned parallel to the most effective wind direction, leaving behind distinctive, elongated ridges called **yardangs**. Yardangs can range from meters to kilometers in length and up to many meters in height. Abrasion is concentrated on the windward end of each yardang, with deflation operating on the leeward portion. A wind-sculpted formation in Snow Canyon, Utah, and one on windblown Bear Island in the Norwegian Sea (74° N) are shown in Figure 12.3.

On Earth, some yardangs are large enough to be detected on satellite imagery. The Ica Valley of southern Peru contains yardangs reaching 100 m (330 ft) in height and several kilometers in length, and yardangs in the Lūt

(a)

(b)

FIGURE 12.3 A yardang.
(a) A small wind-sculpted rock formation in Snow Canyon outside St. George, Utah. (b) Consistent winds shape this exposed rock on Bear Island. The rock sits on a surface of patterned ground and periglacial features (see Chapter 14). [Photos by Bobbé Christopherson.]

Desert of Iran attain 150 m (490 ft) in height. The Sphinx in Egypt perhaps partially formed as a yardang, suggesting a head and body to the ancients. Some scientists think this shape led them to complete the bulk of the sculpture artificially with masonry.

Eolian Transportation

As mentioned in Chapter 4, atmospheric circulation is capable of transporting fine material such as volcanic debris worldwide within days. The distance that wind is capable of transporting particles varies greatly with particle size. Wind exerts a drag or frictional pull on surface materials. Only the finest dust particles travel significant distances (Figure 12.4). The finer material suspended in a *dust storm* is lifted much higher than the coarser particles of a *sand storm*, which may be lifted only about 2 m (6.5 ft). People living in areas of frequent dust storms face the infiltration of very fine particles into their homes and businesses through even the smallest cracks. (Figure 2.25 illustrates such dust storms in Nevada and over the eastern Mediterranean Sea.)

Deflation and wind transport produced the American Dust Bowl catastrophe in the 1930s. (See News Report 12.1.) Overgrazing and poor agricultural practices left soil vulnerable. The transported dust darkened the skies of midwestern cities and the lives of millions.

Through processes of weathering, erosion, and transportation, mineral grains are removed from parent rock

and redistributed. In Chapter 10, Figure 10.4c, you can see the relationship between the composition and color of the sandstone in the background and the derived sandy surface in the foreground. Wind action is not significant in the weathering process that frees individual grains of sand from the parent rock, but it is active in relocating the weathered grains.

The term *saltation* was used in Chapter 11 to describe the movement of particles by water. The term also describes the wind transport of grains along the ground—grains usually larger than 0.2 mm (0.008 in.). Skipping and bouncing actions account for about 80% of wind transport of particles (Figure 12.5). Compared with fluvial transport, in which saltation is accomplished by hydraulic lift, eolian saltation is executed by aerodynamic lift, elastic bounce, and impact (compare Figures 12.5 and 11.12).

Saltating particles crash into other particles, knocking them both loose and forward. This action causes **surface creep**, which is the sliding and rolling of particles too large for saltation. Surface creep affects about 20% of the material being transported. Once in motion, particles can continue moving at lower wind velocities. In a desert or along a beach, you can hear the myriad saltating grains of sand produce a slight hissing sound, almost like steam escaping, as they bounce along and collide with surface particles.

Sand erosion and transport from a beach may be slowed by conservation measures such as the introduction of stabilizing native plants, the use of fences, and the restriction of pedestrian traffic to walkways (Figure 12.6).

(a)

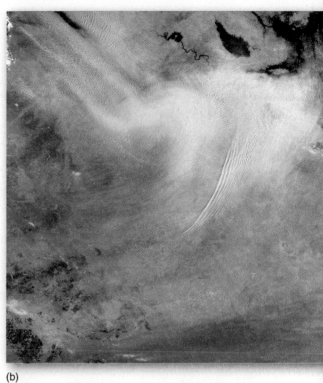

(b)

FIGURE 12.4 Dust storm in Iceland and Iraq.
(a) A dust storm is blowing off the north coast of Iceland over the Melrakkasletta Peninsula into the Greenland Sea, September 17, 2008. The source area is near the northern edge of the Vatnajökull Ice Cap and is a fine dust called loess that is produced as the glaciers grind the rock over which they flow. (b) A sandstorm moves across the Tigris and Euphrates rivers in Iraq on September 14, 2008. Dust is picked up as the storm moves along. Lake Tharthar in the image is about 120 km north of Baghdad. [Images courtesy of (a) *Terra* MODIS Rapid Response Team, GSFC/NASA; (b) *Envisat* MERIS sensor, European Space Agency.]

NEWS REPORT 12.1

The Dust Bowl

Deflation and wind transport of loess soils produced a catastrophe in the American Great Plains in the 1930s—the Dust Bowl. More than a century of overgrazing and intensive agriculture left soil susceptible to drought and eolian processes. The deflation of many centimeters of soil occurred in southern Nebraska, Kansas, Oklahoma, Texas, and eastern Colorado, and even in southern Canada and northern Mexico.

Fine sediments were lifted by winds, forming severe dust storms. The transported dust darkened the skies of Midwestern cities and drifted over farmland. Streetlights were left on throughout the day in Kansas City, St. Louis, and other Midwestern cities and towns. Such episodes can devastate economies, cause tremendous loss of topsoil, and even bury farmsteads. For more on the Dust Bowl, see http://www.pbs.org/wgbh/amex/dustbowl/. During the past decade Southeastern Australia experienced severe dust storms that included consequences similar to those of the American Dust Bowl.

Scientists are seeking answers as to why these events occur. Tree-ring analysis shows that intense droughts have happened about twice a century over the past 400 years. One such drought in the thirteenth century in what is now called Nebraska appears to have been as intense as that in the 1930s; however, it lasted 38 years. The Dust Bowl weather conditions seem to correlate with cool Pacific Ocean sea-surface temperatures (SST) and warm Atlantic Ocean SST. Chapter 4 discusses the Pacific Decadal Oscillation (PDO) and its relation to drought occurrences in the United States. Chapter 6 describes types of drought and the connections between drought and climate change.

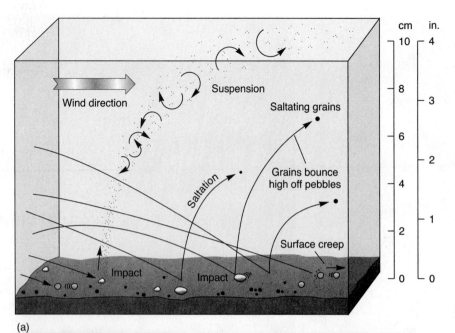

(a)

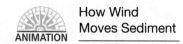

How Wind
Moves Sediment

(b)

FIGURE 12.5 Wind moves sand.
(a) Eolian suspension, saltation, and surface
creep are mechanisms of sediment transportation. (b) Sand grains saltating along the
surface in the Stovepipe Wells dune field,
Death Valley. [Photo by author.]

Eolian Depositional Landforms

The smallest features shaped by individual saltating grains
are *ripples* (Figure 12.7). Ripples form in crests and
troughs positioned transversely (at a right angle) to the
direction of the wind. The length of time particles are airborne influences their formation patterns. Eolian ripples
are different from fluvial ripples because the impact of
saltating grains is slight in water.

A common misconception is that sand covers most
deserts. Instead, desert pavements predominate across
most subtropical arid landscapes; sand covers only about
10% of desert areas. Transient ridges or hills of sand
grains form dunes. A **dune** is a wind-sculpted accumulation of sand. An extensive area of dunes, such as that
found in North Africa, is characteristic of an **erg desert**,
which means **sand sea**.

The Grand Erg Oriental in the central Sahara
exceeds 1200 m (4000 ft) in thickness and covers 192,000
km^2 (75,000 mi^2). This sand sea has been active for
more than 1.3 million years, with average dune heights of

120 m (400 ft). The Sahara Marzūq sand sea shown in
Figure 12.8a is wider than 300 km (185 mi). Similar sand
seas, such as the Grand Ar Rub'al Khālī Erg, are active in
Saudi Arabia and in the Taklimakan Desert. Eolian
processes are also at work on the Martian surface, forming
similar parallel ridges along the floor of Melas Chasma in
Figure 12.8b. The Sand Hills region of central Nebraska
represents a dune-covered region that was active in the
past (pictured in Figure 12.8c and noted on the map in
Figure 12.12).

Dune Movement and Form   Dune fields, whether
in arid regions or along coastlines, tend to migrate in
the direction of effective, sand-transporting winds. In
this regard, stronger seasonal winds or winds from a
passing storm may prove more effective than average
prevailing winds. When saltating sand grains encounter
small patches of sand, their kinetic energy (motion) is
dissipated and they accumulate. As height increases
above 30 cm (12 in.), a *slipface* and characteristic dune
features form.

FIGURE 12.7 Sand ripples.
Myriad sand ripple patterns may become lithified into fixed patterns in rock. The area in the photo is approximately 1 m (3.3 ft) wide. [Photo by author.]

FIGURE 12.6 Preventing sand transport.
(a) Further erosion and transport of coastal dunes can be controlled by stabilizing strategies, planting native plants, and confining pedestrian traffic to walkways, as at Popham Beach State Park, Maine. (b) Tufts of grass are planted to stabilize this Atlantic City, New Jersey, beach and a fence has been installed to prevent human foot traffic. [Photos by Bobbé Christopherson.]

By studying the dune in Figure 12.9 you can see that winds characteristically create a gently sloping *windward side* (stoss side) and a more steeply sloped slipface on the *leeward side*. A dune usually is asymmetrical in one or more directions. The angle of a slipface is the angle at which loose material is stable—its *angle of repose*. Thus, the constant flow of new material makes a slipface a type of *avalanche slope*. As sand moves over the crest of the dune to the brink, it builds up and avalanches as the slipface continually adjusts, seeking its angle of repose (usually 30°–34°). In this way, a dune migrates downwind with the effective wind, as suggested by the successive dune profiles in Figure 12.9.

Dunes that move actively are called *free dunes* and reflect most dynamically the interaction between fluid atmospheric winds and moving sand. However, because the sand is moving close to the ground, it may encounter an obstruction, such as stabilizing vegetation or a rock outcrop, resulting in a *tied dune*, or one that is fixed in place. The ever-changing form of these eolian deposits is part of their beauty, eloquently described by one author: "I see hills and hollows of sand like rising and falling waves. Now at midmorning, they appear paper white. At dawn they were fog gray. This evening they will be eggshell brown."* Their many wind-shaped styles make dune classification difficult. We can simplify dune forms into three classes—*crescentic*, *linear*, and *star* dunes (Figure 12.10).

Crescentic dunes are crescent-shaped ridges of sand that form in response to a fairly unidirectional wind pattern. The crescentic group is most common, with related forms including *barchan dunes* (limited sand), *transverse dunes* (abundant sand), *parabolic dunes* (vegetation controlled), and *barchanoid ridges* (rows of coalesced barchans).

Linear dunes generally form long parallel ridges separated by sheets of sand or bare ground. Linear dunes characteristically are much longer than they are wide; some exceed 100 km (60 mi) in length. Winds producing these dunes are principally bidirectional, so the slipface alternates from side to side. Related forms include *longitudinal dunes* and the *seif* (Arabic for "sword"), which is a sharper, narrower linear dune with a more sinuous crest.

*J. E. Bowers, *Seasons of the Wind* (Flagstaff, AZ: Northland Press, 1985), p. 1.

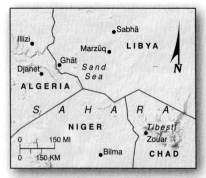

(a)

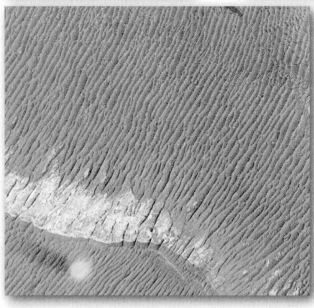

(b)

(c)

FIGURE 12.8 A sand sea and sand hills.
(a) The Sahara Marzūq, an erg desert that dominates southwestern Libya. Effective northwesterly winds shape the pattern and direction of the transverse and barchanoid (series of connected barchans) dunes. This sand sea exceeds 300 km (185 mi) across. (b) A similar pattern of dunes appears on the Martian surface in the southern area of Melas Chasma in Valles Marineris. Note the dust devil at the lower left. The area covered is about 2 km (1.2 mi) wide. (c) Although not active for at least 600 years, these densely packed barchan dunes, fixed (stabilized) by vegetation, represent a large region of sand dunes. Sand Hills, central Nebraska, sit atop the High Plains Aquifer (see Figure 9.1.1), where sand and silt deposits derived from glaciated regions to the north and the Rockies to the west accumulated. [(a) *Terra* image courtesy of MODIS Land Rapid Response Team, NASA/GSFC, November 9, 2001; (b) *Mars Global Surveyor*, Mars Orbiter Camera image courtesy of NASA/JPL/Malin Space Science Systems, July 11, 1999; (c) photo by Bobbé Christopherson.]

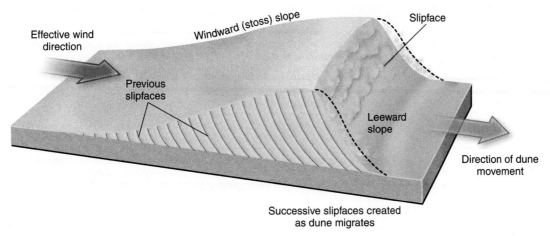

ANIMATION
Formation of
Cross Bedding

FIGURE 12.9 Dune cross section. Cross section of a dune, showing the pattern of successive slipfaces as the dune migrates in the direction of the effective wind.

Star dunes are the mountainous giants of the sandy desert. They form in response to complicated, changing wind patterns and have multiple slipfaces. They are pinwheel-shaped, with several radiating arms rising and joining to form a common central peak. The best examples of star dunes are in the Sahara and the Namib Desert, where they approach 200 m (650 ft) in height (Figure 12.11, p. 412).

These same dune-forming principles and terms (for example, dune, barchan, slipface) apply to snow-covered landscapes. *Snow dunes* form as wind deposits snow in drifts. In semiarid farming areas, fences and tall stubble are left in the fields to capture drifting snow, which contributes significantly to soil moisture on melting.

Figure 12.12a correlates active sand regions with deserts (tropical, continental interior, and coastal). Active sand dunes cover only about 10% of all continental land between 30° N and 30° S. Also noted on the map are dune fields in humid climates: along coastal Oregon, the south shore of Lake Michigan (Figure 12.12b), along the Gulf and Atlantic coastlines, in Europe, and elsewhere. The remarkable coastal desert sands of Namibia are pictured in Figure 12.12c.

Loess Deposits

Approximately 15,000 years ago, in several episodes, Pleistocene glaciers retreated in many parts of the world, leaving behind large glacial outwash deposits of fine-grained clays and silts (<0.06 mm, or 0.0023 in.). These materials were blown great distances by the wind and redeposited in unstratified, homogeneous deposits, named **loess** by peasants working along the Rhine River Valley in Germany.

No specific landforms were created; instead, loess covered existing landforms with a thick blanket of material that assumed the general topography of the existing landscape. Because of its own binding strength, loess weathers and erodes into steep bluffs, or vertical faces. At Xi'an, Shaanxi, in China, a loess wall was excavated for dwelling space (Figure 12.13a, p. 413). When a bank is cut into a loess deposit, it generally will stand vertically, although it can fail if saturated (Figure 12.13b).

Figure 12.14 shows the worldwide distribution of loess deposits. Significant accumulations of loess throughout the Mississippi and Missouri valleys form continuous deposits 15–30 m (50–100 ft) thick. Loess deposits also occur in eastern Washington state and Idaho. The hills and gullies in the Loess Hills of Iowa in Figure 12.14b demonstrate the erosion potential of these soils. The soils in these regions are fertile, for loess deposits are well drained, deep, and have excellent moisture retention. Loess deposits also cover much of Ukraine, central Europe, China, the Pampas-Patagonia regions of Argentina, and lowland New Zealand. These soils derived from loess are some of Earth's "bread basket" farming regions. Transport of loess soils occurred in the catastrophic Dust Bowl in the 1930s in the United States (see News Report 12.1).

In Europe and North America, loess originates mainly from glacial sources. The vast deposits of loess in China, covering more than 300,000 km² (116,000 mi²), are thought to have come from windblown desert sediment rather than glacial sources. Accumulations in the Loess Plateau of China exceed 300 m (1000 ft) in thickness, forming some complex weathered badlands and some good agricultural land. These windblown deposits are interwoven with much of Chinese history. New research has found that plumes of windblown dust from African deserts have moved across the Atlantic Ocean to enrich soils of the Amazon rain forest of South America, the southeastern United States, and the Caribbean islands.

Overview of Desert Landscapes

Arid and semiarid dry climates cover as much as 35% of all land, constituting the largest single climatic region on Earth. See Figures 7.4 and 7.5 for the location of these *arid deserts* and *semiarid steppe* climate regions and the world biome map in Chapter 16 for the distribution of these desert environments. And Earth's deserts are expanding, as we discuss shortly. Now let us look at the link between climate and Earth's deserts.

Class	Type	Description
Crescentic	Barchan	Crescent-shaped dune with horns pointed downwind. Winds are constant with little directional variability. Limited sand available. Only one slipface. Can be scattered over bare rock or desert pavement or commonly in dune fields.
	Transverse	Asymmetrical ridge, transverse to wind direction (right angle). Only one slipface. Results from relatively ineffective wind and abundant sand supply.
	Parabolic	Role of anchoring vegetation important. Open end faces upwind with U-shaped "blowout" and arms anchored by vegetation. Multiple slipfaces, partially stabilized.
	Barchanoid ridge	A wavy, asymmetrical dune ridge aligned transverse to effective winds. Formed from coalesced barchans; look like connected crescents in rows with open areas between them.

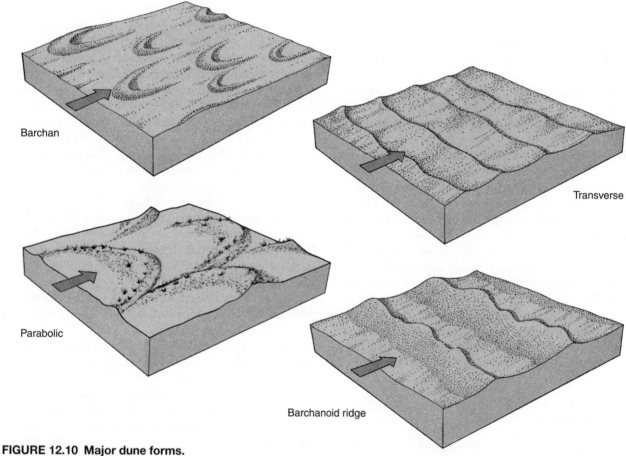

FIGURE 12.10 Major dune forms.
Three major dune classes include crescentic, linear, and star dune. Arrows show wind direction.
[Adapted from E. D. McKee, *A Study of Global Sand Seas*, US Geological Survey Professional Paper 1052 (Washington, DC: 1979).]

Desert Climates

Physical conditions explain the spatial distribution of these dry lands: subtropical high-pressure cells between 15° and 35° N and S (see Figures 4.10 and 4.12), rain shadows on the lee side of mountain ranges (see Figure 5.30), and areas at great distance from moisture-bearing air masses, such as central Asia. Figure 12.15 (p. 414) portrays this distribution according to the climate classification used in this text, with photographs and an image of four major desert regions. These areas pos-

sess unique landscapes created by the interaction of intermittent precipitation events, weathering processes, and wind. Rugged, hard-edged desert landscapes of cliffs and slopes contrast sharply with the vegetation-covered, rounded and smoothed slopes characteristic of humid regions.

Desert environments experience high sensible heat conditions and intense ground heating. Such areas receive a high input of insolation through generally clear skies and experience high radiative heat losses at night. A typical desert water balance shows high potential evapotranspiration (moisture

Class	Type	Description
Linear	Longitudinal	Long, slightly sinuous, ridge-shaped dune, aligned parallel with the wind direction; two slipfaces. Average 100 m high and 100 km long, and the "draa" at the extreme in size can reach to 400 m high. Results from strong effective winds varying in one direction.
	Seif	Arabic word for "sword"; a more sinuous crest and shorter than longitudinal dunes. Rounded toward upwind direction and pointed downwind. (Not illustrated.)
Star dune		The giant of dunes. Pyramidal or star-shaped with three or more sinuous, radiating arms extending outward from a central peak. Slipfaces in multiple directions. Results from effective winds shifting in all directions. Tend to form isolated mounds in high effective winds and connected sinuous arms in low effective winds.
Other	Dome	Circular or elliptical mound with no slipface. Can be modified into barchanoid forms; vegetation can play role in coastal locales.
	Reversing	Asymmetrical ridge form intermediate between star dune and transverse dune. Wind variability can alter shape between forms.

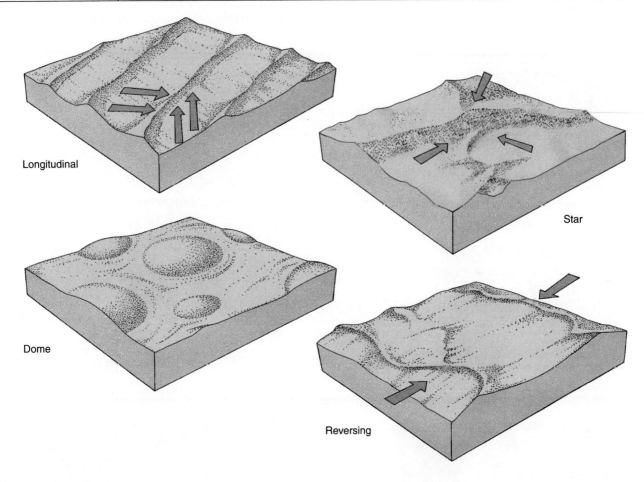

Longitudinal

Star

Dome

Reversing

demand), low precipitation (moisture supply), and prolonged seasonal moisture deficits (see, for example, Figure 6.11 for Phoenix, Arizona). Fluvial processes in the desert feature intermittent running water, with hard, poorly vegetated desert pavement yielding high runoff during rainstorms—the infamous flash flood.

Desert Fluvial Processes

Precipitation events in a desert may be rare—indeed, a year or two apart—but when they do occur, a dry streambed fills with a **flash flood** torrent. Such channels

may fill in a few minutes and surge briefly during and after a storm. Depending on the region, such a dry streambed is known as a **wash**, an *arroyo* (Spanish), or a *wadi* (Arabic). A desert highway that crosses a wash usually is posted to warn drivers not to proceed if rain is in the vicinity, for a flash flood can suddenly sweep away anything in its path.

When washes fill with surging flash floodwaters, a unique set of ecological relationships quickly develops. Crashing rocks and boulders break open seeds. The seeds respond to the timely moisture and germinate. Other plants and animals also spring into brief life cycles as the

FIGURE 12.11 Mountains of the desert.
Star dune in the Namib Desert of Namibia in southwestern
Africa. [Photo by Comstock, Inc.]

water irrigates their limited habitats. Please see the desert comparison photos in Focus Study 7.1 for a look at the great contrasts fostered by moisture.

At times of intense rainfall, remarkable scenes fill the desert. Figure 12.16 shows two photographs made just 1 month apart in a sand dune field in Death Valley, California. The rainfall event that occurred produced 2.57 cm (1.01 in.) of precipitation in 1 day, in a place that receives only 4.6 cm (1.83 in.) in an average year. The runoff flowed for hours and then collected in low spots on the hard, underlying clay surfaces. The water was quickly consumed by the high evaporation demand, so that in just a month these short-lived watercourses were dry and covered with accumulations of alluvial materials.

The 1985 precipitation event just described was exceeded by another one on August 15, 2004, when several

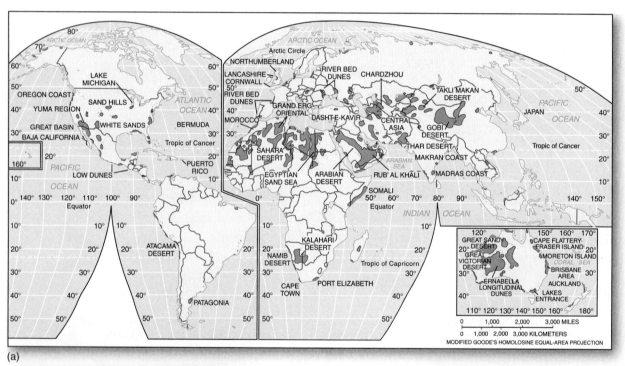

(a)

(b)

(c)

FIGURE 12.12 Sandy regions of the world.
(a) Worldwide distribution of active and stable sand regions. (b) Sand dunes along the shore of Lake Michigan in Indiana Dunes State Park in Indiana. (c) Sandy area in the Namib Desert, Namibia, Africa.
[(a) After R. E. Snead, *Atlas of World Physical Features*, p. 134, © 1972 by John Wiley & Sons. Adapted by permission of John Wiley & Sons, Inc. Photos by (b) Bobbé Christopherson; (c) Nigel J. Dennis/Photo Researchers, Inc.]

(a)

(b)

FIGURE 12.13 Example of loess deposits.
(a) A loess formation in Xi'an, Shaanxi, China, has a strong enough structure to be excavated for dwelling rooms. (b) A loess bluff in western Iowa. [Photos by (a) Betty Crowell; (b) Bobbé Christopherson.]

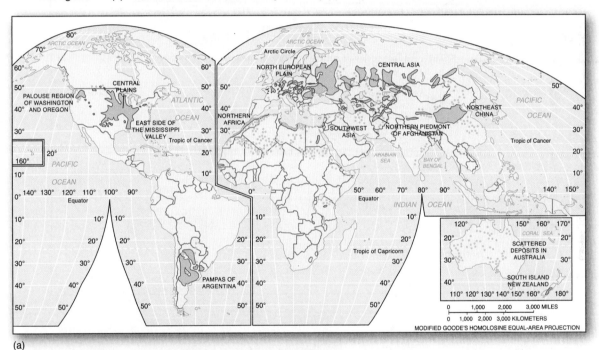
(a)

(b)

FIGURE 12.14 Worldwide loess deposits.
(a) Dots represent small, scattered loess formations. (b) The Loess Hills of western Iowa date from 159,000 to 12,500 years ago. The loess deposits reach heights of about 61 m (200 ft) above the nearby prairie farmlands and range north and south more than 322 km (200 mi). Only China has deposits that exceed these dimensions. [(a) After R. E. Snead, *Atlas of World Physical Features*, p. 138, copyright © 1972 by John Wiley & Sons. Adapted by permission of John Wiley & Sons, Inc.; (b) photos by Bobbé Christopherson.]

413

(a) Sonoran

(b) Taklimakan

(c) Atacama

(d) Kalahari

FIGURE 12.15 The world's dry regions.
Worldwide distribution of arid lands (*arid desert* climates) and semiarid lands (*semiarid steppe* climates). (a) Sonoran Desert in the American Southwest. (b) Taklimakan Desert in central Asia. (c) Atacama Desert in subtropical Chile near Baquedano. (d) Kalahari Desert in south-central Africa. [Photos by (a) author; (b) October 27, 2001, *Terra* image courtesy of MODIS Land Rapid Response Team, NASA/GSFC; (c) Jacques Jangoux/Photo Researchers, Inc.; and (d) Nigel J. Dennis/Photo Researchers, Inc.]

(a)   (b)

FIGURE 12.16 An improbable river in Death Valley.
The Stovepipe Wells dune field of Death Valley, California: (a) the day after a 2.57-cm (1.01-in.)
rainfall and (b) 1 month later (identical location). [Photos by author.]

thunderstorm cells stalled over the mountains along the
east side of Death Valley. Rainfall totals reached 6.35 cm
(2.5 in.) in little more than an hour. Cars were swept away
from the Furnace Creek Inn parking lot, about 4.8 km
(3 mi) of Highways 190 and 178 washed out, and there
were two fatalities in the Zabriskie Point area. One annex
building had a 3-m- (10-ft-) high water mark on it. Again,
water is sparse in the deserts, yet it is the major erosional
force—sometimes dramatically so. For recent information
and maps of Death Valley, see **http://www.nps.gov/deva/**.

As runoff water evaporates, salt crusts may be left
behind on the desert floor. An intermittently wet and dry
low area in a region of closed drainage is a **playa**, the site of
an *ephemeral lake* when water is present. Accompanying our
earlier discussion of evaporites, Figure 8.10 shows such a
playa in Death Valley, covered with salt precipitate just
1 month after the record rainfall event mentioned here.

Permanent lakes and continuously flowing rivers are
uncommon features in the desert, although the Nile River
and the Colorado River are notable exceptions. Both
these rivers are *exotic streams*, with their headwaters in a
wetter region and the majority of their channels in a
desert region. Focus Study 12.1 describes the Colorado
River and the problem of its overuse.

Alluvial Fans   In arid climates, a prominent landform is an
alluvial fan, which occurs at the mouth of a canyon where it
exits into a valley. Flowing water produces the fan as the
water loses velocity when it leaves the constricted channel of
the canyon. It drops layer upon layer of sediment along the
base of the mountain block. Water then flows over the surface of the fan and produces a braided drainage pattern,
shifting from channel to channel with each precipitation
event (Figure 12.17). A continuous apron, or **bajada** (Spanish for "slope"), may form if individual alluvial fans coalesce
into one sloping surface (see Figure 12.23). Fan formation
of any sort is reduced in humid climates because perennial
streams constantly carry away much of the alluvium.

An interesting aspect of an alluvial fan is the natural
sorting of materials by size. Near the mouth of the canyon
at the apex of the fan, coarser materials are deposited, grading slowly to pebbles and finer gravels with distance out
from the mouth. Next, deposits of sands and silts drop, with
the finest clays and salts carried in suspension and solution
all the way to the valley floor. Minerals dissolved in solution
accumulate there as evaporite deposits. Figure 12.18 shows
the edge of a playa and active dune field, so there is a natural
sorting of materials by grain size, with sands dropped in the
dunes, silts and clays moving farther toward the playa, and
the finest and dissolved materials in the playa.

Well-developed alluvial fans also can be a major source of
groundwater. Some cities—San Bernardino, California, for
example—are built on alluvial fans and extract their municipal water supplies from them. However, because water
resources in an alluvial fan are recharged from surface supplies and these fans are in arid regions, groundwater mining
and overdraft beyond recharge rates are common. In other
parts of the world, such water-bearing alluvial fans are known
as *qanat* (Iran), *karex* (Pakistan), and *foggara* (Western Sahara).

Desert Landscapes

Contrary to popular belief, deserts are not wastelands, for
they abound in specially adapted plants and animals.
Moreover, the limited vegetation, intermittent rainfall,
intense insolation, and distant vistas produce starkly beautiful landscapes. And all deserts are not the same: For
example, North American deserts have more vegetation
cover than do the generally barren Saharan expanses.

Deserts occur worldwide as topographic plains, such as
the Great Sandy and Simpson Deserts of Australia, the Arabian and Kalahari Deserts of Africa, and portions of the
extensive Taklimakan Desert in the central Tarim Basin of
China. Also, deserts are found in mountainous regions: interior Asia, from Iran to Pakistan, and in China and Mongolia. In South America, lying between the ocean and the
Andes, is the rugged Atacama Desert.

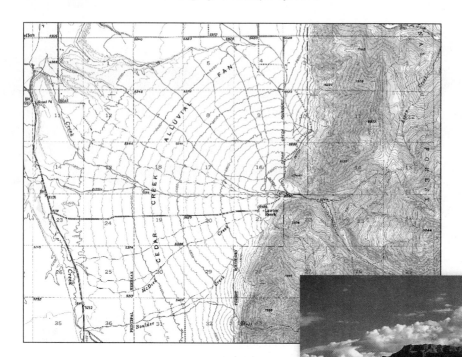

FIGURE 12.17 An alluvial fan.
The photo shows an alluvial fan in a desert landscape.
A topographic map shows the Cedar Creek alluvial
fan. (Topographic map is the Ennis Quadrangle,
15-minute series, scale 1:62,500, contour
interval = 40 ft; latitude/longitude coordinates for
mouth of canyon are 45°2' N 111°35' W.) [Photo by
Bobbé Christopherson; USGS map.]

**FIGURE 12.18 Sand and playa—natural sorting
of materials.**
Playa of dissolved and finer materials (top of photo); sand
dunes grading to coarser particles in the lower area along the
mountains demonstrate the natural sorting that occurs in desert
landscapes. [Photo by Bobbé Christopherson.]

The shimmering heat waves and related mirage
effects in the desert are products of light refraction
through layers of air that develop a temperature gradient
near the hot ground. The book *Desert Solitaire* captures
the desert's enchantment:

> Around noon the heat waves begin flowing upward
> from the expanses of sand and bare rock. They shim-
> mer like transparent, filmy veils between my sanctu-
> ary in the shade and all the sun-dazzled world
> beyond. Objects and forms viewed through this
> tremulous flow appear somewhat displaced or dis-
> torted . . . the great Balanced Rock floats a few inches
> above its pedestal, supported by a layer of superheated
> air. The buttes, pinnacles, and fins in the windows area
> bend and undulate beyond the middle ground like a
> painted backdrop stirred by a draft of air.*

Resistant horizontal rock strata in buttes, pinnacles,
and mesas of arid landscapes exhibit differential weathering. Removal of the less-resistant rock strata produces

*E. Abbey, *Desert Solitaire* (New York: McGraw-Hill, 1968), p. 154.
Copyright © 1968 by Edward Abbey.

(a)

(b)

FIGURE 12.19 A balanced rock—differential weathering.
(a) Balanced Rock in Arches National Park, Utah, where writer/naturalist Edward Abbey (quoted in the text) worked as a ranger years before it became a park. The overall feature is 39 m (128 ft) tall and composed of Entrada sandstone. The balance-rock portion is 17 m (55 ft) tall and weighs 3255 metric (3577 short) tons. (b) Spider Rock towers 244 m (800 ft) above the canyon floor of Canyon de Chelly, Arizona—a dramatic monolith; note the lighter color in the capstone rock. [Photos by author.]

unusual desert sculptures—arches, pedestals, towers, and delicately balanced rocks (Figure 12.19). Specifically, the upper layers of sandstone along the top of an arch or butte are more resistant to weathering and protect the sandstone rock beneath (another example is in Figure 10.1).

Desert landscapes are places where stark erosional remnants stand above the surrounding terrain as knobs, hills, or "island mountains." Such a bare, exposed rock, called an *inselberg*, or island mountain, is exemplified by Uluru (Ayers) Rock in Australia (Figure 12.20).

(text continued on page 422)

FIGURE 12.20 Australian landmark.
Uluru (Ayers) Rock in Northern Territory, Australia, is an isolated mass of weathered rock. The formation is 348 m (1145 ft) high, 2.5 km long, and 1.6 km wide (1.5 by 1.0 mi). Sacred to Aboriginal peoples, this natural monument in the desert is protected in Uluru National Park, established in 1950. [Photo by Porterfield/Chickering.]

FOCUS STUDY 12.1

The Colorado River: A System Out of Balance

An exotic stream has headwaters in a humid region of water surpluses, then flows mostly through arid lands for the rest of its journey to the sea. Exotic streams have few incoming tributaries. Consequently, an exotic stream has a discharge pattern that is opposite that of a typical stream: Instead of discharge increasing downstream, it decreases. The Nile and Colorado Rivers are prominent examples. These natural conditions are further affected by the human impacts of water withdrawals and reservoir construction with high evaporation rates.

In the case of the Nile, the East African mountains and plateaus provide a humid source area. The Nile first rises in remote headwaters as the Kagera River in the eastern portion of the Lake Plateau country of East Africa. On its way to Lake Victoria, it forms the partial boundary of Tanzania, Rwanda, and Uganda. The Nile itself then flows from the lake and continues on its 6650-km (4132-mi) course to its mouth on the Mediterranean Sea near Cairo.

The Colorado River Basin

The Colorado rises in Rocky Mountain National Park, Colorado (Figure 12.1.1a), and flows almost 2317 km (1440 mi). The river ends as a trickle of water that disappears in the sand, kilometers short of its former mouth in the Gulf of California.

Orographic precipitation totaling 102 cm (40 in.) per year (mostly snow) falls in the Rockies, feeding the Colorado headwaters. But at Yuma, Arizona, near the river's end, annual precipitation is a scant 8.9 cm (3.5 in.), an extremely small amount when compared with the annual potential evapotranspiration demand in the Yuma region of 140 cm (55 in.).

From its source region, the Colorado River quickly leaves the humid Rockies and spills out into the arid desert of western Colorado and eastern Utah. At Grand Junction, Colorado, near the Utah border, annual precipitation is only 20 cm (8 in.; Figure 12.1.1b). After carving its way through the intricate labyrinth of canyonlands in Utah, the river enters Lake Powell, behind Glen Canyon Dam (Figure 12.1.1c). The Colorado then flows through the Grand Canyon chasm, formed by its own erosive power.

West of the Grand Canyon, the river turns southward, tracing its final 644 km (400 mi) as the Arizona–California border. Along this stretch of the river we find: Hoover Dam, just east of Las Vegas (Figure 12.1.1d); Davis Dam (Figure 12.1.1e), built to control the releases from Hoover; Parker Dam for the water needs of Los Angeles; three more dams for irrigation water (Figure 12.1.1f)— Palo Verde, Imperial, and Laguna— to bring water to irrigated fields; and, finally, Morelos Dam at the Mexican border (Figure 12.1.1g). Mexico owns the end of the river and whatever water is left, although the river no longer reaches its mouth in the Gulf of California (Figure 12.1.1i). Note on the image the agricultural landscape differences separated by the California–Mexico border.

Figure 12.1.1j shows the annual water discharge and suspended sediment load for the Colorado River at Yuma, Arizona, from 1905 to 1964. The completion of Hoover Dam in the 1930s dramatically reduced suspended sediment. The addition of Glen Canyon Dam upstream from Hoover Dam in 1963 further reduced stream flows. In fact, Lake Powell, which formed upstream behind the artificial base level of Glen Canyon Dam, is forecast to fill with sediment if it remains operational this century.

Overall, the drainage basin encompasses 641,025 km^2 (247,500 mi^2) of mountain, basin-and-range, plateau, canyon, and desert landscapes in parts of seven states and two countries. A discussion of the Colorado River is included in this chapter because of its crucial part in the history of the Southwest and its role in the future of this drought-plagued arid and semiarid region.

Dividing Up the Colorado's Dammed Water

John Wesley Powell (1834–1902), the first person on record to successfully navigate the Colorado River through the Grand Canyon, was the first director of the US Bureau of Ethnology and later director of the US Geological Survey (1881–1892). Powell perceived that the challenge of the West was too great for individual efforts and believed that solutions to

FIGURE 12.1.1 The Colorado River Basin.
The Colorado River Basin, showing division of the upper and lower basins near Lees Ferry in northern Arizona. (a) Headwaters of the Colorado River near Mount Richthofen in the Colorado Rockies. (b) The river near Moab, Utah. (c) Glen Canyon Dam, a regulatory, administrative facility near Lees Ferry, Arizona. (d) Hoover Dam spillways in rare operation during 1983 floods. (e) Davis Dam in full release during flood. (f) Irrigated fresh-cut flowers in the Imperial Valley; refrigerated trucks and a cargo jet at a local airport await the shipment. (g) Morelos Dam at the Mexican border is the final stop as the river dwindles to a mere canal. (h) Central Arizona Project aqueduct west of Phoenix. (i) The Colorado River stops short of its former delta in the Gulf of California. Note the ecological contrast between the California and Mexico sides of the border. (j) Dam construction affects river discharge and sediment yields. [Photos by (a) and (f) Bobbé Christopherson; (b, c, d, e, g) author; (h) Tom Bean/DRK Photo; (i) August 25, 2002, *Terra* image courtesy of MODIS Rapid Response Team, NASA/GSFC; and (j) data from USGS, 1985, *National Water Summary 1984*, Water Supply Paper 2275 (Washington, DC: Government Printing Office, p. 55.]

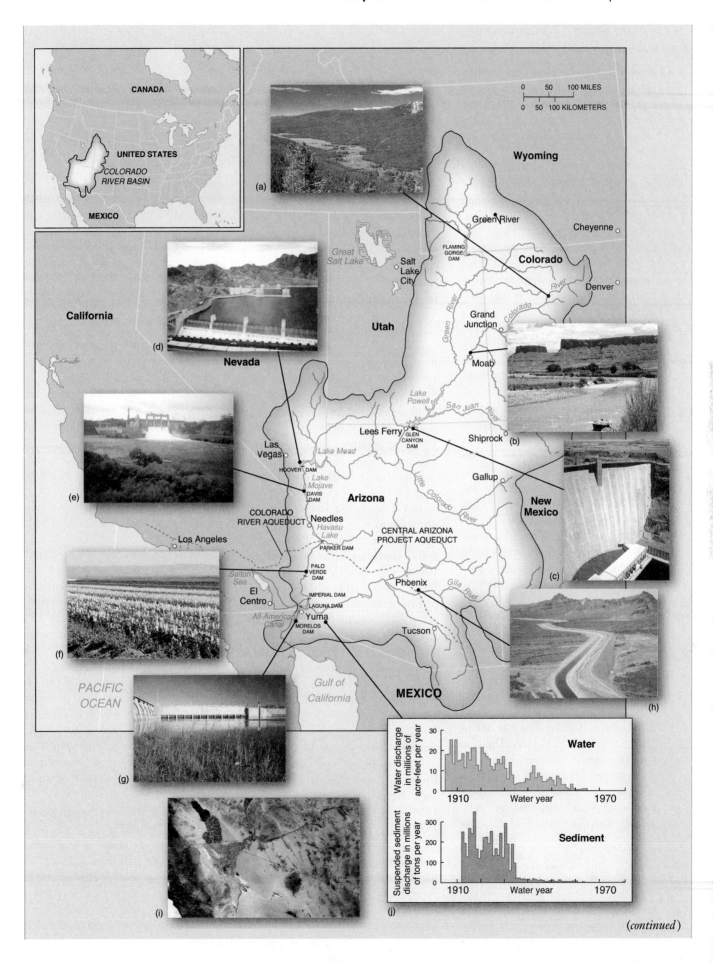

CANADA

UNITED STATES

COLORADO
RIVER BASIN

MEXICO

0   50   100 MILES

0   50   100 KILOMETERS

Wyoming

(a)

Green River

Cheyenne

FLAMING
GORGE
DAM

Colorado

Great
Salt Lake

Salt
Lake
City

River

Denver

Grand
Junction

Colorado

California

Utah

Moab

(b)

Lake
Powell

San Juan   River

Nevada

(d)

Lees Ferry

GLEN
CANYON
DAM

Shiprock

Las
Vegas

Lake Mead

Gallup

(c)

(e)

Lake
Mojave

DAVIS
DAM

HOOVER DAM

Lake
Mead

Arizona

Little Colorado

River

New
Mexico

COLORADO
RIVER AQUEDUCT

Needles

Havasu
Lake

CENTRAL ARIZONA
PROJECT AQUEDUCT

Los Angeles

PARKER DAM

(f)

Salton
Sea

PALO
VERDE
DAM

IMPERIAL DAM

El
Centro

LAGUNA DAM

All-American
Canal

Yuma

MORELOS
DAM

Phoenix

Gila River

Tucson

(h)

PACIFIC
OCEAN

Gulf of
California

MEXICO

(g)

(i)

Water discharge
in millions of
acre-feet per year

30

20

10

Water

1910   Water year   1970

Suspended sediment
discharge in millions
of tons per year

300

200

100

Sediment

1910   Water year   1970

(j)

(continued)

Focus Study 12.1 *(continued)*

problems such as water availability could be met only through private cooperative efforts. His 1878 study (reprinted 1962), *Report of the Lands of the Arid Region of the United States*, is a conservation landmark.

Today, Powell probably would be skeptical of the intervention by government agencies in building large-scale reclamation projects. Lake Powell is named after him, despite his probable opposition were he alive. An anecdote in Wallace Stegner's book *Beyond the Hundredth Meridian*[*] relates that, at an 1893 international irrigation conference held in Los Angeles, Powell observed development-minded delegates bragging that the entire West could be conquered and reclaimed from nature, that "rain will certainly follow the plow." Powell spoke against that sentiment: "I tell you, gentlemen, you are piling up a heritage of conflict and litigation over water rights, for there is not sufficient water to supply the land."[*] He was booed from the hall. But history has shown Powell to be correct.

The Colorado River Compact was signed by six of the seven basin states in 1923. (The seventh, Arizona, signed in 1944, the same year the Mexican Water Treaty was signed.) With this compact, the Colorado River basin was divided into an upper basin and a lower basin, arbitrarily separated for administrative purposes at Lees Ferry near the Utah–Arizona border (noted on the map in Figure 12.1.1). Congress adopted the Boulder Canyon Act in 1928, authorizing Hoover Dam as the first major reclamation project on the river. Also authorized was the All-American Canal into the Imperial Valley, which required an additional dam (Figure 12.1.2). Los Angeles then began its project to bring Colorado River water 390 km

(240 mi) from still another dam and reservoir to the city.

Shortly after Hoover Dam was finished and downstream enterprises were thus offered flood protection, the other projects were quickly completed. There are now eight major dams on the river and many irrigation works. The latest effort to redistribute Colorado River water is the Central Arizona Project, which carries water to the Phoenix area (Figure 12.1.1h).

Highly Variable River Flows

The flaw in all this planning and water distribution is that exotic streamflows are highly variable, and the Colorado is no exception. In 1917, the discharge measured at Lees Ferry totaled 24 million acre-feet (maf), whereas in 1934 it dropped by nearly 80%, to only 5.03 maf. In 1977, the discharge dropped again to 5.02 maf, but in 1984 it rose to an all-time high of 24.5 maf. Yet, in contrast, from 2000 to 2007 river discharge fell to exceptionally low levels, with 2002 dropping to a record low of 3.1 maf, and 2003 to 6.4 maf. The 2002 to 2007 water years marked the lowest 6-year average flow in the total record. Please revisit the chapter-opening pair of photos for Chapter 6 showing the comparison of Lake Mead water levels at Hoover Dam in July 1983 and September 2007 as evidence of this variability. By August 2008, lake levels dropped more to 336.8 m (1105 ft), which is 38.2 m (125.3 ft) below the 375-m capacity.

The average flows between 1906 and 1930 were almost 18 maf a year. As a planning basis for the Colorado River Compact, the government used average river discharges from 1914 up to the Compact signing in 1923, an exceptionally high average of 18.8 maf. That amount was perceived as more than enough for the upper and lower basins, each to receive 7.5 maf, and, later, for Mexico to receive 1.5 maf in the 1944 Mexican Water Treaty.

FIGURE 12.1.2 Colorado water in a canal.
All-American canal brings water from the Colorado River to the irrigated lands of the Salton Trough: the Imperial and Coachella valleys. Only a small portion of this canal is lined, losing water to seepage in transit. [Photo by Bobbé Christopherson.]

We might question whether proper long-range planning should rely on the providence of high variability. Tree-ring analyses of past climates have disclosed that the only other time Colorado discharges were at the high 1914–1923 level was between A.D. 1606 and 1625. The dependable flows of the river have been consistently overestimated. This shortfall problem is shown in an estimated budget of 20 million acre-feet for the river (Table 12.1.1): Clearly, the situation is out of balance, for there is not enough discharge to meet budgeted demands.

Presently, the seven states ideally want rights that total as high as 25.0 maf. When added to the guarantee for Mexico, this comes up to 26.5 maf of water wants. And six states share one opinion: California's right to the water must be limited to a court-ordered 4.4 maf, which it exceeded until California's access to surplus Colorado River water ceased in 2002. At present, there is no river surplus during the ongoing western drought.

Water Loss at Glen Canyon Dam and Lake Powell

Glen Canyon Dam began water impoundment (Lake Powell) in 1963, 27.4 km (17 mi) north of the basin division point at Lees Ferry. The advancing water flooded the deep, fluted, inner gorges of many canyons, including

Table 12.1.1 Estimated Colorado River Budget through 2007

Water Demand	Quantity (maf)[a]
Upper Basin (7.5), Lower Basin (7.5)[b]	15.0
Central Arizona Project (rising to 2.8 maf)	1.0
Mexican allotment (1944 Treaty)	1.5
Evaporation from reservoirs	1.5
Bank storage at Lake Powell	0.5
Phreatophytic losses (water-demanding plants)	0.5
Budgeted total demand	**20.0 maf**

Average Flows at Lees Ferry	
1906–1930	17.70 maf
1930–2003 average flow of the river	14.10 maf
1990–2006	11.34 maf
2000	11.06 maf
2001	10.75 maf
2002	3.1 maf
2003	6.4 maf
2004	5.6 maf
2005	8.3 maf
2006	8.4 maf
2007	8.0 maf

A water year runs from October 1 to September 30.
[a]1 million acre-feet = 325,872 gallons; 1.24 million liters; 43,560 ft³.
[b]In-basin consumptive uses 75% agricultural.
Source: Bureau of Reclamation states and states of Arizona, California, Nevada.

the Glen, Navajo, Labyrinth, and Cathedral. Glen Canyon Dam's primary purpose, according to the Bureau of Reclamation, is to regulate flows between the upper and lower basins. An additional benefit is the production of hydroelectric power, which is sold at a low wholesale rate to utilities across the Southwest. Also, there is a growing recreation and tourism industry around Lake Powell, as previously inaccessible desert scenery now is reachable by boat.

Many thought that Lake Mead, behind Hoover Dam, could serve the primary administrative function of flow regulation, especially considering the serious water-loss problems with the Glen Canyon Dam. Porous Navajo sandstone underlies most of the Lake Powell reservoir at Glen Canyon Dam. This sandstone absorbs an estimated 0.5 maf of the river's overall annual discharge as bank storage. The higher the lake level rises, the greater the loss into the sandstone.

Secondly, Lake Powell is an open body of water in an arid desert, where hot, dry winds accelerate evaporative losses. Another 0.5 maf of the Colorado's overall discharge is lost annually from the Lake Powell reservoir in this manner; again, the higher the lake level the greater the evaporative loss. Thirdly, now-permanent sandbars and banks stabilized along the regulated river, allowing water-demanding plants called phreatophytes to establish and extract an additional 0.5 maf of the river flow. Therefore, if Lake Powell ever returns to capacity, some 1.5 maf, or 13.2% of the overall river discharge (1990–2006 average) or 24% of average overall discharge of the low flows since 2002, would be lost from the system budget.

The Western Drought and Beyond

The severity of the western drought from 1999 to 2008, in its ninth year, surpasses anything in the historical

record and is approaching the driest in the record according to tree-ring analyses, despite a slightly wetter 2005 water year. Satellites can spot the symptoms of this western drought: reduced snow pack in the Rockies, reduced soil moisture, the drought-stressed conditions of vegetation, the prevalence of wildfires, and lowered reservoir levels.

Since A.D. 1226, nine droughts lasting 10 to 20 years and four lasting more than 20 years have occurred. Whether the current drought will be of this duration or be more like the three droughts of the twentieth century that ran 4 to 11 years is unknown. However, this is the first drought to occur in the Colorado River system in the presence of a human demand for water that is increasing at such record levels.

Water inflow to Lake Powell is running less than one-third of the 30-year average (1961–1990)—with 2002 at just 25%. System reservoirs are at record lows. The production of hydroelectric power has fallen to 50% of capacity from these dams. The system is in an official "drought condition." There is speculation that Glen Canyon Dam might cease operations before the end of the decade because of these low-water conditions, leading some citizen groups to suggest decommissioning, possible removal, and restoration of the canyons to predevelopment status.

The worst appears yet to come. As of 2007, the total system stands at less than 50% of capacity and engineers estimate that it will take 13 normal winters of snowpack in the Rockies and basin-wide precipitation to "reset" the system. This recovery is much slower than previous drought recoveries because of the rapid increase in water demand across the Southwest. Arizona's population doubled between 1990 and 2007, reaching 6.2 million. Las Vegas went from 368,000 in 1985 to 1.8 million by 2007 (490% increase). Present probabilities for the period after 2016 put the chances of surplus

(continued)

Focus Study 12.1 *(continued)*

discharge occurring in any given year at only 20%. Imagine what the population in the region will be after 2016 if no steps are taken to control growth.

Chapter 6 describes drought and the changing conditions of an expanding subtropical high-pressure dry zone with climate change. A portion of a quote from Chapter 6 explains the severity of the drought: "The projected future climate of intensified aridity in the Southwest is caused by . . . a poleward expansion of the subtropical dry zones. The drying of subtropical land areas . . . imminent or already underway is unlike any climate state we have seen in the instrumental record. . . . "*

The Scripps Institution of Oceanography forecasted in a February 2008 report a 50% probability that Lake Mead, on the Colorado River behind Hoover Dam, could be completely dry by 2021 and without

*Richard Seager et al., "Model projections of an imminent transition to a more arid climate in southwestern North America," *Science* 316, no. 5828 (May 25, 2007): 1184.

enough water to turn turbines by 2017; see http://scrippsnews.ucsd.edu/Releases/?releaseID=876. Such is the state of this drought.

Government stopgap measures came to a head in December 2007 when the Department of Interior signed a historic "Record of Decision," which was a form of consensus among the stakeholders (see the URL listed in the "NetWork and Critical Thinking Tools"). The "Decision" agreement sets the rules for dealing with shortages during the drought and where water deliveries will be cut; establishes the drought as a basis for planning in times of water shortages; guides the operation rules for Lake Mead and Lake Powell for better coordination during the risk of drought; sets rules for distributing any surpluses should they ever again arise; and encourages implementation of water conservation initiatives.

Conservation (using less water) and efficiency (using water more effectively) are not yet being stressed as ways to reduce the tremendous demand for water in the region;

perhaps this "Record of Decision" will lead to serious conservation and efficiency strategies. For instance, southern Nevada has launched a campaign to replace lawns with drought-tolerant xeroscaping—desert landscaping—but with 45,000 new hotel rooms under construction, we might question the effectiveness of overall planning.

Limiting or halting further metropolitan construction and population growth in the Colorado River Basin are not part of the present strategies to lower demand. Despite the conditions along the river, we still seem to be "supply" focused in our approach and not focusing on "demand." Remember, it is the tremendous increase in the demand for this water that stands in the way of resetting the Colorado River system in the foreseeable future.

We might wonder what John Wesley Powell would think if he were alive today to witness such errant attempts to control the mighty and variable Colorado. He foretold such a "heritage of conflict and litigation."

Differential weathering removes all surrounding rock and leaves enormous buttes as residuals on the landscape. If you imagine a line intersecting the tops of the Mitten Buttes shown in Figure 12.21, you can gain some idea of the quantity of material that has been removed. These buttes exceed 300 m (1000 ft) in height, similar to the Chrysler Building in New York City or First Canadian Place in Toronto. Totem Pole is a monolithic rock formation that appeared in Figure 1.9; can you find it in the aerial view of Monument Valley in Figure 12.21d?

In a desert area, weak surface material may weather to a complex, rugged, low topography, or **badland**, probably so named because it offered little economic value and was difficult to traverse in nineteenth-century wagons. The Badlands region of the Dakotas is of this form, as are portions of Death Valley, California.

Sand dunes that existed in some ancient deserts lithified, forming sandstone structures that bear the imprint of cross-stratification. When such a dune was accumulating, sand cascaded down its slipface and distinct bedding planes (layers) were established that remained after the dune lithified (Figure 12.22). Ripple marks, animal tracks, and fossils also are found preserved in these sandstones, which originally were eolian-deposited sand dunes.

Basin and Range Province A *province* is a large region that shares several geologic or physiographic traits. The **Basin and Range Province** of the western United States consists of alternating basins and mountain ranges that lie in the rain shadow of mountains to the west. Thus, the province has a dry climate, few permanent streams, and *interior drainage patterns*—drainage basins that lack any outlet to the ocean (see Figure 11.5). The Basin and Range Province—almost 800,000 km² (300,000 mi²)—was a major barrier to early settlers in their migration westward (Figure 12.23a,b). The desert climate and north–south-trending mountain ranges were harsh challenges.

As the North American plate moved westward, it overrode former oceanic crust and hot spots at such a rapid pace that slabs of subducted material literally were run over. This motion stretched the crust by tensional forces, creating a landscape fractured by many faults. The present landscape consists of nearly parallel sequences of *horsts* (upward-faulted blocks, which are the "ranges") and *grabens* (downward-faulted blocks, which are the "basins" or valleys).

Figure 12.23c illustrates this pattern. Note the **bolson**, a slope-and-basin area between the crests of two adjacent ridges in a dry region of interior drainage. The figure also identifies a *playa* (central salt pan) and a *bajada* (coalesced

(a)

FIGURE 12.21 Monument Valley landscape.
(a) Mitten Buttes, Merrick Butte, and rainbow in Monument Valley, Navajo Tribal Park, along the Utah–Arizona border. (b) A schematic of the tremendous removal of material by weathering, erosion, and transport. (c) Aerial view of this landscape that has appeared in so many movies. (d) View just east of (c) showing the residual spires of the Yie Bi Chei rock formation that ends with Totem Pole, 137 m (450 ft) tall and only 10 m wide. [Photos by (a) author; (c) and (d) Bobbé Christopherson.]

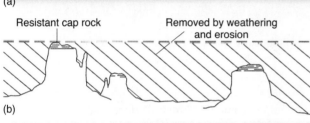

(b)

(c)

(d)

ANIMATION

Formation of
Cross Bedding

FIGURE 12.22 Cross bedding in sedimentary rocks.
The bedding pattern, called cross stratification, in these sandstone rocks tells us about conditions that were established in the dunes before lithification (hardening into rock). [Photo by Bobbé Christopherson.]

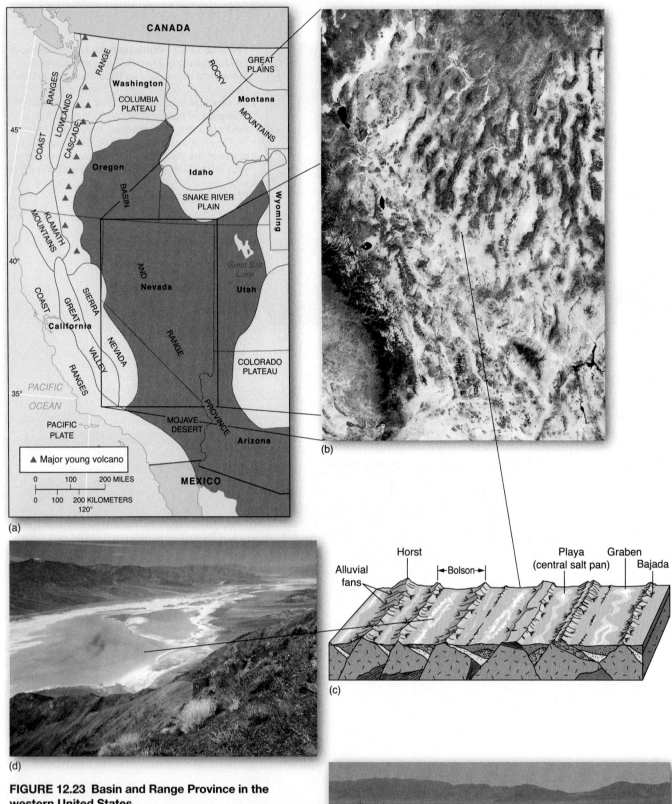

FIGURE 12.23 Basin and Range Province in the western United States.

(a) Map and (b) *Landsat* image of the area. Recent scientific discoveries demonstrate that this province extends south through northern and central Mexico. (c) A bolson in the mountainous desert landscape of the Basin and Range Province. Parallel normal faults produce a series of ranges and basins. (d) Death Valley features a central playa, parallel mountain ranges, alluvial fans, and bajada along the base of the ranges. (e) "The Loneliest Road in America," Highway US 50, slashes westward through the province.

[(b) Image from NASA; photos by (d) author; (e) Bobbé Christopherson.]

424

alluvial fans). A *pediment* is an area of bedrock that is covered with a thin veneer, or coating, of alluvium. It is an erosional surface, as opposed to the depositional surface of the bajada, and is created as a mountain front retreats through weathering and erosion. John McPhee captured the feel of this desert province in his book *Basin and Range*:

> Supreme over all is silence. Discounting the cry of the occasional bird, the wailing of a pack of coyotes, silence—a great spatial silence—is pure in the Basin and Range. It is a soundless immensity with mountains in it. You stand . . . and look up at a high mountain front, and turn your head and look fifty miles down the valley, and there is utter silence.*

Basin-and-range relief is abrupt, and rock structures are angular and rugged. As the ranges erode, transported materials accumulate to great depths in the basins, gradually producing extensive desert plains. The basin's elevation averages between 1220 and 1525 m (4000–5000 ft) above sea level, with mountain crests rising higher by some 915 to 1525 m (3000–5000 ft). Death Valley is an extreme example (Figure 12.23d); its lowest basin has an elevation of −86 m (−282 ft). However, to the west of the valley, the Panamint Range rises to 3368 m (11,050 ft) at Telescope Peak—over 3 vertical kilometers (2 mi) of desert relief.

When you traverse US Highway 50 across Nevada, you cross five passes of over 1950 m (6400 ft) and numerous

*J. McPhee, *Basin and Range* (New York: Farrar, Straus, Giroux, 1981), p. 46.

basins. Throughout the drive you are reminded where you are by prideful signs that plainly state: "The Loneliest Road in America." It is difficult to imagine crossing this topography with wagons and oxen in the nineteenth century (Figure 12.23e).

Vast arid and semiarid lands remain an enigma on the water planet. They challenge our technology to access water, our courage, and personal need for water. Yet, these lands hold a mysterious fascination, perhaps because they are so lacking in the moisture that infuses our lives.

Desertification

We are witnessing an unwanted expansion of Earth's desert lands in a process known as **desertification**. This now is a worldwide phenomenon along the margins of arid and semiarid lands. Desertification is due principally to poor agricultural practices (overgrazing and agricultural activities that abuse soil structure and fertility), improper soil-moisture management, erosion and salinization, deforestation, and the ongoing global climatic change that is shifting temperature and precipitation patterns.

The southward expansion of Saharan conditions through portions of the *Sahel region* has left many African peoples on land that no longer experiences the rainfall of just two decades ago. Other regions at risk of desertification stretch from Asia and central Australia to portions of North and South America. The United Nations estimates that degraded lands have covered some 800 million hectares (2 billion acres) since 1930; many millions of additional hectares are added each year (Figure 12.24).

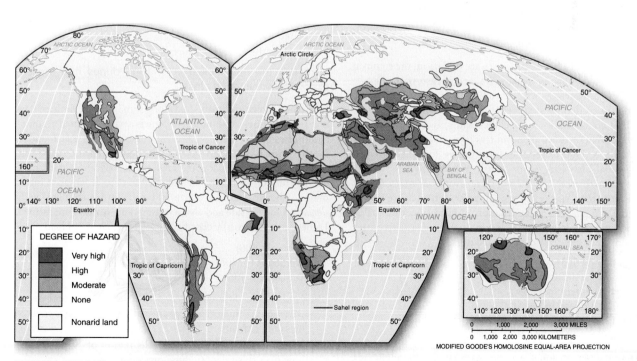

FIGURE 12.24 The desertification hazard.
Worldwide desertification estimates by the United Nations. The climate-stressed Sahel region is outlined on the map. [Data from UN Food and Agricultural Organization (FAO), World Meteorological Organization (WMO), United Nations Educational, Scientific, and Cultural Organization (UNESCO), Nairobi, Kenya.]

Urgently needed is an improved database for a more accurate accounting of the problem and a better understanding of what is occurring.

Figure 12.24 is drawn from a map prepared for a UN Conference on Desertification; although dated, the patterns persist to this day. Desertification areas are ranked: A moderate hazard area has an average 10%–25% drop in agricultural productivity; a high hazard area has a 25%–50% drop; and a very high hazard area has more than a 50% decrease. Because human activities and economies, especially unwise grazing practices, appear to be the major cause of desertification, solutions to slow the process are readily available but progress is slow. The

severity of this problem is magnified by the poverty in many of the affected regions, for the poor lack the capital to change practices and implement conservation strategies.

One of the successes coming from the 1992 Earth Summit in Rio de Janeiro was an initiative for empowering a Convention to Combat Desertification, which began operations in 1994 and is still active. International conferences are convened every other year and many programs are underway. For more on these global arid lands, see the UN site at http://www.unccd.int/main.php; and for the UNDPs Dryland Development Centre work, see http://www.undp.org/drylands/.

Summary and Review—Wind Processes and Desert Landscapes

■ *Characterize* the unique work accomplished by wind and eolian processes.

The movement of the atmosphere in response to pressure differences produces wind. Wind is a geomorphic agent of erosion, transportation, and deposition. **Eolian** processes modify and move sand accumulations along coastal beaches and deserts. Wind's ability to move materials is small compared with that of water and ice.

eolian (p. 402)

1. Who was Ralph Bagnold? What was his contribution to eolian studies?
2. Explain the term *eolian* and its application in this chapter. How would you characterize the ability of wind to move material?

■ *Describe* eolian erosion, including deflation, abrasion, and the resultant landforms.

Two principal wind-erosion processes are **deflation**, the removal and lifting of individual loose particles, and **abrasion**, the "sandblasting" of rock surfaces with particles captured in the air. Fine materials are eroded by wind deflation and moving water, leaving behind concentrations of pebbles and gravel called **desert pavement**. Wherever wind encounters loose sediment, deflation may remove enough material to form basins, forming **blowout depressions**. They range from small indentations less than a meter wide up to areas hundreds of meters wide and many meters deep. Rocks that bear evidence of eolian erosion are called **ventifacts**. On a larger scale, deflation and abrasion are capable of streamlining rock structures, leaving behind distinctive rock formations or elongated ridges as **yardangs**.

deflation (p. 402)
abrasion (p. 402)
desert pavement (p. 402)
blowout depressions (p. 403)
ventifacts (p. 403)
yardangs (p. 403)

3. Describe the erosional processes associated with moving air.
4. Explain deflation and provide an overview of each hypothesis on the processes that produce desert pavement surfaces.
5. How are ventifacts and yardangs formed by the wind?

■ *Describe* eolian transportation and *explain* saltation and surface creep.

Wind exerts a drag or frictional pull on surface particles until they become airborne. Only the finest dust particles travel significant distances, so the finer material suspended in a dust storm is lifted much higher than the coarser particles of a sand storm. Saltating particles crash into other particles, knocking them both loose and forward. The motion of **surface creep** slides and rolls particles too large for saltation.

surface creep (p. 404)

6. Differentiate between a dust storm and a sand storm. How might military activities increase the probability of available sediment for transport?
7. What is the difference between eolian saltation and fluvial saltation?
8. Explain the concept of surface creep.

■ *Identify* the major classes of sand dunes and *present* examples within each class.

In arid and semiarid climates and along some coastlines where sand is available, **dunes** accumulate. A dune is a wind-sculpted accumulation of sand. An extensive area of dunes, such as that found in North Africa, is characteristic of an **erg desert**, or **sand sea**. When saltating sand grains encounter small patches of sand, their kinetic energy (motion) is dissipated and they start to accumulate into a dune. As height increases above 30 cm (12 in.), a steeply sloping *slipface* on the lee side and characteristic dune features are formed. Dune forms are broadly classified as *crescentic*, *linear*, and *star*.

dunes (p. 406)
erg desert (p. 406)
sand sea (p. 406)

9. Explain the difference between an erg desert and desert pavement. Which type is a sand sea? Are all deserts covered by sand? Explain.
10. What occurred in the Sand Hills region of Nebraska and how does this relate to dune processes?
11. What are the three classes of dune forms? Describe the basic types of dunes within each class. What do you think is the major shaping force for sand dunes?

12. Which form of dune is the mountain giant of the desert? What are the characteristic wind patterns that produce such dunes?

■ *Define* loess deposits and their origins, locations, and landforms.

Eolian-transported materials contribute to soil formation in distant places. Windblown **loess** deposits occur worldwide and can develop into good agricultural soils. These fine-grained clays and silts are moved by the wind many kilometers, where they are redeposited in unstratified, homogeneous deposits. The binding strength of loess causes it to weather and erode in steep bluffs, or vertical faces.

Significant accumulations throughout the Mississippi and Missouri valleys form continuous deposits 15–30 m (50–100 ft) thick. Loess deposits also occur in eastern Washington state, Idaho, Iowa, much of Ukraine, central Europe, China, the Pampas-Patagonia regions of Argentina, and lowland New Zealand.

loess (p. 409)

13. How are loess materials generated? What form do they assume when deposited?

14. Name a few examples of significant loess deposits on Earth.

■ *Portray* desert landscapes and *locate* these regions on a world map.

Dry and semiarid climates occupy about 35% of Earth's land surface. The spatial distribution of these dry lands is related to subtropical high-pressure cells between 15° and 35° N and S, to rain shadows on the lee side of mountain ranges, or to areas at great distance from moisture-bearing air masses, such as central Asia.

Water events are rare, yet running water is still the major erosional agent in deserts. Although rare, when precipitation events do occur, a dry streambed may fill with a torrent called a **flash flood**. Depending on the region, such a dry streambed is known as a **wash**, an *arroyo* (Spanish), or a *wadi* (Arabic). As runoff water evaporates, salt crusts may be left behind on the desert floor. An intermittently wet and dry low area in a region of closed drainage is a **playa**, site of an *ephemeral lake* when water is present.

In arid climates, a prominent landform is the **alluvial fan** at the mouth of a canyon where it exits into a valley. The fan is produced by flowing water that abruptly loses velocity as it leaves the constricted channel of the canyon and deposits a layer of sediment along the mountain block. A continuous apron, or **bajada**, may form if individual alluvial fans coalesce. In a dry region, weak surface materials may weather to a complex, rugged, low topography, or **badland**. A *province* is a large region that is characterized by several geologic or physiographic traits. The **Basin and Range Province** of the western United States consists of alternating basins and mountain ranges. A slope-and-basin area between the crests of two adjacent ridges in a dry region of interior drainage is termed a **bolson**. **Desertification** is the process that leads to unwanted expansion of the Earth's desert lands.

flash flood (p. 411)
wash (p. 411)
playa (p. 415)
alluvial fan (p. 415)
bajada (p. 415)
badland (p. 422)
Basin and Range Province (p. 422)
bolson (p. 422)
desertification (p. 425)

15. Characterize desert energy and water-balance regimes. What are the significant patterns of occurrence for the world's arid landscapes?

16. How would you describe the water budget of the Colorado River? What was the basis for agreements regarding distribution of the river's water? Why has thinking about the river's discharge been so optimistic? Overview current conditions of the Colorado River system in light of the ongoing drought and increasing water demands.

17. Describe a desert bolson from crest to crest. Draw a simple sketch with the components of the landscape labeled.

18. Where is the Basin and Range Province? Briefly describe its appearance and character.

19. What is meant by desertification? Using the map in Figures 12.24 and 12.15, and the text description, locate several of the regions affected by desert expansion.

Network and Critical Thinking Tools

A. "Water is the major erosion and transport medium in the desert." Respond to this quotation. How is it possible for this to be true when deserts are so dry? What factors have you learned from this chapter that prove this statement true?

B. Where are the nearest eolian features (coastal, lakeshore, desert dunes, or loess deposits) to your present location? Which causative factors discussed in this chapter explain the features you identified? Is it possible to actually visit the site or to virtually visit the site on the Internet? Explain.

C. Relative to the Colorado River system, define, compare, and contrast what you think is meant by a "supply strategy" or a "demand strategy" in dealing with the present Colorado River water budget (Table 12.1.1 in Focus Study 12.1). Which strategy seems focused on centralized or decentralized solutions? What political and economic forces do you think are at play in planning for the river? Who are the stakeholders? With your analysis, briefly speculate on what strategies you think would be most effective in dealing with this situation.

D. A historic decision was signed in December 2007, the "Record of Decision," that attempts to deal with the Colorado River shortages. See this document at http://www.usbr .gov/lc/region/programs/strategies/RecordofDecision.pdf and read through the "Introduction." What do you find?

Breakers, sandy beach, and weathered rock outcrops near Los Cabos at the southern tip of the state of Baja California Sur. The Pacific Ocean is in the distance. This scene is at 22° N latitude and is the same beach as in the photo of the pen stuck in the sand in Figure 2.11 to demonstrate a subsolar point. *[Photo by Bobbé Christopherson.]*

The Oceans, Coastal Processes, and Landforms

KEY LEARNING CONCEPTS

After reading the chapter, you should be able to:

- **Describe** the chemical composition of seawater and the physical structure of the ocean.

- **Identify** the components of the coastal environment and **list** the physical inputs to the coastal system, including tides and mean sea level.

- **Describe** wave motion at sea and near shore and **explain** coastal straightening as a product of wave refraction.

- **Identify** characteristic coastal erosional and depositional landforms.

- **Describe** barrier islands and their hazards and **relate** them to human settlement.

- **Assess** living coastal environments: corals, wetlands, salt marshes, and mangroves.

- **Construct** an environmentally sensitive model for settlement and land use along the coast.

STUDENT LEARNING CENTER TOOLS

The *Elemental Geosystems* Student Learning Center provides on-line resources for this chapter by clicking on the book cover at **www.mygeoscienceplace.com**. Some highlights:

- Work with satellite and bathymetric imagery in identifying Earth's oceans and seas.

- A Career Link features a geographer who works as a research network coordinator in coastal management and climate change impacts.

- Five animations present tidal cycles, wave motion and wave refraction, beach drift, coastal erosion, and the operation of coastal stabilization structures.

- Watch a satellite loop of Hurricane Isabel, which struck Cape Hatteras in 2003.

- More than 15 additional URLs along with Quick Links to all the URLs in the chapter.

Walk along a shoreline and you witness the dramatic interaction of Earth's vast oceanic, atmospheric, and lithospheric systems. At times, the ocean attacks the coast in a stormy rage of erosive power; at other times, the moist sea breeze, salty mist, and repetitive motion of the water are gentle and calming. Few have captured the confrontation between land and sea as well as Rachel Carson:

> The edge of the sea is a strange and beautiful place. All through the long history of Earth it has been an area of unrest where waves have broken heavily against the land, where the tides have pressed forward over the continents, receded, and then returned. For no two successive days is the shoreline precisely the same. Not only do the tides advance and retreat in their eternal rhythms, but the level of the sea itself is never at rest. It rises or falls as the glaciers melt or grow, as the floors of the deep ocean basins shift under its increasing load of sediments, or as the Earth's crust along the continental margins warps up or down in adjustment to strain and tension. Today a little more land may belong to the sea, tomorrow a little less. Always the edge of the sea remains an elusive and indefinable boundary.*

Commerce and access to sea routes, fishing, and tourism prompt many people to settle near the ocean. A scientific assessment estimates that about 40% of Earth's population lives within 100 km (62 mi) of coastlines. In the United States, about 50% of the people live in areas designated as *coastal* (this includes the Great Lakes).

Therefore, an understanding of coastal processes and landforms is important to nearly half the world's population. And because these processes along coastlines often produce dramatic change, they are essential to consider in planning and development. A World Resources Institute study found the world's coastlines at some risk of loss through erosion and rising sea level or disruption from pollution. The National Ocean Service coordinates many scientific activities related to the ocean. You find information about these activities at http://www.nos.noaa.gov/.

The ocean is a vast ecosystem, intricately linked to life on the planet and to life-sustaining systems in the atmosphere, the hydrosphere, and the lithosphere. In the recent *Atlas of the Oceans—The Deep Frontier*, Jean-Michel Cousteau commented on "The Future of the Ocean":

> Today, we are coming to better appreciate the extent to which our actions affect an ecosystem—and the people who depend on it—thousands of miles away. The reef fisherman in Fiji is not undone by the local poacher, but by global warming intensified by the driving of a car in downtown Toronto, Canada. Yet these connections are not all bad news. The web of

interdependence is built with strands of responsibility and hope. Our ever expanding ability to communicate across borders and oceans is helping to drive a truly global dialogue about the planet's most pressing environmental challenges.[†]

In this chapter: We begin the chapter with a brief look at our global oceans and seas—a decade ago we celebrated The International Year of the Ocean with all United Nations countries (see http://www.yoto98.noaa .gov/). The physical and chemical properties of the ocean distinguish it from the waters of the continent. Coverage includes discussions of tides, waves, coastal erosional and depositional landforms, beaches, barrier islands, and organic processes, including corals, wetlands, salt marshes, and mangroves. A systems framework of specific inputs (components and driving forces), actions (movements and processes), and outputs (results and consequences) organizes our discussion of coastal processes. We conclude with a look at the considerable human impact on coastal environments.

Global Oceans and Seas

The ocean is one of Earth's last great scientific frontiers and is of great interest to geographers. Remote sensing from orbiting spacecraft, aircraft, surface vessels, a network of buoys, and submersibles is providing a wealth of data and a new understanding of the oceanic system. The pattern of sea-surface temperatures is presented in Figure 3.22 and ocean currents in Figure 4.21. The world's oceans, their area, volume, and depth, are listed in a table in Figure 13.1 along with the locations of oceans and major seas.

Chemical Composition of Seawater

Water is the "universal solvent," dissolving at least 57 of the 92 elements found in nature. In fact, most natural elements and the compounds they form are found in the seas as dissolved solids, or *solutes*. Thus, seawater is a solution, and the concentration of dissolved solids is **salinity**.

The oceans are a remarkably homogeneous mixture. The ratio of individual salts does not change, despite minor fluctuations in overall salinity. In 1874 the British HMS *Challenger* sailed around the world, taking surface and depth measurements and collecting samples of seawater. Analyses of those samples first demonstrated the uniform composition of seawater.

Ocean Chemistry Ocean chemistry is a result of complex exchanges among seawater, the atmosphere, minerals, bottom sediments, and living organisms. In addition, significant flows of mineral-rich water enter the ocean through hydrothermal (hot water) vents in the ocean floor. These vents are "black smokers," known for the dense, black, mineral-laden water that spews from them

---

*"The Marginal World," in *The Edge of the Sea* by Rachel Carson. © 1955 by Rachel Carson, © renewed 1983 by R. Christie (Boston: Houghton Mifflin), p. 11.

[†]*Atlas of the Oceans—The Deep Frontier* by Sylvia Earle. © 2001 National Geographic Society, text © 2001 by Sylvia Earle, p. 171.

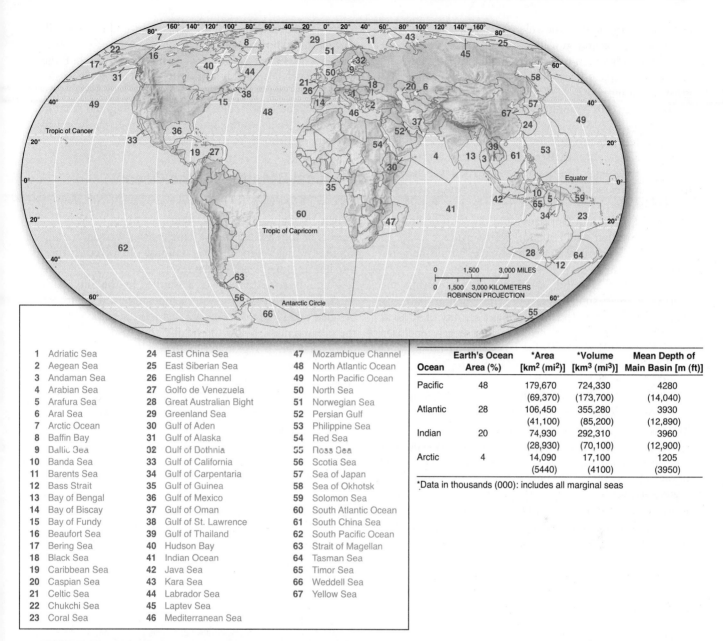

FIGURE 13.1 Principal oceans and seas of the world.
A sea is generally smaller than an ocean and is near a landmass; sometimes the term refers to a
large, inland, salty body of water. A profile of Earth's four oceans is in the inset table.

1	Adriatic Sea	24	East China Sea	47	Mozambique Channel
2	Aegean Sea	25	East Siberian Sea	48	North Atlantic Ocean
3	Andaman Sea	26	English Channel	49	North Pacific Ocean
4	Arabian Sea	27	Golfo de Venezuela	50	North Sea
5	Arafura Sea	28	Great Australian Bight	51	Norwegian Sea
6	Aral Sea	29	Greenland Sea	52	Persian Gulf
7	Arctic Ocean	30	Gulf of Aden	53	Philippine Sea
8	Baffin Bay	31	Gulf of Alaska	54	Red Sea
9	Baltic Sea	32	Gulf of Bothnia	55	Ross Sea
10	Banda Sea	33	Gulf of California	56	Scotia Sea
11	Barents Sea	34	Gulf of Carpentaria	57	Sea of Japan
12	Bass Strait	35	Gulf of Guinea	58	Sea of Okhotsk
13	Bay of Bengal	36	Gulf of Mexico	59	Solomon Sea
14	Bay of Biscay	37	Gulf of Oman	60	South Atlantic Ocean
15	Bay of Fundy	38	Gulf of St. Lawrence	61	South China Sea
16	Beaufort Sea	39	Gulf of Thailand	62	South Pacific Ocean
17	Bering Sea	40	Hudson Bay	63	Strait of Magellan
18	Black Sea	41	Indian Ocean	64	Tasman Sea
19	Caribbean Sea	42	Java Sea	65	Timor Sea
20	Caspian Sea	43	Kara Sea	66	Weddell Sea
21	Celtic Sea	44	Labrador Sea	67	Yellow Sea
22	Chukchi Sea	45	Laptev Sea		
23	Coral Sea	46	Mediterranean Sea		

Ocean	Earth's Ocean Area (%)	*Area [km² (mi²)]	*Volume [km³ (mi³)]	Mean Depth of Main Basin [m (ft)]
Pacific	48	179,670 (69,370)	724,330 (173,700)	4280 (14,040)
Atlantic	28	106,450 (41,100)	355,280 (85,200)	3930 (12,890)
Indian	20	74,930 (28,930)	292,310 (70,100)	3960 (12,900)
Arctic	4	14,090 (5440)	17,100 (4100)	1205 (3950)

*Data in thousands (000): includes all marginal seas

(see Figure 8.11b). The uniformity of seawater results from complementary chemical reactions and continuous mixing—after all, the ocean basins interconnect and water circulates among them.

Seven elements account for more than 99% of the dissolved solids in seawater. They are (with their ionic form): chlorine (as chloride, Cl^-), sodium (as Na^+), magnesium (as Mg^{2+}), sulfur (as sulfate, SO_4^{2-}), calcium (as Ca^{2+}), potassium (as K^+), and bromine (as bromide, Br^-). Seawater also contains dissolved gases (such as carbon dioxide, nitrogen, and oxygen), suspended and dissolved organic matter, and a multitude of trace elements.

Commercially, only sodium chloride (common table salt), magnesium, and bromine are extracted in any significant amount from the ocean. Future mining of minerals

from the seafloor is technically feasible, although it remains uneconomical.

Average Salinity: 35‰   There are several ways to express ocean salinity (dissolved solids by volume) in seawater, using the worldwide average value:

- 3.5% (% parts per hundred)
- 35,000 ppm (parts per million)
- 35,000 mg per liter
- 35 g/kg
- 35‰ (‰ parts per thousand), the most common notation

Salinity worldwide normally varies between 34‰ and 37‰; variations are attributable to atmospheric conditions above the water and to the volume of freshwater inflows. In equatorial water, precipitation is great throughout the year, diluting salinity values to slightly lower than average. In subtropical oceans—where evaporation rates are greatest because of the influence of hot, dry, subtropical high-pressure cells—salinity is more concentrated, increasing to 36‰.

The term **brine** applies to water that exceeds the average of 35‰ salinity. **Brackish** describes water that is less than 35‰ salts. In general, oceans are lower in salinity near landmasses because of freshwater runoff and river discharges. Extreme examples include the Baltic Sea (north of Poland and Germany) and the Gulf of Bothnia (between Sweden and Finland), which average 10‰ or less salinity because of heavy freshwater runoff and low evaporation rates.

On the other hand, the Sargasso Sea, within the North Atlantic subtropical gyre (circulation pattern), averages 38‰. The Persian Gulf has a salinity of 40‰ as a result of high evaporation rates in a nearly enclosed basin. Deep pockets, or "brine lakes," along the floor of the Red Sea and the Mediterranean Sea register up to a salty 225‰.

The ocean reflects conditions in Earth's environment. As an example, the high-latitude oceans have been freshening over the past decade owing to glacial ice melt, as mentioned earlier. Another change underway is oceanic acidification, taking place as the ocean absorbs carbon dioxide from the atmosphere, where it is increasing rapidly. Hydrolysis forms carbonic acid in the seawater and a lowering of the ocean pH—an acidification. A more acid ocean will cause certain marine organisms such as corals and some plankton to have difficulty maintaining external calcium carbonate structures. pH could decrease by –0.4 to –0.5 units this century; the oceans' average pH today is 8.2. Oceanic biodiversity and food webs will change in unknown ways.

The dissolved solids remain in the ocean, but the water recycles endlessly through the hydrologic cycle, driven by energy from the Sun. The water you drink today may have water molecules in it that not long ago were in the Pacific Ocean, in the Yangtze River, in groundwater in Sweden, or airborne in the clouds over Peru.

Physical Structure of the Ocean

The basic physical structure of the ocean is layered, as shown in Figure 13.2. The figure also graphs four key aspects of the ocean, each of which varies with increasing depth: average temperature, salinity, dissolved carbon dioxide, and dissolved oxygen level.

The ocean's surface layer is warmed by the Sun and is wind-driven. Variations in water temperature and solutes are blended rapidly in a *mixing zone* that represents only 2% of the oceanic mass. Below the mixing zone is the *thermocline transition zone*, a region more than a kilometer deep of decreasing temperature gradient that lacks the motion of the surface. Friction at these depths dampens the effect of surface currents. In addition, colder water temperatures at the lower margin tend to inhibit any convective movements.

From a depth of 1–1.5 km (0.6–0.9 mi) to the ocean floor, temperature and salinity values are quite uniform. Temperatures in this *deep cold zone* are near 0°C (32°F). Water in the deep cold zone does not freeze, however, because of its salinity and intense pressures at those depths; seawater freezes at about –2°C (28.4°F) at the surface. The coldest water is at the bottom, except near the poles, where the coldest water may be near or at the surface. Now let us shift to the edge of the sea and examine Earth's coastlines.

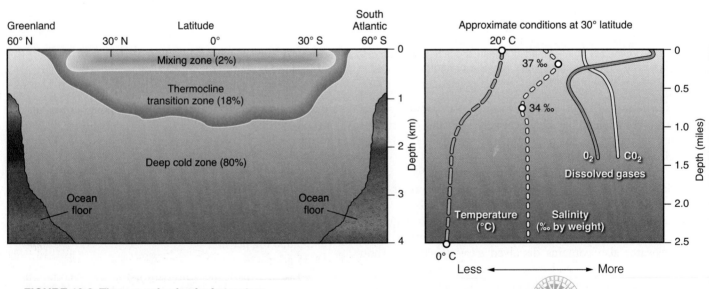

FIGURE 13.2 The ocean's physical structure.
Schematic of average physical structure observed throughout the ocean's vertical profile as sampled along a line from Greenland to the South Atlantic. Temperature, salinity, and dissolved gases are plotted by depth.

ANIMATION
Midlatitude Productivity

Coastal System Components

Earth's surface features, such as mountains and crustal plates, were formed over millions of years. However, most of Earth's coastlines are relatively new, existing in their present state as the setting for continuous change. The land, ocean, atmosphere, Sun, and Moon interact to produce tides, currents, waves, erosional features, and depositional features along the continental margins.

Inputs to the coastal environment include many elements we have already discussed:

- *Solar energy* input drives the atmosphere and the hydrosphere. A conversion of insolation to kinetic energy produces prevailing winds, weather systems, and climate.
- *Atmospheric winds*, in turn, generate ocean currents and waves, key inputs to the coastal environment.
- *Climatic regimes*, which result from insolation and moisture, strongly influence coastal geomorphic processes.
- *The nature of coastal rock* is important in determining rates of erosion and sediment production.
- *Human activities* are an increasingly significant input to coastal change.

All these inputs occur within the ever-present influence of gravity's pull, not only from Earth but also from the Moon and the Sun. Gravity provides the potential energy of position for materials in motion and produces the tides. A dynamic equilibrium among all these components produces coastline features of infinite variety and beauty.

The Coastal Environment and Sea Level

The coastal environment is the **littoral zone**. (*Littoral* comes from the Latin word for "shore.") The littoral zone spans some land as well as water. Landward, it extends to the highest water line that occurs on shore during a storm. Seaward, it extends to the point at which storm waves can no longer move sediments on the seafloor (usually at depths of approximately 60 m, or 200 ft). The specific contact line between the sea and the land is the *shoreline*, and adjacent land is considered the *coast*. Figure 13.3 illustrates the littoral zone and includes specific components discussed later in the chapter.

Because the level of the ocean varies, the littoral zone naturally shifts position from time to time. A rise in sea level causes submergence of land, whereas a drop in sea level produces coastline emergence. Uplift and subsidence of the land also cause changes in the littoral zone.

Sea level is a relative term and reflects changes in both water quantity and land elevations. Because sea level is not a straightforward measurement, there exists no international system to determine exact sea level. The Global Sea Level Observing System (GLOSS) is an international group actively working on sea-level issues and is part of the larger

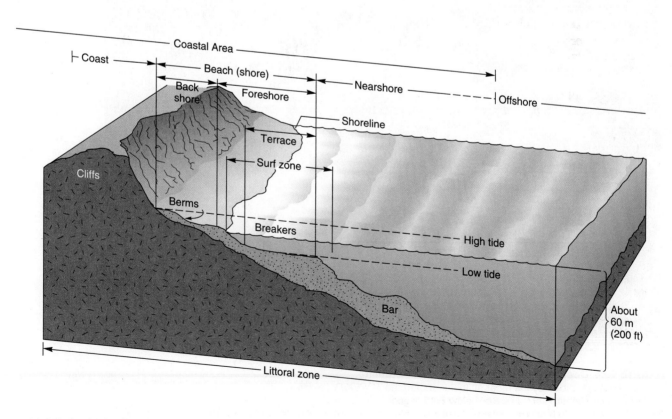

FIGURE 13.3 The littoral zone.
The littoral zone includes the coast, beach, and nearshore environments.

Permanent Service for Mean Sea Level, which you can find at http://www.pol.ac.uk/psmsl/programmes/.

Mean sea level (MSL) is a value based on average tidal levels recorded hourly at a given site over a period of at least 19 years, which is one full lunar tidal cycle. Mean sea level varies because of ocean currents and waves, tides, air temperature and pressure differences, ocean temperature variations, slight variations in Earth's gravity, and changes in oceanic volume.

At present, the overall mean sea level for the United States is calculated at approximately 40 locations along the coastal margins of the continent. These sites are being upgraded with new equipment in the Next Generation Water Level Measurement System, specifically along the US and Canadian Atlantic coasts, Bermuda, and the Hawaiian Islands. The *NAVSTAR* satellites that make up the Global Positioning System (GPS) make possible the correlation of data within a network of ground- and ocean-based measurements. See News Report 13.1 for more information about sea level and the present rise.

Remote-sensing technology augments these measurements, including the *TOPEX/Poseidon* satellite launched in 1992 and still in service (see http://topex-www.jpl.nasa.gov/). This satellite has two radar altimeters that measure changes in mean sea level at any one location every 10 days between 66° N and 66° S latitudes. These measurements are made to an astonishing precision of 4.2 cm (1.7 in., Figure 13.4). Spectacular *TOPEX/Poseidon*

portraits of El Niño and La Niña in the Pacific are in Focus Study 7.1. Another ocean surface topography satellite launched in December 2001, named *Jason-1*, is extending the science of determining mean sea level, ocean topography, atmosphere–ocean interactions, and ocean circulation. *Jason-2* launched in 2008.

Coastal System Actions

The coastal system is the scene of complex tidal fluctuation, winds, waves, ocean currents, and the occasional impact of storms. These forces shape landforms ranging from gentle beaches to steep cliffs, and they sustain delicate ecosystems.

Tides

Tides are complex daily oscillations in sea level, ranging worldwide from barely noticeable to several meters. They are experienced to varying degrees along every ocean shore around the world. As tides flood (rise) and ebb (fall), the daily migration of the shoreline landward and seaward causes significant changes that affect sediment erosion and transportation.

Tides are important in human activities, including navigation, fishing, and recreation. Tides are especially important to ships because the entrance to many ports is limited by shallow water, and thus high tide is required for

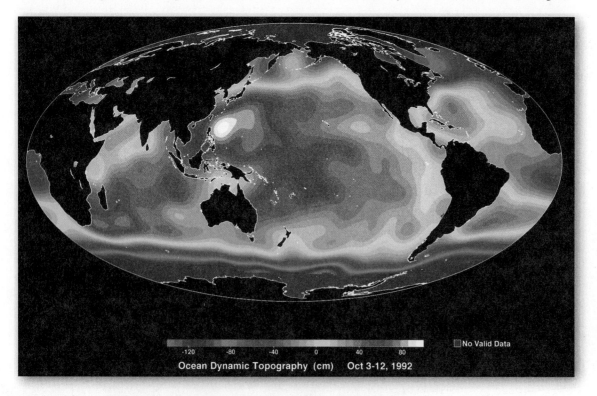

FIGURE 13.4 Ocean topography as revealed by satellite.
Sea-level data recorded by the radar altimeter aboard the *TOPEX/Poseidon* satellite. The color scale is given in centimeters above or below Earth's geoid. The overall relief portrayed in the image is about 2 m (6.6 ft). The maximum sea level is in the western Pacific Ocean (white). The minimum is around Antarctica (blue and purple). [Image from JPL and NASA/GSFC. *TOPEX/Poseidon* is a joint US–French Mission.]

NEWS REPORT 13.1

Sea Level Variations and the Present MSL Increase

Sea level varies along the extent of North American shorelines. The mean sea level (MSL) of the US Gulf Coast is about 25 cm (10 in.) higher than that of Florida's east coast, which is the lowest in North America. MSL rises northward along the eastern coast, to 38 cm (15 in.) higher in Maine than in Florida. Along the US western coast, MSL is higher than Florida's by about 58 cm (23 in.) in San Diego and by about 86 cm (34 in.) in Oregon.

Overall, North America's Pacific coast MSL averages about 66 cm (26 in.) higher than on the Atlantic coast. MSL is affected by differences in ocean currents, air pressure and wind patterns, water density, and water temperature.

Over the long term, sea-level fluctuations expose a range of coastal landforms to tidal and wave processes. As average global temperatures cycle through cold or warm climatic spells, the quantity of ice locked up in the ice sheets of Antarctica and Greenland and in hundreds of mountain glaciers can increase or decrease and result in sea-level changes accordingly. At the peak of the Pleistocene glaciation, about 18,000 B.P. (years before the present), sea level was about 130 m (430 ft) lower than it is today. On the other hand, if Antarctica and Greenland ever became

ice-free (ice sheets fully melted), sea level would rise at least 65 m (215 ft) worldwide.

Just 100 years ago sea level was 38 cm (15 in.) lower along the coast of southern Florida. Venice, Italy, has experienced a rise of 25 cm (10 in.) since 1890. During the last century, sea level rose 10–20 cm (4–8 in.), a rate 10 times higher than the average rate during the last 3000 years. The present sea-level rise is spatially uneven; for instance, the rate along the coast of Argentina is nearly 10 times the rate along the coast of France. For rates in the United States, see http://tidesandcurrents.noaa.gov/sltrends/sltrends.shtml.

Given these trends and the predicted climatic change, sea level will continue to rise and be potentially devastating for many coastal locations. A rise of only 0.3 m (1 ft) would cause shorelines worldwide to move inland an average of 30 m (100 ft). This elevated sea level would inundate some valuable real estate along coastlines all over the world. Some 20,000 km² (7800 mi²) of land along North American shores alone would be drowned, at a staggering loss of more than a trillion dollars. A 95-cm (3.1-ft) sea-level rise could inundate 15% of Egypt's arable land, 17% of Bangladesh, and many island countries and communities.

The 2007 IPCC forecast for global mean sea-level rise this century, given regional variations, is a range from 0.18 to 0.59 m (7.1 to 23.2 in.). Unfortunately, the new 2006–2008 measurements of Greenland's ice loss were not included in the forecasts, and data appear to suggest an acceleration of ice melt and rising sea level. A fair working estimate for this century now stands at a 1.0 to 1.2 m (3.28 to 3.94 ft) rise in MSL.

The University of Arizona's Environmental Studies Laboratory prepared a dramatic map series using USGS DEM data for elevations. The scientists plotted 1-m to 6-m increases in sea level and the amount of coastal inundation (Figure 13.1.1). As you examine the three maps, consider what adjustments will be required. What is your assessment of southern Louisiana? Certainly any calculations on the costs of slowing increasing greenhouse gas emissions should be balanced against damage estimates from such coastal inundation.

Despite any uncertainty, planning should start now along coastlines worldwide because preventive strategies are cheaper than recovery costs from possible destruction. Insurance underwriters have begun the process by refusing coverage for shoreline properties vulnerable to rising sea level.

FIGURE 13.1.1 Beach Impact of a 1-m (3.28-ft) rise in sea level.
Along three portions of US coastline, scientists at the University of Arizona using USGS digital elevation data produced these maps to show a 1-m sea-level rise and its impact. [Maps prepared by Weiss and Overpeck, Environmental Studies Laboratory, Department of Geosciences, University of Arizona. Used by permission.]

passage. Tall-masted ships may need a low tide to clear overhead bridges. Tides also exist in large lakes, but because the tidal range is small, they are difficult to distinguish from changes caused by wind. Lake Superior, for instance, has a tidal variation of only about 5 cm (2 in.).

Causes of Tides Earth's relation to the Sun and the Moon and the reasons for the seasons are described in Chapter 2. These astronomical relationships also produce the pattern of tides, the complex daily oscillations in sea level experienced to varying degrees around the world. Figure 13.5 illustrates the relationship among the Moon, the Sun, and Earth and the generation of variable tidal bulges on opposite sides of the planet. Also, tides are influenced by the size, depth, and topography of ocean basins, by latitude, and by shoreline shape.

The gravitational pull of the Moon tugs on Earth's atmosphere, oceans, and lithosphere. The same is true for the Sun, to a lesser extent. Earth's solid and fluid surfaces all experience some stretching as a result of this gravitational pull. The stretching raises large *tidal bulges* in the atmosphere (which we can't see), smaller tidal bulges in the ocean, and very slight bulges in Earth's rigid crust. Our concern here is the tidal bulges in the ocean.

Gravity and inertia are essential elements in understanding tides. *Gravity* is the force of attraction between two bodies. *Inertia* is the tendency of objects to stay still if motionless or to keep moving in the same direction if in motion. The gravitational effect on the side of Earth facing the Moon or Sun is greater than that experienced by the far side, where inertial forces are slightly greater. In effect, from this inertial point of view, as the nearside water and Earth are drawn toward the Moon and Sun, the farside water is left behind because of the slightly weaker gravitational pull. This arrangement produces the two opposing tidal bulges on opposite sides of Earth.

Tides appear to move in and out along the shoreline, but they do not actually do so. Instead, Earth's surface rotates into and out of the relatively "fixed" tidal bulges as Earth changes its position in relation to the Moon and Sun. Every 24 hours and 50 minutes, any given point on Earth rotates through two bulges as a direct result of this rotational positioning. Thus, every day, most coastal locations experience two high (rising) tides, known as *flood tides*, and two low (falling) tides, known as *ebb tides*. The difference between consecutive high and low tides is considered the *tidal range*.

Spring and Neap Tides When Earth, Sun, and the Moon are aligned as shown in Figure 13.5a and b, their combined gravitational pull creates a greater *spring tide* (meaning to "spring forth," not the season). When positioned as in 13.5c and d, their weaker pulls create a lesser *neap tide* (*neap* means "without the power of advancing").

Locally, tides are influenced by other factors, causing a great variety of tidal ranges. For example, some locations may experience almost no difference between high and low tides. The highest tides occur where open water is

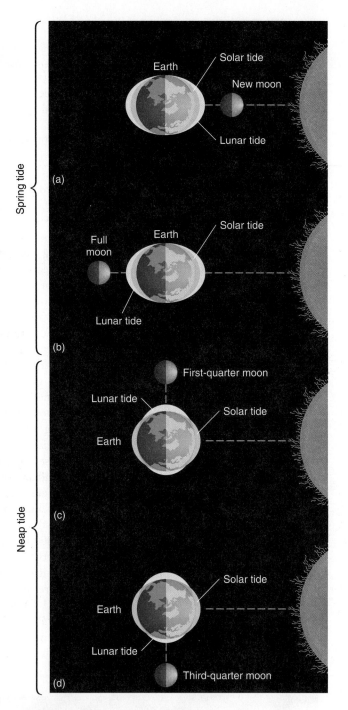

FIGURE 13.5 The cause of tides.
Gravitational relations of Sun, Moon, and Earth combine to produce spring tides (a, b) and neap tides (c, d). (Tides are greatly exaggerated for illustration.)

ANIMATION | Monthly Tidal Cycles

forced into partially enclosed gulfs or bays. The Bay of Fundy in Nova Scotia records the greatest tidal range on Earth, a difference of 16 m (52.5 ft) (Figure 13.6a and b). The comparison in 13.6c shows a smaller tidal range along England's east coast. (For more on tides and tide prediction, and a complete listing of links, see http://ocean.peterbrueggeman.com/tidepredict.html.)

FIGURE 13.6 Tidal range and tidal power.
Tidal range is great in some bays and estuaries, such as in Hall's Harbour along the Bay of Fundy at flood tide (a) and ebb tide (b). (c) Tidal variation of about 3 m (9.8 ft) in the harbor at Bamburgh, Northumbria, England, over several hours. (d) The flow of water between high and low tide is ideal for turning turbines and generating electricity, as is done at the Annapolis Tidal Generating Station, near the Bay of Fundy in Nova Scotia, in operation since 1984. [All photos by Bobbé Christopherson.]

Tidal Power   The fact that sea level changes daily with the tides suggests an opportunity: Could these predictable flows be harnessed to produce electricity? The answer is yes, given the right conditions. The bay or estuary under consideration must have a narrow entrance suitable for the construction of a dam with gates and locks, and it must experience a tidal range of flood and ebb tides large enough to turn turbines, at least a 5-m (16-ft) range.

Only 30 locations in the world are suited for tidal power generation. At present, only 3 are producing electricity. Two are outside North America—a 4-megawatt-capacity station in Russia in operation since 1968 (at Kislaya-Guba Bay on the White Sea) and a facility in France operating since 1967 (on the Rance River estuary on the Brittany coast). The tides in the Rance estuary fluctuate up to 13 m (43 ft), and power production has

been almost continuous there, providing an electrical-generating capacity of a moderate 240 megawatts (about 20% of the capacity of Hoover Dam).

The third area is the Bay of Fundy in Nova Scotia, Canada. At one of several favorable sites on the bay, the Annapolis Tidal Generating Station was built in 1984. Nova Scotia Power Incorporated operates this 20-megawatt plant (Figure 13.6d). According to the Canadian government, tidal power generation at ideal sites is economically competitive with fossil-fuel plants.

Another kind of tidal power taking form in Norway involves sea-floor devices that harness the motion of coastal currents. Windmill-like turbines are set in motion by the tides and currents, as are paddle-driven turbines. Electrical production using this method began in 2003.

Waves

Wind friction on the surface of the ocean generates undulations of water; these are **waves**. They travel in *wave trains*, or groups of waves. On a small scale, a moving boat creates a wake, which consists of waves; at a larger scale, storms around the world generate large groups of wave trains. Notable is the wind wake produced by the presence of the Hawaiian Islands, traceable westward across the Pacific Ocean surface for 3000 km (1865 mi). A stormy area at sea is a *generating region* for these waves, which radiate outward in all directions. Intricate patterns of waves traveling in all directions crisscross the ocean. The waves seen along a coast may be the product of a storm center thousands of kilometers away.

Regular patterns of smooth, rounded waves are **swells**—these are the mature undulations of the open ocean. As waves leave the generating region, wave energy continues to run in these swells, which can range from small ripples to very large flat-crested waves. A deep-water wave leaving a generating region tends to extend its wavelength many meters (see Figure 13.7, noting crest, trough, wavelength, and wave height).

Water within a wave in the open ocean is not really migrating but is transferring energy from molecule to molecule through the water in simple cyclic undulations, or *waves of transition* (Figure 13.7). Individual water particles move forward only slightly, forming a vertically circular pattern. The diameter of the paths formed by the orbiting water particles decreases with depth. As a deep-ocean wave approaches the shoreline and enters shallower water (10–20 m, or 30–65 ft), the water particle orbits are vertically reduced. The restriction causes more elliptical (flattened) orbits to form near the bottom. This change from circular to elliptical orbits of the water particles slows the entire wave, although more waves continue arriving. As the peak of each wave rises, a point is reached when the wave height exceeds its vertical stability and the wave falls into a characteristic **breaker**, crashing onto the beach (Figure 13.7b).

In a breaker the orbital motion of transition gives way to form *waves of translation*, in which energy and water both move forward toward shore as water cascades down from the wave crest. Both the energy of arriving waves and the slope of the shore determine wave style: Plunging breakers indicate a steep bottom profile, whereas spilling breakers indicate a gentle, shallow bottom profile.

When the backwash of water flows to the ocean from the beach in a concentrated column, usually at a right angle to the line of breakers, it is a *rip current*. A person caught in one of these can be swept offshore, but usually only a short distance. These brief, short torrents of water can be dangerous (Figure 13.7c).

As various wave trains move along in the open sea, they interact by *interference*. These interfering waves sometimes align so that the wave crests and troughs from one wave train are in phase with those of another. When this in-phase condition occurs, the height of the waves is increased, sometimes dramatically. Occasionally, resulting "killer waves" or "sleeper waves" can sweep in unannounced and overtake unsuspecting victims. Signs along portions of the California, Oregon, Washington, and British Columbia coastline warn beachcombers to watch for "killer waves." On the other hand, out-of-phase wave trains will dampen wave energy at the shore. Varying patterns of *wave interference* produce the changing beat of the surf along a beach and indicate actions in far-distant areas of the ocean.

Wave Refraction   Generally, wave action results in coastal straightening. As waves approach an irregular coast, they bend around *headlands*, which are protruding landforms generally composed of resistant rocks (Figure 13.8). The submarine topography refracts, or bends, approaching waves into patterns that focus energy around headlands and dissipate energy in coves and the submerged coastal valleys between them. Thus, headlands represent a specific focus of wave attack along a coastline. This **wave refraction** (wave bending) along a coastline redistributes wave energy, so that different sections of the coastline are subject to variations in erosion potential.

When waves approach a coast, they usually arrive at some angle other than parallel to it (Figure 13.9). As the waves enter shallow water, they bend and generate a current parallel to the coast, zigzagging in the prevalent direction of the incoming waves. Think of the end of the wave nearer the shore slowing, while the part of the wave farther from shore in deeper water is moving faster. This **longshore current**, or *littoral current*, depends on wind direction and resultant wave direction. A longshore current generates only in the surf zone and works in combination with wave action to transport large amounts of sand, gravel, sediment, and debris along the shore as *longshore drift*, or **littoral drift**, the more comprehensive term.

A process of *beach drift* moves particles on the beach along in a shifting back and forth between water and land with each swash and backwash of surf. Individual sediment grains trace arched paths along the beach. These dislodged materials are available for transport and eventual deposition in coves and inlets and can represent a significant volume.

Tsunami, or Seismic Sea Wave   A type of wave that can greatly influence coastlines is the **tsunami**. Tsunami is Japanese for "harbor wave," named for its devastating effect in harbors. Tsunami are referred to incorrectly as "tidal waves"—they have no relation to the tides. Sudden, sharp motions in the seafloor, caused by earthquakes, submarine landslides, or eruptions of undersea volcanoes, produce tsunami. Thus, they properly are *seismic sea waves*.

A large undersea disturbance usually generates a solitary wave of great wavelength. Tsunami exceed 100 km (60 mi) in wavelength (crest to crest) but are only a meter (3 ft) or so in height. They travel at great speeds in deep-ocean

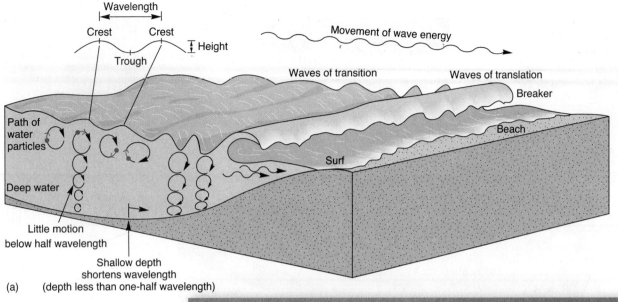

ANIMATION Wave Motion/Wave Refraction

FIGURE 13.7 Wave formation and breakers.
(a) The orbiting tracks of water particles change from circular motions and swells in deep water (waves of transition) to more elliptical orbits near the bottom in shallow water (waves of translation). (b) Cascades of waves attack the shore along Baja California, Mexico. (c) A dangerous rip current interrupts approaching breakers. Note the churned-up water where the rip current enters the surf. [Photos by Bobbé Christopherson.]

water—velocities of 600–800 kmph (375–500 mph) are not uncommon—but often pass unnoticed on the open sea because their great wavelength makes the slow rise and fall of water hard to observe.

As a tsunami approaches a coast, the shallow water forces the wavelength to shorten. As a result, the wave height may increase up to 15 m (50 ft) or more. Such a wave has the potential for coastal devastation, resulting in property damage and death. For example, in September 1992 the citizens of Casares, Nicaragua, were surprised by a 12-m (39-ft) tsunami that took 270 lives, and another hit Indonesia and killed 1000. A 1998 Papua New Guinea

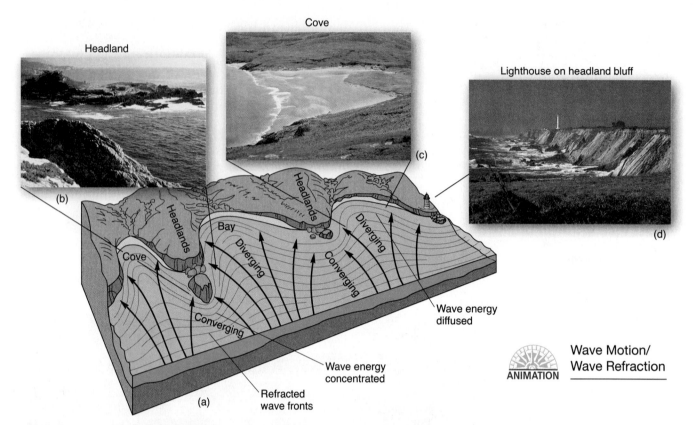

Wave Motion/
Wave Refraction

ANIMATION

FIGURE 13.8 Coastal straightening.
(a) The process of coastal straightening is brought about by wave refraction. (b) Wave energy is concentrated as it converges on headlands and (c) is diffused as it diverges in coves and bays. (d) Headlands are frequent sites for lighthouses such as the light at Point Arena, California. [Photos by (b) author; (c) and (d) Bobbé Christopherson.]

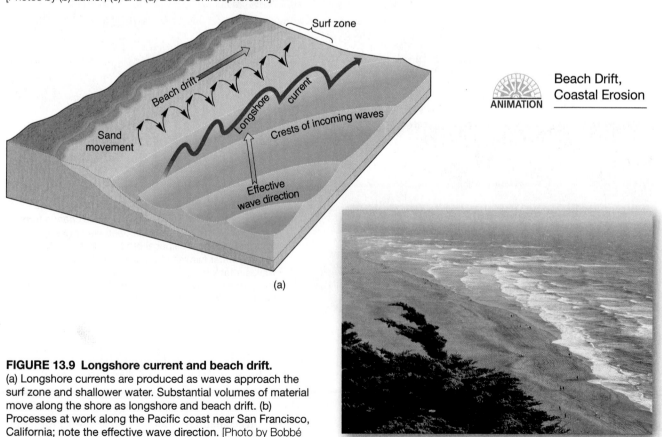

Beach Drift,
Coastal Erosion

ANIMATION

FIGURE 13.9 Longshore current and beach drift.
(a) Longshore currents are produced as waves approach the surf zone and shallower water. Substantial volumes of material move along the shore as longshore and beach drift. (b) Processes at work along the Pacific coast near San Francisco, California; note the effective wave direction. [Photo by Bobbé Christopherson.]

tsunami, launched by a massive undersea landslide of some 4 km³ (1 mi³), killed 2000. Earthquakes triggered all these waves. During the twentieth century there were 141 damaging tsunami and perhaps 900 smaller ones according to one estimate, with a total death toll of about 70,000. News Report 13.2 offers some details of the deadly Indian Ocean tsunami that hit in December 2004.

Hawai'i is vulnerable to tsunami because of its position in the open central Pacific, surrounded by the ring of fire that outlines the Pacific Basin. As an example, a tsunami produced near the Philippines would reach Hawai'i in 10 hours. The US Army Corps of Engineers has reported 41 damaging tsunami in Hawai'i during the past 142 years—statistically, 1 every 3.5 years. Forecasters watch the unstable Kīlauea region of the southeast coast of Hawai'i, where the possible collapse of a 40-km by 20-km by 2000-m (25-mi × 12-mi × 6560-ft) thick portion of relatively new and unstable crustal basalt could

affect the Pacific Basin. In the Atlantic Basin, there is continuing threat from volcanic island collapses, such as the closely watched Cumbre Vieja volcano in the Canary Islands.

Because tsunami travel such great distances so quickly and are undetectable in the open ocean, accurate forecasts are difficult and occurrences are often unexpected. A warning system is now in operation for nations surrounding the Pacific, where the majority of tsunami occur. A warning is issued whenever seismic stations detect a significant quake or landslide underwater, where it might generate a tsunami. A new tsunami hazard mitigation network began operating in 2004 with the deployment of fifteen pressure sensors on the Pacific Ocean floor with accompanying surface buoys—this is the Deep-ocean Assessment and Reporting of Tsunamis (DART).

No warning system was in place in the Indian Ocean as it is in the Pacific to detect the December 2004 tsunami

NEWS REPORT 13.2

Quake and Tsunami Strike!

On December 26, 2004, a magnitude 9.3 earthquake (fourth largest in a century) struck off the west coast of northern Sumatra, triggering a massive tsunami across the Indian Ocean—the Sumatra–Andaman earthquake and tsunami. The earthquake trigger was the continuing underthrust of the Indo-Australian plate beneath the Burma plate along the Sunda Trench subduction zone. Imagine the island of Sumatra springing up about 13.7 m (45 ft) in

elevation! You can find this oceanic trench on the Chapter 9 opening map of the ocean floor, stretching along the coast of Indonesia in the eastern Indian Ocean Basin.

Energy from this wave actually traveled several times around the global ocean basins before settling down. With Earth's mid-ocean ridge mountain chain acting as a guide, related large waves arrived at the shores of Nova Scotia, Antarctica, and Peru.

Figure 13.2.1 shows two satellite radar images and the extent of the tsunami, one at 2:05 hours and the other at 7:10 hours after the event. Geophysicists even picked up slight changes in Earth's rotation speed (slowed 3 microseconds) and a shift in the North Pole location by 2.5 cm (1 in.) in response to this event. Total deaths from the quake and tsunami exceeded 170,000, and a final count may never be known. Recovery will take years.

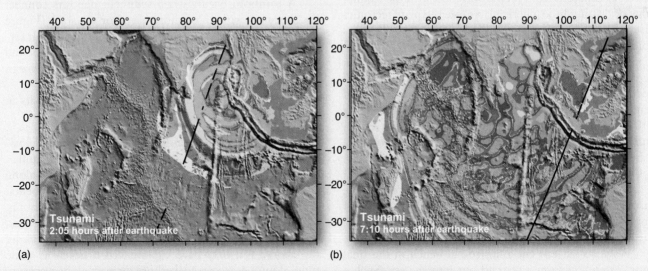

FIGURE 13.2.1 Satellites track tsunami waves across the Indian Ocean, 2004.
(a) A *Jason-1* image 2 hours after the magnitude 9.3 earthquake showing wave heights of 60 cm (24 in.) radiating outward from the epicenter off Sumatra. (b) The *GFO-satellite* radar captured the wave patterns 7 hours and 10 minutes after the earthquake. [Images courtesy of NOAA, Laboratory for Satellite Altimetry.]

disaster. NOAA provided images from four Earth-orbiting radar satellites, including *Jason-1*, *GFO* (*Geosat Follow-On*), *Envisat*, and *TOPEX/Poseidon*, that clearly showed the tsunami wave energy traveling across the Indian Ocean (News Report 13.2, Figure 13.1.1). The challenge to the world community is to apply this technology so that early warnings are possible in all vulnerable oceans, such as the Indian.

Warnings always should be heeded, despite the many false alarms, for the causes lie hidden beneath the ocean and are difficult to monitor in any consistent manner, although remote-sensing capabilities are improving accuracy. (For the tsunami research program, see **http://nctr.pmel .noaa.gov/**. The tsunami home page is at **http://www. ess.washington.edu/tsunami/index.html**.)

Coastal System Outputs

Coastlines are active portions of continents, with energy and sediment continuously delivered to a narrow environment. The action of tides, currents, wind, waves, and changing sea level produces a variety of erosional and depositional landforms. We look first at erosional coastlines such as the US West Coast, then at depositional coastlines as found along the East and Gulf coasts. In this era of rising sea level, coastlines are becoming more dynamic.

Erosional Coastal Processes and Landforms

The active margins of the Pacific Ocean along the North and South American continents are characteristic coastlines affected by erosional landform processes. *Erosional coastlines* tend to be rugged, of high relief, and tectonically active, as expected from their association with the leading edge of drifting lithospheric plates (see the plate tectonics discussion in Chapter 8). Figure 13.10 presents features commonly observed along an erosional coast.

Sea cliffs form by the undercutting action of the sea. As indentations are produced at water level, such a cliff becomes notched, leading to subsequent collapse and retreat of the cliff. Other erosional forms evolve along cliff-dominated coastlines, including *sea caves*, *sea arches*, and *sea stacks*. As erosion continues, arches may collapse, leaving isolated stacks in the water (Figure 13.10e). The coasts of England and Oregon are prime examples of such erosional landscapes. Figure 13.11 shows *Stac an Armin*, in the St. Kilda Archipelago, 165 km (103 mi) west of northern Scotland in the Atlantic Ocean. This sea stack, the highest in the British Isles, is 191 m (627 ft) tall.

Wave action can cut a horizontal bench in the tidal zone, extending from a sea cliff out into the sea. Such a structure is called a **wave-cut platform** or *wave-cut terrace*. If the relationship between the land and sea level changes over time, multiple platforms or terraces may rise like stair steps back from the coast. These marine terraces are remarkable indicators of an emerging coastline, with some terraces more than 365 m (1200 ft) above sea level. A tectonically active region, such as the California coast, has many examples of multiple wave-cut platforms (Figure 13.10f) and the many hazards associated with building there (Figure 13.10d).

Depositional Coastal Processes and Landforms

Depositional coasts generally are located along land of gentle relief, where sediments are available from many sources. Such is the case with the Atlantic and Gulf coastal plains of the United States, which lie along the relatively passive, trailing edge of the North American lithospheric plate. Erosional processes and inundation, particularly during storm activity, influence these depositional coasts.

Characteristic wave- and current-deposited landforms are illustrated in Figure 13.12. A **barrier spit** consists of material deposited in a long ridge extending out from a coast attached at one end; it partially crosses and blocks the mouth of a bay. Classic examples include Sandy Hook, New Jersey (south of New York City), and Cape Cod, Massachusetts. Figure 13.12 shows barrier spits forming partway across the mouths of the Little Sur River in California (Figure 13.12a) and Prion Bay in Southwestern National Park in Tasmania (Figure 13.12b).

A spit becomes a **bay barrier**, sometimes referred to as a *baymouth bar*, if it completely cuts off the bay from the ocean and forms an inland **lagoon**. Spits and barriers are made up of materials that have been eroded and transported by *littoral drift* (beach and longshore drift combined). For much sediment to accumulate, offshore currents must be weak. Tidal flats and salt marshes are characteristic low-relief features wherever tidal influence is greater than wave action.

A **tombolo** occurs when sediment deposits connect the shoreline with an offshore island or sea stack (Figure 13.12c) by accumulating on an underwater wave-built terrace. News Report 13.3 (p. 447) examines isostatic rebound of the crust that causes such shoreline features to rise in relation to sea level.

We next examine these transient coastal deposits in more detail. Let's go to the beach.

Beaches   Of all the features associated with a depositional coastline, beaches probably are the most familiar. Beaches vary in type and permanence, especially along coastlines dominated by wave action. Technically, a **beach** is that place along a coast where sediment is in motion, deposited by waves and currents. Material from the land temporarily resides there while it is in active transit along the shore. You may have experienced a beach at some time, along a seacoast, a lakeshore, or even along a stream. Perhaps you have even built your own "landforms" in the sand, only to see them washed away by the waves.

The beach zone ranges, on average, from 5 m (16 ft) above high tide to 10 m (33 ft) below low tide (see

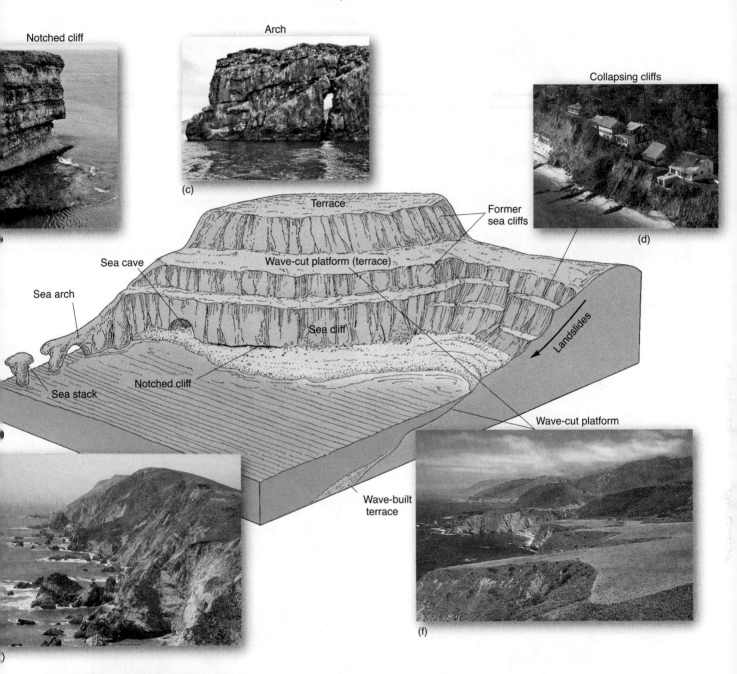

FIGURE 13.10 Erosional coastal features.
(a) Characteristic coastal erosional landforms: caves, (b) notched cliff, (c) sea arch, (d) collapsing cliffs, (e) sea stacks (see Fig. 13.11), and (f) wave-cut platforms near Bixby Bridge along the Cabrillo Highway. [Photos by (d) Lowell Georgia/Photo Researchers, Inc.; (e) author; (b), (c), and (f) Bobbé Christopherson.]

Figure 13.3). The specific definition varies greatly along individual shorelines. Worldwide, sands of quartz (SiO_2) dominate beaches because it resists weathering and therefore remains after other minerals are removed. In volcanic areas, beaches are formed from wave-processed lava. Hawai'i and Iceland feature examples of these black-sand beaches.

Many beaches, such as those in southern France and western Italy, are of pebbles and cobbles—a type of *shingle beach* (shingle meaning beach gravel and shells), shown in Figure 13.12d. Some shores have no beaches at all; scrambling across boulders and rocks may be the only way to

move along the coast. The coast of Maine and portions of the Atlantic provinces of Canada are classic examples, composed of resistant granite rock that is scenically rugged but forms few beaches.

A beach acts to stabilize a shoreline by absorbing wave energy, as is evident by the amount of material that is in almost constant motion (see "sand movement" in Figure 13.9). Some beaches are stable, whereas others cycle seasonally. Protected areas along a coastline tend to accumulate sand. It accumulates during the summer, is moved offshore by winter storm waves, forming a submerged bar, and is redeposited onshore the following summer.

Changes in coastal sediment transport can disrupt human activities—beaches are lost, harbors closed, and coastal highways and beach houses can be inundated with sediment. Consequently, various strategies are employed to interrupt longshore currents and beach drift. The goal of intervention is usually to halt sand accumulation or to force accumulation in a desired way through construction of engineered structures—"hard" shoreline protection.

Figure 13.13 illustrates common approaches: a *jetty* to block material from harbor entrances; a *groin* to slow drift action along the coast; and a *breakwater* to create a zone of still water near the coastline. However, interrupting

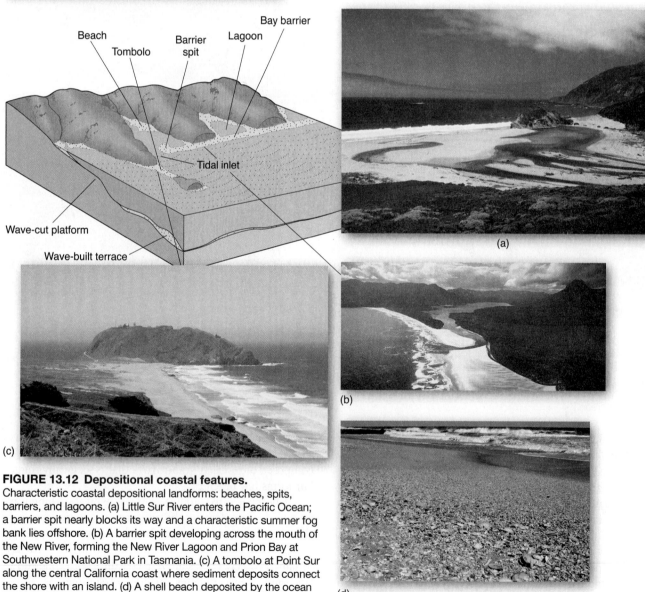

FIGURE 13.12 Depositional coastal features.
Characteristic coastal depositional landforms: beaches, spits, barriers, and lagoons. (a) Little Sur River enters the Pacific Ocean; a barrier spit nearly blocks its way and a characteristic summer fog bank lies offshore. (b) A barrier spit developing across the mouth of the New River, forming the New River Lagoon and Prion Bay at Southwestern National Park in Tasmania. (c) A tombolo at Point Sur along the central California coast where sediment deposits connect the shore with an island. (d) A shell beach deposited by the ocean along the New Jersey shore. [Photos by (b) Reg Morrison/Auscape International Pty. Ltd.; (a), (c), and (d) Bobbé Christopherson.]

444

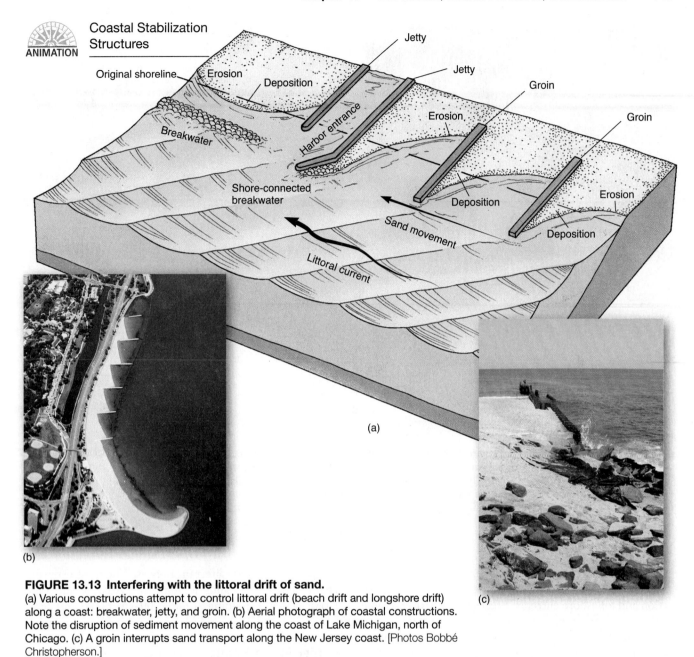

ANIMATION
Coastal Stabilization
Structures

FIGURE 13.13 Interfering with the littoral drift of sand.
(a) Various constructions attempt to control littoral drift (beach drift and longshore drift) along a coast: breakwater, jetty, and groin. (b) Aerial photograph of coastal constructions. Note the disruption of sediment movement along the coast of Lake Michigan, north of Chicago. (c) A groin interrupts sand transport along the New Jersey coast. [Photos Bobbé Christopherson.]

the coastal drift, which naturally replenishes beaches, may lead to unwanted changes in sediment distribution downcurrent. Careful planning and impact assessment should be part of any strategy for preserving or altering a beach.

Beach nourishment refers to the artificial placement of sand along a beach. Through such efforts, a beach that normally experiences a net loss of sediment will instead show a net gain. In contrast to hard structures, this hauling of sand to replenish a beach is considered "soft" shoreline protection; see News Report 13.4 (p. 450). Years of human effort and expense to build beaches can be erased by a single storm, as occurred when Hurricane Camille completely eliminated offshore islands along the Gulf Coast in 1969. Much sand was lost to Hurricane Georges in 1998 and also when Hurricanes Charley, Frances, Ivan, and Jeanne in 2004, Dennis, Katrina, Rita, and Wilma in 2005,

and Gustav and Ike in 2008, which devastated coastal beaches along the Gulf and Atlantic coasts.

Barrier Beach Formations Barrier chains are long, narrow, depositional features, generally of sand, that form offshore roughly parallel to the coast. Common forms are **barrier beaches** or the broader, more extensive landform, **barrier islands**. Tidal variation in the area usually is moderate to low, with adequate sediment supplies coming from nearby coastal plains. Figure 13.14 illustrates the many features of barrier chains, using North Carolina's famed Outer Banks as an example, including Cape Hatteras, across Pamlico Sound from the mainland. The area presently is designated as one of ten national seashore reserves supervised by the National Park Service.

On the landward side of a barrier formation are tidal flats, marshes, swamps, lagoons, coastal dunes, and beaches.

FIGURE 13.14 Barrier island chain.
(a) *Landsat* image of a barrier island chain along the North Carolina coast. You can see key depositional forms: spit, island, beach, lagoon, and inlet. A sound is a large inlet of the ocean; Pamlico Sound is an ideal example. (b) The path, partially a parking lot, along which the Cape Hatteras lighthouse was moved inland in 1999 to safer ground. (c) Cape Hatteras light in its new location 488 m farther inland. [(a) *Terra* MODIS image courtesy of NASA/GSFC; photos (b) and (c) by Bobbé Christopherson.]

Barrier beaches appear to adjust to sea level and may naturally shift position from time to time in response to wave action and longshore currents. Breaks in the barrier, visible in the image, are *inlets*, connecting the bay with the ocean. The name "barrier" is appropriate, for they take the brunt of storm energy and actually act as protection for the mainland. Because of the continuing loss occurring to a barrier island, the famous Cape Hatteras lighthouse was moved inland in 1999 to safer ground (Figure 13.14b). Its new position is about 488 m (1600 ft) from the ocean, or approximately the distance it was from the water in 1870 when it was built—that much sand was lost back to the sea. (See http://www.ncsu.edu/coast/chl/ for details of the extraordinary effort to save this landmark.)

However, vulnerable settlements fill the Outer Banks and Cape Hatteras and are repeatedly slammed by tropical storms—Hurricanes Emily, 1993; Dennis, Floyd, and Irene, 1999; Isabel in 2003; Alex and Bonnie in 2004; Ophelia in 2005; and Hanna in 2008—causing damage and beach erosion.

Barrier beaches and islands are common worldwide, lying offshore of nearly 10% of Earth's coastlines. Examples are found offshore of Africa, India (east coast), Sri Lanka, Australia, the north slope of Alaska, and the shores of the Baltic and Mediterranean Seas. The most extensive chain of barrier islands is along the Atlantic and Gulf Coast states, extending some 5000 km (3100 mi) from Long Island to Texas and Mexico.

Barrier Island Hazards Because many barrier islands are barely above sea level and seem to be migrating landward, they are an unwise choice for homesites or commercial building. Nonetheless, they are a common choice, even though they take the brunt of storm energy. The hazard represented by the settlement of barrier islands was made graphically clear when Hurricane Hugo assaulted South Carolina in 1989. The storm attacked the Grand Strand barrier islands off the northern half of South Carolina's coastline, affecting most of the Charleston and the South Strand portion of the islands. The storm made landfall as the worst hurricane to strike there in 35 years—beachfront houses, barrier-island developments, and millions of tons of sand were swept away.

The barrier islands off the Louisiana shore are disappearing. They are affected by subsidence through compaction of Mississippi delta sediments, a changing sea level that is rising at 1 cm (0.4 in.) per year in the region, and an increase in tropical storm intensity. In 1998, Hurricane

NEWS REPORT 13.3

A Rebounding Shoreline and Coastal Features

We learned in Chapter 8, Figure 8.4, that the lighter crust rests on denser materials beneath—the principle of buoyancy applies. Where the load is greater, owing to glaciers, sediment, or mountains, the crust tends to sink, or ride lower. When a glacier melts, for instance, the crust rides higher with the decreased load, in a recovery uplift known as *isostatic rebound*. Thus, the entire crust is in a constant state of compensating adjustment.

As Earth systems came out of the last ice age and the massive glaciers and continental ice retreated, the unburdened crust began its rebound, or *postglacial uplift*. The last continental ice retreated about 6000 years ago. The floor of Hudson Bay, Canada, for example, is rising at about 15 mm a year. Recently, as we see in Chapter 14, southeastern Alaska is experiencing an accelerated rise beyond the postglacial uplift associated with the last ice age—a rebound rate of 36 mm/year, discussed in Chapter 8. This new uplift is thought to be a result of the substantial losses of glacial ice related to increasing temperatures across Alaska.

In the Svalbard Archipelago in the Arctic Ocean north of Norway, along Duve Fjord on Nordaustlandet Island, the land is rising at approximately 2.5 m (8.3 ft) per century. Figure 13.3.1 shows landscape features caused by this phenomenon. The former shoreline appears in successive terraces, or raised beaches. Where a depositional tombolo had formed, the neck of land connecting to the offshore island is now stranded above sea level (Figure 13.3.1b).

(a)

(b)

FIGURE 13.3.1 A rebounding coast.
Along Duve Fjord, Nordaustlandet Island, Arctic Ocean: (a) Each of the curved terraces of cobbles along this coast is a former shoreline, formed as the island isostatically rebounded following the retreat of the glaciers. (b) A raised tombolo on the uplifting coast. [Photos by Bobbé Christopherson.]

Georges destroyed large tracts of the Chandeleur Islands, located 30 to 40 km (19 to 25 mi) from the Louisiana and Mississippi Gulf Coast, leaving the mainland with reduced protection (Figure 13.15). Louisiana's increasingly exposed wetlands are disappearing at rates of 65 km² (25 mi²) per year. Hurricane Katrina (2005) alone removed this amount of wetlands and swept away most of the remaining Chandeleur Islands (Figure 13.15c, d). A USGS coastal researcher, Abby Sallenger, said when viewing the scene after Katrina, "I've never seen it this bad, the sand is just gone." The regional office of the USGS predicts that in a few decades the barrier islands may be gone.

Figure 13.16a is an aerial photo of development along the Texas Gulf Coast, along the Bolivar Peninsula, a barrier spit near Galveston. Portions of the barrier island chain along the Texas coast are experiencing beach erosion rates averaging 2 m (6.5 ft) per year as sea level continues to rise. Hurricane Ike (September 13, 2008) struck Galveston Island, Bolivar Peninsula, and the Texas–Louisiana coast—landfall was in the area pictured in 13.16a. The coastal barrier island urbanization was devastated as billions of dollars were swept away along with more than three dozen lives, and hundreds remain missing. Surveys found 80% to 95% of homes were destroyed. The NOAA aerial photo in 13.16c is a valuable tool that helps local governments assess damage and organize recovery.

Biological Processes: Coral Formations

Not all coastlines form due to purely physical processes. Some form as the result of organic processes, such as coral growth. A **coral** is a simple marine animal with a

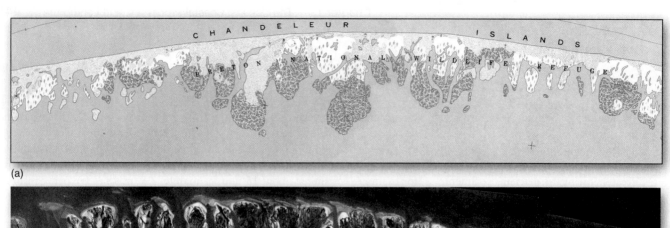

(a)

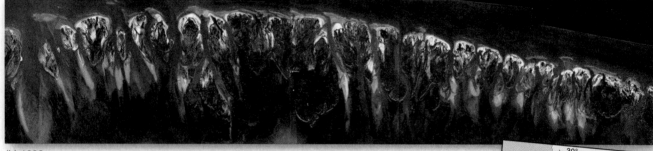

(b) 1988

(c) September 16, 2005

 Hurricane Isabel 9/6–9/19/04

SATELLITE

 Hurricane Georges

SATELLITE

FIGURE 13.15 Hurricanes Georges (1998) and Katrina (2005) take their toll.
Hurricane Georges eroded huge amounts of sand from the Chandeleur Islands, just off the Louisiana and Mississippi Gulf coasts. Compare the *before* topographic map (a) with the *after* aerial photo composite (b). (c) Little of the barrier-island chain is visible after Katrina's passage removed some 80% of the sand. (d) Aerial photo comparison between 2001 and two days after Katrina shows the devastation in the Northern Chandeleur Islands. The yellow arrow points to the same location in each photo. [(a) Topographic map provided by USGS; (b) photo by Aerial Data Service, Earth Imaging; (c) and (d) USGS Hurricane Impact Studies aerial photos—see *Sound Waves Newsletter*, September 2005.]

July 17, 2001

August 31, 2005

(d)

FIGURE 13.16 Coastal barrier formations devastated along the Texas Gulf coast.
(a) Bolivar Peninsula, a barrier spit, stands between Texas and the Gulf of Mexico. Note the
Intercoastal Waterway for shipping along Bolivar Peninsula. The scene is between Crystal Beach
and Gilcrist in the distance where the spit narrows. (b) Hurricane Ike moves toward this coast
September 12, 2008. The red dot notes the location of photo (a). (c) Devastation of Crystal Beach,
Texas, on the Bolivar Peninsula. Some seawater remains ashore in pools amid the sand dumped
by the surge; debris from former homes and businesses is everywhere. See before and after
photo comparison of Bolivar Peninsula devastation on p. 459. [(a) Photo by Bobbé Christopherson;
(b) *Terra* image courtesy of MODIS Direct Broadcast System, SSEC/NASA; (c) NOAA aerial photo
courtesy NOAA.]

small, cylindrical, saclike body called a polyp; it is related
to other marine invertebrates, such as anemones and jelly-
fish. Corals secrete calcium carbonate ($CaCO_3$) from the
lower part of their bodies, forming a hard external skeleton. Corals function in a *symbiotic* (mutually helpful) relationship with algae—that is, they live in close association
with the algae, and each depends on the other for survival.
Algae photosynthesize some of the food for the coral, and
in turn the corals provide some nutrients and shelter for
the algae.

Coral reefs are the most diverse marine ecosystems.
Preliminary estimates of coral species range to upward
of 9 million worldwide, yet, as in most ecosystems in
water or on land, biodiversity is declining in these
communities.

Figure 13.17 shows the distribution of currently living
coral formations. Corals thrive in warm tropical oceans, so
the difference in ocean temperature between the western
coasts and eastern coasts of continents is critical to their
distribution. Western coastal waters tend to be cooler,
thereby discouraging coral activity, whereas eastern coastal
currents are warmer and thus enhance coral growth.

Living colonial corals range in distribution from
about 30° N to 30° S and occupy a very specific ecological
zone: 10–55 m (30–180 ft) depth, 27‰–40‰ (parts per
thousand) salinity, and 18° to 29°C (64°–85°F) water
temperature. However, an interesting exception to this
are unique species of cool corals that exist in cold, dark
water and deep ocean, at temperatures as low as 4°C
(39°F) and in depths to 2000 m (6562 ft), beyond the

NEWS REPORT 13.4

Engineers Nourish a Beach

The city of Miami, Florida, and surrounding Dade County have spent almost $70 million since the 1970s in a continuing effort to rebuild their beaches. Sand is transported to the replenishment area from a source area. To maintain a 200-m-wide beach (660 ft), planners determine net sand loss per year and set a schedule for replenishment. In Miami Beach, an 8-year replenishment cycle is maintained. During Hurricane Andrew, in 1992, the replenished Miami Beach is thought to have prevented millions of dollars in shoreline structural damage. The cost of beach replenishment following the 2004 and 2005 hurricane onslaught and sand loss is still undetermined. Following Hurricane Wilma in 2005, massive replenishment of sands along Cancún, Mexico, tourist beaches took place. This work is already eroding away from active tropical storms and surge.

Unforeseen environmental impact may accompany the addition of sand to a beach, especially if the sand is from an unmatched source, in terms of its ecological traits. If the new sands do not physically and chemically match the existing varieties, disruption of coastal marine life is possible. The US Army Corps of Engineers, which operates the Miami replenishment program, is running out of "borrowing areas" for sand that

matches the natural sand of the beach. A proposal to haul a different type of sand from the Bahamas is being studied as to possible environmental consequences.

In contrast to the term *hard structures*, this hauling of sand to replenish a beach is considered "soft" shoreline protection (Figure 13.4.1). Enormous energy and material must be committed to counteract the relentless energy that nature expends along the coast. The hurricanes that hit Florida in 2004 and Louisiana in 2005, and

Louisiana and Texas in 2008, removed enormous quantities of sand from beaches and barrier islands. Each loss makes the coast that much more vulnerable. Nearly $2 billion has been spent from New York to Florida through 2005 to replenish beaches, with costs over the next 10 years forecasted to rise to $5.5 billion more. For more on beach nourishment, see **http://www.csc.noaa.gov/beachnourishment/** and **http://www.brynmawr.edu/geology/geomorph/beachnourishmentinfo.html**.

FIGURE 13.4.1 Beach sand replenishment.
Efforts to create protection for the beach and former primary dune must be continuous. Barrier islands are a dynamic system of rapid change; increasing storm activity and rising sea level place these coastal environments at risk. Note the sign to help keep the sand in place. One bad storm can wipe out these efforts. [Photo by Bobbé Christopherson.]

---

expected range. Scientists study these remarkable oddities that do not rely on algae but harvest nutrients from plankton and particulate matter.

Corals require clear, sediment-free water and consequently do not locate near the mouths of sediment-charged

freshwater streams. For example, note the lack of these structures along the US Gulf Coast.

Coral Reefs Most corals are colonial and their skeletons accumulate in enormous structures. Through many

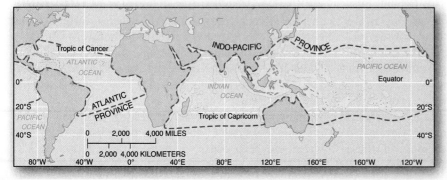

FIGURE 13.17 Worldwide distribution of living coral formations.
Yellow areas include prolific reef growth and atoll formation. The red dotted line marks the geographical limits of coral activity. Colonial corals range in distribution from about 30° N to 30° S. [After J. L. Davies, *Geographical Variation in Coastal Development* (Essex, England: Longman House, 1973). Adapted by permission.]

generations, live corals near the ocean's surface build on the foundation of older coral skeletons, which in turn may rest upon a volcanic seamount or some other submarine feature built up from the ocean floor. *Coral reefs* form by this process. Thus, a coral reef is a biologically derived sedimentary rock. It can assume one of several distinctive shapes.

In 1842 Charles Darwin presented a hypothesis for the evolution of reef formation: As reefs develop around a volcanic island and the island itself gradually subsides, equilibrium is maintained between the subsidence of the island and the upward growth of coral. This generally accepted idea is shown in Figure 13.18 with specific examples of each reef stage: *fringing reefs* (platforms of surrounding coral rock), *barrier reefs* (forming enclosed lagoons), and *atolls* (circular, ring-shaped).

Coral Bleaching A troubling phenomenon is occurring among corals around the world as normally colorful corals turn stark white by expelling their own nutrient-supplying, colorful algae (red-brown to green), a phenomenon known as *bleaching*. Exactly why the corals eject their symbiotic partner is unknown, for without algae the corals die. Scientists are tracking this unprecedented bleaching and dying worldwide. Locations in the Caribbean Sea and the Indian Ocean, as well as off the shores of Australia, Indonesia, Japan, Kenya, Florida, Texas, and Hawai'i, are experiencing this phenomenon.

Possible causes include local pollution, disease, sedimentation, and changes in salinity. Another cause is the 1 to 2 C° (1.8 to 3.6 F°) warming of sea-surface temperatures, as stimulated by greenhouse warming of the atmosphere. In a report, the *Status of Coral Reefs of the World: 2000* from the Global Coral Reef Monitoring Network, warmer water was found to be a greater threat to corals than local pollution or other environmental problems (see http://www.coris.noaa.gov/).

Coral bleaching is continuing all over the world as average ocean temperatures climb higher, thus linking the issue of climate change to the health of all living coral formations. By the end of 2000, approximately 30% of reefs were lost, especially following the record El Niño event of 1998. As

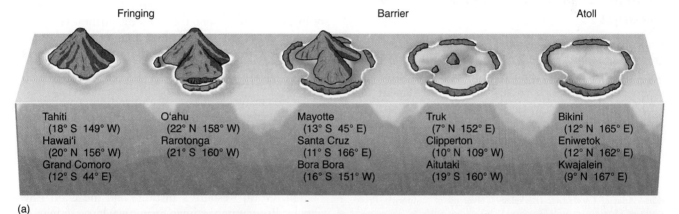

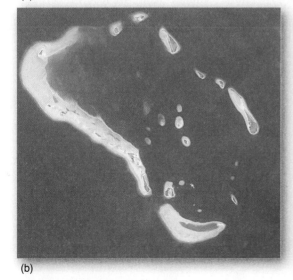

FIGURE 13.18 Coral formations.
(a) Common coral formations in a sequence of reef growth formed around a subsiding volcanic island: fringing reefs, barrier reefs, and an atoll. (b) Satellite image of a portion of the Maldive Islands, in the Indian Ocean (5° N 75° E). (c) Aerial photograph of the atolls in Bora Bora, Society Islands (16° S 152° W). [(a) After D. R. Stoddart, *The Geographical Magazine* 63 (1971): 610; (b) *Landsat-7* image courtesy of NASA; (c) photo by Harvey Lloyd/Stock Market.]

sea-surface temperatures continue to rise, forecasts state that coral losses will continue. For more information and Internet links, see http://www.usgs.gov/coralreef.html and the Coral Reef Monitoring Network at http://www.coral.noaa.gov/gcrmn/. The *World Atlas of Coral Reefs* summarized in 2001:

> Humans are thus bringing new pressures to bear on the world's coral reefs and driving more profound changes, more rapidly, than any natural impact has ever done. Overfishing has become so widespread that there are few, if any, reefs in the world which are not threatened. . . . From onshore a much greater suite of damaging activities is taking place. Often remote from reefs, deforestation, urban development, and intensive agriculture are now producing vast quantities of sediments and pollutants which are pouring into the sea and rapidly degrading coral reefs. . . . A further specter overshadowing the world of coral reefs is that of global climate change.*

Wetlands, Salt Marshes, and Mangrove Swamps

Some coastal areas have great *biological productivity* (plant growth, spawning grounds for fish, shellfish, and other organisms) stemming from trapped organic matter and sediments. Such a rich coastal marsh environment can greatly outproduce a wheat field in raw vegetation per acre. These wetland ecosystems are quite fragile and are threatened by human development.

*M. D. Spalding, C. Ravilious, and E. P. Green, and UNEP/WCMC, *World Atlas of Coral Reefs* (Berkeley: University of California Press, 2001), p. 11.

Wetlands are saturated with water enough of the time to support *hydrophytic vegetation* (plants that grow in water or very wet soil). Wetlands usually occur on poorly drained soils. Geographically, they occur not only along coastlines but also as bogs (peatlands with high water tables), as potholes in prairie lands, as cypress swamps (with standing or gently flowing water), as river bottomlands and floodplains, and as arctic and subarctic environments that experience permafrost during the year.

Coastal Wetlands   Coastal wetlands are of two general types—salt marshes and mangrove swamps. In the Northern Hemisphere, **salt marshes** tend to form north of the 30th parallel, whereas **mangrove swamps** form equatorward of that point. Freezing conditions control the survival of mangrove seedlings and thus dictate their distribution. Roughly the same latitudinal limits apply in the Southern Hemisphere.

Salt marshes usually form in estuaries and behind barrier beaches and spits. An accumulation of mud produces a site for the growth of *halophytic* (salt-tolerant) plants. This vegetation then traps additional alluvial sediments and adds to the salt marsh area. Because salt marshes are in the intertidal zone (between the farthest reaches of high and low tides), sinuous, branching channels are produced as tidal waters flood into and ebb from the marsh (Figure 13.19).

Sediment accumulation on tropical coastlines provides the site for mangrove trees, shrubs, and other vegetation. The prop roots of the mangrove are constantly finding new anchorages (Figure 13.20b, c). They are visible above the waterline but reach below the water surface, providing a habitat for a multitude of specialized life forms. Mangrove swamps often secure enough material to form islands.

The World Resources Institute and the U.N. Environment Programme estimate mangrove losses at between

FIGURE 13.19 Coastal salt marsh.
Salt marshes are productive ecosystems commonly occurring poleward of 30° latitude in both hemispheres. This is Gearheart Marsh, part of the Arcata Marsh system on the Pacific coast in northern California. [Photo by Bobbé Christopherson.]

FIGURE 13.20 Mangroves.
Mangroves tend to grow equatorward of 30° latitude. (a) Mangroves along the East Alligator
River (12° S latitude) in Kakadu National Park, Northern Territory, Australia. (b) Mangroves retain
sediments and can form anchors for island formations, seen here at sunset in "Ding" Darling
National Wildlife Refuge, Sanibel Island, Florida. (c) Mangroves in the Florida Keys. [Photos by
(a) Belinda Wright/DRK Photo; (b) Bobbé Christopherson; (c) author.]

40% (example, Cameroon and Indonesia) and nearly 80% (example, Bangladesh and the Philippines) since preagricultural times. Deliberate removal was a common practice by many governments because of false fears of disease or pestilence in these swamplands. Today, clearing for commercial and residential development is common.

Human Impact on Coastal Environments

The development of modern societies is closely linked to estuaries, wetlands, barrier beaches, and coastlines. Estuaries are important sites of human settlement because they provide natural harbors, a food source, and convenient sewage and waste disposal. Society depends on daily tidal flushing of estuaries to dilute pollution. Thus, the estuarine and coastal environment is vulnerable to abuse

and destruction if development is not carefully planned. We are approaching a time when most barrier islands will be developed and occupied (Figure 13.21).

Barrier islands and coastal beaches do migrate over time. Houses that were more than a mile from the sea in the Hamptons along the southeastern shore of Long Island, New York, are now within only 30 m (100 ft) of it (Figure 13.22). The time remaining for these homes can be quickly shortened by a single hurricane or by a sequence of storms with the intensity of the series that occurred in 1962 or the severe storm that struck the US and Canadian Atlantic coast in December 1992.

Despite our understanding of beach and barrier-island migration, the effect of storms, and warnings from scientists and government agencies, coastal development proceeds. Society behaves as though beaches and barrier islands are stable, fixed features, or as though they can be engineered to be permanent. Experience shows us that

(a)

FIGURE 13.21 Barrier island subdivisions.
The town of Barnegat Light abruptly stops where Long Beach Island is protected in Barnegat Lighthouse State Park, New Jersey. The protected area is almost entirely composed of replenished sand. [Photo by Bobbé Christopherson.]

severe erosion generally cannot be prevented. We witness the degradation of the cliffs of southern California, the shores of the Great Lakes, the New Jersey shore, the Gulf Coast, the devastation of coastal South Carolina by Hurricane Hugo, and the destruction brought by five recent hurricanes to North Carolina, by five hurricanes to the Gulf Coast, and by the nine hurricanes that hit in Florida that tore at the coast between 2004 and 2008. Shoreline planning is the topic of Focus Study 13.1.

A reasonable conclusion seems to be that the key to protective environmental planning and zoning is to *allocate responsibility and cost in the event of a disaster*. An ideal system places a hazard tax on land, based on assessed risk, and restricts the government's responsibility to fund reconstruction or an individual's right to reconstruct on frequently damaged sites. Comprehensive mapping of erosion-hazard areas would help avoid the ever-increasing costs from recurring disasters.

(b)

FIGURE 13.22 Beach erosion dooms houses in Massachusetts and New York.
(a) Two homes on stilts sit atop a primary berm made of cobbles. The ocean is just beyond the houses along this Massachusetts shore. (b) Placing houses on stilts along the shore, here on Long Island, offers little protection from an attacking storm surge. [Photos by (a) Bobbé Christopherson; (b) Mark Wexler/Woodfin Camp & Associates.]

FOCUS STUDY 13.1

An Environmental Approach to Shoreline Planning

Coastlines are places of wonderful opportunity. They also are zones of specific constraints. Poor understanding of this resource and a lack of environmental analysis often go hand in hand. Ecologist and landscape architect Ian McHarg, in his book *Design with Nature*, discusses the New Jersey shore. He shows how proper understanding of a coastal environment could have avoided problems from major storms and coastal development. Much of what is presented here

applies to coastal areas elsewhere. Figure 13.1.1 illustrates the New Jersey shore from ocean to back bay. Let us walk across this landscape and discover how it should be treated under ideal conditions.

Beaches and Dunes—Where to Build?

We begin our walk at the water's edge and proceed across the beach. Sand beaches are the primary natural defense against the ocean; they act as

bulwarks against the pounding of a stormy sea. The shoreline tolerates recreation but not construction, because of its shifting, changing nature during storms, daily tidal fluctuations, and the potential effects of rising sea level. Beaches are susceptible to pollution and require environmental protection to control nearshore dumping of dangerous materials.

Walking inland in Figure 13.1.1, we encounter the primary dune. The

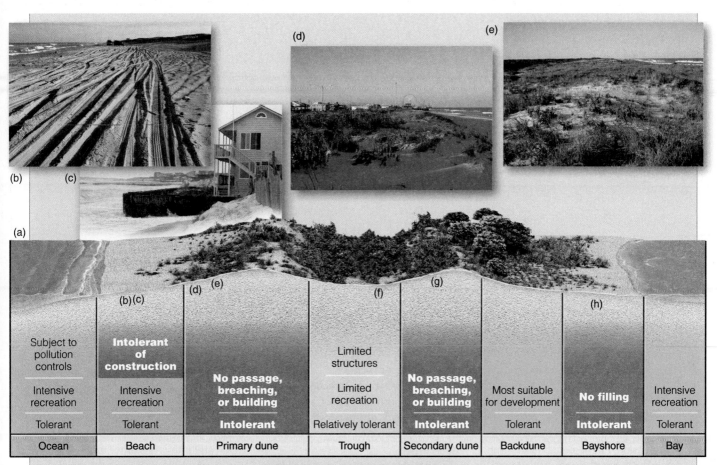

Subject to pollution controls	Intolerant of construction		No passage, breaching, or building	Limited structures	No passage, breaching, or building	Most suitable for development	No filling	Intensive recreation
Intensive recreation	Intensive recreation			Limited recreation				
Tolerant	Tolerant		Intolerant	Relatively tolerant	Intolerant	Tolerant	Intolerant	Tolerant
Ocean	Beach		Primary dune	Trough	Secondary dune	Backdune	Bayshore	Bay

FIGURE 13.1.1 Coastal environment and planning.

(a) Coastal environment—a planning perspective from ocean to bay along the New Jersey shore. Note the placement of each letter on the illustration to identify photo location. (b) Light trucks and SUVs drive along the beach in the protected Island Beach State Park, whereas they are banned from the stretches of beach that are developed with houses and business. (c) Houses once distant from the surf are hit by waves. Sand is brought in to strengthen defenses. (d) Development of an amusement park along the primary dunes challenges the natural setting. (e) The primary dune in a somewhat natural state. (f) An undeveloped trough behind the primary dune. (g) The former trough, paved and developed, barely above high tide levels. (h) Development and commerce in the bayshore and bay. [(a) After Design with Nature by Ian McHarg. Copyright © 1969 by Ian L. McHarg. Adapted by permission. All photos by Bobbé Christopherson.]

primary dune is even more sensitive than the beach. It is fragile, easily disturbed, and vulnerable to erosion; it cannot tolerate the passage of people trekking to the beach. Some vegetation is present, but it is easily disturbed and has only a small effect on sand stabilization. Primary dunes are like human-made dikes in the Netherlands: They are the next defense against the sea, so disturbance by development or heavy traffic should be limited. Carefully controlled access points to the beach should be designated, and restricted access should be enforced, for even foot traffic can cause destruction. (See the beach stabilization example in Figure 12.6.)

The trough behind the primary dune is relatively tolerant of limited recreation and building. The plants that affix themselves to the surface send

(continued)

Focus Study 13.1 *(continued)*

FIGURE 13.1.2 **Vulnerable coastal settlements along Cape Cod.**
Settlements along Beach Point on Cape Cod are vulnerable to sea-level rise, storm surges, and saltwater intrusion into groundwater systems. Houses are barely above sea level and use septic systems. Since the National Seashore was set aside, urbanization has developed in a patchwork in and around protected areas, making the Cape a complex of different jurisdictions, land usage, and plans. [Photo by Bobbé Christopherson.]

roots down to fresh groundwater reserves and anchor the sand in the process. Thus, if construction should inhibit the surface recharge of that water supply, the natural protective ground cover could die and destabilize the environment. Or subsequent saltwater intrusion might contaminate well water for human use. Clearly, groundwater resources and the location of recharge aquifers must be considered in planning. The continuing use of septic sewage systems in the trough is not sustainable in these environments.

Behind the trough is the secondary dune, another line of defense against the sea. It, too, is tolerant of some use yet is vulnerable to destruction. Next is the backdune, more suitable for development than any zone between it and the sea. Further inland are the bayshore and the bay, where no dredging or filling and only limited dumping of treated wastes and toxics should be permitted. This zone is tolerant to intensive recreation. In reality, the opposite of this careful assessment and planning prevails along most coastlines.

Cape Cod Vulnerability
Cape Cod, Massachusetts, a narrow coastal sandy peninsula composed of glacial outwash and morainal deposits, is one of the National Seashore reserves under the National Park Service. The Cape Cod National Seashore was

established in 1961. The Cape is subject to extensive human impacts and aggressive coastal processes. Over the years, population growth encroached across unprotected portions of the Cape, leaving a patchwork of interwoven private and public protected lands, marshes, and wetlands.

The urbanization outside the reserve is filled with houses and businesses barely above sea level (Figure 13.1.2). Most of the Cape has only septic systems for waste disposal with leach fields near sea level. These systems and most structures will be increasingly vulnerable to future sea-level rise. See http://pubs.usgs.gov/of/2002/of02-233/ for a report on the concept of coastal vulnerability along Cape Cod. Scientists are working to better understand this dynamic environment. Proper planning for the present and the future is made difficult as various stakeholders defend their positions.

A Scientific View and a Political Reality
Before an intensive government study of the New Jersey shore was completed in 1962, no analysis of coastal hazards had been done outside of academic circles. What is common knowledge to geographers, botanists, biologists, and ecologists in the classroom and laboratory still has not filtered through to the general planning and political

processes. As a result, on the New Jersey shore and along much of the Atlantic and Gulf coasts, improper development of the fragile coastal zone (primary dunes, trough, and secondary dunes) led to extensive destruction during storms in 1962, 1992, 2004, 2005, and numerous other times.

Scientists estimate that the coastlines will continue their retreat, in some cases tens of meters in a few decades. Such estimates include the entire East Coast around to the Gulf Coast. Society should reconcile ecology and economics if these coastal environments are to be sustained.

South Carolina enacted the Beach Management Act of 1988 (modified in 1990) to apply some of McHarg's principles. In its first 2 years, more than 70 lawsuits protested the act as an invalid seizure of private property without compensation. Similar measures in other states have faced the same difficult path.

Implementation of any planning process is problematic; political pressure is intense, and results are mixed. Ian McHarg voices an optimistic hope: "May it be that these simple ecological lessons will become known and incorporated into ordinance [law] so that people can continue to enjoy the special delights of life by the sea."*

*Ian McHarg, *Design with Nature*. © 1969 by Ian L. McHarg. (Bantam Doubleday Dell Publishing Group, Inc.), p. 17.

Summary and Review—The Oceans, Coastal Processes, and Landforms

■ *Describe* the chemical composition of seawater and the physical structure of the ocean.

Water is the "universal solvent," dissolving at least 57 of the 92 elements found in nature. Most natural elements and the compounds they form are found in the seas as dissolved solids. Seawater is a solution, and the concentration of dissolved solids is **salinity**. **Brine** exceeds the average 35‰ (parts per thousand) salinity; **brackish** applies to water that is less than 35‰. The ocean is divided by depth into a narrow mixing zone at the surface, a thermocline transition zone, and the deep cold zone.

> salinity (p. 430)
> brine (p. 432)
> brackish (p. 432)

1. Describe the salinity of seawater: its composition, amount, and distribution.
2. Review the table of ocean information in Figure 13.1. Locate each of the four oceans and the Southern Ocean on a map.
3. What are the three general zones relative to physical structure within the ocean? Characterize each by temperature, salinity, dissolved oxygen, and dissolved carbon dioxide.

■ *Identify* the components of the coastal environment and *list* the physical inputs to the coastal system, including tides and mean sea level.

The coastal environment is called the **littoral zone** and exists where the tide-driven, wave-driven sea confronts the land. Inputs to the coastal environment include solar energy, wind and weather, ocean currents and waves, climatic variation, and the nature of coastal rock. **Mean sea level (MSL)** is based on average tidal levels recorded hourly at a given site over many years. MSL varies spatially because of ocean currents and waves, tidal variations, air temperature and pressure differences, ocean temperature variations, slight variations in Earth's gravity, and changes in oceanic volume.

Tides are complex daily oscillations in sea level, ranging worldwide from barely noticeable to many meters. Tides are produced by the gravitational pull of both the Moon and the Sun. Most coastal locations experience two high (rising) *flood tides* and two low (falling) *ebb tides* every day. The difference between consecutive high and low tides is the tidal range. *Spring tides* exhibit the greatest tidal range, when the Moon and Sun are either in conjunction or opposition. *Neap tides* produce a lesser tidal range.

> littoral zone (p. 433)
> mean sea level (MSL) (p. 434)
> tides (p. 434)

4. What are the key terms used to describe the coastal environment?
5. Define mean sea level. How is this value determined? Is it constant or variable around the world? Explain.
6. What interacting forces generate the pattern of tides?
7. What characteristic tides are expected during a new Moon or a full Moon? During the first-quarter and third-quarter phases of the Moon? What is meant by a flood tide? An ebb tide?

8. Is tidal power being used anywhere to generate electricity? Explain briefly how such a plant would utilize the tides to produce electricity. Are there any sites in North America? Where are they?

■ *Describe* wave motion at sea and near shore and *explain* coastal straightening as a product of wave refraction.

Friction between moving air (wind) and the ocean surface generates undulations of water that we call **waves**. Wave energy in the open sea travels through water, but the water itself stays in place. Regular patterns of smooth, rounded waves, called **swells**, are the mature undulations of the open ocean. Near shore, the restricted depth of water slows the wave, forming *waves of translation*, in which both energy and water actually move toward shore. As the crest of each wave rises, the wave falls into a characteristic **breaker**.

Wave refraction redistributes wave energy so that different sections of the coastline vary in erosion potential. Headlands are eroded, whereas coves and bays receive materials, with the long-term effect of straightening the coast. As waves approach a shore at an angle, refraction produces a **longshore current** of water moving parallel to the shore, producing the *longshore drift*, which moves sand, sediment, and gravel and assorted materials along the beach as *beach drift*. **Littoral drift** is transport of materials along the shore; a more comprehensive term, it combines *beach drift* and *longshore drift*.

A **tsunami** is a seismic sea wave triggered by an undersea landslide or earthquake. It travels at great speeds in the open sea and gains height as it comes ashore, posing a coastal hazard.

> waves (p. 438)
> swell (p. 438)
> breaker (p. 438)
> wave refraction (p. 438)
> longshore current (p. 438)
> littoral drift (p. 438)
> tsunami (p. 438)

9. What is a wave? How are waves generated, and how do they travel across the ocean? Does the water travel with the wave? Discuss the process of wave formation and transmission.
10. Describe the refraction process that occurs when waves reach an irregular coastline. How is the coastline straightened?
11. Define the components of beach drift and the longshore current and longshore drift.
12. Explain how a seismic sea wave attains such tremendous velocities. Why is it given a Japanese name?

■ *Identify* characteristic coastal erosional and depositional landforms.

An *erosional coast* features wave action that cuts a horizontal bench in the tidal zone, extending from a sea cliff out into the sea. Erosional features include *stacks*, *arches*, and *cliffs*. Such a structure is a **wave-cut platform**, or *wave-cut terrace*. In contrast, *depositional coasts* generally are located along land of gentle relief, where depositional sediments are available from many sources. Characteristic landforms deposited by waves and currents are a **barrier spit** (material deposited in a long ridge extending out from a coast); a **bay barrier**, or *baymouth bar* (a spit that cuts off the bay from the ocean and forms an inland **lagoon**); a **tombolo** (where sediment deposits connect the shoreline with an offshore island or sea stack); and a **beach** (temporary land along the shore where sediment is in motion, deposited by waves and currents). A beach helps to stabilize the shoreline, although it may be unstable seasonally.

> wave-cut platform (p. 442)
> barrier spit (p. 442)
> bay barrier (p. 442)
> lagoon (p. 442)
> tombolo (p. 442)
> beach (p. 442)

13. What is meant by an erosional coast? What are the expected features of such a coast?

14. What is meant by a depositional coast? What are the expected features of such a coast?

15. How do people attempt to modify littoral drift? What strategies are used? What are the positive and negative impacts of these actions?

16. Describe a beach—its form, composition, function, and evolution.

17. What success has Miami had with beach replenishment? Is it a practical strategy?

■ *Describe* barrier islands and their hazards and *relate* them to human settlement.

Barrier chains are long, narrow, depositional features, generally of sand, that form offshore roughly parallel to the coast. Common forms are **barrier beaches** and the broader, more extensive **barrier islands**. Barrier formations are transient coastal features, constantly on the move, and they are a poor, but common, choice for development.

> barrier beach (p. 445)
> barrier island (p. 445)

18. On the basis of the information in the text and any other sources at your disposal, do you think barrier islands and beaches should be used for development? If so, under what conditions? If not, why not?

19. After Hurricane Hazel destroyed the Grand Strand off South Carolina in 1954, settlements were rebuilt, only to be hit by Hurricane Hugo 35 years later, in 1989. After the storm events along the Gulf Coast in 2005 and the devastation of Hurricane Ike along the Texas coast in 2008, what

do you think the future holds—relocate, rebuild, abandon? Why does this type of recurring damage happen to human populations? Propose some ways that these events could have been avoided.

■ *Assess* living coastal environments: corals, wetlands, salt marshes, and mangroves.

A **coral** is a simple marine invertebrate that forms a hard, calcified, external skeleton. Over generations, corals accumulate in large reef structures. Corals live in a *symbiotic* (mutually helpful) relationship with algae; each is dependent on the other for survival.

    Wetlands are lands saturated with water that support specific plants adapted to wet conditions. They occur along coastlands and inland in bogs, swamps, and river bottomlands. Coastal wetlands form as **salt marshes** poleward of the 30th parallel in each hemisphere and as **mangrove swamps** equatorward of these parallels.

> coral (p. 447)
> wetlands (p. 452)
> salt marsh (p. 452)
> mangrove swamp (p. 452)

20. How are corals able to construct reefs and islands?

21. Describe a trend in corals that is troubling scientists, and discuss some possible causes.

22. Why are the coastal wetlands poleward of 30° N and S latitude different from those that are equatorward? Describe the differences.

■ *Construct* an environmentally sensitive model for settlement and land use along the coast.

Coastlines are zones of specific constraints. Poor understanding of this resource and a lack of environmental analysis often go hand in hand in producing frequent disastrous disruption of coastal ecosystems and property losses. Society must reconcile ecology and economics if these coastal environments are to be sustained.

23. Describe the condition of the Chandeleur Islands in the Gulf of Mexico before and after Hurricane Georges in 1998 and Katrina in 2005 (Figure 13.15). What does this tell us about the fate of other barrier islands?

24. What type of environmental analysis is needed for rational development and growth in a region like the New Jersey shore? Examine, analyze, and explain the photographs in Focus Study 13.1, Figure 13.1.1. Evaluate South Carolina's approach to coastal hazards and protection.

NetWork and Critical Thinking Tools

A. This chapter concludes with the following statement:

> A reasonable conclusion seems to be that the key to protective environmental planning and zoning is to *allocate responsibility and cost in the event of a disaster*. An ideal system places a hazard tax on land, based on assessed risk, and restricts the government's responsibility to fund reconstruction or an

individual's right to reconstruct on frequently damaged sites. Comprehensive mapping of erosion-hazard areas would help avoid the ever-increasing costs from recurring disasters.

What do you think about this as a policy statement? How would you approach implementing such a strategy? In what way could you use geographic information systems (GIS), as described in Chapter 1, to survey, assess, list owners, and follow taxation status for a vulnerable stretch of coastline?

B. Under "Destinations" in Chapter 13 of the *Elemental Geosystems* Student Learning Center, there is a link called "Coral Reefs." Sample some of the buttons on this page. Do you find any information about the damage to and bleaching of

coral reefs reported in 1998 or 1999? Which places in the world were affected? Are there some suspect causes presented in "Coral Bleaching Hot Spots" or any of the "Global Coral Reef Alliance" references? Do you find any current status reports concerning coral bleaching?

September 15, 2008

Bolivar Peninsula, TX before and after Hurricane Ike. [USGS photo comparisons, September 9 and 15, 2008.]

A dramatic moment at the 14th July Glacier (Fjortende Julibreen), west coast of Spitsbergen Island near 79° N, June 22, 2006. An active glacial front calves blocks of glacial ice as it moves forward into a body of water. Five seconds after the main photo, the camera caught an enormous calving, shown in the inset. Wave energy quickly spread across the bay, producing breakers. The glacial front is approximately 35 m (115 ft) high, yet when the ice plunges into the water heaving up and down to establish equilibrium, it reaches a point where it is 1/7th exposed and 6/7th submerged—an iceberg. *[Photos by Bobbé Christopherson.]*

Chapter

14

Glacial and Periglacial Landscapes

KEY LEARNING CONCEPTS

After reading the chapter, you should be able to:

- **Differentiate** between alpine and continental glaciers and **describe** their principal features.

- **Describe** the process of glacial ice formation and **portray** the mechanics of glacial movement.

- **Describe** characteristic erosional and depositional landforms created by alpine glaciation and continental glaciation.

- **Analyze** the spatial distribution of periglacial processes and **describe** several unique landforms and topographic features related to permafrost and frozen ground phenomena.

- **Explain** the Pleistocene ice-age epoch and related glacials and interglacials and **describe** some of the methods used to study paleoclimatology.

STUDENT LEARNING CENTER TOOLS

The *Elemental Geosystems* Student Learning Center provides on-line resources for this chapter by clicking on the book cover at **www.mygeoscienceplace.com**. Some highlights:

▶ Read Career Links about an avalanche scientist in the Rockies and an environmental geochemist in Churchill, Manitoba, who both work with snow and ice and the cryosphere.

▶ Case Studies highlight "Ice Cores and High-Elevation Climatology" and present a 14-month diary kept by those stationed at the South Pole, "Living at the South Pole."

▶ Two animations offer glacial budgets and ice flows within glaciers, bringing these concepts to life.

▶ Two narrated photo galleries offer scenes of East Greenland and polar bears, and an author video clip shows you "Mom and Cubs on the Pack Ice."

▶ More than 15 additional URLs along with Quick Links to all the URLs in the chapter are included.

A large measure of the freshwater on Earth is frozen, with the bulk of that ice sitting restlessly in just two places—Greenland and Antarctica. The remaining ice covers various mountains and fills some alpine valleys. More than 29 million km³ (7 million mi³) of water is tied up as ice, or about 77% of all freshwater. These deposits of ice provide an extensive frozen record of Earth's climatic history over the past several million years and perhaps some clues to its climatic future. This is Earth's **cryosphere**, the portion of the hydrosphere and groundwater that is perennially frozen, generally at high latitudes and elevations.

Worldwide, mountain glacial ice is melting at rates that exceed anything in the ice record. Later in this chapter the loss of Alaska's glacial ice is documented along with its contribution to rising sea levels. In 2007, the Arctic Ocean sea ice reduced to its smallest extent in the past century, as regional temperatures set records of more than 5 C° (9 F°) above normal. The ice loss in 2008 was second to this record. Earth's cryosphere is in a state of dramatic change, as worldwide glacial and polar ice retreats. In the European Alps alone some 75% of the glaciers have receded in the past 50 years, losing more than 50% of the ice mass accumulated since 1850.

In this chapter: We focus on Earth's extensive ice deposits—their formation and movement and the ways in which they produce various erosional and depositional landforms. Glaciers, transient landforms themselves, leave in their wake a variety of landscape features. The fate of glaciers is intricately tied to change in global temperature, which ultimately concerns us all. We discuss the methods used to decipher past climates—the science of *paleoclimatology*—and the clues for understanding future climate patterns. Exciting discoveries are being made from the study of ice cores taken from Earth's two ice sheets, one dating back 800,000 years from the present.

We examine the cold, near-glacial world of permafrost and periglacial processes. Approximately 25% of Earth's land area is subject to freezing conditions and frost action characteristic of periglacial regions. These areas near former and existing glaciers are reminders of the last ice age. The chapter ends at Earth's poles, the Arctic Ocean and Antarctica. Higher latitudes are experiencing greater increases in average temperatures than are lower latitudes.

Rivers of Ice

A **glacier** is a large mass of ice, resting on land or floating shelflike in the sea adjacent to land. Glaciers are not frozen lakes or groundwater ice but form by the continual accumulation of snow that recrystallizes into an ice mass. Glaciers move under the pressure of their own great weight and the pull of gravity. In fact, they move slowly in streamlike patterns as "rivers" of ice. Tributary glaciers flow into a compound glacier and merge alongside one another by extending and thinning rather than by blending, as do rivers. Each tributary maintains its own patterns of transported debris, as you can see in the image and photo in Figure 14.1.

(a)

(b)

(c)

FIGURE 14.1 Rivers and sheets of ice.
(a) Alpine glaciers merge from adjoining glacial valleys in the northeast region of Ellesmere Island (80° N 75° W) in the Canadian Arctic. (b) Note how each lateral moraine (dark streaks of rock debris) remains identifiable as they combine in this valley glacier. How many moraines can you count? Note the medial moraine flowing from the upper-left corner of the photo.
(c) These peaks, known as nunataks, rise above the Greenland Ice Sheet. There are no clouds in the photo, for this is all ice at approximately 2500 m (8200 ft) elevation—an accumulation of ice perhaps 100,000 years in the making. [(a) *Terra* ASTER image courtesy of University of Alberta, NASA/GSFC/ERSDAC/JAROS, and the US/Japan ASTER Science Team, July 31, 2000; (b) and (c) photos by Bobbé Christopherson.]

Today, these slowly flowing rivers and sheets of ice cover about 11% of Earth's land area. During colder episodes in the past as much as 30% of continental land was covered by glacial ice. Through these "ice ages," below-freezing temperatures prevailed at lower latitudes more than they do today, allowing snow to accumulate year after year.

Glaciers form in areas of permanent snow, both at high latitudes and high elevations. A *snow line* is the lowest elevation where snow remains year-round—specifically, the lowest line where winter snow accumulation persists throughout the summer. Glaciers exist on some high mountains along the equator, such as in the Andes Mountains of South America and on Mount Kilimanjaro in Tanzania. In equatorial mountains, the snow line is around 5000 m (16,400 ft); on midlatitude mountains, such as the European Alps, snow lines average 2700 m (8850 ft); and in southern Greenland snow lines are down to 600 m (1970 ft) or less. (For Internet links to Global Land Ice Measurements from Space, go to http://www.glims.org/, which includes an inventory of world glaciers, or the National Snow and Ice Data Center at http://nsidc.org/. *Landsat*-7 images are listed at http://www.emporia.edu/earthsci/gage/glacier7.htm.)

Glaciers are as varied as the landscape itself; however, they fall within two general groups, based on their form, size, and flow characteristics: alpine glaciers and continental glaciers.

Alpine Glaciers

With few exceptions, a glacier in a mountain range is an **alpine glacier**, or *mountain glacier*. The *snowfield* that continually feeds the glacier with new snow is at a higher elevation. The name comes from the Alps of central Europe, where such glaciers abound. Alpine glaciers form in several subtypes. One prominent type is a *valley glacier*, an ice mass confined within a valley that originally formed by stream action. Such glaciers range in length from only 100 m (325 ft) to over 100 km (60 mi). In Figure 14.1a and b, how many valley glaciers join the main glaciers?

In Figure 14.2, at least a dozen valley glaciers are identifiable in the high-altitude photograph of the Eldridge and Ruth Glaciers. They fill valleys as they flow from source areas near Mount McKinley.

As a valley glacier flows slowly downhill, the mountains, canyons, and river valleys beneath its mass are profoundly altered by its erosive passage. Some of the debris created by the glacier's excavation is transported in and on the ice, visible as dark streaks of material being carried for deposition elsewhere.

Most alpine glaciers originate in a mountain snowfield that is confined in a bowl-shaped recess. This scooped-out erosional landform at the head of a valley is a **cirque**, and a glacier that forms in a cirque is a *cirque glacier*. Several cirque glaciers may jointly feed a valley glacier (see Figure 14.5d, p. 466). Numerous cirques and cirque glaciers appear in the high-altitude photo in Figure 14.2.

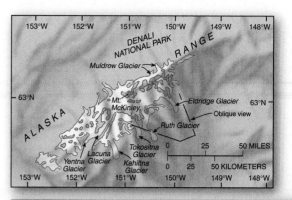

FIGURE 14.2 Glaciers in south-central Alaska.
Oblique infrared (false-color) photo of Eldridge and Ruth Glaciers, with Mount McKinley at upper left, in the Alaska Range of Denali National Park. Photo made at 18,300 m (60,000 ft). [Alaska High Altitude Aerial Photography from EROS Data Center, USGS.]

Wherever several valley glaciers pour out of their confining valleys and coalesce at the base of a mountain range, a *piedmont glacier* is formed and spreads freely over the nearby lowlands. A *tidal glacier*, such as the Columbia Glacier on Prince William Sound in Alaska, ends in the sea, *calving* (breaking off) to form floating ice, known as **icebergs** (Figure 14.3). Icebergs usually form wherever glaciers meet an ocean, bay, or fjord. Icebergs are inherently unstable, as their center of gravity shifts with melting and further breaking. Figure 14.3c shows such adjustments; note the former waterline, melt channels, and delicately sculpted underwater surfaces, far above water level. In Figure 14.3d, a magnificent ice tower brings to mind the ratio of 1/7 (14%) exposed and 6/7 (86%) submerged—these proportions vary depending on the age of the ice and air content.

Continental Glaciers

On a larger scale than individual alpine glaciers, a continuous mass of ice is known as a **continental glacier** and in its most extensive form is an **ice sheet**. Most glacial ice exists in the snow-covered ice sheets that blanket 80% of Greenland and 90% of Antarctica. Antarctica alone has 91% of all the glacial ice on the planet. A portion of the Greenland Ice Sheet is shown in Figure 14.1c.

(a)

(b)

(c)

(d)

FIGURE 14.3 Piedmont glacier and ice calving into the sea.
(a) When glaciers reach the sea, large pieces begin calving off, forming icebergs, such as these massive 30-m- (98-ft-) high blocks ready to fall at Neko Harbor, Antarctica. (b) A tidewater glacier calving numerous icebergs into the ocean in southern Greenland. (c) A weathered iceberg off Antarctica; note the former waterline marking where the berg floated in the water when it was greater in mass. (d) This iceberg tower in East Greenland is about 60 m (200 ft) tall, so imagine its keel extending some 360 m (1180 ft) below the surface. [All photos by Bobbé Christopherson.]

The Antarctic and Greenland ice sheets represent such an enormous mass that large regions of crust within each landmass are isostatically depressed (under the weight of the ice) more than 2000 m (6500 ft) below sea level. Each ice sheet is more than 3000 m (9800 ft) deep, burying all but the highest peaks.

Ice caps and *ice fields* are two additional types of continuous ice cover associated with mountain locations. An **ice cap** is roughly circular and covers an area of less than 50,000 km² (19,300 mi²); it completely buries the underlying landscape. The volcanic island of Iceland features several ice caps, such as the Vatnajökull Ice Cap (Figure 14.4a). Iceland's Grímsvötn Volcano lies beneath the ice cap and

erupted in 1996, producing large quantities of melted glacial water and floods, a flow called *jökulhlaup* by Icelanders.

An **ice field** extends in a characteristic elongated pattern in a mountainous region, with ridges and peaks visible above the buried terrain. Figure 14.4b shows the southern Patagonian ice field in the Andes, one of Earth's largest, stretching 360 km (224 mi) in length between 46° S and 51° S latitude and up to 90 km (56 mi) in width. An ice field is not extensive enough to form the characteristic dome of an ice cap.

Glacial Processes

A glacier is a dynamic body, moving relentlessly downslope at rates that vary within its mass, excavating the landscape through which it flows. The mass is dense ice that forms from snow and water through a complex process over time. A glacier's *mass budget* consists of net gains or losses of this glacial ice, which determine whether the glacier expands or

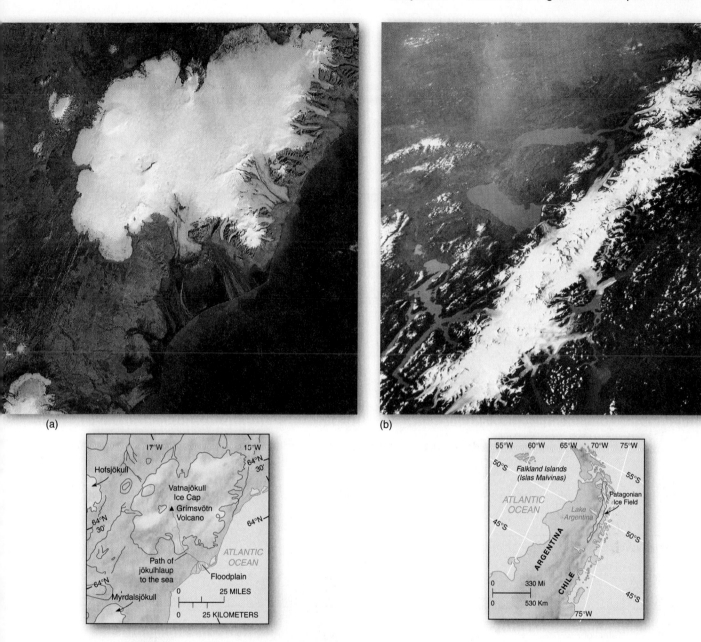

FIGURE 14.4 Ice cap and ice field.
(a) The Vatnajökull ice cap in southeastern Iceland (*jökull* means "ice cap" in Danish). Note the location of the Grímsvötn Volcano on the map and the path of the floods to the sea. (b) The southern Patagonian ice field of Argentina. [(a) *Landsat* image from NASA; (b) photo by Cosmonauts G. M. Greshko and Yu V. Romanenko, *Salyut 6*.]

retreats. Let us now look at glacial ice formation, mass balance, movement, and erosion before we discuss the fascinating landforms produced by these processes.

Formation of Glacial Ice

Dense ice comprises a glacier, formed from snow and water through a process of compaction, recrystallization, and growth. The essential input to a glacier is precipitation that accumulates in a *snowfield*, a glacier's accumulation zone (Figure 14.5a and c). Snowfields usually are at the highest elevation of an ice sheet, ice cap, or head of a valley glacier—usually in a cirque. Highland snow accumulation sometimes is the product of orographic processes, noted in

Chapter 5. Avalanches from surrounding mountain slopes can add to the snowfield.

As the snow accumulation deepens in sedimentary-like layers, the increasing thickness results in increased weight and pressure on underlying portions. Rain and summer snowmelt then contribute water, which stimulates further melting, and that meltwater seeps down into the snowfield and refreezes.

The slow transformation into glacial ice begins with snow that survives the summer into the following winter. Air spaces among ice crystals are pressed out as snow packs to a greater density. The ice crystals recrystallize and consolidate under pressure. In a transition step to glacial ice, snow becomes **firn**, which has a compact, granular texture.

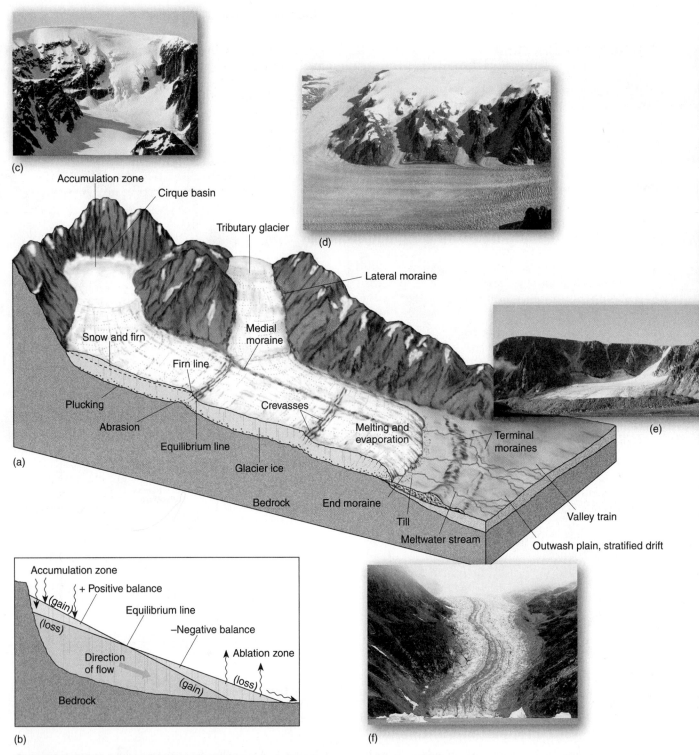

(c)

(d)

Accumulation zone

Cirque basin

Tributary glacier

Lateral moraine

Snow and firn

Medial moraine

Firn line

Plucking

Crevasses

Abrasion

Melting and evaporation

Terminal moraines

(e)

Equilibrium line

Glacier ice

Bedrock

End moraine

Till

Valley train

Meltwater stream

Outwash plain, stratified drift

(a)

Accumulation zone

+ Positive balance

(gain)

(loss)

Equilibrium line

−Negative balance

Direction of flow

Ablation zone

(gain)

(loss)

Bedrock

(b)

(f)

FIGURE 14.5 A retreating alpine glacier and mass balance.
(a) Cross section of a typical retreating alpine glacier. (b) Annual mass balance of a glacial system, showing how the relation between accumulation and ablation controls the location of the equilibrium line. (c) The cirque accumulation zone for a glacier in Antarctica. (d) Four tributary glaciers flow into a compound valley glacier in Greenland. (e) A terminal moraine marks the farthest advance of this glacier on Nordaustlandet Island, Arctic Ocean. (f) Moraine rock and debris stain this glacier in Greenland as it advances to the sea. [All photos by Bobbé Christopherson.]

ANIMATION

Budget of a Glacier, Mass Balance, Flow of Ice Within a Glacier

As this process continues, many years pass before dense glacial ice is produced. Formation of **glacial ice** is analogous to formation of metamorphic rock: Sediments (snow and firn) are pressured and recrystallized into a dense metamorphic rock (glacial ice). In Antarctica, glacial ice formation may take 1000 years because of the dryness of the climate (minimal snow input), whereas in more moist climates this time is reduced to just a few years because of the volume of new snow constantly being added to the system.

Glacial Mass Balance

A glacier is an open system, with *inputs* of snow and *outputs* of ice, meltwater, and water vapor. In the *accumulation zone* at its upper end, snowfall and other moisture feed a glacier (Figure 14.5a). This area ends at the **firn line**, indicating where the winter snow and ice accumulation survived the summer melting season.

Toward a glacier's lower end, it is wasted (reduced) through several processes: melting on the surface, internally, and at its base; ice removal by deflation (wind); the calving of ice blocks; and sublimation (recall from Chapter 5 that this is the direct evaporation of ice). Collectively, these losses are **ablation**.

The zone where accumulation gain balances ablation loss is the **equilibrium line** (Figure 14.5b). This area of a glacier generally coincides with the firn line. A glacier achieves *positive net balance* of mass—grows larger—during cold periods with adequate precipitation. In warmer times, the equilibrium line migrates up-glacier, and the glacier retreats—grows smaller—because of its *negative net balance*. Internally, gravity continues to move a glacier forward even though its lower terminus might be in retreat owing to ablation. The mass-balance losses from the South Cascade glacier and Alaskan glaciers provide two examples in News Report 14.1.

Highways of ice that flow from accumulation areas high in the mountains are marked by trails of transported debris called *moraines* (Figure 14.5e and f). A *lateral moraine* accumulates along the sides. A *medial moraine* forms down the middle when two glaciers merge and their lateral moraines combine, as we saw in Figure 14.1.

Glacial Movement

We generally think of ice as those small, brittle cubes from the freezer, but glacial ice is quite different. In fact, glacial ice behaves in a plastic (pliable) manner, for it distorts and flows in its lower portions in response to weight and pressure from above and the degree of slope below. In contrast, the glacier's upper portions are quite brittle. Rates of flow range from almost nothing to a kilometer or

NEWS REPORT 14.1

South Cascade and Alaskan Glaciers Lose Mass

As part of a global trend, glacial mass budgets are running in the negative, on average, with significant ice losses in Alaska, the Andes, the European Alps, and the Himalayas. The net mass balance of the South Cascade Glacier in Washington state demonstrated significant losses between 1955 and 2006. In just 1 year (September 1991 to October 1992), the terminus of the glacier retreated 38 m (125 ft), resulting in major changes to the surface and sides of the glacier, with more than a 2% loss of the glacier's mass. The net accumulation and net wastage are illustrated in Figure 14.1.1.

(continued)

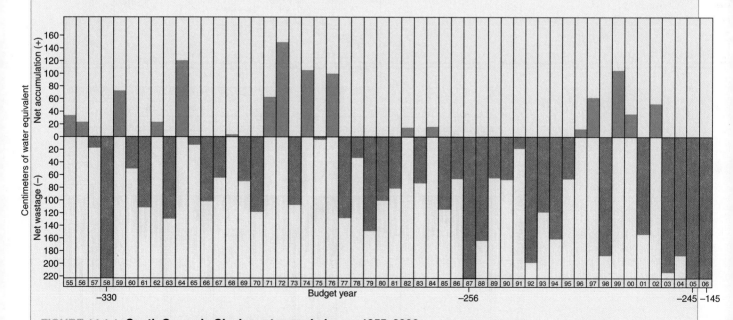

FIGURE 14.1.1 South Cascade Glacier net mass balance, 1955–2006.
A negative net mass balance has dominated this shrinking glacier since 1955. Data in centimeters (2.54 cm = 1 in.). [Data from R. M. Krimmel, *Water, Ice, Meteorological, and Speed Measurements at South Cascade Glacier, Washington, 2002 Balance Year.* USGS Water Resources Report, Tacoma, Washington, personal communication, and http://ak.water.usgs.gov/glaciology/all_bmg/3glacier_balance.htm.]

The annual mean of the 1955–2006 net mass balance for the South Cascade Glacier is –55 cm. The heavy snowfall in the Northwest associated with the La Niña episode is reflected in the positive mass balance for 1999 and 2000. Even though the graph shows a slight positive mass balance for 2002, the glacier's terminus retreated 4 m (13.1 ft) during that year. The years 2003–2006 showed significant losses as regional temperatures increased. The glacier has retreated every year in the record except 1972. Figure 14.1.2 is a photo comparison between 1979 and 2003.

The reasons for these losses are not fully understood. Increasing average air temperature and decreasing precipitation are the leading possible causes.

A comparison of the trend of this glacier's mass balance with that of others in the world shows that temperature changes are causing widespread reductions in middle- and lower-elevation glacial ice. The present wastage (ice loss) from alpine glaciers worldwide is thought to contribute over 25% to the rise in sea level.

Research covering almost 50 years for 67 glaciers in Alaska was published in 2002. The study estimated volume changes in these glaciers, which were

FIGURE 14.1.2 **Photo comparison of 24 years at South Cascade.** Photos visually remind us of the tremendous quantities of ice that have ablated from South Cascade Glacier. [Photos courtesy of USGS Water Resources for Alaska, Glacier and Snow Program.]

FIGURE 14.1.3 **Negative mass balances for Alaska's glaciers.**
(a) Sixty-seven Alaskan glaciers divided into seven regions: 55 in Alaska, 11 crossing the Alaskan-Canadian border, and one in the Yukon. (b) Rate of change (meters per year) in glacierwide thickness for 67 glaciers in the earlier period (1950s to 1995, solid brown bars) and 28 glaciers in the recent period (1995 to 2001, red bars). Note the grouping of glaciers into seven regions noted on the map of Alaska in (a). [Adapted by permission of the AAAS from Anthony A. Arendt et al., Figures 1 and 3, "Rapid wastage of Alaska glaciers and their contribution to rising sea level," *Science* 297 (July 19, 2002): 382–86.]

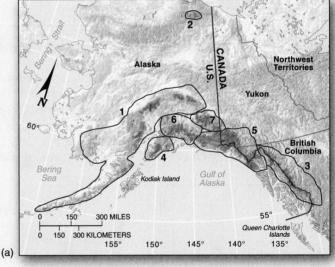

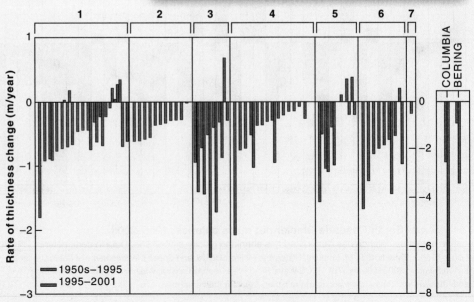

grouped into seven regions, shown in Figure 14.1.3a. The rates of glacier-wide thickness loss for 67 glaciers between the mid-1950s and 1995 and for 28 glaciers between 1995 and 2001 are in Figure 14.1.3b.

A total volume change was estimated to be −52 ± 15 km^3 per year (−12.5 ± 3.6 mi^3). About 9% of global sea-level rise is coming from this Alaskan meltdown, and the recent isostatic rebound of the crust in this region is a result of this glacial ice loss, as discussed in News Report 8.3.

The reasons for these losses are complex. Thinning and wastage are not solely a result of climatic warming; however, these changes appear to be initiated by negative mass balances due to the warming. The individual dynamics of each glacier must be considered. Relative to the contribution of the Alaskan meltdown to sea-level rise, the scientists summarized:

Compared with estimated inputs from the Greenland ice sheet

and other sources, Alaskan glaciers have, over the past 50 years, made the largest single glaciological contribution to rising sea level yet measured. . . . the different rates of thinning observed in the various Alaskan regions may be important in characterizing patterns of climate change.*

*A. A. Arendt et al., "Rapid wastage of Alaska glaciers and their contribution to rising sea level," *Science* 297 (July 19, 2002): 382–86.

---

two per year on a steep slope. The rate of accumulation of snow in the formation area is critical to the speed.

Glaciers are not rigid blocks that simply slide downhill. The greatest movement within a valley glacier occurs *internally*, below the rigid surface layer, which fractures as the underlying zone moves forward (Figure 14.6a). At the same time, the base creeps and slides along, varying its speed with temperature and the presence of any lubricating water beneath the ice. This *basal slip* usually is much less rapid than the internal plastic flow of the glacier, so the upper portion of the glacier flows ahead of the lower portion.

Unevenness in the landscape beneath the ice may cause the pressure to vary, melting some of the basal ice by compression at one moment, only to have it refreeze later. This process is *ice regelation*, meaning to refreeze, or re-gel. Such melting/refreezing action incorporates rock debris into the glacier. Consequently, the basal ice layer, which can extend tens of meters above the base of the glacier, has a much greater debris content than the ice above.

A flowing glacier can develop vertical cracks known as **crevasses** (Figure 14.6b, c, and d). These crevasses result from friction with valley walls, or tension from stretching as the glacier passes over convex slopes, or compression as the glacier passes over concave slopes. Traversing a glacier, whether an alpine glacier or an ice sheet, is dangerous because a thin veneer of snow sometimes masks the presence of a crevasse.

Glacier Surges   Although glaciers flow plastically and predictably most of the time, some will lurch forward with little or no warning in a **glacial surge**. This is not quite as abrupt as it sounds; in glacial terms, a surge can be tens of meters per day. The Jakobshavn Glacier in Greenland, for example, is known to move between 7 and 12 km (4.3 and 7.5 mi) a year.

The doubling of ice-mass loss from Greenland between 1996 and 2005 means that glaciers are surging overall. In southern Greenland in 2007, outlet glaciers were averaging flow rates of 22.8 m (75 ft) per year. This is significant because in 1999 the average flow rate was

1.8 m (6 ft) per year. In Greenland, scientists determined that 2006–2007 experienced more days of melting snow at higher elevations than average. The fear is that the meltwater works its way to the basal layer, lubricating underlying *soft beds* of clay.

The exact cause of such a glacier surge is under study. Some surge events result from a buildup of water pressure under the glacier, sometimes enough to actually float the glacier slightly, detaching it from its bed during the surge. As a surge begins, ice quakes are detectable and ice faults are visible. Surges can occur in dry conditions as well, as the glacier plucks (picks up) rock from its bed and moves forward. Another cause of glacier surges is the presence of a water-saturated layer of sediment *soft bed* beneath the glacier. This is a deformable layer that cannot resist the tremendous sheer stress produced by the moving ice of the glacier. Scientists examining cores taken from several ice streams now accelerating through the West Antarctic Ice Sheet think they have confirmed this cause of glacial surges—although water pressure is still important.

Glacial Erosion   Glacial erosion is similar to a large excavation project, with the glacier hauling debris from one site and depositing it in another. The passing glacier mechanically plucks rock material and carries it away. Rock pieces actually freeze to the basal layers of the glacier in this *glacial plucking* process. Once embedded in the ice, this debris scours and sandpapers the landscape as the glacier moves—a process of **abrasion**. Abrasion and gouging produce a smooth surface on exposed rock, which shines with *glacial polish* when the glacier retreats. Larger rocks in the glacier act much like chisels, working the underlying surface to produce glacial striations (scratches) parallel to the flow direction (Figure 14.7).

Glacial Landforms

Glacial erosion and deposition produce unique landforms. You might expect all glaciers to create the same landforms, but this is not so. Alpine and continental glaciers

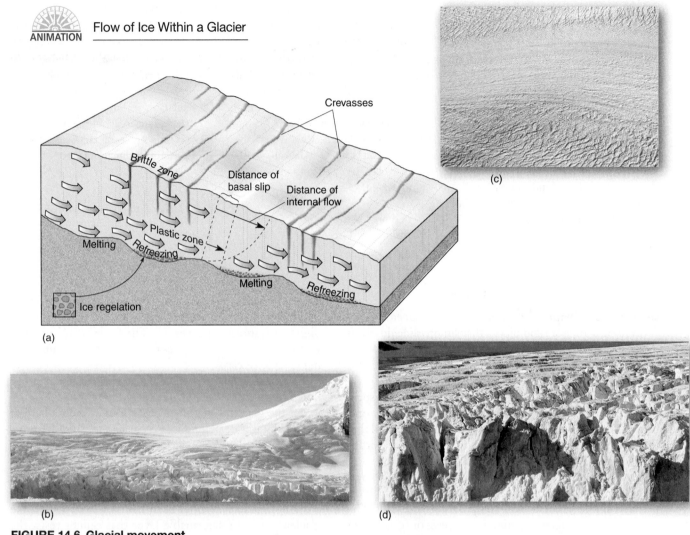

ANIMATION Flow of Ice Within a Glacier

FIGURE 14.6 Glacial movement.
(a) Cross section of a glacier, showing its forward motion and brittle cracking at the surface and flow along its basal layer.
(b) Surface crevasses and cracks are evidence of a glacier's forward motion on Yalour Island, Antarctica. (c) Crevasses frame this ice stream flowing off the Greenland Ice Sheet, evidence of great stress forces at work in the ice. (d) Crevasses mark the jumbled surface of this glacier in Hornsund, Spitsbergen, Arctic Ocean. [All photos by Bobbé Christopherson.]

FIGURE 14.7 Glacial sandpapering polishes rock.
Glacial polish and striations are evidence of glacial abrasion and erosion. The polished, marked surface is seen sheltered beneath a glacial erratic—rock left behind by a retreating glacier. [Photo by Bobbé Christopherson.]

generate their own characteristic landscapes. We look first at erosional landforms created by alpine glaciers, then depositional landforms. Finally, we examine the results of continental glaciation.

Erosional Landforms Created by Alpine Glaciation

Alpine glaciers create spectacular, dramatic landforms that bring to mind the Canadian Rockies, the Swiss Alps, or vaulted Himalayan peaks. Geomorphologist William Morris Davis characterized the stages of a valley glacier in a set of drawings published in 1906 and redrawn in Figure 14.8. Study of these figures reveals ice as the sculptor:

- In 14.8a, you see typical stream-cut valleys as they exist before glaciation. Note the prominent **V**-shape.
- In 14.8b, you see the same landscape during subsequent glaciation. Glacial erosion and transport are actively removing much of the regolith (weathered

bedrock) and the soils that covered the stream valley landscape. As the cirque walls erode away, sharp ridges form, dividing adjacent cirque basins. These **arêtes** ("knife-edge" in French) become the sawtooth, serrated ridges in glaciated mountains. Two eroding cirques may reduce an arête to a saddlelike depression or pass, a **col**. A **horn** (pyramidal peak) results when several cirque glaciers gouge an individual mountain summit from all sides. The most famous horn is the Matterhorn in the Swiss Alps, but many others occur worldwide. A *bergschrund* forms when a crevasse or wide crack opens along the headwall of a glacier, most visible in summer when covering snow is gone.

• In 14.8c, you see the same landscape at a time of warmer climate when the ice has retreated. The glaciated valleys now are **U**-shaped, greatly changed from their previous stream-cut **V** form. You can see the steep sides and the straightened course of the valleys. Physical weathering from the freeze–thaw cycle has loosened rock along the steep cliffs, where it has fallen to form *talus slopes* along the valley sides.

A small mountain lake, especially one that collects in a cirque basin behind risers of rock material, is a **tarn** (Figure 14.8c). Small, circular, stair-stepped lakes in a series are called **paternoster lakes**, because they look like a string of religious beads forming in individual rock basins aligned down the course of a glaciated valley. Such lakes may have formed by the differing resistance of rock to glacial processes or by the damming effect of glacial deposits.

The valleys carved by tributary glaciers are left stranded high above the glaciated valley floor following the removal of previous slopes and stream courses by the glacier. These *hanging valleys* are the sites of spectacular waterfalls as streams plunge down the steep cliffs. In Figure 14.9 see how many of the erosional forms from Figure 14.8 (arêtes, col, horn, cirque, cirque glacier, sawtooth ridge, **U**-shaped valleys, erratics, tarn, and truncated spurs, among others) you can identify in this photograph of glaciers at work.

Where a glacial trough intersects the ocean, the glacier can erode the landscape below sea level. As the glacier retreats, the trough floods and forms a deep **fjord** in which the sea extends inland, filling the lower reaches of the steep-sided valley. The fjord may be flooded further by a rising sea level or by changes in the elevation of the coastal region. All along the glaciated coast of Alaska, glaciers now are in retreat, thus opening many new fjords that previously were blocked by ice. Coastlines with notable fjords include those of Greenland, Chile, South Island of New Zealand, Alaska, British Columbia, and Norway (in Figure 14.10).

Depositional Landforms Created by Alpine Glaciation

You see how glaciers excavate tremendous amounts of material and create fascinating landforms in the process. Glaciers also produce another set of landforms when they melt and deposit their eroded and transported debris cargo. Accumulated debris marks its former margins at the toe where the glacier retreats. This is the end, or *terminus*, of the glacier (Figure 14.5a and e). **Glacial drift** is the general term for both unsorted and sorted glacial deposits. Direct ice deposits are unstratified and unsorted and are known as **till**. Sediments deposited by glacial meltwater are sorted and form **stratified drift**.

Moraine is the name for specific landforms produced by the deposition of these sediments. A **lateral moraine** forms along each side of a glacier. If two glaciers with lateral moraines join, a **medial moraine** may form as lateral moraines merge. Can you identify these elements on the active glacier in Figure 14.11a? Also, revisit Figure 14.1b (p. 474). Along St. Jonsfjorden, Spitsbergen (78° 30′ N), the view from the lateral moraine reveals the retreating Gaffelbreen (Gaffel glacier) that left this unsorted deposit behind (Figure 14.11b). The large terminal moraine deposit (Figure 14.11c) forms an island, Isispynten, in the Arctic. Just 9 years earlier this was a peninsula at the edge of an ice cap. The ice cap rapidly retreated more than a kilometer in this time frame. Note the unsorted material in the moraine.

Eroded debris that is dropped at the glacier's farthest extent is a *terminal moraine*. However, there also may be *end moraines*, formed wherever a glacier pauses after reaching a new equilibrium. Lakes may form behind terminal and end moraines following a glacier's retreat, with the moraine acting as a dam. If a glacier is in retreat, individual deposits are *recessional moraines*. A deposition of till generally spread across a surface is a *ground moraine*, or *till plain*, and may hide the former landscape. Such plains are found in portions of the Midwest.

All these types of till are unsorted and unstratified debris. In contrast, streams of glacial meltwater can deposit sorted and stratified drift beyond a terminal moraine. This type of meltwater-deposited material down-valley from a glacier forms a *valley train deposit*. Peyto Glacier in Alberta produces such a valley train that continues into Peyto Lake (Figure 14.12, p. 474). Distributary stream channels appear braided across its surface. The picture also shows the milky meltwater laden with finely ground "rock flour." Glaciers produce meltwater at all times, not just when they are retreating.

Erosional and Depositional Features of Continental Glaciation

The extent of the most recent continental glaciation in North America and Europe, 18,000 years ago, is portrayed in Figure 14.24. Because continental glaciers form across broad, open landscapes under different circumstances than alpine glaciers, the intricately carved alpine features, lateral moraines, and medial moraines all are lacking in continental glaciation.

Figure 14.13 (p. 475) illustrates some of the most common erosional and depositional features associated with the retreat of a continental glacier. A **till plain** forms

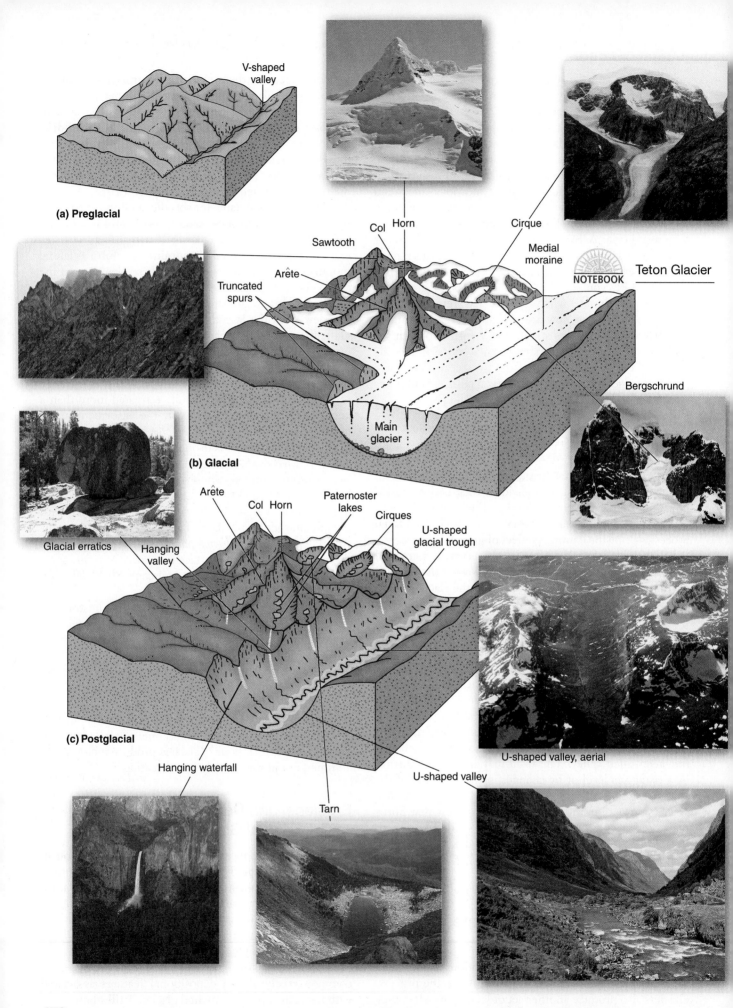

(a) Preglacial

V-shaped valley

(b) Glacial

Sawtooth

Col   Horn

Arête

Truncated spurs

Main glacier

Cirque

Medial moraine

NOTEBOOK   Teton Glacier

Bergschrund

Glacial erratics

(c) Postglacial

Arête

Col   Horn

Paternoster lakes

Cirques

U-shaped glacial trough

Hanging valley

Hanging waterfall

Tarn

U-shaped valley

U-shaped valley, aerial

FIGURE 14.9 Erosional features of alpine glaciation.
Many of the erosional glacial features appear in this scene from southern Greenland. How many of the erosional features illustrated in Figure 14.8 can you find in the photo? [Photo by Bobbé Christopherson.]

FIGURE 14.10 Norwegian fjord.
The beautiful Fjœrfjorden, Norway, where an inlet of the sea fills a U-shaped, glacially carved valley. [Photo by Bobbé Christopherson.]

FIGURE 14.8 The geomorphic handiwork of alpine glaciers.
(a) A preglacial landscape with V-shaped stream-cut valleys. (b) The same landscape filled with valley glaciers. Note inset photos of a *horn*, *cirque basin*, *sawtooth ridge*, and *bergschrund*. (c) When the glaciers retreat, the new landscape is unveiled. Note inset photos of a *hanging valley* and waterfall, *tarn* (lake), *glacial erratics*, and characteristic U-shaped glacial valley (trough; aerial and ground views). [After W. M. Davis, in Tarbuck, Lutgens, and Pinzke, *Applications and Investigations in Earth Science* (New York: Macmillan, an imprint of Prentice Hall, Inc., 1994), p. 85. Photos: waterfall by author; all other photos by Bobbé Christopherson.]

behind an end moraine, featuring *unstratified* coarse till, low, rolling relief, and deranged drainage patterns (Figure 14.13a; see Figure 11.9g). Beyond the morainal deposits, **outwash plains** of *stratified drift* feature stream channels that are meltwater-fed, braided, and overloaded with debris deposited across the landscape.

Figure 14.13b shows a sinuously curving, narrow ridge of coarse sand and gravel called an **esker**. An esker forms along the channel of a meltwater stream that flows beneath a glacier, in an ice tunnel, or between ice walls beneath the glacier. As a glacier retreats, the steep-sided esker is left behind in a pattern roughly parallel to the path of the glacier. The ridge may not be continuous, and in places may even appear branched, following the course set by the subglacial watercourse. Commercially valuable deposits of sand and gravel are quarried from some eskers.

Sometimes an isolated block of ice, perhaps more than a kilometer (0.6 mi) across, persists in a ground moraine, an outwash plain, or a valley floor after a glacier retreats. Perhaps 20 to 30 years are required for it to melt. In the interim, sediment continues to accumulate around the melting ice block. When the block finally melts, it leaves behind a steep-sided hole that frequently fills with water. This feature is a **kettle** (Figure 14.13c). Thoreau's Walden Pond (quotation in Chapter 5) is such a kettle.

Another feature of outwash plains is a **kame**—a small hill, knob, or mound of poorly sorted sand and gravel deposited directly by water, by ice in crevasses, or in ice-caused indentations in the ground surface (Figure 14.13d). Kames also can be found in deltaic forms and in terraces along valley walls.

Glacial action also forms streamlined hills—one erosional, a roche moutonnée, and the other depositional, a drumlin. A **roche moutonnée** ("sheep rock" in French) is an asymmetrical hill of exposed bedrock. Its gently sloping

(a)

(b)

(c)

FIGURE 14.11 Depositional features of alpine glaciation.
(a) As these tributary glaciers merge, their lateral moraines combine to form a medial moraine in eastern Greenland. (b) Retreating ice leaves behind a substantial lateral moraine; the retreating Gaffelbreen (Gaffel Glacier) is in the background. (c) A terminal moraine forms this island separated from its ice-cap glacier by more than a kilometer of ocean at Isispynten Island, Arctic Ocean. [All photos by Bobbé Christopherson.]

FIGURE 14.12 A valley train deposit.
Peyto Glacier in Alberta, Canada. Note valley train, braided stream, and milky-colored glacial meltwater. [Photo by author.]

upstream side (stoss side) was polished smooth by glacial action, whereas its downstream side (lee side) is abrupt and steep where the glacier plucked rock pieces (Figure 14.14).

A **drumlin** is deposited till that was streamlined in the direction of continental ice movement, blunt end upstream and tapered end downstream. Multiple drumlins (called *swarms*) occur across the landscape in portions of New York

and Wisconsin, among other areas. Sometimes their shape is that of an elongated teaspoon bowl, lying face down. Drumlins may attain lengths of 100–5000 m (330 ft to 3.1 mi) and heights up to 200 m (650 ft). Figure 14.15 shows a portion of a topographic map for the area south of Williamson, New York, which experienced continental glaciation. In studying the map, can you identify the numerous drumlins? In what direction do you think the continental glaciers moved across this region?

Periglacial Landscapes

In 1909, Polish geologist W. Lozinski coined the term **periglacial** to describe frost weathering and freeze–thaw rock shattering in the Carpathian Mountains. These periglacial regions occupy over 20% of Earth's land surface (Figure 14.16). Periglacial landscapes are either near-permanent ice or at high elevation and are seasonally snow-free. Under these conditions, a unique set of periglacial processes operate, including permafrost, frost action, and ground ice.

Climatologically, these regions are in *subarctic* and *polar* climates (especially the *tundra* climate region). Such climates occur either at high latitude (tundra and boreal forest environments) or high elevation in lower-latitude

FIGURE 14.13 Continental glacier depositional features.
Common depositional landforms produced by glaciers. (a) Deranged drainage tells us continental glaciers advanced and retreated across the prairies of central Saskatchewan, Canada. (b) An esker through farmland near Dahlen, North Dakota. (c) Walden Pond is a kettle surrounded by mixed forest near Boston, Massachusetts. (d) A kame covered by a woodlot near Campbellsport, Wisconsin. [Illustration redrawn from R. M. Busch, ed., *Laboratory Manual in Physical Geology*, 3rd ed. (New York: Macmillan, an imprint of Prentice Hall, Inc., 1993), p. 188. Photos (a) and (c) by Bobbé Christopherson; (b) and (d) Tom Bean/DRK.]

mountains (alpine environments). Processes that relate to physical weathering and mass movement (Chapter 10), climate (Chapter 7), and soil (Chapter 15) dominate these periglacial regions.

Geography of Permafrost

When soil or rock temperatures remain below 0°C (32°F) for at least 2 years, a condition of **permafrost** develops. An area that has permafrost but is not covered by glaciers is considered periglacial. Note that this criterion is based solely on temperature and not on how much water is present. Other than high latitude and low temperatures, two other factors contribute to permafrost: the presence of fossil permafrost from previous ice-age conditions and the insulating effect of snow cover or vegetation that inhibits heat loss.

Continuous and Discontinuous Zones Permafrost regions are divided into two general categories, continuous and discontinuous, that merge along a general transition zone. *Continuous permafrost* is a perennial region of the most severe cold (purple areas in Figure 14.16). Continuous permafrost affects all surfaces except those beneath deep lakes or rivers and may exceed 1000 m (over 3000 ft) in depth, averaging approximately 400 m (1300 ft).

Discontinuous permafrost occurs in unconnected patches that gradually merge poleward toward the continuous permafrost zone. Equatorward, along lower-latitude margins, permafrost becomes scattered or sporadic until it gradually disappears (dark blue area on the map). Permafrost affects as much as 50% of Canada and 80% of Alaska. In central Eurasia, the effects of continentality and elevation produce discontinuous permafrost that extends equatorward to the 50th parallel. In addition to these two types, zones of high-elevation alpine permafrost extend to lower latitudes, as shown on the map.

Behavior of Permafrost We looked at the spatial distribution of permafrost; let us now examine how permafrost behaves. Figure 14.17 is a stylized cross section from approximately 75° N to 55° N, using the three sites located on the map in Figure 14.16. The **active layer** is the zone of seasonally frozen ground that exists between the subsurface permafrost layer and the ground surface.

(a)

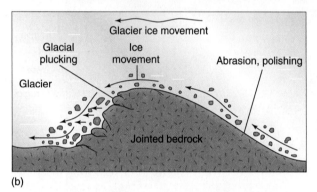
(b)

FIGURE 14.14 Glacially eroded streamlined rock.
Roche moutonnée, as exemplified by Lembert Dome in the Tuolumne Meadows area of Yosemite National Park, California. [Photo by author.]

FIGURE 14.15 Glacially deposited streamlined features.
Topographic map south of Williamson, New York, features numerous drumlins (7.5-minute series quadrangle map, originally produced at a 1:24,000 scale, 10-ft contour interval). [USGS map courtesy of US Geological Survey.]

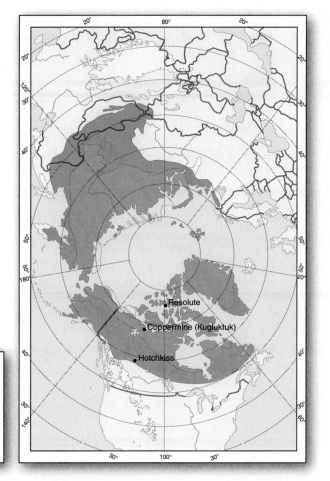

FIGURE 14.16 Permafrost distribution.
Distribution of permafrost in the Northern Hemisphere. Alpine permafrost is noted except for small alpine occurrences in Hawai'i, Mexico, Europe, and Japan. Subsea permafrost occurs in the ground beneath the Arctic Ocean along the margins of the continents noted. Note the towns of Resolute and Coppermine, now Kugluktuk in Nunavut, and Hotchkiss in Alberta. A cross section of the permafrost beneath these towns is shown in Figure 14.17.
[Adapted from Troy L. Péwé, "Alpine permafrost in the contiguous United States: A review," *Arctic and Alpine Research* 15, no. 2 (May 1983): 146. © University of Colorado. Used by permission.]

Subsea permafrost

Continuous permafrost

Discontinuous permafrost

Alpine permafrost

The active layer is subjected to consistent daily and seasonal freeze-thaw cycles. This cyclic melting of the active layer affects as little as 10 cm (4 in.) depth in the north (Ellesmere Island, 78° N), up to 2 m (6.6 ft) in the southern margins (55° N) of the periglacial region, and 15 m (50 ft) in the alpine permafrost of the Colorado Rockies (40° N).

The depth and thickness of the active layer and permafrost zone change slowly in response to climatic change. Higher temperatures degrade (reduce) permafrost and increase the thickness of the active layer; lower temperatures gradually aggrade (increase) permafrost depth and reduce active-layer thickness. Although somewhat sluggish in response, the active layer is a dynamic open system driven by energy gains and losses in the subsurface environment. As you might expect, most permafrost exists in disequilibrium with environmental conditions and therefore actively adjusts to inconstant climatic conditions.

With the incredibly warm temperatures recorded in the Canadian, Alaskan, and Siberian Arctic since 1990, more disruption of surfaces is occurring—leading to highway, railway, and building damage. This warming is causing a large release of carbon from peat-rich thawed ground that in turn influences the global greenhouse—a real-time positive feedback mechanism. In Siberia, many lakes have disappeared in the discontinuous permafrost region as subsurface drainage opens, yet new lakes have formed in the continuous region as thawed soils become waterlogged. In Canada, hundreds of lakes have disappeared

simply from excessive evaporation into the warm air. These trends are measurable from satellite imagery.

A *talik* is unfrozen ground that may occur above, below, or within a body of discontinuous permafrost or beneath a water body in the continuous region. Taliks occur beneath deep lakes and may extend to bedrock and noncryotic (unfrozen) soil beneath large, deep lakes (Figure 14.17). Taliks form connections between the active layer and groundwater, whereas in continuous permafrost groundwater is essentially cut off from surface water. In this way, permafrost disrupts aquifers and taliks, leading to water-supply problems.

Ground Ice and Frozen-Ground Phenomena

In regions of permafrost, subsurface water that is frozen is **ground ice**. The moisture content of areas with ground ice may vary from nearly none in regions of drier permafrost to almost 100% in saturated soils.

Frost-Action Processes Frozen water in the soil initiates *frost action*. The 9% expansion of water as it freezes produces strong mechanical forces that fracture rock and disrupt soil at and below the surface. If sufficient water freezes, the saturated soil and rocks are subject to *frost heaving* (vertical movement) and *frost thrusting* (horizontal motions). Boulders and slabs of rock are thrust to the surface. Layers of soil (soil horizons) may appear disrupted

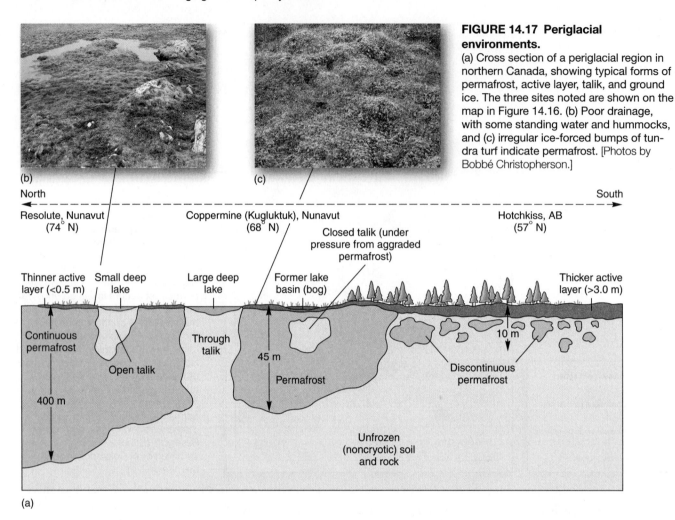

(b)

(c)

FIGURE 14.17 Periglacial environments.
(a) Cross section of a periglacial region in northern Canada, showing typical forms of permafrost, active layer, talik, and ground ice. The three sites noted are shown on the map in Figure 14.16. (b) Poor drainage, with some standing water and hummocks, and (c) irregular ice-forced bumps of tundra turf indicate permafrost. [Photos by Bobbé Christopherson.]

(a)

as if stirred or churned. Frost-action processes also produce a contraction in soil and rock volume, opening up cracks for ice wedges to form.

An *ice wedge* develops when water enters a crack in the permafrost and freezes. Thermal contraction in ice-rich soil forms a tapered crack—wider at the top, narrowing toward the bottom. Repeated seasonal freezing and melting progressively expand the wedge, which may widen from a few millimeters to 5–6 m and up to 30 m (100 ft) in depth (Figure 14.18). Widening may be small each year, but after many years the wedge can become significant.

Frost action forces expansion and contraction that results in the transport of stones and boulders. As the water-ice volume changes and an ice wedge deepens, coarser particles move toward the surface. An area with a system of ground ice and frost action develops sorted and unsorted accumulations of rock at the surface that become organized forms, known as **patterned ground**. Patterned ground may include polygons of sorted rocks that coalesce into stone polygon nets (Figure 14.19a, b, and c). Various terms are in use to describe such ice-wedge and stone polygon forms: nets, circles, vegetation-covered hummocks, and stripes (elongated polygons formed on a hillslope, Figure 14.19d).

Fascinating research has identified unique synergies at work to form patterned ground. Expansion and contraction of frost action results in the movement of soil particles,

stones, and small boulders. During this freeze–thaw process is when self-organization begins. Imagine a process where stones move toward stone domains (stone-rich areas) and soil particles move toward soil domains (soil-rich areas). Patterned ground emerges that may take centuries to form. Added to this is the role of slope angles—greater slopes produce striped patterns (Figure 14.19e), whereas lesser slopes result in sorted polygons (Figure 14.19a, b, and c). Figure 14.19d shows stone islands forming where stone concentration decreases. These polygons and circles are complex forms resulting from interacting forces in the frozen ground. Also a factor is the amount of stone confinement that occurs: more circular forms when sorting is dominant; more polygons when confinement of stone domains is dominant. This means that such landscapes are self-making, self-maintaining, and self-organizing—all traits of living systems, making this an interesting path for future research in Earth systems. (See M. A. Kessler and B. T. Warner, "Self-organization of patterned ground," *Science* 17, v. 299, no. 5605 (January 2003): 354–55; 380–83.)

Such polygon nets in patterned ground provide vivid evidence of frozen subsurface water on Mars (Figure 14.19f). In this 1999 image, the *Mars Global Surveyor* captured a region of these features on the Martian northern plains. To date, some 600 different sites on Mars demonstrate such patterned ground, and in 2008 the

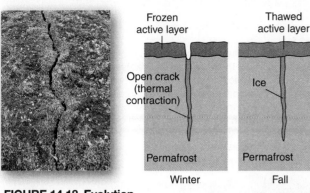

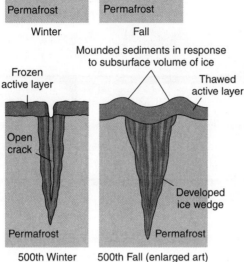

FIGURE 14.18 Evolution of an ice wedge.
(a) Sequential illustration of ice-wedge formation beginning with the crack in the tundra in the photo. (b) An ice wedge and ground ice in northern Canada. [(a) Illustration adapted from A. H. Lachenbruch, "Mechanics of thermal contraction and ice-wedge polygons in permafrost," *Geological Society of America Bulletin Special Paper 70* (1962). Photos by (a) Bobbé Christopherson; (b) H. M. French.]

Frozen active layer / Thawed active layer

Open crack (thermal contraction) / Ice

Permafrost / Permafrost

Winter / Fall

Mounded sediments in response to subsurface volume of ice

Frozen active layer / Thawed active layer

Open crack

Developed ice wedge

Permafrost / Permafrost

500th Winter / 500th Fall (enlarged art)

(a)

(b)

Phoenix Lander set down in the middle of a Martian arctic plain covered with polygonal forms. The craft dug into the Martian surface and found much ground (water) ice.

Hillslope Processes Soil drainage is poor in areas of permafrost and ground ice. When the surface thaws seasonally, the upper layer of soil and regolith saturates with soil moisture, and the whole layer flows if the landscape is sloped. Such soil flows are in general called *solifluction*; in the presence of ground ice, the more specific term *gelifluction* applies (Figure 14.20).

The cumulative effect of this landflow can be an overall flattening of a rolling landscape, with sagging surfaces and scalloped and lobed patterns in downslope soil movements. Periglacial mass movement processes are related to slope dynamics and processes discussed in Chapter 10.

Humans and Periglacial Landscapes

Human populations face various difficulties in areas that experience permafrost and frozen-ground phenomena. Because thawed ground above the permafrost zone frequently shifts, highways and rail lines become warped and twisted and fail, and utility poles are tilted. In addition, any building placed directly on frozen ground literally will melt itself into the defrosting soil (Figure 14.21).

Construction in periglacial regions dictates placing structures above the ground to allow air circulation beneath. This airflow allows the ground to cycle through its normal annual temperature pattern. Utilities such as water and sewer lines must be built aboveground in "utilidors" to protect them from freezing and thawing ground (Figure 14.22). For this reason, the Alaskan oil pipeline was constructed

above ground on racks for 675 km of its 1285-km (420 mi of its 800-mi) length to avoid melting the frozen ground, causing shifting that could rupture the line. Earthquakes remain a threat, as happened in November 2002—see Figure 9.1b for a photo of this pipeline with broken support brackets.

The effects of global warming are already showing up in Siberia, where the active layer is now more than twice the depth it was in the past. Buildings constructed with shallow pilings for support will certainly fail if this trend continues. In an era of global change, the regions of permafrost provide further climatic indicators.

The Pleistocene Ice-Age Epoch

Imagine almost a third of Earth's land surface buried beneath ice sheets and glaciers thousands of meters thick—most of Canada, the northern Midwest, England, and northern Europe, including many mountain ranges. This is how it was at the height of the Pleistocene Epoch of the late Cenozoic Era.

The Pleistocene is thought to have begun about 1.65 million years ago and is one of the more prolonged cold periods in Earth's history. It featured not just one glacial advance and retreat, but at least 18 expansions of ice over Europe and North America, each obliterating and confusing the evidence from the one before. Apparently, glaciation can take about 90,000 years, whereas deglaciation is rapid, requiring less than about 10,000 years to melt away the accumulation.

The term **ice age**, or *glacial age*, is applied to any such extended episode of cold (not a single cold spell). An ice age is an episode of generally cold climate that includes

FIGURE 14.19 Patterned ground phenomena.
(a) Patterned ground in Beacon Valley of the McMurdo Dry Valleys of East Antarctica. (b) Polygons and circles (about a meter across) near Duvefjord, Nordaustlandet Island, Arctic Ocean. (c) Polygons and circles Spitsbergen Island. (d) Intermediate organization in progress on Bear Island (74° 30′ N). (e) Patterned ground in stripes, near Sundbukta, Barents Island. (f) Polygons in the Martian northern plains; each cell averages a little more than 100 m across (300 ft). [Photos by (a) writer Joan Myers; (b), (c), (d), and (e) Bobbé Christopherson; and (f) image from the *Mars Global Surveyor*, Mars Orbiter Camera, courtesy of NASA/JPL/Malin Space Science Systems, May 1999.]

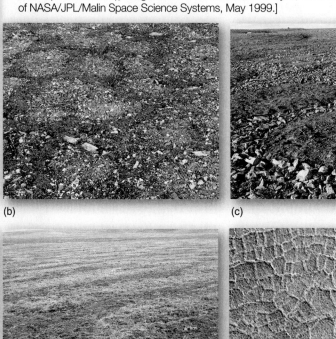

(a)

(b)

(c)

(d)

(e)

(f)

one or more *glacials*, interrupted by warm spells known as *interglacials*. Each glacial and interglacial is given a name that is usually based on the location where evidence of the episode is prominent—for example, "Wisconsinan glacial."

Modern research techniques to understand past climates include the examination of ancient ratios of oxygen isotopes, depths of coral growth in the tropics, analysis of ocean and lake sediments worldwide, and analysis of the latest ice cores from Greenland and Antarctica, including the

Dome C ice core that plunges the record back 800,000 years. These techniques have opened the way for a new chronology and understanding of past climates and for gaining perspective on the present warming trends. The record is clear and correlates across all proxy methods: Present levels of carbon dioxide (CO_2) and methane (CH_4) are the highest in 800,000 years, as shown in Chapter 7. Earth systems have no experience in dealing with levels this high during this time span, and these concentrations are increasing.

FIGURE 14.20 Soil flowage in periglacial environments.
Gelifluction lobes of soil underlain by frozen ground in permafrost conditions, on a hillside near Sundbukta, Barents Island, Arctic Ocean. [Photo by Bobbé Christopherson.]

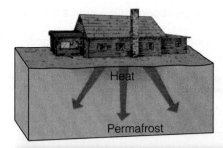

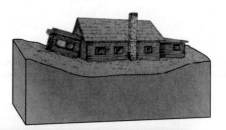

FIGURE 14.21 Permafrost melting and structure collapse.
Building failure due to the melting of permafrost south of Fairbanks, Alaska. [Adapted from US Geological Survey. Photo by Steve McCutcheon; illustration based on US Geological Survey pamphlet "Permafrost" by Louis L. Ray.]

Glaciologists currently recognize the Illinoian glacial and Wisconsinan glacial interval, with the Sangamon interglacial between them. These events span the 300,000-year interval prior to our present Holocene Epoch (Figure 14.23).

The chart shows that the Illinoian glacial actually consisted of two glacials occurring during Marine (oxygen) Isotope Stages (MIS) 6 and 8, as did the Wisconsinan (MIS 2 and 4), which are dated at 10,000 to 35,000 years ago. The oxygen isotope glacial/interglacial stages on the chart are numbered back to MIS 23 at approximately 900,000 years ago. To overcome local bias in core records, investigators correlate oxygen isotope data with other indicators worldwide. Note the label on the chart in the figure at MIS 20.2 for the Dome C record of 800,000 years.

The polar map projection in Figure 14.24a illustrates the continental ice sheets over Canada, the United States, Europe, and Asia some 18,000 years ago. The ice sheets were more than 2 km (1.2 mi) in thickness. In North America, the Ohio and Missouri River systems mark the southern terminus of continuous ice. The edge of an ice sheet is not even; instead, it expands and retreats in ice lobes. Such lobes were positioned where the Great Lakes are today and contributed to the formation of these lakes

Pipeline
1.2 m diameter
Average height
1.5 to 3.0 m

(b)

FIGURE 14.22 Special structures for permafrost.
(a) Proper construction in periglacial environments requires raising of buildings above ground and running water and sewage lines in elevated "utilidors," here in Barentsburg, a Russian settlement on Spitsbergen Island. (b) Supporting the Trans-Alaska oil pipeline on racks protects the permafrost from heat in the oil (compare with the damaged pipeline in Figure 9.1b). [Photos by (a) Bobbé Christopherson; (b) Galen Rowell/Mountain Light Photography, Inc.]

(a)

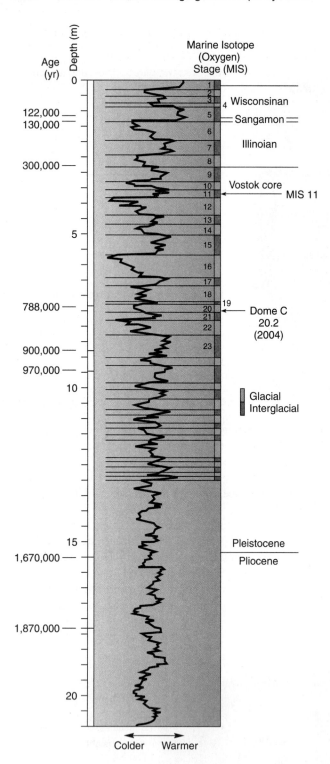

FIGURE 14.23 Temperature record of the past 2 million years.
Pleistocene temperatures, determined by oxygen isotope fluctuations in fossil planktonic foraminifera (tiny marine organisms having a calcareous shell) from deep-sea cores. Twenty-three Marine Isotope Stages (MIS) cover 900,000 years, with names assigned for the past 300,000 years. [After N. J. Shackelton and N. D. Opdyke, *Oxygen-Isotope and Paleomagnetic Stratigraphy of Pacific Core V28-239, Late Pliocene to Latest Pleistocene.* Geological Society of America Memoir 145. ©1976 by the GSA. Adapted by permission.]

landscapes is important, for it allows us to understand *paleoclimatology* (the study of past climates) and discover the mechanisms that produce ice ages and long-term climatic change. (For web links on glaciers and Pleistocene geology, see **http://research.umbc.edu/~miller/geog111/glacierlinks.htm.**)

Paleolakes

Figure 14.25 portrays the West dotted with large lakes 12,000–30,000 years ago. Except for the Great Salt Lake in Utah (a remnant of the former Lake Bonneville noted on the map) and a few smaller lakes, only dry basins, ancient shorelines, and lake sediments remain today. These ancient lakes are **paleolakes**. The three photos in Figure 14.25 show how these paleolake sites look today: the Bonneville Salt Flats in Utah, Mono Lake and its tufa towers in California, and Sevier Dry Lake, Utah.

The term *pluvial* (Latin "rain") describes this period of wetter conditions, principally during the Pleistocene epoch. During pluvial periods, lake levels increased in arid regions. The drier intervals between pluvials are called *interpluvials*. Interpluvials are marked with **lacustrine deposits**—lake sediments that form terraces along former shorelines.

Early researchers attempted to correlate pluvial and glacial ages, given their coincidence during the Pleistocene. However, few sites actually demonstrate such a simple relationship. For example, in the western United States, the estimated volume of melted ice is only a small portion of the water volume of pluvial lakes of the region. Also, these lakes tend to predate glacial times and correlate instead with periods of wetter climate, or periods thought to have had lower evaporation rates. In North America, the two largest late-Pleistocene paleolakes were in the Basin and Range Province of the West. Figure 14.25 portrays these— Lake Bonneville and Lake Lahontan—and other paleolakes at their highest levels. Pluvial lakes also occurred in Mexico, South America, Africa, Asia, and Australia.

The Great Salt Lake, near Salt Lake City, Utah, and the Bonneville Salt Flats in western Utah are remnants of Lake Bonneville, which at its greatest extent covered more than 50,000 km² (19,500 mi²) and reached depths of 300 m (1000 ft), spilling over into the Snake River drainage to the north. Today, the area is a closed basin with no drainage except an artificial outlet to the west where during floods excess water from the Great Salt Lake is pumped to a basin called Newfoundland Evaporation Basin.

when the ice retreated from their gouged, isostatically (weighted down by ice) depressed basins.

As these ice sheets and alpine glaciers retreated (see map in Figure 14.24c), they exposed a drastically altered landscape: the rocky soils of New England, the polished and scarred surfaces of Canada's Atlantic Provinces, the sharp crests of the Sawtooth Range and Tetons of Idaho and Wyoming, the scenery of the Canadian Rockies and the Sierra Nevada, the Great Lakes of the United States and Canada, and much more. Study of these glaciated

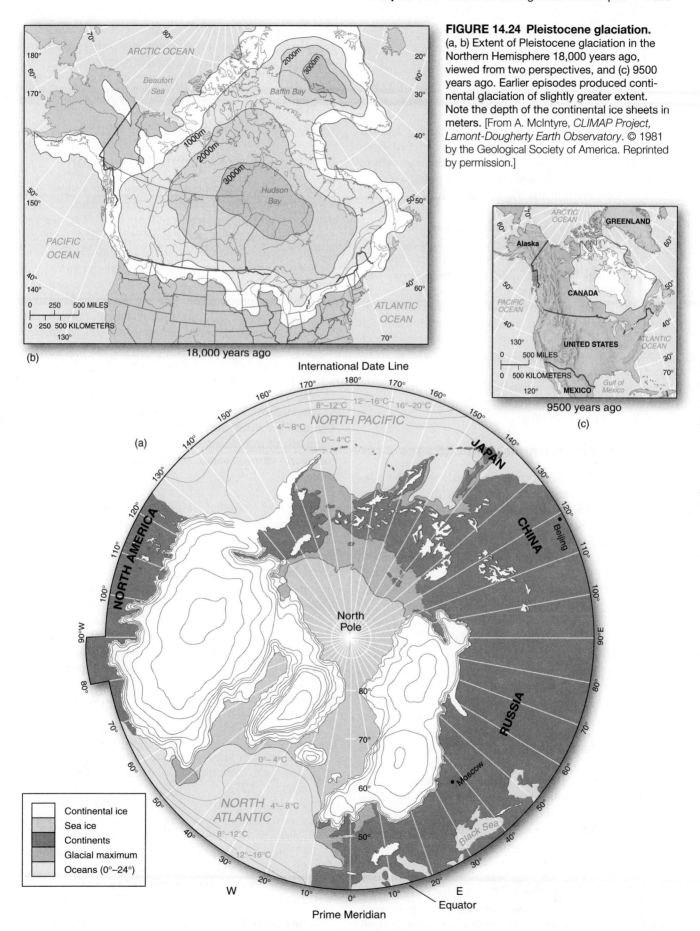

FIGURE 14.24 Pleistocene glaciation.
(a, b) Extent of Pleistocene glaciation in the Northern Hemisphere 18,000 years ago, viewed from two perspectives, and (c) 9500 years ago. Earlier episodes produced continental glaciation of slightly greater extent. Note the depth of the continental ice sheets in meters. [From A. McIntyre, *CLIMAP Project, Lamont-Dougherty Earth Observatory*. © 1981 by the Geological Society of America. Reprinted by permission.]

(b) 18,000 years ago

(c) 9500 years ago

(a) International Date Line

Prime Meridian

Legend:
- Continental ice
- Sea ice
- Continents
- Glacial maximum
- Oceans (0°–24°)

Deciphering Past Climates: Paleoclimatology

We observe glacials and interglacials because Earth's climate has fluctuated in and out of cold and warm ages. Documentation for this fluctuation now is traced in ice cores from Greenland and Antarctica, in layered deposits of silts and clays, in the extensive pollen record from ancient plants, and in the relation of past coral growth to sea level. One especially interesting fact is emerging from these studies: We humans (*Homo erectus* and *H. sapiens* of the last 1.9 million years) have never experienced Earth's normal (more moderate, less extreme) climate, most characteristic of Earth's entire 4.6-billion-year span.

Apparently, Earth's climates slowly fluctuated until the past 1.2 billion years, when temperature patterns with cycles of 200–300 million years became more pronounced. The most recent cold episode was the Pleistocene Epoch, which began in earnest 1.65 million years ago and through which we may still be progressing. The Holocene Epoch began approximately 10,000 years ago, when average temperatures abruptly increased 6 C° (11 F°). The period we live in may represent an end to the Pleistocene, or it may be merely a mild interglacial. Figure 14.26 details the climatic record of the past 160,000 years.

Medieval Warm Period and Little Ice Age

In A.D. 1001, Leif Eriksson inadvertently ventured onto the North American continent, perhaps the first European to do so. He and his fellow Vikings were favored by a medieval warming episode as they sailed the less-frozen North Atlantic to settle Iceland and Greenland.

The mild climatic episode that lasted from about A.D. 800 to 1200 is known as the *Medieval Warm Period*. During the warmth, grape vineyards were planted far into England some 500 km (310 mi) north of present-day commercial plantings. Oats and barley were planted in Iceland, and wheat was planted as far north as Trondheim, Norway. The shift to warmer, wetter weather influenced migration and settlement northward in North America, Europe, and Asia.

However, from approximately 1200–1350 through 1800–1900, a *Little Ice Age* took place. Parts of the North Atlantic froze, and expanding glaciers blocked many key mountain passes in Europe. Snowlines in Europe lowered about 200 m (650 ft) in the coldest years. The Greenland colonies were deserted. Cropping patterns changed and northern forests declined, along with human population in those regions. In the winter of 1779–80, New York's Hudson and East Rivers and the entire Upper Bay froze over. People walked and hauled heavy loads across the ice between Staten and Manhattan Islands!

But the Little Ice Age was not consistently cold throughout its 700-year reign. Reading the record of the Greenland ice cores, scientists found many mild years among the harsh. More accurately, this was a time of rapid, short-term climate fluctuations that lasted only decades.

Ice cores drilled in Greenland revealed a record of annual snow and ice accumulation that, when correlated with other aspects of the core sample, was indicative of air temperature. Figure 14.26c presents this record since A.D. 500. The warmth of the Medieval Warm Period is evident, whereas the Little Ice Age appears mixed with colder conditions around 1200, 1500, and after about 1800.

Mechanisms of Climate Fluctuation

What mechanisms cause short-term fluctuations? And why is Earth pulsing through long-term climatic changes that span several hundred million years? The ice-age concept is being researched and debated with an unprecedented intensity for three principal reasons: (1) Continuous ice cores from Greenland and Antarctica are providing a new, detailed record of weather and climate patterns, volcanic eruptions, and trends in the biosphere (discussed in Focus Study 14.1); (2) to understand present and future climate change and to refine general circulation models, we must understand the natural variability of the atmosphere and climate; (3) anthropogenic global warming, and its relation to ice ages, is a major concern.

Because past occurrences of low temperature appear to have followed a pattern, researchers looked for causes that also are cyclic in nature. They identified a complicated mix of interacting variables that appear to influence long-term climatic trends. Let us take a look at several of them.

Climate and Celestial Relations As our Solar System revolves around the distant center of the Milky Way, it crosses the plane of the galaxy approximately every 32 million years. At that time, Earth's plane of the ecliptic aligns parallel to the galaxy's plane, and we pass through regions in space of increased interstellar dust and gas, which may have some climatic effect.

Milutin Milankovitch (1879–1954), a Yugoslavian astronomer who studied Earth-Sun orbital relations, developed other possible astronomical factors. Milankovitch wondered whether the development of an ice age relates to seasonal astronomical factors—Earth's revolution around the Sun, rotation, and tilt (Figure 14.27). In summary:

- Earth's elliptical orbit about the Sun is not constant. The shape of the ellipse varies by more than 17.7 million km (11 million mi) during a 100,000-year cycle, from nearly circular to an extreme ellipse (Figure 14.27a).
- Earth's axis "wobbles" through a 26,000-year cycle, in a movement much like that of a spinning top winding down. Earth's wobble is *precession*. As you can see in Figure 14.27b, precession changes the orientation of hemispheres and landmasses to the Sun.
- Earth's present axial tilt of 23.5° varies from 22° to 24° during a 41,000-year period (Figure 14.27c).

Milankovitch calculated, without the aid of today's computers, that the interaction of these Earth-Sun relations

(a)

(b) Bonneville Salt Flats

(c) Mono Lake, tufa towers

(d) Sevier Dry Lake

FIGURE 14.25 Paleolakes in the western United States.
Paleolakes of the western United States at their greatest extent 12,000 to 30,000 years ago, a
recent pluvial period. Lake Lahontan and Lake Bonneville were the largest. The Great Salt Lake
and Sevier Dry Lake in Utah are remnants of Lake Bonneville; Mono Lake in California is what
remains of pluvial Lake Russell. [After R. F. Flint, *Glacial and Pleistocene Geology*. © 1957 by John
Wiley & Sons, Inc. Adapted by permission. Photos by (b) author; (c) and (d) Bobbé Christopherson.]

creates a 96,000-year climatic cycle. His glaciation model
assumes that changes in astronomical relations affect the
amounts of insolation received.

Milankovitch died in 1954, his ideas still not accepted
by a skeptical scientific community. Now, in the era of
computers, remote-sensing satellites, and worldwide
efforts to decipher past climates, Milankovitch's valuable
work has received some degree of confirmation. A
roughly 100,000-year climatic cycle is confirmed in such
diverse places as ice cores in Greenland and the accumulation of sediment in Lake Baykal, Siberia.

Climate and Solar Variability If the Sun significantly
varied its output over the years, as some other stars do,
that variation would seem a convenient and plausible
cause of ice-age timing. However, lack of evidence that
the Sun's radiation output varies significantly over long
cycles argues against this hypothesis. Nonetheless,
inquiry about the Sun's variability continues. As we saw in
Chapter 7, in the IPCC *Fourth Assessment Report* in Figure

7.29, solar variability is not correlated to temperature
increases occurring in the atmosphere and oceans this
past century.

Climate and Tectonics Major glaciations also can be
associated with plate tectonics because some landmasses
migrated to higher, cooler latitudes. Chapters 8 and 9
explain that the shape and orientation of landmasses and
ocean basins changed greatly during Earth's history. Continental plates drifted from equatorial locations to the
polar regions and vice versa, thus exposing the land to a
gradual change in climate. Gondwana (the southern half
of Pangaea) experienced extensive glaciation that left its
mark on the rocks of parts of present-day Africa, South
America, India, Antarctica, and Australia. Landforms in
the Sahara, for example, bear the markings of even earlier
glacial activity. These markings are partly explained by
the fact that portions of Africa were centered near the
South Pole during the Ordovician Period, 465 million
years ago (see Figure 8.15a).

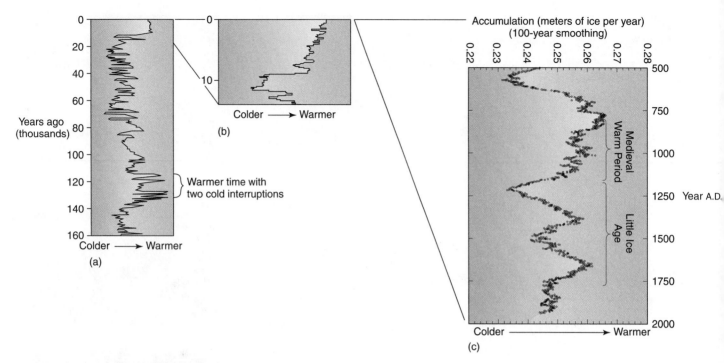

FIGURE 14.26 **Recent climates determined from an ice core.**
(a) Temperature patterns during the past 160,000 years, back to MIS 6. Note the two cold spells that interrupted the earlier interglacial (between 115,000–135,000 years ago). (b) Higher resolution of the last 12,000 years. The cold period known as the "Younger Dryas" intensified at the beginning of this record. Warming was underway by 11,700 years ago and abruptly began to increase by the Holocene. (c) The ice-core record for the past 1500 years. Note the Medieval Warm Period and consistent temperatures as compared with the chaotic record of the Little Ice Age. [(a) and (b) Courtesy of the Greenland Ice Core Project (GRIP); (c) from D. A. Meese et al., "The accumulation record from the GISP-2 core as an indicator of climate change throughout the Holocene," *Science* 266 (December 9, 1994): 1681, © AAAS.]

Episodes of mountain building over the past billion years forced mountain summits above the snowline, where snow remains after the summer melt. Mountain chains influence downwind weather patterns and jet-stream circulation, which in turn guides weather systems. More dust was present during glacial periods, suggesting drier weather and more extensive deserts beyond the frozen regions.

Climate and Atmospheric Factors Some events alter the atmosphere and produce climate change. A volcanic eruption might produce lower temperatures for a year or two. The lower temperatures could initiate a buildup of long-term snow cover at high latitudes. These high-albedo snow surfaces then would reflect more insolation away from Earth, to further enhance cooling in a positive feedback system. The eruption of Mount Pinatubo in the Philippines in 1991 is closely studied, for the eruption caused a temporary cooling and possibly other climatic effects.

The fluctuation of atmospheric greenhouse gases can trigger higher or lower temperatures. Each interglacial, or warmer interval, appears to be correlated to higher levels of carbon dioxide. During May 2008, atmospheric

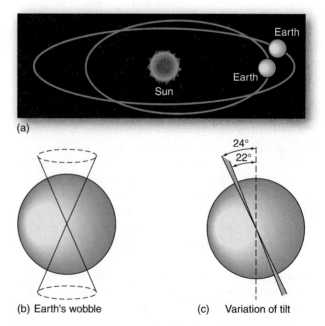

FIGURE 14.27 **Astronomical factors that may affect broad climatic cycles.**
(a) Earth's elliptical orbit varies widely during a 100,000-year cycle, stretching out to an extreme ellipse. (b) Earth's 26,000-year axial wobble. (c) Variation in Earth's axial tilt every 41,000 years.

FOCUS STUDY 14.1

GRIP, GISP-2, and Dome-C: Boring Ice for Exciting History

The Greenland Ice Core Project (GRIP) was launched in 1989. A site was selected near the summit of the Greenland Ice Sheet at 3200 m (10,500 ft) so that the maximum thickness of ice history would be accessed (Figure 14.1.1). After 3 years, the drills hit bedrock 3030 m (9940 ft) below the site—or, in terms of time, 250,000 years into the past. The core is 10 cm (4 in.) in diameter.

In 1990, about 32 km (20 mi) west of the summit, the Greenland Ice Sheet Project (GISP-2) began to bore back through time (see Figure 14.1.1). GISP-2 reached bedrock in 1993. This core is slightly larger in diameter, at 13.2 cm (5.2 in.), and collects about twice the data as GRIP. The existence of a second core helped scientists compensate for any folds or disturbed sections they encountered below 2700 m, or about 115,000 years, in the first core.

What is being discovered from a 3030-m ice core? Locked into the core is a record of past precipitation and air bubbles of past atmospheres, which indicate ancient gas concentrations. Of special interest are the greenhouse gases, carbon dioxide and methane. Chemical and physical properties of the atmosphere and the snow that accumulated each year are frozen in place.

Pollutants are locked into the ice-core record. For example, during cold periods, high concentrations of dust were present, brought by winds from distant dry lands. An invaluable record of past volcanic eruptions is included in the layers, as if on a calendar. Even the exact beginning of the Bronze Age is recorded in the ice core—about 3000 B.C. When the Greeks and later the Romans began smelting copper, they produced ash and smoke that the winds carried to this distant place. The presence of ammonia indicates ancient forest fires at lower latitudes. And, as an important analog of past temperatures on

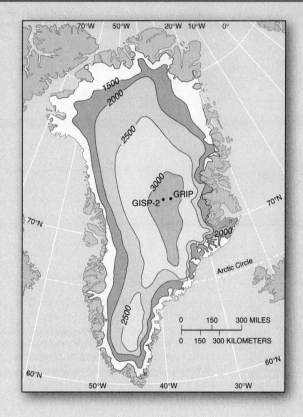

FIGURE 14.1.1
Greenland ice-core locations.
GRIP is at 72.5° N 37.5° W. GISP-2 is at 72.6° N 38.5° W. [Map courtesy of GISP 2 Science Management Office.]

the ice-sheet surface when each snowfall occurred, the ratio between stable forms of oxygen is measured.

Dome-C Ice Core
On the East Antarctic Plateau, 75.1° S, 123.3° E, approximately 900 km (560 mi) from the coast and 1750 km (1087 mi) from the South Pole, sits Dome C. This is the point on the Antarctic continent where the ice sheet is the thickest (see Figure 14.28c for location, and Figure 14.1.2). Here the ice is 3309-m

(10,856-ft) deep. The mean annual surface temperature is −54.5°C (−66.1°F); however, the science crews experience temperatures that range from −50°C when they arrive to −25°C by midsummer. Dome C is 560 km (348 mi) from the Vostok base, where the previous coring record of 400,000 years was set. The Dome-C 8-year project is part of a 10-country European Project for Ice Coring in Antarctica (EPICA).

The mechanical drill bit used for coring is 10 cm wide and provides a

FIGURE 14.1.2 Dome C station.
An overview of the Dome C scientific station, with its gathering of tents and semipermanent structures. [Photo courtesy of British Antarctic Survey.]

(continued)

Focus Study 14.1 *(continued)*

(a)

(b)

FIGURE 14.1.3 Dome C ice-core analysis.
(a) Dr. Eric Wolff of the British Antarctic Survey carefully inspects one of the ice-core segments. A quarter-section of each ice core is kept at Dome C in case there is an accident in transport back to labs in Europe. (b) Light shining through a thin section from the ice core; Earth's ancient atmosphere is trapped in the ice awaiting analysis. [Both photos courtesy of British Antarctic Survey, http://www.antarctica.ac.uk.]

high-resolution record of ancient atmospheric gases, ash from volcanic eruptions, and materials from other atmospheric events. By December 2004, coring to a 3270.20-m (10,729-ft) depth has brought up 800,000 years of Earth's past climate history (Figure 14.1.3). In Figure 14.23, this is Marine Isotope Stage (MIS) 20.2. This remarkable record contains eight glacial and interglacial cycles. Scientists found the Dome-C record affirms the Vostok findings and correlates perfectly with the deep-sea core of oxygen isotope fluctuations in foraminifera shells (microfossils) from the Atlantic Ocean.

Confirmed is the finding that the present concentration of carbon dioxide, methane, and nitrous oxide in the atmosphere is the highest it has been in the past 800,000 years and that these changing levels have marched in step with higher and lower temperatures throughout this time span.

Go to Figure 14.23 and find MIS 11 for an additional item of importance at Dome C. This warm spell at MIS 11, from 425,000 to 395,000 years ago, is perfectly recorded in the Dome-C core and is of interest because CO_2 in the atmosphere during MIS 11 was similar to our preindustrial level of 280 ppm. Earth's orbital alignment was similar to that of today as well. The MIS 11 analogy means that a return to the next glacial period is perhaps 16,000 years in the future. Understanding the past helps decode future trends. The challenge for physical geographers is to put these data through spatial analysis and find the linkages to other Earth systems and to human economies and societies.

carbon dioxide reached 389 ppm—higher than at any time in the past 800,000 years, principally owing to anthropogenic (human-created) sources. CO_2 ranged between 100 ppm and 300 ppm over the span of the Dome C ice-core record. With carbon dioxide emissions rising at 3.0% per year since 2000, such increases in atmospheric concentrations is of utmost concern.

Climate and Oceanic Circulation Finally, oceanic circulation patterns have changed. For example, the Isthmus of Panama formed about 3 million years ago and effectively separated the circulation of the Atlantic and Pacific Oceans. Changes in ocean basin configuration, surface temperatures, and salinity and in upwelling and downwelling rates affect air mass formation and air temperature.

Our understanding of Earth's climate—past, present, and future—is unfolding. We are learning that climate is a multicyclic system controlled by an interacting set of cooling and warming processes, all founded on celestial relations, tectonic factors, atmospheric variables, and changes in oceanic circulation. In the era of industrialization and coal-fired electrical production and combustion of fossil fuels, we add human activity as dominant in the climate-change mix.

Arctic and Antarctic Regions

Climatologists use environmental criteria to define the Arctic and the Antarctic regions. The *Arctic region* is defined in Figure 14.28 by the green line, which is the 10°C (50°F) isotherm for July (the Northern Hemisphere summer). This line coincides with the visible tree

FIGURE 14.28 The Arctic and Antarctic regions.
(a) Note the 10°C (50°F) isotherm in midsummer, which designates the Arctic region, dominated by rapidly retreating pack ice (b). (c) The Antarctic convergence designates the Antarctic region. Arrows on the ice sheet show the general direction of ice movement on Antarctica. Note the location of the Dome C and Vostok bases, sites where important ice cores have been removed. (d) Mountains, glacial ice, and a bergy bit along the Danco Coast, Antarctic Peninsula. [Photos (b) and (d) by Bobbé Christopherson.]

Geographic Scenes:
East Greenland

High Latitude
Connection Videos

line—the boundary between the northern forests and tundra. (For more information on polar region ice, see the National Ice Center at http://www.natice.noaa.gov/ and the Canadian Ice Service at http://ice-glaces.ec.gc.ca/app/WsvPageDsp.cfm.)

The *Antarctic convergence* defines the *Antarctic region*, a narrow zone that extends around the continent as a boundary between colder antarctic water and warmer water at lower latitudes. This boundary follows roughly the 10°C

(50°F) isotherm for February (Southern Hemisphere summer) and is located near 60° S latitude. The Antarctic region of the Southern Ocean surrounding the continent, when covered with sea ice, represents an area greater than North America, Greenland, and Western Europe combined!

Antarctica is a continent-sized landmass and therefore is much colder overall than the Arctic, which is an ocean. In simplest terms, Antarctica can be thought of as a continent covered by a single enormous glacier,

(a)

FIGURE 14.29 Disintegrating ice shelves, Antarctic and Arctic coastlines.
(a) 2003 image of a broken B-15 iceberg which calved off the Ross Ice Shelf, near Roosevelt Island, Antarctica, in 2000; originally twice the size of Delaware. (b) Disintegration and retreat of Larsen-B ice shelf between January 31 and March 7, 2002. Note the meltponds in the January image. (c) Loss of ice shelves along Ellesmere Island, Canada; the former positions of six are noted on August 29, 2008. (d) On the north coast of Greenland, some 30 km² of the 80-km-long tidewater portion of Petermann Glacier disintegrated in this July 25, 2008 image. Meltponds and a large crack opening up all point to more near-term losses. [All images Terra MODIS courtesy MODIS Rapid Response Team, GSFC/NASA.]

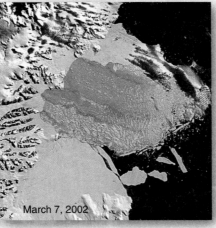

January 31, 2002   March 7, 2002

(b)

(d)

(c)

although it contains distinct regions, such as the East Antarctic and West Antarctic ice sheets, which respond differently to slight climatic variations.

Ice-sheet edges that enter coastal bays form extensive ice shelves, with sharp ice cliffs rising up to 30 m (100 ft) above the sea. Large tabular islands of ice form when sections of the shelves break off and move out to sea. These ice islands can be very large; several in the late 1980s and again in 1995, 1998, 2000, and 2002 exceeded the area of Rhode Island, and one twice the area of Delaware (see Figure 14.29a).

Changes Are Underway in the Polar Regions

Review the "Global Climate Change" section in Chapter 7, where a portrait of high-latitude temperatures and ongoing physical changes to pack ice, ice shelves, and glaciers is presented. News Report 14.2 looks at meltpond occurrence as an indicator of changing surface energy budgets.

The Arctic Ocean is covered by two kinds of ice: *floating sea ice* (frozen seawater) and *glacier ice* (frozen

NEWS REPORT 14.2

Increase in Meltponds Indicates Changing Surface Energy Budgets

An increase in meltponds across the Arctic Ocean indicates changing surface energy conditions, for these meltponds represent positive feedback—they are darker, absorb more insolation, and become warmer, which melts more ice, making more meltponds, etc. (Figure 1.5, Chapter 1). The *Landsat*-7 satellite in tandem with aircraft equipped with video cameras spotted this increase in *meltpond* occurrence on glaciers, icebergs, ice shelves, and the Greenland Ice Sheet.

Figure 14.2.1a is a scene from west Greenland in June 2003, using a sensor aboard the *Terra* satellite. Examine the bare rock along the eastern margin, indicating higher snow levels. The image shows rapidly increasing numbers of meltponds (small blue dots across the ice) and areas of water-saturated ice in the *melt zone*, advancing 400% in the 2-year interval since 2001. You can see the meltponds and melt streams in southwest Greenland in Figure 14.2.1b (June 2008).

The streams sometimes melt through the ice sheet, forming a *moulin*, or drainage channel, that works its way to the base of the glacier (Figure 14.2.1c). The glacial water can flow through the basal layers of fine clays, thus lubricating and causing acceleration of glacial flow rates; even hydrostatically lifting the ice is possible. Sometimes these moulin channels exit from the glacier's face and flow into the sea (Figure 14.2.1d).

Meltponds provide evidence of warming trends. The action of such melting is evident in the potholes mottling a glacier in southern Greenland (Figure 14.2.2a) and the actual meltponds on the ice sheet (Figure 14.2.2b, c). If you look back at the Larsen-B ice shelf breakup in Figure 14.29b, you can see the presence of meltponds as the ice decayed.

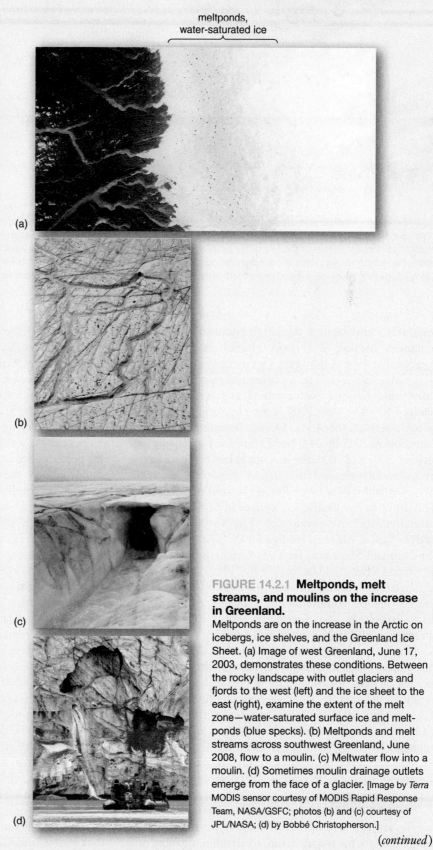

meltponds, water-saturated ice

(a)

(b)

(c)

(d)

FIGURE 14.2.1 **Meltponds, melt streams, and moulins on the increase in Greenland.**
Meltponds are on the increase in the Arctic on icebergs, ice shelves, and the Greenland Ice Sheet. (a) Image of west Greenland, June 17, 2003, demonstrates these conditions. Between the rocky landscape with outlet glaciers and fjords to the west (left) and the ice sheet to the east (right), examine the extent of the melt zone—water-saturated surface ice and meltponds (blue specks). (b) Meltponds and melt streams across southwest Greenland, June 2008, flow to a moulin. (c) Meltwater flow into a moulin. (d) Sometimes moulin drainage outlets emerge from the face of a glacier. [Image by *Terra* MODIS sensor courtesy of MODIS Rapid Response Team, NASA/GSFC; photos (b) and (c) courtesy of JPL/NASA; (d) by Bobbé Christopherson.]

(continued)

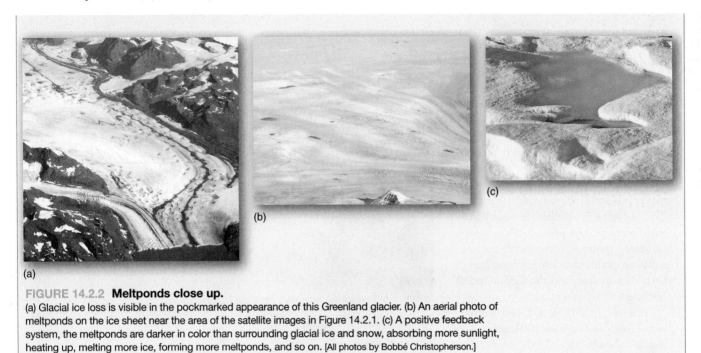

FIGURE 14.2.2 Meltponds close up.
(a) Glacial ice loss is visible in the pockmarked appearance of this Greenland glacier. (b) An aerial photo of meltponds on the ice sheet near the area of the satellite images in Figure 14.2.1. (c) A positive feedback system, the meltponds are darker in color than surrounding glacial ice and snow, absorbing more sunlight, heating up, melting more ice, forming more meltponds, and so on. [All photos by Bobbé Christopherson.]

freshwater). This ice pack thins in the summer months and sometimes breaks up (Figure 14.28b). A 1999 study reported that 43% of the Arctic Ocean ice has melted since 1970, owing to regional temperature increases. An increase in meltponds indicates such warming, as described in News Report 14.2. The year 2007 broke the record for lowest sea-ice extent in the Arctic Ocean, beating the previous record low in 2005 by 24% (see Figure 3.1.1 in Chapter 3). Between 2005 and 2007, the increase in loss of sea ice in the Arctic Ocean exceeded the area of California (424,000 km²). The fabled Northwest Passage across the Arctic from the Atlantic to the Pacific was ice-free in September 2007, as the Arctic ice continued to melt. The Northeast Passage, north of Russia, has been ice-free for the past several years. The National Snow and Ice Data Center forecasted that the Arctic Ocean could be ice-free by 2012.

By late fall 2003, dramatic news spread through the climate science community that the Ward Hunt Ice Shelf—largest in the Arctic, on the north coast of Ellesmere Island in Nunavut—was breaking up. This was shocking news, for the shelf had been stable for at least 4500 years. What scientists observed was that after three decades of mass losses, the ice shelf reached a systems threshold and rapidly broke up between 2000 and 2003. More ice loss at Ward Hunt and across the Ellesmere Island coast continued in 2008 with losses to the Serson, Peterson, Milne, Ayles, and Markham Ice Shelves (Figure 14.29c). The Petermann Glacier on the north coast of Greenland, dotted with meltponds, began cracking up in 2008 as well (Figure 14.29d).

Since 1993, six ice shelves have disintegrated in Antarctica. More than 8000 km² (3090 mi²) of ice shelf is gone, changing maps, freeing up islands to circumnavigation, and creating thousands of icebergs. The Larsen Ice Shelf, along the east coast of the Antarctic Peninsula, has been retreating slowly for years. Larsen-A suddenly disintegrated in 1995. In only 35 days in early 2002, Larsen-B collapsed into icebergs (Figure 14.29b). Larsen-B was at least 11,000 years old, meaning this collapse was farther south than any previously during the Holocene. Larsen-C, the next segment to the south, is losing mass on its underside since the water temperature is warmer by 0.65 C° (1.17 F°) than the melting point for ice at 300-m depth—warmth is melting the shelf from both the ocean and atmosphere sides.

Farther south along the Antarctic Peninsula in 2008, the Wilkins Ice Shelf began breaking up. This ice loss is likely a result of the 2.5 C° (4.5 F°) temperature increase in the peninsula region during the last 50 years. In response to the increasing warmth, the Antarctic Peninsula is supporting previously unseen vegetation growth, reduced sea ice, and disruption of penguin feeding, nesting, and fledging activities.

The Pine Island Glacier flows as an outlet glacier from the West Antarctic Ice Sheet (WAIS) to the Amundsen Sea (noted on Figure 14.28c). As discussed in Chapter 7, this glacier accelerated 3.5% in flow rate between April 2001 and April 2003. The Pine Island Glacier is accelerating toward the ocean and calving at a greater rate than previously recorded along its 40-km-wide ice shelf. Scientists determined that the acceleration could be sensed through the ice in the glacier, far back into the WAIS. This is similar to a discovery following the Larsen-B ice-shelf breakup in 2002, whose flow-rate changes were detected in the ice it held back.

Also, as happened with the Larsen-B breakup, warmer-than-average ocean water is melting the ice shelf holding back Pine Island Glacier. With its ice-shelf buttress changing, flows of grounded ice upstream in the glacier and the

WAIS itself, which are not yet displacing ocean water, are moving toward the sea. The impact on sea level is a scientific concern because present prediction capabilities and models of glacial behavior are still evolving. Meanwhile ice loss from Antarctica increased from 112 gigatonnes/year loss in 1996 to 196 gigatonnes/year loss in 2006—a 75% increase in an accelerated rate (gigatonne, or Gt, equals 1 billion metric tones; multiply by 1.1 for a US short ton).

The fact that Antarctica is a place so remote from civilization makes it an excellent laboratory for sampling past and present human and natural variables that are transported by atmospheric and oceanic circulation. High altitude, winter cold and darkness, and distance from pollution sources make the polar region an ideal location for certain astronomical and atmospheric observations (Figure 14.30).

FIGURE 14.30 Scientific base at the South Pole.
An aerial view of the Amundsen-Scott South Pole Station in January 2008. This station is at an elevation of 2835 m (9301 ft). The geodesic dome (left), now decommissioned, opened in 1975 and is 50 m wide by 16 m high (165 ft by 52 ft). To the left you see the elevated station completed in 2006, which stands on pylons 3 m (10 ft) above the surface to accommodate snow accumulation and protect the ice underneath from melting. Below the modules is the ceremonial South Pole with a semicircle of Antarctic Treaty-signing national flags; to its right is the geographic South Pole (brass pole marker, 90° S; see inset photo). More than 50 scientists and support people work through the winter (February to October) and 150 personnel, or more, research and work there in the brief summer. [Photos courtesy of Ethan Dicks, US Antarctic Program, National Science Foundation.]

Summary and Review—Glacial and Periglacial Landscapes

■ *Differentiate* between alpine and continental glaciers and *describe* their principal features.

More than 77% of Earth's freshwater is frozen. Ice covers about 11% of Earth's surface, and periglacial features occupy another 20% of ice-free but cold-dominated landscapes. Earth's **cryosphere** is the portion of the hydrosphere and ground that is perennially frozen, generally at high latitudes and elevations. A **glacier** is a mass of ice sitting on land or floating as an ice shelf in the ocean next to land. Glaciers form in areas of permanent snow. A *snowline* is the lowest elevation where snow occurs year-round; its altitude varies by latitude—higher near the equator, lower poleward.

A glacier in a mountain range is an **alpine glacier**. If confined within a valley, it is termed a *valley glacier*. The area of origin is a snowfield, usually in a bowl-shaped erosional landform called a **cirque**. Where alpine glaciers flow down to the sea, they calve and form **icebergs**. A **continental glacier** is a continuous mass of ice on land. Its most extensive form is an **ice sheet**; a smaller, roughly circular form is an **ice cap**; and the least extensive form, usually in mountains, is an **ice field**.

cryosphere (p. 462)
glacier (p. 462)
alpine glacier (p. 463)
cirque (p. 463)
iceberg (p. 463)
continental glacier (p. 463)
ice sheet (p. 463)
ice cap (p. 464)
ice field (p. 464)

1. Describe the location of most freshwater on Earth today. What is the cryosphere?

2. What is a glacier? What is implied about existing climate patterns in a glacial region?

3. Differentiate between an alpine glacier and a continental glacier.

4. Name the three types of continental glaciers. What is the basis for dividing continental glaciers into types? Which type covers Antarctica?

■ *Describe* the process of glacial ice formation and *portray* the mechanics of glacial movement.

Snow becomes glacial ice through accumulation, increasing thickness, pressure on underlying layers, and recrystallization. Snow progresses through transitional steps from **firn** (compact, granular) to a denser **glacial ice** after many years.

A glacier is an open system with inputs and outputs that can be analyzed through observation of the growth and wasting of the glacier itself. A **firn line** is the lower extent of a fresh snow-covered area. A glacier is fed by snowfall and is wasted by **ablation** (losses from its upper and lower surfaces and along its margins). Accumulation and ablation achieve a mass balance in each glacier. The zone where accumulation gain balances ablation loss is the **equilibrium line**.

As a glacier moves downhill, vertical **crevasses** (cracks and openings) may develop. Sometimes a glacier will move rapidly in a **glacier surge**. The presence of water along the basal layer appears to be important in glacial movements. As a glacier moves, it plucks rock pieces and debris, incorporating them into the ice, and this debris scours and sandpapers underlying rock through **abrasion**.

firn (p. 465)
glacial ice (p. 466)
firn line (p. 467)
ablation (p. 467)
equilibrium line (p. 467)
crevasses (p. 469)
glacier surge (p. 469)
abrasion (p. 469)

5. Trace the evolution of glacial ice from fresh fallen snow.
6. What is meant by glacial mass balance? What are the basic inputs and outputs underlying that balance?
7. What is meant by a glacier surge? What do scientists think produces surging episodes?

■ *Describe* characteristic erosional and depositional land-forms created by alpine glaciation and continental glaciation.

Extensive valley glaciers have profoundly reshaped mountains worldwide, carving **V**-shaped stream valleys into **U**-shaped glaciated valleys, producing many distinctive erosional and depositional landforms. As cirque walls erode away, sharp **arêtes** (sawtooth, serrated ridges) form, dividing adjacent cirque basins. Two eroding cirques may reduce an arête to a saddlelike **col**. A **horn** results when several cirque glaciers gouge an individual mountain summit from all sides, forming a pyramidal peak. An ice-carved rock basin left as a glacier retreats may fill with water to form a **tarn**; **paternoster lakes** are a string of tarns separated by moraines. Where a glacial valley trough joins the ocean, and the glacier retreats, the sea extends inland to form a **fjord**.

All glacial deposits, whether ice-borne or meltwater-borne, constitute **glacial drift**. Direct deposits from ice, called **till**, are unstratified and unsorted. Glacial meltwater deposits are sorted and are called **stratified drift**. A specific landform produced by the deposition of drift is a **moraine**. A **lateral moraine** forms along each side of a glacier; merging glaciers with lateral moraines form a **medial moraine**; and eroded debris dropped at the glacier's terminus is a *terminal moraine*.

Continental glaciation leaves different features than does alpine glaciation. A **till plain** forms behind end moraines, featuring unstratified coarse till, low and rolling relief, and deranged drainage. Beyond the morainal deposits, **outwash plains** of

stratified drift feature stream channels that are meltwater-fed, braided, and overloaded with debris that is sorted and deposited across the landscape. An **esker** is a sinuously curving, narrow ridge of coarse sand and gravel that forms along the channel of a meltwater stream beneath a glacier. An isolated block of ice left by a retreating glacier becomes surrounded by debris; when the block finally melts, it leaves a steep-sided **kettle**. A **kame** is a small hill, knob, or mound of poorly sorted sand and gravel that is deposited directly by water or by ice in crevasses.

Glacial action may form two types of streamlined hills: the erosional **roche moutonnée** is an asymmetrical hill of exposed bedrock, gently sloping upstream and abruptly sloping down-stream; the depositional **drumlin** is deposited till, streamlined in the direction of continental ice movement (blunt end upstream and tapered end downstream).

arête (p. 471)
col (p. 471)
horn (p. 471)
tarn (p. 471)
paternoster lakes (p. 471)
fjord (p. 471)
glacial drift (p. 471)
till (p. 471)
stratified drift (p. 471)
moraine (p. 471)
lateral moraine (p. 471)
medial moraine (p. 471)
till plain (p. 471)
outwash plain (p. 473)
esker (p. 473)
kettle (p. 473)
kame (p. 473)
roche moutonnée (p. 473)
drumlin (p. 474)

8. How does a glacier accomplish erosion?
9. Describe the evolution of a **V**-shaped stream valley to a **U**-shaped glaciated valley. What features are visible after the glacier retreats?
10. How is an iceberg generated?
11. Differentiate between two forms of glacial drift—till and outwash.
12. What is a morainal deposit? What specific moraines are created by alpine and continental glaciers?
13. What are some common depositional features encountered in a till plain?
14. Contrast a roche moutonnée and a drumlin regarding appearance, orientation, and the way each forms.

■ *Analyze* the spatial distribution of periglacial processes and *describe* several unique landforms and topographic features related to permafrost and frozen ground phenomena.

The term **periglacial** describes cold-climate processes, land-forms, and topographic features that exist along the margins of glaciers, past and present. Periglacial regions occupy over 20% of Earth's land surface. These areas are either near-permanent ice or are at high elevation, and the ground is seasonally snow-free. When soil or rock temperatures remain below 0°C (32°F) for at least 2 years, **permafrost** ("permanent frost") develops. An area of permafrost that is not covered by glaciers is considered periglacial. Note that this criterion is based solely on temperature and has nothing to do with how much or how little water is present. The **active layer** is the zone of seasonally frozen ground that exists between the subsurface permafrost layer and the ground surface. In regions of permafrost, frozen subsurface water forms **ground ice**. **Patterned ground** forms in the periglacial environment where freezing and thawing of the ground create polygonal forms of arranged rocks at the surface; these forms can be circles, polygons, stripes, nets, and steps.

periglacial (p. 474)
permafrost (p. 475)
active layer (p. 476)
ground ice (p. 477)
patterned ground (p. 478)

15. In terms of climatic types, describe the areas on Earth where periglacial landscapes occur. Include both higher-latitude and higher-altitude climate types.

16. Define two types of permafrost and differentiate their occurrence on Earth. What are the characteristics of each?

17. Describe the active zone in permafrost regions and relate the degree of development to specific latitudes.

18. What is a talik? Where might you expect to find taliks and to what depth do they occur?

19. What is the difference between permafrost and ground ice?

20. Describe the role of frost action in the formation of various landforms in the periglacial region.

21. Relate some of the specific problems humans encounter in developing periglacial landscapes.

■ *Explain* the Pleistocene ice-age epoch and related glacials and interglacials and *describe* some of the methods used to study paleoclimatology.

The study of Earth's past climates is the science of *paleoclimatology*. An **ice age** is any extended episode of cold. The late Cenozoic Era has featured pronounced ice-age conditions in an epoch called the Pleistocene. During this time, alpine and continental glaciers covered about 30% of Earth's land area in at least 18 glacials, punctuated by interglacials of milder weather. Beyond the ice, **paleolakes** formed because of wetter conditions. Evidence of ice-age conditions is gathered from ice cores drilled in Greenland and Antarctica, from ocean sediments, from coral growth in relation to past sea levels, and from rock.

The apparent pattern followed by these low-temperature episodes indicates cyclic causes. A complicated mix of interacting variables appears to influence long-term climatic trends: celestial relations, solar variability, tectonic factors, atmospheric variables, and oceanic circulation. **Lacustrine deposits** are lake sediments that form terraces along former shorelines.

ice age (p. 479)
paleolake (p. 482)
lacustrine deposits (p. 482)

22. What is paleoclimatology? Describe Earth's past climatic patterns. Are we experiencing a normal climate pattern in this era, or have scientists noticed any significant climate trends?

23. Define an ice age. When was the most recent? Explain *glacial* and *interglacial* in your answer.

24. Summarize what science discovered about the causes of ice ages by listing and explaining at least four possible factors in climate change.

25. Describe the role of ice cores in deciphering past climates. What record do they preserve? Where were they drilled?

26. Explain the relationship between the criteria defining the Arctic and Antarctic regions. Is there any coincidence between these criteria and the distribution of Northern Hemisphere forests on the continents?

27. What is the latest news from the analysis of the Dome C ice core as described in the text? Review Chapter 7 information as you prepare your thoughts on Dome C.

28. What is the significance of having the Dome C ice core include Marine Isotope Stage 11 (MIS 11) shown on Figure 14.23, relative to understanding present climates?

29. Briefly characterize changes in the cryosphere, specifically in the Arctic and Antarctic regions, that scientists are tracking.

30. What happened to the ice shelves along the Ellesmere Island coast in northern Canada? To Petermann Glacier in northern Greenland? What is thought to be the reason for these accelerating losses?

NetWork and Critical Thinking Tools

A. After reading my summary in the "Living at the South Pole" diary, on the *Elemental Geosystems* Student Learning Center site, imagine duty there for yourself. Of the people who winter over at the station, some serve as scientists, technicians, and support staff. Remember, the last airplane leaves mid-February and the first airplane lands mid-October—such is the isolation, which is unimaginable. What do you see as the positives and negatives to such service? How would you combat the elements? The isolation? The cold and dark winter conditions? During the 1999 winter, a doctor who was working at the South Pole station detected her own breast cancer. An unusual aerial winter drop of medications was delivered so she could self-administer chemotherapy. (In addition, see http://www.southpolestation.com/ and the US Antarctic Program at http://www.usap.gov/.)

B. Speculate on the relationship between the Alaskan meltdown discussed in this chapter and the occurrence of earthquakes, such as the magnitude 7.9 quake near Denali National Park, Alaska, the largest on Earth during 2002. The rapid melting of glacial ice is causing an isostatic rebound of the landscape (see Figure 8.4). Vertical motion in Alaska is averaging 36 mm (1.42 in.) per year of uplift—the fastest rate anywhere on Earth. Scientists see a tie between the rebounding crust and loss of glacial ice weight over the past 150 years and especially over the past 30 years. Hypothetically, how might isostatic rebound, in response to the unloading of glacial ice, produce strain along faults?

C. Analyze and compare the initial results from ice-core drilling efforts in Greenland and Antarctica described in this chapter (Focus Study 14.1 and text) with the type of ice cores described in the Student Learning Center Case Study that are taken from high-elevation tropical glaciers. What do they tell us about past climates? How far back do they go in time? Has any evidence of human activity on Earth been found in the cores?

Part IV   Biogeography Systems

Bison females and calves graze on mixed grasses and wildflowers in the Big Basin Prairie Reserve, some 735 hectares (1820 acres), 65 km (40 mi) south of Dodge City, Kansas. Males are kept in a separate section of the reserve to control breeding in this free-roaming wild herd of 60 bison. Note the bison are shedding wooly winter coats for a cooler summer hide. In the prairie ecology, birds use the hair for nest-building material. Bison (*Bison bison*), or American buffalo, were reduced to only a few hundred in the late nineteenth century out of a former population estimated in the tens of millions. Today nearly 500,000 bison exist in public and private managed herds, with some 40,000 harvested a year for their meat. Big Basin is in the Great Plains of history and legend, semiarid, expansive, and challenging. Walter Prescott Webb accurately said: "Historically the buffalo had more influence on man than all other Plains animals combined. It was life, food, raiment, and shelter to the Indians. The buffalo and the Plains Indians lived together, and together passed away." *[Photo by Bobbé Christopherson.]*

E arth is the home of the only known biosphere in the Solar System—a unique, complex, and interactive system of abiotic (nonliving) and biotic (living) components working together to sustain a tremendous diversity of life. Energy enters the system through conversion of solar energy by means of photosynthesis in the leaves of plants. Soil is the essential link among the lithosphere, plants, and the rest of Earth's physical systems. Thus, soil helps sustain life.

Life is organized into a feeding hierarchy from producers to consumers, ending with decomposers. Taken together, the soils, plants, animals, and all abiotic components produce aquatic and terrestrial ecosystems, generally grouped together in various biomes. Today we face crucial issues, principally the preservation of the diversity of life in the biosphere and the survival of the biosphere itself. Patterns of land and ocean temperatures, precipitation, weather phenomena, and stratospheric ozone, among many elements, are changing as global climate systems shift. The resilience of the biosphere, as we know it, is being tested in a real-time, one-time experiment. These important issues of biogeography are considered in PART IV.

Freshly plowed Spodosols, a soil type in the Soil Taxonomy (Podzolic soils in the Canadian System), near Lakeville, Nova Scotia, inland from the Bay of Fundy. Formed beneath coniferous forests, the land is cleared for crops and apple orchards (visible in the distance). This fertile silt-loam soil under a cool, moist climate is being prepared for planting vegetables. *[Photo by Bobbé Christopherson.]*

The Geography of Soils

KEY LEARNING CONCEPTS

After reading the chapter, you should be able to:

- **Define** soil and soil science and **describe** a pedon, polypedon, and typical soil profile.

- **Describe** soil properties of color, texture, structure, consistence, porosity, and soil moisture.

- **Explain** basic soil chemistry, including cation-exchange capacity, and **relate** these concepts to soil fertility.

- **Evaluate** principal soil formation factors, including the human element.

- **Describe** the 12 soil orders of the Soil Taxonomy classification system and **explain** their general occurrence.

STUDENT LEARNING CENTER TOOLS

The *Elemental Geosystems* Student Learning Center provides on-line resources for this chapter by clicking on the book cover at **www.mygeoscienceplace.com**. Once you are in our Learning Center, bookmark the URL address. For this chapter find reviewing tools, eBook links, self-tests and quizzes, critical thinking items, figure animations, photo galleries, and satellite loops. Some highlights:

- ▶ In Thinking Spatially, work through exercises with specific figures from the chapter for better understanding.

- ▶ Study and interact with an animation demonstrating soil ion exchange, soil particles, and soil water.

- ▶ Review soil science with four styles of questions and your interactive responses.

- ▶ More than 15 additional URLs along with Quick Links to all the URLs in the chapter are included.

Earth's landscape generally is covered with soil. **Soil** is a dynamic natural material composed of fine particles in which plants grow, and it contains both mineral and organic matter. If you have ever planted a garden, tended a houseplant, or been concerned about famine, this chapter will interest you. A knowledge of soil is at the heart of agriculture and food production.

You kneel down and scoop a handful of prairie soil, compressing it and breaking it apart with your fingers. You are holding a historical object—one that bears the legacy of the last 15,000 years or more. This lump of soil contains information about the last ice age and intervening warm intervals, about distinct and distant source materials, and about several physical processes. We are using and abusing this legacy at rates much faster than it formed. Soils do not reproduce, nor can they be recreated.

Soil science is interdisciplinary, involving physics, chemistry, biology, mineralogy, hydrology, taxonomy, climatology, and cartography. Physical geographers are interested in the spatial patterns formed by soil types and the physical factors that interact to produce them. As an integrative science, physical geography is well suited for this task. *Pedology* deals with the origin, classification, distribution, and description of soils (*ped* is from the Greek *pedon*, meaning "soil" or "earth"). *Edaphology* specifically focuses on the study of soil as a medium for sustaining the growth of higher plants (*edaphos* means "soil" or "ground").

Soil science deals with a complex substance whose characteristics vary from kilometer to kilometer, and even centimeter to centimeter. In many locales, an *agricultural extension service* can provide specific information and perform a detailed analysis of local soils. Soil surveys and local soil maps are available for most counties in the United States and for the Canadian provinces. Your local phonebook may list the US Department of Agriculture, Natural Resources Conservation Service (http://www.nrcs.usda.gov/), or Agriculture Canada's Soil Information System (http://sis.agr.gc.ca/cansis/intro.html); or you may contact the appropriate department at a local college or university. See the National Soil Survey Center's site at http://soils.usda.gov/, and more information on Soil Taxonomy at http://www.metla.fi/info/vlib/soils/old.htm and http://soils.ag.uidaho.edu/soilorders/.

In this chapter: The geography of soils deals spatially with a complex substance, the characteristics of which vary across the landscape. We begin with soil characteristics and the basic soil-sampling and soil-mapping units. The soil profile is a dynamic structure, mixing and exchanging materials and moisture across its horizons. Properties of soil include color, texture, structure, consistence, porosity, moisture, and chemistry—all integrating to form soil types. The chapter discusses both natural and human factors that affect soil formation. A global concern exists over the loss of soils to erosion, mistreatment, and conversion to other uses. The chapter concludes with a brief examination of the Soil Taxonomy, the 12 principal soil orders, and their spatial distribution.

Soil Characteristics

Classifying soils is similar to classifying climates because both involve interacting variables. Before we look at soil classification, let us examine the physical properties that distinguish soils as they develop through time, in response to climate, relief, and topography.

Soil Profiles

Just as a book should not be judged by its cover, so soils should not be evaluated at the surface layer only. Instead, a soil profile should extend from the surface to the deepest extent of plant roots, or to where regolith or bedrock is encountered. Such a profile, a **pedon**, is conceptually a hexagonal column from 1 to 10 m^2 in surface area (Figure 15.1). *The pedon is the basic sampling unit used in soil surveys.*

Many pedons together in one area make up a polypedon, which has distinctive characteristics differentiating it from surrounding polypedons. A **polypedon** is an essential soil individual, constituting an identifiable *series* of soils in an area. *The polypedon is the soil unit used in preparing local soil maps.*

Soil Horizons

Each layer exposed in a pedon is a **soil horizon**. A horizon is roughly parallel to the pedon's surface and has characteristics distinctly different from horizons directly above or below. The boundary between horizons usually is visible when viewed in profile, such as along a road cut, using the properties of color, texture, structure, consistence (meaning soil consistency), porosity, and the presence or absence of certain minerals, moisture, and chemical processes (Figure 15.2). Soil horizons are the building blocks of soil classification.

At the top of the soil profile is the *O (organic) horizon*, named for its organic composition, derived from plant and animal litter deposited on the surface and transformed into humus. **Humus** is a mixture of decomposed organic materials and is usually dark in color. Microorganisms work busily on this organic debris, performing a portion of the *humification* (humus-making) process. The O horizon is 20%–30% or more organic matter, important because of its water-absorbing ability and its nutrients.

At the bottom of the soil profile is the *R (rock) horizon*, consisting of either unconsolidated (loose) material or consolidated bedrock. The A, E, B, and C horizons mark important mineral strata between O and R. These middle layers are composed of sand, silt, clay, and other weathered by-products.

In the *A horizon*, the presence of humus and clay particles is particularly important, for they provide essential chemical links between soil nutrients and plants. The A horizon usually is darker and richer in organic content than lower horizons. It grades into the *E horizon*, made up

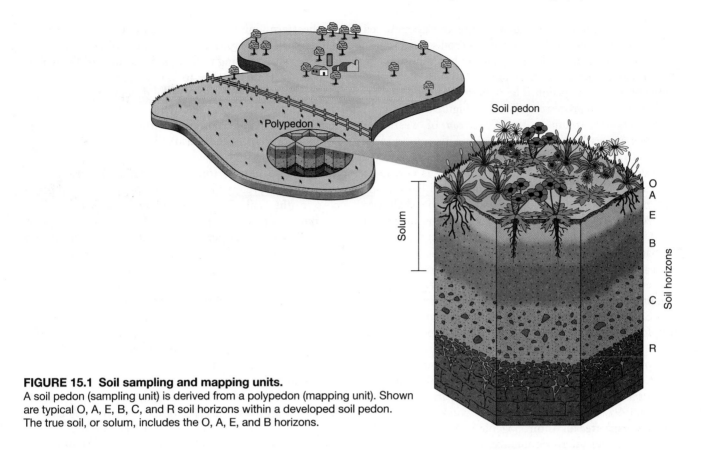

FIGURE 15.1 Soil sampling and mapping units.
A soil pedon (sampling unit) is derived from a polypedon (mapping unit). Shown are typical O, A, E, B, C, and R soil horizons within a developed soil pedon. The true soil, or solum, includes the O, A, E, and B horizons.

of coarse sand, silt, and resistant minerals. From the E horizon, clays and oxides of aluminum and iron are leached, removed by water, and carried to lower horizons with water as it percolates through the soil. **Eluviation** is the process in which water rinses upper horizons and removes fine particles and minerals; thus the designation E for this horizon. As precipitation increases, so does the rate of eluviation.

In contrast to the A and E horizons, *B horizons* accumulate clays, aluminum, and iron. **Illuviation** dominates the B horizons, a depositional process. (Eluviation is an erosional removal process; illuviation is depositional.) These horizons may exhibit reddish or yellowish hues because of the illuviated presence of mineral and organic oxides. In the humid tropics, these layers often develop to some depth.

Note that some materials occurring in the B horizon may have formed in place from weathering processes rather than arriving there by *translocation*, or migration. Likewise, clay losses in an A horizon may be caused by destructive processes and not eluviation. Research to

FIGURE 15.2 A typical soil profile.
A Mollisol pedon found in southeastern South Dakota. The parent material is till, and the soil is well drained. The dark O and A horizons above the #1 transition into an E horizon. Distinct carbonate nodules are visible in the lower B and upper C horizons. [Photo from Marbut Collection, Soil Science Society of America, Inc.]

better understand erosion and deposition of clays between soil horizons is one of the challenges in soil science.

The combination of the A and E horizons and the B horizon is designated the **solum**, considered the true definable soil of the pedon. The A, E, and B horizons experience active soil processes (labeled on Figure 15.1).

Below the solum is the *C horizon* of weathered bedrock or weathered parent material. This zone is identified as *regolith* (although the term sometimes is used to include the solum as well). The C horizon is not much affected by soil operations in the solum and lies outside the biological influences experienced in the shallower horizons. Plant roots and soil microorganisms are rare in the C horizon. It lacks clay concentrations and generally is made up of carbonates, gypsum, or soluble salts, or of iron and silica, which form cemented soil structures. In dry climates, calcium carbonate commonly forms the cementing material of these hardened layers.

Soil scientists using the US classification system employ letter suffixes to further designate special conditions within each soil horizon. A few examples include Ap (A horizon has been plowed), Bt (B horizon is of illuviated clay), Bf (permafrost or frozen soil), and Bh (illuviated humus as a dark coating on sand and silt particles).

In 1995, an international organization—the International Committee on Anthropogenic Soils (ICOMANTH, http://clic.cses.vt.edu/icomanth/)—worked on classification of anthropogenic soils, those with major properties created by human activities. The designation M is used in the NRCS *Keys to Soil Taxonomy*, 10th edition (2006), to indicate the presence of horizontal artifacts such as concrete and asphalt pavement and plastic, rubber, and textile liners placed by humans for some practical purpose. An M tells us that a subsoil layer is present that limits root growth. This innovation assists soil science in assessing construction or septic system monitoring, among other considerations.

Soil Properties

Soils are complex and varied, as this section reveals. Observing a real soil profile will help you identify color, texture, structure, and other soil properties. A good opportunity to observe soil profiles is at a construction site or excavation, perhaps on your campus, or at a road cut along a highway. The USDA Natural Resources Conservation Service *Soil Survey Manual* (US Department of Agriculture Handbook No. 18, October 1993) presents information on all soil properties. The NRCS publication *Soil Survey Laboratory Methods Manual* (US Department of Agriculture Soil Survey Investigations Report No. 42, v. 4.0, November 2004) details specific methods and practices in conducting soil analysis.

Soil Color Color is important, for it sometimes suggests composition and chemical makeup. If you look at exposed soil, color may be the most obvious trait. Among the many possible hues are the red and yellow soils of the southeastern United States, which are high in ferric (iron) oxides; the black-prairie, richly organic soils of portions of the US grain-growing regions and Ukraine; or the white-to-pale soils attributable to the presence of aluminum oxides and silicates. Color can be deceptive too, for soils of high humus content are often dark, yet clays of warm-temperate and tropical regions with less than 3% organic content are some of the world's blackest soils.

Soil Texture Soil texture, perhaps a soil's most permanent attribute, refers to the size and organization of particles in the soil. Individual mineral particles are *soil separates*. Particles range from the finest clays, to silts, to coarser sands, to larger pebbles and gravels. (Sands are graded from coarse, to medium, to fine, down to 0.05 mm; silt to 0.002 mm; and clay at less than 0.002 mm.)

Figure 15.3 is a soil triangle, showing the relation of sand, silt, and clay concentrations distributed along each side. Each corner of the triangle represents a soil consisting solely of the particle size noted (rarely are true soils composed of a single separate). Every soil on Earth is defined somewhere in this triangle.

Figure 15.3 includes the common designation **loam**, which is a balanced mixture of sand, silt, and clay beneficial to plant growth. Because of its water-holding characteristics and ease of cultivation, farmers consider a sandy loam with clay content below 30% (lower left) ideal. Soil texture is important in determining water-retention and water-transmission traits.

Consider a soil type in Indiana called *Miami silt loam*. Samples from this soil type are plotted on the soil texture triangle as points 1, 2, and 3. A sample taken near the surface in the A horizon is recorded at point 1, in the B horizon at point 2, and in the C horizon at point 3. Textural analyses of these samples are summarized in pie diagrams and a table in Figure 15.3. Note that silt dominates the surface, clay the B horizon, and sand the C horizon. The *Soil Survey Manual* presents guidelines for estimating soil texture by feel, a relatively accurate method when used by an experienced person. However, laboratory methods using graduated sieves and separation by mechanical analysis in water allow more precise measurements.

Soil Structure *Soil structure* refers to the *arrangement* of soil particles, whereas *soil texture* describes the size of soil particles. Structure can partially modify the effects of soil texture. The term *ped* describes an individual unit of soil particles; it is a tiny natural lump or cluster of particles held together. The shape of these peds determines which of the structural types the soil exhibits (Figure 15.4).

Peds separate from each other along zones of weakness, creating voids, or pores, that are important for moisture storage and drainage. More rounded peds have more pore space and greater permeability, and therefore are more productive for plant growth than coarse, blocky, prismatic, or platy peds, despite comparable fertility. Terms used to describe soil structure include *fine*, *medium*, or *coarse*. Adhesion among peds ranges from weak to strong.

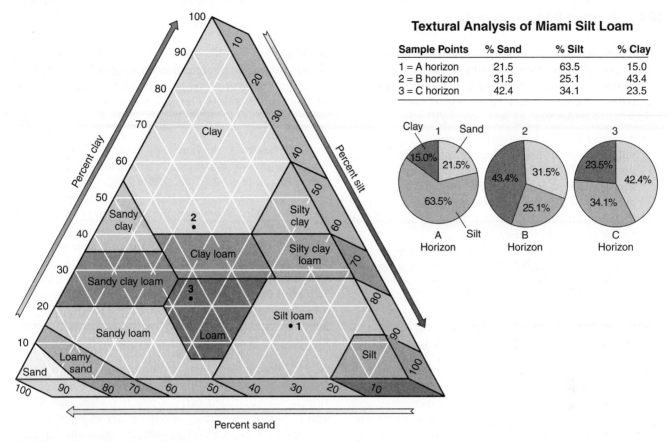

Textural Analysis of Miami Silt Loam

Sample Points	% Sand	% Silt	% Clay
1 = A horizon	21.5	63.5	15.0
2 = B horizon	31.5	25.1	43.4
3 = C horizon	42.4	34.1	23.5

FIGURE 15.3 Soil texture triangle.
Measures of the ratio of clay, silt, and sand determine soil texture. As an example, points 1 (horizon A), 2 (horizon B), and 3 (horizon C) designate samples taken at three different horizons in the Miami silt loam in Indiana. Note that silt dominates the surface, clay the B horizon, and sand the C horizon. [After US Department of Agriculture, Natural Resources Conservation Service, *Soil Survey Manual*, Agricultural Handbook No. 18 (Washington, DC: US Government Printing Office, 1993), p. 138.]

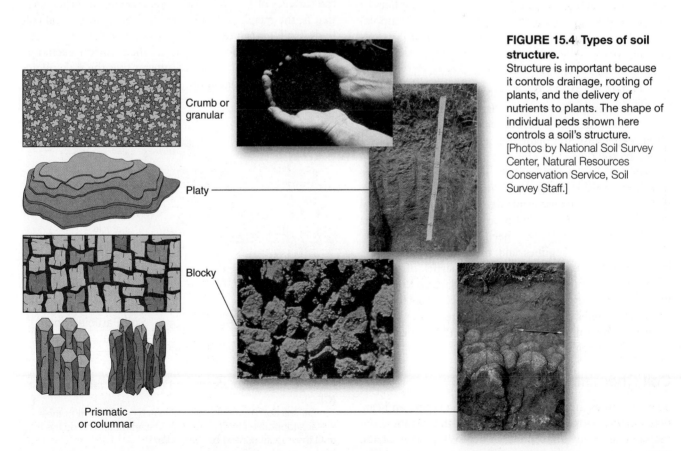

FIGURE 15.4 Types of soil structure.
Structure is important because it controls drainage, rooting of plants, and the delivery of nutrients to plants. The shape of individual peds shown here controls a soil's structure. [Photos by National Soil Survey Center, Natural Resources Conservation Service, Soil Survey Staff.]

Soil Consistence In soil science, the term *consistence* is used to describe the consistency of soil particles. It refers to cohesion in soil, a product of texture and structure. Consistence reflects a soil's resistance to breaking and manipulation under varying moisture conditions.

- A *wet soil* is sticky between the thumb and forefinger, ranging from a little adherence to either finger, to sticking to both fingers, to stretching when the fingers are moved apart. *Plasticity*, the quality of being moldable, is roughly measured by rolling a piece of soil between your fingers and thumb to see whether it rolls into a thin strand.
- A *moist soil* is filled to about half of field capacity (the usable water capacity of soil), and its consistence grades from loose (noncoherent), to *friable* (easily pulverized), to firm (not crushable between thumb and forefinger).
- A *dry soil* is typically brittle and rigid, with consistence ranging from loose, to soft, to hard, to extremely hard.

Soil Porosity Soil porosity, permeability, and moisture storage are discussed in Chapter 6. Pores in the soil horizon control the flow of water, its intake and drainage, and air ventilation. Important porosity factors are *pore size*, *continuity* (whether the pores are interconnected), *structure* (whether pores are spherical, irregular, or tubular), *orientation* (whether pore spaces are vertical, horizontal, or random), and *pore location* (whether pores are within or between soil peds).

Porosity is improved by the biotic actions of plant roots, animal activity such as the tunneling of gophers or worms, and human intervention through soil manipulation (plowing, adding humus or sand, or planting soil-building crops). Much of a farmer's soil preparation work is done to improve soil porosity.

Soil Moisture Reviewing Figures 6.8 and 6.9 will help you understand this section. Plants operate most efficiently when the soil is at *field capacity*, that is, maximum water availability for plant use. Field capacity is determined by soil type. The effective rooting depth of a plant determines the amount of soil moisture to which the plant's roots are exposed. If soil moisture is drawn below field capacity, plants use available water inefficiently until the plant reaches its wilting point, beyond which plants cannot extract the water they need.

Soil moisture regimes and their associated climate types shape the biotic and abiotic properties of the soil more than any other factor. The US Natural Resources Conservation Service Soil Taxonomy recognizes five soil-moisture regimes based on Thornthwaite's water-balance principles, ranging from constantly wet ("Aquic regime") to dry ("Xeric regime").

Soil Chemistry

The soil atmosphere is mostly nitrogen, oxygen, and carbon dioxide. Nitrogen concentrations in the soil are about the same as in the atmosphere. There is less oxygen and more carbon dioxide in soil than in the atmosphere because of ongoing respiration processes.

Water present in soil pores is the *soil solution*. It is the medium in the soil for chemical reactions. Soil solution is critical to plants as their source of nutrients and is the foundation of *soil fertility*. Carbon dioxide and various organic materials combine with water to produce carbonic and organic acids, respectively. These acids are then active in soil processes, as are dissolved alkalis and salts.

To understand how the soil solution behaves, let us go through a quick chemistry review. An *ion* is an atom, or group of atoms, that carries an electrical charge (examples: Na^+, Cl^-, HCO_3^-). An ion has either a positive charge or a negative charge. For example, when NaCl (sodium chloride) dissolves in solution, it separates into two ions: Na^+, a *cation* (positively charged ion), and Cl^-, an *anion* (negatively charged ion). Some ions in soil carry single charges, whereas others carry double or even triple charges (e.g., sulfate, SO_4^{2-}; and aluminum, Al^{3+}).

Soil Colloids and Mineral Ions Soil colloids retain ions in soil. **Soil colloids** are tiny particles of clay and organic material that carry a negative electrical charge and consequently attract any positively charged ions in the soil (Figure 15.5). The positive ions, many metallic, are critical to plant growth. If it were not for the soil colloids, the positive ions would be leached away by the soil solution and thus would be unavailable to plant roots.

Individual clay colloids are thin and platelike, with parallel surfaces that are negatively charged. Cations attach to the surfaces of the colloids by **adsorption** (not *absorption*); that is, the metallic cations are adsorbed onto the soil colloids. Colloids can exchange cations between their surfaces and the soil solution because of their **cation-exchange capacity (CEC)**. A high CEC means that the soil colloids can store or exchange more cations from the soil solution,

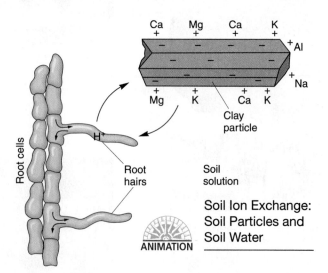

FIGURE 15.5 Soil colloids and cation-exchange capacity (CEC).
Typical soil colloids retain mineral ions by adsorption to their surface (opposite charges attract). This attraction holds the ions until they are absorbed by root hairs.

an indication of good soil fertility (unless there is a complicating factor, such as a soil that is too acid).

Soil fertility is the ability of soil to sustain plants. Soil has fertility when it contains organic substances and clay minerals that absorb water and certain elements needed by plants. Billions of dollars are expended to create fertile soil conditions, yet the future of Earth's most fertile soils is threatened because soil erosion is increasing worldwide.

Soil Acidity and Alkalinity A soil solution may contain significant hydrogen ions (H^+), cations that stimulate acid formation. The result is an *acid soil*. In contrast, a soil high in base cations (calcium, magnesium, potassium, and sodium) is a *basic* or *alkaline soil*. Acidity and alkalinity are expressed on the pH scale (Figure 15.6).

Pure water is nearly *neutral*, with a pH of 7.0. Readings below 7.0 represent increasing acidity. Readings above 7.0 represent increasing alkalinity. Acidity usually is strong at 5.0 or lower on the pH scale, whereas 10.0 or above is strongly alkaline.

Today the major contributor to soil acidity is acid precipitation (rain, snow, fog). Acid rain actually has been measured below pH 2.0—an incredibly low value for natural precipitation—as acidic as lemon juice. Because most crops are sensitive to specific pH levels, acid soils below pH 6.0 require treatment to raise the pH. The pH is elevated by the addition of bases in the form of minerals that are rich in base cations, usually lime (calcium carbonate, $CaCO_3$).

Soil Formation Factors and Management

Three soil-forming factors are important: Two natural factors are *dynamic* (climatic and biologic) and *passive* (parent material, topography and relief, and time), and the *human* component is a third. These three factors work together as a system for soil formation. Chapter 10 discusses the role of parent material in providing weathered minerals to form soils.

Natural Factors Physical and chemical weathering of rocks in the upper lithosphere provides the raw mineral ingredients for soil formation. These rocks supply the parent materials, and their composition, texture, and chemical nature help determine the type of soil that forms. The principal weathered by-products in soil are clay minerals. Climate types correlate closely with soil types worldwide. The moisture, evaporation, and temperature regimes of climate determine the chemical reactions, organic activity, and eluviation rates of soils. Not only is the present climate important but many soils exhibit the imprint of past climates, sometimes over thousands of years. Most notable is the effect of glaciations. Among other contributions, glaciation produced the loess soil materials that have been windblown thousands of kilometers to their present locations (discussed and mapped in Chapter 12).

Vegetation, animal, and bacterial activity determine the organic content of soil, along with all that is living in soil—algae, fungi, worms, and insects. The chemical makeup of the vegetation contributes to the acidity or alkalinity of the soil solution. For example, broadleaf trees tend to increase alkalinity, whereas needleleaf trees tend to produce higher acidity. Thus, when civilization moves into new areas and alters the natural vegetation by logging or plowing, the affected soils are likewise altered, often permanently.

Topography also affects soil formation, mainly through slope and orientation. Slopes that are too steep cannot have full soil development because gravity and erosional processes remove materials. Slopes that are nearly level inhibit soil drainage. In the Northern Hemisphere, a south-facing slope is warmer because it receives direct sunlight. North-facing slopes are colder, causing slower snowmelt and lower evaporation rates, which result in more moisture for plants than is available on south-facing slopes. Temperature affects water-balance relationships.

All of the identified natural factors in soil development—parent material, climate, biological activity, landforms and topography—require *time* to operate. This brings us to the human factor in soil processes.

The Human Factor Human intervention has a major impact on soils. Millennia ago, farmers in most cultures learned to plant slopes "on the contour"—to make rows or mounds around a slope at the same elevation, not vertically up and down the slope. Planting on the contour prevents water from flowing straight down the slope and thus reduces soil erosion. It was common to plant and harvest a floodplain but to live on higher ground nearby. Floods were celebrated as blessings that brought water, nutrients, and more soil to the land. Society is drifting away from these commonsense strategies.

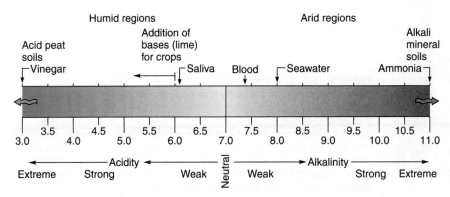

FIGURE 15.6 pH scale.
Soil pH scale, which measures acidity (lower pH) and alkalinity (higher pH). The complete pH scale ranges between 0 and 14.

A few centimeters' thickness of prime farmland soil may require *500 years* to mature. Yet, this same thickness is being lost annually through soil erosion when the soil-holding vegetation is removed and the land is plowed regardless of topography. Flood control structures block sediments and nutrients from replenishing floodplain soils, leading to additional soil losses. Over the same period, exposed soils may be completely leached of needed cations, thereby losing their fertility.

Unlike living species, soils do not reproduce nor can they be re-created. Some 35% of farmlands are losing soil faster than it can form—a loss exceeding 23 billion metric tonnes (25 billion tons) per year. Soil depletion and loss are at record levels from Iowa to China, Peru to Ethiopia, the Middle East to the Americas. The impact on society is potentially disastrous as population and food demands increase.

Soil erosion can be compensated for in the short run by using more fertilizer, increasing irrigation, and by planting higher-yielding strains. But the potential yield from prime agricultural land will drop by as much as 20% over the next 20 years if only moderate erosion continues. One study tabulated the market value of lost nutrients and other variables in the most comprehensive soil-erosion study to date. The sum of direct damage (to agricultural land) and indirect damage (to streams, society's infrastructure, and human health) was estimated at more than $25 billion a year in the United States and hundreds of billions of dollars worldwide. (Of course, this is a controversial assessment in the agricultural industry.) The cost to bring erosion under control in the United States is estimated at approximately $8.5 billion, or about 30 cents on every dollar of damage and loss (Figure 15.7). A recent article by scientist David Montgomery well summarizes our predicament:

> Recent compilations of data from around the world show that soil erosion under conventional agriculture exceeds both rates of soil production and geological erosion rates by several orders of magnitude. Consequently, modern agriculture—and therefore global society—faces a fundamental question over the upcoming centuries. Can an agricultural system capable of feeding a growing population safeguard both soil fertility and the soil itself?*

*D. R. Montgomery, "Is agriculture eroding civilization's foundation?" *GSA Today*, October 2007, p. 4.

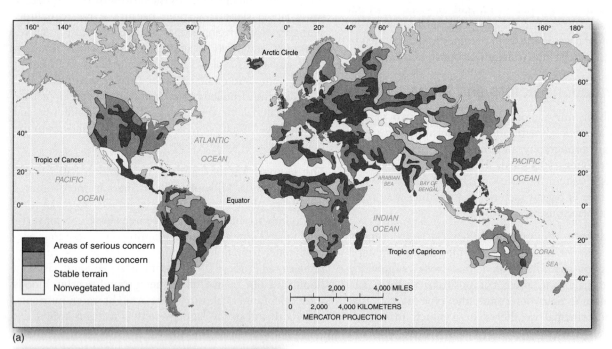

(a)

(b)

FIGURE 15.7 Soil degradation.
(a) Approximately 1.2 billion hectares (3.0 billion acres) of Earth's soils suffer some degree of degradation through erosion caused by human misuse and abuse. (b) An example of soil loss through sheet erosion and gully erosion on a northwestern Iowa farm. One millimeter of soil lost from an acre weighs about 5 tons. [(a) *A Global Assessment of Soil Degradation*, adapted from UNEP, International Soil Reference and Information Centre, "Map of Status of Human-Induced Soil Degradation," Sheet 2, Nairobi, Kenya, 1990; (b) photo by USDA/NRCS.]

Soil Classification

Classification of soils is complicated by the variety of interactions that create thousands of distinct soils—well over 15,000 soil types in the United States and Canada alone. Not surprisingly, a number of different classification systems are in use worldwide. The United States, Canada, the United Kingdom, Germany, Australia, Russia, and the United Nations Food and Agricultural Organization (FAO) have their own soil classification systems. Each system reflects the environment of its country.

Soil Taxonomy

The US soil classification system was published in 1975 (*Soil Taxonomy—A Basic System of Soil Classification for Making and Interpreting Soil Surveys*) and revised in a second edition in 1999. Soil scientists refer to it as **Soil Taxonomy**. Over the years, various revisions and clarifications in the system were published in *Keys to the Soil Taxonomy*, now in its 10th edition (2006, http://soils.usda .gov/technical/classification/tax_keys/), which includes all the revisions to the 1975 *Soil Taxonomy*. Much of the information in this chapter is derived from these two keystone publications. Two soil orders were added to the original 10: Andisols (volcanic soils) in 1990 and Gelisols (cold and frozen soils) in 1998.

Soil properties and morphology (appearance, form, and structure) actually seen in the field are key to the Soil Taxonomy system. Thus, it is open to addition, change, and modification as the sampling database grows. The system recognizes the importance of interactions between humans and soils and the changes that humans have introduced, both purposely and inadvertently.

The classification system divides soils into six categories, creating a hierarchical sorting system. Each *soil series* (the smallest, most detailed category) ideally includes only one polypedon but may include portions continuous with adjoining polypedons in the field. In sequence from smallest to largest categories, including the number of occurrences within each, the Soil Taxonomy recognizes: *soil series* (15,000), *soil families* (6,000), *soil subgroups* (1,200), *soil great groups* (230), *soil suborders* (47), and *soil orders* (12).

Pedogenic Regimes Before the Soil Taxonomy system, *pedogenic regimes* were used. These regimes keyed specific soil-forming processes to climatic regions. Although each pedogenic process may be active in several soil orders and in different climates, we discuss them within the soil order where they commonly occur. Such climate-based regimes are convenient for relating climate and soil processes. However, *the Soil Taxonomy system recognizes the great uncertainty and inconsistency in basing soil classification on such climatic variables*. The principal pedogenic regimes are

- **laterization**: a leaching process in humid and warm climates, discussed with Oxisols and Figure 15.10;
- **salinization**: a process that concentrates salts in soils in climates with excessive potential evapotranspiration (moisture demand) rates, discussed with Aridisols;
- **calcification**: a process that produces an illuviated accumulation of calcium carbonates in continental climates, discussed with Mollisols and Aridisols and Figure 15.17;
- **podzolization**: a process of soil acidification associated with forest soils in cool climates, discussed with Spodosols and Figure 15.20c;
- **gleization**: a process that includes an accumulation of humus and a thick, water-saturated gray layer of clay beneath, usually in cold, wet climates and poor drainage conditions.

Diagnostic Soil Horizons

To identify specific soil types within the Soil Taxonomy, the US Natural Resources Conservation Service describes *diagnostic horizons*. A diagnostic horizon reflects a distinctive physical property (color, texture, structure, consistence, porosity, moisture) or a dominant soil process (discussed with the soil types).

In the solum, two diagnostic horizons may be identified:

- The **epipedon** (literally, "over the soil") is the diagnostic horizon at the surface. It is visibly darkened by organic matter and sometimes is leached of minerals. Excluded from the epipedon are alluvial deposits, eolian deposits, and cultivated areas, because soil-forming processes have lacked the time to erase these relatively short-lived characteristics.
- The **diagnostic subsurface horizon** is located below the surface at varying depths. It may include part of the A or B horizon or both.

The presence or absence of either diagnostic horizon usually distinguishes a soil for classification. The Natural Resources Conservation Service publications identify many different types within each diagnostic horizon using a myriad of terminology.

The 12 Soil Orders of the Soil Taxonomy

At the heart of the Soil Taxonomy are 12 general soil orders, which are described in Table 15.1. Their worldwide distribution is shown in Figure 15.8, with an inset enlargement of most of North America. Please consult this table and map as you read the following descriptions. Because the taxonomy evaluates each soil order on its own characteristics, there is *no priority* to the classification. You will find the 12 orders loosely arranged here by latitude, beginning along the equator as is done in Chapter 7, "Climate Systems and Climate Change," and Chapter 16, "Ecosystems and Biomes."

Oxisols The persistent moisture, intense temperature, and uniform daylength of equatorial latitudes greatly affect soils. These generally old landscapes are deeply

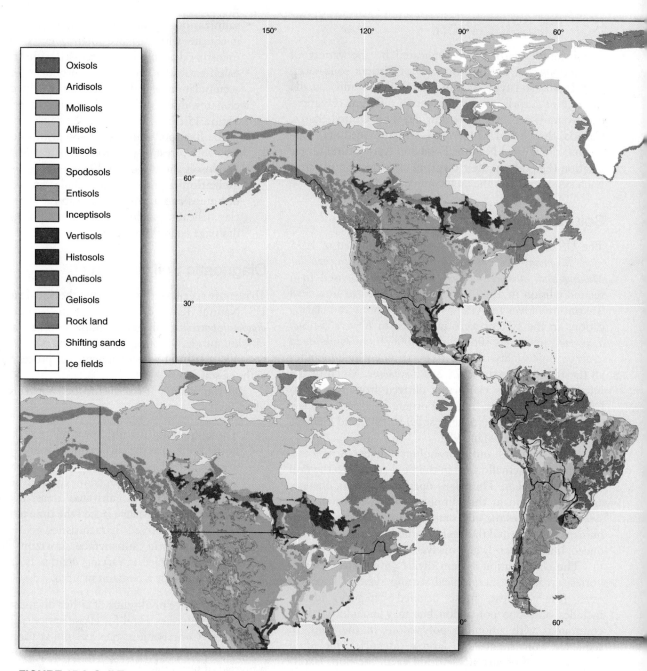

FIGURE 15.8 Soil Taxonomy.
Worldwide distribution of the 12 soil orders, with North America enlarged in inset map. [Adapted from maps prepared by World Soil Resources Staff, Natural Resources Conservation Service, USDA, 1999, 2006.]

developed and have greatly altered minerals (Figure 15.9). These soils, the **Oxisols** (tropical soils), have a distinctive horizon with a mixture of iron and aluminum oxides. Related vegetation is the luxuriant and diverse tropical and equatorial rain forest.

Heavy precipitation leaches soluble minerals and soil constituents (such as silica) from the A horizon. Typical Oxisols are reddish and yellowish from the iron and aluminum oxides left behind, with a weathered claylike texture, sometimes in a granular soil structure that is easily broken apart. The high degree of eluviation removes basic cations and colloidal material to lower illuviated horizons. Thus, Oxisols are low in CEC (cation-exchange capacity) and fertility, except in regions augmented by alluvial (water-borne sediment) or volcanic materials. Figure 15.10 illustrates *laterization*, the leaching process that operates in well-drained soils in warm, humid, tropical and subtropical climates.

To have the lush rain forests in the same regions as soils poor in inorganic nutrients seems an irony. However, this forest system relies on the recycling of nutrients from soil organic matter to sustain fertility, although this nutrient recycling ability is quickly lost when the ecosystem is disturbed.

The subsurface diagnostic horizon in an Oxisol is highly weathered, containing iron and aluminum oxides,

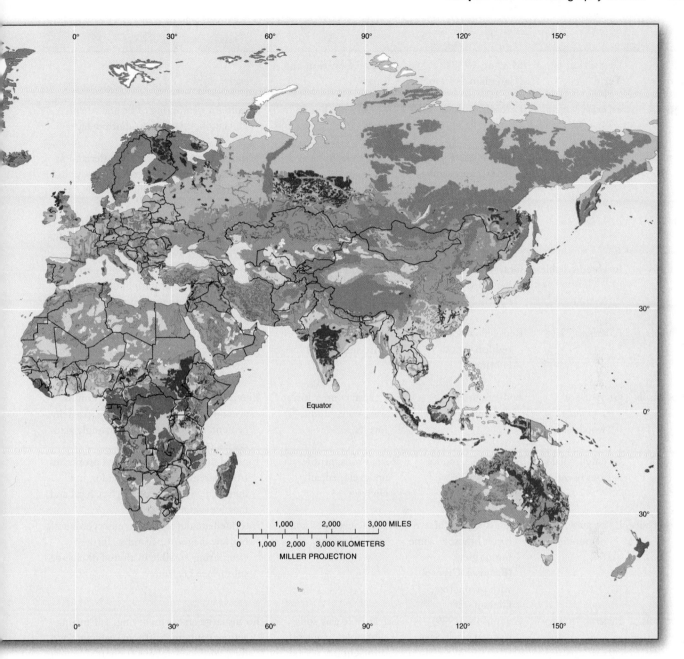

at least 30 cm (12 in.) thick and within 2 m (6.5 ft) of the surface (Figure 15.9). If these horizons are subjected to repeated wetting and drying, a *hardpan* (hardened soil layer) develops, called a **plinthite** (from the Greek *plinthos*, meaning "brick"). This form of soil, also called a laterite, can be quarried in blocks and used as a building material (Figure 15.11).

Simple agricultural activities can be conducted with care in these soils. Early *slash-and-burn* shifting cultivation practices were adapted to these soil conditions and formed a unique style of crop rotation. The scenario went like this: People in the tropics cut down (slashed) and burned the rain forest in small tracts and then cultivated

the land with stick and hoe. After several years, the soil lost fertility and the people shifted cultivation to another tract and repeated the process. After many years of moving from tract to tract, the group returned to the first patch to begin the cycle again. This practice protected the limited fertility of the soils somewhat, allowing periods of recovery.

The invasion of foreign plantation interests, development by local governments, vastly increased population pressures, and conversion of vast forest tracts to pasturage disrupted this orderly land rotation. Permanent tracts of cleared land, taken out of the former rotation mode, put tremendous pressure on the remaining forest and brought

Table 15.1	Soil Taxonomy Soil Orders			
Order	Derivation of Term	Marbut, 1938* (Canadian System**)	General Location and Climate	Description
Oxisols	Fr. *oxide*, "oxide" Gr. *oxide*, "acid or sharp"	Latosols, lateritic soils	Tropical soils; hot, humid areas	Maximum weathering of Fe and Al and eluviation, continuous plinthite layer
Aridisols	L. *aridos*, "dry"	Reddish desert, gray desert, sierozems	Desert soils; hot, dry areas	Limited alteration of parent material, low climate activity, light color, low humus content, subsurface illuviation of carbonates
Mollisols	L. *mollis*, "soft"	Chestnut, chernozem (Chernozemic)	Grassland soils; subhumid, semiarid lands	Noticeably dark with organic material; humus rich; base saturation; high, friable surface with well-structured horizons
Alfisols	Invented syllable	Gray-brown podzolic, degraded chernozem (Luvisol)	Moderately weathered forest soils; humid temperate forests	B horizon high in clays, moderate to high degree of base saturation, illuviated clay accumulates, no pronounced color change with depth
Ultisols	L. *ultimus*, "last"	Red-yellow podzolic, reddish yellow lateritic	Highly weathered forest soils; subtropical forests	Similar to Alfisols, B horizon high in clays, generally low amount of base saturation, strong weathering in subsurface horizons, redder than Alfisols
Spodosols	Gr. *spodos* or L. *spodus*, "wood ash"	Podzols, brown podzolic (Podzol)	Northern conifer forest soils; cool, humid forests	Illuvial B horizon of Fe/Al clays, humus accumulates, without structure, partially cemented, highly leached, strongly acid, coarse texture of low bases
Entisols	Invented syllable from *recent*	Azonal soils, tundra	Recent soils, profile undeveloped, all climates	Limited development, inherited properties from parent material, pale color, low humus, few specific properties, hard and massive when dry
Inceptisols	L. *inceptum*, "beginning"	Ando, subarctic brown forest lithosols, some humic gleys (Brunisol, Cryosol with permafrost, Gleysol wet)	Weakly developed soils; humid regions	Intermediate development, embryonic soils but few diagnostic features, further weathering possible in altered or changed subsurface horizons
Vertisols	L. *verto*, "to turn"	Grumusols, (1949) tropical black clays	Expandable clay soils; subtropics, tropics; sufficient dry period	Forms large cracks on drying, self-mixing action, contains >30% in swelling clays, light color, low humus content
Histosols	Gr. *histos*, "tissue"	Peat, muck, bog (Organic)	Organic soils, wet places	Peat or bog, >20% organic matter, much with clay >40 cm thick, surface, organic layers, no diagnostic horizons
Andisols	L. *ando*, "volcanic ash"	—	Areas affected by frequent volcanic activity (formerly within Inceptisols and Entisols)	Volcanic parent materials, particularly ash and volcanic glass; weathering and mineral transformation important; high CEC and organic content, generally fertile
Gelisols	L. *gelatio*, "freezing"	Formerly Inceptisols and Entisols (Cryosols, some Brunisols)	High latitudes in Northern Hemisphere, southern limits near tree line	Permafrost within 100 cm of the soil surface; evidence of cryoturbation (frost churning) and/or an active layer; patterned-ground

*C. F. Marbut, USDA soil scientist, developed the first American system of soil classification in the 1930s, first published in the USDA *Yearbook of Agriculture* in 1938.

**For comparison, soil orders from the Canadian System of Soil Classification (CSSC), 3rd edition, last revised in 1998.

(a)

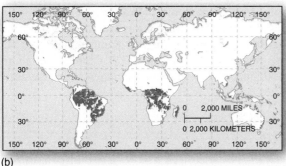

(b)

FIGURE 15.9 Oxisols.
(a) Deeply weathered Oxisol profile in central Puerto Rico.
(b) General map of worldwide distribution. [(a) Photo from the Marbut Collection, Soil Science Society of America.]

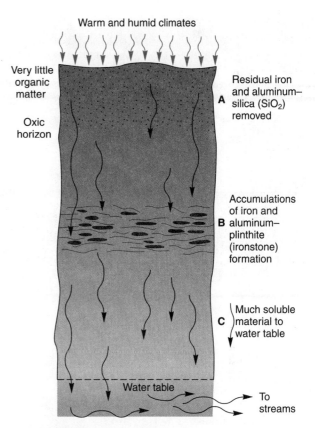

FIGURE 15.10 Laterization.
Laterization process that is characteristic of Oxisols in tropical and subtropical climatic regimes.

FIGURE 15.11 Oxisols are used for building materials.
Plinthite blocks are quarried for building materials in India.
[Photos by Henry D. Foth; inset from NRCS/USDA.]

disastrous consequences. When Oxisols are disturbed, soil loss can exceed a thousand tons per square kilometer per year, not to mention the greatly increased extinction rate of plant and animal species that accompanies such soil depletion and rain forest destruction. The regions dominated by the Oxisols and rain forests are rightfully the focus of much worldwide environmental attention.

Aridisols The largest single soil order occurs in the world's dry regions. **Aridisols** (desert soils) occupy approximately 19% of Earth's land surface and some 12% of the land in the United States (Figure 15.8 world map). A pale soil color near the surface is diagnostic (Figure 15.12a).

Not surprisingly for desert soils, the water balance in Aridisol regions is marked by periods of soil moisture deficit and generally inadequate soil moisture for plant growth. High potential evapotranspiration (moisture demand) and

(a)

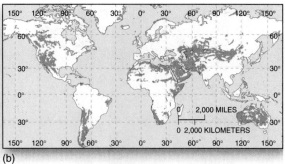

(b)

FIGURE 15.12 Aridisols.
(a) Soil profile from central Arizona. (b) General map showing worldwide distribution of these desert soils. [Photo from the Marbut Collection, Soil Science Society of America.]

early 1980s by contaminated agricultural drainage. Focus Study 15.1 elaborates on the Kesterson tragedy and the challenge of farming semiarid lands.

Mollisols Mollisols (grassland soils) are some of Earth's most significant agricultural soils. The dominant diagnostic horizon is a dark, organic surface layer some 25 cm (10 in.) thick (Figure 15.14). Mollisols are soft, even when dry, with granular or crumbly peds, loosely arranged. These humus-rich organic soils are high in basic cations (calcium, magnesium, and potassium) and have a high CEC (therefore, high fertility). In terms of water balance, these soils are intermediate in moisture between humid and arid soils.

Soils of the steppes and prairies of the world belong to this soil group: the North American Great Plains, the Palouse of Washington state, the Pampas of Argentina, and the region from Manchuria in China through to Europe (Figure 15.15). Agriculture ranges from large-scale

low precipitation produce very shallow soil horizons. Usually there is no period greater than 3 months when the soils have adequate moisture. Aridisols also lack organic matter of any consequence. Although the soils are not leached often because of the low precipitation, they are leached easily when exposed to excessive water, for they lack a significant colloidal structure.

Salinization is a soil process that occurs in Aridisols. Salinization results from high potential evapotranspiration rates in the deserts and semiarid regions of the world. Salts dissolved in soil water are brought to surface horizons and deposited as surface water evaporates. These deposits will kill plants when the salts accumulate near the root zone.

Obviously, salinization complicates farming in Aridisols. The introduction of irrigation water may either waterlog poorly drained soils or lead to salinization. Nonetheless, vegetation does grow where soils are well drained and kept low in salt content. If large capital investments are made in water, drainage, and fertilizers, Aridisols possess much agricultural potential (Figure 15.13).

In the Nile and Indus River valleys, for example, Aridisols are intensively farmed with a careful balance of these environmental factors, although thousands of acres of once-productive land, not so carefully treated, now sit idle and salt-encrusted. In California, the Kesterson Wildlife Refuge was reduced to a toxic waste dump in the

FIGURE 15.13 Agriculture in arid lands.
Rows of Aloe plants in irrigated Aridisols in the Imperial Valley. The irrigation water used in the Imperial Valley of southeastern California is derived from the dwindling Colorado River and is the subject of much tension and political action between agricultural interests and the cities of southern California. [Photo by Bobbé Christopherson.]

Selenium Concentration in Western Soils

Irrigated agriculture has increased greatly since 1800, when only 8 million hectares (about 20 million acres) were irrigated worldwide. Today, approximately 255 million hectares (about 630 million acres) are irrigated, and this figure is on the increase (USDA, FAO). Representing about 16% of Earth's agricultural land, irrigated land accounts for nearly 36% of the harvest.

Two related problems common in irrigated lands are salinization and waterlogging, especially in arid lands that are poorly drained. In many areas, production has decreased and even ended because of salt buildup in the soils. Examples include areas along the Tigris and Euphrates Rivers, the Indus River valley, sections of South America and Africa, and the western United States.

Irrigation in the West
About 95% of the irrigated acreage in the United States lies west of the 98th meridian. This region is increasingly troubled with salinization and waterlogging problems. In addition, at least nine sites in the West, particularly California's western San Joaquin Valley, are experiencing related contamination of a more serious nature—increasing selenium concentrations. Toxic effects of selenium were reported during the 1980s in some domestic animals grazing on grasses grown in selenium-rich soils in the Great Plains. In California, as parent materials weathered, selenium-rich alluvium washed into the semiarid valley, forming the soils that needed only irrigation water to become productive.

Drainage of agricultural wastewater poses a particular problem in semiarid and arid lands, where river discharge is inadequate to dilute and remove field runoff. One solution to prevent salt accumulations and waterlogging is to place field drains beneath the soil to collect gravitational water from fields that have been purposely overwatered to keep salts away from the effective rooting depth of the crops. But agricultural drain water must go somewhere, and for the San Joaquin Valley of central

FIGURE 15.1.1 Fields drain into canals.
Soil drainage canal collects contaminated water from field drains. [Photo by author.]

California this problem triggered a 15-year controversy (Figure 15.1.1).

Death of Kesterson
Central California's potential outlets for agricultural runoff are to the ocean, or San Francisco Bay, or the Central Valley. But all these suggested destinations failed to pass environmental impact assessments under Environmental Policy Act requirements. Nonetheless, about 80 miles of a drain were finished by the late 1970s, even though no formal plan was in place—a drain was built with no outlet. Large-scale irrigation continued, supplying the field-drainage tile outlets with salty, selenium-laden runoff that made its way to the Kesterson National Wildlife Refuge in the northern portion of the San Joaquin Valley east of San Francisco. The unfinished drain abruptly stopped at the boundary to the refuge.

The selenium-tainted drainage took only 3 years to destroy the wildlife refuge, which was officially declared a contaminated toxic waste site (Figure 15.1.2). Aquatic life forms (e.g., marsh plants, plankton, and insects) had taken in the selenium, which then made its way up the food chain and into the diets of other life forms in the refuge. According to US Fish and Wildlife Service scientists, the toxicity moved through the food chain and genetically damaged and killed wildlife, including all varieties of birds that nested at Kesterson; approximately 90% of the exposed birds perished or were injured. Because this wildlife refuge was a major

migration flyway and stopover point for birds from throughout the Western Hemisphere, the destruction of this refuge also violated several multinational wildlife protection treaties.

The field drains were sealed and removed in 1986, following a court order that forced the federal government to uphold existing laws. Irrigation water then immediately began backing up in the corporate farmlands, producing both waterlogging and selenium contamination.

Since 1985, more than 0.6 million hectares (1.5 million acres) of irrigated Aridisols and Alfisols have gone out of production in California, marking the end of several decades of irrigated farming in climatically marginal lands. Severe cutbacks in irrigated acreage no doubt will continue, underscoring the need to preserve prime farmlands in wetter regions.

Frustrated agricultural interests asked the federal government to finish the drain, either to San Francisco Bay or to the ocean. However, neither option appears capable of passing an environmental impact analysis. Another strategy is to allow irrigated lands to start pumping again into the former wildlife refuge because it is now a declared toxic site anyway! There are nine such threatened sites in the West; Kesterson was simply the first of these to fail from contamination. Such damage to a wildlife refuge presents a real warning to human populations. Remember where we are—at the top food chain.

FIGURE 15.1.2 Soil contamination in the wildlife refuge.
Salt encrusts soil and plants in portions of the contaminated Kesterson National Wildlife Refuge. [Photo by Gary R. Zahm.]

(a)

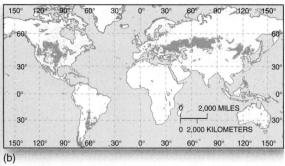

(b)

FIGURE 15.14 Mollisols.
(a) Profile from central Iowa and (b) general map of worldwide distribution. [(a) Photo from Marbut Collection, Soil Science Society of America, Inc.]

(a)

(b)

FIGURE 15.15 Fertile Mollisols in the Palouse and Midwest.
(a) Mollisols of eastern Washington state form the basis of a productive agricultural landscape, here filled with wheat ready to harvest. Wheat, when added to maize (corn) and rice, provides more than 50% of the world's grain supply. (b) Light rains grace the fertile soils of southwestern Minnesota on a warm, humid Midwestern afternoon; fields are planted in feed grains. Former prairies have sustained a productive agriculture since the early nineteenth century. [Photos by Bobbé Christopherson.]

commercial grain farming to grazing along the drier portions of the soil order. With fertilization or soil-building practices, high crop yields are common. The "fertile triangle" of Ukraine, Russia, and western portions of the former Soviet Union is of this soil type.

In North America, the Great Plains straddle the 98th meridian, which is coincident with the 51-cm (20-in.) isohyet of annual precipitation—wetter to the east and drier to the west. The Mollisols here mark the historic division between the short- and tall-grass prairies (Figure 15.16).

Calcification is a soil process characteristic of some Mollisols and adjoining marginal areas of Aridisols. Calcification is the accumulation of calcium carbonate or magnesium carbonate in the B and C horizons. Calcification by calcium carbonate ($CaCO_3$) forms a diagnostic subsurface horizon that is thickest along the boundary between dry and humid climates (Figure 15.17). When cemented or hardened, these deposits are called *caliche*, or *kunkur*; they occur in widespread soil formations.

Alfisols Alfisols (moderately weathered forest soils) are the most widespread of the soil orders, extending from near the equator to high latitudes. Representative Alfisol areas include Boromo and Burkina (interior western Africa); Fort Nelson, British Columbia; the states near the Great Lakes; and the valleys of central California. Most Alfisols have a pale, grayish brown-to-reddish epipedon

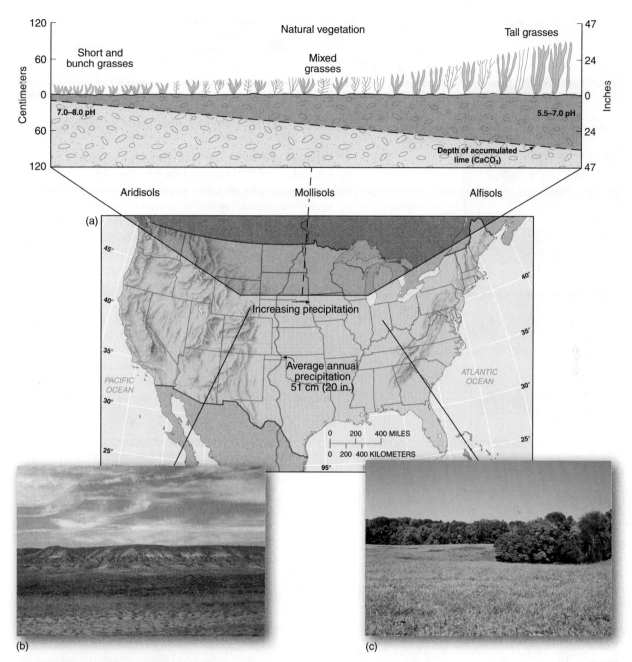

FIGURE 15.16 Soils of the Midwest.
(a) Aridisols (to the west), Mollisols (central), and Alfisols (to the east)—a soil continuum in the north-central United States and southern Canadian prairies. Graduated changes that occur in soil pH and the depth of accumulated lime are shown. (b) Bunch grasses and shallow soils of Wyoming. (c) Farmlands and Alfisols south of Bedford, Indiana, planted in soy beans. [(a) Illustration adapted from N. C. Brady, *The Nature and Properties of Soils*, 10th ed., © 1990 by Macmillan Publishing Company, adapted by permission; photos by (b) author; (c) Bobbé Christopherson.]

and are considered moist versions of the Mollisol soil group. Moderate eluviation is present, as well as a subsurface horizon of illuviated white clays because of increased precipitation (Figure 15.18).

Alfisols have moderate-to-high reserves of basic cations and are fertile. However, productivity depends on moisture and temperature. Alfisols usually are supplemented by a moderate application of lime and fertilizers in areas of active agriculture. Some of the best farmland in

the United States stretches from Illinois, Wisconsin, and eastern Minnesota through Indiana, Michigan, and Ohio to Pennsylvania and New York. This land produces grains and hay and supports a dairy industry. These naturally productive soils are farmed intensively.

The Xeralfs, another Alfisol subgroup, correlate with the moist-winter, dry-summer pattern of the Mediterranean climate. These soils are farmed for subtropical fruits, nuts, and specialty crops that can grow in only a few

POTET equal to or
greater than PRECIP

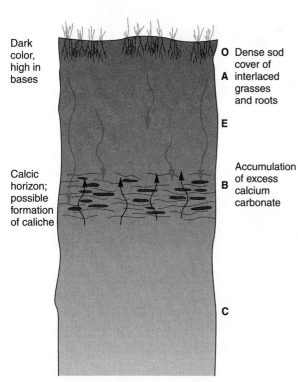

Dark
color,
high in
bases

O Dense sod
cover of
A interlaced
grasses
and roots

E

Calcic
horizon;
possible
formation
of caliche

B Accumulation
of excess
calcium
carbonate

C

FIGURE 15.17 Calcification in soil.
Calcification process in Aridisol/Mollisol soils in climatic regimes
that have a potential evapotranspiration equal to or greater than
precipitation.

locales worldwide, for example, California grapes, olives,
citrus, artichokes, almonds, and figs.

Ultisols Farther south in the United States are the
Ultisols, highly weathered forest soils. An Alfisol might
degenerate into an Ultisol, given time and exposure to
increased weathering under moist conditions. These soils
tend to be reddish because of residual iron and aluminum
oxides in the A horizon (Figure 15.19).

The increased precipitation in Ultisol regions means
greater mineral alteration, more eluvial leaching, and
therefore a lower level of basic cations, leading to infer-
tility. Certain agricultural practices and the effect of soil-
damaging crops such as cotton and tobacco reduce
fertility, depleting nitrogen and exposing soil to erosion.
However, these soils respond well if subjected to good
management—for example, crop rotation that restores
nitrogen and cultivation practices that prevent sheetwash
and soil erosion. Figure 15.19c shows peanut plantings

(a)

(b)

(c)

FIGURE 15.18 Alfisols.
(a) Soil profile from central California. (b) General map showing worldwide distribution of these
moderately weathered forest soils. (c) Cultivated Alfisols farmland near Lompoc, California.
[Photos from (a) the Marbut Collection, Soil Science Society of America, Inc.; (c) by Bobbé
Christopherson.]

FIGURE 15.19 Ultisols.
(a) General map showing worldwide distribution of these highly weathered forest soils. (b) A type of Ultisol in central Georgia, bearing its characteristic dark reddish color, is planted with pecan trees. (c) A type of Ultisol planted with rows of peanuts, near Plains in west-central Georgia, shows its characteristic reddish color. [Photos by Bobbé Christopherson.]

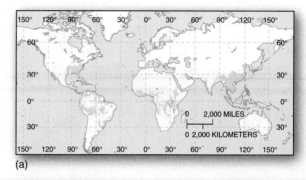

(a)

(b)

(c)

that assist in nitrogen restoration. Much needs to be done to achieve sustainable management of these soils.

Spodosols   The **Spodosols** (northern coniferous forest soils) occur generally to the north and east of the Alfisols. They are in cold and forested moist regimes (*humid-continental, mild-summer* climates) in northern North America and Eurasia, Denmark, the Netherlands, and southern England. Because there are no comparable climates in the Southern Hemisphere, this soil type is not identified there. Spodosols form from sandy parent materials, shaded under evergreen forests of spruce, fir, and pine. Spodosols with more moderate properties form under mixed or deciduous forests (Figure 15.20).

Spodosols lack humus and clay in the A horizons. An eluviated white horizon, sandy and leached of clays and irons, lies in the A horizon instead and overlies a horizon of illuviated organic matter and iron and aluminum oxides. The surface horizon receives organic litter from base-poor, acid-rich evergreen trees, which contribute to acid accumulations in the soil. The low pH (acidic) soil solution effectively removes clays, iron, and aluminum, passing to the upper diagnostic horizon. An ashen gray color is common in these subarctic forest soils and is characteristic of a formation process called *podzolization* (Figure 15.20c). In the Canadian system, Spodosols fall within the Podzolic Great Group, as seen in the temperate rain forests of Vancouver Island, British Columbia

(Figure 15.20d), and the rich forest soils of south-central Norway in Figure 15.20e.

When agriculture is attempted, the low base-cation content of Spodosols requires the addition of nitrogen, phosphate, and potash (potassium carbonate), and perhaps crop rotation as well. A soil amendment such as limestone can significantly increase crop production by raising the pH of these acidic soils. For example, the yields of several crops (corn, oats, wheat, and hay) grown in specific Spodosols in New York State increased up to a third with the application of limestone in a 6-year rotation program.

Entisols   The **Entisols** (recent, undeveloped soils) lack vertical development of horizons. The presence of Entisols is not climate-dependent, for they occur in many climates worldwide. Entisols are true soils that have not had sufficient time to generate the usual horizons.

Entisols generally are poor agricultural soils, although those formed from river silt deposits are quite fertile. The conditions that inhibited complete development also prevented adequate fertility—too much or too little water, poor structure, and insufficient accumulation of weathered nutrients. Active slopes, alluvium-filled floodplains, poorly drained tundra, tidal mud flats, dune sands and erg (sandy) deserts, and plains of glacial outwash all are characteristic regions with these soils. Figure 15.21 shows an Entisol in a desert climate where shales formed the parent material.

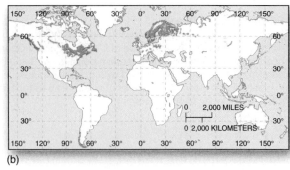

(b)

(a)

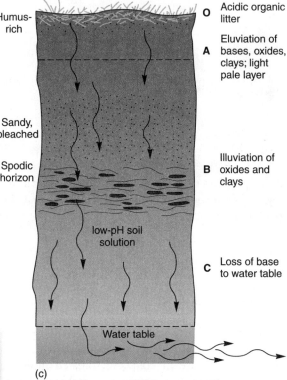

Cool and moist
climate

Humus-
rich

Sandy,
bleached

Spodic
horizon

low-pH soil
solution

Water table

O Acidic organic
litter

A Eluviation of
bases, oxides,
clays; light
pale layer

B Illuviation of
oxides and
clays

C Loss of base
to water table

(c)

(d)

(e)

FIGURE 15.20 Spodosols.
(a) Profile from northern New York and (b) general map of worldwide distribution; (c) podzolization
process, typical in cool and moist climatic regimes. (d) Characteristic temperate forest and
Spodosols in the cool, moist climate of central Vancouver Island. (e) Spodosols and farms
growing feed grains in central Norway along a fjord. [Photos from (a) Marbut Collection, Soil
Science Society of America, Inc.; (d) and (e) by Bobbé Christopherson.]

FIGURE 15.21 Entisols.
A characteristic Entisol: a young, poorly developed soil, lacking mature horizons, forming from a shale parent material in the Anza-Borrego desert, California. [Photo by Bobbé Christopherson.]

Inceptisols **Inceptisols** (weakly developed soils) are inherently infertile. They are young soils, although more developed than the Entisols. Inceptisols include a wide variety of different soils, all having in common a lack of maturity with evidence of weathering just beginning. Inceptisols are associated with moist soil regimes and are regarded as eluvial because they demonstrate a loss of soil constituents throughout their profile but retain some weatherable minerals. This soil group has no distinct illuvial horizons. Inceptisols include glacially derived till and outwash materials from New York down through the Appalachians and alluvium on the Mekong and Ganges floodplains.

Gelisols The newest addition to the Soil Taxonomy are the **Gelisols** (cold and frozen soils) and their three suborders that represent inclusion of high-latitude (Canada, Alaska, Russia, Arctic Ocean islands, and the Antarctic Peninsula) and high-elevation (mountain) soil conditions (Figure 15.22a). Temperatures in these regions are at or below 0°C (32°F), making soil development a slow process and disturbances of the soil long-lasting. Gelisols can develop organic diagnostic horizons because cold temperatures slow decomposition of materials (Figure 15.22b and c). Characteristic vegetation is that of tundra, such as lichens, mosses, sedges, and other plants adapted to the harsh cold.

Gelisols are subject to *cryoturbation* (frost churning and mixing in soil) in the freeze-thaw cycle in the active layer of periglacial regions (see Chapter 14). This process disrupts soil horizons, pulling organic material to lower layers and lifting rocky C-horizon material to the surface. Patterned-ground phenomena are possible under such conditions.

As we saw in Chapter 14, periglacial processes occupy about 20% of Earth's land surface, including permafrost (frozen ground) that underlies some 13% of these lands.

FIGURE 15.22 Gelisols.
(a) General map of worldwide distribution. (b) Gelisols and tundra green in the brief summer season on Spitsbergen Island. (c) Turning over a clod exposes the fibrous organic content and slow decomposition on this arctic island location. [Photos by Bobbé Christopherson.]

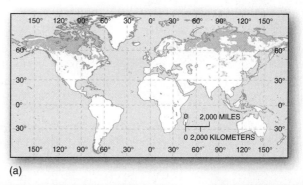

(a)

(b)

(c)

Previously these soils were included in the Inceptisols, Entisols, and Histosols soil orders. In the Canadian System of Soil Classification, these are included in the *Crysolic* soil order.

Andisols Areas of volcanic activity feature **Andisols** (volcanic parent materials). These soils formerly were classified as Inceptisols and Entisols, but in 1990 they were placed in this new order. Andisols are derived from volcanic ash and volcanic glass. Previous soil horizons frequently are buried by ejecta from repeated volcanic eruptions. Andisols are unique in their mineral content and in their recharge by recurring eruptions.

Weathering and mineral transformations are important in this soil order. Volcanic glass weathers readily into a clay colloid and oxides of aluminum and iron. Andisols feature a high CEC and high water-holding ability and develop moderate fertility. The fertile Andisol fields of Hawai'i produce sugar cane, pineapple, and other important cash crops (Figure 15.23a). Andisol distribution is

(a)

(b)

FIGURE 15.23 Andisols in agricultural production.
(a) Hawaiian landscape, trees, macadamia nut orchard, and fertile Andisols. (b) Pasture and feed grains for sheep raising grow in Andisols in southwestern Iceland. [Photos by Bobbé Christopherson.]

small in areal extent; however, such soils are locally important around the volcanic "ring of fire" of the Pacific Rim. In Iceland, feed crops and pasture land of fertile Andisols sustain a large sheep industry. The annual fall roundup is a time when many ranchers combine forces over several weeks to bring in the herds (Figure 15.23b).

Vertisols **Vertisols** (expandable clay soils) are heavy clay soils. They contain more than 30% swelling clays (which swell significantly when they absorb water) and are located in regions experiencing highly variable soil moisture balances through the seasons. These soils occur in areas of subhumid-to-semiarid moisture and moderate-to-high temperature. Vertisols frequently form under savanna and grassland vegetation in tropical and subtropical climates and are sometimes associated with a distinct dry season following a wet season. Although widespread, individual Vertisol units are limited in extent.

Vertisol clays are black when wet (not because of organics, but because of specific mineral content) and range to brown and dark gray. These deep clays swell when moistened and shrink when dried. In the process, vertical cracks form, which widen and deepen as the soil dries, producing cracks 2–3 cm (0.8–1.2 in.) wide and up to 40 cm (16 in.) deep. Loose material falls into these cracks, only to disappear when the soil again expands and the cracks close. After many such cycles, soil contents tend to invert or mix vertically, bringing lower horizons to the surface (Figure 15.24).

Despite the fact that clay soils are plastic and heavy when wet, with little available soil moisture for plants, Vertisols are high in bases and nutrients and thus are some of the better farming soils wherever they occur, for example, in a narrow zone along the coastal plain of Texas (Figure 15.24c) and in a section along the Deccan region of India. Vertisols often are planted with grain sorghums, corn, and cotton.

Histosols **Histosols** (organic soils) are formed from accumulations of thick organic matter. In the midlatitudes, when conditions are right, beds of former lakes may turn into Histosols with water gradually being replaced by organic material, forming a bog and layers of peat. (Lake succession and bog/marsh formation are discussed in Chapter 16.) Histosols also form in small, poorly drained depressions such as a bog (Figure 15.25a). This material can be cut, baled, and sold as a soil amendment.

Peat is cut by hand with a spade into blocks, which are then set out to dry. In Figure 15.25c, note the fibrous texture of the sphagnum moss growing on the surface and the darkening layers with depth in the soil profile as the peat is compressed and chemically altered. Such beds can be more than 2 m thick. Once dried, the peat blocks burn hot and smoky. Peat is the first stage in the natural formation of lignite, an intermediate step toward coal. Imagine such soils forming in plant-lush swamp environments in the Carboniferous Period (359 to 299 m.y.a.), only to go through coalification to become coal deposits.

(a)

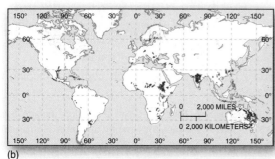

(b)

(c)

FIGURE 15.24 Vertisols.
(a) Profile in the Lajas Valley of Puerto Rico and (b) general map of worldwide distribution. (c) Vertisols in the Texas coastal plain, northeast of Palacios near the Tres Palacios River, planted with a commercial sorghum crop. Note the dark soil color indicative of Vertisols, wet and shiny from the rains of Tropical Storm Allison, June 2001. [Photos (a) from Marbut Collection, Soil Science Society of America, Inc.; and (c) by Bobbé Christopherson.]

(a)

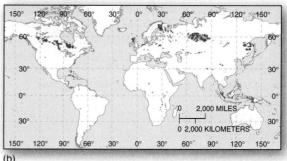

(b)

(c)

FIGURE 15.25 Histosols.
(a) A Histosol bog in coastal Maine, near Popham Beach State Park. (b) General map of worldwide distribution. (c) Sphagnum peat soil profile on Mainland Island, in the Orkneys, north of Scotland. Note the drying peat blocks in the inset photo. [Photos by Bobbé Christopherson.]

Summary and Review—The Geography of Soils

■ *Define* soil and soil science and *describe* a pedon, polypedon, and typical soil profile.

Soil is the portion of the land surface in which plants can grow. It is a dynamic natural body composed of fine materials and contains both mineral and organic matter. **Soil science** is the interdisciplinary study of soils involving physics, chemistry, biology, mineralogy, hydrology, taxonomy, climatology, and cartography. *Pedology* deals with the origin, classification, distribution, and description of soil. *Edaphology* specifically focuses on the study of soil as a medium for sustaining the growth of plants.

The basic sampling unit used in soil surveys is the **pedon**. The **polypedon** is the soil unit used to prepare local soil maps and may contain many pedons. Each discernible layer in an exposed pedon is a **soil horizon**. The horizons are designated as follows: O (contains **humus**, a complex mixture of decomposed and synthesized organic materials); A (rich in humus and clay, darker); E (zone of **eluviation**, the removal of fine particles and minerals by water); B (zone of **illuviation**, the deposition of clays and minerals translocated from elsewhere); C (regolith, weathered bedrock); and R (bedrock). Soil horizons A, E, and B experience the most active soil processes and together are designated the **solum**.

soil (p. 500)
soil science (p. 500)
pedon (p. 500)
polypedon (p. 500)
soil horizon (p. 500)
humus (p. 500)
eluviation (p. 501)
illuviation (p. 501)
solum (p. 502)

1. Soils provide the foundation for animal and plant life and therefore are critical to Earth's ecosystems. Why is this true?
2. What are the differences among soil science, pedology, and edaphology?
3. Define *polypedon* and *pedon*, the basic units of soil.
4. Characterize the principal aspects of each soil horizon. Where does the main accumulation of organic material occur? Where does humus form? Explain the difference between the eluviated layer and the illuviated layer. Which horizons constitute the solum?

■ *Describe* soil properties of color, texture, structure, consistence, porosity, and soil moisture.

We use several physical properties to classify soils. *Soil color* suggests composition and chemical makeup. *Soil texture* refers to the size of individual mineral particles and the proportion of different sizes. For example, **loam** is a balanced mixture of sand, silt, and clay. *Soil structure* refers to the arrangement of soil *peds*, which are the smallest natural cluster of particles in a soil. The cohesion of soil particles to each other is called *soil consistence*. Soil porosity refers to the size, alignment, shape, and location of spaces in the soil. *Soil moisture* refers to water in the soil pores and its availability to plants.

loam (p. 502)

5. How can soil color be an indication of soil qualities? Give a couple of examples.
6. Define a *soil separate*. What are the various sizes of particles in soil? What is loam? Why is loam regarded so highly by agriculturists?
7. What is a quick, hands-on method for determining soil consistence?
8. Summarize the role of soil moisture in mature soils.

■ *Explain* basic soil chemistry, including cation-exchange capacity, and *relate* these concepts to soil fertility.

Particles of clay and organic material form negatively charged **soil colloids** that attract and retain positively charged mineral ions in the soil. **Adsorption** is a process whereby cations attach to soil colloids. The capacity to exchange ions between colloids and roots is the **cation-exchange capacity (CEC)**. CEC is a measure of **soil fertility**, the ability of soil to sustain plants. Fertile soil contains organic substances and clay minerals that absorb water and retain certain elements needed by plants.

soil colloids (p. 504)
adsorption (p. 504)
cation-exchange capacity (CEC) (p. 504)
soil fertility (p. 505)

9. What are soil colloids? How are they related to cations and anions in the soil? Explain cation-exchange capacity.
10. What is meant by the concept of soil fertility?

■ *Evaluate* principal soil formation factors, including the human element.

Environmental factors that affect soil formation include parent materials, climate, vegetation, topography, and time. Human influence is having great impact on Earth's prime soils. Mismanagement, destruction, and conversion to other uses threaten essential soil fertility for agriculture. Much soil loss is preventable through the application of known technologies, improved agricultural practices, and government policies.

11. Briefly describe the contribution of the following factors and their effect on soil formation: parent material, climate, vegetation, landforms, time, and humans.
12. Explain some of the details that support the concern over loss of our most fertile soils. What cost estimates have been placed on soil erosion?

■ *Describe* the 12 soil orders of the Soil Taxonomy classification system and *explain* their general occurrence.

The **Soil Taxonomy** classification system is used in the United States and is built around an analysis of various diagnostic horizons and 12 soil orders as actually seen in the field. The system divides soils into six hierarchical categories: series, families, subgroups, great groups, suborders, and orders.

Specific soil-forming processes keyed to climatic regions (not a basis for classification) are *pedogenic regimes*: **laterization** (leaching in warm and humid climates), **salinization** (collection of salt residues in surface horizons in hot, dry climates), **calcification** (accumulation of carbonates in the B and C horizons in drier continental climates), **podzolization** (soil acidification in forest soils in cool climates), and **gleization** (humus and clay accumulation in cold, wet climates with poor drainage).

The Soil Taxonomy system uses two diagnostic horizons to identify soil: the **epipedon**, or the surface soil, and the **diagnostic subsurface horizon**, or the soil below the surface at various depths. (For an overview and definition of the 12 soil orders, please refer to Table 15.1, "Soil Taxonomy Soil Orders.") The 12 soil orders are **Oxisols** (tropical soils, including a hardpan **plinthite**, a building material), **Aridisols** (desert soils), **Mollisols** (grassland soils), **Alfisols** (moderately weathered, temperate forest soils), **Ultisols** (highly weathered, subtropical forest soils), **Spodosols** (northern coniferous forest soils), **Entisols** (recent, undeveloped soils), **Inceptisols** (weakly developed, humid-region soils), **Gelisols** (cold soils underlain by permafrost), **Andisols** (soils formed from volcanic materials), **Vertisols** (expandable clay soils), and **Histosols** (organic soils).

Soil Taxonomy (p. 507)
laterization (p. 507)
salinization (p. 507)
calcification (p. 507)
podzolization (p. 507)
gleization (p. 507)
epipedon (p. 507)
diagnostic subsurface horizon (p. 507)
Oxisols (p. 508)
plinthite (p. 509)
Aridisols (p. 511)
Mollisols (p. 512)
Alfisols (p. 514)
Ultisols (p. 516)
Spodosols (p. 517)
Entisols (p. 517)
Inceptisols (p. 519)
Gelisols (p. 519)
Andisols (p. 520)
Vertisols (p. 520)
Histosols (p. 520)

13. Summarize what led soil scientists to develop the Soil Taxonomy classification system.

14. What is the basis of the Soil Taxonomy system? How many orders, suborders, great groups, subgroups, families, and soil series are there?

15. Define *epipedon* and *subsurface diagnostic horizon*. Give a simple example of each.

16. Locate each soil order on the world map and on the US map as you give a general description of it.

17. How was slash-and-burn shifting cultivation, as practiced in the past, a form of crop and soil rotation and conservation of soil properties?

18. Describe the salinization process in arid and semiarid soils. What associated soil horizons develop?

19. Which of the soil orders are associated with Earth's most productive agricultural areas?

20. What is the significance to plants of the 51-cm (20-in.) isohyet in the Midwest relative to soils, pH, and lime content?

21. Describe the podzolization process associated with northern coniferous forest soils. What characteristics are associated with the surface horizons? What strategies might enhance these soils?

22. What former Inceptisols now form a new soil order named in 1998? Describe these soils as to location, nature, and formation processes. Why do you think they were separated into their own order?

23. Why has a selenium contamination problem arisen in western US soils? Explain the impact of agricultural practices and tell why you think this is or is not a serious problem.

24. Examine the peat cutting and drying shown in Figure 15.25c. Explain the soil-forming processes at work to produce these organic soils. How are the peat blocks used? Into which Soil Taxonomy soil order do these soils fall?

NetWork and Critical Thinking Tools

A. Select a small soil sample from your campus or near where you live. Using the sections in this chapter on soil characteristics, properties, and formation, describe this sample as completely as possible, within these constraints. Using the general soil map and any other sources available, are you able to roughly place this sample in one of the soil orders?

B. At our Student Learning Center, "Destinations," locate an agency that provides information about soils and advice for soil management. Is there a place where you can have soil tested in your region?

C. Refer to "Critical Thinking" at our Student Learning Center, item 1. Please complete the analysis of the three photographs: deciduous forests of eastern North America, evergreen forests, and arctic tundra environments. What soil order would you expect in these regions? Briefly list and explain some of each soil type's qualities as described in the chapter.

Fields of rapeseed (*Brassica napus*) are in bloom on coastal farms, northeastern Scotland. Grown for vegetable oil (known as the modern canola oil), animal feed, and increasingly as a source for biofuel, rapeseed outproduces soy beans or corn in oil volume per crop. The abundant nectar produced makes it popular with honey bees and honey producers. *[Photo by Bobbé Christopherson.]*

Ecosystems and Biomes

KEY LEARNING CONCEPTS

After reading the chapter, you should be able to:

- **Define** ecology, biogeography, and the ecosystem, community, habitat, and niche concepts.

- **Explain** photosynthesis and respiration and **derive** net photosynthesis and the world pattern of net primary productivity.

- **List** abiotic and biotic ecosystem components and **relate** those components to ecosystem operations and trophic relationships.

- **Define** succession and **outline** the stages of general ecological succession in both terrestrial and aquatic ecosystems. **Relate** biodiversity and evolution to these concepts.

- **Define** the concepts of terrestrial ecosystem, biome, ecotone, and formation classes.

- **Describe** 10 major terrestrial biomes and **locate** them on a world map.

- **Relate** human impacts, real and potential, to several of the biomes.

STUDENT LEARNING CENTER TOOLS

The *Elemental Geosystems* Student Learning Center provides on-line resources for this chapter by clicking on the book cover at **www.mygeoscienceplace.com**. Some highlights:

▶ Use four animations to work through the carbon cycle, nitrogen cycle, mid-latitude productivity, and building a food web in an ecosystem.

▶ See three narrated photo galleries featuring scenes from East Greenland, high-latitude animals, and an exploration of polar bears.

▶ Read two Career Links featuring a professor of biogeography and her fascinating work in paleolimnology, and a professor studying the Bay of Fundy, salt marsh sedimentation, and ecosystem health.

▶ More than 30 additional URLs along with Quick Links to all the URLs in the chapter are included.

Diversity is an impressive feature of the living Earth. The diversity of organisms is a response to the interaction of the atmosphere, hydrosphere, and lithosphere, which produce a variety of conditions within which the biosphere exists. The diversity of life also results from the intricate interplay of living organisms themselves. We, as part of this vast natural complex, seek to find our place and understand the feelings nature stimulates within us.

> The physical beauty of nature is certainly among its most powerful appeals to the human animal. The complexity of the aesthetic response is suggested by its wide-ranging expression from the contours of a mountain landscape to the ambient colors of a setting Sun to the fleeting vitality of a breaching whale. Each exerts a powerful aesthetic impact on most people, often accompanied by feelings of awe at the extraordinary physical appeal and beauty of the natural world.*

The biosphere, the sphere of life and organic activity, extends from the ocean floor to about 8 km (5 mi) into the atmosphere. The biosphere includes myriad ecosystems from simple to complex, each operating within general spatial boundaries. An **ecosystem** is a self-regulating association of living plants and animals and their nonliving physical environment. In an ecosystem, a change in one component causes changes in others as systems adjust to new conditions. Earth's surface environment itself is the largest ecosystem within the natural boundary of the atmosphere. Natural ecosystems are open systems for both energy and matter, with almost all ecosystem boundaries functioning as transition zones rather than as sharp demarcations.

Ecology is the study of the relationships between organisms and their environment and among the various ecosystems in the biosphere. The word *ecology*, developed by German naturalist Ernst Haeckel in 1869, is derived from the Greek *oikos* ("household," or "place to live") and *logos* ("study of"). **Biogeography**, essentially a spatial ecology, is the study of the distribution of plants and animals, the diverse spatial patterns they create, and the physical and biological processes, past and present, that produce this distribution across Earth.

The degree to which modern society understands ecosystems will help determine our success as a species and the long-term survival of a habitable Earth:

> The time is ripe to step up and expand current efforts to understand the great interlocking systems of air, water, and minerals nourishing the Earth. . . . Moreover, without vigorous action toward that goal, nations will be seriously handicapped in trying to cope with proven and suspected threats to ecosystems and to human health and welfare resulting from alterations in the cycles of carbon, nitrogen, phosphorus, sulfur, and

related materials. . . . Society depends upon this life-support system of planet Earth.[†]

Here we are, 30 years after Gilbert White's statement, reading in the UNEP assessment *Global Environmental Outlook Geo4* (October 25, 2007, *Summary for Policy Makers*, p. 12):

> Biodiversity decline and loss of ecosystem services continue to be a major threat to future development. The reduction in distribution and functioning of land, fresh water, and marine biodiversity is more rapid than at any time in human history. Ecosystems such as forests, wetlands, and drylands are being transformed and, in some cases, irreversibly degraded. Rates of species extinction are increasing. The great majority of well-studied species, including commercially important fish stocks, are declining in distribution or abundance or both. Genetic diversity of agricultural and other species is widely considered to be in decline.

What was learned in all these years? The essential challenge for us in geographic education is in the above topic sentence, "The degree to which modern society understands ecosystems. . . ." We must work to better understand these "Ecosystems and Biomes."

Earth's most influential biotic (living) agents are the humans. This is not arrogance; it is fact, because we powerfully influence every ecosystem on Earth. From the time humans first developed agriculture, the raising of animals, and the use of fire, we began a process that has led to our dominance over Earth's physical systems.

In this chapter: We explore ecosystems and the community, habitat, and niche concepts. Plants are the essential living component in the biosphere, translating solar energy into usable forms to energize life. We examine the role of nonliving systems, including biogeochemical cycles. We cover the organization of living ecosystems along complex food chains and webs. The biodiversity of organisms is seen as a product of biological evolution over the past 3.6+ billion years. Topics include ecosystem stability and resilience, the effects of global change on ecosystems and rates of succession, and consideration of global biodiversity. We explore Earth's major terrestrial ecosystems, conveniently grouped into 10 biomes. Table 16.2 brings together nearly all the information about Earth's physical systems presented through the pages of this text. Through biomes we synthesize the physical geography of place and region.

Ecosystem Components and Cycles

An ecosystem is a complex of many variables, all functioning independently yet in concert, with complicated flows of energy and matter. The key is the linkages and interconnections within and between systems (Figure 16.1). The

---

*S. R. Kellert, "The biological basis for human values of nature," in S. R. Kellert and E. O. Wilson, eds., *The Biophilia Hypothesis* (Washington, DC: Island Press, 1993), p. 49.

[†]G. F. White and M. K. Tolba, *Global Life Support Systems*, United Nations Environment Programme Information, No. 47 (Nairobi, Kenya: UN, 1979), p. 1.

(a)

b)

FIGURE 16.1 The web of life.
(a) Study the spider web as you read the following quotation:

> Life devours itself: everything that eats is itself eaten; everything that can be eaten is eaten; every chemical that is made by life can be broken down by life; all the sunlight that can be used is used. . . . The web of life has so many threads that a few can be broken without making it all unravel, and if this were not so, life could not have survived the normal accidents of weather and time, but still the snapping of each thread makes the whole web shudder, and weakens it. . . . You can never do just one thing: the effects of what you do in the world will always spread out like ripples in a pond.

(b) Abandoned Black-browed Albatross nests indicate a change in government fishing regulations in the surrounding ocean—
". . . snapping each thread makes the whole web shudder. . . ."
[(a) Photo by author; quotation from Friends of the Earth and Amory Lovins, The United Nations Stockholm Conference, *Only One Earth* (London: Earth Island Limited, 1972), p. 20. (b) Photos by Bobbé Christopherson.]

decrease in the Black-browed Albatross (*Diomedea melanophris*) colony evident from the empty nests in Figure 16.1b was the result of changes in regulations in the fishing industry around the Falkland Islands. An increase in quotas for the squid harvest, a primary food source for the albatross, led to the reduction in bird population.

An ecosystem includes both *biotic* (living—producers, consumers, and detritivores) and *abiotic* (nonliving—gaseous and sedimentary nutrient cycles) components. Nearly all depend on the input of solar energy; the few limited ecosystems that exist in dark caves or on the ocean floor depend on chemical reactions (chemosynthesis). Figure 16.2 illustrates the essential elements of an ecosystem.

Communities

A convenient biotic subdivision within an ecosystem is a **community**, which is formed by interactions among populations of living animals and plants. Therefore, an ecosystem is the interaction of many communities with the abiotic physical components of their environment. A community is identified in several ways—by its physical appearance, the number of species and the abundance of each, the complex patterns of their interdependence, and the trophic (feeding) structure of the community.

For example, in a forest ecosystem, a specific community may exist on the forest floor while another community functions in the canopy of leaves high above. Similarly, within a lake ecosystem, the plants and animals that flourish in the bottom sediments form one community, whereas those near the surface form another. Not all forests appear in familiar structure, as the dwarf willow forest in Greenland demonstrates (Figure 16.2c).

Within a community are two important concepts: habitat and niche. **Habitat** is the specific physical location of an organism, the type of environment in which it resides or is biologically adapted to live. In terms of physical and natural factors, most species have specific habitat requirements with definite limits and a specific regimen of sustaining nutrients.

Niche (French *nicher*, "to nest") refers to the function, or occupation, of a life form within a given community. It is the way an organism obtains and sustains the physical, chemical, and biological factors it needs to survive. An individual species must satisfy several aspects in its niche. Among these are a *habitat* niche, a *trophic* (food) niche, and a *reproductive* niche. For example, the Red-winged Blackbird (*Agelaius phoeniceus*) occurs throughout the United States and most of Canada in habitats of meadow, pastureland, and marsh where it nests. Its

FIGURE 16.2 Abiotic and biotic components of ecosystems.
(a) Solar energy is the input that drives the biosphere. Heat energy and biomass are the outputs
from the biosphere. (b) Abiotic and biotic ingredients operate together to sustain this subtropical
swamp (low, waterlogged ground) and basking turtles at Juniper Springs Recreation Area, Florida.
(c) Fall colors highlight this dwarf willow forest in Greenland. Musk oxen (*Ovibos moschatus*) are
foraging for food and drinking meltwater. The inset photo shows a 13-year-old willow, dated by
the small annual growth segments below the leaf. Some dwarf willows can live over 300 years.
[Photos by Bobbé Christopherson.]

trophic niche is weed seeds and cultivated seed crops throughout the year, adding insects to its diet during the nesting season.

Similar habitats produce comparable niches. In a stable community, no niche is left unfilled. The *competitive exclusion principle* states that no two species can occupy the same niche (food or space) successfully in a stable community. Thus, closely related species are separated spatially from one another. In other words, each species operates to reduce competition. This strategy in turn leads to greater diversity as species shift and adapt to fill each niche. Figure 16.3 shows 11 different community environments and representative life forms.

Some species are *symbiotic*, an arrangement where two or more species exist together in an overlapping relationship. One type of symbiosis, *mutualism*, occurs when each

FIGURE 16.3 Plants and animals work to fit specific niches.
(a) Elephant heads (*Pedicularis groenlandica*), a wildflower, above 1800 m (6000 ft) in wet meadows. (b) A western yellow-bellied racer (*Coluber constrictor*) lives in prairies and feeds on insects, lizards, and mice. (c) A female polar bear (*Ursus maritimus*) with two 8-month-old cubs in the Arctic Ocean. (d) A Svalbard reindeer (*Rangifer tarandus platyrhynchus*) on Spitsbergen Island, Arctic Ocean, losing velvet from his antlers. (e) Killdeer chicks (*Charadrius vociferus*) begin life in an exposed rocky nest. (f) Coral mushroom (*Ramaria* sp.), a fungal decomposer, at work on organic matter. (g) A large barrel cactus flourishes in the rocky soils of the desert. (h) A leopard seal (*Hydrurga leptonyx*) resting on a small iceberg near the Antarctic Peninsula; the blood stains are from a recent meal. (i) An alligator (*Alligator mississippiensis*) hides in a mangrove swamp. (j) Tricolored Heron (*Egretta tricolor*) in wetlands. (k) Five kinds of lichen and sea thrift flower in Scotland. [Photos (a, c, d, f, h, i, j, k) by Bobbé Christopherson; (b, e, g) by author.]

(a) Elephant heads

(b) Western yellow-bellied racer

(c) Polar bear and cubs

(d) Svalbard reindeer

(e) Killdeer chicks

(f) Coral mushroom

(g) Desert scene, barrel cactus

(h) Leopard seal

(i) Alligator in swamp

(j) Tricolor heron

(k) Lichens and flower

organism benefits and is sustained over an extended period by the relation. For example, lichen (pronounced "liken") is made up of algae and fungi living together. The alga is the producer and food source for the fungus, and the fungus provides structure and physical support. Their mutualism allows the two to occupy a niche in which neither could survive alone. Lichen developed from an earlier parasitic relationship in which the fungi broke into the alga cells. Today, the two organisms have evolved into a supportive harmony and symbiotic relationship (Figure 16.3k). The partnership of corals and algae discussed in Chapter 13 is another example of mutualism in a symbiotic relationship.

By contrast, another form of symbiosis is a *parasitic* relationship, which may eventually kill the host, thus destroying the parasite's own niche and habitat. An example is mistletoe (*Phoradendron* sp.), which lives on and may kill various kinds of trees. Some scientists are questioning whether our human society and the physical systems of Earth constitute a global-scale symbiotic relationship: mutualistic (sustainable) or parasitic (nonsustainable).

Plants: The Essential Biotic Component

Plants are the critical biotic link between life and solar energy. *Ultimately, the fate of the biosphere rests on the success of plants and their ability to capture sunlight and turn it into food.*

Land plants (and animals) became common about 430 million years ago, according to fossilized remains. **Vascular plants** developed conductive tissues and true roots for internal transport of fluid and nutrients. (*Vascular* is from a Latin word for "vessel-bearing," referring to the conducting cells.) Presently, about 270,000 species of plants are known to exist, and most are vascular. Many more species have yet to be identified. They represent an untapped resource base. Only about 20 species of plants provide 90% of the world's food; just three—wheat, maize (corn), and rice—comprise half of the food supply.

Leaf Activity   Leaves are solar-powered chemical factories. Flows of carbon dioxide, water, light, and oxygen enter and exit the surface of each leaf (see Figure 1.4). They flow in and out through small pores called **stomata** (singular: stoma). Each stoma is surrounded by small guard cells that open and close the pore, depending on the plant's needs at the moment. Veins in each leaf connect to the stems and branches of the plant and thus to its main circulation system. The veins bring in water and nutrients and carry away the sugars produced by photosynthesis.

Water that moves through a plant exits the leaves through the stomata and evaporates from leaf surfaces, thereby assisting heat regulation within the plant. As water evaporates from the leaves, a pressure deficit is created that allows atmospheric pressure to push water up through the plant in the same manner that a soda straw works. We can only imagine the complex operation of a 100-m (330-ft) tree.

Photosynthesis and Respiration   Under the influence of certain wavelengths of visible light, **photosynthesis** unites carbon dioxide and hydrogen (derived from water in the plant). The process releases oxygen and produces energy-rich organic material. The name of this process is descriptive: *photo-* refers to sunlight, and *-synthesis* describes the combining reaction of materials within plant leaves.

The largest concentration of light-responsive, photosynthetic structures (known as *organelles*) in a leaf rests below the leaf's upper layers. These organelle units within cells are the *chloroplasts*, and within each resides a green, light-sensitive pigment called **chlorophyll**. Within this pigment, light stimulates photochemistry. Consequently, competition for light is a dominant factor in the formation of plant communities. This competition is expressed in the height, orientation, distribution, and structure of plants.

Photosynthesis essentially follows this equation:

$$6CO_2 + 6H_2O + Light \rightarrow C_6H_{12}O_6 + 6O_2$$

(carbon dioxide)   (water)   (solar energy)   (glucose, carbohydrate)   (oxygen)

From the equation, you can see that photosynthesis removes carbon (in the form of CO_2) from Earth's atmosphere (Figure 16.4). Carbohydrates, the organic result of the photosynthetic process, are combinations of carbon, hydrogen, and oxygen. They form simple sugars, such as glucose ($C_6H_{12}O_6$). Plants use glucose to build starches, which are more complex carbohydrates and the principal food stored in plants. *Primary productivity* refers to the rate at which energy is stored in such organic substances.

Plants not only store energy; they also must consume some of this energy by converting carbohydrates for their other operations. Thus, **respiration** is essentially a reverse of the photosynthetic process:

$$C_6H_{12}O_6 + 6O_2 \rightarrow 6CO_2 + 6H_2O + energy$$

(glucose, carbohydrate)   (oxygen)   (carbon dioxide)   (water)   (solar energy)

In respiration, plants oxidize carbohydrates, releasing carbon dioxide, water, and energy as heat. The *compensation point* is the break-even point between the production and consumption of organic material. Each leaf must operate on the production side of the compensation point, or else the plant eliminates it—something each of us has no doubt experienced with a houseplant that received inadequate water or light. Figure 16.4 presents a simple schematic of this process, which yields plant growth. The difference between photosynthetic production and respiration loss is the *net photosynthesis*.

Net Primary Productivity   The net photosynthesis for an entire plant community is its **net primary productivity**. This is the amount of stored chemical energy (biomass) that the community generates for the ecosystem. **Biomass** is the net dry weight of organic material; it is biomass that feeds the food chain.

FIGURE 16.4 How plants live and grow.
The balance between photosynthesis and respiration determines net photosynthesis and plant growth. [Temperate rain forest photo by Bobbé Christopherson.]

Net primary productivity is mapped in terms of *fixed carbon per square meter per year.* ("Fixed" means chemically bound into plant tissues.) Study the map and satellite images in Figure 16.5, and you can see that on land, net primary production tends to be highest in the tropics and decreases toward higher latitudes. Even though deserts receive high amounts of solar radiation, other controlling factors are more important, namely, water availability and soil conditions. But precipitation also affects productivity, as evidenced on the map by the correlations of abundant precipitation with high productivity (tropics) and reduced precipitation with low productivity (subtropical deserts).

Differing nutrient levels limit productivity in the oceans. Regions with nutrient-rich upwelling currents generally are the most productive (off western coastlines). The map shows that tropical oceans and areas of subtropical high pressure are quite low in productivity.

In temperate and high latitudes, the rate at which carbon dioxide is fixed by vegetation varies seasonally. The rate increases in spring and summer as plants flourish with increasing solar input and, in some areas, more available (nonfrozen) water, and decreases in late fall and winter. Rates in the tropics are high throughout the year, and turnover in the photosynthesis-respiration cycle is faster, exceeding by many times the rates experienced in a desert environment or in the far northern limits of the tundra. A lush hectare (2.47 acres) of sugar cane in the tropics might fix 45 metric tons (50 tons) of carbon in a year, whereas desert plants in an equivalent area might achieve only 1% of this amount.

Table 16.1 lists various ecosystems, their net primary productivity per year, and an estimate of net total biomass worldwide. World net primary productivity is estimated at 170 billion metric tons (189 billion tons) of dry organic matter per year. Compare the various ecosystems, especially cultivated land, with the natural communities.

Abiotic Ecosystem Components

Critical in each ecosystem are the flow of energy and the cycling of nutrients and water. Nonliving abiotic components set the stage for ecosystem operations.

Light, Temperature, Water, and Climate Solar energy powers ecosystems, so the pattern of solar energy receipt is crucial. Solar energy enters an ecosystem by way of photosynthesis, and heat energy is dissipated from the system at many points. Of the total energy intercepted at Earth's surface and available for work, only about 1.0% is actually fixed by photosynthesis as carbohydrates in plants.

The duration of Sun exposure is the *photoperiod.* Along the equator, days are essentially 12 hours long year-round; however, with increasing distance from the equator, seasonal effects become pronounced. Plants have adapted their flowering and seed germination to seasonal changes in insolation. Some seeds germinate only when daylength reaches a certain number of hours. A plant that responds in the opposite manner is the poinsettia (*Euphorbia pulcherrima*), which requires at least 2 months of 14-hour nights to start flowering.

Monarch butterflies (*Danaus plexippus*) migrate some 3600 km (2235 mi) from eastern North America to wintering forests in Mexico. The butterflies navigate using an internal time-compensated Sun compass to maintain their southwestward track while they compensate for time-of-day Sun changes. Researchers found receptors in their eyes sensitive to ultraviolet light that play the key role in accurate navigation. For Monarchs, ultraviolet light is an important abiotic component.

Other components are important to ecosystem processes. Air and soil temperatures determine the rates at which chemical reactions proceed. Significant temperature factors are seasonal variation and duration and the pattern of minimum and maximum temperatures.

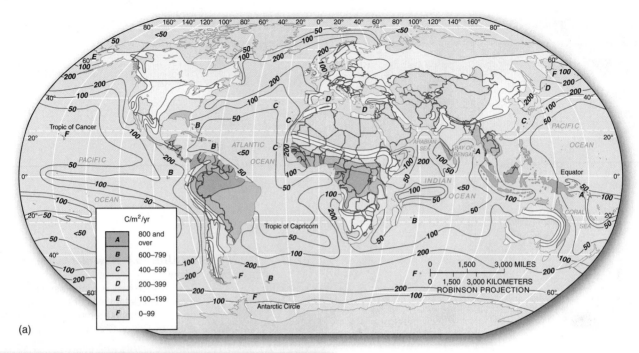

(a)

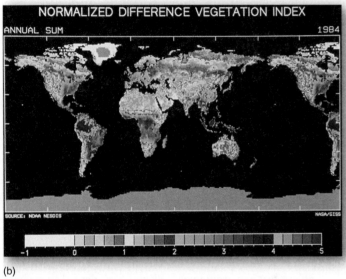

(b)

FIGURE 16.5 Net primary productivity.
(a) Worldwide net primary productivity in grams of carbon per square meter per year (approximate values). (b) Normalized difference vegetation index. False coloration indicates bare ground in browns, dense vegetation in blues. [(a) After D. E. Reichle, *Analysis of Temperate Forest Ecosystems* (Heidelberg, Germany: Springer-Verlag, 1970). Adapted by permission; (b) Goddard Space Flight Center and NOAA.]

Operations of the hydrologic cycle and water availability depend on precipitation and evaporation rates and their seasonal distribution. Water quality—its mineral content, salinity, and levels of pollution and toxicity—is important. Also, regional climates affect the pattern of vegetation and ultimately influence soil development. All of these factors work together to form the limits for ecosystems in a given location.

Figure 16.6 illustrates the general relationship among temperature, precipitation, and vegetation. In general terms, can you identify the characteristic vegetation type and related temperature and moisture relationship that fit the area of your home or school?

Life Zones Alexander von Humboldt (1769–1859), an explorer, geographer, and scientist, deduced that plants and animals recur in related groupings wherever conditions in the abiotic environment are similar. After several years of study in the Andes Mountains of Peru, he described a distinct relationship between elevation and plant communities, his **life zone** concept. As he climbed the mountains, he noticed that the experience was similar to that of traveling away from the equator toward higher latitudes (Figure 16.7).

This zonation of plants with altitude is noticeable on any trip from lower valleys to higher elevations. Each life zone possesses its own temperature, precipitation, and insolation relationships and therefore its own biotic communities.

Elemental Cycles

The most abundant natural elements in living matter are hydrogen (H), oxygen (O), and carbon (C). Together, these elements make up more than 99% of Earth's biomass; in fact, all life (organic molecules) contains hydrogen and

Table 16.1	Net Primary Productivity and Plant Biomass on Earth			
Ecosystem	Area (10^6 km^2)[a]	Net Primary Productivity per Unit Area (g/m^2/yr)[b]		World Net Biomass (10^9/tons/yr)[c]
		Normal Range	Mean	
Tropical rain forest	17.0	1000–3500	2200	37.4
Tropical seasonal forest	7.5	1000–2500	1600	12.0
Temperate evergreen forest	5.0	600–2500	1300	6.5
Temperate deciduous forest	7.0	600–2500	1200	8.4
Boreal forest	12.0	400–2000	800	9.6
Woodland and shrubland	8.5	250–1200	700	6.0
Savanna	15.0	200–2000	900	13.5
Temperate grassland	9.0	200–1500	600	5.4
Tundra and alpine region	8.0	10–400	140	1.1
Desert and semidesert scrub	18.0	10–250	90	1.6
Extreme desert, rock, sand, ice	24.0	0–10	3	0.07
Cultivated land	*14.0*	*100–3500*	*650*	*9.1*
Swamp and marsh	2.0	800–3500	2000	4.0
Lake and stream	2.0	100–1500	250	0.5
Total continental	**149.0**	—	773	**115.17**
Open ocean	332.0	2–400	125	41.5
Upwelling zones	0.4	400–1000	500	0.2
Continental shelf	26.6	200–600	360	9.6
Algal beds and reefs	0.6	500–4000	2500	1.6
Estuaries	1.4	200–3500	1500	2.1
Total marine	**361.0**	—	152	**55.0**
Grand total	**510.0**	—	333	**170.17**

Source: R. H. Whittaker, *Communities and Ecosystems* (Heidelberg: Springer, 1975), p. 224. Reprinted by permission.

[a]1 km^2 = 0.38 mi^2.

[b]1 g per m^2 = 8.92 lb per acre.

[c]1 metric ton (10^6 g) = 1.1023 tons.

carbon. In addition, nitrogen (N), calcium (Ca), potassium (K), magnesium (Mg), sulfur (S), and phosphorus (P) are essential *nutrients*, elements necessary for the growth and development of a living organism.

Several key chemical cycles function in nature. Oxygen, carbon, and nitrogen all have *gaseous cycles*, part of which are in the atmosphere. Other elements have *sedimentary cycles*, which principally involve the mineral and solid phases (major elements that have these cycles include phosphorus, calcium, and sulfur). Some elements combine gaseous and sedimentary cycles. These gaseous and sedimentary recycling processes are the **biogeochemical cycles**, so called because they involve chemical reactions in both living and nonliving systems.

Oxygen and Carbon Cycles We consider oxygen and carbon together because they are so closely intertwined through photosynthesis and respiration (Figure 16.8, p. 536). The atmosphere is the principal reserve of available oxygen. Larger reserves of oxygen exist in Earth's crust, but they are unavailable, for they are chemically bound to other elements.

As for carbon, the greatest pool of it is in the ocean—about 42,900 billion metric tonnes. (Metric tonnes times 1.10 equals short tons, or here, 47,190 billion tons.) However, all this carbon is bound chemically in carbon dioxide, calcium carbonate, and other compounds. The ocean absorbs carbon dioxide through photosynthesis by phytoplankton. On land, carbon is stored in certain carbonate minerals, such as limestone ($CaCO_3$).

The atmosphere, which is the integrating link in the cycle, contains only about 700 billion tons of carbon (as carbon dioxide) at any moment. This is far less carbon than is stored in fossil fuels and oil shales (13,200 billion metric tons, as hydrocarbon molecules) or in living and dead organic matter (2500 billion metric tons, as carbohydrate molecules). Respiration of plants and animals, volcanic activity, and fossil-fuel combustion by industry and transportation add carbon dioxide to the atmosphere.

The carbon dumped into the atmosphere by human activity constitutes a vast geochemical experiment, using the real-time atmosphere as a laboratory. Annually, we are adding carbon to the atmospheric pool in an amount 400% greater than 1950 levels. Previous to 2000, carbon

(e)

(i)

(h)

(g)

(d)

Polar and Alpine

Moist tundra

Dry tundra

Pine-Spruce-Fir forest

Cool

Cool desert

Deciduous forest

Temperature

Temperate

Short grass

Tall grass

Temperate desert

Tropical forest

Grasslands

Hot

Hot desert

Wet

Precipitation

Dry

(a)

(c)

(b)

(f)

FIGURE 16.6 Temperature and precipitation affect ecosystems.
(a) Climate controls (wetness, dryness, warmth, cold) and ecosystem types; (b) subtropical desert near Death Valley, California; (c) Sonoran desert of Arizona; (d) cold desert of north-central Nevada; (e) dry tundra of East Greenland; (f) El Yunque tropical rain forest of Puerto Rico; (g) deciduous trees in fall colors in Ohio; (h) needleleaf forest near Independence Pass, Colorado; and (i) moist tundra of Spitsbergen, Arctic Ocean. [Photos (c, d) by author; (f) photo by Tom Bean; (b, e, g, h, i) by Bobbé Christopherson.]

emissions averaged an increase of 1.1% per year; however, since 2000, emissions have risen to a 3.0% per year rate of increase. This is a significant acceleration.

The Nitrogen Cycle   The *nitrogen cycle* is the major constituent of the atmosphere; it is 78.084% of each breath we take. Also, nitrogen is important in the makeup of organic molecules, especially proteins, and therefore is essential to living processes. A simplified view of the nitrogen cycle is portrayed in Figure 16.9. This vast atmospheric reservoir is not directly accessible to most organisms.

Nitrogen-fixing bacteria, which live principally in the soil and are associated with the roots of certain plants, are the key link to life. The roots of legumes such as clover, alfalfa, soybeans, peas, beans, and peanuts have such bacteria. Bacteria colonies reside in nodules on the legume roots and chemically combine the nitrogen from the air in the form of nitrates (NO_3) and ammonia (NH_3). Plants use these chemically bound forms of nitrogen to produce their own organic matter. Anyone or anything feeding on the plants thus ingests the nitrogen. Finally, the nitrogen in the organic wastes of the consuming organisms is freed by denitrifying bacteria, which recycle it to the atmosphere.

FIGURE 16.7 Vertical and latitudinal zonation of plant communities.
(a) Progression of plant community life zones with changing elevation or latitude. (b) Alpine tundra ecosystem in the Colorado Rockies. (c) The timberline for a needleleaf forest in the Rockies. The forest line marks the highest continuous forest, whereas the tree line above it marks the zone in which no trees grow. [Photos by Bobbé Christopherson.]

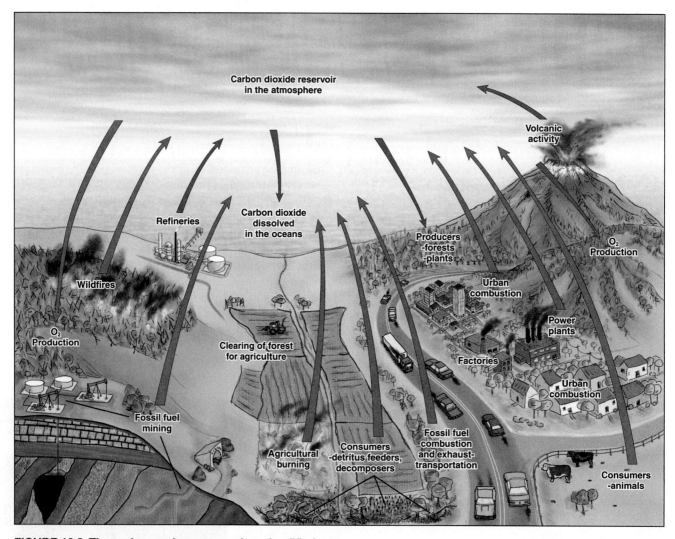

FIGURE 16.8 The carbon and oxygen cycles, simplified.
Carbon is fixed through photosynthesis and begins its passage through the ecosystem. Respiration by living organisms, burning of forests and grasslands, and the combustion of fossil fuels release carbon to the atmosphere. These cycles are greatly influenced by human activities.

ANIMATION   The Global Carbon Cycle

To improve agricultural yields, many farmers use synthetic inorganic fertilizers, as opposed to soil-building organic fertilizers (manure and compost). Inorganic fertilizers are chemically produced through artificial nitrogen fixation at factories. Humans presently fix more nitrogen as synthetic fertilizer per year than all terrestrial sources combined—and the present production of synthetic fertilizers now is doubling every 8 years; some 1.82 million tons (2 million tons) are produced per week, worldwide. The crossover point when anthropogenic sources of fixed nitrogen exceeded the normal range of naturally fixed nitrogen was 1970.

The surplus of usable nitrogen accumulates in Earth's ecosystems. This excess nitrogen load begins a water pollution process that feeds an excessive growth of algae and phytoplankton, increases biochemical oxygen demand, diminishes dissolved oxygen reserves, and eventually disrupts the aquatic ecosystem.

Figure 16.10 maps and images the core region of the dead zone, a region of oxygen-depleted (hypoxia) water off the coast of Louisiana in the Gulf of Mexico. In 2008, researchers at Texas A&M University reported dead-zone conditions stretched along the entire Texas Gulf Coast and extended in a discontinuous spread some 32 km from shore.

The Mississippi River carries agricultural fertilizers, farm sewage, and other wastes to the Gulf, causing huge spring blooms of phytoplankton, a greening of primary productivity. The Mississippi drainage system handles the runoff for 41% of the continental United States. By summer, the biological oxygen demand of bacteria feeding on the decay exceeds the dissolved oxygen; hypoxia develops, killing any fish that venture into the area. These low-oxygen conditions act as a limiting factor on marine life. From 2002 on, the dead zone has expanded to more than 22,000 km² (8500 mi²). The agricultural, feedlot, and fertilizer industries dispute the connection between their nutrient input and the dead zone.

Following Hurricane Katrina in 2005, this region was overwhelmed by the discharge carrying all the sewage, chemical toxins, oil spills, household pesticides, animal

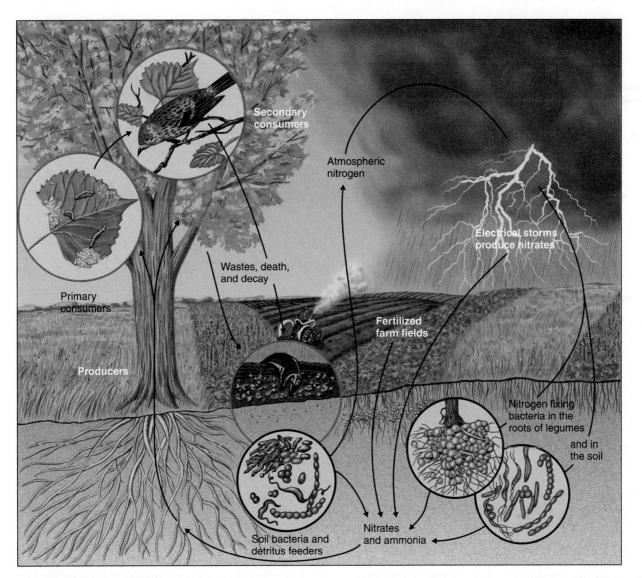

FIGURE 16.9 The nitrogen cycle.
The atmosphere is the essential reservoir of gaseous nitrogen. Atmospheric nitrogen gas is chemically fixed by bacteria to produce ammonia. Lightning and forest fires produce nitrates, and fossil-fuel combustion forms nitrogen compounds that are washed from the atmosphere by precipitation. Plants absorb nitrogen compounds and produce organic material. [Adapted from Gerald Audesirk and Teresa Audesirk, *Biology, Life On Earth*, 4th ed., Figure 45-10, p. 903. Copyright © 1996, by permission of Prentice-Hall, Inc., Upper Saddle River, N.J.]

The Nitrogen Cycle

wastes and carcasses, and nutrients. Full disclosure of the contents of the "muck" has yet to be made and total volume will never be known. The plume, now mostly dissipated, from this event tracked around the Florida peninsula and northward up the Eastern seaboard (see suggested plot in Figure 16.10a).

Similar coastal dead zones occur as the result of nutrient outflows from more than 400 river systems worldwide affecting almost 250,000 km² (94,600 mi²) of offshore oceans and seas. Climate change and related oceanic warming create water-column stratification with oxygen depletion; therefore if nutrient loads continue to increase and warming continues, the area afflicted by dead-zone conditions will no doubt continue to grow exponentially. Many think this disruption of dissolved oxygen in coastal

ecosystems is a major environmental problem, probably the most widespread of human-forced impacts on marine environments.

Limiting Factors

The term **limiting factor** identifies the one physical or chemical abiotic component that most inhibits biotic operations through its lack or excess. In most ecosystems, precipitation is the limiting factor. However, temperature, light levels, and soil nutrients certainly affect vegetation patterns. Here are a few examples:

- Low temperatures limit plant growth at high elevations or high latitudes.
- Lack of water limits growth in a desert.

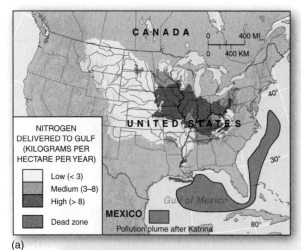

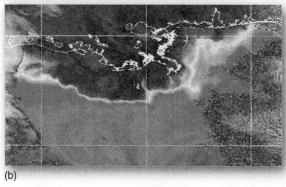

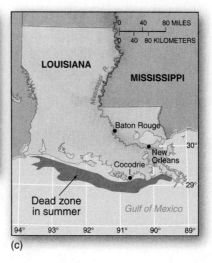

FIGURE 16.10 The dead zone.
(a) Nitrogen originates from agricultural activities in the upstream watershed that feeds the Mississippi River system. Additional shading shows the estimated path of post-Katrina discharge, now mostly dissipated. (b) The flush of nutrients from the Mississippi River drainage basin enriches the offshore waters in the Gulf of Mexico, causing huge phytoplankton blooms in early spring (red areas). (c) By late summer, oxygen-depleted (hypoxic) waters dominate what is known as the *dead zone*, each year expanding to cover a new record region. [(a) Adapted from R. B. Alexander, R. A. Smith, and G. E. Schwarz, "Effect of stream channel size on the delivery of nitrogen to the Gulf of Mexico," *Nature* (February 17, 2000): 761, as presented in *Environment* (Jan./Feb. 2001): 16. (b) *SeaWiFS* image from the Goddard Space Flight Center, NASA, courtesy of Omnimage, Dulles, Virginia.]

- Excess water limits growth in a bog.
- Changes in salinity levels limit aquatic ecosystems.
- Lack of iron in ocean surface environments limits photosynthetic production.
- Low phosphorus content of eastern US soils limits plant growth.
- General lack of active chlorophyll above 6100 m (20,000 ft) limits primary productivity.

Each organism possesses a range of tolerance for each limiting factor in its environment. Limiting factors are illustrated vividly in Figure 16.11, which shows the geographic range for two tree species and two bird species. The coast redwood (*Sequoia sempervirens*) is limited to a narrow section of the Coast Ranges in California, covering barely 9500 km² (less than 4000 mi²) and concentrated in areas that receive necessary summer advection fog. The red maple (*Acer rubrum*), on the other hand, thrives over a large area under varying conditions of moisture and temperature, thus demonstrating a broader tolerance to environmental variations.

The Mallard Duck (*Anas platyrhynchos*) and the Snail Kite (*Rostrhamus sociabilis*) also demonstrate a variation in tolerance and range (Figure 16.11b). The Mallard, a generalist, feeds from widely diverse sources, is easily domesticated, and is found throughout most of North America in at least one season of the year. By contrast, the Snail Kite is a specialist that feeds only on one specific type of snail. This single food source, then, is its limiting factor.

Biotic Ecosystem Operations

The abiotic components of energy, atmosphere, water, weather, climate, and minerals make up the life support for the biotic components of each ecosystem. The flow of energy, cycling of nutrients, and trophic (feeding) relations determine the nature of an ecosystem. As energy cascades through this process-flow system, it is constantly replenished by the Sun. But nutrients and minerals cannot be replenished from an external source, so they constantly cycle within each ecosystem and through the biosphere in general (Figure 16.12). Let us examine these biotic operations.

Producers, Consumers, and Detritivores Organisms that are capable of using carbon dioxide as their sole source of carbon are *autotrophs*, or **producers**. They chemically fix carbon through photosynthesis. Organisms that depend on autotrophs for their carbon are heterotrophs, or **consumers**. Plants are the essential producers in an

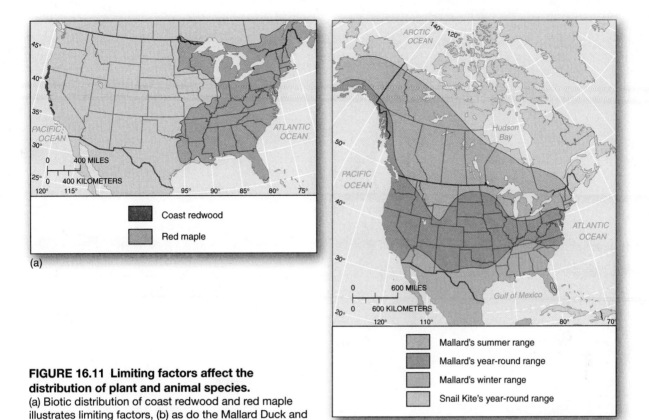

FIGURE 16.11 Limiting factors affect the distribution of plant and animal species.
(a) Biotic distribution of coast redwood and red maple illustrates limiting factors, (b) as do the Mallard Duck and Snail Kite distribution across North America.

(b)

ecosystem—capturing light and generating heat energy and converting it to chemical energy, incorporating carbon, forming new plant tissue and biomass, and freeing oxygen.

From the producers, which manufacture their own food, energy flows through the system along a **food chain** circuit, reaching consumers and eventually detritivores and decomposers. Ecosystems generally are structured in a **food web**, a complex network of interconnected food chains. In a food web, consumers participate in several different food chains. Organisms that share the same basic foods are said to be at the same *trophic* (feeding, nutrition) *level*.

In Figure 16.12b a bearded seal (*Erignathus barbatus*), with body fat exceeding 30%, rests on a bergy bit, or small iceberg, in Arctic waters where it feeds on fish and clams. Figure 16.12c shows a male polar bear (*Ursus maritimus*) hauling its bearded seal kill onto the ice. Polar bears are the dominant Arctic predator, a marine mammal that depends on the presense of sea ice. In 16.12d, on the pack ice many kilometers from land, we see a mother and her 6-month-old cubs-of-the-year feeding on a seal catch. Even though the cubs still nurse they share the kill with mom. The polar bear consumes most of the seal except for the bones and intestines, which are quickly eaten by birds scavenging the leftovers. Note the Glaucous Gulls and Ivory Gulls in the photo.

As an example of a trophic food web, look at the one based on krill (Figure 16.13). Krill is a shrimplike crustacean about 7.5 cm (3 in.) long that is a major food for a

diverse group of organisms, including whales, fish, seabirds, seals, and squid in the Antarctic region. All these organisms participate in numerous other food chains as well, some consuming and some being consumed. Because krill are a protein-rich, plentiful food, commercial factory ships, such as those from Japan and Russia, hunt krill. The annual harvest currently approaches a million tons, principally as feed for chickens and livestock and as protein for human consumption. Many of the Antarctic-dwelling birds depend on krill and on fish that eat krill.

The impact on the food web of further increases in the krill harvest is uncertain, for little is known about the krill reproduction rate. Their long life span (greater than 8 years) and slow growth rate limit the ultimate extent of the resource. In addition, the possible effect on krill of increased ultraviolet radiation resulting from the seasonal "hole" in the ozone layer above Antarctica is under investigation.

Climate change is breaking down this Antarctic food web. More than half the penguin population, including Emperors, will face population declines, if not ultimate extinction, due to climate change in their habitats. Declining sea-ice conditions and food web productivity are happening and food stress is underway. Adélie penguins (lower left in 16.13b) are most vulnerable, and forecasts place their loss as high as a 75% decline in population in 50 years or less. They have declined more than 50% in numbers since 1980.

Another example of a complex food web comes from eastern North America and a temperate deciduous forest. As with the krill food web, the diagram is simplified from

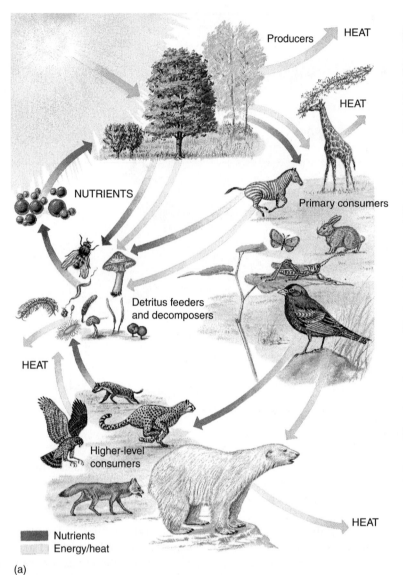

NUTRIENTS

Producers

HEAT

HEAT

Primary consumers

Detritus feeders
and decomposers

HEAT

Higher-level
consumers

HEAT

Nutrients
Energy/heat

(a)

(b)

(c)

(d)

ANIMATION Building a Food Web

FIGURE 16.12 Energy, nutrient, and food pathways in the environment.
(a) The flow of energy, cycling of nutrients, and trophic (feeding) relationships portrayed for a generalized ecosystem. The operation is fueled by radiant energy supplied by sunlight and first captured by the plants. (b) A bearded seal on a bergy bit in the Arctic Ocean. (c) A solitary male polar bear drags its seal feast onto pack ice. (d) A mother polar bear and cubs eat dinner on an iceberg after mom caught a seal. [From T. Audesirk and G. Audesirk, *Biology: Life on Earth,* 4th ed., Figure 45-1, p. 892. © 1996 Prentice Hall Inc. Used by permission. Photos (b, c, and d) by Bobbé Christopherson.]

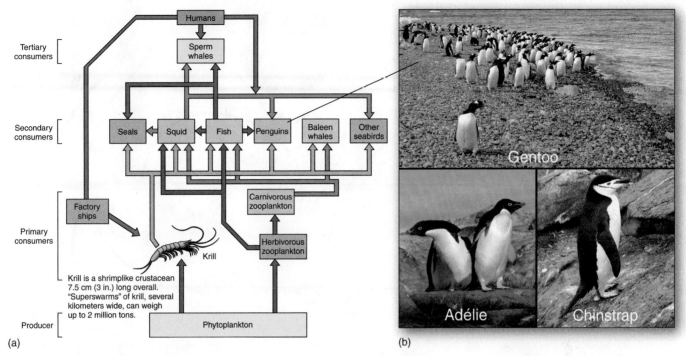

(a)   (b)

FIGURE 16.13 An Antarctic food web.
(a) The food web of the krill in Antarctic waters, from phytoplankton producers (bottom) through various consumers. Phytoplankton begin this chain by using solar energy in photosynthesis. The krill, along with other organisms, feed on the phytoplankton. Krill in turn are fed upon by the next trophic level. (b) Gentoo penguins (*Pygoscelis papua*) on a rocky shore in Antarctica watch the water for leopard seal predators. Adélie penguins (*Pygoscelis adeliae*) in Antarctica depend on the ice as a base for feeding and are the most threatened. Chinstrap penguins (*Pygoscelis antarctica*) show us that we mainly identify penguins by differences in black and white patterns on their heads. [Photos by Bobbé Christopherson.]

 Building a Food Web

 Midlatitude Productivity

the actual complexity of nature (Figure 16.14). From our discussion, find the primary producers and then locate the primary, secondary, and tertiary consumers in this community. Do you find the detritivores (worms are shown) playing a role as well? Bacteria act as decomposers in the system; do you find any other decomposers in the illustration?

Primary consumers feed on producers. Because producers are always plants, the primary consumer is a **herbivore**, or plant eater. A **carnivore** is a *secondary consumer* and primarily eats meat. A *tertiary consumer* eats primary and secondary consumers and is the "top carnivore" in the food chain; an example is the sperm whale in the krill food web. A consumer that feeds on both producers (plants) and consumers (meat) is an **omnivore**—a category occupied by humans, among others.

Any assessment of world food resources depends on the level of consumer being studied. Using humans as an example, many can be fed from the amount of wheat harvested from an acre of land. An acre of land (0.41 hectare) produces about 810 kg (1800 pounds) of grain. However, if herbivores, such as cattle, eat that grain, only 82 kg (180 pounds) of biomass is produced, which can feed far fewer people (Figure 16.15).

In terms of energy, *only about 10% of the calories in plant matter survive from the primary to the secondary trophic level.* When humans, or any carnivore or omnivore, consume meat, there is a loss of biomass to the system. More energy is lost to the environment at each progressive step in the food chain. Today, approximately half of the cultivated acreage in the United States and Canada is planted for animal consumption—beef and dairy cattle, hogs, chickens, and turkeys. You can see that an omnivorous diet—the human dietary pattern—is quite expensive in terms of biomass and energy.

Detritivores (detritus feeders and decomposers) are the final link in the endless chain. Detritivores renew the entire system by releasing simple inorganic compounds and nutrients with the breaking down of organic materials. Detritus refers to all the dead organic debris, remains, fallen leaves, and wastes that living processes leave. *Detritus feeders*—worms, mites, termites, centipedes, snails, crabs, and even vultures, among others—work like an army to consume detritus and excrete nutrients that fuel an ecosystem. **Decomposers** are primarily bacteria and fungi that digest organic debris outside their bodies and absorb and release nutrients in the process. This metabolic work of microbial decomposers produces the rotting that breaks down detritus. Detritus feeders and decomposers, although different in operation, have a similar function in an ecosystem.

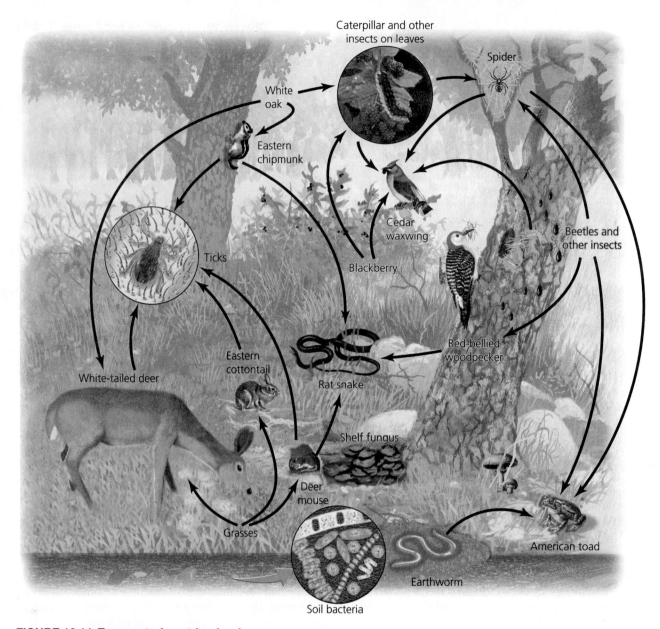

FIGURE 16.14 Temperate forest food webs.
A simplified illustration of food webs in a temperate forest. Identify primary producers and primary, secondary, and tertiary consumers. Featured are strong interactions, but many weak interactions not illustrated hold such natural communities together, affirming the hypothesis that complexity is stabilizing and not weakening the community. [Jay Withgott and Scott Brennan, *Environment, The Science Behind the Stories* (San Francisco: Benjamin Cummings, © 2007), Fig. 6.11, p. 159. Used by permission.]

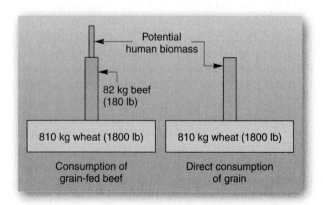

FIGURE 16.15 Pyramids show us efficiency and inefficiency.
Biomass pyramids illustrating the difference in efficiency between direct and indirect consumption of grain.

Ecosystems, Evolution, and Succession

Far from being static, Earth's ecosystems are dynamic (vigorous and energetic) and ever changing from the beginning of life on the planet. Over time, communities of plants and animals have adapted, evolved, and in turn shaped their environments to produce great diversity. A constant interplay exists between increasing growth in a community and decreasing growth caused by resistance factors that create limits and disruptions (Figure 16.16). Each ecosystem is constantly adjusting to changing conditions and disturbances in the struggle to survive. The concept of change is key to understanding ecosystem stability.

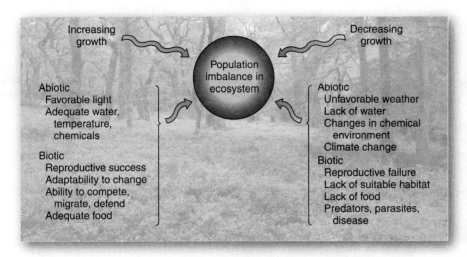

FIGURE 16.16 Factors controlling the population balance of an ecosystem.
Potential growth factors and resistance factors ultimately determine the population of an ecosystem. [Background photo of Scottish forest by Bobbé Christopherson.]

For most of the last century, scientists thought that an undisturbed ecosystem—whether forest, grassland, aquatic, or other—would progress to a stage of equilibrium, a stable point, with maximum chemical element storage and biomass. Modern research has determined, however, that ecosystems do not progress to some static conclusion. In other words, there is no "balance of nature"; instead, think of a "constant interplay of physical, chemical, and living factors in a dynamic equilibrium."

Biological Evolution Delivers Biodiversity

A critical aspect of ecosystem stability and vitality is **biodiversity**, or species richness of life on Earth (a combination of *bio*logical and *diversity*). Biodiversity includes the number of different species and quantity of each species, the genetic diversity in species (number of genetic characteristics), and ecosystem and habitat diversity. A working estimate says that 13.6 million species presently populate Earth; however, scientists have classified fewer than 2 million of these.

The origin of this diversity is embodied in the *theory of evolution*. (The definition of a *scientific theory* is given in Focus Study 1.1.) A theory is based on many extensively tested hypotheses and represents a broad general principle and a powerful tool with which to understand nature and the general laws in operation across the planet and the Universe.

Evolution states that single-cell organisms adapted, modified, and passed along inherited changes to multicellular organisms. The genetic makeup of successive generations was shaped by environmental factors, physiological functions, and behaviors that created a greater rate of survival and reproduction, traits that were passed along—literally, "survival of the species that reproduces the most." A species refers to a population that can reproduce sexually and produce viable offspring. By definition, this means reproductive isolation from other species.

Each inherited trait is guided by a *gene* that is a segment of the primary genetic material called *DNA* (deoxyribonucleic acid) that resides in the nest of chromosomes in every cell nucleus. These traits are passed to successive generations, especially those traits that are most successful at maximizing niches different from those of other species or traits that help the species adapt to environmental changes. Such differential reproduction and adaptation passes along successful genes or groups of genes in a process of genetic favoritism known as *natural selection*. Evolution follows generation after generation and tracks the passage of inherited characteristics that were successful—the failures passed into extinction. Thus, it is not an understatement to say that today's humans are the result of billions of years of positive change and affirmative natural selection.

New mixes in the *gene pool*, or all genes possessed by individuals in a given population, involve mutations. Mutation is a process that occurs when a random action, or perhaps an error as the DNA reproduces, forms new genetic material and sets new traits into the inherited stream. Geography comes into play, as spatial effects are important. A species may disperse through migration to different environments or may be separated by natural *vicariance* (fragmentation of the environment) events that establish natural barriers. Such isolation allows natural selection to evolve new sets of genes. Imagine the migrations across ice bridges or land connections at times of low sea level, or the slowly evolving rise of a mountain range, or the formation of an isthmus such as Panama connecting continents and separating oceans. To restate, the physical and chemical evolution of Earth's systems is closely linked to the biological evolution of life.

The deciphering of complete species *genomes*, or the total genetic constituents of an organism, is a scientific triumph and is disclosing a complex history and geography of each species—where species originated, how species are dispersed, what diseases and traumas were of consequence. Analysis of the human genome has indicated our origins and provided a road map of how we got to the present.

Ecosystem Stability and Diversity In an ecosystem the tendency of species populations to remain stable

(a)

(b)

FIGURE 16.17 Poor practices disrupt stable forest communities.
(a) An example of clear-cut timber harvesting that has devastated a stable forest community and produced drastic changes in microclimatic conditions. Full exposure to sunlight and wind is different from former shade conditions. With few riparian buffer strips of uncut forest along streams, sedimentation clogs streams as soils erode. (b) A logged pine forest and disruptive logging road. About 10% of the Northwest's old-growth forests remain, as identified through satellite images and GIS analysis. [Photos (a) by Bobbé Christopherson; (b) by author.]

(*inertial stability*) does not necessarily foster the ability to recover from change (*resilience*). Several of the dramatic asteroid impact episodes that triggered partial extinctions over the past 440 million years tested communities of plants and animals and their resilience. Think of resilience as an ecosystem's ability to recover from disturbance, its ability to absorb disturbance up to some threshold point before it snaps to a different set of relations, thus altering some ecological progression.

Examples of communities with high inertial stability include a redwood forest, a pine forest at a high elevation, and a tropical rain forest near the equator. In contrast, a midlatitude grassland is low in stability; yet, when burned, its resilience is high because the community recovers rapidly.

In the context of ecosystem stability and resilience, consider the elimination of a full section of forest ecosystem, on private land in southern Oregon, south-southeast of Crater Lake, shown in Figure 16.17. Note the high road density, the inadequate buffers along watercourses to prevent sedimentation of streams, the drag lines and skid trails to each pickup spot that pockmark the landscape. On nearby US Forest Service land, the rules are clear: a full 95-m (312-ft) tree buffer is left on either side of a stream, road densities are minimized, and harvesting regimes that include partial removals, commercial thinning, and group selection harvests, among others, are all in contrast to the devastation on private lands in the photo. Scientists determined that for sustainable forests, the sole objective should never be the complete removal of the timber.

Greater biodiversity in an ecosystem results in greater stability and greater productivity. In other words, the

strength of biodiversity is that it spreads risk over the entire community, because several food sources exist at each trophic level. Therefore, biodiversity matters. Consider more than just the diversity of species; think of all the positive interactions of species embodied in the *facilitation concept*. Facilitation describes how a diverse community manages nutrient dynamics among the populations, and pest resistance, and resilience to events such as fire. News Report 16.1 presents exciting confirmation of biodiversity.

The present loss of biodiversity is irreversible, and yet, here we are in the midst of the biosphere's sixth major extinction episode. Extinction is final no matter how complex the organism or how long it has lived since it evolved. E. O. Wilson, the famed biologist, stated:

Biological diversity—the full sweep from ecosystems to species within ecosystems, thence to genes within species—is in trouble. Mass extinctions are commonplace, especially in tropical regions where most of the biodiversity occurs. Among the more recent are more than half the exclusively freshwater fishes of peninsular Asia, half of the fourteen birds of the Philippines island of Cebu, and more than ninety plant species growing on a single mountain ridge in Ecuador. In the United States an estimated 1 percent of all species have been extinguished; another 32 percent are imperiled.*

*E. O. Wilson, *Consilience, The Unity of Knowledge* (New York: Knopf, 1998), p. 292.

NEWS REPORT 16.1

Experimental Prairies Confirm the Importance of Biodiversity

Little of the original midlatitude prairies remains. Only a few widespread areas of unaltered prairie are protected. Iowa was once 80% prairie, and now 99.9% of the prairie is gone, converted to other uses. One such plot of 240 acres (97 ha) is in Howard County, Iowa, south of the Minnesota border. The Hayden Prairie State Preserve sits quietly, almost missed by passing cars as they kick up pale dust from the gravel county road (Figure 16.1.1).

Field experiments confirmed an important scientific assumption: that greater biological diversity in an ecosystem leads to greater stability, productivity, and soil nutrient use within that ecosystem. For instance, during a drought, some species of plants will be damaged. In a diverse ecosystem, however, other species with deeper roots and better water obtaining ability will thrive.

Exciting research on natural prairie ecosystems and biodiversity, the role of fire, and climate-change impacts is ongoing at Konza Prairie LTER (Long-Term Ecological Research) in

the Flint Hills of northeastern Kansas. Kansas State University and the Nature Conservancy operate this native tallgrass prairie research station (see http://www.konza.ksu.edu/).

To better understand prairies and biodiversity, ecologist David Tilman at the University of Minnesota tracked the operation of 147 grassland plots, each 3 m (9.8 ft) square. Species diversity was carefully controlled on each plot (Figure 16.1.2). Plots were sown with different numbers of native North American prairie plant seeds and cared for by a team of 50 people. The results demonstrated that the plots with a more diverse plant community were able to retain and use nutrients more efficiently than the plots with less diversity. This efficiency reduced the loss of soil nitrogen through leaching, thus increasing soil fertility. Biodiversity improved the nutrient dynamic in the prairie ecosystem.

Greater plant diversity led to both higher productivity and better resource utilization. Also, total plant cover was found to increase with

FIGURE 16.1.2 **Biodiversity experiment.**
Experimental plots are planted with native North American prairie grasses. A total of 147 plots were sown with random seed selections. Experimental results confirm many assumptions regarding the value of rich biodiversity. [Photo by David Tilman/University of Minnesota, Twin Cities Campus.]

species richness. Tilman and his colleagues affirm:

> This extends the earlier results to the field, providing direct evidence that the current rapid loss of species on Earth, and management practices that decrease local biodiversity, threaten ecosystem productivity and the sustainability of nutrient cycling. Observational, laboratory, and now field experimental evidence supports the hypothesis that biodiversity influences ecosystem productivity, sustainability, and stability.*

FIGURE 16.1.1 **One of the few remaining tracts of prairie.**
The Hayden Prairie State Preserve in Howard County, Iowa, is 240 acres of tallgrass prairie named after Dr. Ada Hayden, a former botany professor at Iowa State University. Grasses and shooting star flowers (*Dodecatheon* sp.), and a blocked former road and trails, mark this pristine place in Iowa.
[Photos by Bobbé Christopherson.]

*D. Tilman, D. Wedin, and J. Knops, "Productivity and sustainability influenced by biodiversity in grassland ecosystems," *Nature* 379 (February 22, 1996): 720. Also see the summary in *Science* 271 (March 15, 1996): 1497.

Agricultural Ecosystems Humans simplify communities by eliminating biodiversity, and in this way we place more ecosystems at risk of harmful change and perhaps failure. An artificially produced monoculture community, such as a field of wheat, is singularly vulnerable to failure owing to weather or attack from insects or plant disease. In some regions, simply planting multiple crops brings more stability to the ecosystem. This is an important principle of sustainable agriculture, a movement to make farming more ecologically compatible.

A modern agricultural ecosystem is vulnerable to failure not only because of its lack of ecological diversity but also because it creates an enormous demand for energy, chemical pesticides and herbicides, artificial fertilizer, and irrigation water. The practice of harvesting and removing biomass from the land interrupts the cycling of materials into the soil (Figure 16.18). This net loss of nutrients must be artificially replenished. The type of plantation agriculture in Figure 16.18b is depleting soil nutrients over time; this farming practice requires subsidies of chemical fertilizers and herbicides to eliminate competing vegetation.

Climate Change Obviously, the distribution of plant species is affected by climate change, though many species have survived wide climate swings in the past. Consider the beginning of the Tertiary Period, 75 million years ago. Warm, humid conditions and tropical forests dominated the land to southern Canada, pines grew in the Arctic, and deserts were few. Then, between 15 and 50 million years ago, deserts began developing in the southwestern United States. Mountain-building processes created higher elevations, causing rain-shadow aridity and affecting plant distribution.

Recall, too, that the movement of Earth's tectonic plates created climate changes important in the evolution and distribution of plants and animals (see Figure 8.15). Europe and North America were joined in Pangaea and positioned near the equator, where vast swamps formed (the site of coal deposits today). The southern mass of Gondwana was extensively glaciated as it drifted at high latitudes in the Southern Hemisphere. Glaciation left matching glacial scars and specific plant and animal distributions across South America, Africa, India, and Australia. The diverse and majestic dinosaurs dispersed with the drifting continents.

FIGURE 16.18 Agricultural ecosystems.
(a) The Sacramento Valley of California is one of the most intensively farmed regions of the world. Field crops, including irrigated rice, an irrigation canal, and the Sacramento River, are in the scene. (b) A pine tree plantation (farm) in Georgia, principally for the production of wood pulp—the forest ecosystem is reduced to efficient rows for clear-cutting when mature. (c) Modern mechanized farming and a freshly mowed field of oats near Dokka, Norway. (d) Highly fragmented, fertile agricultural land along the north German plain. Note the lack of sprawl from each German city, as agricultural lands are protected. [All photos by Bobbé Christopherson.]

The key question for our future is: As temperature patterns change, how fast can plants either adapt to new conditions or migrate through succession (location change) to remain within their shifting specific habitats? Satellite observations indicate rapid changes in vegetation patterns in northern latitudes since 1980. Global warming is reducing spring snow cover, allowing spring greening to occur earlier, up to 3 weeks sooner in some high-latitude locations. Likewise, the onset of fall is delayed to a later date than previously experienced.

Nina Leopold Bradley studied the Wisconsin setting her father Aldo Leopold made famous in his book *A Sand County Almanac* (Oxford University Press, 1949). He kept detailed records of all things dealing with nature and seasonal change on his tract of land. Her findings show that more than one-third of animal and plant species start spring activities 4 weeks earlier than they did 50 years earlier.

Adaptation to conditions is key to evolution. Through mutation and natural selection, species have either adapted or failed to adapt to changing environmental conditions over millions of years. Current global change is occurring rapidly, at the rate of decades instead of millions of years. Thus, we see die-out and succession of different species along disadvantageous habitat margins. The displaced species may colonize new regions made more hospitable by climate change. Also shifting are current agricultural climates that produced wheat, corn, soybeans, and other commodities. Society will have to adapt to crops growing in different places.

A study completed by biologist Margaret Davis on North American forests suggests that trees will have to respond quickly if temperatures increase. Changes in the climate inhabited by certain species could shift 100–400 km (60–250 mi) during the next 100 years. Some species, such as the sugar maple, are migrating northward, disappearing from the United States except for Maine and moving into eastern Ontario and Québec. Davis prepared a map showing the possible impact of increasing temperatures on the distribution of beech and hemlock trees (Figure 16.19).

> Changes in the geographical distributions of plant and animal species in response to future greenhouse warming threaten to reduce biotic diversity. . . . The risk posed by CO_2-induced warming depends on the distances that regions of suitable climate are displaced northward [in the Northern Hemisphere] and on the rate of displacement. . . . If the change occurs too rapidly for colonization of newly available regions, population sizes may fall to critical levels, and extinction will occur.*

*M. B. Davis and C. Zabinski, "Changes in the Geographical Range Resulting from Greenhouse Warming: Effects on Biodiversity in Forests," in R. L. Peters and T. E. Lovejoy, eds., *Global Warming and Biological Diversity* (New Haven, CT: Yale University Press, 1992), p. 297.

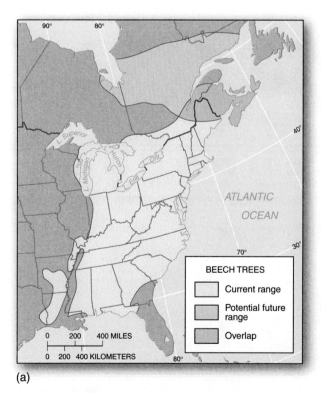

(a)

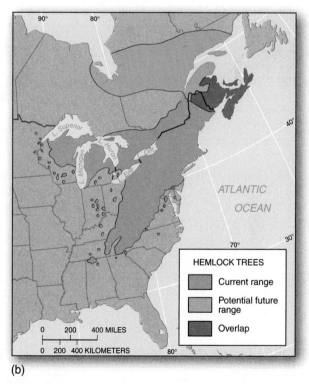

(b)

FIGURE 16.19 Present and predicted distribution of two tree species.
(a) Beech trees and (b) hemlock trees in North America. The potential future range shown for each reflects expected climate change resulting from a doubling of carbon dioxide. These maps are based on forecasts from the Goddard Fluid Dynamics Laboratory (GFDL) general circulation model.
[After M. B. Davis and C. Zabinski, "Changes in the Geographical Range Resulting from Greenhouse Warming: Effects on Biodiversity in Forests," in R. L. Peters and T. E. Lovejoy, eds., *Global Warming and Biological Diversity* (New Haven: Yale University Press, 1992), p. 301.]

Another sign of the impact of climate change on ecosystems is the updating of planting and gardening guides so that they better fit actual changing climatic patterns. A comparison of a 2006 Hardiness Zone map with the USDA 1990 map illustrates such change in a dramatic way. Each zone is based on the low temperature reached, with the zones set at 10° increments—a zone 3 has a low of −30°F to −40°F (−34°C to −40°C), a zone 4 increases by 10 F° (5.6 C°), and so forth. Based on NOAA and National Climatic Data center input from 5000 stations, the Arbor Day Foundation produced the new plant hardiness map. Figure 16.20 presents both the 1990 and 2006 maps and a map that portrays the differences, or zone changes, that occurred between the two. For instance Ohio, Indiana, and Illinois shifted from Zone 5 to the warmer Zone 6.

Ecological Succession

Ecological succession occurs when more complex communities replace older, usually simpler, communities of plants and animals. Each successive community of species modifies the physical environment in a manner suitable for the establishment of a later community of species. Changes apparently move toward a more stable and mature condition, although disturbances are common.

Traditionally, it was assumed that plants and animals eventually formed a *climax community*—a stable, self-sustaining, and symbiotically functioning community with balanced birth, growth, and death—but scientists think differently now. Contemporary conservation biology, biogeography, and ecology assume nature to be in constant adaptation and nonequilibrium.

Rather than thinking of an ecosystem as a uniform set of communities, think of an ecosystem as a patchwork mosaic of habitats—each striving to achieve an optimal range and low environmental stress. *Patch dynamics* is the study of disturbed portions of habitats. Within an ecosystem, individual patches may arise only to fail later. The complexity of succession is in the interactions of structure and function among patches. Most ecosystems are actually made up of patches of former landscapes.

Given the complexity of natural ecosystems, there may be several stages, or a *polyclimax* condition, with adjoining ecosystems, or patches within ecosystems, at different stages in the same environment. Mature communities are properly thought of as being in dynamic equilibrium. Or at times, these communities may be out of phase and in nonequilibrium with the immediate physical environment because of the usual lag time in their adjustment.

Succession often requires an initiating disturbance. External examples include windstorm, severe flooding, a

Find your home town/city and campus town/city on this "Zone Change" map and on the 1990 and 2006 maps. What zone changes do you note for each location?
Home: _____
Campus: _____

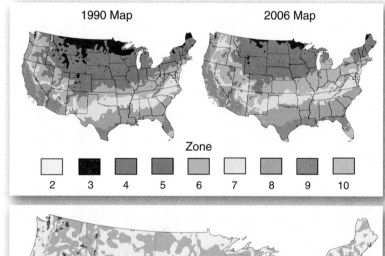

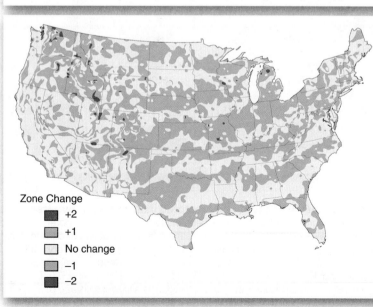

FIGURE 16.20 Climate change impacts hardiness zones.
The Arbor Day Foundation revised the 1990 USDA Hardiness Zone map used to guide farmers and gardeners. Climate changes necessitated the 2006 revision. Compare the two maps, then check the differences map. Locate your home and your college towns. Did these places get reclassified?
[Maps courtesy of Arbor Day Foundation and USDA; see
http://www.arborday.org/media/map_change.cfm.]

(a)

(b)

(c)

(d)

FIGURE 16.21 Disturbances in ecosystems alter succession patterns.
(a) River flooding. (b) Succession progresses in a burned forest, perhaps by chance, as species with an adaptive edge succeed, such as fireweed (*Epilobium* sp.). (c) Unwise practices cause extensive soil erosion on a hog farm in Iowa; a disrupted prairie landscape. (d) Drought-stressed Coulter pines are attacked by bark beetles, altering relationships in the ecosystem. [Photos (a), (b), and (d) by Bobbé Christopherson; photo (c) by USDA Natural Resources Conservation Service.]

volcanic eruption, a devastating wildfire, an agricultural practice such as prolonged overgrazing, climate change, or an insect infestation (Figure 16.21). When existing organisms are disturbed or removed, new communities can emerge. At such times of nonequilibrium transition, the interrelationships among species produce elements of chance, and species having an adaptive edge will succeed in the competitive struggle for light, water, nutrients, space, time, reproduction, and survival.

In Figure 16.21d you see dead and dying Coulter pines (*Pinus coulteri*) in southern California mountains. These drought-stressed trees can die within a week of infestation by opportunistic bark beetles. Thus, the succession of plant and animal communities is an intricate process with many interactive variables, both in space and time and from internal and external processes.

Land and water experience different forms of succession. We first look at terrestrial succession, which is characterized by competition for sunlight, and then at aquatic succession, characterized by progressive changes in nutrient levels. Succession is a dynamic process of unpredictable outcomes—the basis of *dynamic ecology*.

Terrestrial Succession An area of bare rock or soil with no vestige of a former community can be a site for **primary succession**, the beginning of development of an ecosystem.

Examples are any new surface created by mass movement of land, areas exposed by a retreating glacier, cooled lava flows, or lands disturbed by surface mining, clear-cut logging, land development, or volcanic eruption. Illustrating primary succession are plants taking hold on the lava flows of Hawai'i in Figure 16.22. In terrestrial ecosystems, succession begins

FIGURE 16.22 Primary succession.
Succession and recovery of pioneer species as ferns take hold on new lava flows coming from Kīlauea volcano, Hawai'i.
[Photo by Bobbé Christopherson.]

with early species that form a **pioneer community**. We first see lichens, mosses, and ferns growing on bare rock. These early inhabitants prepare the way for further succession to grasses, shrubs, and trees.

More common is **secondary succession**, which begins if the vestiges of a previously functioning community are present. An area where the natural community has been destroyed or disturbed, but where the underlying soil remains intact, may experience secondary succession.

As succession progresses, soil develops and a different set of plants and animals with different niche requirements may adapt. Further niche expansion follows as the community matures. Succession, which is really the developmental process forming new communities, is a set of interactions with sometimes unpredictable outcomes, steered by sometimes random events that trigger a *threshold* leap to a new set of relations as in a dynamic ecology.

Examples of secondary succession are most of the areas affected by the Mount St. Helens eruption and blast in 1980 (Figure 16.23). Some soils, young trees, and plants were protected under ash and snow, so community development began almost immediately after the event. About 38,450 hectares (95,000 acres) of trees were blown down and burned. As the photos show, the effects of the eruption were devastating and lingering. Of course, the areas completely destroyed near the Mount St. Helens volcano or those buried beneath the massive landslide north of the mountain became areas for primary succession.

Wildfire and Fire Ecology Fire and its effects are one of Earth's significant ecosystem and economic processes. In the United States during 2000, about 7.0 million acres, more than double the annual average, burned in wildfires—an incredible 84,210 fires. In subsequent years, US wildfires consumed the following: 2001, 3.3 million acres; 2002, 6.7 million acres; 2003, 3.2 million acres; 2005, 8.2 million acres (4.4 million acres of this in Alaska); 2006, 9.4 million acres; and 2007, 8.2 million acres (see National Interagency Fire Center at http://www.nifc.gov/fire_info/nfn.htm).

Over the past 50 years, the role of fire in ecosystems has been the subject of much scientific research and experimentation. Today, fire is recognized as a natural component of most ecosystems and not the enemy of nature that it was once thought. In fact, in many forests, undergrowth and surface litter is purposely burned in controlled "cool fires" to remove fuel and prevent catastrophic and destructive "hot fires." Forestry experts have learned that when fire-prevention strategies are rigidly followed, it can lead to the abundant undergrowth accumulation that fuels major fires.

Modern society's demand for fire prevention to protect property goes back to European forestry of the 1800s. Fire prevention became an article of faith for forest managers in North America. But, in studies of the longleaf pine forest that stretches in a wide band from the Atlantic coastal plain to Texas, fire was discovered to be an integral part of regrowth following lumbering. In fact, seed dispersal of some pine species, such as the knobcone pine, does not occur unless assisted by a forest fire. Heat from the fire opens the cones, releasing seeds so they can fall to the ground for germination. Also, these fire-disturbed areas quickly recover with protein-rich woody growth, young plants, and a stimulated seed production that provides abundant food for animals.

The science of **fire ecology** imitates nature by recognizing fire as a dynamic ingredient in community succession. Controlled ground fires, deliberately set to prevent accumulation of forest undergrowth, now are widely regarded as a wise forest management practice and are used across the country. After the highly visible Yellowstone National Park fires, an outcry was heard from forestry and recreational interests (Figure 16.24c). Critics called fire ecology practices the government's "let it burn" policy.

In its final report on the 1988 Yellowstone fires, a government interagency task force concluded, "... an attempt to exclude fire from these lands leads to major unnatural changes in vegetation ... as well as creating fuel accumulation that can lead to uncontrollable, sometimes very damaging wildfire." Thus, federal land managers and scientists reaffirmed their stand that fire ecology is a fundamentally sound concept.

An increasingly serious problem is emerging as people seek to live in the wildlands near cities. Urban development has encroached on forests, and the added suburban landscaping has created new fire hazards. Uncontrolled wildfires now can destroy homes and threaten public safety. People living in these areas demand fire protection that actually causes undergrowth to accumulate and worsens the risk. In 2003 and again in 2007, wildfires devastated communities in southern California, especially where development was in the "wilds" of chaparral country. In 2007, fire destroyed more than 2000 homes, and over 500,000 acres burned in two dozen wildfires (see Figure 2.24a for a satellite view of the southern California inferno). Strong Santa Ana winds (see Chapter 4 for description) added the blowtorch effect, taking embers miles from their ignition source.

The connection between the record wildfires in the American West since 2000 and climate change was reported in *Science* in 2006:

> Higher spring and summer temperatures and earlier snowmelt are extending the wildfire season and increasing the intensity of wildfires in the western US. . . . Here, we show that large wildfire activity increased suddenly and markedly in the mid-1980s, with higher large-wildfire frequency. Longer wildfire durations, and longer wildfire seasons. . . . strongly associated with increased spring and summer temperatures and an earlier spring snowmelt.*

---

*A. L. Westerling, H. G. Hidalgo, et al., "Warming and earlier spring increase Western US forest wildfire activity," *Science* 313, no. 5789 (August 2006): 927, 940–43.

Landsat-7 8/22/1999

FIGURE 16.23 The pace of change in the region of Mount St. Helens.
(a) The area north of the volcano before the eruption (1979) and (b) after the eruption (1980) experienced profound change in all ecosystems. A series of four photo comparisons made in 1983 and 1999 of matching scenes illustrate the slow recovery of secondary succession: (c) and (d) at Meta Lake; (e) Spirit Lake panorama view (note the mat of floating logs after 19 years); (f) detail of the destroyed forest community. [(a) and all 1983 photos by author; (b) and all 1999 photos by Bobbé Christopherson; *Landsat-7* 8/22/99 satellite image courtesy of NASA/GSFC.]

FIGURE 16.24 Wildfire strikes and recovery, Bitterroot and Yellowstone.
(a) The drama of wildfire unfolds in the Bitterroot National Forest in Montana, August 2000, as elk
stand in the Bitterroot River attempting to survive the conflagration. (b) Fire-adapted chaparral
vegetation a few months after a wildfire, as recovery begins. Note the stump sprouts emerging.
(c) Yellowstone National Park in northwestern Wyoming: 1988 wildfire progresses up a slope;
(d) trees and shrubs return to Yellowstone with secondary succession. [Photos by (a) John
McColgan, BLM Alaska Fire Service; (b) and (d) by Bobbé Christopherson; and (c) Joe Peaco/National
Park Service.]

Aquatic Succession   Lakes, ponds, and oxbows exhibit
another form of ecological succession. A lake experiences
successional stages as it fills with nutrients and sediment
and as aquatic plants take root and grow, capturing more
sediment and adding organic debris to the system (Figure
16.25a). This gradual enrichment of water bodies is
known as **eutrophication**, from the Greek *eutrophos*,
meaning "well nourished."

For example, in moist climates, a floating mat of vegetation grows outward from the shore to form a bog
(Figure 16.25b). Cattails and other marsh plants become
established, and partially decomposed organic material
accumulates in the basin, with additional vegetation bordering the remaining lake surface. A meadow may form as
water is displaced by the peat bog; willow trees follow, and
perhaps cottonwood trees. Eventually, the lake may
evolve into a forest community.

Now that we have examined ecosystem operations, let
us look at terrestrial ecosystem patterns—Earth's terrestrial biomes.

Earth's Major Terrestrial Biomes

A **terrestrial ecosystem** is a self-regulating association of
plants and animals and their abiotic environment characterized by specific plant formations. In their growth, form,
and distribution, plants reflect Earth's physical systems: its
energy patterns; atmospheric composition; temperature
and winds; air masses; water quantity, quality, and seasonal
timing; soils; regional climates; geomorphic processes; and
global climate change. We know from history that the
effect of climate change on ecosystems will be significant
and that understanding current changes underway is critical. This is a challenge for physical geography and the

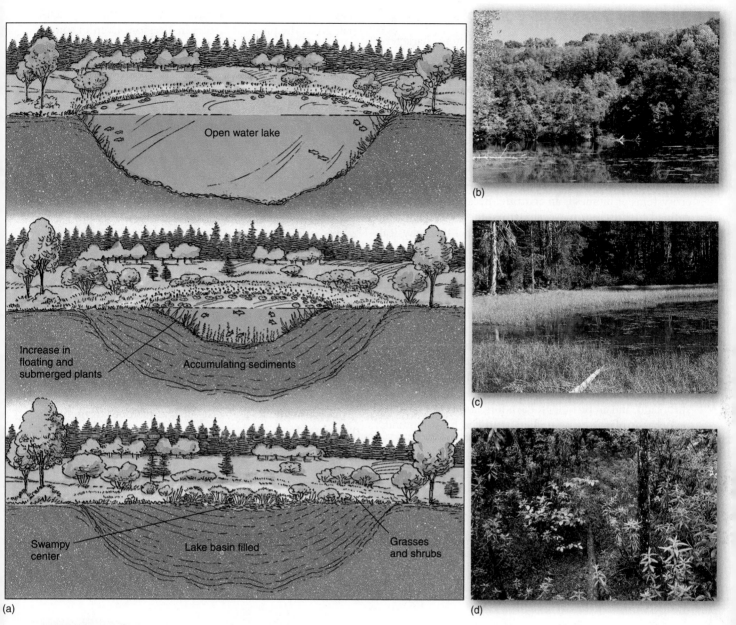

FIGURE 16.25 Lake-bog-meadow succession.
(a) What begins as a lake gradually fills with organic and inorganic sediments, which successively shrink the area of the pond. A bog forms, then a marshy area, and finally a meadow completes the successional stages. (b) Spring Mill Lake, Indiana. (c) Aquatic succession in a mountain lake. (d) The Richmond Nature Park and its Lulu Island Bog was ocean just 8000 years ago. Frasier River sediment created the mudflats and set the stage for the evolution of this sphagnum bog with acidic soils that "quakes" when you walk on it. Plants in the bog include moss, shore pine, hemlock, blueberry, salal shrubs, and Labrador tea, among others. [Photos by Bobbé Christopherson.]

other geospatial sciences—to provide information that might be applied in some ways to benefit the future of natural systems that sustain living systems.

An assessment was given in UNEP's 2007 *Global Environmental Outlook–GEO4 Executive Summary* (p. 4):

> Human life and all other species depend on healthy ecosystems. But current biodiversity changes, the fastest in human history, mean losses are restricting future development options. About 60% of the ecosystem services that have been previously assessed are degraded or used unsustainably. Species are becoming extinct at rates which are 100 times faster than the rate shown in the fossil record, because of land-use changes, habitat loss, overexploitation of resources, pollution, and the spread of invasive alien species.

Humans now have become the most powerful biotic agent on Earth, influencing all ecosystems on a planetary scale. With this in mind, let's explore Earth's major terrestrial ecosystems, conveniently grouped into 10 biomes.

Terrestrial Ecosystem Concepts

A **biome** is a large, stable, terrestrial ecosystem characterized by specific plant and animal communities. Each biome usually is named for its *dominant vegetation*. We can generalize Earth's wide-ranging plant species into these broad biomes: *forest*, *savanna*, *grassland*, *shrubland*, *desert*, and *tundra*. Because plant distributions respond to environmental conditions and reflect variations in climate and soil, the biome-related world climate map (Figure 7.5) is a helpful reference for this chapter.

A boundary transition zone between adjoining ecosystems or biomes is an **ecotone**. Because ecotones are defined by different physical factors, they vary in width. Climatic ecotones usually are more gradual than physical ecotones, because differences in soil or topography sometimes form abrupt boundaries. An ecotone between prairies and northern forests may occupy many kilometers of land. The ecotone is an area of tension, as similar species of plants and animals compete for the resource base.

These general biomes are divided into more specific **formation classes**, vegetation units that refer to the dominant plants in a terrestrial ecosystem. Examples are equatorial rain forest, northern needleleaf forest, Mediterranean shrubland, and arctic tundra. Each formation class includes numerous plant communities, and each community includes innumerable plant habitats. Within those habitats, Earth's diversity is expressed in 270,000 plant species. Despite this intricate complexity, we can generalize Earth's numerous formation classes into 10 global terrestrial biome regions.

Few natural communities of plants and animals remain; most biomes have been greatly altered by human intervention. Thus, the "natural vegetation" identified on many biome maps reflects ideal mature vegetation potential, given the physical factors in a region. Even in somewhat remote, pristine Norway, the former needleleaf and montane forest today is a mix of second- and third-growth forests, farmlands, and altered landscapes (Figure 16.26). Even though human practices have greatly altered these ideal forms, it is valuable to study the natural (undisturbed) biomes to better understand the natural environment and to assess the extent of human-caused alteration.

In addition, knowing the ideal in a region helps us approximate natural vegetation in the plants we introduce. In the United States and Canada, we humans are perpetuating a type of transition community, somewhere between grassland and a forest. We plant trees and lawns and then must invest energy, water, and capital to sustain such artificial modifications of nature. Some irreversible damage might occur when plants, animals, and organisms are knowingly or accidentally brought into an ecosystem or biome in which they are not native. These exotic species pose an increasing concern to scientists and society—a subject of News Report 16.2.

The global distribution of Earth's 10 major terrestrial biomes, based on vegetation formation classes, is portrayed in Figure 16.27. Table 16.2 describes each biome on the map and summarizes other pertinent information from throughout this text—a compilation from many chapters, for Earth's biomes are a synthesis of physical geography.

Equatorial and Tropical Rain Forest

Earth is girdled with a lush biome—the **equatorial and tropical rain forest**. In a climate of consistent daylength (12 hours), high insolation, and average annual

FIGURE 16.26 Needleleaf forest landscape modified by human activity.
Forests and farms sweep down to meet the long, narrow Vagavatnet Lake near the town of Vagamo in south-central Norway, at 62° N latitude. [Photo by Bobbé Christopherson.]

NEWS REPORT 16.2

Alien Invaders of Exotic Species

The title above sounds like that of a cult science fiction movie; instead, it refers to a real problem in the integrity of many ecosystems, both aquatic and terrestrial. Animal and plant species brought or somehow getting into biomes outside their native home is an important subject in biogeography and invasion biology. Niche takeovers by nonnative species can prove damaging to otherwise healthy ecosystems. These intruders are known as *exotic species*, or *alien* or *nonnative species*.

The Institute for Biological Invasions at the University of Tennessee (http://invasions.bio.utk.edu/bio_invasions/index.html) is dedicated to the study of these invasive plants, animals, and other organisms. Some states post warning signs at their borders (Figure 16.2.1a) and others offer web sites, such as Minnesota's "Prohibited and Noxious Plants by Scientific Name" (http://www.dnr.state.mn.us/invasives/index.html).

The African "killer bees" are frequently in the headlines, as are brown tree snakes in Guam, zebra mussels in the Great Lakes and Mississippi River, and Eurasian cheat-grass in the Utah desert, to name but a few examples. Brought from Africa to the central coast of Brazil in 1957 to increase honey production, killer bees now have interbred with native bees and range to southern California, Arizona, New Mexico, Texas, and Puerto Rico. More than 1000 people have died from their attacks, due to the dangers of their mass response to disturbances and the allergic reaction some people have to even one bee sting. This is a classic case of geographical diffusion of an exotic species.

Probably 90% of exotic species fail when they try to move into established niches in a community. The 10% that succeed prove damaging to as much as one-fifth of ecosystems they invade, as demonstrated by three examples.

Purple loosestrife (*Lythrum salicaria*) was introduced from Europe in the 1800s as a desired ornamental and had some medicinal applications. The plant's seeds also arrived in ships that used soil for ballast. A hardy perennial, it got loose and invaded wetlands across the eastern portions of the United States and Canada, through the upper Midwest, and as far west as Vancouver Island, British Columbia, replacing plants on which native wildlife depended (Figure 16.2.1b). The plant is known to infect drier landscapes as well and poses a potential threat to agriculture.

The infamous kudzu (*Pueraria montana*) was brought from Japan in 1876 for cattle feed, erosion control, and as an ornamental. It has spread to east Texas, southern Pennsylvania, and across the South into Florida. The plant can grow 0.3 m (1 ft) on a hot, humid summer day and overtake forests, homes, or anything in its path (Figure 16.2.1c). Serious effort is

(a)

(b)

(c)

(d)

FIGURE 16.2.1 **Exotic species.**
(a) A sign in Wallowa County, Washington state, pleads for awareness and help in this agricultural country, listing noxious (nonnative, often problematic) weeds at the county line. (b) Invasive purple loosestrife near Lake Michigan and the Indiana Dunes National Lakeshore, Indiana (foreground and left center). (c) Kudzu overruns pasture and forest in western Georgia. (d) Gorse (yellow shrub), intended for barriers between pastures, is loose across the landscape in West Point, Falkland Islands.
[Photos by Bobbé Christopherson.]

(continued)

News Report 16.2 *(continued)*

going into finding some practical use for the prodigious plant, from pulp (treeless paper), to cooking recipes, to the telling of new kudzu jokes.

Gorse was introduced into the timber-poor Falkland Islands as a barrier plant, like a substitute fence.

Its thorny stickers and dense growth were meant to keep grazing sheep and a few cattle under control. The plant got loose and spread beyond this intended use (Figure 16.2.1d).

In learning about biomes, we find out about native plant, animal,

and organism locations and the unique place each species occupies. We must avoid knowingly or perhaps unconsciously adding to an escalating problem by bringing exotic species into nonnative sites.

temperatures around 25°C (77°F), plant and animal populations have responded with the most diverse body of life on the planet. The Amazon region, also called the *selva*, is the largest tract of equatorial and tropical rain forest. Rain forests also cover equatorial regions of Africa, Indonesia,

the margins of Madagascar and Southeast Asia, the Pacific coast of Ecuador and Colombia, and the east coast of Central America, with small discontinuous patches elsewhere (find on Figure 16.27). The cloud forests of western Venezuela are such tracts of rain forest at high

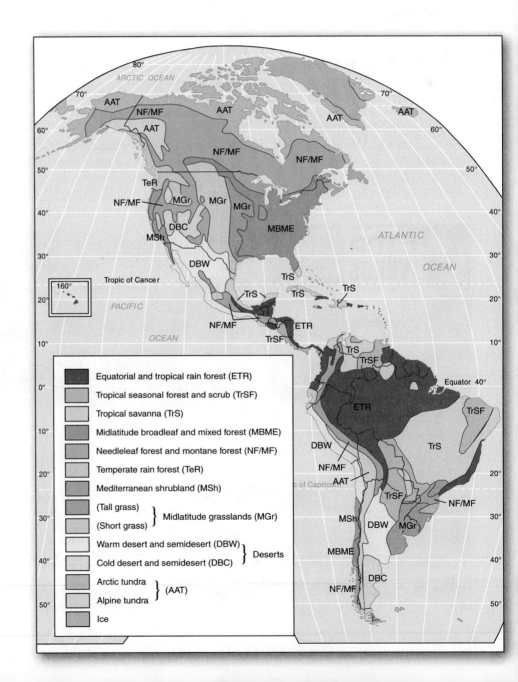

FIGURE 16.27 The 10 major global terrestrial biomes (idealized).

elevation, perpetuated by high humidity and cloud cover. Undisturbed tracts of rain forest are rare.

Rain forests feature ecological niches distributed vertically rather than horizontally, because of the competition for light. The canopy is filled with a rich variety of plants and animals. *Lianas* (vines) stretch from tree to tree, entwining them with cords that can reach 20 cm (8 in.) in diameter. *Epiphytes* flourish there, too: such plants as orchids, bromeliads, and ferns that live entirely above the ground, supported physically but not nutritionally by the structures of other plants. Windless conditions on the forest floor make pollination difficult—insects, other animals, and self-pollination predominate.

The rain forest canopy forms three levels—see Figure 16.28. The upper level is not continuous but features tall trees whose high crowns rise above the middle canopy. The middle canopy is the most continuous (Figure 16.28b), its broad leaves blocking much of the light and creating a darkened *understory* area and forest floor. The lower level is composed of seedlings, ferns, bamboo, and the like, leaving the litter-strewn ground level in deep shade and fairly open.

Aerial photographs of a rain forest, or views along river banks covered by dense vegetation, aided by the false Hollywood-movie imagery of the jungle, make it difficult to imagine the shadowy environment of the actual rain forest floor (Figure 16.29a). The forest floor receives only about 1% of the sunlight arriving at the canopy. Together, the constant moisture, rotting fruit and moldy odors, strings of thin roots and vines dropping down from above, windless air, and echoing sounds of life in the trees create a unique environment.

The smooth, slender trunks of rain forest trees are covered with thin bark and buttressed by large wall-like

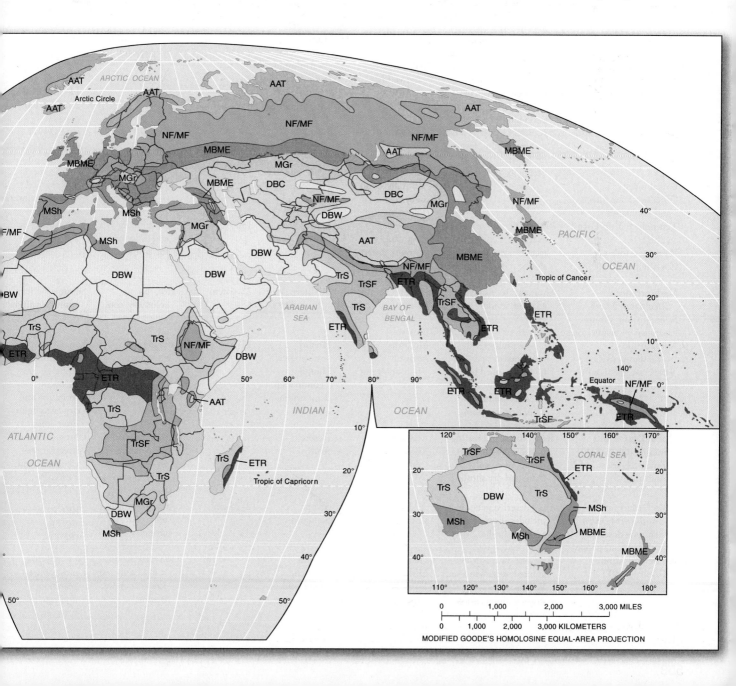

Table 16.2 Major Terrestrial Biomes and Their Characteristics

Biome and Ecosystems (map symbol)	Vegetation Characteristics	Soil Orders (Soil Taxonomy and CSSC)	Climate Type	Annual Precipitation Range	Temperature Patterns	Water Balance
Equatorial and Tropical Rain Forest (ETR) Evergreen broadleaf forest Selva	Leaf canopy thick and continuous; broadleaf evergreen trees, vines (lianas), epiphytes, tree ferns, palms	Oxisols Ultisols (on well-drained uplands)	Tropical	180–400 cm (>6 cm/mo)	Always warm (21–30°C; avg 25°C)	Surpluses all year
Tropical Seasonal Forest and Scrub (TrSF) Tropical monsoon forest Tropical deciduous forest Scrub woodland and thorn forest	Transitional between rain forest and grasslands; broadleaf, some deciduous trees; open parkland to dense undergrowth; acacias and other thorn trees in open growth	Oxisols Ultisols Vertisols (in India) Some Alfisols	Tropical monsoon savanna	130–200 cm (<40 rainy days during 4 driest months)	Variable, always warm (>18°C)	Seasonal surpluses and deficits
Tropical Savanna (TrS) Tropical grassland Thorn tree scrub Thorn woodland	Transitional between seasonal forests, rain forests, and semiarid tropical steppes and desert; trees with flattened crowns, clumped grasses, and bush thickets; fire association	Alfisols (dry: Ultalfs) Ultisols Oxisols	Tropical savanna	9–150 cm, seasonal	No cold-weather limitations	Tends toward deficits, therefore fire- and drought-susceptible
Midlatitude Broadleaf and Mixed Forest (MBME) Temperate broadleaf Midlatitude deciduous Temperate needleleaf	Mixed broadleaf and needleleaf trees; deciduous broadleaf, losing leaves in winter; southern and eastern evergreen pines demonstrate fire association	Ultisols Some Alfisols (Podzols, red and yellow)	Humid subtropical warm summer Humid continental warm summer	75–150 cm	Temperate with cold season	Seasonal pattern with summer maximum PRECIP and POTET (PET); no irrigation needed
Needleleaf Forest and Montane Forest (NF/MF) Taiga Boreal forest Other montane forests and Highlands	Needleleaf conifers, mostly evergreen pine, spruce, fir; Russian larch, a deciduous needleleaf	Spodosols Histosols Inceptisols Alfisols (Boralfs: cold) (Gleysols) (Podzols)	Subarctic Humid continental cool summer	30–100 cm	Short summer, cold winter	Low POTET (PET), moderate PRECIP, moist soils, some waterlogged and frozen in winter; no deficits
Temperate Rain Forest (TeR) West coast forest Coast redwoods (US)	Narrow margin of lush evergreen and deciduous trees on windward slopes; redwoods, tallest trees on Earth	Spodosols Inceptisols (mountainous environs) (Podzols)	Marine west coast	150–500 cm	Mild summer and mild winter for latitude	Large surpluses and runoff
Mediterranean Shrubland (MSh) Sclerophyllous shrubs Australian eucalyptus forest	Short shrubs, drought adapted, tending to grassy woodlands and chaparral	Alfisols (Xeralfs) Mollisols (Xerolls) (Luvisols)	Mediterranean dry summer	25–65 cm	Hot, dry summers, cool winters	Summer deficits, winter surpluses
Midlatitude Grasslands (MGr) Temperate grassland Sclerophyllous shrub	Tallgrass prairies and shortgrass steppes, highly modified by human activity; major areas of commercial grain farming; plains, pampas, and veld	Mollisols Aridisols (Chernozemic)	Humid subtropical Humid continental hot summer	25–75 cm	Temperate continental regimes	Soil moisture utilization and recharge balanced; irrigation and dry farming in drier areas
Warm Desert and Semidesert (DBW) Subtropical desert and scrubland	Bare ground graduating into xerophytic plants including succulents, cacti, and dry shrubs	Aridisols Entisols (sand dunes)	Desert Arid	<2 cm	Average annual temperature, around 18°C, highest temperatures on Earth	Chronic deficits, irregular precipitation events, PRECIP <½ POTET (PET)
Cold Desert and Semidesert (DBC) Midlatitude desert, scrubland, and steppe	Cold desert vegetation includes short grass and dry shrubs	Aridisols Entisols	Steppe Semiarid	2–25 cm	Average annual temperature around 18°C	PRECIP >½ POTET (PET)
Arctic and Alpine Tundra (AAT)	Treeless; dwarf shrubs, stunted sedges, mosses, lichens, and short grasses; alpine, grass meadows	Gelisols Histosols Entisols (permafrost) (Organic) (Cryosols)	Tundra Subarctic very cold	15–180 cm	Warmest months <10°C, only 2 or 3 months above freezing	Not applicable most of the year, poor drainage in summer
Ice			Ice sheet, ice cap			

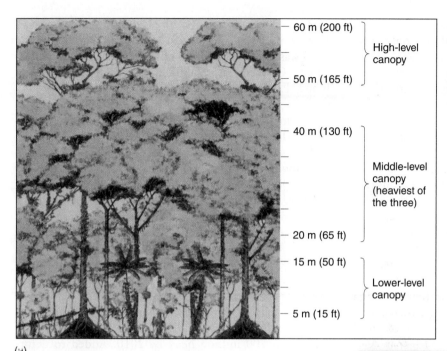

(a)

FIGURE 16.28 Rain forest—equatorial and tropical.
(a) The three levels of a rain forest canopy.
(b) Undisturbed rain forest on the Peninsula de Osa in Costa Rica. Note the few high-level trees that extend beyond the dominant middle-level canopy. [Photo by Barbara Cushman Rowell/Mountain Light Photography, Inc.]

(b)

(a)

(b)

FIGURE 16.29 The rain forest.
(a) Equatorial rain forest is thick along the Amazon River where light breaks through to the surface, producing a rich gallery of vegetation. (b) The rain forest floor in Corcovado, Costa Rica, with typical buttressed trees and lianas. [Photos by (a) Wolfgang Kaehler; (b) Frank S. Balthis.]

flanks that grow out from the trees to brace the trunks (Figure 16.29b). These buttresses form angular open enclosures, a ready habitat for various animals. There are usually no branches for at least the lower two-thirds of the tree trunks.

Varieties of trees include mahogany, ebony, and rosewood. Logging is difficult because individual species are widely scattered; a species may occur only once or twice per square kilometer. Selective cutting is required for species-specific logging, whereas pulpwood production takes everything. Conversion of the forest to pasture usually is accomplished by setting destructive fires.

Rain forests represent approximately half of Earth's remaining forests, occupying about 7% of the total land area worldwide. This biome is stable in its natural state, as a result of the long-term residence of these continental plates near equatorial latitudes and their escape from glaciation. The soils, principally Oxisols, are essentially infertile, yet they support rich vegetation. The trees have adapted to these soil conditions with root systems able to capture nutrients from litter decay at the soil surface. With much investment in fertilizers, pesticides, and machinery, these soils can be productive.

The animal and insect life of the rain forest is diverse, ranging from small decomposers (bacteria) working the surface to many animals living exclusively in the upper stories of the trees. These tree dwellers are referred to as *arboreal*, from the Latin for "tree," and include sloths, monkeys, lemurs, parrots, and snakes. Beautiful birds of many colors, tree frogs, lizards, bats, and a rich insect life that includes over 500 species of butterflies are found in rain forests. Surface animals include pigs (bushpigs and the giant forest hog in Africa, wild boar, bearded pig in Asia, and peccary in South America), species of small antelopes (bovids), and mammal predators (tiger in Asia, jaguar in South America, and leopard in Africa and Asia).

Deforestation of the Tropics More than half of Earth's original rain forest is gone, cleared for pasture, timber, fuel wood, and farming. Worldwide, an area nearly the size of Wisconsin is lost each year and about a third more is disrupted by selective cutting of canopy trees, where damage occurs along the *edges* of newly deforested areas (Figure 16.30).

When orbiting astronauts look down on the rain forests at night, they see thousands of human-set fires. During the day, the lower atmosphere in these regions is choked with smoke. These fires are used to clear land for agriculture, which is intended to feed the domestic population as well as to produce cash exports of beef, rubber, coffee, and other commodities. Because of the poor soil fertility, the cleared lands are quickly exhausted under intensive farming and are then generally abandoned in favor of newly burned and cleared lands, unless the lands are artificially maintained.

Figure 16.30a, b, c shows satellite false-color images of a portion of western Brazil called Rondônia, recorded in June 1975 (*Landsat-2*), August 1986 (*Landsat-5*), and June 1992 (*Landsat-4*), and in (d) a true-color June 2001 image (*Terra*, MODIS). These images give a sense of the rain forest destruction in progress. You can clearly see encroachment along new roads branching from highway BR364. The *edges* of every road and cleared area represent a significant portion of the species habitat disturbance, population dynamic changes, and carbon losses to the atmosphere—significant impact occurs along these edges next to tracts of clear-cut logging. The hotter, drier, and windier edge conditions penetrate the forest up to 100 m. Figure 16.30 shows a tract of former rain forest that has just been burned to begin the clearing and road-building process.

Between 1972 and 1999 the losses totaled more than the area of the north-central states of Minnesota, Iowa, Missouri, North and South Dakota, Nebraska, and Kansas, noted on the map in Figure 16.30f. The loss in 2000 was 19,836 km² (7660 mi²), an area roughly equivalent to Connecticut, Rhode Island, and Delaware combined. Estimates based on satellite imagery and ground surveys place the losses for the 3 years 2001–2003 at a staggering 75,545 km² (29,170 mi²)—or an area equal to slightly less than the states of Vermont, New Hampshire, and Massachusetts combined. The losses in 2004–2005 are about half the area of Maine. Estimates for 2006–2008, some 40,532 km² (15,650 mi²), are equivalent to the rest of Maine in area. In 1970 only about 1% of the Brazilian Amazon had been deforested, yet here we see forest area more than that of France lost since 1972, more than 20% of the original forest.

Brazil is a major beef exporter and consumer of its own rain forest resource base. This is a highly charged issue, as can be seen by the growth of Brazil's cattle industry that uses these new pasture lands. Cattle herds will have increased from 25 million head (1990) to a forecasted 200 million head by 2010. Added to deforestation, the Brazilian rain forest is experiencing several years of severe drought, the worst in more than 100 years.

The United Nations Food and Agricultural Organization (FAO) estimates that if this destruction to rain forests continues unabated, they will be completely removed by about A.D. 2050. By continent, forest losses are estimated at more than 50% in Africa, over 40% in Asia, and 40% in Central and South America. (Among many references, see Tropical Rainforest Coalition at **http://www.rainforest. org/**, World Resources Institute publications on forests at **http://www.globalforestwatch.org/english/index.htm**, the Rainforest Information Centre at **http://www. rainforestinfo.org.au/welcome.htm**, and the Rainforest Action Network at **http://www.ran.org/**.)

Focus Study 16.1 looks more closely at efforts to curb increasing rates of species extinction and loss of Earth's biodiversity, much of which is directly attributable to the loss of rain forests.

Tropical Seasonal Forest and Scrub

A varied biome on the margins of the rain forest is the **tropical seasonal forest and scrub**, which occupies regions of low and erratic rainfall. The shifting intertropical convergence zone (ITCZ) brings precipitation with the seasonally shifting high Sun and dryness with the low Sun, producing a seasonal pattern of moisture deficits, some leaf loss, and dry-season flowering. The term *semideciduous* applies to some of the broadleaf trees that lose their leaves during the dry season.

Areas of this biome have fewer than 40 rainy days during their 4 consecutive driest months, yet heavy monsoon downpours characterize their summers (see Figure 4.20). The climates *tropical monsoon* and *tropical savanna* apply to these transitional communities between rain forests and tropical grasslands.

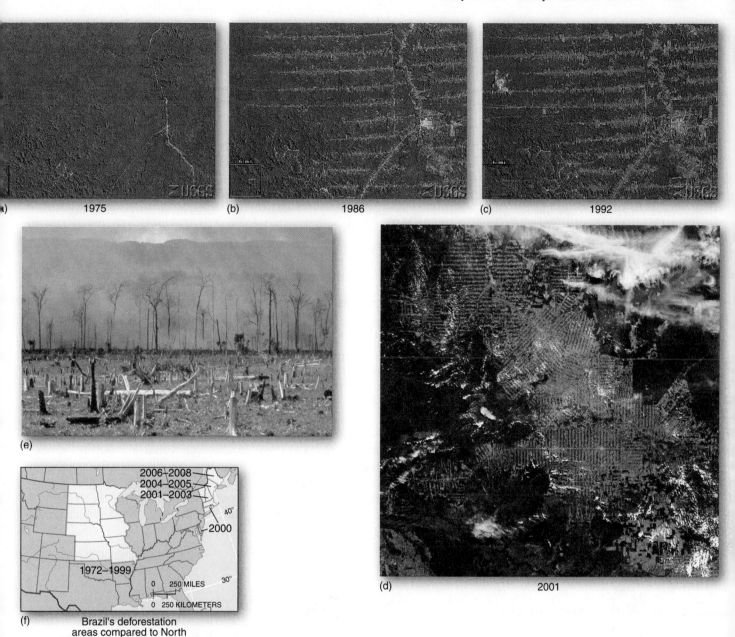

FIGURE 16.30 Amazon rain forest losses over 36 years.
Satellite images of a portion of western Brazil called Rondônia, recorded in false color in (a) June 1975 (*Landsat 2*), (b) August 1986 (*Landsat 5*), (c) June 1992 (*Landsat 4*), and in true color in (d) June 1, 2001 (the area of the other three images is in the upper left). Highway BR364, the main artery in the region, passes through the towns of Jaru and Ji-Paraná near the center of image (d) at about 11° S 62° W. The branching pattern of feeder roads encroaches on the rain forest. (e) Surface view of burning rain forest. (f) Deforested areas in Brazil, cleared between the years 1972 and 2000, 2001–2003, the record-setting 2004–2005, and 2006–2008, are compared to North America. [*Landsat* images courtesy of NASA/USGS EROS Data Center; *Terra* image courtesy of NASA/MODIS Land Rapid Response Team; (e) photo by George Holton/Photo Researchers, Inc.]

Portraying such a varied biome is difficult. In many areas, humans disturb the natural biome so that the savanna grassland adjoins the rain forest directly. The biome does include a gradation from wetter to drier areas: monsoonal forests, to open woodlands and scrub woodland, to thorn forests, to drought-resistant scrub species (Figure 16.31). An area of transitional tropical deciduous forest surrounds an area of savanna in southeastern Brazil and portions of Paraguay.

The monsoonal forests average 15 m (50 ft) high with no continuous canopy of leaves, graduating into open orchardlike parkland with grassy openings or into areas choked by dense undergrowth. In more open tracts, a common tree is the acacia, with its flat top and usually

(*text continued on page 564*)

FOCUS STUDY 16.1

Biodiversity and Biosphere Reserves

When the first European settlers landed on the Hawaiian Islands in the late 1700s, 43 species of birds were counted. Today, 15 of those species are extinct, and 19 more are threatened or endangered, with only one-fifth of the original species populations relatively healthy. In most of Hawai'i there are no native species below 1220-m (4000-ft) elevation because of an introduced avian virus. Humans have great impact on animals, plants, and global biodiversity. We are facing a loss of genetic diversity that may be unparalleled in Earth's history, even compared with the major extinctions over the geologic record.

As more is learned about Earth's ecosystems and their related communities, more is known of their value to civilization and our interdependence with them. Natural ecosystems are a major source of new foods, new chemicals, new medicines, and specialty woods, and, of course, they are indicators of a healthy, functioning biosphere.

International efforts are under way to study and preserve specific segments of the biosphere: among others, the UN Environment Programme, World Resources Institute, The World Conservation Union, Natural World Heritage Sites, Wetlands of International Importance, and the Nature Conservancy. An important part of the effort is the setting aside of biosphere reserves and the focus provided by the World Conservation Monitoring Centre and its IUCN Red List of global endangered species. (See IUCN Red List at http://www.iucnredlist.org/, the IUCN at http://www.iucn.org/, the endangered species home page of the Fish and Wildlife Service at http://www.fws.gov/endangered/, and the World Wildlife Fund at http://www.panda.org/.)

Species Threatened—Examples

The International Union for the Conservation of Nature (IUCN) in 2006 listed the polar bear (*Ursus maritimus*) as *vulnerable* to extinction, a "threatened" category on their Endangered Species List—the *Red Book*. In spring 2007, during the 90-day comment period concerning the bears' status, the Fish and Wildlife Service received more than 600,000 letters concerning the inclusion of polar bears under the US Endangered Species Act. Research released by the USGS in September 2007 predicted that with the loss of Alaskan, Canadian, and Russian sea-ice habitat, some two-thirds of the world's polar bears will die off by 2050. The remaining 7500 bears will be struggling (Figure 16.1.1).

Following an April 2008 court-ordered deadline, the Bush administration listed the polar bear as "threatened" on the Endangered Species List on May 14, 2008, although an accompanying "signing statement" reserved the right to explore and develop oil, gas, and mineral resources in critical habitats. On August 4, the state of Alaska filed a lawsuit against the US Fish and Wildlife Service to remove the listing and cease references to global warming. Meanwhile, 2007 broke the record for the lowest extent of sea ice, and enough ice melted in 2008 to place it second in the record. The polar bears, seals, walruses, beluga whales, and other species of the Arctic face increasing difficulty in obtaining food and rearing their young.

Black rhinos (*Diceros bicornis*) and white rhinos (*Ceratotherium simum*) in Africa exemplify species in jeopardy. Rhinos once grazed over much of the savanna grasslands and woodlands. Today, they survive only in protected districts in heavily guarded sanctuaries and are threatened even there (Figure 16.1.2). Rhinoceros horn sells for US $29,000 per kilogram as an aphrodisiac (a false use). These large land mammals are nearing extinction and will survive only as a dwindling zoo population. The limited genetic pool that remains complicates further reproduction. In 1998, 2599 black rhinos remained, 96% fewer than the

FIGURE 16.1.1 The polar bear.
A male polar bear, scar on his snout, seems to be looking directly at us. One study predicts that 67% of the polar bear population of 23,000 will be gone by 2050 because of climate change and loss of sea ice—that's 7590 bears remaining across the Arctic region. [Photo by Bobbé Christopherson.]

NOTEBOOK

Polar Bear Photo Gallery
Polar Bear Movie - Mom and Cubs on the Pack Ice

70,000 in 1960. Populations have remained statistically stable through the last decade, and this count is maintained by the guarding of about 50% of the herd by South Africa. A slow recovery is underway as the number of black rhinos rose to 3600 in 2003 (*source*: IUCN/SSC/ African Rhino Specialist Group). This is not the case with the western black rhino, a subspecies, which is down to its last few individuals.

There were 11 white rhinos (northern subspecies) in existence in 1984, increasing to an estimated 29 by 1998 (Congo 25, Côte d'Ivoire 4), although the IUNC suggests this number might actually have fallen to only 5 to 10 rhinos left in 2005. Political unrest in Congo and the area of the Garamba National Park has led to the deaths of an unknown number of these few. The southern white rhino population reached 11,000 by 2002.

Table 16.1.1 summarizes the numbers of known and estimated species on Earth. Scientists have classified only 1.75 million species of plants and animals of an estimated 13.6 million overall; this figure represents an increase in what scientists once thought to be the diversity of life on Earth. And those yet-to-be-discovered species represent a potential future resource for society. The wide range of estimates places the expected species count between a low of 3.6 million and a high of 111.7 million. Estimates of annual species loss range between 1000 and 30,000, although this range might be conservative. The possibility exists that over half of Earth's present species could be extinct within the next 100 years.

The Question of Medicine and Food

Wheat, maize (corn), and rice—just three grains—fulfill about 50% of our planetary food demands. About 7000 plant species have been gathered for food throughout human history, but over 30,000 plant species have edible parts. Undiscovered potential food

FIGURE 16.1.2 The rhinoceros in Africa.
Black rhino and young in Tanzania, escorted by Oxpecker birds. These rhinos are quite nearsighted (they can see clearly only up to 10 m); the birds act as the rhino's early-warning system for disturbances in the distance. [Photo by Stephen J. Krasemann/ DRK Photo.]

resources are in nature waiting to be found and developed. Biodiversity, if preserved, provides us a potential cushion for all future food needs, but only if species are identified, inventoried, and protected. The same is true for potential pharmaceuticals.

Nature's biodiversity is like a full medicine cabinet. Since 1959, 25% of all prescription drugs were originally derived from plants. In preliminary surveys, 3000 plants are identified as having anticancer properties. The rosy periwinkle (*Catharanthus roseus*)

Table 16.1.1   Known and Estimated Species on Earth

Categories of Living Organisms	Number of Known Species	Estimated Number of Species		Working Estimate (000)	Accuracy
		High (000)	Low (000)		
Viruses	4,000	1,000	50	400	Very poor
Bacteria	4,000	3,000	50	1000	Very poor
Fungi	72,000	27,000	200	1500	Moderate
Protozoa	40,000	200	60	200	Very poor
Algae	40,000	1,000	150	400	Very poor
Plants	270,000	500	300	320	Good
Nematodes	25,000	1,000	100	400	Poor
Arthropods					
Crustaceans	40,000	200	75	150	Moderate
Arachnids	75,000	1,000	300	750	Moderate
Insects	950,000	100,000	2,000	8,000	Moderate
Mollusks	70,000	200	100	200	Moderate
Chordates	45,000	55	50	50	Good
Others	115,000	800	200	250	Moderate
Total	**1,750,000**	**111,655**	**3,635**	**13,620**	**Very poor**

Source: United Nations Environment Programme, *Global Biodiversity Assessment* (Cambridge: Cambridge University Press, 1995), Table 3.1–2, p. 118, used by permission.

(continued)

Focus Study 16.1 *(continued)*

of Madagascar contains two alkaloids that combat two forms of cancer. (Alkaloids are compounds found in certain plants that help the plant defend against insects and are potentially significant as human medicines; examples are atropine, quinine, and morphine.) Yet less than 3% of flowering plants have been examined for alkaloid content. It defies common sense to throw away the medicine cabinet before we even open the door to see what is inside.

Biosphere Reserves

Formal natural reserves are a possible strategy for slowing the loss of biodiversity and protecting this resource base. Setting up such a biosphere reserve involves principles of *island biogeography*. Island communities are special places for study because of their spatial isolation and the relatively small number of species present. Islands resemble natural experiments, because the impact of individual factors, such as civilization, can be more easily assessed on islands than over larger continental areas.

Studies of islands also can assist in the study of mainland ecosystems, for in many ways a park or biosphere reserve, surrounded by modified areas and artificial boundaries, is like an island. Indeed, a biosphere reserve is conceived as an ecological island in the midst of change. The intent is to establish a core in which genetic

material is protected from outside disturbances, surrounded by a buffer zone that is, in turn, surrounded by a transition zone and experimental research areas. An important variable to consider in setting aside a biosphere reserve is any change that might occur in temperature and precipitation patterns as a result of global change. A carefully considered reserve could end up outside its natural range as ecotones (climatic/ecosystem boundaries) shift.

Scientists predict that new, undisturbed reserves will not be possible in a decade, because pristine areas will be gone. The biosphere reserve program is coordinated by the Man and the Biosphere (MAB) Programme of UNESCO (http://www.unesco.org/mab/). More than 480 biosphere reserves are now operated voluntarily in 100 countries.

Not all protected areas are ideal bioregional entities. Some are simply imposed on existing park space and some remain in the planning stage, although they are officially designated. Some of the best biosphere reserve examples have been in operation since the late 1970s and range from the struggling Everglades National Park in Florida, to the Changbai Reserve in China, to the Tai Forest on the Côte d'Ivoire (Ivory Coast), to the Charlevoix Biosphere Reserve in Québec, Canada. Added to these efforts is the work of the Nature Conservancy, which

acquires land for preservation as a valuable part of the reserve (http://nature.org/).

The ultimate goal, about half achieved, is to establish at least one reserve in each of the 194 distinctive biogeographic communities presently identified. UNESCO and the World Conservation Monitoring Centre compile the United Nations' list of national parks and protected areas. Presently in these biogeographic communities, 6930 areas covering 657 million hectares (1.62 billion acres) are under some protective form.

The preservation of species diversity is a problem that must today be confronted by one species, *Homo sapiens*.... The diversity of species is worth preserving because it represents a wealth of knowledge that cannot be replaced. Moreover, today's extinctions are unlike those in previous eras, in which long periods of recovery could follow extinctions. The present situation is an inexorably irreversible...that more energy be devoted to planning and priorities and less to emotionalism and indignation [on all sides].*

*D. E. Koshland, Jr., "Preserving biodiversity," *Science* 253 (August 16, 1991): 717.

---

thorny stems. These trees have branches that look like an upside-down umbrella (spreading and open skyward), as do trees in the tropical savanna.

Local names are given to these communities: the *caatinga* of the Bahia State of northeastern Brazil, the *chaco* area of Paraguay and northern Argentina, the *brigalow* scrub of Australia, and the *dornveld* of southern Africa. The world map in Figure 16.27 shows this biome in Africa, extending from eastern Angola through Zambia to Tanzania; in southeast Asia and portions of India, from interior Myanmar through northeastern Thailand; and in parts of Indonesia.

The trees throughout most of this biome make poor lumber, but some, especially teak, may be valuable for fine cabinetry. In addition, some of the plants with dry-season

adaptations produce usable waxes and gums, such as carnauba and palm-hard waxes. Animal life includes the koalas and cockatoos of Australia and the elephants, large cats, rodents, and ground-dwelling birds in other occurrences of this biome.

Tropical Savanna

Large expanses of grassland interrupted by trees and shrubs describe the **tropical savanna**. This is a transitional biome between the tropical forests and semiarid tropical steppes and deserts. The savanna biome also includes treeless tracts of grasslands. The trees of the savanna woodlands are characteristically flat-topped.

(a)

(b)

FIGURE 16.31 Kenyan landscapes.
Two views of the open thorn forest and savanna near and in
the Samburu Reserve, Kenya, representing a broad ecotone
between the two biomes. [(a) Photo by Gael Summer; (b) photo
by Stephen J. Krasemann/DRK Photo.]

Savannas covered more than 40% of Earth's land surface before human intervention but were especially modified by human-caused fire. Fires occur annually
throughout the biome. The timing of these fires is important. Early in the dry season they are beneficial and
increase tree cover; if late in the season, they are very hot
and kill trees and seeds. Savanna trees are adapted to resist
the "cooler" fires.

Elephant grasses averaging 5 m (16 ft) high and forests
once penetrated much farther into the dry regions, for they
are known to survive there when protected. Savanna grasslands are much richer in humus than the wetter tropics and
are better drained, thereby providing a better base for agriculture and grazing. Sorghums, wheat, and groundnuts
(peanuts) are common commodities.

Tropical savannas receive their precipitation during
less than 6 months of the year, when they are influenced
by the ITCZ. The rest of the year they are under the drier
influence of shifting, subtropical, high-pressure cells.

Savanna shrubs and trees are frequently *xerophytic*, or
drought resistant, with various adaptations to protect
them from the dryness: small, thick leaves, rough bark, or
waxy leaf surfaces.

Africa has the largest region of this biome, including
the famous Serengeti Plains and the Sahel region. Sections of Australia, India, and South America also are part
of the savanna biome. Some of the local names for these
lands include the *Llanos* in Venezuela, stretching along the
coast and inland east of Lake Maricaibo and the Andes;
the *Campo Cerrado* of Brazil and Guiana; and the *Pantanal*
of southwestern Brazil.

Particularly in Africa, savannas are the home of large
land mammals (zebra, giraffe, buffalo, gazelle, wildebeest,
antelope, rhinoceros, elephant). They graze on savanna
grasses (Figure 16.32) or feed upon the grazers themselves

(a)

(b)

FIGURE 16.32 Animals and plants of the Serengeti Plains.
(a) Savanna landscape of the Serengeti Plains, with wildebeest,
zebras, and thorn forest. (b) Near Samburu, Kenya, Grevy's
zebra and reticulated giraffe forage. [Photos by (a) Stephen F.
Cunha; (b) Galen Rowell/Mountain Light Photography, Inc.]

(lion, cheetah). Birds include the Ostrich, Martial Eagle (largest of all eagles), and Secretary Bird. Many species of venomous snakes as well as the crocodile are present in this biome.

However, in our lifetime we may see the reduction of these animal herds to zoo stock only because of poaching and habitat losses. The loss of the rhino is discussed in Focus Study 16.1. Establishment of large tracts of savanna as biosphere reserves is critical for the preservation of this biome and its associated fauna.

Midlatitude Broadleaf and Mixed Forest

Moist continental climates support mixed broadleaf and needleleaf forest in areas of warm-to-hot summers and cool-to-cold winters. The **midlatitude broadleaf and mixed forest** biome includes several distinct communities in North America, Europe, and Asia. Relatively lush evergreen broadleaf forests occur along the Gulf of Mexico. Northward are mixed deciduous and evergreen needleleaf stands associated with sandy soils and burning. When areas are given fire protection, broadleaf trees quickly take over.

Pines predominate in the southeastern and Atlantic coastal plains. Into New England and westward in a narrow belt to the Great Lakes, white and red pines and eastern hemlock are the principal evergreens, mixed with deciduous varieties of oak, beech, hickory, maple, elm, chestnut, and many others.

These mixed stands contain valuable timber, but their distribution has been greatly altered by human activity. Native stands of white pine in Michigan and Minnesota were removed before 1910 (Figure 16.33a); only later reforestation sustains their presence today. In northern China, these forests have almost disappeared as a result of centuries of occupation. Deforestation in China, as elsewhere, was principally for agricultural purposes, as well as for construction materials and fuel.

The midlatitude broadleaf and mixed-forest biome is quite consistent in appearance from continent to continent; at one time it represented the principal vegetation of the *humid subtropical hot-summer* climates *marine west coast*, and *cool-summer, winter-drought* climatic regions of North America, Europe, and Asia. To the north of this biome, poorer soils and colder climates favor stands of coniferous trees and a gradual transition to the needleleaf forests.

A wide assortment of mammals, birds, reptiles, and amphibians is distributed throughout this biome. Representative animals (some migratory) include red fox, southern flying squirrel, opossum, bear, and a great variety of birds, including Tanager and Cardinal. A white-tailed deer is characteristic of the mixed forest (Figure 16.33b).

As with all terrestrial biomes, climate change is causing ecotone boundaries to shift. The Hardiness Zone planting maps presented earlier in Figure 16.20 illustrated the changes over just 16 years. The climate forecast models predict these changes. For example, Figure 16.34 illustrates the "migration" of Illinois this century based on changes in temperature and precipitation, summer

(a)

(b)

FIGURE 16.33 Mixed forest community.
(a) Pioneer Mothers' Memorial Forest near Paoli, Indiana, is 36 hectares (88 acres) of old-growth forest, virtually undisturbed since around 1816, featuring black walnut, white oak, yellow poplar, white ash, and beech trees, among others. (b) A young white-tailed deer grazes in the undergrowth of a mixed forest. [Photos by (a) Bobbé Christopherson; (b) John Shaw/Tom Stack & Associates.]

conditions in particular. By 2030, Illinois' summers will be similar to the Missouri–Arkansas border; by 2090, Illinois' climate will be similar to that of east Texas. In turn, east Texas will be like north-central Mexico.

Simply stated, a physical geography text written 20 to 70 years from now will have to redo the biome maps to reflect these climatic shifts. Imagine the change underway for farm crops, pollinating insects, bird populations, cities and energy-use patterns, and water resources.

Needleleaf Forest and Montane Forest

Stretching from the east coast of Canada and the Atlantic provinces westward to Alaska and continuing from Siberia across the entire extent of Russia to the European Plain is the **needleleaf forest** biome, also called the **boreal forest**

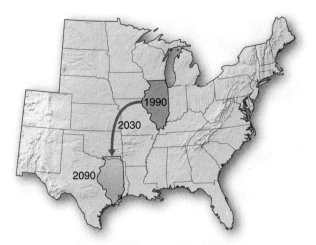

FIGURE 16.34 Climatic shifts impact the Midwest.
The relative future position of Illinois in response to climatic shifts in temperature and precipitation by 2090. [After Canadian General Circulation Model (CGCM2) and *Climate Change Impacts on the United States*, US Global Change Research Program, 2000.]

(Figures 16.35 and 16.36). A more open form of boreal forest, transitional to arctic and subarctic regions, is the **taiga**. The Southern Hemisphere, lacking *humid microthermal* (at least one month averaging below freezing) climates except in mountainous locales, has no such

biome. However, *montane forests* of needleleaf trees exist worldwide at high elevation.

Boreal forests of pine, spruce, and fir occupy most of the subarctic climates on Earth that are dominated by trees. Although these forests are similar in formation, individual species vary between North America and Eurasia. Certain regions of the northern needleleaf biome experience permafrost (discussed in Chapter 14). When coupled with rocky and poorly developed soils, these conditions generally limit the existence of trees to those with shallow root systems.

The Sierra Nevada, Rocky Mountains, Alps, and Himalayas have similar forest communities occurring at lower latitudes. Douglas and white fir grow in the western mountains in the United States and Canada. Economically, these forests are important for lumbering, with saw timber occurring in the southern margins of the biome and pulpwood throughout the middle and northern portions. Present logging practices and whether these yields are sustainable are issues of increasing controversy.

Representative fauna in this biome include wolf, elk, moose (the largest in the deer family), bear, lynx, beaver, wolverine, marten, small rodents, and migratory birds during the brief summer season (Figure 16.36). Birds include hawks and eagles, several species of grouse, Pine Grosbeak, Clark's Nutcracker, and several owls. About 50 species of insects particularly adapted to the presence of coniferous trees inhabit the biome.

Temperate Rain Forest

The **temperate rain forest** biome is identified by its lush forests at middle and high latitudes and occurs only along narrow margins of the Pacific Northwest in North America (Figure 16.37). This biome contrasts with the diversity of the equatorial and tropical rain forest in that only a few species make up the bulk of the trees. The rain forest of the Olympic Peninsula in Washington state is a mixture of

FIGURE 16.35 Boreal needleleaf forest of Canada.
Boreal means "northern." [Photo by author.]

FIGURE 16.36 Animals of the northern needleleaf forest.
A male elk (*Cervus Canadensis*) in a snowstorm looking for any available vegetation to eat among needleleaf trees during winter. [Photo by Steven K. Huhtala.]

(b)

(a)

FIGURE 16.37 Temperate rain forest.
(a) The temperate rain forest of Macmillan Provincial Park, central Vancouver Island, British Columbia—an old-growth Douglas fir forest, with western red cedar and hemlock, sword and bracken ferns, and hanging mosses. (b)The Northern Spotted Owl is characteristic of the temperate rain forest. [Photos by (a) Bobbé Christopherson; (b) Greg Vaughn/Tony Stone Images.]

broadleaf and needleleaf trees, huge ferns, and thick undergrowth. Precipitation approaching 400 cm (160 in.) per year on the western slopes, moderate air temperatures, summer fog, and an overall maritime influence produce this moist, lush vegetation community. Animals include bear, badger, deer, wild pig, wolf, bobcat, and fox. The trees are home to numerous bird species (Figure 16.37b).

The tallest trees in the world occur in this biome—the coastal redwoods (*Sequoia sempervirens*). Their distribution is shown on the map in Figure 16.11a. These trees can exceed 1500 years in age and typically range in height from 60 to 90 m (200–300 ft), with some exceeding 100 m (330 ft). Virgin stands of other representative trees—Douglas fir, spruce, cedar, and hemlock—have been reduced to a few remaining valleys in Oregon and Washington, less than 10% of the original forest.

The failing ecology of these forest ecosystems requires a change in policies, one of economic-ecological synthesis rather than continuing conflict and forest losses—the forests can't all be cut down, nor can they all be preserved.

Mediterranean Shrubland

The **Mediterranean shrubland** biome occupies those regions poleward of the shifting subtropical high-pressure cells. As those cells shift poleward with the high Sun, they cut off available storm systems and moisture. Their stable high-pressure presence produces the characteristic dry summer climate—*Mediterranean dry-summer*—and establishes conditions conducive to fire. Plant ecologists think that this biome is well adapted to frequent fires, for many of its characteristically deep-rooted plants have the ability to resprout from their roots after a fire (review Figure 16.24b).

The dominant shrub formations that occupy these regions are stunted and tough in their ability to withstand hot-summer drought. The vegetation is *sclerophyllous* (from

sclero or "hard" and *phyllos* for "leaf"); it averages a meter or two in height and has deep, well-developed roots, leathery leaves, and uneven low branches.

Typically, the vegetation varies between woody shrubs covering more than 50% of the ground and grassy woodlands with 25%–60% coverage. In California, the Spanish word *chaparro* for "scrubby evergreen" gives us the word **chaparral** for this vegetation type (Figure 16.38). A counterpart to chaparral in the Mediterranean region, *maquis*, includes live and cork oak trees (source of cork), as well as pine and olive trees. In Chile, such a region is the *mattoral*; in southwest Australia, *mallee scrub*. Of course, in Australia the bulk of the eucalyptus species are sclerophyllous in form and structure in whichever climate it occurs.

As described in Chapter 7, Mediterranean climates are important in commercial agriculture for subtropical fruits, vegetables, and nuts, with many food types produced only in this biome (for example, artichokes, olives, almonds). Larger animals, such as different types of deer, are grazers and browsers, with coyote, wolf, and bobcat as predators. Many rodents, other small animals, and a variety of birds also proliferate.

Midlatitude Grasslands

Of all the natural biomes, the **midlatitude grasslands** are most modified by human activity. Here are the world's "breadbaskets"—regions of grain and livestock production (Figure 16.39). In these regions, the only naturally occurring trees were deciduous broadleafs along streams and other limited sites. These regions are called grasslands because of the original predominance of grasslike plants:

In this study of vegetation, attention has been devoted to grass because grass is the dominant feature of the Plains and is at the same time an index

FIGURE 16.38 Mediterranean chaparral.
Mixed chaparral vegetation associated with the Mediterranean dry-summer climate in southern California. [Photo by Bobbé Christopherson.]

to their history. Grass is the visible feature which distinguishes the Plains from the desert. Grass grows, has its natural habitat, in the transition area between timber and desert. . . . The history of the Plains is the history of the grasslands.*

In North America, tall-grass prairies once rose to heights of 2 m (6.5 ft) and extended westward to about the 98th meridian, with short-grass prairies farther west. The 98th meridian is roughly the location of the 51-cm (20-in.) isohyet, with wetter conditions to the east and drier to the west (see Figure 15.16). The deep sod of those grasslands posed problems for the first settlers, as did the climate. The self-scouring steel plow, introduced in 1837 by John Deere, allowed the interlaced grass sod to be bro-

*From *The Great Plains*, ©1931 by Walter Prescott Webb, published by Ginn and Company, Needham Heights, MA., p. 32.

ken apart, freeing the soils for agriculture. Other inventions were critical to opening this region and solving its unique spatial problems: barbed wire (the fencing material for a treeless prairie), well-drilling techniques developed by Pennsylvania oil drillers but used for water wells, windmills for pumping, and railroads to conquer the distances.

Few patches of the original prairies (tall grasslands) or steppes (short grasslands) remain within this biome. For prairies alone, the reduction of natural vegetation went from 100 million hectares (247 million acres) down to a few areas of several hundred hectares each. Read the story of the bison in the caption to the PART IV opening photo. The Tallgrass Prairie National Preserve is a 4410-hectare (10,894-acre) section of prairie located in the Flint Hills of Kansas. This is the only national park unit preserving this vestige of the former grasslands that covered much of the North American Great Plains. A state-designated 10-hectare (25-acre) remnant of the "prairie" is located 8 km

FIGURE 16.39 Farming in the grasslands of North America.
Cattle, wheat, corn, sorghum grain, and sunflowers are produced in Kansas. Grain farms are placed next to massive feedlots near Dodge City, Kansas in this aerial view. This feedlot and slaughterhouse processes 6000 cattle a day, 6 days a week. Overall, Kansas processing facilities process 8 million head a year. [Photo by Bobbé Christopherson.]

north of Ames, Iowa; the Hayden Prairie Preserve, also in Iowa, and the Konza Prairie research reserve station in Kansas are discussed in News Report 16.1. Some of these patches of original grassland have never felt the plow.

Outside North America, the Pampas of Argentina and Uruguay and the grasslands of Ukraine are characteristic midlatitude grassland biomes. In each region of the world where these grasslands have occurred, human development of them has been critical to territorial expansion.

This biome is the home of large grazing animals, including deer, antelope, pronghorn sheep, and bison (the almost complete annihilation of the latter is part of American history). Grasshoppers and other insects feed on the grasses and crops as well, and gophers, prairie dogs, ground squirrels, Turkey Vultures, grouse, and prairie chickens are on the land. Predators include coyote, the nearly extinct black-footed ferret, badger, and birds of prey, including hawks, eagles, and owls.

Deserts

Earth's **desert biomes** cover more than one-third of its land area. In Chapter 7, we examined desert climates, and in Chapter 12, desert landscapes. On a planet with such a rich biosphere, the deserts stand out as unique regions of fascinating adaptations for survival. Much as a group of humans in the desert might behave with short supplies, plant communities also compete for water and site advantage. Some desert plants, called *ephemerals*, wait years for a rainfall event, at which time their seeds quickly germinate and the plants develop, flower, and produce new seeds, which then rest again until the next rainfall event. The seeds of some xerophytic species open only when fractured by the tumbling, churning action of flash floods cascading down a desert arroyo, and such an event produces the moisture that a germinating seed needs.

Perennial desert plants employ other adaptive features to cope with the desert, such as long, deep tap roots (for example, the mesquite); succulence (thick, fleshy, water-holding tissue such as that of cacti); spreading root systems to maximize water capture; waxy coatings and fine hairs on leaves to retard water loss; leafless conditions during dry periods (for example, palo verde and ocotillo); reflective surfaces to reduce leaf temperatures; and tissue that tastes bad to discourage herbivores.

The faunas of both warm and cold deserts are limited by the extreme conditions and include few resident large animals. Exceptions are the camel, which can lose up to 30% of its body weight in water without suffering (for humans, a 10%–12% loss is dangerous), and the desert bighorn sheep in scattered populations in inaccessible mountains and in places such as the Grand Canyon, along rocky outcrops and cliffs but not at the rim. In an effort to reestablish bighorn populations, several states are doing transplant operations to bring the animals to their former ranges.

Some representative desert animals are the ringtail cat, kangaroo rat, lizards, scorpions, and snakes. Most of these animals are quite secretive and become active only

at night, when temperatures are lower. In addition, various birds have adapted to desert plants and other available food sources—for example, Roadrunners, Thrashers, Ravens, Wrens, Hawks, Grouse, and Nighthawks.

We are witnessing an unwanted expansion of the desert biome (discussed in Chapter 12 and Figure 12.24). This process, known as *desertification*, is now a worldwide phenomenon along the margins of semiarid and arid lands. Desertification is due principally to poor agricultural practices (overgrazing and agricultural activities that abuse soil structure and fertility), improper soil-moisture management, erosion and salinization, deforestation, and the ongoing climatic change.

Earth's deserts are subdivided into desert and semidesert associations, to distinguish those with expanses of bare ground from those covered by xerophytic plants of various types. The two broad associations are further separated into warm deserts, principally tropical and subtropical, and cold deserts, principally midlatitude.

Warm Desert and Semidesert Earth's **warm desert and semidesert** biomes are caused by the presence of dry air and low precipitation from subtropical high-pressure cells. These areas are very dry, as evidenced by the Atacama Desert of northern Chile, where only a minute amount of rain has ever been recorded—a 30-year annual average of only 0.05 cm (0.02 in.)! As in the Atacama, the true deserts of Earth are under the influence of the descending, drying, and stable air of high-pressure systems from 8 to 12 months of the year. A sampling of deserts and a worldwide distribution map is in Figure 12.15.

A few of the subtropical deserts—such as those in Chile, Western Sahara, and Namibia—are right on the seacoast and are influenced by cool offshore ocean currents. As a result, these true deserts experience summer fog that mists the plant and animal populations with needed moisture.

Vegetation ranges from almost none in the arid deserts to numerous xerophytic shrubs, succulents, and thorn tree forms. The lower Sonoran Desert of southern Arizona is a warm desert (Figure 16.40). This desert landscape features the unique saguaro cactus (*Carnegiea gigantea*), which grows to many meters in height and up to 200 years in age if undisturbed. First blooms do not appear until it is 50 to 75 years old.

The equatorward margin of the subtropical high-pressure cell is a region of transition to savanna, thorn tree, scrub woodland, and tropical seasonal forest. Poleward of the warm deserts, the subtropical cells shift to produce the Mediterranean dry-summer regime along west coasts and may grade into cool deserts elsewhere.

Cold Desert and Semidesert The **cold desert and semidesert** biomes tend to occur at higher latitudes, where seasonal shifting of the subtropical high is of some influence less than 6 months of the year. Specifically, interior locations are dry because of their distance from moisture sources or their location in rain-shadow areas on the lee side of mountain ranges such as the Sierra

FIGURE 16.40 Sonoran desert scene.
Lower Sonoran desert west of Tucson, Arizona (32.3° N latitude; elevation 900 m, or 2950 ft). Note adaptive features of the plants pictured: tall, fleshy, saguaro cactus; thorny, sparsely leafed ocotillo (right); barrel and prickly pear cactus, among others. [Photo by author.]

Nevada of the western United States, the Himalayas in Asia, and the Andes in South America. The combination of interior location and rain-shadow positioning produces the cold deserts of the Great Basin of western North America.

Winter snows occur in the cold deserts, but are generally light. Summers are hot, with highs from 30°C to 40°C (86°F to 104°F). Nighttime lows, even in the summer, can cool 10 C° to 20 C° (18 F° to 36 F°) from the daytime high. The dryness, generally clear skies, and sparse vegetation lead to high radiative heat loss and cool evenings. Many areas of these cold deserts that are covered by sagebrush and scrub vegetation were actually dry short-grass regions in the past, before extensive grazing altered their appearance. The look of the deserts in the upper Great Basin is the result of more than a century of such cultural practices (see Figure 7.24b).

Arctic and Alpine Tundra

The **arctic tundra** is found in the extreme northern area of North America and Russia, bordering on the Arctic Ocean and generally north of the 10°C (50°F) isotherm for the warmest month. Daylength varies greatly throughout the year, seasonally changing from almost continuous day to continuous night. The region, except for a few portions of Alaska and Siberia, was covered by ice during all of the Pleistocene glaciations. In terms of climate change, these regions have been warming at more than twice the rate of the rest of the planet over the past few decades; see related discussions and photos in Chapters 7 (Figure 7.20) and 14 (Figures 14.17–14.22). The impact of such heating on plants and animals, permafrost, First Nation peoples and their towns, glacial and sea ice, and decreasing ocean salinity are topics of active scientific inquiry and the International Polar Year for science and research (ending March 2009).

Winters in this biome, a *tundra* climate classification, are long and cold; cool summers are brief. Intensely cold continental polar and arctic air masses and stable highpressure anticyclones dominate winter. A growing season of sorts lasts only 60–80 days, and even then frosts can occur at any time. Vegetation is fragile in this flat, treeless world; soils are poorly developed periglacial surfaces that are underlain by permafrost. In the summer months only the surface horizons thaw, thus producing a mucky surface of poor drainage. Roots can penetrate only to the depth of thawed ground, usually about a meter (3 ft). In these periglacial regions the surface is shaped by freeze-thaw cycles that create the frozen ground phenomena discussed in Chapter 14.

Tundra vegetation is low, ground-level herbaceous plants such as sedges, mosses, arctic meadow grass, and snow lichen, and some woody species such as dwarf willow (Figure 16.41a, and b, and see Figure 16.2c). Owing to the short growing season, some perennials form flower buds one summer and open them for pollination the next. The tundra supports musk ox, caribou, reindeer, rabbit, ptarmigans, lemmings, and other small rodents (important food for the larger carnivores), wolf, fox, weasel, Snowy Owl, polar bears, and, of course, mosquitoes. The tundra is an important breeding ground for geese, swans, and other waterfowl.

Alpine tundra is similar to arctic tundra but can occur at lower latitudes because it is associated with high elevation. This biome usually is described as above the timberline—that elevation above which trees cannot grow. Timberlines increase in elevation equatorward in both hemispheres. Alpine tundra communities occur in the Andes near the equator, the White Mountains of California, the Rockies, the Alps, and Mount Kilimanjaro of equatorial Africa, as well as mountains from the Middle East to Asia. Alpine meadows feature grasses and stunted shrubs, such as willows and heaths (Figure 16.41c). Characteristic fauna include mountain goats, bighorn sheep, elk, and voles.

 High Latitude Animals Photo Gallery
NOTEBOOK

 Geographic Scenes
East Greenland Photo Gallery
NOTEBOOK

(a)

(c)

(b)

FIGURE 16.41 Arctic and alpine tundra.
(a) Tundra on the Kamchatka Peninsula, Russia. The Uzon Caldera is in the center-background mountain range. (b) Lush by tundra standards, flowers, grasses, mosses, and dwarf willow flourish in the cold climate of the Arctic. (c) An alpine tundra and mountain goats near Mount Evans, Colorado, at 3660 m (12,000 ft). [Photos (a) by Wolfgang Kaehler; (b) and (c) by Bobbé Christopherson.]

Summary and Review—Ecosystems and Biomes

■ *Define* ecology, biogeography, and the ecosystem, community, habitat, and niche concepts.

Earth's biosphere is unique in the Solar System; its ecosystems are the core of life. **Ecosystems** are self-sustaining associations of living plants and animals and their nonliving physical environment. **Ecology** is the study of the relationships between organisms and their environment and among the various ecosystems in the biosphere. **Biogeography** is the study of the distribution of plants and animals and the diverse spatial patterns they create. A **community** is formed by the interactions among populations of living animals and plants at a particular time. Within a community, a **habitat** is the specific physical location of an organism— its address. A **niche** is the function or operation of a life form within a given community—its profession.

ecosystem (p. 526)
ecology (p. 526)
biogeography (p. 526)

community (p. 527)
habitat (p. 527)
niche (p. 527)

1. What is the relationship between the biosphere and an ecosystem? Define *ecosystem* and give some examples.

2. What does biogeography include? Describe its relationship to ecology.

3. Briefly summarize what ecosystem operations imply about the complexity of life.

4. Define a community within an ecosystem.

5. What do the concepts of habitat and niche involve? Relate them to some specific plant and animal communities. Do you have a "habitat" and "niche" in your life? Describe.

6. Describe symbiotic and parasitic relationships in nature. Draw an analogy between these relationships and human societies on our planet. Explain.

■ *Explain* photosynthesis and respiration and *derive* net photosynthesis and the world pattern of net primary productivity.

As plants evolved, the **vascular plants** developed conductive tissues and true roots for internal transport of fluid and nutrients. *Vascular* is from a Latin word for "vessel-bearing," referring to the conducting cells. **Stomata** on the underside of leaves are the portals through which the plant interacts with the atmosphere and hydrosphere. Plants (primary producers) perform **photosynthesis** as sunlight stimulates a light-sensitive pigment called **chlorophyll**. This process produces the food sugars and oxygen that drive biological processes. **Respiration** is essentially the reverse of photosynthesis and is the way the plant derives energy by oxidizing carbohydrates. **Net primary productivity** is the net photosynthesis (photosynthesis minus respiration) of an entire community. This stored chemical energy is the **biomass** that the community generates, or the net dry weight of organic material.

vascular plants (p. 530)
stomata (p. 530)
photosynthesis (p. 530)
chlorophyll (p. 530)
respiration (p. 530)
net primary productivity (p. 530)
biomass (p. 530)

7. How many plant species are there on Earth?
8. How do plants function to link the Sun's energy to living organisms? What is formed within the light-responsive cells of plants?
9. Compare photosynthesis and respiration and the derivation of net photosynthesis. What is the importance of knowing the net primary productivity of an ecosystem and how much biomass an ecosystem has accumulated?
10. Briefly describe the global pattern of net primary productivity.

■ *List* abiotic and biotic ecosystem components and *relate* those components to ecosystem operations and trophic relationships.

Light, temperature, water, and the cycling of gases and nutrients constitute the life-supporting abiotic components of ecosystems. Elevation and latitudinal position on Earth create a variety of physical environments. The zonation of plants with altitude is called **life zones** and is visible as you travel between different elevations.

Life is sustained by **biogeochemical cycles** through which circulate the gases and sedimentary materials (nutrients) necessary for growth and development of living organisms. The environment may inhibit biotic operations, either through lack or excess. These **limiting factors** may be physical or chemical in nature.

Trophic refers to the feeding and nutrition relations in an ecosystem. They represent the flow of energy and cycling of nutrients. **Producers**, which fix the carbon they need from carbon dioxide, are the plants, including phytoplankton in aquatic ecosystems. **Consumers**, generally animals, including zooplankton in aquatic ecosystems, depend on the producers as their carbon source. Energy flows from the producers through the system along a circuit called the **food chain**. Within ecosystems, the feeding interrelationships are complex and arranged in a **food web**.

The primary consumer is an **herbivore**, or plant eater. A **carnivore** (meat eater) is a secondary consumer. A consumer that eats both producers and consumers is called an **omnivore**—a category occupied by humans. **Detritivores** (detritus feeders and decomposers) renew the entire system by releasing simple

inorganic compounds and nutrients by breaking down organic materials. *Detritus feeders* include worms, mites, termites, centipedes, snails, crabs, and even vultures, among others. **Decomposers**, primarily bacteria and fungi, digest organic debris outside their bodies and absorb and release nutrients in the process. This metabolic work of microbial decomposers produces the rotting that breaks down detritus.

life zone (p. 532)
biogeochemical cycles (p. 533)
limiting factor (p. 537)
producers (p. 538)
consumers (p. 538)
food chain (p. 539)
food web (p. 539)
herbivore (p. 541)
carnivore (p. 541)
omnivore (p. 541)
detritivores (p. 541)
decomposers (p. 541)

11. What are the principal abiotic components in terrestrial ecosystems?
12. Describe what Alexander von Humboldt found that led him to propose the life-zone concept. What are life zones? Explain the interaction among altitude, latitude, and the types of communities that develop.
13. What are biogeochemical cycles? Describe several of the essential cycles.
14. What is a limiting factor? How does it function to control the spatial distribution of plant and animal species?
15. What roles do producers and consumers play in an ecosystem? What are the detritivores? Describe their "job" in an ecosystem.
16. Describe the relationship among producers, consumers, decomposers, and detritus feeders in an ecosystem. What is the trophic nature of an ecosystem? What is the place of humans in a trophic system?

■ *Define* succession and *outline* the stages of general ecological succession in both terrestrial and aquatic ecosystems. Relate biodiversity and evolution to these concepts.

A critical aspect of ecosystem stability is **biodiversity**, or species richness of life (a combination of *biological* and *diversity*). The more diverse the species population (both in number of different species and quantity of each species), the species genetic diversity (number of genetic characteristics), and ecosystem and habitat diversity, the better the risk is spread over the entire community. The greater the biodiversity within an ecosystem, the more stable and resilient it is, and the more productive it will be.

Evolution states that single-cell organisms adapted, modified, and passed along inherited changes to multicellular organisms. The genetic makeup of successive generations was shaped by environmental factors, physiological functions, and behaviors that created a greater rate of survival and reproduction that were passed along through natural selection.

Biogeographers trace communities across the ages, considering plate tectonics and ancient dispersal of plants and animals. Past glacial and interglacial climatic episodes have created a long-term succession. Climate change is forcing accelerated succession in ecosystems.

Ecological succession occurs when more complex communities replace older, usually simpler, communities of plants and animals. An area of bare rock or soil with no trace of a former community can be a site for **primary succession**. The initial community that occupies an area of early succession is a **pioneer community**. **Secondary succession** begins in an area that has a vestige of a previously functioning community in place. Rather than progressing smoothly to a definable stable point, ecosystems tend to operate in a dynamic condition, with succeeding communities overlapping in time and space.

Wildfire is one external factor that disrupts a successional community. The science of **fire ecology** has emerged in an effort to understand the natural role of fire in ecosystem maintenance and succession. Aquatic ecosystems also experience community succession, as exemplified by the eutrophication of a lake ecosystem. A natural process in which lakes receive nutrients and sediment and become enriched is **eutrophication**, a natural aging of water bodies.

> biodiversity (p. 543)
> evolution (p. 543)
> ecological succession (p. 548)
> primary succession (p. 549)
> pioneer community (p. 550)
> secondary succession (p. 550)
> fire ecology (p. 550)
> eutrophication (p. 552)

17. Referring back to Figure 1.1.1 in Focus Study 1.1, define the scientific method. Construct a simple model to show the progressive stages that lead to the development of a theory; use the theory of evolution as an example.

18. In what ways did the process of evolution and natural selection contribute to Earth's biodiversity?

19. What is meant by ecosystem stability?

20. How does ecological succession proceed? What are the relationships between existing communities and new, invading communities?

21. Discuss the concept of fire ecology in the context of the Yellowstone National Park fires of 1988. What were the findings of the government task force?

22. Summarize the process of succession in a body of water. What is meant by cultural eutrophication?

■ *Define* the concepts of terrestrial ecosystem, biome, ecotone, and formation classes.

Earth is the only planet in the Solar System with a biosphere. A **terrestrial ecosystem** is a self-sustaining association of plants and animals and their abiotic environment that is characterized by specific plant formation classes. A **biome** is a large, stable ecosystem characterized by specific plant and animal communities. Biomes carry the name of the dominant vegetation because it is the most easily identified feature: forest, savanna, grassland, shrubland, desert, tundra. A boundary transition zone adjoining ecosystems is an **ecotone**. Biomes are divided into more specific vegetation units called **formation classes**. The structure and appearance of the vegetation units are described, for example, as rain forest, needleleaf forest, Mediterranean shrubland, and arctic tundra. Specific life-form designations include trees, lianas, shrubs, herbs, bryophytes, epiphytes (plants growing above the ground on other plants), and thallophytes (lacking leaves, stems, or roots, including bacteria, fungi, algae, and lichens).

> terrestrial ecosystem (p. 552)
> biome (p. 554)

> ecotone (p. 554)
> formation class (p. 554)

23. Describe a transition zone between two ecosystems. How wide is an ecotone? Explain.

24. Define *biome*. What is the basis of the designation?

25. Distinguish between formation classes and life-form designations as a basis for spatial classification.

■ *Describe* 10 major terrestrial biomes and *locate* them on a world map.

Biomes are Earth's major terrestrial ecosystems, each named for its dominant plant community. The 10 major biomes are generalized from numerous formation classes that describe vegetation. Ideally, a biome represents a mature community of natural vegetation. In reality, few undisturbed biomes exist in the world, for most have been modified by human activity. Many of Earth's plant and animal communities are experiencing an accelerated rate of change that could produce dramatic alterations within our lifetime.

For an overview of Earth's 10 major terrestrial biomes and their vegetation characteristics, soil orders, climate designation, annual precipitation range, temperature patterns, and water balance characteristics, please review Table 16.2.

> equatorial and tropical rain forest (p. 554)
> tropical seasonal forest and scrub (p. 560)
> tropical savanna (p. 564)
> midlatitude broadleaf and mixed forest (p. 566)
> needleleaf forest (p. 566)
> boreal forest (p. 566)
> taiga (p. 567)
> temperate rain forest (p. 567)
> Mediterranean shrubland (p. 568)
> chaparral (p. 568)
> midlatitude grasslands (p. 568)
> desert biome (p. 570)
> warm desert and semidesert (p. 570)
> cold desert and semidesert (p. 570)
> arctic tundra (p. 571)

26. Using the integrative chart in Table 16.2 and the world map in Figure 16.27, select any two biomes and study the correlation of vegetation characteristics, soil, moisture, and climate in their spatial distribution. Then contrast the two using each characteristic.

27. Describe the equatorial and tropical rain forests. Why is the rain forest floor somewhat clear of plant growth? Why are logging activities for specific species so difficult there?

28. What issues surround the deforestation of the rain forest? What is the impact of these losses on the rest of the biosphere? What new threat to the rain forest has emerged?

29. What do *caatinga, chaco, brigalow,* and *dornveld* refer to? Explain.

30. Describe the role of fire or fire ecology in the tropical savanna biome and the midlatitude broadleaf and mixed forest biome.

31. Why does the northern needleleaf forest biome not exist in the Southern Hemisphere? Where is this biome located in the Northern Hemisphere, and what is its relationship to climate type?

32. In which biome do we find Earth's tallest trees? Small, stunted plants, lichens, and mosses dominate which biome?

33. What type of vegetation predominates in the Mediterranean dry-summer climates? Describe the adaptations necessary for these plants to survive.

34. What is the significance of the 98th meridian in terms of North American grasslands? What types of inventions were necessary for humans to cope with the grasslands?

35. Describe some of the unique adaptations found in a desert biome.

36. What is desertification (review Chapter 12 and this chapter)? Explain its impact.

37. What physical weathering processes are specifically related to the tundra biome? What types of plants and animals are found there?

■ *Relate* human impacts, real and potential, to several of the biomes.

The equatorial and tropical rain forest biome is undergoing rapid deforestation. Because the rain forest is Earth's most diverse biome and is important to the climate system, such losses are creating great concern among citizens, scientists, and nations. Efforts are under way worldwide to set aside and protect remaining representative sites within most of Earth's principal biomes. The Man and the Biosphere (MAB) Programme of UNESCO coordinates these biosphere reserve programs. More than 480 such biosphere reserves are now operated voluntarily in 100 countries.

38. What is the relationship between island biogeography and biosphere reserves? Describe a biosphere reserve. What are the goals?

39. Compare the map in Figure 16.27 with the global climate map in Chapter 7, Figure 7.5. Select several regions. What correlations can you make between them?

40. As an example of shifting climate impacts, we tracked temperature and precipitation conditions for Illinois in Figures 16.34. What impacts do you think such climate change will have in the United States and in other countries? See Critical Thinking item "B."

NetWork and Critical Thinking Tools

A. This chapter states, "Some scientists are questioning whether our human society and the physical systems of Earth constitute a global-scale symbiotic relationship: mutualistic (sustainable) or parasitic (nonsustainable)." Referring to the definition of these terms, what is your response to the statement? How do you equate our planetary economic system with the need to sustain life-supporting natural systems? Mutualism? Parasitism?

B. Given the information presented in this chapter about deforestation in the tropics, assess the present situation for yourself. What are the main issues? What natural assets are at stake? Natural resources? What is the perspective of the less-developed countries that possess most of the rain forest? What is the perspective of the more-developed countries and their transnational corporations and millions of environmentally active citizens? What kind of action plan would you develop to accommodate all parties? At our Student Learning Center, under Chapter 16, "Destinations," sample rain forest sources to augment your conclusions.

C. Using Figures 16.27 (biomes), Figure 7.2 (precipitation), Figure 7.5 (climates), and Figure 5.25 (air masses), and the printed graphic scales on these four maps, consider the following hypothetical. Assume a northward climatic shift in the United States and Canada of 500 km (310 mi). Describe in your analysis conditions through the Midwest from Texas to the prairie provinces of Canada. Describe conditions from New York, through New England, and into the Maritime provinces. What economic dislocations and relocations do you envision? How might a GIS (geographic information system) program help in your analysis?

Tourists in raincoats stand under the mist from the Canadian Falls as the Niagara River plunges 57 m (187 ft) over the Niagara Escarpment. Note how the power of the falling water undercuts the cliff, which will eventually produce further collapse and obvious changes to the human presence. This nickpoint in the stream gradient has migrated upstream toward Lake Erie 11 km (7 mi) during the past 12,000 years. *[Photo by Bobbé Christopherson.]*

17

Earth and the Human Denominator

KEY LEARNING CONCEPTS

After reading the chapter, you should be able to:

- **Determine** an answer to Carl Sagan's question, "Who speaks for Earth?"
- **Describe** the growth in human population and **speculate** on possible future trends.
- **Analyze** "An Oily Bird" and **relate** your analysis to energy consumption patterns in the United States and Canada.
- **Recall** climate change and intense weather that is occurring and **relate** them to physical geography/Earth systems science.
- **List** and **discuss** the 12 paradigms for the twenty-first century.
- **Appraise** your place in the biosphere, **calculate** your personal Earth *footprint*, and **realize** your physical identity as an inhabitant of Earth.

STUDENT LEARNING CENTER TOOLS

The *Elemental Geosystems* Student Learning Center provides on-line resources for this chapter by clicking on the book at **www.mygeoscienceplace.com**. Some highlights:

▶ Complete an end-of-the-course overview to assess your experience, views, and feelings at this time.

▶ More than 10 additional URLs along with Quick Links to all the URLs in the chapter are included.

During my space flight, I came to appreciate my profound connection to the home planet and the process of life evolving in our special corner of the Universe, and I grasped that I was part of a vast and mysterious dance whose outcome will be determined largely by human values and actions.*

*Rusty Sweickart, "Our backs against the bomb, our eyes on the stars," *Discovery*, July 1987, p. 62.

Earth can be observed from profound vantage points, as this astronaut experienced on the 1969 *Apollo IX* mission. Since November 2000, the International Space Station (ISS) has operated from another vantage point at an orbital altitude of about 400 km (250 mi), in the upper reaches of Earth's thermosphere (see Figure 2.18). At 27,686 kmph (17,165 mph), each orbit takes 91.6 min, making 15.7 orbits a day. Just imagine—a 16-nation coalition built this scientific workstation in orbit. The ISS is 44.5 m long, 78 m wide, and 27.5 m tall (256 ft by 146 ft by 90 ft) and it is not yet finished (Figure 17.1). The 219-metric-ton (482,385-lb) complex has had 167 (through 2008) astronauts aboard in 392 m³ (14,000 ft³) of pressurized work space.

Several hundred scientific investigations and experiments are completed or underway to help us better understand Earth and life systems through research in the unique space environment. All life-support supplies must be brought from Earth. Water recycling and purification techniques provide the bulk of the water needed. Oxygen is electrolytically derived from water and supplemented by compressed oxygen. Enormous solar-panel arrays generate electricity. Without plants, carbon dioxide must be scrubbed from the air the astronauts breathe.

If completed as planned, the ISS will weigh 455 metric tons (1 million lb) and have the cabin volume of a 747 jumbo jet. This seems like an important fulfillment of our human nature: to question, to explore, to discover, to test the limits of the known, and to push beyond to the unknown. (See http://www.nasa.gov/mission_pages/station/main/index.html for more information and updates.)

Our vantage point in this book is that of physical geography. We examine Earth's many systems: its energy, atmosphere, winds, ocean currents, water, weather, climate, endogenic and exogenic systems, soils, ecosystems, and biomes. These systems sustain us as do the ISS systems that sustain the astronauts. This exploration has led us to an examination of the planet's most abundant large animal, *Homo sapiens*.

We stand in the first decade of the twenty-first century. The twenty-first century will be an adventure for the global society, historically unparalleled in experimenting with Earth's life-supporting systems. What preparations and "future thinking" are we doing to understand all that is to occur?

In his 1980 book and public television series *Cosmos*, the late astronomer Carl Sagan asked:

> What account would we give of our stewardship of the planet Earth? We have heard the rationales offered by the nuclear superpowers. We know who speaks for the nations. But who speaks for the human species? Who speaks for Earth?*

Indeed, who does speak for Earth? Perhaps we physical geographers, and other scientists who have studied Earth and know the operations of the global ecosystem, should speak for Earth. However, some might say that questions of technology, environmental politics, and future thinking

FIGURE 17.1 Our orbital perspective—Earth's Space Station.
The International Space Station (ISS) as it looked in August 2007. The ISS began scientific operations in November 2000. The station orbits Earth in 91.6 min, 15.7 times a day. [Photo of ISS courtesy of NASA.]

belong outside of science, and that our job is merely to learn how Earth's processes work and to leave the spokesperson's role to others. Biologist-ecologist Marston Bates offered insight on this question:

> Then we came to humans and their place in this system of life. We could have left humans out, playing the ecological game of "let's pretend humans don't exist." But this seems as unfair as the corresponding game of the economists, "let's pretend nature doesn't exist." The economy of nature and the ecology of humans are inseparable and attempts to separate them are more than misleading, they are dangerous. Human destiny is tied to nature's destiny and the arrogance of the engineering mind does not change this. Humans may be a very peculiar animal, but they are still a part of the system of nature.†

A fact of life is that Earth's more-developed countries (MDCs), through their economic dominance, speak for the billions who live in less-developed countries (LDCs). The reality is that the US defense budget alone is greater than the gross domestic product (GDP) of many countries. The scale of disparity on our home planet is difficult to comprehend.

The fate of traditional modes of life may rest in some distant financial capital. But economics aside, the reality is that the remote lands of Siberia are linked by Earth systems to the Pampas of Argentina, to the Great Plains in North America, and to those harvesting grain by hand in the remote Pamirs of Tajikistan (Figure 17.2).

To understand these linkages among Earth's myriad systems is our quest in *Elemental Geosystems* (see Figure 1.9). We explored the atmosphere, hydrosphere, lithosphere, and biosphere, synthesizing operating systems into the web

*C. Sagan, *Cosmos* (New York: Random House, 1980), p. 329.

†M. Bates, *The Forest and the Sea* (New York: Random House, 1960), p. 247.

FIGURE 17.2 Lands distant from the global centers of power.
These lands were depopulated under Stalin when they were part of the former USSR. Only in the late twentieth century or so have people returned to this traditional rural landscape in the Pamirs of Tajikistan, beginning life again, so distant from the more-developed countries. [Photo by Stephen F. Cunha.]

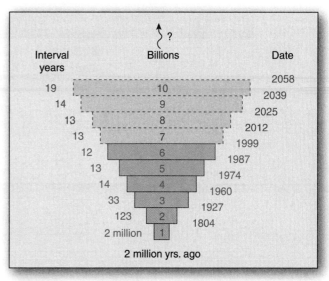

FIGURE 17.3 Human population growth.
The number of years required for the human population to add 1 billion to the count grows smaller and smaller. The new century should mark a slowing of this history of growth if policy actions are taken. Note the population forecasts for the next half-century.

of life. We now see global linkages among physical and living systems and how actions in one place can affect change elsewhere.

In this sense, Earth can be compared to a spaceship. In the same way the members of a crew aboard the ISS are linked to each other's lives and survival, we too are each connected through the operation of planetary systems. There is a growing international environmental awareness in the public sector that is prodding governments into action. People, when informed, generally favor environmental protection and public health progress over economic interests.

The Human Count and the Future

Because human influence is so pervasive, we consider the totality of our impact the *human denominator*—thus, the chapter title. Just as the denominator in a fraction tells how many parts a whole is divided into, so the growing human population, the increasing demand for resources, and the resulting rise in planetary impact must be divided into Earth's relatively fixed resource base.

The human population of Earth passed 6 billion in August 1999 and added another 705 million by mid-2008; more people are alive today than at any previous point in the planet's long history, unevenly arranged in 193 countries and numerous colonies. In a year, about 80 million more people are added—this is 219,000 a day, or a new US population every 3.8 years. Over the span of human history these billion-mark milestones are occurring at closer intervals (Figure 17.3).

Thirty-seven percent of Earth's 6.705 billion people (mid-2008) live in just two countries (China, 21%, and India, 16%—2.47 billion people combined). Moreover, we

are a young planetary population, with some 28% of those now alive under the age of 15. (For population data, go to the Population Reference Bureau, http://www.prb.org; and for the count, see US Census Bureau *PopClock Projection*: http://www.census.gov/cgi-bin/popclock).

The present annual global *natural increase* in population is 1.2%—the annual difference between the *crude birth rate* (per thousand population) and *crude death rate* (per thousand population). This natural increase, if unchanged, will produce a 39% increase in world population by 2050. Growth is not uniformly distributed. *Virtually all new population growth is in the less-developed countries* (LDCs) that now possess 82%, or 5.5 billion, of the total population; the MDCs have the other 18%, or 1.2 billion. Amazingly, 53% of the world's people live on less than $2 (US) per day. Table 17.1 illustrates the natural increase and the differences in growth found in the MDCs and LDCs.

If you consider only the population count, the MDCs do not have a population growth problem. In fact, Japan and some European countries are actually declining in growth or are near replacement levels. However, people in the developed world produce a greater impact on the planet per person and therefore constitute a population-impact crisis. An equation expresses this concept:

$$\text{Planetary impact } (I) = P \cdot A \cdot T$$

where P is *population* A is *affluence*, or consumption per person; and T is *technology*, or the level of environmental impact per unit of production. The United States and Canada, with about 5% of the world's population, produce more than 23% ($15.1 trillion in 2007) of the world's annual gross domestic product and use more than double the energy per capita of Europeans, more than 7 times

Table 17.1 Planetary Natural Increase Rates				
	Birth Rate	Death Rate	Natural Increase (%)	Population Growth 2008–2050 (%)
World—2008	**21**	**8**	**1.2**	39%
More developed	**12**	**10**	**0.2**	5%
United States	14	8	0.6	44%
Japan	9	9	0.0	−25%
United Kingdom	13	9	0.3	26%
Less developed	**23**	**8**	**1.5**	47%
LDC exclu. China	**26**	**9**	**1.8**	59%
Mexico	20	5	1.6	22%
Nigeria	43	18	2.5	91%
India	24	8	1.6	53%

Source: World Population Data Sheet 2008, Population Reference Bureau, Washington, DC, http://www.prb.org.

that of Latin Americans, 10 times that of Asians, and 20 times that of Africans.

Therefore, the vantage point of people in the MDCs on the state of Earth systems, natural resources, and sustainability of current practices is critical. Earth systems science is here to provide a high level of planetary monitoring and analysis to assist this process.

The concept of an individual's "footprint" has emerged—ecological footprint, carbon footprint, lifestyle footprint. These consider what your affluence and technology, the *A* and the *T* in the equation, cost planetary systems. Footprint assessment is grossly simplified, but it can give you an idea of your impact and even an estimate of how many planets it would take to sustain that lifestyle and economy if everyone lived like you. Your author has taken action to neutralize the carbon footprint on science expeditions and for the production and printing of these textbooks using sustainable-forestry green paper for *Elemental Geosystems*. There are many Internet links to try for checking your footprint; here is a sample:

http://www.earthday.net/search/node/footprint
http://www.ecologicalfootprint.org/
http://www.carbonfootprint.com/calculator.html
https://www.econeutral.com/carboncalculator.html
http://www.ucsusa.org/publications/greentips/
  whats-your-carb.html

The MDCs not only should effect change in their own planetary impact but also provide direction for the LDCs, which have a right to move along the road of progress and improve their living conditions. Let us examine this idea of global impact.

An Oily Bird

At first glance, the chain of events that exposes wildlife to oil contamination seems to stem from a technological problem. An oil tanker splits open at sea and releases its petroleum cargo, which is moved by ocean currents toward shore, where it coats coastal waters, beaches, and animals. In response, concerned citizens mobilize and try to save as much of the spoiled environment as possible (Figure 17.4). But the real problem goes far beyond the physical facts of the spill.

In Prince William Sound off the southern coast of Alaska, in clear weather and calm seas, the *Exxon Valdez*, a single-hulled supertanker operated by Exxon Corporation, struck a reef in 1989. The tanker spilled 42 million liters (11 million gallons) of oil. It took only 12 hours for the *Exxon Valdez* to empty its contents, yet a complete cleanup was impossible, and costs and private claims exceeded $15 billion. Scientists are still finding damage and residual oil spill. Eventually, more than 2400 km (about 1500 mi) of sensitive coastline was ruined for years

FIGURE 17.4 An oily bird.
A Western Grebe contaminated with oil from the *Exxon Valdez* tanker accident in Prince William Sound, Alaska. This oily bird is the result of a long chain of events and mistakes. [Photo by Geoffrey Orth/© Sipa Press.]

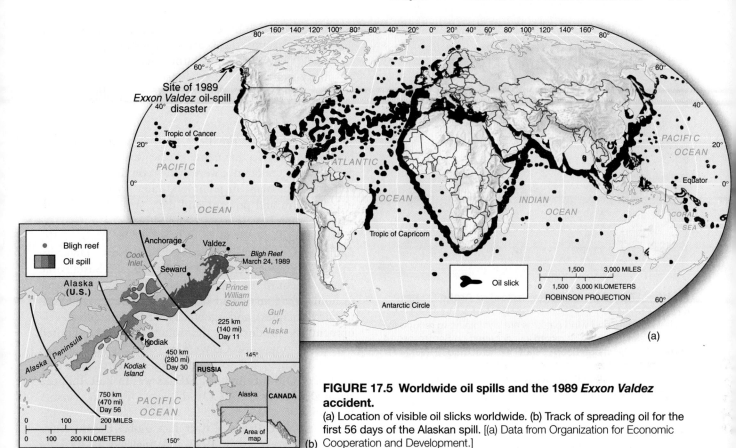

FIGURE 17.5 Worldwide oil spills and the 1989 *Exxon Valdez* accident.
(a) Location of visible oil slicks worldwide. (b) Track of spreading oil for the first 56 days of the Alaskan spill. [(a) Data from Organization for Economic Cooperation and Development.]

to come, affecting three national parks and eight other protected areas (Figure 17.5).

The death toll of animals was massive: At least 5000 sea otters died, or about 30% of resident otters; about 300,000 birds and uncounted fish, shellfish, plants, and aquatic microorganisms also perished. Sublethal effects, namely mutations, now are appearing in fish. The Pacific herring is still in significant decline, as are the harbor seals; other species are in recovery, such as the Bald Eagle and Common Murre. More than two decades later, oil remains in mudflats and marsh soils and still can be scooped from beneath rocks along shorelines.

The immediate problem of cleaning oil off a bird symbolizes a national and international concern with far-reaching spatial significance. And while we search for answers, oil slicks continue their contamination. An average of 27 oil-spill accidents occur each day, 10,000 a year worldwide, ranging from a few disastrous spills to numerous small ones. There were 50 spills equal to the *Exxon Valdez* or larger since 1970. In addition to oceanic oil spills, people improperly dispose of crankcase oil from their automobiles in a volume that annually exceeds all of these tanker spills.

The immediate effect of global oil spills on wildlife is contamination and death. But the issues involved are much bigger than dead birds. Let's ask some fundamental questions about the *Exxon Valdez* accident:

- Why was the oil tanker there in the first place?
- Why do unsafe, single-hulled tankers still traverse the ocean (double-hull tankers will not be standard until 2020)?
- Why is petroleum imported into the continental United States from Alaska in such enormous quantity? Is the demand for petroleum products based on real need and the operation of efficient systems?
- The US demand for oil is higher per capita than the demand of any other country. Efficiency and mileage ratings in the US transportation sector again fell for the 2005 fleet of cars and trucks, with inefficient SUVs (light trucks) leading in sales. Why?

The spike in oil and gasoline prices in 2007–2008 coincided with a record decrease in miles driven, with few difficulties. This demonstrates the effectiveness of individual conservation and efficiency strategies that reduce demand. Strangely, the word *conservation* was not used in media or political coverage. The task of physical geography is to analyze all the spatial aspects of these events in the environment and the ironies that are symbolized by the oily bird and our linkages to it.

With the shrinking of Arctic Ocean permanent and seasonal sea ice, countries and corporations are lining up to exploit the open water and resources (Figure 17.6). Freighters and tankers are being designed wider than the

FIGURE 17.6 2007 record low extent of Arctic Ocean sea ice.
Global warming and wind patterns are combining to reduce the Arctic Ocean sea ice. The record low occurred on September 14, 2007. What the image doesn't show is the degraded thickness of the multiyear ice. Not only did the atmosphere record increased warmth but the Arctic Ocean warmed as well.
[Images courtesy of Scientific Visualization Studio, GSFC/NASA.]

Panama or Suez canals for shipping a decade or two from now through the newly ice-free Northwest and Northeast passages. The exhaust particulates, soot, smoke, and wastewater discharges will damage the fragile Arctic environment. These pollutants will darken the remaining ice shelves and Greenland's ice sheet, forcing lower albedos, thus increasing insolation absorption and accelerating the melting of remaining ice inventories. Think of the impact of physical geography's perspective on this situation using elements from the chapters you have studied. What should be on the planning table for discussion—unrestrained commerce or albedo-forcing impacts on ice melt? Global sea-level rise is occurring at a rate faster than forecast and is the subject of increasing international concern.

The Need for International Cooperation

The idea for a global meeting on the environment was put forward at the 1972 U.N. Conference on the Human Environment held in Stockholm. The U.N. General Assembly in 1987 achieved a landmark in global planning by agreeing to hold an Earth Summit, the first one occurring in 1992. *Our Common Future* set the tone:

> The Earth is one but the world is not. We all depend on one biosphere for sustaining our lives. Yet each community, each country, strives for survival and prosperity with little regard for its impact on others. Some consume the Earth's resources at a rate that would leave little for future generations. Others, many more in number, consume far too little and live with the prospect of hunger, squalor, disease, and early death.*

The setting for the 1992 Earth Summit in Rio de Janeiro was ironic, for many of the very problems discussed at the conference were evident on the city's streets, with their abundant air pollution, water pollution, toxins, noise, wealth and poverty, and daily struggle for health and education.

Five agreements were written at the summit: the Climate Change Framework; Biological Diversity Treaty; the Management, Conservation, and Sustainable Development of All Types of Forests; the Earth Charter (a nonbinding statement of 27 environmental and economic principles); and Agenda 21 for Sustainable Development. A product of the summit was the United Nations Framework Convention on Climate Change (FCCC), which led to a series of Conference of the Parties (COP) meetings—COP–13 was in Bali, Indonesia in December 2007; COP–14 was in Poznañ, Poland, in 2008.

The Kyoto Protocol to reduce global carbon emissions was agreed to in 1997 and finalized in 2001, in Marrakech, Morocco, at COP-7. Following Russian ratification, the Kyoto Protocol and Rulebook became international law in March 2005, without United States or Australian participation. Australia has since ratified the treaty.

Asking whether the Earth Summit "succeeded" or "failed" is the wrong question. The occurrence of this largest-ever official gathering is a remarkable accomplishment. From the Earth Summit emerged a new organization—the U.N. Commission on Sustainable Development—to oversee the promises made in the five agreements. This momentum lead to a second Earth Summit in 2002 (http://www.earthsummit2002.org/) in Johannesburg, South Africa, that completed an agenda including climate change, freshwater, gender issues, global public goods, HIV/AIDS, sustainable finance, and the five Rio conventions.

A clear conclusion is that members of society must work to move the solutions for environmental and developmental problems off the bench and into play. We need international cooperation to consider our symbiotic relations with each other and with Earth's resilient, yet fragile, life-support systems.

A critical corollary to these international efforts is the linkage of academic disciplines. A positive step in that direction is the Earth systems science approach, embodied in physical geography, that synthesizes

---

*World Commission on Environment and Development, *Our Common Future* (Oxford, UK: Oxford University Press, 1987), p. 27.

(a)

(b)

FIGURE 17.7 Polar bears in peril.
(a) An underweight male polar bear walks slowly along the shore of Hudson Bay in mid-November. He is hungry and is scouting for any sign of pack ice, sniffing the air for any hint of seals. The ice is late again, as the open waters of Hudson Bay fill his view. The bear has not eaten since he left the ice in early July. (b) In the Barents Sea, along the edge of the pack ice at 81.5° N, we spot two 8-month-old cubs-of-the-year. The sea ice is breaking up and disappearing, as we watch the cubs try to follow their mom across the open leads. [Photos by Bobbé Christopherson.]

content from across the disciplines to create a holistic perspective. Exciting progress toward an integrated understanding of Earth's physical and biological systems is in progress.

Following decades of research, in 2006 the International Union for the Conservation of Nature (IUCN) listed the polar bear (*Ursus maritimus*) as *vulnerable* to extinction, a "threatened" category on its Endangered Species List—the *Red Book* (see http://www.iucnredlist .org/). The dramatic loss of pack ice caused by human-forced warming in the circumarctic region is key to what is destroying the bears' habitat. As described in Chapter 16, the US Fish and Wildlife Service added polar bears to its Endangered Species List in May 2008, although shortly after the declaration the agency was sued by the state of Alaska to cease the listing.

A September 2007 report from the USGS concluded that fully two-thirds of the polar bears will disappear by 2050; out of the present population of 23,000, that would leave about 7300 bears across the Arctic Region. The loss of sea ice, shown in Figure 17.6, is the root cause of this disaster. Polar bears depend on sea ice for hunting and rearing cubs. Seals depend on the ice as well and the ice itself is habitat for the food web in the Arctic (Figure 17.7).

In the Arctic, these 180–680 kg (400–1500 lb) marine mammals are our unexpected miner's canary. The full meaning of this metaphor—polar bear as canary—unfolds as climate changes across the globe. Their vulnerability to extinction is a warning of difficulties ahead for society if our present course remains unchanged.

Twelve Paradigms for the Twenty-first Century

As we conclude, it seems appropriate to consider a brief list of the dominant themes and patterns of concern for the twenty-first century. I hope these *paradigms* provide a useful framework for brainstorming and discussing the critical issues that will affect us.

Twelve Paradigms for the Twenty-first Century

1. Population increases in the less-developed countries
2. Planetary impact per person (on the biosphere and resources; $I = P \cdot A \cdot T$)
3. Feeding the world's population
4. Global climate change (temperatures, sea level, weather and climate patterns, disease, animal and plant extinctions, and biomes shifting across latitudes)
5. Status of women and children (health, welfare, rights)
6. Global and national disparities of wealth and resource allocation
7. Energy supplies and energy demands; renewables and demand management
8. Loss of biodiversity (habitats, genetic wealth, and species richness)
9. Pollution of air, surface water (quality and quantity), groundwater, oceans, and land
10. The persistence of wilderness (biosphere reserves and biodiversity hot spots)
11. Globalization versus cultural diversity
12. Conflict resolution

FIGURE 17.8 **Earth made the cover of** *Time.*
Global concerns about the environment prompted *Time* magazine to deviate from its 60-year tradition of naming a prominent
citizen as its person of the year, instead naming Earth the
"Planet of the Year." The magazine devoted 33 pages to Earth's
physical and human geography. Importantly, *Time* also offered
positive policy strategies for consideration—an interesting
time capsule for comparison with real events over the years.
[January 2, 1989 issue, Copyright © 1989 The *Time* Inc. Magazine
Company. Reprinted by permission.]

Who Speaks for Earth?

Almost two decades ago, *Time* magazine gave Earth
an interesting honor and today that issue is a time capsule of activist suggestions (Figure 17.8). We wonder if that
Earth awareness is to be a victim of short-lived fame, or if it
fits into the global scheme of things today. Is there a follow-
up in pop culture to making the cover of *Time*?

Geographic awareness and education is an increasingly positive force on Earth. The National Geographic
Society conducts an annual National Geography Bee to
promote geography education and global awareness to
millions of 6th- to 8th-graders, their parents, and teachers. The National Geographic World Championship is
now an annual event (Figure 17.9). There are presently
60 geographic alliances in 47 states, which coordinate
geographic education among teachers and students at
all levels: K–12, community college, college, and university. People are learning more about Earth-human
relations.

Ideological and ethical differences do remain within society. Edward O. Wilson, the respected biologist, addressed
this dichotomy:

FIGURE 17.9 **National Geographic World**
Championship.
Students from across the globe meet every 2 years for the
National Geographic World Championship, sponsored by the
National Geographic Society, with Alex Trebek of *Jeopardy!* as
moderator. Here Alex congratulates finalist teams from three
countries. The contest is an indicator of growing international
geographic awareness. [Photo by O. Louis Mazzatenta, National
Geographic Society.]

The evidence of swift environmental change calls
for an ethic uncoupled from other systems of belief.
Those committed by religion to believe that life was
put on Earth in one divine stroke will recognize
that we are destroying the Creation; and those who
perceive biodiversity to be the product of blind
evolution will agree.... Defenders of both premises
seem destined to gravitate toward the same position
on conservation.... For what, in the final analysis, is
morality but the command of conscience seasoned
by a rational examination of consequences?... An
enduring environmental ethic will aim to preserve
not only the health and freedom of our species,
but access to the world in which the human spirit was
born.*

In an ideal sense, we hold all of Earth's systems in
common. Now, what about Sagan's question to us? The
late Carl Sagan asked, "Who speaks for Earth?" He
answered with this perspective:

We have begun to contemplate our origins: starstuff
pondering the stars; organized assemblages of ten
billion billion billion atoms considering the evolution of atoms; tracing the long journey by which,
here at least, consciousness arose. Our loyalties are
to the species and the planet. We speak for Earth.
Our obligation to survive is owed not just to

*E. O. Wilson, *The Diversity of Life* (Cambridge, MA: Harvard University Press, 1992), p. 351.

FIGURE 17.10 View of Earth from Saturn.
The *Cassini–Huygens* spacecraft was on the far side of Saturn so the Sun backlit the rings and atmosphere. The surprise catch was a glimpse of Earth as a little, pale blue dot, some 1.23 billion km (791 million mi) distant. It was September 19, 2006, on Earth. [Image courtesy of NASA/JPL/ Space Science Institute.]

ourselves but also to that Cosmos, ancient and vast, from which we spring.[†]

The *Cassini–Huygens* spacecraft on its Saturn mission provided us a dramatic perspective of our Home

[†]C. Sagan, *Cosmos* (New York: Random House, 1980), p. 345.

Planet. In September 2006, JPL scientists directed the craft so that Saturn was backlit by the Sun, giving a view of the rings that we never see from Earth. The small pale dot to the left of Saturn (see arrow), just inside the bright ring, is us! All that we studied in the text and that we are trying to understand is on that small, blue dot (Figure 17.10).

May we all perceive our spatial importance within Earth's ecosystems and do our part to maintain a life-supporting and sustaining Earth for ourselves and countless generations in the future.

NetWork and Critical Thinking Tools

1. What part do you think technology, politics, and thinking about the future should play in science courses?
2. Assess population growth issues: the count, the impact per person, and future projection. What strategies do you see as important?
3. According to the discussion in the chapter, what worldwide factors led to the *Exxon Valdez* accident? Describe the complexity of that event from a global perspective. In your analysis, examine both supply-side and demand-side issues, as well as environmental and strategic factors.
4. Relate the content of the various chapters in this text to the integrative Earth systems science concept. Which chapters help you to better understand Earth-human relations and human impacts?
5. What do you see as the lessons learned from the recent devastating climate and weather events described in this text? How might some of these lessons be applied to reducing the impact of future catastrophes?
6. After examining the list of 12 paradigm issues for the twenty-first century, suggest items that need to be added to the list, omitted from the list, or expanded in coverage. Rearrange the list items as needed to match your concerns.
7. Using several of the URLs offered, calculate your global footprint for energy and carbon.
8. This chapter states that we already know many of the solutions to the problems we face. Why do you think these solutions are not being implemented at a faster pace?
9. Who speaks for Earth?

Maps in This Text and Topographic Maps

Maps Used in This Text

Elemental Geosystems uses several map projections: Goode's homolosine, Robinson, and Miller cylindrical, among others. Each was chosen to best present specific types of data. **Goode's homolosine projection** is an interrupted world map designed in 1923 by Dr. J. Paul Goode of the University of Chicago. Rand McNally's *Goode's Atlas* first used it in 1925. Goode's homolosine equal-area projection (Figure A.1) is a combination of two oval projections (*homo*lographic and *sin*usoidal projections).

Two equal-area projections are cut and pasted together to improve the rendering of landmass shapes. A *sinusoidal projection* is used between 40° N and 40° S latitudes. Its central meridian is a straight line; all other meridians are drawn as sinusoidal curves (based on sine-wave curves) and parallels are evenly spaced. A *Mollweide projection*, also called a *homolographic projection*, is used from 40° N to the North Pole and from 40° S to the South Pole. Its central meridian is a straight line; all other meridians are drawn as elliptical arcs and parallels are unequally spaced—farther apart at the equator, closer together poleward. This technique of combining two projections preserves areal size relationships, making the projection excellent for mapping spatial distributions when interruptions of oceans or continents do not pose a problem.

We use Goode's homolosine projection throughout this book. Examples include the world climate map and smaller climate type maps in Chapter 7, topographic regions and continental shield maps (Figures 9.3 and 9.4), world sand regions and loess deposits (Figures 12.12 and 12.14), and the terrestrial biomes map in Figure 16.27.

Another projection we use is the **Robinson projection**, designed by Arthur Robinson in 1963 (Figure A.2). This projection is neither equal area nor true shape, but is a compromise between the two. The North and South Poles appear as lines slightly more than half the length of the equator; thus higher latitudes are exaggerated less than on other oval and cylindrical projections. Some of the Robinson maps employed include the latitudinal geographic zones map in Chapter 1 (Figure 1.13), daily net radiation map (Figure 2.10), the world temperature range map in Chapter 3 (Figure 3.29), the maps of lithospheric plates of crust and volcanoes and earthquakes in Chapter 8 (Figures 8.16 and 8.18), and the global oil spills map in Chapter 17 (Figure 17.6).

Another compromise map, the **Miller cylindrical projection**, is used in this text (Figure A.3). Examples of this projection include the world time zone map (Figure 1.17), global temperature maps in Chapter 3 (Figures 3.24 and 3.27), two global pressure maps in Figure 4.10, and the global soil maps and soil order maps in Chapter 15 (Figure 15.8). This projection is neither true shape nor true area but is a compromise that avoids the severe scale distortion of the Mercator. The Miller projection frequently appears in world atlases. The American Geographical Society presented Osborn Miller's map projection in 1942.

Mapping, Quadrangles, and Topographic Maps

The westward expansion across the vast North American continent demanded a land survey for the creation of accurate maps. Maps were needed to subdivide the land and to guide travel,

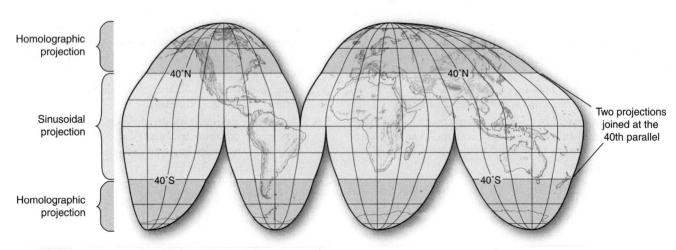

FIGURE A.1 Goode's homolosine projection.
An equal-area map. [Copyright by the University of Chicago. Used by permission of the University of Chicago Press.]

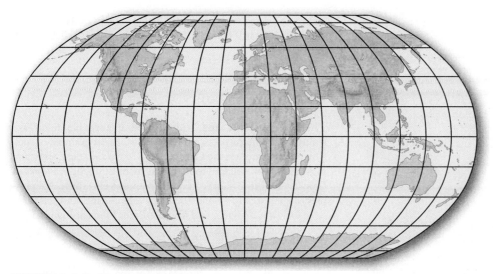

FIGURE A.2 Robinson projection.
A compromise projection between equal area and true shape. [Developed by Arthur H. Robinson, 1963.]

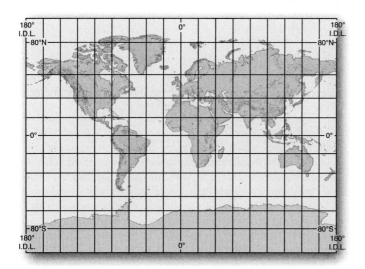

FIGURE A.3 Miller cylindrical projection.
A compromise map projection between equal area and true shape. [Developed by Osborn M. Miller, American Geographical Society, 1942.]

exploration, settlement, and transportation. In 1785 the Public Lands Survey System began surveying and mapping government land in the United States. In 1836 the Clerk of Surveys in the Land Office of the Department of the Interior directed public-land surveys. The Bureau of Land Management replaced this Land Office in 1946. The actual preparation and recording of survey information fell to the US Geological Survey (USGS), also a branch of the Department of the Interior (see http://www.usgs.gov/pubprod/maps.html).

In Canada, National Resources Canada conducts the national mapping program. Canadian mapping includes base maps, thematic maps, aeronautical charts, federal topographic maps, and the *National Atlas of Canada*, now in its fifth edition (see http://atlas.nrcan.gc.ca/).

Quadrangle Maps

The USGS depicts survey information on quadrangle maps, so called because they are rectangular maps with four corner angles. The angles are junctures of parallels of latitude and meridians of longitude rather than political boundaries. These quadrangle maps utilize the Albers equal-area projection, from the conic class of map projections.

The accuracy of conformality (shape) and scale of this base map is improved by the use of not one but two standard parallels. (Remember from Chapter 1 that standard lines are where the projection cone touches the globe's surface, producing greatest accuracy.) For the conterminous United States (the "lower 48"), these parallels are 29.5° N and 45.5° N latitude (noted on the Alber's projection shown in Figure 1.21c). The standard parallels shift for conic projections of Alaska (55° N and 65° N) and for Hawai'i (8° N and 18° N).

Because a single map of the United States at 1:24,000 scale would be more than 200 m wide (more than 600 ft), some system had to be devised for dividing the map into manageable size. Thus, a quadrangle system using latitude and longitude coordinates was developed. Note that these maps are not perfect rectangles, because meridians converge toward the poles. The width of quadrangles narrows noticeably as you move north (poleward).

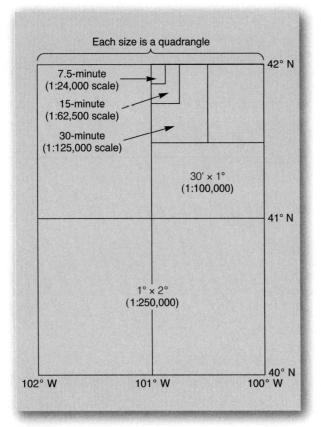

Each size is a quadrangle

7.5-minute
(1:24,000 scale)

15-minute
(1:62,500 scale)

30-minute
(1:125,000 scale)

42° N

30' × 1°
(1:100,000)

41° N

1° × 2°
(1:250,000)

40° N

102° W   101° W   100° W

FIGURE A.4 Quadrangle system of maps used by the USGS.

Quadrangle maps are published in different series, covering different amounts of Earth's surface at different scales. You see in Figure A.4 that each series is referred to by its angular dimensions, which range from 1° to 2° (1:250,000 scale) to 7.5' × 7.5' (1:24,000 scale). A map that is one-half a degree on each side is called a "30-minute quadrangle" (30'), and a map one-fourth of a degree on each side is a "15-minute quadrangle" (15'; this was the USGS standard size from 1910 to 1950). A map that is one-eighth of a degree on each side is a 7.5-minute (7.5') quadrangle, the most widely produced of all USGS topographic maps and the standard since 1950. The progression toward more detailed maps and a larger-scale map standard through the years reflects the continuing refinement of geographic data and new mapping technologies.

The USGS National Mapping Program completed coverage of the entire country (except Alaska) on 7.5-minute maps (1 in. to 2000 ft, a large scale). It takes 53,838 separate 7.5-minute quadrangles to cover the lower 48 states, Hawai'i, and the US territories. A series of smaller-scale, more general, 15-minute topographic maps offer Alaskan coverage.

In the United States, most quadrangle maps remain in English units of feet and miles. The eventual changeover to the metric system requires revision of the units used on all maps, with the 1:24,000 scale eventually changing to a scale of 1:25,000. However, after completing only a few metric quads, the USGS halted the program in 1991. In Canada, the entire country is mapped at a scale of 1:250,000, using metric units (1.0 cm to 2.5 km). About half the country also is mapped at 1:50,000 (1.0 cm to 0.50 km).

Topographic Maps

The most popular and widely used quadrangle maps are **topographic maps** prepared by the USGS. An example of such a map is a portion of the Cumberland, Maryland, quad shown in Figure A.5. You will find topographic maps throughout *Elemental Geosystems* because they portray landscapes so effectively. As examples, see Figures 10.15 and 10.16, karst landscapes and sinkholes near Orleans, Indiana, and Winter Park, Florida; Figure 11.8, river drainage patterns; Figure 11.21, river meander scars; Figure 12.17, an alluvial fan in Montana; and Figure 14.15, drumlins in New York.

A **planimetric map** shows the horizontal position (latitude/longitude) of boundaries, land-use aspects, bodies of water, and economic and cultural features. A highway map is a common example of a planimetric map.

A topographic map adds a vertical component to show topography (configuration of the land surface), including slope and relief (the vertical difference in local landscape elevation). These fine details are shown through the use of elevation contour lines (Figure A.6). A *contour line* connects all points at the same elevation. Elevations are shown above or below a vertical datum, or reference level, which usually is mean sea level. The contour interval is the vertical distance in elevation between two adjacent contour lines (20 ft, or 6.1 m, in Figure A.6b).

The topographic map in Figure A.6b shows a hypothetical landscape, demonstrating how contour lines and intervals depict slope and relief, which are the three-dimensional aspects of terrain. The pattern of lines and the spacing between them indicate slope. The steeper a slope or cliff, the closer together the contour lines appear—in the figure, note the narrowly spaced contours that represent the cliffs to the left of the highway. A wider spacing of these contour lines portrays a more gradual slope, as you can see from the widely spaced lines on the beach and to the right of the river valley.

In Figure A.7 are the standard symbols commonly used on these topographic maps. These symbols and the colors used are standard on all USGS topographic maps: black for human constructions, blue for water features, brown for relief features and contours, pink for urbanized areas, and green for woodlands, orchards, brush, and the like.

The margins of a topographic map contain a wealth of information about its concept and content. In the margins, you find the quadrangle name, names of adjoining quads, quad series and type, position in the latitude-longitude and other coordinate systems, title, legend, magnetic declination (alignment of magnetic north) and compass information, datum plane, symbols used for roads and trails, the dates and history of the survey of that particular quad, and more.

Topographic maps may be purchased directly from the USGS or Centre for Topographic Information, NRC (http://maps.nrcan.gc.ca/). Many state geological survey offices, national and state park headquarters, outfitters, sports shops, and bookstores also sell topographic maps to assist people in planning their outdoor activities.

FIGURE A.5 An example of a topographic map from the Appalachians.
Cumberland, Md., Pa., W.Va. 7.5-minute quadrangle topographic map prepared by the USGS. Note the water gap through Haystack Mountain.

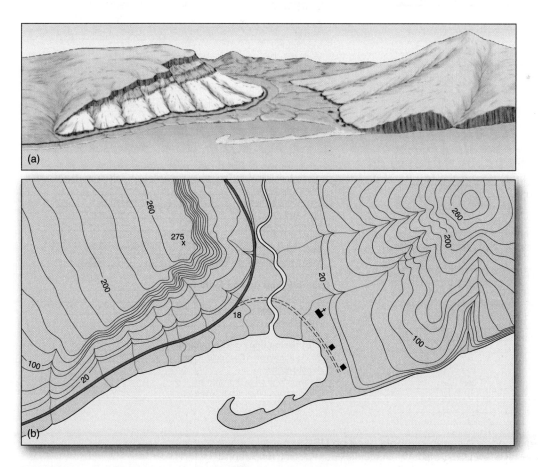

FIGURE A.6 Topographic map of a hypothetical landscape.
(a) Perspective view of a hypothetical landscape. (b) Depiction of that landscape on a topographic map. The contour interval on the map is 20 feet (6.1 m). [After the US Geological Survey.]

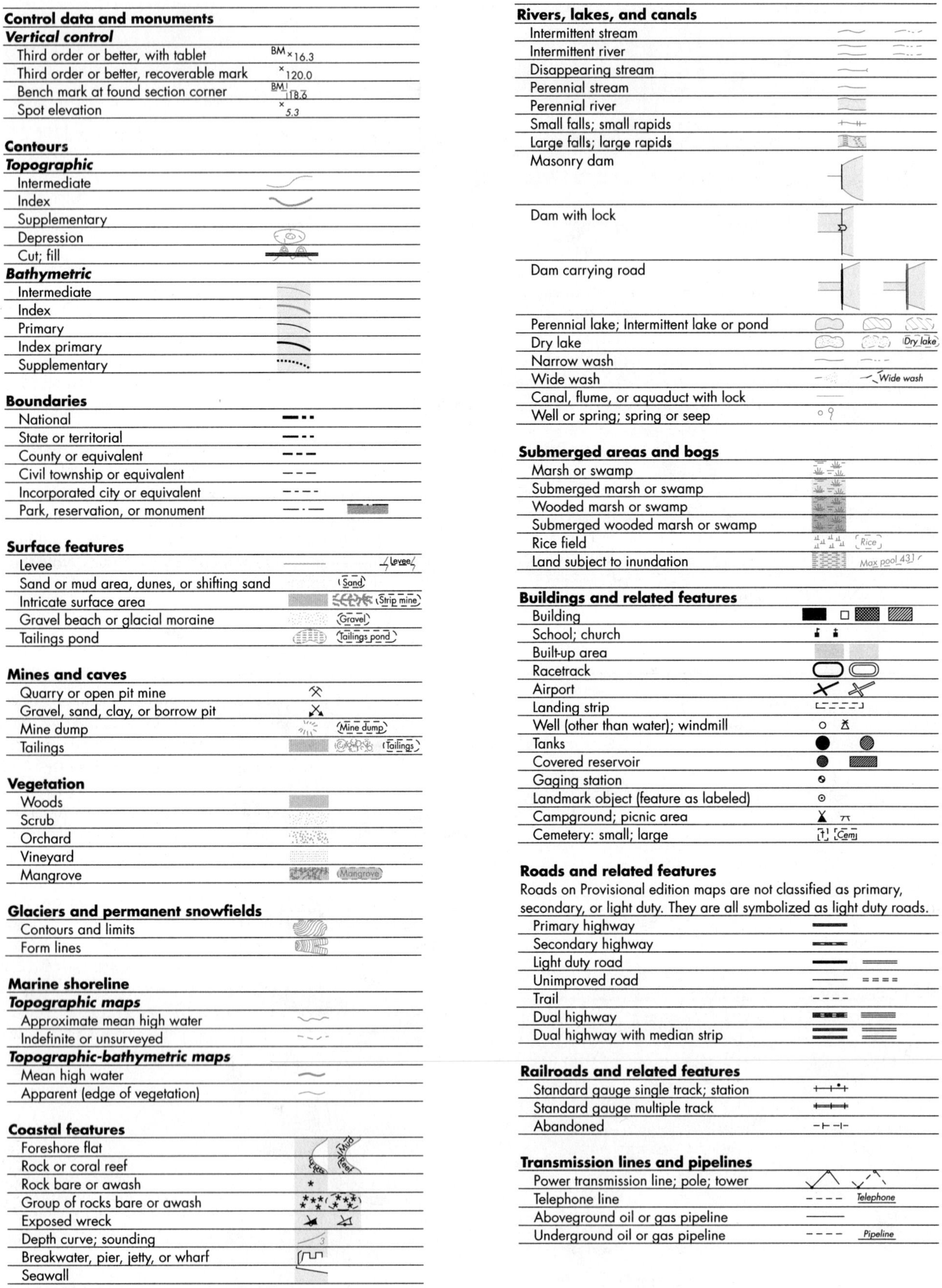

Control data and monuments
Vertical control

Third order or better, with tablet	BM ×16.3
Third order or better, recoverable mark	× 120.0
Bench mark at found section corner	BM! 18.3
Spot elevation	× 5.3

Contours
Topographic

Intermediate	
Index	
Supplementary	
Depression	
Cut; fill	

Bathymetric

Intermediate	
Index	
Primary	
Index primary	
Supplementary	

Boundaries

National	
State or territorial	
County or equivalent	
Civil township or equivalent	
Incorporated city or equivalent	
Park, reservation, or monument	

Surface features

Levee	Levee
Sand or mud area, dunes, or shifting sand	Sand
Intricate surface area	Strip mine
Gravel beach or glacial moraine	Gravel
Tailings pond	Tailings pond

Mines and caves

Quarry or open pit mine	
Gravel, sand, clay, or borrow pit	
Mine dump	Mine dump
Tailings	Tailings

Vegetation

Woods	
Scrub	
Orchard	
Vineyard	
Mangrove	Mangrove

Glaciers and permanent snowfields

Contours and limits	
Form lines	

Marine shoreline
Topographic maps

Approximate mean high water	
Indefinite or unsurveyed	

Topographic-bathymetric maps

Mean high water	
Apparent (edge of vegetation)	

Coastal features

Foreshore flat	
Rock or coral reef	
Rock bare or awash	
Group of rocks bare or awash	
Exposed wreck	
Depth curve; sounding	
Breakwater, pier, jetty, or wharf	
Seawall	

Rivers, lakes, and canals

Intermittent stream	
Intermittent river	
Disappearing stream	
Perennial stream	
Perennial river	
Small falls; small rapids	
Large falls; large rapids	
Masonry dam	
Dam with lock	
Dam carrying road	
Perennial lake; Intermittent lake or pond	
Dry lake	Dry lake
Narrow wash	
Wide wash	Wide wash
Canal, flume, or aquaduct with lock	
Well or spring; spring or seep	

Submerged areas and bogs

Marsh or swamp	
Submerged marsh or swamp	
Wooded marsh or swamp	
Submerged wooded marsh or swamp	
Rice field	Rice
Land subject to inundation	Max pool 431

Buildings and related features

Building	
School; church	
Built-up area	
Racetrack	
Airport	
Landing strip	
Well (other than water); windmill	
Tanks	
Covered reservoir	
Gaging station	
Landmark object (feature as labeled)	
Campground; picnic area	
Cemetery: small; large	Cem

Roads and related features
Roads on Provisional edition maps are not classified as primary, secondary, or light duty. They are all symbolized as light duty roads.

Primary highway	
Secondary highway	
Light duty road	
Unimproved road	
Trail	
Dual highway	
Dual highway with median strip	

Railroads and related features

Standard gauge single track; station	
Standard gauge multiple track	
Abandoned	

Transmission lines and pipelines

Power transmission line; pole; tower	
Telephone line	Telephone
Aboveground oil or gas pipeline	
Underground oil or gas pipeline	Pipeline

FIGURE A.7 Standardized topographic map symbols used on USGS maps.
English units still prevail, although a few USGS maps are in metric. [From USGS, Topographic Maps, 1969.]

The Köppen Climate Classification System

The Köppen climate classification system was designed by Wladimir Köppen (1846–1940), a German climatologist and botanist, and is widely used owing to its ease of comprehension. The basis of any empirical classification system is the choice of criteria used to draw lines on a map to designate different climates. Köppen-Geiger climate classification uses *average monthly temperatures*, *average monthly precipitation*, and *total annual precipitation* to devise its spatial categories and boundaries. But we must remember that boundaries really are transition zones of gradual change. The trends and overall patterns of boundary lines are more important than their precise placement, especially with the small scales generally used on world maps. Take a few minutes and examine the Köppen system, the criteria, and the considerations for each of the principal climate categories.

The modified Köppen-Geiger system has its drawbacks. It does not consider winds, temperature extremes, precipitation intensity, quantity of sunshine, cloud cover, or net radiation. Yet the system is important because its correlations with the actual world are reasonable and the input data are standardized and readily available.

Köppen's Climatic Designations

Figure 7.4 in Chapter 7 shows the distribution of each of Köppen's six climate classifications on the land. This generalized map shows the spatial pattern of climate. The Köppen system uses capital letters (A, B, C, D, E, H) to designate climatic categories from the equator to the poles. The guidelines for each of these categories are in the margin of Figure B.1.

Five of the climate classifications are based on thermal criteria:

A Tropical (equatorial regions)
C Mesothermal (Mediterranean, humid subtropical, marine west coast regions)
D Microthermal (humid continental, subarctic regions)
E Polar (high latitudes and polar regions)
H Highland (compared to lowlands at the same latitude, highlands have lower temperatures—recall the normal lapse rate—and more efficient precipitation due to lower moisture demand)

Only one climate classification is based on moisture as well:

B Dry (deserts and semiarid steppes)

Within each climate classification additional lowercase letters are used to signify temperature and moisture conditions.

For example, in a tropical rain forest *Af* climate, the *A* tells us that the average coolest month is above 18°C (64.4°F, average for the month), and the *f* indicates that the weather is constantly wet, with the driest month receiving at least 6 cm (2.4 in.) of precipitation. (The designation *f* is from the German *feucht*, for moist.) As you can see on the climate map, the tropical rain forest Af climate dominates along the equator and equatorial rain forest.

In a *Dfa* climate, the *D* means that the average warmest month is above 10°C (50°F), with at least one month falling below 0°C (32°F); the *f* indicates that at least 3 cm (1.2 in.) of precipitation falls during every month; and the *a* indicates a warmest summer month averaging above 22°C (71.6°F). Thus, a *Dfa* climate is a *humid-continental*, *hot-summer* climate in the microthermal category.

What's in a Boundary?

Originally, Köppen proposed that the isotherm boundary between mesothermal *C* and microthermal *D* climates be a coldest month of −3°C (26.6°F) or lower. That might be an accurate criterion for Europe, but for conditions in North America, the 0°C isotherm is considered more appropriate. The difference between the 0°C and −3°C isotherms for January covers an area about the width of the state of Ohio. Remember, these isotherm lines are really transition zones and do not mean abrupt change from one temperature to another.

A line denoting at least one month below freezing runs from New York City roughly along the Ohio River, trending westward until it meets the dry climates in the southeastern corner of Colorado. The line marking −3°C as the coldest month runs farther north along Lake Erie and the southern tip of Lake Michigan. From year to year, the position of the 0°C isotherm for January can shift back and forth several hundred kilometers as weather conditions vary. The map in Figure B.1 uses the 0°C isotherm for the C–D climate boundary.

Köppen Guidelines

The Köppen guidelines and map portrayal are in Figure B.1. First, check the guidelines for a climate type, then the color legend for the subdivisions of the type, and then check out the distribution of that climate on the map. You may want to compare this with the climate map in Figure 7.5 that presents causal elements that produce these climates.

FIGURE B.1 World climates and their guidelines according to the Köppen classification system.

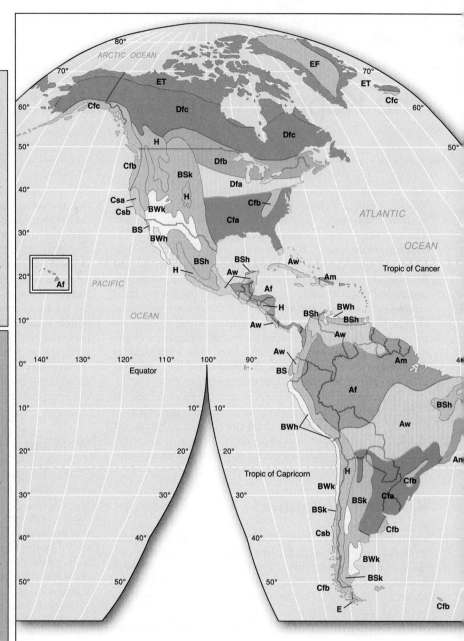

Köppen Guidelines
Tropical Climates — A

Consistently warm with all months averaging above 18°C (64.4°F); annual water supply exceeds water demand.

Af — Tropical rain forest:
  f = All months receive precipitation in excess of 6 cm (2.4 in.).

Am — Tropical monsoon:
  m = A marked short dry season with 1 or more months receiving less than 6 cm (2.4 in.) precipitation; an otherwise excessively wet rainy season. ITCZ 6–12 months dominant.

Aw — Tropical savanna:
  w = Summer wet season, winter dry season; ITCZ dominant 6 months or less, winter water-balance deficits.

Mesothermal Climates — C

Warmest month above 10°C (50°F); coldest month above 0°C (32°F) but below 18°C (64.4°F); seasonal climates.

Cfa, Cwa — Humid subtropical:
  a = Hot summer; warmest month above 22°C (71.6°F).
  f = Year-round precipitation.
  w = Winter drought, summer wettest month 10 times more precipitation than driest winter month.

Cfb, Cfc — Marine west coast, mild-to-cool summer:
  f = Receives year-round precipitation.
  b = Warmest month below 22°C (71.6°F) with 4 months above 10°C.
  c = 1–3 months above 10°C.

Csa, Csb — Mediterranean summer dry:
  s = Pronounced summer drought with 70% of precipitation in winter.
  a = Hot summer with warmest month above 22°C (71.6°F).
  b = Mild summer; warmest month below 22°C.

Microthermal Climates — D

Warmest month above 10°C (50°F); coldest month below 0°C (32°F); cool temperature-to-cold conditions; snow climates. In Southern Hemisphere, occurs only in highland climates.

Dfa, Dwa — Humid continental:
  a = Hot summer; warmest month above 22°C (71.6°F).
  f = Year-round precipitation.
  w = Winter drought.

Dfb, Dwb — Humid continental:
  b = Mild summer; warmest month below 22°C (71.6°F).
  f = Year-round precipitation.
  w = Winter drought.

Dfc, Dwc, Dwb — Subarctic:
Cool summers, cold winters.
  f = Year-round precipitation.
  w = Winter drought.
  c = 1–4 months above 10°C
  d = Coldest month below –38°C (–36.4°F), in Siberia only.

Dry Arid and Semiarid Climates — B

Potential evapotranspiration* (natural moisture demand) exceeds precipitation (natural moisture supply) in all B climates. Subdivisions based on precipitation timing and amount and mean annual temperature.

Earth's arid climates.

BWh — Hot low-latitude desert

BWk — Cold midlatitude desert
  BW = Precipitation less than 1/2 natural moisture demand.
  h = Mean annual temperature >18°C (64.4°F).
  k = Mean annual temperature <18°C.

Earth's semiarid climates.

BSh — Hot low-latitude steppe

BSk — Cold midlatitude steppe
  BS = Precipitation more than 1/2 natural moisture demand but not equal to it.
  h = Mean annual temperature >18°C.
  k = Mean annual temperature <18°C.

Polar Climates — E

Warmest month below 10°C (50°F); always cold; ice climates.

ET — Tundra:
  Warmest month 0–10°C (32–50°F); precipitation exceeds small potential evapotranspiration demand*; snow cover 8–10 months.

EF — Ice cap:
  Warmest month below 0°C (32°F); precipitation exceeds a very small potential evapotranspiration demand; the polar regions.

EM — Polar marine:
  All months above –7°C (20°F), warmest month above 0°C; annual temperature range <17 C° (30 F°).

*Potential evapotranspiration = the amount of water that would evaporate or transpire if it were available — the natural moisture demand in an environment; see Chapter 0

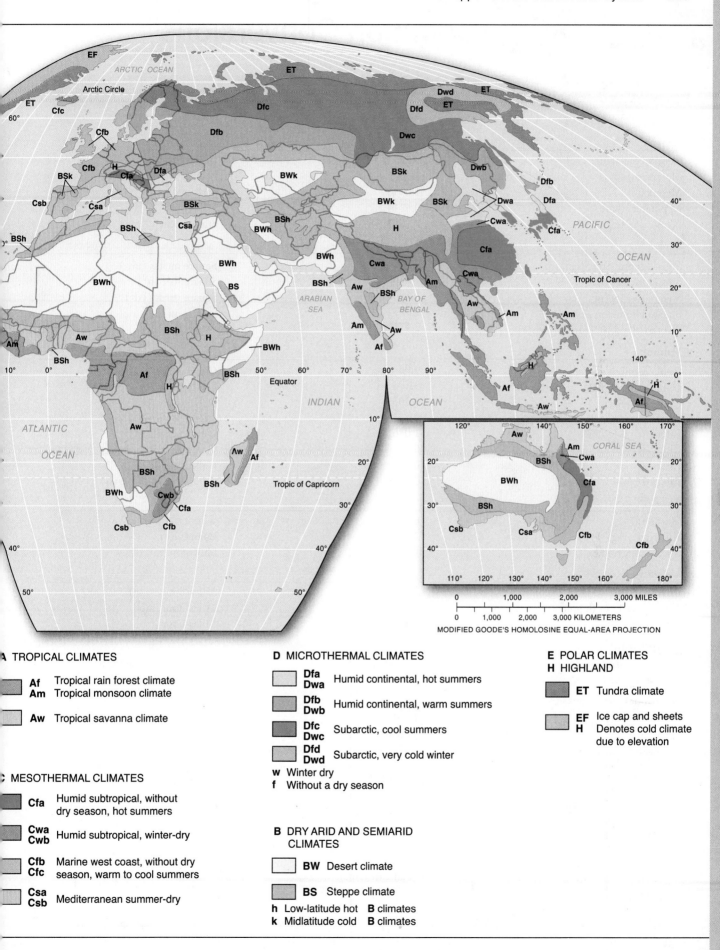

A TROPICAL CLIMATES

- **Af** Tropical rain forest climate
- **Am** Tropical monsoon climate
- **Aw** Tropical savanna climate

C MESOTHERMAL CLIMATES

- **Cfa** Humid subtropical, without dry season, hot summers
- **Cwa Cwb** Humid subtropical, winter-dry
- **Cfb Cfc** Marine west coast, without dry season, warm to cool summers
- **Csa Csb** Mediterranean summer-dry

D MICROTHERMAL CLIMATES

- **Dfa Dwa** Humid continental, hot summers
- **Dfb Dwb** Humid continental, warm summers
- **Dfc Dwc** Subarctic, cool summers
- **Dfd Dwd** Subarctic, very cold winter

- **w** Winter dry
- **f** Without a dry season

B DRY ARID AND SEMIARID CLIMATES

- **BW** Desert climate
- **BS** Steppe climate
- **h** Low-latitude hot **B** climates
- **k** Midlatitude cold **B** climates

E POLAR CLIMATES
H HIGHLAND

- **ET** Tundra climate
- **EF** Ice cap and sheets
- **H** Denotes cold climate due to elevation

Common Conversions

Metric to English

Metric Measure	Multiply by	English Equivalent
Length		
Centimeters (cm)	0.3937	Inches (in.)
Meters (m)	3.2808	Feet (ft)
Meters (m)	1.0936	Yards (yd)
Kilometers (km)	0.6214	Miles (mi)
Nautical mile	1.15	Statute mile
Area		
Square centimeters (cm^2)	0.155	Square inches ($in.^2$)
Square meters (m^2)	10.7639	Square feet (ft^2)
Square meters (m^2)	1.1960	Square yards (yd^2)
Square kilometers (km^2)	0.3831	Square miles (mi^2)
Hectare (ha) ($10,000\,m^2$)	2.4710	Acres (a)
Volume		
Cubic centimeters (cm^3)	0.06	Cubic inches ($in.^3$)
Cubic meters (m^3)	35.30	Cubic feet (ft^3)
Cubic meters (m^3)	1.3079	Cubic yards (yd^3)
Cubic kilometers (km^3)	0.24	Cubic miles (mi^3)
Liters (l)	1.0567	Quarts (qt), US
Liters (l)	0.88	Quarts (qt), Imperial
Liters (l)	0.26	Gallons (gal), US
Liters (l)	0.22	Gallons (gal), Imperial
Mass		
Grams (g)	0.03527	Ounces (oz)
Kilograms (kg)	2.2046	Pounds (lb)
Metric ton (tonne) (t)	1.10	Short ton (tn), US
Velocity		
Meters/second (mps)	2.24	Miles/hour (mph)
Kilometers/hour (kmph)	0.62	Miles/hour (mph)
Knots (kn) (nautical mph)	1.15	Miles/hour (mph)
Temperature		
Degrees Celsius (°C)	1.80 (then add 32)	Degrees Fahrenheit (°F)
Celsius degree (C°)	1.80	Fahrenheit degree (F°)
Additional water measurements:		
Gallon (Imperial)	1.201	Gallon (US)
Gallons (gal)	0.000003	Acre-feet

1 cubic foot per second per day = 86,400 cubic feet = 1.98 acre-feet

ADDITIONAL ENERGY AND POWER MEASUREMENTS

$1\,watt\,(W) = 1\,joule/s$

$1\,joule = 0.239\,calorie$

$1\,calorie = 4.186\,joules$

$1\,W/m^2 = 0.001433\,cal/min$

$697.8\,W/m^2 = 1\,cal/cm^2min^{-1}$

$1\,W/m^2 = 2.064\,cal/cm^2\,day^{-1}$

$1\,W/m^2 = 61.91\,cal/cm^2\,month^{-1}$

$1\,W/m^2 = 753.4\,cal/cm^2\,year^{-1}$

$100\,W/m^2 = 75\,kcal/cm^2\,year^{-1}$

Solar constant:

$1372\,W/m^2$

$2\,cal/cm^2min^{-1}$

English to Metric

English Measure	Multiply by	Metric Equivalent
Length		
Inches (in.)	2.54	Centimeters (cm)
Feet (ft)	0.3048	Meters (m)
Yards (yd)	0.9144	Meters (m)
Miles (mi)	1.6094	Kilometers (km)
Statute mile	0.8684	Nautical mile
Area		
Square inches (in.^{2})	6.45	Square centimeters (cm^2)
Square feet (ft^2)	0.0929	Square meters (m^2)
Square yards (yd^2)	0.8361	Square meters (m^2)
Square miles (mi^2)	2.5900	Square kilometers (km^2)
Acres (a)	0.4047	Hectare (ha)
Volume		
Cubic inches (in.^{3})	16.39	Cubic centimeters (cm^3)
Cubic feet (ft^3)	0.028	Cubic meters (m^3)
Cubic yards (yd^3)	0.765	Cubic meters (m^3)
Cubic miles (mi^3)	4.17	Cubic kilometers (km^3)
Quarts (qt), US	0.9463	Liters (l)
Quarts (qt), Imperial	1.14	Liters (l)
Gallons (gal), US	3.8	Liters (l)
Gallons (gal), Imperial	4.55	Liters (l)
Mass		
Ounces (oz)	28.3495	Grams (g)
Pounds (lb)	0.4536	Kilograms (kg)
Short ton (tn), US	0.91	Metric ton (tonne) (t)
Velocity		
Miles/hour (mph)	0.448	Meters/second (mps)
Miles/hour (mph)	1.6094	Kilometers/hour (kmph)
Miles/hour (mph)	0.8684	Knots (kn) (nautical mph)
Temperature		
Degrees Fahrenheit (°F)	0.556 (after subtracting 32)	Degrees Celsius (°C)
Fahrenheit degree (F°)	0.556	Celsius degree (C°)
Additional water measurements:		
Gallon (US)	0.833	Gallons (Imperial)
Acre-feet	325,872	Gallons (gal)

Additional Notation

Multiples	Prefixes	
$1{,}000{,}000{,}000 = 10^9$	giga	G
$1{,}000{,}000 = 10^6$	mega	M
$1{,}000 = 10^3$	kilo	k
$100 = 10^2$	hecto	h
$10 = 10^1$	deka	da
$1 = 10^0$		
$0.1 = 10^{-1}$	deci	d
$0.01 = 10^{-2}$	centi	c
$0.001 = 10^{-3}$	milli	m
$0.000001 = 10^{-6}$	micro	μ

Glossary

The chapter in which each term appears is **boldfaced** in parentheses and is followed by a specific definition relevant to the key term's usage in the chapter. Cross-reference terms appear in *italics*.

Aa (9) Basaltic lava that is rough, jagged, and clinkery, with sharp edges. This texture is caused by the loss of trapped gases, a slow flow, and the development of a thick skin that cracks into the jagged surface.

Abiotic (1) Nonliving; Earth's nonliving systems of energy and materials.

Ablation (14) Loss of glacial ice through melting, sublimation, wind removal by deflation, or the calving off of blocks of ice.

Abrasion (11, 12, 14) Mechanical wearing and erosion of bedrock accomplished by the rolling and grinding of particles and rocks carried in a stream, moved by wind in a "sandblasting" action, or imbedded in glacial ice.

Absorption (3) Assimilation and conversion of radiation from one form to another in a medium. In the process, the temperature of the absorbing surface is raised, thereby affecting the rate and quality of radiation from that surface.

Active layer (14) A zone of seasonally frozen ground that exists between the subsurface permafrost layer and the ground surface. The active layer is subject to consistent daily and seasonal freeze-thaw cycles (see *permafrost, periglacial*).

Actual evapotranspiration (6) Actual amount of evaporation and transpiration that occurs (ACTET); derived in the water-balance equation by subtracting the deficit (DEFIC) from potential evapotranspiration (POTET).

Adiabatic (5) Pertaining to the heating and cooling of a descending or ascending parcel of air through compression and expansion, without any exchange of heat between the parcel and the surrounding environment.

Adsorption (15) The process whereby cations become attached to soil colloids, or the adhesion of gas molecules and ions to solid surfaces with which they come into contact.

Advection (3) Horizontal movement of air or water from one place to another (compare with *convection*).

Advection fog (5) Active condensation formed when warm, moist air moves laterally over cooler water or land surfaces, causing the lower layers of the overlying air to be chilled to the dew-point temperature.

Aggradation (11) The general building up of land surface because of deposition of material; opposite of *degradation*. When the sediment load of a stream exceeds the stream's capacity, the stream channel is filled through this process.

Air mass (5) A distinctive, homogeneous body of air in terms of temperature and humidity that takes on the moisture and temperature characteristics of its source region.

Air pressure (2) Pressure produced by the motion, size, and number of gas molecules and exerted on surfaces in contact with the air. Normal sea-level pressure, as measured by the height of a column of mercury (Hg), is expressed as 1013.2 millibars, 760 mm of Hg, or 29.92 inches of Hg, or 101.32 kPa (kilopascals) as used in Canada. Air pressure can be measured with mercury or aneroid barometers (see listing for both).

Albedo (3) The reflective quality of a surface, expressed as the percentage of reflected insolation to incoming insolation; a function of surface color, angle of incidence, and surface texture.

Alfisols (15) A soil order in the Soil Taxonomy. Moderately weathered forest soils that are moist versions of Mollisols, with productivity dependent on specific patterns of moisture and temperature; rich in organics; most wide-ranging of the soil orders in the Soil Taxonomy classification.

Alluvial fan (12) Fan-shaped fluvial landform at the mouth of a canyon; generally occurs in arid landscapes where streams are intermittent.

Alluvial terraces (11) Level areas that appear as topographic steps above the stream; created by a stream as it scours with renewed downcutting into its floodplain; composed of unconsolidated alluvium (see *alluvium*).

Alluvium (11) General descriptive term for clay, silt, and sand, transported by running water and deposited in sorted or semi-sorted sediment on a floodplain, delta, or streambed.

Alpine glacier (14) A glacier confined in a mountain valley or walled basin, consisting of three subtypes: valley glacier (within a valley), piedmont glacier (coalesced at the base of a mountain, spreading freely over nearby lowlands), and outlet glacier (flowing outward from a continental glacier).

Altitude (2) The angular distance between the horizon (a horizontal plane) and the Sun (or any point in the sky).

Altocumulus (5) Middle-level, puffy clouds that occur in several forms: patchy rows, wave patterns, a "mackerel sky," or lens-shaped "lenticular" clouds.

Andisols (15) A soil order in the Soil Taxonomy; derived from volcanic parent materials in areas of volcanic activity; a new order created in 1990 of soils previously considered under Inceptisols and Entisols.

Anemometer (4) A device to measure wind velocity.

Aneroid barometer (4) A device to measure air pressure using a partially emptied, sealed cell (see *air pressure*).

Antarctic high (4) A consistent high-pressure region centered over Antarctica; source region for an intense polar air mass that is dry and associated with the lowest temperatures on Earth.

Anthropogenic atmosphere (2) Earth's next atmosphere, so named because humans appear to be the principal causative agent.

Anticline (9) Upfolded strata in which layers slope away from the axis of the fold, or central ridge (compare *syncline*).

Anticyclone (4) A dynamically or thermally caused area of high atmospheric pressure with descending and diverging air flows that rotate clockwise in the Northern Hemisphere and counterclockwise in the Southern Hemisphere (compare *cyclone*).

Aphelion (2) The most distant point in Earth's elliptical orbit about the Sun; reached on July 4 at a distance of 152,083,000 km (94.5 million mi); variable over a 100,000-year cycle (compare *perihelion*).

Aquatic ecosystem (16) An association of plants and animals and their nonliving environment in a water setting.

Aquiclude (6) A body of rock that does not conduct water in usable amounts; an impermeable layer.

Aquifer (6) Rock strata permeable to groundwater flow.

Aquifer recharge area (6) The surface area where water enters an aquifer to recharge the water-bearing strata in a groundwater system.

Arctic tundra (16) A biome in the northernmost portions of North America, Europe, and Russia, featuring low, ground-level herbaceous plants as well as some woody plants.

Arête (14) A sharp ridge that divides two cirque basins; means "fish bone" in French. Arêtes form sawtooth and serrated ridges in glaciated mountains.

Aridisols (15) The largest soil order in areal extent in the Soil Taxonomy; typical of dry (arid and semiarid) climates; low in organic matter and dominated by calcification and salinization.

Artesian water (6) Pressurized groundwater that rises in a well or a rock structure above the local water table; may flow out onto the ground.

Asthenosphere (8) Region of the upper mantle just below the lithosphere known as the plastic layer; the least rigid portion of Earth's interior; shatters if struck yet flows under extreme heat and pressure.

Atmosphere (1) The thin veil of gases surrounding Earth that forms a protective boundary between outer space and the biosphere; generally considered to extend to about a 480-km (300-mi) elevation from Earth's surface.

Auroras (2) A spectacular glowing light display in the ionosphere, stimulated by the interaction of the solar wind with oxygen and nitrogen gases at high latitudes; called aurora borealis (Northern Hemisphere) and aurora australis (Southern Hemisphere).

Autumnal (September) equinox (2) The time around September 22–23 when the Sun's declination crosses the equatorial parallel; all places on Earth experience days and nights of equal length. The Sun rises at the South Pole and sets at the North Pole (compare *vernal equinox*).

Available water (6) The portion of capillary water that is accessible to plant roots; usable water held in soil-moisture storage.

Axial parallelism (2) Earth's axis is parallel to itself throughout the year; the North Pole points to near Polaris.

Badland (12)   Rugged topography, usually of relatively low, varied relief and barren of vegetation; associated with arid and semiarid regions and rocks with low resistance to weathering.

Bajada (12)   A continuous apron of coalesced alluvial fans, formed along the base of mountains in arid climates; presents a gently rolling surface from fan to fan (see *alluvial fan*).

Barrier beach (13)   Narrow, long, depositional feature, generally composed of sand, that forms offshore and is roughly parallel to the coast; may appear as barrier islands and long chains of barrier beaches.

Barrier island (13)   Generally, a broadened barrier beach.

Barrier spit (13)   A depositional form that develops when transported sand or gravel in a barrier beach or island is deposited in long ridges that are attached at one end to the mainland and partially cross the mouth of a bay.

Basalt (8)   A common extrusive igneous rock whose mafic composition is fine-grained; comprising the bulk of the ocean floor crust, lava flows, and volcanic forms.

Base level (11)   A hypothetical level below which a stream cannot erode its valley, and thus the lowest operative level for denudation processes; in an absolute sense, represented by sea level extending back under the landscape.

Base load (11)   Those coarser materials dragged along a streambed by a stream's traction (compare *suspended load*; see *traction*).

Basin and Range Province (12)   A region of dry climates, few permanent streams, and interior drainage patterns in the western United States; composed of a sequence of *horsts* and *grabens*.

Batholith (8)   The largest plutonic form exposed at the surface; an irregular intrusive mass ($>100\,\mathrm{km}^2$; $>40\,\mathrm{mi}^2$) that invades crustal rocks, cooling slowly so that crystals develop (see *pluton*).

Bay barrier (13)   An extensive sand spit that encloses a bay, cutting it off completely from the ocean and forming a lagoon; produced by littoral drift and wave action; sometimes referred to as a baymouth bar (see *barrier spit*, *lagoon*).

Beach (13)   The portion of the coastline where an accumulation of sediment is in motion.

Bed load (11)   Coarse materials that are dragged along the bed of a stream by traction or by the rolling and bouncing motion of saltation; involves particles too large to remain in suspension (see *traction*, *saltation*).

Bedrock (10)   The rock of Earth's crust that is below the soil and basically unweathered; such solid crust sometimes is exposed as an outcrop.

Biodiversity (16)   A principle of ecology and biogeography: The more diverse the species population in an ecosystem (in number of species, quantity of members in each species, and genetic content), the more risk is spread over the entire community, which results in greater overall stability, greater productivity, and increased use of soil nutrients, as compared to a monoculture of no diversity.

Biogeochemical cycles (16)   The various circuits of flowing elements and materials (carbon, oxygen, nitrogen, phosphorus, water) that combine Earth's biotic (living) and abiotic (nonliving) systems; the cycling of materials is continuous and renewed through the biosphere and the life processes.

Biogeography (16)   The study of the distribution of plants and animals and related ecosystems; the geographical relationships with related environments over time.

Biomass (16)   The total mass of living organisms on Earth or per unit area of a landscape; also, the weight of the living organisms in an ecosystem.

Biome (16)   A large terrestrial ecosystem characterized by specific plant communities and formations; usually named after the predominant vegetation in the region (see *terrestrial ecosystem*).

Biosphere (1)   That area where the atmosphere, lithosphere, and hydrosphere function together to form the context within which life exists; an intricate web that connects all organisms with their physical environment; sometimes called *ecosphere*.

Biotic (1)   Living; Earth's living system of organisms.

Blowout depressions (12)   Eolian erosion whereby deflation forms a basin in areas of loose sediment. Size may range up to hundreds of meters (see *deflation*).

Bolson (12)   The slope and basin area between the crests of two adjacent ridges in a dry region.

Boreal forest (16)   See *needleleaf forest*.

Brackish (13)   Descriptive of seawater with a salinity of less than 35‰; for example, the Baltic Sea (contrast *brine*).

Braided stream (11)   A stream that becomes a maze of interconnected channels laced with excess sediments. Braiding often occurs with a reduction of discharge that affects a stream's transportation ability or an increase in sediment load.

Breaker (13)   The point where a wave's height exceeds its vertical stability and the wave breaks as it approaches the shore.

Brine (13)   Seawater with a salinity of more than 35‰; for example, the Persian Gulf (contrast *brackish*).

Calcification (15)   The illuviated accumulation of calcium carbonate or magnesium carbonate in the B and C soil horizons.

Caldera (9)   An interior sunken portion of a composite volcanic crater; usually steep-sided and circular, sometimes containing a lake; also found in conjunction with shield volcanoes.

Canadian System of Soil Classification (15)   A classification system for all soils presently recognized in Canada and adapted to Canada's particular forest, tundra, prairie, frozen ground, and colder climates. The CSSC is organized at five levels of generalization with 10 soil orders (order, great group, subgroup, family, and series), with each level referred to as a category of classification. The system is described in *The Canadian Soil Classification System*, 3rd edition, published by Agriculture Canada, 1998 (compare *Soil Taxonomy*; see *Gelisols*).

Capillary water (6)   Soil moisture, most of which is accessible to plant roots; held in the soil by the water's surface tension and cohesive forces between water and soil (see also *available water*, *field capacity*, and *wilting point*).

Carbonation (10)   A process of chemical weathering by a weak carbonic acid (water and carbon dioxide) that reacts with many minerals containing calcium, magnesium, potassium, and sodium—especially limestone—transforming them into carbonates.

Carbon monoxide (2)   An odorless, colorless, tasteless combination of carbon and oxygen produced by the incomplete combustion of fossil fuels or other carbon-containing substances; toxicity to humans is due to its affinity for hemoglobin, displacing oxygen in the bloodstream; CO.

Carnivore (16)   A secondary consumer that principally eats meat for sustenance. The top carnivore in a food chain is considered a tertiary consumer (compare *herbivore*).

Cartography (1)   The making of maps and charts; a specialized science and art that blends aspects of geography, engineering, mathematics, graphics, computer science, and artistic specialties.

Cation-exchange capacity (CEC) (15)   The ability of soil colloids to exchange cations between their surfaces and the soil solution; a measured potential that indicates soil fertility (see *soil colloid*, *soil fertility*).

Chaparral (16)   Dominant shrub formations of Mediterranean dry-summer climates; characterized by (sclerophyllous) scrub and short, stunted, and tough forests; derived from the Spanish *chapparo*; specific to California (see *Mediterranean shrubland*).

Chemical weathering (10)   Decomposition and decay of the constituent minerals in rock through chemical alteration of those minerals. Water is essential, with rates keyed to temperature and precipitation values. Chemical reactions are active at microsites even in dry climates. Processes include *hydration*, *hydrolysis*, *oxidation*, *carbonation*, and *solution*.

Chlorofluorocarbon compounds (CFCs) (2)   A manufactured molecule (polymer) containing chlorine, fluorine, and carbon; inert and possessing remarkable heat properties; also known as halogens. After slow transport to the stratospheric ozone layer, CFCs react with ultraviolet radiation, freeing chlorine atoms that act as a catalyst to produce reactions that destroy ozone; manufacture banned by international treaties.

Chlorophyll (16)   A light-sensitive pigment that resides within chloroplast (organelle) bodies in leaf cells of plants; the basis of photosynthesis.

Cinder cone (9)   A landform of tephra and scoria, usually small and cone-shaped and generally not more than 450 m (1500 ft) in height, with a truncated top.

Circle of illumination (2)   The division between lightness and darkness on Earth; a day-night great circle.

Cirque (14)   A scooped-out, amphitheater-shaped basin at the head of an alpine glacier valley; an erosional landform.

Cirrus (5)   Wispy filaments of ice-crystal clouds that occur above 6000 m (20,000 ft); they appear in a variety of forms, from feathery hair-like fibers to veils of fused sheets.

Classification (7)   The process of ordering or grouping data or phenomena in related classes; results in a regular distribution of information; a taxonomy.

Climate (7) The consistent, long-term behavior of weather over time, including its variability; in contrast to *weather*, which is the condition of the atmosphere at any given place and time.

Climatic regions (7) Areas of similar climate, which contain characteristic regional weather and air mass patterns.

Climatology (7) A scientific study of climate and climatic patterns and the consistent behavior of weather and weather variability and extremes over time in one place or region, including the effects of climate change on human society and culture.

Climographs (7) Graphs that plot daily, monthly, or annual temperature and precipitation values for a selected station; may also include additional weather information.

Closed system (1) A system that is shut off from the surrounding environment so that it is entirely self-contained in terms of energy and materials; Earth is a closed material system (see *open system*).

Cloud (5) An aggradation of moisture droplets and ice crystals that are suspended in air and are great enough in volume and density to be visible; basic forms include stratiform, cumuliform, and cirroform.

Cloud-albedo forcing (3) An increase in albedo (the reflectivity of a surface) caused by clouds due to their reflection of incoming insolation.

Cloud-greenhouse forcing (3) An increase in greenhouse warming caused by clouds because they can act like insulation, trapping longwave (infrared) radiation.

Col (14) Formed by two headward-eroding cirques that reduce an arête (ridge crest) to form a high pass or saddlelike narrow depression.

Cold desert and semidesert (16) A type of desert biome found at higher latitudes than warm deserts. Interior location and rain shadow locations produce these cold deserts in North America.

Cold front (5) The leading edge of a cold air mass; identified on a weather map as a line marked with a series of triangular spikes pointing in the direction of frontal movement (compare *warm front*).

Community (16) A convenient biotic subdivision within an ecosystem; formed by interacting populations of animals and plants in an area.

Composite volcano (9) A volcano formed by a sequence of explosive volcanic eruptions; steep-sided, conical in shape; sometimes referred to as a stratovolcano, although composite is the preferred term (compare *shield volcano*).

Condensation nuclei (5) Necessary microscopic particles on which water vapor condenses to form moisture droplets; can be sea salts, dust, soot, or ash.

Conduction (3) The slow molecule-to-molecule transfer of heat from warmer to cooler portions through a medium.

Cone of depression (6) The depressed shape of the water table around a well after active pumping from an aquifer. The water table adjacent to the well is drawn down during the process of water removal.

Confined aquifer (6) An aquifer that is bounded above and below by impermeable layers of rock or sediment.

Consumers (16) Organisms in an ecosystem that depend on producers (autotrophs—organisms capable of using carbon dioxide as their sole source of carbon) for their source of nutrients (compare *producer*).

Continental divide (11) A ridge or elevated area that determines the drainage pattern of drainage basins; specifically, that ridge in North America that separates drainage to the Pacific in the west from drainage to the Atlantic and Gulf in the east and to Hudson Bay and the Arctic Ocean in the north.

Continental effect (3) A qualitative designation applied to regions that lack the temperature-moderating effects of the sea and that exhibit a greater range between minimum and maximum temperatures, both daily and annually (see *marine effect, land-water heating differences*).

Continental glacier (14) A continuous mass of unconfined ice, covering at least $50,000 \text{ km}^2$ ($19,500 \text{ mi}^2$); most extensive as ice sheets covering Greenland and Antarctica.

Continental landmasses (9) The broadest category of landforms, including those masses of crust that reside above or near sea level and the adjoining undersea continental shelves along the coastline.

Continental shield (9) Generally old, low-elevation heartland regions of continental crust; various cratons (granitic cores) and ancient mountains exposed at the surface.

Contour lines (Appendix A) Isolines on a topographic map that connect all points at the same elevation relative to a reference elevation called the vertical datum.

Convection (3) Vertical transfer of heat from one place to another through the actual physical movement of air; involves a strong vertical motion.

Convectional lifting (5) Air passing over warm surfaces gains buoyancy and lifts, initiating adiabatic processes.

Convergent lifting (5) Air flows in conflict force lifting and displacement of air upward, initiating adiabatic processes.

Coordinated Universal Time (UTC) (1) The official reference time in all countries, formerly known as Greenwich Mean Time; now measured by six primary standard atomic clocks whose time calculations are collected in Paris, France, at the International Bureau of Weights and Measures (BIPM); the legal reference for time in all countries and broadcast worldwide.

Coral (13) A simple, cylindrical marine animal with a saclike body that secretes calcium carbonate to form a hard external skeleton and landforms called reefs; lives symbiotically with nutrient-producing algae; presently in a worldwide state of decline due to bleaching (loss of algae).

Core (8) The deepest inner portion of Earth, representing one-third of its entire mass; differentiated into two zones—a solid iron inner core surrounded by a dense, molten, fluid metallic iron outer core.

Coriolis force (4) The apparent deflection of moving objects on Earth from a straight path, in relationship to the differential speed of rotation at varying latitudes. Deflection is to the right in the Northern Hemisphere and to the left in the Southern Hemisphere; it produces a maximum effect at the poles and zero effect along the equator.

Crater (9) A circular surface depression formed by volcanism (accumulation, collapse, or explosion); usually located at a volcanic vent or pipe; can be at the summit or on the flank of a volcano.

Crevasses (14) Vertical cracks that develop in a glacier as a result of friction between valley walls, or tension forces of extension on convex slopes, or compression forces on concave slopes.

Crust (8) Earth's outer shell of crystalline surface rock, ranging from 5 to 60 km (3 to 38 mi) in thickness from oceanic crust to mountain ranges.

Cryosphere (14) The portion of the hydrosphere and ground that is perennially frozen, generally at high latitudes and elevations.

Cumulonimbus (5) A towering, precipitation-producing cumulus cloud that is vertically developed across altitudes associated with other clouds; frequently associated with lightning and thunder and thus sometimes termed a thunderhead.

Cumulus (5) Bright and puffy cumuliform clouds up to 2000 m in altitude (6500 ft).

Cyclone (4) A dynamically or thermally caused area of low atmospheric pressure with converging and ascending air flows; rotates counterclockwise in the Northern Hemisphere and clockwise in the Southern Hemisphere (compare *anticyclone*; see *midlatitude cyclone, tropical cyclone*).

Daylength (2) Duration of exposure to insolation, varying during the year depending on latitude; an important aspect of seasonality.

Daylight saving time (1) The setting ahead of time 1 hour in the spring and the setting back by 1 hour in the fall in the Northern Hemisphere. In the United States and Canada, time is set ahead on the second Sunday in March and set back on the first Sunday in November—except in Hawai'i, Arizona, and Saskatchewan, which exempt themselves.

Debris avalanche (10) A mass of falling and tumbling rock, debris, and soil; can be dangerous because of the tremendous velocities achieved by the onrushing materials.

Declination (2) The latitude that receives direct overhead (perpendicular) insolation on a particular day; migrates annually through 47° of latitude between the Tropics of Cancer (23.5° N) and Capricorn (23.5° S).

Decomposer (16) Bacteria and fungi that digest organic debris outside their bodies and absorb and release nutrients in an ecosystem (see *detritivores*).

Deficit (6) In a water balance, the amount of unmet, or unsatisfied, potential evapotranspiration (DEFIC).

Deflation (12) A process of wind erosion that removes and lifts individual particles, literally blowing away unconsolidated, dry, or noncohesive sediments (see *blowout depression*).

Delta (11) A depositional plain formed where a river enters a lake or an ocean; named after the triangular shape of the Greek letter delta, Δ.

Denudation (10) A general term that refers to all processes that cause degradation of the landscape: weathering, mass movement, erosion, and transport.

Deposition (11)   The process whereby weathered, wasted, and transported sediments, deposited by air, water, or ice, are laid down.

Derechos (5)   Strong linear winds in excess of 26 m/s (58 mph) associated with thunderstorms and bands of showers crossing a region.

Desert biome (16)   Arid landscapes of uniquely adapted dry-climate plants and animals.

Desertification (12)   The expansion of deserts worldwide, related principally to poor agricultural practices (overgrazing and inappropriate agricultural practices), improper soil-moisture management, erosion and salinization, deforestation, and the ongoing climatic change; an unwanted semipermanent invasion into neighboring biomes.

Desert pavement (12)   In arid landscapes, a surface formed when wind deflation and sheet flow remove smaller particles, leaving residual pebbles and gravels to concentrate at the surface; resembles a cobblestone street (see *deflation*, *sheet flow*).

Detritivores (16)   Detritus feeders and decomposers that consume, digest, and destroy organic wastes and debris. *Detritus feeders*—worms, mites, termites, centipedes, snails, crabs, and even vultures, among others—consume detritus and excrete nutrients and simple inorganic compounds that fuel an ecosystem (compare with *decomposers*).

Dew-point temperature (5)   The temperature at which a given mass of air becomes saturated, absorbing all the water it can. Any further cooling or addition of water vapor results in active condensation.

Diagnostic subsurface horizon (15)   A soil horizon that originates below the epipedon at varying depths; may be part of the A and B horizons; important in soil description as part of the Soil Taxonomy.

Differential weathering (10)   The effect of different resistance in rock, coupled with variations in the intensity of physical and chemical weathering.

Diffuse radiation (3)   The downward component of scattered incoming insolation from clouds and the atmosphere.

Discharge (11)   The measured volume of flow in a river that passes by a given cross section of a stream in a given unit of time; expressed in cubic meters per second or cubic feet per second.

Dissolved load (11)   Materials carried in a stream in chemical solution derived from minerals such as limestone and dolomite or from soluble salts.

Downwelling current (4)   An area of the sea where a convergence or accumulation of water thrusts excess water downward, occurring, for example, at the western end of the equatorial current or along the margins of Antarctica (compare *upwelling currents*).

Drainage basin (11)   The basic spatial geomorphic unit of a river system; distinguished from a neighboring basin by ridges and highlands that form a divide.

Drainage pattern (11)   A geometric arrangement of streams in a region determined by slope, differing rock resistance to weathering and erosion, climatic and hydrologic variability, and structural controls of the landscape.

Drawdown (6)   See *cone of depression*.

Drumlin (14)   A depositional landform related to glaciation that is composed of till and is streamlined in the direction of continental ice movement—blunt end upstream and tapered end downstream with a rounded summit.

Dry adiabatic rate (DAR) (5)   The rate at which a parcel of air that is less than saturated cools (if ascending) or heats (if descending); a rate of 10 C° per 1000 m (5.5 F° per 1000 ft) (see also *adiabatic*).

Dunes (12)   A depositional feature of sand grains deposited in transient mounds, ridges, and hills; extensive areas of sand dunes are called sand seas.

Dust dome (3)   A dome of airborne pollution associated with every major city; may be blown by winds into elongated plumes downwind from the city.

Dynamic equilibrium model (10)   The balancing act between tectonic uplift and reduction rates of erosion, and between the resistance of crust materials and the work of denudation processes. Landscapes evidence ongoing adaptation to rock structure, climate, local relief, and elevation.

Earthquake (9)   A sharp release of energy that produces shaking in Earth's crust at the moment of rupture along a fault or in association with volcanic activity. The moment magnitude scale expresses earthquake magnitude; the Mercalli scale describes earthquake intensity (see *moment magnitude scale*).

Earth systems science (1)   An emerging science of Earth as a complete, systematic entity. The study of an integrated set of chemical, biological, and physical systems that include the processes of a whole Earth system and of planetary change resulting from system operations; includes a desire for a more quantitative understanding among components, rather than qualitative description.

Ecological succession (16)   The process whereby different and usually more complex assemblages of plants and animals replace older and usually simpler communities. Communities are in a constant state of change as each species adapts to conditions; ecosystems do not exhibit a stable point or successional climax condition as previously thought (refer to *primary succession*, *secondary succession*).

Ecology (16)   The science that studies the interrelationships among organisms and their environment and among various ecosystems.

Ecosphere (1)   Another name for the *biosphere*.

Ecosystem (16)   A self-regulating association of living plants, animals, and their nonliving physical and chemical environment.

Ecotone (16)   A boundary transition zone between adjoining ecosystems that may vary in width and represent areas of tension as similar species of plants and animals compete for the resources (see *ecosystem*).

Effusive eruption (9)   An eruption characterized by low viscosity, basaltic magma, and readily escaping low-gas content. Lava pours forth onto the surface with relatively small explosions and little tephra; tends to form shield volcanoes (see *shield volcano*, *lava*, *pyroclastic*; compare *explosive eruption*).

Elastic-rebound theory (9)   A concept describing the faulting process, in which the two sides of a fault appear locked despite the motion of adjoining pieces of crust, but with accumulating strain they rupture suddenly, snapping to new positions relative to each other, generating an earthquake.

Electromagnetic spectrum (2)   All the radiant energy produced by the Sun placed in an ordered range, divided according to wavelengths (see *wavelength*).

Eluviation (15)   The downward removal of finer particles and minerals from the upper horizons of soil; an erosional process within a soil body (compare *illuviation*).

Empirical classification (7)   A climate classification based on weather statistics or other data; used to determine general climate categories (compare *genetic classification*).

Endogenic system (8)   The system internal to Earth, driven by radioactive heat derived from sources within the planet. In response, the surface is fractured, mountain building occurs, and earthquakes and volcanoes are activated (compare *exogenic system*).

Entisols (15)   A soil order in the Soil Taxonomy that specifically lacks vertical development of horizons; usually young or undeveloped and found in active slopes, alluvial-filled floodplains, and poorly drained tundra.

Environmental lapse rate (2, 5)   The actual lapse rate in the lower atmosphere at any particular time under local weather conditions; may deviate above or below the average normal lapse rate of 6.4 C° per 1000 m (3.5 F° per 1000 ft); compare *normal lapse rate*.

Eolian (12)   Caused by wind; refers to the erosion, transportation, and deposition of materials; spelled *aeolian* in some countries.

Epicenter (9)   The surface area directly above the subsurface focus where movement along the fault plane was initiated. Shock waves radiate outward from the epicenter area.

Epipedon (15)   The diagnostic soil horizon that forms at the surface, not to be confused with the A horizon; may include all or part of the illuviated B horizon.

Equal area (1)   A trait of a map projection; indicates the equivalence of all areas on the surface of the map, although shape is distorted.

Equatorial and tropical rain forest (16)   A lush biome of tall, broadleaf evergreen trees and diverse plants and animals. The dense canopy of leaves usually is arranged in three levels.

Equatorial low-pressure trough (4)   A thermally caused low-pressure area that almost girdles Earth, with air converging and ascending all along its extent; also called the intertropical convergence zone (ITCZ).

Equilibrium line (14)   The area of a glacier where accumulation (gain) and ablation (loss) are balanced.

Erg desert (12)   Sandy deserts, or areas where sand is so extensive that it constitutes a *sand sea*.

Erosion (11)   Denudation by wind, water, and ice, which dislodges, dissolves, or removes surface material.

Esker (14)   A sinuously curving, narrow deposit of coarse gravel that forms along a meltwater stream channel, developing in a tunnel beneath the glacier.

Estuary (11) The point at which the mouth of a river enters the sea and freshwater and seawater are mixed; a place where tides ebb and flow.

Eustasy (5) Refers to worldwide changes in sea level that are not related to movements of land but rather to a rise and fall in the volume of water in the oceans.

Eutrophication (16) A natural process in which lakes receive nutrients and sediment and become enriched; the gradual filling and natural aging of water bodies.

Evaporation (6) The movement of free water molecules away from a wet surface into air that is less than saturated; the phase change of water to water vapor.

Evaporation fog (5) A fog formed when cold air flows over the warm surface of a lake, ocean, or other body of water as the water molecules evaporate from the water surface into the cold, overlying air; also known as a steam fog or sea smoke.

Evaporation pan (6) A weather instrument; a standardized pan from which evaporation occurs, with water automatically replaced and measured; an evaporimeter.

Evaporite (8) Chemical sediments formed from inorganic sources when water evaporates and leaves behind a residue of salts previously in solution.

Evapotranspiration (6) The merging of evaporation and transpiration water loss into one term (see *potential evapotranspiration* and *actual evapotranspiration*).

Evolution (16) A theory that single-cell organisms adapted, modified, and passed along inherited changes to multicellular organisms. The genetic makeup of successive generations is shaped by environmental factors, physiological functions, and behaviors that created a greater rate of survival and reproduction and were passed along through natural selection.

Exfoliation dome (10) A dome-shaped feature of weathering produced by the response of granite to the overburden removal process, which relieves pressure from the rock. Layers of rock slough (sluff) off in slabs or shells in a sheeting process.

Exogenic system (8) The external surface system, powered by insolation, that energizes air, water, and ice and sets them in motion, under the influence of gravity; includes all processes of landmass denudation (compare *endogenic system*).

Exosphere (2) An extremely rarefied outer atmospheric halo beyond the thermopause at an altitude of 480 km (300 mi); probably composed of hydrogen and helium atoms, with some oxygen atoms and nitrogen molecules present near the *thermopause*.

Exotic stream (11) A river that rises in a humid region and flows through an arid region, with discharge decreasing toward the mouth; for example, the Nile River and the Colorado River.

Explosive eruption (9) A violent and unpredictable eruption—the result of magma that is thicker, stickier (more viscous), and higher in gas content and silica than that of an effusive eruption; tends to form blockages within a volcano; produces composite volcanic landforms (see *composite volcano*; compare *effusive eruption*).

Faulting (9) The process whereby displacement and fracturing occur between two portions of Earth's crust; usually associated with earthquake activity.

Feedback loop (1) A portion of the system output that cycles back as an information input, causing changes that guide further system operations (see *negative feedback* and *positive feedback*).

Field capacity (6) Water held in the soil by hydrogen bonding against the pull of gravity, remaining after water drains from the larger pore spaces; the available water for plants (see *available water, capillary water*).

Fire ecology (16) Recognition of fire as a dynamic ingredient in community succession. Controlled fires secure plant reproduction and prevent the accumulation of forest litter and brush; widely regarded as a wise forest management practice.

Firn (14) Snow of a granular texture that is transitional in the slow transformation of snow to glacial ice; snow that has persisted through a summer season in the zone of accumulation.

Firn line (14) The snowline that is visible on the surface of a glacier, where winter snows survive the summer ablation season; analogous to a snowline on land (see *ablation, equilibrium line*).

Fjord (14) Along a coast, a drowned glaciated valley, or glacial trough, that is filled by the sea.

Flash flood (12) A sudden and short-lived torrent of water that exceeds the capacity of a stream channel; associated with desert and semiarid washes.

Flood (11) A high water level that overflows the natural (or artificial) banks along any portion of a stream.

Floodplain (11) A low-lying area near a stream channel, subject to recurrent flooding; alluvial deposits generally mask underlying rock.

Fluvial (11) Stream-related processes; from the Latin *fluvius* for "river" or "running water."

Fog (5) A cloud, generally stratiform, in contact with the ground, with visibility usually restricted to less than 1 km (3300 ft).

Folding (9) A process that bends and deforms beds of various rocks subjected to compressional forces.

Food chain (16) The circuit along which energy flows from producers, who manufacture their own food, to consumers; a one-directional flow of chemical energy, ending with detritivores.

Food web (16) A complex network of interconnected food chains.

Formation class (16) That portion of a biome that concerns the plant communities only, subdivided by size, shape, and structure of the dominant vegetation.

Friction force (4) The effect of drag by the wind as it moves across a surface; may be operative through 500 m (1640 ft) of altitude. Surface friction slows the velocity of the wind and therefore reduces the effectiveness of the Coriolis force.

Frost action (10) A powerful mechanical force produced as water volume expands 9% as it freezes and can exceed the tensional strength of rock, breaking it.

Funnel cloud (5) The visible swirl extending from the bottom side of a cloud, which may or may not develop into a tornado. A tornado is a funnel cloud that has extended all the way to the ground (see *tornado*).

Fusion (2) The process of forcibly joining, under extreme temperature and pressure, positively charged hydrogen and helium nuclei; occurs naturally in thermonuclear reactions within stars, such as our Sun.

Gelisols (15) A new soil order in the Soil Taxonomy, added in 1998, describing cold and frozen soils at high latitudes or high elevations; characteristic tundra vegetation (see *Canadian System of Soil Classification*).

General circulation model (GCM) (7) Complex computer-based climate models that produce generalizations about reality and predictive forecasts of future weather and climate conditions.

Genetic classification (7) A type of classification that uses causative factors to determine climatic regions; for example, an analysis of the effect of interacting air masses (compare *empirical classification*).

Geodesy (1) The science that determines Earth's shape and size through surveys, mathematical means, and remote sensing (see *geoid*).

Geographic information system (GIS) (1) A computer-based data-processing tool or methodology used for gathering, manipulating, and analyzing geographic information to produce a holistic, interactive analysis.

Geography (1) The science that studies the interdependence among geographic areas, natural systems, processes, society, and cultural activities over space—a spatial science. The five themes of geographic education include: location, place, movement, regions, and human-Earth relationships.

Geoid (1) A word that describes Earth's shape; literally, "the shape of Earth is Earth-shaped." A theoretical surface at sea level that extends through the continents, it deviates from a perfect sphere.

Geologic cycle (8) A general term characterizing the vast cycling (hydrologic, tectonic, and rock) in and on the lithosphere. It encompasses the hydrologic cycle, tectonic cycle, and rock cycle.

Geologic time scale (8) A listing of eras, periods, and epochs that span Earth's history; reflects the relative relationship of various layers of rock strata and the absolute dates as determined by scientific methods such as radioactive isotopic dating.

Geomagnetic reversal (8) With an uneven regularity the magnetic field fades to zero, then phases back to full strength, but with the magnetic poles reversed. Reversals have occurred nine times during the past 4 million years.

Geomorphic threshold (10)   The threshold up to which landforms change before lurching into a new set of relationships, with rapid realignments of landscape materials and slopes.

Geomorphology (10)   The science that analyzes and describes the origin, evolution, form, classification, and spatial distribution of landforms.

Geostrophic wind (4)   Winds moving between pressure areas along paths that are parallel to the isobars. In the upper troposphere, the pressure gradient force equals the Coriolis force so that the amount of deflection is proportional to air movement (see *isobar, pressure gradient force, Coriolis force*).

Geothermal energy (8)   The energy potential of boiling steam produced by subsurface magma in near contact with groundwater; active examples include locations in Iceland, New Zealand, Italy, and northern California.

Glacial drift (14)   The general term for all glacial deposits, both unsorted (*till*) and sorted (*stratified drift*).

Glacial ice (14)   A hardened form of ice, very dense in comparison to normal snow or *firn*, that is under pressure within a glacier and is capable of downhill movement.

Glacial surge (14)   The rapid, lurching movement of a glacier.

Glacier (14)   A large mass of perennial ice resting on land or floating shelflike in the sea adjacent to the land; formed from the accumulation and recrystallization of snow, which then flows slowly under the pressure of its own weight and the pull of gravity.

Gleization (15)   A process of humus and clay accumulation in cold, wet climates with poor drainage.

Global dimming (3)   The decline in sunlight reaching Earth's surface owing to pollution, aerosols, and clouds.

Global Positioning System (GPS) (1)   A handheld instrument that accurately determines latitude, longitude, and elevation by calibrating radio signals from satellites.

Goode's homolosine projection (Appendix A)   An equal-area projection developed in 1925 by Dr. Paul Goode that combines two oval projections: the holographic and the sinusoidal; presented interrupted in the text.

Graben (9)   Pairs or groups of faults that produce downward-faulted blocks; characteristic of the basins of the interior western United States (see *horst* and *Basin and Range Province*).

Graded stream (11)   A condition in a stream of mutual adjustment between the load carried by the stream and the related landscape through which the stream flows, forming a state of dynamic equilibrium among erosion, transported load, deposition, and the stream's capacity.

Gradient (11)   The drop in elevation from a stream's headwaters to its mouth, ideally forming a concave slope.

Granite (8)   A coarse-grained (slow-cooling) intrusive igneous rock of 25% quartz and more than 50% potassium and sodium feldspars; characteristic of the continental crust (compare *basalt*).

Gravitational water (6)   That portion of surplus water that percolates downward from the capillary zone, pulled by gravity to the groundwater zone.

Gravity (2)   The mutual force exerted by the masses of objects that are attracted one to another; produced in an amount proportional to each object's mass.

Great circle (1)   Any circle of circumference drawn on a globe with its center coinciding with the center of the globe. An infinite number of great circles can be drawn, but only one parallel is a great circle—the equator (compare *small circle*).

Greenhouse effect (3)   The process whereby radiatively active gases (carbon dioxide, water vapor, methane, and CFCs) absorb insolation and radiate the energy at longer wavelengths, which are retained, delaying the loss of infrared wavelengths to space. Thus, the lower troposphere is warmed through the radiation of infrared wavelengths. The approximate similarity between this process and that of a greenhouse explains the name.

Greenwich Mean Time (GMT) (1)   Former world standard time, now known as Coordinated Universal Time (UTC) (see *Coordinated Universal Time*).

Ground ice (14)   Subsurface water that is frozen in regions of permafrost. The moisture content of areas with ground ice may vary from nearly absent in regions of drier permafrost to almost 100% in saturated soils.

Groundwater (6)   Water beneath the surface that is beyond the soil-root zone; a major source of potable water.

Groundwater mining (6)   Pumping an aquifer beyond its capacity to flow and recharge; an overuse of the groundwater resource.

Gulf Stream (3)   A strong northward-moving warm current off the east coast of North America that carries its water far into the North Atlantic.

Habitat (16)   That physical location in which an organism is biologically suited to live. Most species have specific habitat parameters, or limits.

Hail (5)   A type of precipitation formed when a raindrop is repeatedly circulated above and below the freezing level in a cloud, with each cycle adding more ice to the hailstone until it becomes too heavy to stay aloft.

Herbivore (16)   The primary consumer in a food chain, which eats plant material formed by a producer that has synthesized organic molecules (compare *carnivore*).

Heterosphere (2)   A zone of the atmosphere above the mesopause, 80 km (50 mi) in altitude, composed of rarified layers of oxygen atoms and nitrogen molecules; includes the ionosphere.

Histosols (15)   A soil order in the Soil Taxonomy; formed from thick accumulations of organic matter, such as beds of former lakes, bogs, and layers of peat.

Homosphere (2)   A zone of the atmosphere from the surface up to 80 km (50 mi) composed of an even mixture of gases, including nitrogen, oxygen, argon, carbon dioxide, and trace gases.

Horn (14)   A pyramidal, sharp-pointed peak that results when several cirque glaciers gouge an individual mountain summit from all sides.

Horst (9)   Upward-faulted blocks produced by pairs or groups of faults; characterized by the mountain ranges of the interior of the western United States (see *graben* and *Basin and Range Province*).

Hot spots (8)   Individual points of upwelling material originating in the asthenosphere and tending to remain fixed relative to migrating plates; some 100 are identified worldwide, in, for example, Yellowstone National Park, Hawai'i, and Iceland.

Human-Earth relationships (1)   One of the oldest themes of geography (the human-land tradition); includes the spatial analysis of settlement patterns, resource utilization and exploitation, hazard perception and planning, and the impact of environmental modification and artificial landscape creation.

Humidity (5)   Water vapor content of the air. The capacity of the air to absorb water vapor is mostly a function of the water vapor temperature and air temperature.

Humus (15)   A mixture of organic debris in soil, worked by consumers and decomposers in the humification process; characteristically formed from plant and animal litter laid down at the surface.

Hurricane (5)   A tropical cyclone that is fully organized and intensified in inward-spiraling rainbands; ranges from 160 to 960 km (100 to 600 mi) in diameter, with wind speeds in excess of 119 kmph (65 knots, or 74 mph); a name used specifically in the Atlantic and eastern Pacific (compare *typhoon*).

Hydration (10)   A chemical weathering process involving water that is added to a mineral, which initiates swelling and stress within the rock, mechanically forcing grains apart, thus physical weathering, as the constituents expand (contrast to *hydrolysis*).

Hydraulic action (11)   The erosive work accomplished by the turbulence of water; causes a squeezing and releasing action in joints in bedrock and is capable of prying and lifting rocks.

Hydrograph (11)   A graph of stream discharge over a period of time (minutes, hours, days, years) at a specific place on a stream. The relationship between stream discharge and precipitation input is illustrated on the graph.

Hydrologic cycle (6)   A simplified model of the flow of water and water vapor from place to place as energy powers system operations. Water flows through the atmosphere; across the land, where it is also stored as ice; and within groundwater.

Hydrology (11)   The science of water, its global circulation, distribution, and properties, specifically water at and below Earth's surface.

Hydrolysis (10)   A chemical weathering process in which minerals chemically combine with water; a decomposition process that causes silicate minerals in rocks to break down and become altered (contrast with *hydration*).

Hydrosphere (1)   An abiotic open system that includes all of Earth's water, including the *cryosphere*, or frozen waters.

Ice age (14)   A cold episode, with accompanying alpine and continental ice accumulations, that has recurred roughly every 200 to 300 million years since the late Precambrian era (1.25 billion years ago); includes the most recent episode during the Pleistocene Ice Age, which began 1.65 million years ago.

Iceberg (14)   Floating ice created by calving ice (a large piece breaking off and floating adrift); a hazard to shipping because the ice is 86% submerged.

Ice cap (14)   A dome-shaped glacier, less extensive than an ice sheet ($<50,000$ km^2, or $<19,300$ mi^2), although it buries mountain peaks and the local landscape.

Ice field (14)   An extensive form of land ice, with mountain ridges and peaks visible above the ice; less extensive than an ice cap or ice sheet.

Ice sheet (14)   An enormous continuous continental glacier. The bulk of glacial ice on Earth covers Antarctica and Greenland in two ice sheets.

Igneous rock (8)   One of the basic rock types; it has solidified and crystallized from a hot molten state (either magma or lava). (Compare *metamorphic rock*, *sedimentary rock*.)

Illuviation (15)   The downward movement and deposition of finer particles and minerals from the upper horizon of the soil; a depositional process. Deposition usually is in the B horizon, where accumulations of clays, aluminum, carbonates, iron, and some humus occur (compare *eluviation*; see *calcification*).

Impermeability (6)   Subsurface structure attribute that obstructs water flows in the groundwater system.

Inceptisols (15)   An order in the Soil Taxonomy; usually young soils that are weakly developed and inherently infertile, although they are more developed than Entisols.

Industrial smog (2)   Air pollution associated with coal-burning industries; may contain sulfur oxides, particulates, carbon dioxide, and exotics.

Infiltration (6)   Water access to subsurface regions of soil-moisture storage through penetration of the soil surface.

Insolation (2)   Incoming solar radiation that is intercepted by Earth.

International Date Line (1)   The 180° meridian; an important corollary to the prime meridian at 0°, Greenwich, London, England; established by a treaty in 1884.

Intertropical convergence zone (ITCZ) (4)   See *equatorial low-pressure trough*.

Ionosphere (2)   A layer in the atmosphere above 80 km (50 mi) altitude where gamma, X-ray, and some ultraviolet radiation is absorbed and converted into infrared, or heat, energy and where the solar wind stimulates the auroras.

Isobar (4)   An isoline connecting all points of equal pressure.

Isostasy (8)   A state of equilibrium formed by the interplay between portions of the lithosphere and the asthenosphere; the crust depresses with weight and recovers with the melting of the ice or removal of the load in isostatic rebound.

Isotherm (3)   An isoline connecting all points of equal atmospheric temperature.

Jet contrails (3)   Condensation trails produced by aircraft exhaust, particulates, and water vapor, can form high cirrus clouds, sometimes called false cirrus clouds.

Jet stream (4)   The most prominent movement in upper-level westerly wind flows; irregular, concentrated, sinuous bands of *geostrophic wind*, traveling at 300 kmph (190 mph).

Joints (10)   Fractures or separations in rock without displacement of the sides; they increase the surface area of rock exposed to weathering processes.

Kame (14)   A depositional feature of glaciation; a small hill of poorly sorted sand and gravel that accumulates in crevasses or in ice-caused indentations in the surface.

Karst topography (10)   Distinctive topography formed in a region of chemically weathered limestone with poorly developed surface drainage and solution features that appear pitted and bumpy; originally named after the Krš Plateau in Slovenia (formerly Yugoslavia).

Katabatic wind (4)   Air drainage from elevated regions, flowing as gravity winds. Layers of air at the surface cool, become denser, and flow downslope; known worldwide by many local names.

Kettle (14)   Formation made when an isolated block of ice persists in a ground moraine, an outwash plain, or a valley floor after a glacier retreats; as the block finally melts, it leaves behind a steep-sided hole that frequently fills with water.

Kinetic energy (2)   The energy of motion in a body derived from the vibration of the body's own movement and stated as temperature.

Köppen-Geiger climate classification (7, Appendix B)   An empirical climate classification system based on average monthly temperature, average monthly precipitation, and total annual precipitation. Capital letters A, B, C, D, E, and H designate climate categories.

Lacustrine deposit (14)   Deposit associated with lake-level fluctuations, for example, benches or terraces marking former shorelines.

Lagoon (13)   A portion of coastal seawater that is virtually cut off from the ocean by a bay barrier or barrier beach; also, the water surrounded and enclosed by an atoll.

Land-sea breeze (4)   Wind along coastlines and adjoining interior areas created by different heating characteristics of land and water surfaces—onshore (landward) breeze in the afternoon and offshore (seaward) breeze at night.

Landslide (10)   A sudden, rapid, downslope movement of a cohesive mass of regolith and/or bedrock in a variety of mass-movement forms under the influence of gravity—form of *mass movement*.

Land-water heating differences (3)   The differences in the way land and water heat as a result of contrasts in transmission, evaporation, mixing, and specific heat capacities. Land surfaces heat and cool faster than water and are characterized as having aspects of continentality, whereas water is regarded as producing a marine influence.

Latent heat (5)   Heat energy that is absorbed or released in the phase change of water and is stored in one of the three states—ice, water, or water vapor; includes the latent heat of melting, freezing, vaporization, evaporation, and condensation.

Latent heat of condensation (5)   The heat energy released in a phase change from water vapor to liquid; 585 calories are released from 1 gram of water vapor that condenses at 20°C (68°F).

Latent heat of evaporation (5)   The heat energy required to change phase from liquid to water vapor; under normal sea-level pressure, 585 calories must be added to 1 gram of water (at 20°C) to achieve a phase change to water vapor.

Latent heat of vaporization (5)   The heat energy absorbed from the environment in a phase change from liquid to water vapor at the boiling point; under normal sea-level pressure, 540 calories must be added to each gram of boiling water to achieve a phase change to water vapor.

Lateral moraine (14)   Debris transported by a glacier that accumulates along the sides of the glacier and is deposited along these margins.

Laterization (15)   A pedogenic process operating in well-drained soils that are found in warm and humid regions; typical of Oxisols. High precipitation values leach soluble minerals and soil particles; soils are usually reddish or yellowish colors.

Latitude (1)   The angular distance measured north or south of the equator from a point at the center of Earth. A line connecting all points of the same latitudinal angle is called a parallel (compare *longitude*).

Lava (8, 9)   Magma that issues from volcanic activity onto the surface; the extrusive rock that results when magma solidifies (see *magma*).

Life zone (16)   An altitudinal zonation of plants and animals that form distinctive communities. Each life zone possesses its own temperature and precipitation relations.

Lightning (5)   Flashes of light caused by tens of millions of volts of electrical charge igniting the air to temperatures of 15,000°C to 30,000°C (see *thunder*).

Limestone (8)   The most common chemical sedimentary rock (nonclastic); lithified calcium carbonate, which is very susceptible to chemical weathering by acids in the environment, including carbonic acid in rainfall.

Limiting factor (16)   The physical or chemical factor that most inhibits (either through lack or excess) biotic processes.

Lithification (8)   The compaction, cementation, and hardening of sediments into sedimentary rock.

Lithosphere (1)   Earth's crust, and that portion of the uppermost mantle directly below the crust, that extends down to 70 km (45 mi). Some use this term to refer to the entire Earth.

Littoral drift (13)   The transport of materials along the shore; a more comprehensive term, it includes *beach drift* and *longshore drift* combined.

Littoral zone (13)   A specific coastal environment; that region between the high-water line during a storm and a depth at which storm waves are unable to move sea-floor sediments.

Loam (15)   A mixture of sand, silt, and clay in almost equal proportions, with no one texture dominant; an ideal agricultural soil.

Location (1)   A basic theme of geography dealing with the absolute and relative position of people, places, and things on Earth's surface.

Loess (12)   Large quantities of fine-grained clays and silts left as glacial outwash deposits and subsequently blown by the wind great distances and redeposited as a generally unstratified, homogeneous blanket of material covering existing landscapes. In China, loess originated from desert lands.

Longitude (1) The angular distance measured east or west of a prime meridian from a point at the center of Earth. A line connecting all points of the same longitude is called a *meridian*.

Longshore current (13) A current that forms parallel to a beach as waves arrive at an angle to the shore; generated in the surf zone by wave action, transporting large amounts of sand and sediment.

Lysimeter (6) A weather instrument for measuring potential and actual evapotranspiration; isolates a portion of a field so that the moisture moving through the plot is measured.

Magma (8) Molten rock from beneath the surface of Earth; fluid, gaseous, under tremendous pressure, and either intruded into country rock or extruded onto the surface as lava (see *lava*).

Magnetosphere (2) Earth's magnetic force field, which is generated by dynamo-like motions within the planet's outer core; it deflects the solar wind toward each pole.

Mangrove swamp (13) Wetland ecosystems between 30° N or S and the equator; tends to form a distinctive community of mangrove plants (compare *salt marsh*).

Mantle (8) An area within the planet representing about 80% of Earth's total volume, with densities increasing with depth; occurs above the core and below the crust; rich in iron and magnesium oxides and silicates.

Map (1, Appendix A) A generalized view of an area, usually some portion of Earth's surface, as seen from above at a greatly reduced size (see *scale* and *map projection*).

Map projection (1, Appendix A) The reduction of a spherical globe onto a flat surface in some orderly and systematic realignment of the latitude and longitude grid.

Marine effect (3) Regions that are dominated by the moderating effects of the ocean and that exhibit a smaller minimum and maximum temperature range than continental stations (see *land water heating differences*).

Mass movement (10) All unit movements of materials propelled by gravity; can range from dry to wet, slow to fast, small to large, and free-falling to gradual or intermittent; a term sometimes used interchangeably with *mass wasting*.

Meandering stream (11) The sinuous, curving pattern common to graded streams, with the outer portion of each curve subjected to the greatest erosive action and the inner portion receiving sediment deposits (see *graded stream*).

Mean sea level (MSL) (13) The average of tidal levels recorded hourly at a given site over a long period of time, which must be at least a full lunar tidal cycle.

Medial moraine (14) Debris transported by a glacier that accumulates down the middle of the glacier when two glaciers merge and their lateral moraines combine; forms a depositional feature following glacial retreat.

Mediterranean shrubland (16) A major biome dominated by Mediterranean dry-summer climates and characterized by scrub and short, stunted, tough (sclerophyllous) forests (see *chaparral*).

Mercator projection (1) A cylindrical, true-shape projection developed by Gerardus Mercator in 1569. All straight lines represent true constant directions, or rhumb lines. It presents false notions of the size (area) of midlatitude and poleward landmasses but presents true compass direction (see *rhumb line*).

Mercury barometer (4) A device that measures air pressure with a column of mercury in a tube that is inserted in a vessel of mercury.

Meridian (1) A line designating an angle of longitude (see *longitude*).

Mesocyclone (5) A large, rotating circulation initiated within a parent cumulonimbus cloud at the mid-troposphere level; generally produces heavy rain, large hail, blustery winds, and lightning; may lead to tornado activity.

Mesosphere (2) The upper region of the homosphere from 50 to 80 km (30 to 50 mi) above the ground; designated by temperature criteria and very low pressures.

Metamorphic rock (8) Existing rock, both igneous and sedimentary, that goes through profound physical and chemical changes under increased pressure and temperature. Constituent mineral structures may exhibit foliated or nonfoliated textures.

Meteorology (5) The scientific study of the atmosphere that includes a study of its physical characteristics and motions; related chemical, physical, and geological processes; the complex linkages of atmospheric systems; and weather forecasting.

Methane (7) A radiatively active gas, CH_4, that participates in the greenhouse effect; derived from the organic processes of burning, digesting, and rotting in the presence of oxygen.

Microclimatology (3) The study of climates at or near Earth's surface, or that height above the Earth's surface where the effects of the surface are no longer of effect.

Midlatitude broadleaf and mixed forest (16) A biome in moist continental climates in areas of warm-to-hot summers and cool-to-cold winters; relatively lush stands of broadleaf forests trend northward into needleleaf evergreen stands.

Midlatitude cyclone (5) An organized area of low pressure, with converging and ascending air flow producing an interaction of air masses; migrates along storm tracks. Such lows or depressions form the dominant weather pattern in the middle and higher latitudes of both hemispheres (see *wave cyclone*).

Midlatitude grassland (16) The major biome most modified by human activity; so named because of the predominance of grasslike plants, although deciduous broadleafs appear along streams and other limited sites; location of the world's breadbaskets of grain and livestock production.

Mid-ocean ridge (8) Submarine mountain ranges that extend more than 65,000 km (40,000 mi) worldwide and average more than 1000 km (620 mi) in width; centered along sea-floor spreading centers (see *sea-floor spreading*).

Milky Way Galaxy (2) A flattened, disk-shaped mass estimated to contain up to 400 billion stars; includes our Solar System.

Miller cylindrical projection (Appendix A) A compromise map projection that avoids the severe distortion of the Mercator projection (see *map projection*).

Mineral (8) An element or combination of elements that form an inorganic natural compound; described by a specific formula and qualities of a specific nature.

Model (1) A simplified version of a system, representing an idealized part of the real world.

Mohorovičić discontinuity (8) The boundary between the crust and the rest of the lithospheric upper mantle, named for the Yugoslavian seismologist Mohorovičić; a zone of sharp material and density contrasts; also called the Moho.

Moist adiabatic rate (MAR) (5) The rate at which a parcel of saturated air cools in ascent; a rate of 6 C° per 1000 m (3.3 F° per 1000 ft). This rate may vary with moisture content and temperature, from 4 C° to 10 C° per 1000 m (2 F° to 6 F° per 1000 ft) (see *adiabatic*; compare *dry adiabatic rate*).

Moisture droplet (5) Initial composition of clouds; each droplet measures approximately 0.002 cm (0.0008 in.) in diameter and is invisible to the human eye.

Mollisols (15) A soil order in the Soil Taxonomy classification that has a humus-rich, organic content high in alkalinity; some of the world's most significant agricultural soils.

Moment magnitude scale (9) An earthquake magnitude scale; considers the amount of fault slippage, the size of the area that ruptured, and the nature of the materials that faulted in estimating the magnitude of an earthquake. Replaces the Richter scale (amplitude magnitude) and is especially valuable in assessing larger-magnitude events.

Monsoon (4) From the Arabic word *mausim*, meaning "season"; refers to an annual cycle of dryness and wetness, with seasonally shifting winds produced by changing atmospheric pressure systems; affects India, Southeast Asia, Indonesia, northern Australia, and portions of Africa.

Moraine (14) Marginal glacial deposits (lateral, medial, terminal, ground) of unsorted and unstratified material.

Mountain-valley breeze (4) A light wind produced as cooler mountain air flows downslope at night and as warmer valley air flows upslope during the day.

Movement (1) A major theme in geography involving migration, communication, and the interaction of people and processes across space.

Natural levees (11) Long, low ridges that occur on either side of a river in a developed floodplain; depositional by-products (coarse gravels and sand) of river-flooding episodes.

Needleleaf forest (16)   Forests of pine, spruce, fir, and larch, stretching from the east coast of Canada westward to Alaska and continuing from Siberia westward across the entire extent of Russia to the European Plain. Called the *taiga* (a Russian word) or the *boreal forest*, it occurs principally in the humid continental and subarctic climates and includes montane forests that may be at lower latitudes at elevation.

Negative feedback (1)   A feedback loop that tends to slow or dampen response in a system, thus promoting self-regulation; far more common than positive feedback in living systems (see *feedback loop, positive feedback*).

Net primary productivity (16)   The net photosynthesis (photosynthesis minus respiration) for a given community; considers all growth and all reduction factors that affect the amount of useful chemical energy (*biomass*) fixed (chemically bound) in an *ecosystem*.

Net radiation (NET R) (3)   The net all-wave radiation available; the final outcome of the radiation balance process between incoming and outgoing shortwave and longwave energy.

Niche (16)   The basic function, or occupation, of a life form within a given community; the way an organism obtains its food, air, and water (compare *habitat*).

Nickpoint (knickpoint) (11)   The point at which the longitudinal profile of a stream is abruptly broken by a change in gradient; for example, a waterfall, rapids, or cascade.

Nimbostratus (5)   A rain-producing, dark, grayish stratiform cloud characterized by gentle drizzles.

Nitrogen dioxide (2)   A reddish-brown choking gas produced in high-temperature combustion engines; can be damaging to human respiratory tracts and to plants; participates in photochemical reactions and acid deposition.

Normal fault (9)   A type of geologic fault in rocks. Tension produces strain that breaks a rock, with one side moving vertically relative to the other side along an inclined fault plane (compare *reverse fault*).

Normal lapse rate (2, 5)   The average rate of temperature decrease with increasing altitude in the lower atmosphere; an average value of 6.4 C° per km, or 1000 m (3.5 F° per 1000 ft).

Occluded front (5)   In a cyclonic circulation, the overrunning of a surface warm front by a cold front and the subsequent lifting of the warm air wedge off the ground; initial precipitation is moderate to heavy.

Ocean basins (9)   The physical containers (a depression in the lithosphere) for Earth's oceans.

Oceanic trench (8)   The deepest single features of Earth's crust; associated with *subduction* zones. The deepest is the Mariana Trench near Guam, which descends to 11,033 m (36,198 ft).

Omnivore (16)   A consumer that feeds on both producers (plants) and consumers (meat)—a classification occupied by humans, among other animals (compare *consumer, producer*).

Open system (1)   A system with inputs and outputs crossing back and forth between the system and the surrounding environment. Earth is an open system in terms of energy (compare *closed system*).

Orogenesis (9)   The process of mountain building that occurs when large-scale compression leads to deformation and uplift of the crust; literally, the birth of mountains.

Orographic lifting (5)   The uplift of migrating air masses in response to the physical presence of a mountain, a topographic barrier. The lifted air cools adiabatically as it moves upslope and may form clouds and produce increased precipitation (see *rain shadow*).

Outgassing (5)   The release of trapped gases from rocks, forced out from within Earth through cracks, fissures, and volcanoes; the terrestrial source of Earth's water.

Outwash plain (14)   Glaciofluvial deposits of stratified drift from meltwater-fed, braided, and overloaded streams; beyond a glacier's morainal deposits.

Oxbow lake (11)   A lake that was formerly part of the channel of a meandering stream, isolated when a stream eroded its outer bank, forming a cutoff through the neck of a looping meander (see *meandering stream*).

Oxidation (10)   A chemical weathering process whereby oxygen oxidizes (combines with) certain metallic elements to form oxides; most familiar as the "rusting" of iron in a rock or soil (Ultisols, Oxisols) that produces a reddish-brown stain of iron oxide (Fe_2O_3).

Oxisols (15)   A soil order in the Soil Taxonomy; tropical soils that are old, deeply developed, and lacking in horizons wherever well drained and are heavily weathered, low in cation-exchange capacity, and low in fertility.

Ozone layer (2)   See *ozonosphere*.

Ozonosphere (2)   A layer of ozone (O_3) occupying the full extent of the stratosphere 20 to 50 km (12 to 30 mi) above the surface; the region of the atmosphere where ultraviolet wavelengths are principally absorbed and converted into heat. Its thinning and depletion relates to human use of CFCs and other compounds. International treaties strive to facilitate recovery and foster protection for the ozonosphere.

Pahoehoe (9)   Basaltic lava that is more fluid than *aa*; forms a thin crust that folds and appears "ropy," like coiled, twisted rope.

Paleoclimatology (7, 14)   The science that studies the climates of past ages.

Paleolake (14)   An ancient lake, such as Lake Bonneville or Lake Lahonton, associated with former wet periods when the lake basins were filled to higher levels than today.

PAN (2)   See *peroxyacetyl nitrates*.

Pangaea (8)   The supercontinent formed by the collision of all continental masses approximately 225 million years ago; named by Wegener (1912) in his continental drift theory (see *plate tectonics*).

Parallel (1)   A line, parallel to the equator, that designates an angle of latitude (see *latitude*).

Parent material (10)   The unconsolidated material, from both organic and mineral sources, that is the basis of soil development.

Particulate matter (PM) (2)   Dust, dirt, soot, salt, sulfate aerosols, fugitive natural particles, or other material particles suspended in air.

Paternoster lake (14)   One of a series of small, circular, stair-stepped lakes, formed in individual rock basins aligned down the course of a glaciated valley; named because they look like a string of rosary (religious) beads.

Pedon (15)   A soil profile extending from the surface to the lowest extent of plant roots or to the depth where regolith or bedrock is encountered, imagined as a hexagonal column; the basic soil sampling unit.

Percolation (6)   The process by which water permeates through to the subsurface environment; vertical water movement through soil or porous rock.

Periglacial (14)   Cold-climate processes, landforms, and topographic features along the margins of glaciers, past and present, that occupy more than 20% of Earth's land surface; including *permafrost, frost action*, and *ground ice*.

Perihelion (2)   That point in Earth's elliptical orbit about the Sun where Earth is closest to the Sun; occurs on January 3 at 147,255,000 km (91,500,000 mi); variable over a 100,000-year cycle (compare *aphelion*).

Permafrost (14)   Forms when soil or rock temperatures remain below 0°C (32°F) for at least 2 years in areas considered periglacial; the criterion is based on temperature and not on whether water is present (see *periglacial*).

Permeablity (6)   The ability of water to flow through soil or rock; a function of the texture and structure of the medium.

Peroxyacetyl nitrates (PAN) (2)   A pollutant formed from photochemical reactions involving nitric oxide (NO) and volatile organic compounds (VOCs). PAN produces no known human health effects but is particularly damaging to plants.

Phase change (5)   The change in phase, or state, between ice, water, and water vapor; involves the absorption or release of latent heat.

Photochemical smog (2)   Air pollution produced by the interaction of ultraviolet light, nitrogen dioxide, and hydrocarbons; produces ozone and PAN through a series of complex photochemical reactions. Automobiles are the major source of the contributive gases.

Photosynthesis (16)   The joining of carbon dioxide and hydrogen in plants under the influence of certain wavelengths of visible light; releases oxygen and produces energy-rich organic material (sugars and starches).

Physical geography (1)   The science concerned with the spatial aspects and interactions of the physical elements and processes that make up the environment: energy, air, water, weather, climate, landforms, soils, animals, plants, microorganisms, and Earth.

Physical weathering (10)   The breaking and disintegrating of rock without any chemical alteration; sometimes referred to as mechanical or fragmentation weathering.

Pioneer community (16)   The initial plant community in an area; usually found on new surfaces or those that have been stripped of life—for example, surfaces created by mass movements of land or land disturbed by human activities.

Place (1)   A major theme in geography focused on the tangible and intangible characteristics that make each location unique.

Plane of the ecliptic (2)   A plane intersecting all the points of Earth's orbit.

Planetesimal hypothesis (2) Early protoplanets formed from the condensing masses of a nebular cloud of dust, gas, and icy comets; a formation process now being observed in other parts of the galaxy.

Planimetric map (Appendix A) A basic map showing the horizontal position of boundaries, land-use activities, and political, economic, and social outlines.

Plate tectonics (8) The conceptual model that encompasses continental drift, sea-floor spreading, and related aspects of crustal movement; accepted as the foundation of crustal tectonic processes.

Plateau basalts (9) An accumulation of horizontal flows formed when lava spreads out from elongated fissures onto the surface in extensive sheets; associated with effusive eruptions; also known as flood basalts (see *basalt*).

Playa (12) An area of salt crust left behind by evaporation on a desert floor, usually in the middle of a bolson or valley; intermittently wet and dry.

Plinthite (15) An ironstone hardpan structure formed in Oxisol subsurfaces or surface horizons that are exposed to repeated wetting and drying sequences; quarried in some areas for building materials.

Pluton (8) A mass of intrusive igneous rock that has cooled slowly in the crust; forms in any size or shape. The largest partially exposed pluton is a *batholith*.

Podzolization (15) A pedogenic process in cool, moist climates; forms a highly leached soil with strong surface acidity because of humus from acid-rich trees.

Point bar (11) The inner portion of a stream meander that receives sediment fill.

Polar easterlies (4) Variable weak, cold, and dry winds moving away from the polar region; an anticyclonic circulation.

Polar front (4) A significant zone of contrast between cold and warm air masses; roughly situated between 50° and 60° N and S latitude.

Polar high-pressure cells (4) A weak, anticyclonic, thermally produced pressure system positioned roughly above each pole.

Polypedon (15) The identifiable soil in an area, with distinctive characteristics differentiating it from surrounding polypedons, forming the basic soil mapping unit; composed of many pedons.

Porosity (6) The total volume of available pore space in soil; a result of the texture and structure of the soil.

Positive feedback (1) Feedback that amplifies or encourages responses in a system (see *negative feedback*, *feedback loop*).

Potential evapotranspiration (6) POTET, or PE; the amount of moisture that would evaporate and transpire if adequate moisture were available; the amount lost under optimum moisture conditions—the moisture demand.

Precipitation (6) Rain, snow, sleet, and hail—the moisture supply (PRECIP).

Pressure gradient force (4) A force that causes air to move from areas of higher barometric pressure to areas of lower barometric pressure due to pressure differences.

Primary succession (16) Succession that occurs among plant species in an area of new surfaces created by mass movement of land, areas exposed by a retreating glacier, cooled lava flows and volcanic eruption landscapes, surface mining and clear-cut logging scars, or an area of sand dunes, with no trace of a former community (see *primary community*).

Prime meridian (1) An arbitrary meridian designated as 0° longitude; the point from which longitude is measured east or west; agreed to in an 1884 treaty.

Process (1) A set of actions and changes that occur in some special order; analysis of processes is central to modern geographic study.

Producers (16) Organisms that are capable of using carbon dioxide as their sole source of carbon, which they chemically fix through photosynthesis to provide their own nourishment; also called autotrophs.

Pyroclastic (9) An explosively ejected rock fragment launched by a volcanic eruption; sometimes described by the more general term *tephra*.

Radiation fog (5) Fog formed by radiative cooling of the surface, especially on clear nights in areas of moist ground; occurs when the air layer directly above the surface is chilled to the dew-point temperature, thereby producing saturated conditions.

Rain gauge (6) A standardized device that catches and measures rainfall.

Rain shadow (5) The area on the leeward slopes of a mountain range, in the shadow of the mountains, where precipitation receipt is greatly reduced compared to that of windward slopes (see *orographic lifting*).

Reflection (3) The portion of arriving energy that returns directly back to space without being converted into heat or performing any work (see *albedo*).

Refraction (3) The bending effect that occurs when insolation enters the atmosphere or another medium; the same process by which a crystal, or prism, disperses the component colors of the light passing through it.

Region (1) A geographic theme that focuses on areas that display unity and internal homogeneity of traits; includes the study of how a region forms, evolves, and interrelates with other regions.

Regolith (10) Partially weathered rock overlying bedrock, whether residual or transported.

Relative humidity (5) The ratio of water vapor actually in the air (content) compared to the maximum water vapor the air is able to hold (capacity) at that temperature; expressed as a percentage (compare *vapor pressure*, *specific humidity*).

Relief (9) Elevation differences in a local landscape; an expression of local height differences of landforms.

Remote sensing (1) Information acquired from a distance, without physical contact with the subject, for example, by photography, orbital imagery, or radar.

Respiration (16) The process by which plants derive energy for their operations—essentially, the reverse of the photosynthetic process—the release of carbon dioxide, water, and heat into the environment (compare *photosynthesis*).

Reverse fault (9) A compression fault, in which one side of the fault moves upward vertically in comparison to the other side (compare *normal fault*).

Revolution (2) The annual orbital movement of Earth about the Sun; determines year length and the length of seasons.

Rhumb line (1) A line of constant compass direction, or constant bearing, which crosses successive meridians at the same angle; appearing as a straight line only on a Mercator projection.

Robinson projection (Appendix A) An oval projection developed by A. Robinson in 1963.

Roche moutonnée (14) A glacial erosion feature; an asymmetrical hill of exposed bedrock displaying a gently sloping upstream side that has been smoothed and polished by a glacier and an abrupt, steep downstream side.

Rock (8) An assemblage of minerals bound together (like granite) or a mass of a single mineral (like rock salt).

Rock cycle (8) A model representing the interrelationships among the three rock-forming processes—igneous, sedimentary, and metamorphic—showing how each can be transformed into another rock type.

Rockfall (10) Free-falling movement of debris from a cliff or steep slope, generally falling straight down or bounding downslope.

Rossby waves (4) Undulating horizontal motions in the upper-air westerly circulation at middle and high latitudes.

Rotation (2) The turning of Earth on its axis, averaging 24 hours in duration and determining the day-night relation; counterclockwise west to east or eastward when viewed from above the North Pole and from above the equator.

Salinity (13) The concentration of natural elements and compounds dissolved in solution, as solutes; measured by weight in parts per thousand (‰) in seawater.

Salinization (15) A pedogenic regime that results from high potential evapotranspiration rates in deserts and semiarid regions. Soil water is drawn to surface horizons, and dissolved salts are deposited as the water evaporates.

Saltation (11, 12) The transport of sand grains (usually larger than 0.2 mm, or 0.008 in.) by stream or wind, which bounce the grains along the ground in asymmetrical paths.

Salt marsh (13) Wetland ecosystem characteristic of latitudes poleward of the 30th parallel (compare *mangrove swamp*).

Sand sea (12) An extensive area of sand and dunes; characteristic of Earth's erg deserts.

Saturation (5) Air that is holding all the water vapor that it can hold at a given temperature.

Scale (1)   The ratio of the distance on a map to that in the real world, expressed as a representative fraction, graphic scale, or written scale.

Scarification (10)   Human-induced mass movements of Earth materials, such as large-scale open-pit mining and strip mining.

Scattering (3)   Deflection and redirection of insolation by atmospheric gases, dust, ice, and water vapor. The shorter the wavelength, the greater the scattering, thus skies in the lower atmosphere are blue.

Scientific method (1)   An approach that uses applied common sense in an organized and objective manner, based on observation, generalization, formulation of a hypothesis, and ultimately the development of a theory.

Sea-floor spreading (8)   As proposed by Hess and Dietz, the mechanism driving the movement of the continents; associated with upwelling flows of magma along the worldwide system of mid-ocean ridges (see *mid-ocean ridge*).

Secondary succession (16)   Succession that occurs among plant species in an area where vestiges of a previously functioning community are present; an area where the natural community has been destroyed or disturbed, but where the underlying soil remains intact.

Sediment (10)   Fine-grained mineral matter that is transported and deposited by air, water, or ice.

Sedimentary rock (8)   One of three basic rock types; formed from the compaction, cementation, and hardening of sediments derived from former rocks (compare *igneous rock, metamorphic rock*).

Seismic waves (8)   The shock waves sent through the planet by an earthquake or underground nuclear test, the transmission varying according to temperature and the density of various layers within the planet and providing indirect diagnostic evidence of Earth's internal structure.

Seismograph (9)   A device that measures seismic waves of energy transmitted throughout Earth's interior or along the crust.

Sensible heat (2)   Heat measured with a thermometer; a measure of the concentration of kinetic energy from molecular motion.

Sheet flow (11)   Water that moves downslope in a thin film as overland flow, not concentrated in channels larger than rills.

Sheeting (10)   A form of weathering associated with fracturing or fragmentation of rock by pressure release; often related to exfoliation processes (see *exfoliation dome*).

Shield volcano (9)   A symmetrical mountain landform built from effusive eruptions that is gently sloped, gradually rising from the surrounding landscape to a summit crater and typical of the Hawaiian Islands (see *effusive eruption;* compare *composite volcano*).

Sinkhole (10)   Nearly circular depression, also known as a doline, created by the weathering of karst landscapes; may collapse through the roof of an underground space (see *karst topography*).

Slopes (10)   Curved, inclined surfaces that bound landforms.

Small circle (1)   Circles on a globe's surface that do not share Earth's center, for example, all parallels other than the equator (compare *great circle*).

Soil (15)   A dynamic natural body made up of fine materials covering Earth's surface in which plants grow; composed of both mineral and organic matter.

Soil colloid (15)   Tiny clay and organic particles in soil that provide chemically active sites for mineral ion adsorption (see *cation-exchange capacity*).

Soil creep (10)   A persistent mass movement of surface soil where individual soil particles are lifted and disturbed by the expansion of soil moisture as it freezes or even by grazing livestock or digging animals.

Soil fertility (15)   The ability of soil to support plant productivity when it contains organic substances and clay minerals that absorb water and certain elemental ions needed by plants (see *cation-exchange capacity*).

Soil horizon (15)   The various layers exposed in a pedon, roughly parallel to the surface and identified as O, A, E, B, and C, residing above bedrock designated R.

Soil-moisture recharge (6)   Water entering available soil storage.

Soil-moisture storage (6)   The retention of moisture within soil, representing a savings account that can accept deposits (soil-moisture recharge) or experience withdrawals (soil-moisture utilization) as conditions change (acronym: Δ STRGE).

Soil-moisture utilization (6)   The extraction of moisture by plants for their needs; inefficiency of withdrawal increases as the soil storage is reduced.

Soil science (15)   Interdisciplinary science of soils. Pedology concerns the origin, classification, distribution, and description of soil. Edaphology focuses on soil as a medium for sustaining higher plants.

Soil Taxonomy (15)   A soil classification system based on observable soil properties actually seen in the field; published in 1975 by the U.S. Soil Conservation Service, revised in 1990 and 1998 by the Natural Resources Conservation Service to include 12 soil orders (compare *Canadian System of Soil Classification*).

Soil-water budget (6)   An accounting system for soil moisture using inputs of precipitation and outputs of evapotranspiration and gravitational water.

Solar constant (2)   The amount of insolation intercepted by Earth on a surface perpendicular to the Sun's rays when Earth is at its average distance from the Sun; a value of 1372 watts per m^2, or 1.968 calories/cm^2 per minute, averaged over the entire globe at the thermopause.

Solar wind (2)   Clouds of ionized (charged) gases emitted by the Sun and traveling in all directions from the Sun's surface. Effects on Earth include auroras, disturbance of radio signals, and possible influences on weather.

Solum (15)   A true soil profile in the *pedon;* ideally, a combination of the A and B horizons.

Spatial (1)   The nature or character of physical space, as in an area; occupying or operating within a space. Geography is a spatial science; spatial analysis is its essential approach.

Spatial analysis (1)   The examination of spatial interactions, patterns, and variations over an area and/or space; a key integrative approach of geography.

Specific heat (3)   The increase of temperature in a material when energy is absorbed; water is said to have a higher specific heat than a comparable volume of soil or rock.

Specific humidity (5)   The mass of water vapor (in grams) per unit mass of air (in kilograms) at any specified temperature. The maximum mass of water vapor that a kilogram of air can hold at any specified temperature is termed its maximum specific humidity (compare *vapor pressure, relative humidity*).

Speed of light (2)   Specifically, 299,792 kmps (186,282 mi per second), covering more than 9.4 trillion km per year (5.9 trillion mi per year)—a distance known as a light-year. At light speed, Earth is 8 minutes and 20 seconds from the Sun.

Spheroidal weathering (10)   A chemical weathering process in which the sharp edges and corners of boulders and rocks are weathered in thin plates, creating a rounded, spheroidal form.

Spodosols (15)   A soil order in the Soil Taxonomy classification that occurs in northern coniferous forests; best developed in cold, moist, forested climates in humid continental or subarctic regions; lacks humus and clay in the A horizons and has high acidity associated with podzolization processes.

Squall line (5)   A zone slightly ahead of a fast-advancing cold front, where wind patterns are rapidly changing and blustery and precipitation is strong.

Stability (5)   The condition of a parcel, whether it remains where it is or changes its initial position. The parcel is stable if it resists displacement upward; it is unstable if it continues to rise.

Steady-state equilibrium (1)   The condition that occurs in a system when rates of input and output are equal and the amounts of energy and stored matter are nearly constant around a stable average.

Stomata (16)   Small openings on the undersides of leaves through which water and gases pass.

Storm surge (5)   Large quantities of seawater pushed inland by the strong winds associated with a tropical cyclone.

Stratified drift (14)   Sediments deposited by glacial meltwater that appear sorted; a specific form of *glacial drift* (compare *till*).

Stratigraphy (8)   An analysis of the sequence, spacing, and spatial distribution of rock strata.

Stratocumulus (5)   A lumpy, grayish, low-level cloud, patchy with sky visible, sometimes present at the end of the day.

Stratosphere (2)   That portion of the *homosphere* that ranges from 20 to 50 km (12.5 to 30 mi) above Earth's surface; temperatures range from −57°C (−70°F) at the tropopause to 0°C (32°F) at the stratopause. The functional *ozonosphere* is within the stratosphere.

Stratus (5)   A stratiform (flat, horizontal) cloud generally below 2000 m (6500 ft).

Strike-slip fault (9)   Horizontal movement along a fault line, that is, movement in the same direction as the fault; also known as a transcurrent fault. Such movement is described as right-lateral or left-lateral, depending on the relative motion observed.

Subduction zone (8)   An area where two plates of crust collide and the denser oceanic crust dives beneath the less dense continental plate, forming deep oceanic trenches and seismically active regions (see *oceanic trench*).

Sublimation (5)   A process in which ice evaporates directly to water vapor or water vapor freezes to ice (deposition).

Subpolar low-pressure cell (4)   A region of low pressure centered approximately at 60° latitude in the North Atlantic near Iceland and in the North Pacific near the Aleutians, as well as in the Southern Hemisphere. Air flow is cyclonic; it weakens in summer and strengthens in winter.

Subsolar point (2)   The only point receiving perpendicular insolation at a given moment—the Sun is directly overhead (see *declination*).

Subtropical high-pressure cells (4)   Dynamic high-pressure areas covering roughly the region from 20° to 35° N and S latitudes; responsible for the hot, dry areas of Earth's arid and semiarid deserts.

Sulfate aerosols (2)   Sulfur compounds in the atmosphere, principally sulfuric acid, which scatter and reflect insolation; principal sources relate to fossil-fuel combustion.

Sulfur dioxide (2)   A colorless gas detected by its pungent odor produced by the combustion of fossil fuels that contain sulfur as an impurity (forming SO_2); can react in the atmosphere to form sulfuric acid, a component of acid deposition.

Summer (June) solstice (2)   The time when the Sun's declination is at the Tropic of Cancer, at 23.5° N latitude; June 20–21 each year (compare to *winter, or December, solstice*).

Sunspots (2)   Magnetic disturbances on the surface of the Sun, occurring in an average 11-year cycle; related flares, prominences, and outbreaks produce surges in solar wind.

Surface creep (12)   A form of *eolian* transport that involves particles too large for *saltation*; a process whereby individual grains are impacted by moving grains and slide and roll.

Surplus (6)   SURPL; the amount of moisture that exceeds potential evapotranspiration; moisture oversupply when soil moisture storage is at field capacity.

Suspended load (11)   Fine particles held in suspension in a stream. The finest particles are not deposited until the stream velocity nears zero.

Swell (13)   Regular patterns of smooth, rounded waves in open water; can range from small ripples to very large waves.

Syncline (9)   A trough in folded strata, with beds that slope toward the axis of the downfold.

System (1)   Any ordered, interrelated set of materials or items existing separate from the environment, or within a boundary; energy transformations and energy and matter storage and retrieval occur within a system.

Taiga (16)   See *needleleaf forest*.

Tarn (14)   A small mountain lake, especially one that collects in a cirque basin behind risers of rock material.

Tectonic processes (8)   Driven by internal energy from within Earth; refers to large-scale movement and deformation of the crust.

Temperate rain forest (16)   A major biome of lush forests at middle and high latitudes; occurs along narrow margins of the Pacific Northwest in North America, among other locations; includes the tallest trees in the world.

Temperature (3)   A measure of sensible heat energy in the atmosphere and other media; indicates the average kinetic energy of individual molecules within the atmosphere.

Temperature inversion (2)   A reversal of the normal decrease of temperature with increasing altitude; can occur anywhere from ground level up to several thousand meters; functions to block atmospheric convection and thereby traps pollutants.

Terranes (9)   A migrating crustal piece, dragged about by processes of mantle convection and plate tectonics. Displaced terranes are distinct in their history, composition, and structure from the continents that accept them.

Terrestrial ecosystem (16)   A self-regulating association characterized by specific plant formations; usually named for the predominant vegetation and known as a biome when large and stable (see *biome*).

Thermal equator (3)   An isoline on an isothermal map that connects all points of highest mean temperature.

Thermohaline circulation (4)   Differences in temperatures and salinity that produce density differences important to the flow of deep, sometimes vertical, currents. Traveling at slower speeds than wind-driven surface currents, the thermohaline circulation hauls larger volumes of water.

Thermopause (2)   A zone approximately 480 km (300 mi) in altitude that serves conceptually as the top of the atmosphere; an altitude used for the determination of the solar constant.

Thermosphere (2)   A region of the heterosphere extending from 80 to 480 km (50 to 300 mi) in altitude; contains the functional ionosphere layer.

Threshold (1)   A point at which a system can no longer maintain its character, so it lurches to a new operational level; places the system in a *metastable equilibrium*, such as a hillside or coastal bluff, which adjusts after a sudden landslide.

Thrust fault (9)   A low-angle fault plane, relative to the horizontal, where an overlying block shifts (thrusts) over an underlying block.

Thunder (5)   The violent expansion of suddenly heated air, created by lightning discharges, sending out shock waves as an audible sonic bang.

Tide (13)   Patterns of daily oscillations in sea level produced by astronomical relationships among the Sun, the Moon, and Earth; experienced in varying degrees around the world.

Till (14)   Direct ice deposits that appear unstratified and unsorted; a specific form of *glacial drift* (compare *stratified drift*).

Till plains (14)   Large, relatively flat plains composed of unsorted glacial deposits behind a terminal or end moraine. Low-rolling relief and unclear drainage patterns are characteristic.

Tombolo (13)   A landform created when coastal sand deposits connect the shoreline with an offshore island outcrop or sea stack.

Topographic map (Appendix A)   A map that portrays physical relief through the use of elevation contour lines that connect all points at the same elevation above or below a vertical datum, such as mean sea level.

Topography (9)   The undulations and configurations that give Earth's surface its texture; the heights and depths of local relief, including both natural and human-made features.

Tornado (5)   An intense, destructive cyclonic rotation, developed in response to extremely low pressure; associated with mesocyclone formation.

Total runoff (6)   Surplus water that flows across a surface toward stream channels; formed by sheet flow, combined with precipitation and subsurface flows into those channels.

Traction (11)   A type of sediment transport that drags coarser materials along the bed of a stream (see *bed load*).

Trade winds (4)   Northeast and southeast winds that converge in the equatorial low-pressure trough, forming the intertropical convergence zone.

Transmission (3)   The passage of shortwave and longwave energy through space, the atmosphere, or water.

Transparency (3)   The quality of a medium (air, water) that allows light to shine through it.

Transpiration (6)   The movement of water vapor out through the pores in leaves, the water being drawn by plant roots from the soil-moisture storage.

Transport (11)   The actual movement of weathered and eroded materials by air, water, and ice.

Tropical cyclone (5)   A cyclonic circulation originating in the tropics, with winds between 30 and 64 knots (39 to 73 mph); characterized by closed isobars, circular organization, and heavy rains (see *hurricane* and *typhoon*).

Tropical savanna (16)   A major biome containing large expanses of grassland interrupted by trees and shrubs; a transitional area between the humid rain forests and tropical seasonal forests and the drier, semiarid tropical steppes and deserts.

Tropical seasonal forest and scrub (16)   A variable biome on the margins of the rain forests, occupying regions of lesser and more erratic rainfall; the site of transitional communities between the rain forests and tropical grasslands.

Tropic of Cancer (2)   The parallel that marks the farthest north the subsolar point migrates during the year; 23.5° north latitude (see *Tropic of Capricorn; summer, or June, solstice*).

Tropic of Capricorn (2)   The parallel that marks the farthest south the subsolar point migrates during the year; 23.5° south latitude (see *Tropic of Cancer; winter or December solstice*).

Troposphere (2)   The home of the biosphere; the lowest layer of the homosphere, containing approximately 90% of the total mass of the atmosphere; extends up to the *tropopause*, defined by a temperature of −57°C (−70°F) occurring at an altitude of 18 km (11 mi) at the equator, 13 km (8 mi) in the middle latitudes, and at lower altitudes near the poles.

True shape (1)   A map property showing the correct configuration of coastlines; a useful trait of conformality for navigational and aeronautical maps, although areal relationships are distorted (see *map projection*; compare *equal area*).

Tsunami (13)   A seismic sea wave, traveling at high speeds across the ocean, formed by sudden and sharp motions in the seafloor, such as seafloor earthquakes, submarine landslides, or eruptions from undersea volcanoes.

Typhoon (5)   A tropical cyclone with inward-spiraling winds in excess of 65 knots (74 mph) that occurs in the western Pacific; same as a hurricane except for location.

Ultisols (15)   A soil order in the Soil Taxonomy, featuring highly weathered forest soils, principally in the humid subtropical climatic classification. Increased weathering and exposure can degenerate an Alfisol into the reddish color and texture of these more humid, tropical soils. Fertility is quickly exhausted when Ultisols are cultivated.

Unconfined aquifer (6)   The zone of saturation in water-bearing rock strata, with no impermeable overburden, and with recharge generally accomplished by water percolating down from above.

Undercut bank (11)   A steep bank formed along the outer portion of a meandering stream, produced by lateral, erosive, undercutting action of a stream (compare *point bar*).

Uniformitarianism (8)   An assumption that physical processes active in the environment today are operating at the same pace and intensity that have characterized them throughout geologic time; proposed by Hutton and Lyell.

Upslope fog (5)   Fog that forms when moist air is forced to higher elevations along a hill or mountain and is thus cooled (compare *valley fog*).

Upwelling current (4)   Ocean currents that cause surface waters to be swept away from the coast by surface divergence or offshore winds. Cool, deep waters, which are generally nutrient-rich, rise to replace the vacating water (compare *downwelling current*).

Urban heat island (3)   Urban microclimates, which are warmer on the average than areas in the surrounding countryside because of various surface characteristics.

Valley fog (5)   The settling of cooler, more dense air in low-lying areas, which produces saturated conditions and fog.

Vapor pressure (5)   That portion of total air pressure that results from water-vapor molecules; expressed in millibars (mb). At a given temperature the maximum capacity of the air is termed its saturation vapor pressure.

Vascular plants (16)   Plants that have developed conductive tissues and true roots for internal transport of fluid and nutrients. *Vascular* is from a Latin word for "vessel-bearing," referring to the conducting cells.

Ventifacts (12)   The individual pieces and pebbles etched and smoothed by eolian erosion—abrasion by windblown particles.

Vernal (March) equinox (2)   The time when the Sun's declination crosses the equatorial parallel and all places on Earth experience days and nights of equal length, around March 20–21 each year. The Sun rises at the North Pole and sets at the South Pole (compare *autumnal equinox*).

Vertisols (15)   A soil order in the Soil Taxonomy that features expandable clay soils composed of more than 30% swelling clays; occurs in regions that experience highly variable soil-moisture balances through the seasons.

Volcano (9)   A landform at the end of a conduit or pipe that rises from below the crust and vents to the surface. Magma rises and collects in a magma chamber deep below, resulting in eruptions that are effusive or explosive, forming the mountain landform.

Warm front (5)   The leading edge of an advancing warm air mass that is unable to push cooler, passive air out of the way; tends to push the cooler, underlying air into a wedge shape; noted on weather maps with a series of rounded knobs placed along the front in the direction of the frontal movement (compare *cold front*).

Wash (12)   An intermittently dry streambed that fills with torrents of water after rare precipitation events in arid lands.

Waterspout (5)   An elongated, funnel-shaped circulation formed when a tornado takes place over water.

Water table (6)   The upper limit of groundwater; that contact point between zones of saturation and aeration in an unconfined aquifer (see *zone of aeration*, *zone of saturation*).

Wave (13)   Undulations of ocean water produced by the conversion of solar energy to wave energy; produced in a generating region or a stormy area of the sea.

Wave-cut platform (13)   A flat, or gently sloping, tablelike bedrock surface that develops in the tidal zone, where wave action cuts a bench that extends from the cliff base out into the sea.

Wave cyclone (5)   An organized area of low pressure, with converging and ascending air flow producing an interaction of air masses; migrates along storm tracks. Such lows or depressions form the dominant weather pattern in the middle and higher latitudes of both hemispheres (see *midlatitude cyclone*).

Wavelength (2)   Measurement of waveform; the actual distance between the crests of successive waves. The number of waves passing a fixed point in 1 second is called the frequency of the wavelength (see *electromagnetic spectrum*).

Wave refraction (13)   A process that concentrates energy on headlands and disperses it in coves and bays; a coastal straightening process.

Weather (5)   The short-term condition of the atmosphere, as compared to *climate*, which reflects long-term atmospheric conditions and extremes. Temperature, air pressure, relative humidity, wind speed and direction, daylength, and Sun angle are important measurable elements that contribute to the weather.

Weathering (10)   The related processes by which surface and subsurface rock disintegrate, or dissolve, or are otherwise broken down. Rocks at or near Earth's surface are exposed to physical, organic, and chemical weathering processes.

Westerlies (4)   The predominant wind-flow pattern from the subtropics to high latitudes in both hemispheres.

Wetland (13)   A narrow, vegetated strip occupying many coastal areas and estuaries worldwide; highly productive ecosystems with an ability to trap organic matter, nutrients, and sediment.

Wilting point (6)   That point in the soil-moisture balance when only hygroscopic water and some bound capillary water remains. Plants wilt and eventually die after prolonged stress from a lack of available water.

Wind (4)   The horizontal movement of air relative to Earth's surface; produced essentially by air pressure differences from place to place; also influenced by the Coriolis force and surface friction. Turbulence adds wind updrafts and downdrafts and a vertical component to this definition.

Wind vane (4)   A weather instrument used to determine wind direction; winds are named for the direction from which they originate.

Winter (December) solstice (2)   That time when the Sun's declination is at the Tropic of Capricorn, at 23.5° S latitude, December 21–22 each year. The day is 24 hours long south of the Antarctic Circle. The night is 24 hours long north of the Arctic Circle (compare to *summer*, or *June, solstice*).

Zone of aeration (6)   A groundwater zone above the water table, which may or may not hold water in pore spaces.

Zone of saturation (6)   A groundwater zone below the water table, in which all pore spaces are filled with water.

Index